Methods in Microbiology
Volume 35

Recent titles in the series

Volume 24 *Techniques for the Study of Mycorrhiza*
JR Norris, DJ Reed and AK Varma

Volume 25 *Immunology of Infection*
SHE Kaufmann and D Kabelitz

Volume 26 *Yeast Gene Analysis*
AJP Brown and MF Tuite

Volume 27 *Bacterial Pathogenesis*
P Williams, J Ketley and GPC Salmond

Volume 28 *Automation*
AG Craig and JD Hoheisel

Volume 29 *Genetic Methods for Diverse Prokaryotes*
MCM Smith and RE Sockett

Volume 30 *Marine Microbiology*
JH Paul

Volume 31 *Molecular Cellular Microbiology*
P Sansonetti and A Zychlinsky

Volume 32 *Immunology of Infection, 2nd edition*
SHE Kaufmann and D Kabelitz

Volume 33 *Functional Microbial Genomics*
B Wren and N Dorrell

Volume 34 *Microbial Imaging*
T Savidge and C Pothoulakis

Methods in Microbiology

Volume 35
Extremophiles

Edited by

Fred A Rainey

Department of Biological Sciences
Louisiana State University
Baton Rouge, LA
USA

and

Aharon Oren

Institute of Life Sciences
The Hebrew University of Jerusalem
Jerusalem
Israel

Amsterdam • Boston • Heidelberg • London • New York
Oxford • Paris • San Diego • San Francisco
Singapore • Sydney • Tokyo

Academic Press is an imprint of Elsevier

ELSEVIER B.V.	ELSEVIER Inc.	ELSEVIER Ltd.	**ELSEVIER Ltd.**
Radarweg 29	525 B Street, Suite 1900	The Boulevard, Langford Lane	**84 Theobalds Road**
P.O. Box 211, 1000 AE	San Diego, CA 92101-4495	Kidlington, Oxford OX5 1GB	**London WC1X 8RR**
Amsterdam, The Netherlands	USA	UK	**UK**

© 2006 Elsevier Ltd. All rights reserved.

This work is protected under copyright by Elsevier Ltd., and the following terms and conditions apply to its use:

Photocopying
Single photocopies of single chapters may be made for personal use as allowed by national copyright laws. Permission of the Publisher and payment of a fee is required for all other photocopying, including multiple or systematic copying, copying for advertising or promotional purposes, resale, and all forms of document delivery. Special rates are available for educational institutions that wish to make photocopies for non-profit educational classroom use.

Permissions may be sought directly from Elsevier's Rights Department in Oxford, UK: phone (+44) 1865 843830, fax (+44) 1865 853333, e-mail: permissions@elsevier.com. Requests may also be completed on-line via the Elsevier homepage (http://www.elsevier.com/locate/permissions).

In the USA, users may clear permissions and make payments through the Copyright Clearance Center, Inc., 222 Rosewood Drive, Danvers, MA 01923, USA; phone: (+1) (978) 7508400, fax: (+1) (978) 7504744, and in the UK through the Copyright Licensing Agency Rapid Clearance Service (CLARCS), 90 Tottenham Court Road, London W1P 0LP, UK; phone: (+44) 20 7631 5555; fax: (+44) 20 7631 5500. Other countries may have a local reprographic rights agency for payments.

Derivative Works
Tables of contents may be reproduced for internal circulation, but permission of the Publisher is required for external resale or distribution of such material. Permission of the Publisher is required for all other derivative works, including compilations and translations.

Electronic Storage or Usage
Permission of the Publisher is required to store or use electronically any material contained in this work, including any chapter or part of a chapter.

Except as outlined above, no part of this work may be reproduced, stored in a retrieval system or transmitted in any form or by any means, electronic, mechanical, photocopying, recording or otherwise, without prior written permission of the Publisher.
Address permissions requests to: Elsevier's Rights Department, at the fax and e-mail addresses noted above.

Notice
No responsibility is assumed by the Publisher for any injury and/or damage to persons or property as a matter of products liability, negligence or otherwise, or from any use or operation of any methods, products, instructions or ideas contained in the material herein. Because of rapid advances in the medical sciences, in particular, independent verification of diagnoses and drug dosages should be made.

First edition 2006

A catalogue record is available from the Library of Congress.

Cover image: "Collage of Extreme Life" by Laura Fukuda

ISBN-10:	0-12-521536-3 (Hardbound – this volume)
ISBN-13:	978-0-12-521536-7 (Hardbound – this volume)
ISBN-10:	0-12-521537-1 (Paperback – this volume)
ISBN-13:	978-0-12-521537-4 (Paperback – this volume)
ISSN:	0580-9517 (Series)

∞ The paper used in this publication meets the requirements of ANSI/NISO Z39.48-1992 (Permanence of Paper).
Printed in Spain.

Working together to grow
libraries in developing countries

www.elsevier.com | www.bookaid.org | www.sabre.org

ELSEVIER   BOOK AID International   Sabre Foundation

Contents

Contributors xi

1. Extremophile Microorganisms and the Methods to Handle Them 1
   F A Rainey and A Oren

Thermophiles

2. *In Situ* Activity Studies in Thermal Environments 29
   NV Pimenov and EA Bonch-Osmolovskaya

3. The Isolation of Thermophiles from Deep-sea Hydrothermal
   Environments 55
   S Nakagawa and K Takai

4. Growth of Hyperthermophilic Microorganisms for Physiological
   and Nutritional Studies 93
   A Godfroy, A Postec and N Raven

5. Toward the Large Scale Cultivation of Hyperthermophiles at
   High-Temperature and High-Pressure 109
   CB Park, BB Boonyaratanakornkit and DS Clark

6. Analysis of Lipids from Extremophilic Bacteria 127
   MS da Costa, MF Nobre and R Wait

7. Membranes of Thermophiles and Other Extremophiles 161
   SV Albers, WN Konings and AJM Driessen

8. Characterization and Quantification of Compatible Solutes in
   (Hyper)thermophilic Microorganisms 173
   H Santos, P Lamosa and N Borges

9. Functional Genomics of the Thermo-Acidophilic Archaeon
   Sulfolobus solfataricus.. 201
   J van der Oost, J Walther, SJJ Brouns, HJG van de Werken,
   APL Snijders, PC Wright, A Andersson, R Bernander
   and WM de Vos

10. Heat Shock Proteins in Hyperthermophiles 233
    FT Robb, HD Shukla and DS Clark

11. Deep-sea Thermococcales and their Genetic Elements: Plasmids and Viruses 253
    D Prieur, G Erauso, D Flament, M Gaillard, C Geslin, M Gonnet, M Le Romancer, S Lucas and P Forterrre

12. Genetic Systems for *Thermus* 279
    B Averhoff

13. Gene Transfer Systems for Obligately Anaerobic Thermophilic Bacteria 309
    MV Tyurin, LR Lynd and J Wiegel

14. Hyperthermophilic Virus–Host Systems: Detection and Isolation 331
    D Prangishvili

15. Preservation of Thermophilic Microorganisms 349
    S Spring

Psychrophiles

16. Handling of Psychrophilic Microorganisms 371
    NJ Russell and DA Cowan

17. Proteins from Psychrophiles 395
    R Cavicchioli, PMG Curmi, KS Siddiqui and T Thomas

Alkaliphiles

18. Cultivation of Aerobic Alkaliphiles 439
    WD Grant

19. Isolation, Cultivation and Characterization of Alkalithermophiles 451
    NM Mesbah and J Wiegel

Acidophiles

20. The Isolation and Study of Acidophilic Microorganisms 471
    E González-Toril, F Gómez, M Malki and R Amils

Halophiles

21. Characterization of Natural Communities of Halophilic Microorganisms 513
    CD Litchfield, M Sikaroodi and PM Gillevet

22. Cultivation of Haloarchaea 535
    D Burns and M Dyall-Smith

23. Extraction of Halophiles from Ancient Crystals 553
    RH Vreeland

24. The Assessment of the Viability of Halophilic Microorganisms in Natural Communities 569
    H Stan-Lotter, S Leuko, A Legat and S Fendrihan

25. Characterization of Lipids of Halophilic Archaea 585
    A Corcelli and S Lobasso

26. Characterization of Organic Compatible Solutes of Halotolerant and Halophilic Microorganisms.. 615
    MF Roberts

27. Genetic Systems for Halophilic Archaea 649
    BR Berquist, JA Müller and S DasSarma

28. The Isolation and Study of Viruses of Halophilic Microorganisms........ 681
    K Porter and ML Dyall-Smith

29. Detection, Quantification and Purification of Halocins: Peptide Antibiotics from Haloarchaeal Extremophiles 703
    RF Shand

30. Storage of Halophilic Bacteria... 719
    JK DiFerdinando and RH Vreeland

Barophiles

31. Handling of Piezophilic Microorganisms 733
    C Kato

Radiation-Resistant Microorganisms

32. Measuring Survival in Microbial Populations Following Exposure to Ionizing Radiation 745
    JM Zimmerman and JR Battista

Strict Anaerobes

33. The Study of Strictly Anaerobic Microorganisms......................... 757
    KR Sowers and JEM Watts

Applications of Extremophiles

34. Applications of Extremophiles: The Industrial Screening of Extremophiles for Valuable Biomolecules 785
    G Ravot, J-M Masson and F Lefèvre

Index 815

Colour plate section between pages 272 and 273

Series Advisors

Gordon Dougan The Wellcome Trust Sanger Institute, Wellcome Trust Genome Campus, Hinxton, Cambridge, CBIO ISA, UK

Graham J Boulnois Schroder Ventures Life Science Advisers (UK) Limited, 71 Kingsway, London WC2B 6ST, UK

Jim Prosser Department of Molecular and Cell Biology, University of Aberdeen, Institute of Medical Sciences, Foresterhill, Aberdeen AB25 2ZD, UK

Ian R Booth Professor of Microbiology, Department of Molecular and Cell Biology, University of Aberdeen, Institute of Medical Sciences, Foresterhill, Aberdeen AB25 2ZD, UK

David A Hodgson Reader in Microbiology, Department of Biological Sciences, University of Warwick, Coventry CV4 7AL, UK

David H Boxer Department of Biochemistry, Medical Sciences Institute, Dundee DD1 4HN, UK

Contributors

SV Albers Molecular Microbiology, University of Groningen, Kerklaan 32, 9753 NN Haren, The Netherlands

R Amils Centro de Biología Molecular "Severo Ochoa", Universidad Autónoma de Madrid, Cantoblanco, Madrid 28049, Spain

A Andersson Department of Biotechnology, KTH – Royal Institute of Technology, Stockholm, Sweden

B Averhoff Johann Wolfgang Goethe-Universität Frankfurt, Campus Riedberg Institute of Microbiology, Marie-Curie-Strasse 9, D-60439 Frankfurt, Germany

JR Battista Department of Biological Sciences, Louisiana State University, 202 Life Sciences Building, Baton Rouge, LA 70803, USA

R Bernander Department of Molecular Evolution, Evolutionary Biology Center, Uppsala University, Uppsala, Sweden

BR Berquist Center of Marine Biotechnology, University of Maryland Biotechnology Institute, 701 East Pratt Street, Suite 236, Columbus Center, Baltimore, MD 21202, USA

EA Bonch-Osmolovskaya Institute of Microbiology, Russian Academy of Sciences, Prospekt 60 Let Oktyabria 7, bldg. 2, 117811 Moscow, Russia

BB Boonyaratanakornkit Department of Chemical Engineering, University of California at Berkeley, 201 Gilman Hall # 1462, Berkeley, CA 94720-1462, USA

N Borges Instituto de Tecnologia Química e Biológica, Universidade Nova de Lisboa, Rua da Quinta Grande 6, Apartado 127, 2780-156 Oeiras, Portugal

SJJ Brouns Laboratory of Microbiology, Wageningen University, Hesselink van Suchtelenweg 4, 6703 CT Wageningen, The Netherlands

D Burns Department of Microbiology and Immunology, University of Melbourne, Royal Parade, 3052 Parkville, Australia

R Cavicchioli School of Microbiology and Immunology, The University of New South Wales, Sydney, NSW 2052, Australia

DS Clark Department of Chemical Engineering, University of California at Berkeley, 201 Gilman Hall # 1462, Berkeley, CA 94720-1462, USA

A Corcelli Dipartimento di Fisiologia generale ed ambientale, Facoltà di Scienze, Università di Bari, Via Amendola 165/a, 70126 Bari, Italy

DA Cowan   Department of Biotechnology, University of the Western Cape, Bellville 7535, Cape Town, South Africa

PMG Curmi   School of Physics, The University of New South Wales, Sydney, NSW 2052, Australia

MS da Costa   Departamento de Bioquímica, Universidade de Coimbra, Apartado 3126, 3004-517 Coimbra, Portugal

S DasSarma   Center of Marine Biotechnology, University of Maryland Biotechnology Institute, 701 East Pratt Street, Suite 236, Columbus Center, Baltimore, MD 21202, USA

WM de Vos   Laboratory of Microbiology, Wageningen University, Hesselink van Suchtelenweg 4, 6703 CT Wageningen, The Netherlands

JK DiFerdinando   Southeast Applied Research Inc, 1423 Candlewood Drive, Fredericksburg, VA 22407, USA

AJM Driessen   Molecular Microbiology, University of Groningen, Kerklaan 30, 9751 NN Haren, The Netherlands

M Dyall-Smith   Department of Microbiology and Immunology, University of Melbourne, 3052 Parkville, Australia

G Erauso   UMR 6539, Université de Brest, Place Nicolas Copernic, 29280 Plouzané, France

S Fendrihan   Institut für Genetik und allgemeine Biologie, Hellburnnerstrasse 34, A-5020 Salzburg, Austria

D Flament   UMR 6539, Université de Brest, Place Nicolas Copernic, 29280 Plouzané, France

P Forterre   Laboratoire de Biologie Moléculaire du Gène chez les Extrémophiles, Département de Microbiologie Fondamentale et Médicale, Institut Pasteur, 25 rue du Dr Roux, 75015 Paris, France

M Gaillard   UMR 6539, Université de Brest, Place Nicolas Copernic, 29280 Plouzané, France

C Geslin   UMR 6539, Université de Brest, Place Nicolas Copernic, 29280 Plouzané, France

PM Gillevet   Department of Environmental Science and Policy, George Mason University, 10900 University Boulevard, MS 5G8, Manassas, VA 20110, USA

A Godfroy   Laboratoire de Microbiologie et Biotechnologie des Extrémophiles, DRV/VP, IFREMER, Technopole de Brest-Iroise, BP70, 29280 Plouzané, France

F Gómez   Centro de Astrobiologia (CSIC-INTA), Torrejón de Ardoz 28850, Spain

M Gonnet   UMR 6539, Université de Brest, Place Nicolas Copernic, 29280 Plouzané, France

E González-Toril   Centro de Astrobiologia (CSIC-INTA), Torrejón de Ardoz 28850, Spain

WD Grant   Department of Microbiology and Immunology, University of Leicester, Medical Sciences Building, PO Box 138, Leicester, LE1 9HN, UK

C Kato   Research Program for Marine Biology and Ecology, Extremobiosphere Research Center, Japan Agency for Marine Science and Technology, 2–15 Natsushima-cho, Yokosuka 237-0061, Kanagawa, Japan

WN Konings   Molecular Microbiology, University of Groningen, Kerklaan 30, 9752 NN Haren, The Netherlands

P Lamosa   Instituto de Tecnologia Química e Biológica, Universidade Nova de Lisboa, Rua da Quinta Grande 6, Apartado 127, 2780-156 Oeiras, Portugal

F Lefèvre   Protéus SA, 70 allée Graham Bell, Parc George Besse, 30000 Nîmes, France

A Legat   Institut für Genetik und allgemeine Biologie, Hellburnnerstrasse 34, A-5020 Salzburg, Austria

S Leuko   Department of Biological Sciences, Australian Centre for Astrobiology, Macquarie University, Sydney, NSW 2109, Australia

CD Litchfield   Department of Environmental Science and Policy, George Mason University, 10900 University Boulevard, MS 5G8, Manassas, VA 20110, USA

S Lobasso   Dipartimento di Fisiologia generale ed ambientale, Facoltà di Scienze, Università di Bari, Via Amendola 165/a, 70126 Bari, Italy

S Lucas   Laboratoire de Biologie Moléculaire du Gène chez les Extrémophiles, Département de Microbiologie Fondamentale et Médicale, Institut Pasteur, 25 rue du Dr Roux, 75015 Paris, France

LR Lynd   MCB/Biochemical Environmental and Metabolic Engineering, Dartmouth College, Cummings 128D, Hanover, NH 03755, USA

M Malki   Centro de Biología Molecular "Severo Ochoa", Universidad Autónoma de Madrid, Cantoblanco, Madrid 28049, Spain

JM Masson   Protéus SA, 70 allée Graham Bell, Parc George Besse, 30000 Nîmes, France

NM Mesbah Department of Microbiology, The University of Georgia, Biological Sciences Building, Athens, GA 30602-2605, USA

JA Müller Center of Marine Biotechnology, University of Maryland Biotechnology Institute, 701 East Pratt Street, Suite 236, Columbus Center, Baltimore, MD 21202, USA

S Nakagawa Deep-Sea Microorganisms Research Group, Japan Marine Science and Technology Center, 2–15 Natsushima-cho, Yokosuka, 237-0061, Japan

MF Nobre Departamento de Bioquímica, Universidade de Coimbra, Apartado 3126, 3004-517 Coimbra, Portugal

A Oren Department of Plant and Environmental Sciences, The Institute of Life Sciences, The Hebrew University of Jerusalem, Edmond Safra Campus – Givat Ram, 91904 Jerusalem, Israel

CB Park Department of Chemical and Materials Engineering, Arizona State University, Tempe, AZ 85287-6006, USA

NV Pimenov Institute of Microbiology, Russian Academy of Sciences, Prospekt 60 Let Oktyabria 7, bldg. 2, 117811 Moscow, Russia

K Porter Department of Microbiology and Immunology, University of Melbourne, 3052 Parkville, Australia

A Postec Laboratoire de Microbiologie et Biotechnologie des Extrémophiles, DRV/VP, IFREMER, Technopole de Brest-Iroise, BP70, 29280 Plouzané, France

D Prangishvili Biologie Moleculaire du Gene chez les Extrémophiles, Institut Pasteur, 25 rue du Dr. Roux, Cedex 15, 75724 Paris, France

D Prieur UMR 6539, Université de Brest, Place Nicolas Copernic, 29280 Plouzané, France

FA Rainey Department of Biological Sciences, Louisiana State University, 202 Life Sciences Building, Baton Rouge, LA 70803, USA

N Raven Health Protection Agency, Porton Down, Salisbury SP4 0JG, UK

G Ravot Protéus SA, 70 allée Graham Bell, Parc George Besse, 30000 Nîmes, France

FT Robb Center of Marine Biotechnology, University of Maryland Biotechnology Institute, 701 East Pratt Street, Suite 236, Columbus Center, Baltimore, MD 21202, USA

MF Roberts Merkert Chemistry Center, Boston College, 2609 Beacon Street, Chestnut Hill, MA 02467, USA

M Le Romancer UMR 6539, Université de Brest, Place Nicolas Copernic, 29280 Plouzané, France

NJ Russell Department of Agricultural Sciences, Imperial College London, Wye Campus, Ashford, Kent TN25 5AH, UK

H Santos Instituto de Tecnologia Química e Biológica, Universidade Nova de Lisboa, Rua da Quinta Grande 6, Apartado 127, 2780-156 Oeiras, Portugal

RF Shand Department of Biological Sciences, Northern Arizona University, PO Box 5640, Flagstaff, AZ 86011, USA

HD Shukla Center of Marine Biotechnology, University of Maryland Biotechnology Institute, 702 East Pratt Street, Suite 236, Columbus Center, Baltimore, MD 21203, USA

KS Siddiqui School of Microbiology and Immunology, The University of New South Wales, Sydney, NSW 2052, Australia

M Sikaroodi Department of Environmental Science and Policy, George Mason University, 10900 University Boulevard, MS 5G8, Manassas, VA 20110, USA

APL Snijders Biological and Environmental Systems Group, Department of Chemical and Process Engineering, Sheffield University, Sheffield, UK

KR Sowers Center of Marine Biotechnology, University of Maryland Biotechnology Institute, 701 East Pratt Street, Columbus Center, Suite 236, Baltimore, MD 21202, USA

S Spring DSMZ-Deutsche Sammlung von Mikroorganismen und Zellkulturen GmbH, Mascheroder Weg 1b, D-38124 Braunschweig, Germany

H Stan-Lotter Institut für Genetik und allgemeine Biologie, Hellburnnerstrasse 34, A-5020 Salzburg, Austria

K Takai Deep-Sea Microorganisms Research Group, Japan Marine Science and Technology Center, 2–15 Natsushima-cho, Yokosuka, 237-0061, Japan

T Thomas School of Microbiology and Immunology, The University of New South Wales, Sydney, NSW 2052, Australia

MV Tyurin MCB/Biochemical Environmental and Metabolic Engineering, Dartmouth College, Cummings 128D, Hanover, NH 03755, USA

HJG van de Werken Laboratory of Microbiology, Wageningen University, Hesselink van Suchtelenweg 4, 6703 CT Wageningen, The Netherlands

J van der Oost Laboratory of Microbiology, Wageningen University, Hesselink van Suchtelenweg 4, 6703 CT Wageningen, The Netherlands

RH Vreeland Department of Biology, University of West Chester, West Chester, PA 19383, USA

R Wait Kennedy Institute of Rheumatology, Imperial College, London, UK

J Walther Laboratory of Microbiology, Wageningen University, Hesselink van Suchtelenweg 4, 6703 CT Wageningen, The Netherlands

JEM Watts Department of Biological Sciences, Towson University, Towson, MD 21252, USA

J Wiegel Department of Microbiology, The University of Georgia, Biological Sciences Building, Athens, GA 30602-2605, USA

PC Wright Biological and Environmental Systems Group, Department of Chemical and Process Engineering, Sheffield University, Sheffield, UK

JM Zimmerman Department of Biological Sciences, Louisiana State University, 202 Life Sciences Building, Baton Rouge, LA 70803, USA

1 Extremophile Microorganisms and the Methods to Handle Them

Fred A Rainey[1] and Aharon Oren[2]

[1] Department of Biological Sciences, 202 Life Sciences Building, Louisiana State University, Baton Rouge, Louisiana 70803, USA; [2] The Institute of Life Sciences, and the Moshe Shilo Minerva Center for Marine Biogeochemistry, The Hebrew University of Jerusalem, Jerusalem 91904, Israel

◆◆◆

CONTENTS

Introduction
The extremophiles – an overview
The mechanisms of extremophile behavior
Methods for studying extremophilic microorganisms

◆◆◆◆◆ INTRODUCTION

"Indeed, many organisms that live in extreme environments have been unfairly neglected, partly because of the difficulty in studying them and obtaining publishable results. Admittedly, it is trying to study microorganisms whose growth media fills the laboratory with steam, or the centrifuge heads with salt, or which grow so slowly that weeks, instead of hours, may be required for experiments and whose genetics are unknown or almost impossible to study. Those who have persisted have found their rewards, both in the satisfaction and leisure for contemplation available to the student of an out-of-the-way field, and in the fascination afforded by the microorganisms themselves and the very clever ways they have found to adapt to such a wide range of environmental conditions."

These words were written more than a quarter of a century ago by the late Donn Kushner in his introductory chapter to one of the first books entirely devoted to the biology of extremophilic microorganisms (Kushner, 1978). They show both the fascination of the microbiologist who studied organisms inhabiting unusual environments, and the specific problems he had to face while working with such microorganisms. Many of these problems are due to the fact that the typical microbiology laboratory is an environment first and foremost reflecting the ideal working conditions for the scientist (a temperature around 20°C, an ambient pressure of 1 atmosphere, radiation levels far lower than those encountered outdoors, etc.) than the ideal conditions of the extremophilic microorganisms he is studying.

Extremophilic microorganisms have fascinated microbiologists from early times. Ferdinand Cohn devoted extensive discussions to the cyanobacteria in hot springs already in the nineteenth century (Cohn, 1897). The study of microbial life at high temperatures made great advancements thanks to the work of Tom Brock and his coworkers in the hot springs of Yellowstone National Park, USA in the 1960s (Brock, 1978), and the search for life in volcanic environments and undersea hot springs pushed the upper temperature limit known to support life up to values undreamt of when Kushner wrote the introduction to the above-mentioned book. The ability of microorganisms to thrive in highly saline environments has also been known for a long time. Thus, Charles Darwin, in his famous report of his travels on H.M.S. Beagle (1839) wrote in his description of a salt lake in Patagonia: "Parts of the lake seen from a short distance appeared of a reddish colour, and this, perhaps, was owing to some infusorial animalcula." Such red extremely salt-requiring "infusorial animalcula", now known as extremely halophilic Archaea of the family *Halobacteriaceae*, have been isolated and studied since the first decades of the twentieth century. Literature surveys on the early studies of acidophilic, alkaliphilic, psychrophilic, barophilic and other types of extremophilic microorganisms will show that the interest in these thought to be unusual life forms is by no means a phenomenon that started in recent years.

However, the last two decades have shown a surge in the interest in the world of microorganisms that live in extreme conditions. More and more extreme environments that could not be accessed earlier because of technical limitations are now being explored, and this has led to the isolation of a wealth of new organisms. Their study not only shows us how life functions at what we consider environmental extremes, but also teaches us much about the nature of life itself and about the possible properties of the first organisms that colonized planet Earth at a time when conditions were far more extreme than those tolerated by most life forms that presently inhabit our planet. A highly relevant question is whether the existing extremophiles have only recently adapted to the unusual environments in which they live, or whether they may be vestiges of ancient types of organisms that evolved under extreme conditions on the early Earth. The study of the extremophiles has therefore important implications for our views on the origin of life. Lastly, the understanding of the limits of life on Earth, as displayed by the extremophiles, may provide us with clues on the possibility of the existence, now or in the past, of similar life elsewhere in the Universe.

A survey of the numbers of new prokaryote species (Archaea and Bacteria combined) described in the years 2002–2004 shows that in these three years more than 70 new thermophiles were named, more than 40 halophiles, about 20 acido- and alkaliphiles, and more than 40 psychro- and/or barophiles (the exact numbers depending on how the different categories of extremophiles are defined), altogether about 12% of the total of 1523 new species names validated during the period. One of the reasons for the renewed interest in the extremophiles is the possibility to exploit these organisms or processes performed by them in economically valuable processes. Notably the thermophiles have yielded a range of enzymes used in biotechnology (Hough and Danson, 1999). The properties of acidophilic metal-leaching microorganisms were used to recover copper and other metals from metal ores long before the nature of the involvement of the bacteria was understood. A famous example of extremophile-based biotechnology is the *Taq* polymerase used for the amplification of DNA in the polymerase chain reaction. This enzyme obtained from the thermophile *Thermus aquaticus*, isolated from a Yellowstone hot spring (Brock and Freeze, 1969), is by no means the only extremophile protein that has found applications in biotechnology and industry. Other examples are the exploitation of exoenzymes of alkaliphilic bacteria in washing powders and other detergents, the (present and predicted future) use of bacteriorhodopsin, the light-driven proton pump of halobacteria, in a range of applications – from holography to computer memories and processors. A recent success story is the industrial production of ectoine (1,4,5,6-tetrahydro-4-pyrimidine carboxylic acid), an organic compatible solute accumulated by many halophilic members of the domain Bacteria to withstand the osmotic pressure of their highly saline environment. This compatible solute is being used both as a stabilizer of enzyme preparations and as an additive to skin treatment cosmetics (Buenger and Driller, 2004). The term "Superbugs", coined a number of years ago by Horikoshi and Grant (1991) for extremophiles with their unusual properties, signifies both the abilities of these organisms to live in unusual environments and the possibilities for their economic exploitation.

Extremophiles are found in all three domains of life: Archaea, Bacteria and Eukarya. When the Archaea (Archaebacteria) were separated from the Bacteria as the third domain of life by Carl Woese in the late 1970s, the impression was gained that the Archaea is the domain of the extremophiles par excellence, and that the group contains nothing but extremophiles (if one considers the mesophilic methanogens as extremophiles as well on the basis of their high sensitivity toward molecular oxygen). Indeed, the microorganisms known to grow at temperatures above 100°C are all Archaea. Both the Crenarchaeota and the Euryarchaeota phyla contain large numbers of thermophiles. Sequencing of 16S rRNA gene clones obtained from Obsidian Pool, one of Yellowstone's hot springs (74–93°C), has even indicated the presence of a new phylum of Archaea: the "Korarchaeota" (Barns *et al.*, 1996). None of the representatives of this phylum have been brought into culture as yet. A fourth proposed phylum of thermophilic Archaea, the "Nanoarchaeota", consists of thermophiles as well. It is currently represented by a single species, designated "Nanoarchaeum equitans", a very small (about 0.4 µm diameter) organism that lives at 90°C as a parasite on another extreme thermophile, *Ignicoccus* (Crenarchaeota) (Huber *et al.*, 2002; Waters *et al.*, 2003). The great stability of the archaeal lipids, having ether bonds rather than the much more labile ester bonds of bacterial and eukaryal lipids, makes the Archaea highly suitable for a thermophilic mode of life.

However, the concept that the most extremely halophilic microorganisms are all Archaea (*Halobacterium* and other representatives of the family *Halobacteriaceae*) was recently challenged with the discovery of *Salinibacter*, a member of the domain Bacteria, which has a salt requirement and tolerance as high as the most halophilic members of the *Halobacteriaceae* (Antón et al., 2002). Thermophiles are also found in the domain Bacteria, the already mentioned *Thermus aquaticus* being a well-known example. Analyses of 16S rRNA gene sequences recovered from Obsidian Pool at Yellowstone have shown that this extremely hot environment harbours representatives of many groups of Bacteria, including species that belong to novel candidate divisions (phyla) within the Bacteria domain that have as yet no cultured representatives (Hugenholtz et al., 1998). It is now becoming more and more clear that Archaea are also widespread in "conventional", non-extreme environments such as seawater, soils etc., and there is increasing evidence that they may form a substantial fraction of the prokaryote biomass in such environments (Karner et al., 2001). Little is known as yet about the properties of these non-extremophilic Archaea, as they still defy the microbiologists' attempts toward their isolation and characterization.

Interesting extremophiles exist in the domain Eukarya as well. To name just a few well-known examples: unicellular green algae of the genus *Dunaliella* grow in salt-saturated environments such as the Dead Sea and the north arm of Great Salt Lake, Utah; one species of this genus, *Dunaliella acidophila*, while not showing prominent halophilic properties, grows only in the pH range 0–3. It has its optimum at pH 1.0 (Pick, 1999), making it one of the most acidophilic organisms known. Another interesting case is *Cyanidium caldarium*, a unicellular thermoacidophilic alga belonging to the Rhodophyta. Its optimal pH for growth is 2–3, it tolerates pH values as low as 0.2, and grows at temperatures up to 57°C (Seckbach, 1994, 1999).

When discussing extremophilic microorganisms, we should not forget the viruses/bacteriophages. Viruses are known to attack halophilic and thermophilic Archaea, halophilic Bacteria, as well as other extremophiles. Such viruses also need to be adapted to the conditions both in the environment (during the extracellular stage) as well as in the cytoplasm of their host – conditions that are often very different from those outside the cell.

The current interest in the world of extremophilic microorganisms is shown by the establishment of the international society "Extremophiles" in 2002 (see www.extremophiles.org), a series of biannual symposia on extremophilic microorganisms from 1996 onwards, the most recent of which were held in Hamburg-Harburg, Germany (2000), Naples, Italy (2002) and Cambridge, Maryland (2004), and specialized international congresses on thermophiles (most recently in Exeter, UK in 2003 and in Gold Coast, Queensland, Australia, 2005) and halophiles (the last held in Ljubljana, Slovenia in 2004). The journal "Extremophiles"

(Springer-Verlag), established in 1997, is dedicated to the publication of articles on all aspects of the biology of microbial life in extreme environments.

◆◆◆◆◆ THE EXTREMOPHILES – AN OVERVIEW

An in-depth review of the world of extremophilic microorganisms and their properties is outside the scope of this volume, which is primarily devoted to practical aspects related to the methods on how to handle the known extremophiles and possibly, how to discover and characterize novel types. The overview given below is thus intended as a general orientation within the diverse world of the extremophilic microorganisms. More information on the properties of the extremophiles can be found in the many books that have been devoted to their biology. Among the multi-author reviews that deal with diverse groups of extremophiles we can name those written or edited by da Costa et al. (1989), Gould and Corry (1980), Heinrich (1976), Herbert and Codd (1986), Horikoshi and Grant (1991, 1998), Kushner (1978), Seckbach (1999, 2000), Shilo (1979), and Wharton (2002). Books specializing in thermophile microbiology include those by Brock (1978), Kristjansson (1991), Reysenbach et al. (2001), and Wiegel and Adams (1998). Monographs on halophiles were presented by Gunde-Cimerman et al. (2005), Oren (2002), Rodriguez-Valera (1988, 1991), Ventosa (2004), and Vreeland and Hochstein (1993). The alkaliphiles were featured in by book by Horikoshi (1999), and the psychrophiles in the monograph on Antarctic microbiology by Vincent (1988). A few useful recent review articles in which different aspects of the life of extremophiles are discussed are those by Kristjansson and Hreggvidsson (1995), Madigan and Marrs (1997), Madigan and Oren (1999), and Rothschild and Mancinelli (2001).

The presently recognized boundaries of environmental parameters that still support life are given in Table 1.1. This table shows our current "Extemophile Book of Records" entries with respect to life at high temperatures, pressures, extremes of pH, salt concentration and ionizing radiation. Many of these extremophiles cannot live under the more moderate conditions preferred by most living organisms. Examples of such specialists with a low level of adaptability to changing environmental conditions are the acidophile *Picrophilus*, which dies after even a short exposure to a pH higher than 4, and the halophile *Halobacterium salinarum*, which lyses as soon as the salt concentration in its surroundings decreases below 100–150 g l^{-1}. Other extremophiles show a much greater flexibility and adaptability to changes in their environment. Good examples are representatives of the halophilic genera *Halomonas* and *Chromohalobacter* (γ-Proteobacteria), which can adapt to life within a wide range of salt concentrations, from near-zero to close to saturation (Oren, 2002; Ventosa et al., 1998).

Even more impressive are those microorganisms that are adapted to more than one form of environmental stress. Such "polyextremophilic"

Table 1.1 "Record-holding" extremophiles and their environmental limits. The table is based on data presented by Ferreira et al. (1997), Madigan (2000), and Seckbach and Oren (2004)

Environmental factor			Organism	Habitat	Phylogenetic affiliation	Tolerance to stress
Temperature		High	*Pyrolobus fumarii*	Hot undersea hydrothermal vents	Crenarchaeota	Maximum 113°C, Optimum 106°C, Minimum 90°C
		Low	*Polaromonas vacuolata*	Sea-ice	Bacteria	Minimum 0°C, Optimum 4°C, Maximum 12°C
pH		Low	*Picrophilus oshimae*	Acidic hot springs	Euryarchaeota	Minimum pH −0.06, Optimum pH 0.7, Maximum pH 4 (is also thermophilic)
		High	*Natronobacterium gregoryi*	Soda lakes	Euryarchaeota	Maximum pH 12, Optimum pH 10, Minimum pH 8.5 (is also halophilic)
Hydrostatic pressure			Strain MT41	Mariana Trench	Bacteria	Maximum >100 MPa, Optimum 70 MPa, Minimum 50 Mpa
Salt concentration			*Halobacterium salinarum*	Salt lakes, salted hides, salted fish	Euryarchaeota	Maximum NaCl saturation, Optimum 250 g l^{-1} salt, Minimum 150 g l^{-1} salt
Ultraviolet and ionizing radiation			*Deinococcus radiodurans*	Isolated from ground meat, radioactive waste, nasal secretion of elephant; true habitats unknown	Bacteria	Resistant to 1.5 kGy gamma radiation and to 1500 J m^{-2} of ultraviolet radiation

organisms, as they were termed by Rothschild and Mancinelli (2001), are quite common, as combinations of different forms of stress are often found in nature. Frequently found combinations are hot + acidic (volcanic areas, hot springs), saline + alkaline (soda lakes), and cold + high pressure (the deep sea). Other combinations are hot + high pressure (deep sea hydrothermal vents) and hot + high pH (certain hot springs). Table 1.2 presents a few examples of such polyextremophiles.

Table I.2 Some examples of "polyextremophilic" microorganisms, adapted to life at a combination of several environmental extremes. Derived from Seckbach and Oren (2004)

Environmental factors	Organism	Habitat	Phylogenetic affiliation	Tolerance to stress
Low pH, high temperature	*Picrophilus oshimae*	Acidic hot springs	Euryarchaeota	Minimum pH –0.06; Maximum temperature 65°C
	Sulfurisphaera	Acid hot spring	Crenarchaeota	Maximum temperature 92°C; Minimum pH 1
	Cyanidium caldarium	Acidic hot springs	Eukarya	Maximum temperature 57°C; Minimum pH 0.2
High pH, high salt concentration	*Natronobacterium gregoryi*	Soda lakes	Archaea - Euryarchaeota	Maximum pH 12; NaCl saturation
High temperature, high pH	*Thermococcus alcaliphilus*	Shallow marine hydrothermal springs	Archaea - Euryarchaeota	Maximum temperature 90°C; Maximum pH 10.5
High temperature, high pressure	*Thermococcus barophilus*	Thermal vent, Mid-Atlantic Ridge	Archaea - Euryarchaeota	Maximum temperature 100°C; Requires 15–17.5 MPa pressure at the highest temperatures
High temperature, high radiation	*Deinococcus geothermalis*	Thermal springs	Bacteria	Maximum temperature 50°C. Resistant to at least 10 kGy gamma radiation
Low temperature, high pressure	All deep-sea bacteria	Deep sea environments	Bacteria	Live at 2–4°C and pressures of 50–110 Mpa

Thermal Environments and Thermophiles

There are different types of thermal environments on Earth that are populated by microorganisms adapted to life at high temperatures. Many of these environments are associated with volcanic activity. Hot springs, geysers, volcanoes and deep-sea hydrothermal vents have all been popular hunting grounds for the search for new types of thermophilic microorganisms. However, thermophiles can also be found in environments such as compost heaps and coal refuse piles, where the high temperature is entirely due to biological activity. Additional thermal environments include the human made domestic and industrial hot water systems and industrial processes at high temperature (e.g. paper processing machinery and food processing facilities). An environment still poorly explored for the presence of prokaryotic thermophilic life is the deep subsurface of planet Earth (Gold, 1992).

As temperature increases, we find progressively fewer groups of organisms that can cope with the stress of high temperatures. The upper limit for eukaryotic life (fungi, algae, protozoa) is about 60°C. The existence of truly thermophilic bacteria was firmly established during Tom Brock's pioneering studies of the hot springs of Yellowstone National Park in the 1960s. A wide variety of microorganisms adapted to life at high temperatures was discovered, including phototrophs (up to about 72–73°C) and heterotrophs (up to 91°C, the temperature at which water boils at the elevation of Yellowstone) (Brock, 1978). At the time, Brock postulated that life may exist as long as there is liquid water. Indeed, the search for thermophiles in undersea volcanic areas in which the boiling point of water exceeds 100°C has shown a wealth of organisms, all belonging to the domain Archaea, that can grow at temperatures higher than those of the boiling point of water at normal atmospheric pressure (Stetter, 1996). Even more challenging environments to search for hyperthermophiles are the deep-sea hydrothermal vents: the anoxic, sulfide and mineral-loaded waters emitted by the so-called "black smoker" chimneys can reach temperatures up to about 350°C.

Thus far the most heat-tolerant, and also most heat-requiring of all documented species of prokaryotes is probably *Pyrolobus fumarii*, a crenarchaeote isolated from the wall of a deep-sea black smoker, which does not grow at temperatures below 90°C, has its optimum at 106°C, and will still grow, albeit at a low rate, at 113°C (Blöchl *et al.*, 1997). There are even reports of an iron-reducing archaeon, not yet named and still incompletely characterized and known only as "Strain 121", that reportedly grows at 121°C, the temperature within an autoclave (Kashefi and Lovley, 2003). It is still unknown what the true upper temperature of life is; it is probably dictated by the limited stability of DNA, RNA, proteins, and other biological molecules.

Molecular ecological studies, based on the sequencing of 16S rRNA genes amplified from the thermal environments of Yellowstone National Park and elsewhere have clearly shown that the true diversity of thermophiles in such high temperature environments is much greater than the diversity of the cultured species. Such thermal areas harbour as yet uncultured organisms that according to their phylogenetic position belong to new families, orders, classes, and even kingdoms that are not yet represented by any cultured species (Barns *et al.*, 1996; Hugenholtz *et al.*, 1998).

Cold Environments and Psychrophiles

Cold environments are not at all rare on our planet. Most of the deeper waters of the oceans that cover more than two-thirds of our planet have temperatures below 5°C. Low temperature environments are the rule in the Arctics and the Antarctics, and large areas of Siberia-Russia, Canada and other northern countries are permanently frozen ("permafrost").

Psychrophilic (cold-loving and even cold-requiring) and psychrotolerant microorganisms have been isolated from all these environments. True psychrophiles are generally defined as such organisms that have their optimum temperature below 15–20°C. The most psychrophilic microorganism described in the literature is probably *Polaromonas vacuolata*, a member of the γ-Proteobacteria isolated from Antarctic marine waters: it grows optimally at 4°C, with a minimum at 0°C and a maximum of 12°C (Irgens *et al.*, 1996). Recent investigations on the microbial community in Antarctic soils have shown that extremely cold environments may harbour surprisingly dense microbial communities: based on the ATP content of the community, between 3×10^6 and 2×10^9 live bacterial cells per gram soil were inferred to be present in soils varying in temperature between -0.5 and $+3.8°C$ (Cowan *et al.*, 2002). Measurements of heterotrophic microbial activity in permafrost soils, based on the assessment of incorporation of ^{14}C-labeled acetate into cellular lipids, estimated doubling times of the community were in the order of 1 day at 5°C, 20 days at $-10°C$, and even at $-20°C$ some activity could be detected (estimated doubling time 160 days) (Rivkina *et al.*, 2000).

Activity of even the most psychrophilic microorganisms depends on the presence of liquid water within the cell: when the cytoplasm becomes frozen, the cell may retain its viability, but metabolic activity is no longer possible. The lowest temperature boundary enabling growth is therefore determined by the freezing point of the intracellular water – a temperature that can be lowered to some extent by the accumulation of organic and inorganic solutes that act as "antifreeze" agents.

Until recently, the world of psychrophilic microorganisms was relatively neglected in comparison with the thermophiles. However, the large number of new species of psychophiles/barophiles described in the past few years shows a renewed interest in the study of life at the lowest possible temperatures.

Acidic Environments and Acidophiles

Environments with extremely low pH values are not very abundant. Acidic environments are often associated with volcanic activity: hot sulfur springs, mud pots, etc. Another type of low pH environments is that caused by microbial activity. Chemoautrophic oxidation of sulfide, elemental sulfur, and other reduced sulfur compounds causes the formation of sulfuric acid which acidifies the medium, and many of the organisms that perform the process have their growth optimum in the low pH range. In mining areas in which sulfur-containing ores are brought to the surface, oxidation of pyrite by chemoautotrophic bacteria often leads to the formation of highly acidic, metal ion-loaded wastewaters. Acidification by such chemoauotrophic sulfur bacteria is exploited in mining operations for the recovery of copper and other precious metals from sulfur-containing ores. Fermentations such as the lactic acid fermentation may also decrease the pH of soils and aquatic systems from which oxygen is excluded. However, the lowest pH levels reached are generally much above those found in environments in which the low pH had been casued by chemoautotrophic activity.

Acidophilic microorganisms can be found in all three domains of life. *Acidithiobacillus thiooxidans* (Bacteria) cannot grow at pH values above 4–6, and growth is still possible down to pH 0.5, (Kelly and Wood, 2000). The pH minimum of the aerobic heterotroph *Picrophilus* species (Euryarchaeota) is even lower: -0.06. With its growth optimum at pH 0.5, it is the most acidophilic organism known. In addition it has thermophilic properties (optimum growth temperature: 60°C). Above pH 4 lysis occurs (Schleper *et al.*, 1995). Also among the Eukarya, acidophiles are found. Thus, *Dunaliella acidophila*, a unicellular green alga isolated from highly acidic waters and soils, is one of the most acidophilic organisms known, as it grows between pH 0 and 3 with an optimum at pH 1 (Pick, 1999). Another extremely acidophilic alga (optimum pH 2–3, minimum pH 0.2) with thermophilic properties as well (optimum 57°C) is *Cyanidium caldarium* (Rhodophyta) (Seckbach, 1994, 1999). A recent culture-independent study of rocks in a Yellowstone National Park hot spring demonstrated a wide diversity of endolithic microorganisms in rock pore water that had a pH of 1.0 (Walker *et al.*, 2005). These included 16S rRNA gene sequences that fell within the radiation of the genus *Mycobacterium*, a group that to date was not considered to be extremophilic in nature. Such studies demonstrate the extensive diversity of extremophiles still to be cultured and their biotechnological potential evaluated.

Acidophiles keep their cytoplasm at near-neutral pH values by means of powerful proton pumps in their cytoplasmic membrane, which has to maintain a proton concentration gradient of four, five, or even more orders of magnitude. The intracellular enzymatic machinery therefore does not require any special adaptations to low pH, and the ability of the cell to withstand (and often require) low pH values is determined by the properties of the cell envelope that separates between the highly acidic environment and the intracellular compartment.

Basic Environments and Alkaliphiles

Alkaliphilic microorganisms are widespread in nature. Bacteria (e.g. species of the genus *Bacillus*) that grow at pH 9–10 while being unable to grow at neutral pH can easily be isolated from soils. High pH conditions are common in the surface layers of productive freshwater lakes when CO_2 becomes depleted during daytime photosynthesis, and many planktonic eukaryotic algae as well as cyanobacteria are adapted to an existence at high pH, at least temporarily. However, most true alkaliphiles known have been isolated from permanently alkaline lakes ("soda lakes") in which the high pH is caused by geological-geochemical rather than by biological processes. Such lakes are found on all continents, examples being Mono Lake, California, Lake Magadi and other East African soda lakes, the Wadi Natrun lakes in Egypt, and soda lakes in China and Tibet. Many of these alkaline lakes are characterized by high salt concentrations as well, so that their inhabitants should have alkaliphilic as well as halophilic properties. Part of the described members of the *Halobacteriaceae*, the family of halophiles par excellence, are obligatory alkaliphilic as well (Oren, 2002). Alkaliphiles are thus found in each of the three domains of life: Archaea, Bacteria, and Eukarya. Alkaliphilic behavior is also not limited to cells with a particular mode of life, so that even the most extremely alkaline lakes appear to support complete cycling of carbon, nitrogen and sulfur (Zavarzin and Zhilina, 2000; Zhilina and Zavarzin, 1994).

Similar to the acidophiles discussed above, the alkaliphiles keep the stressful extreme pH of their surrounding medium out of their cells, and they maintain an intracellular pH close to neutrality. Bioenergetically this is far more problematic than in the case of the acidophiles, as energy transformation in the cell is commonly based on the outward pumping of protons, and the "natural" direction of the pH gradient over the cell membrane is therefore acidic outside and alkaline inside. The alkaliphiles thus maintain an inverted pH gradient. Membrane-linked bioenergetic processes are then based on an exceptionally large membrane potential and/or processes that are linked with the transmembrane transport of sodium ions rather than protons.

Hypersaline Environments and Halophiles

The seas that cover nearly 70% of the surface of planet Earth contain about 35 g l^{-1} dissolved salt. Hypersaline environments are easily formed when seawater dries up in coastal lagoons and salt marshes, as well as in manmade evaporation ponds of saltern systems built to produce common salt by evaporation of seawater. There are also inland saline lakes in which the salt concentration can reach values close to saturation. Well-known examples are the Great Salt Lake, Utah – a lake in which the ionic composition of the salts resembles that of seawater, and the Dead Sea on the border between Israel and Jordan – a lake dominated by magnesium rather than by sodium as the most abundant cation. Furthermore, there are extensive underground deposits of rock salt that originated by the drying up of closed marine basins. All of these environments, as well as others such as saline soils, provide a habitat for salt-adapted microorganisms, obligate halophiles as well as halotolerant types, that can adjust to life over a wide range of salt concentrations.

The pink to red-purple color that characterizes brines approaching NaCl saturation provides direct evidence of the massive numbers of pigmented halophilic microorganisms often found in such environments.

Halophiles adapted to life at salt concentrations up to saturation are found in all three domains of life. *Dunaliella*, a genus of unicellular green algae, is found worldwide in salt lakes and salterns. The archaeal order *Halobacteriales* is entirely composed of highly salt-requiring species, most of which are colored red due to carotenoid pigments as well as by retinal pigments (bacteriorhodopsin, halorhodopsin) (Oren, 2000, 2002). In addition, many phylogenetic branches of the domain Bacteria contain halophilic or highly halotolerant representatives, and these inhabit a variety of hypersaline environments. Physiologically the halophilic world is highly diverse, as it encompasses aerobic and anaerobic heterotrophs, fermentative organisms, sulfate reducers, cyanobacteria, as well as anoxygenic photosynthetic sulfur bacteria (*Halochromatium*, *Halorhodospira*) (Oren, 2002). The recent discovery of *Salinibacter*, a red bacterium affiliated with the *Cytophaga/Flavobacterium/Bacteroides* branch of the Bacteria (Antón et al., 2002), shows that adaptation to the highest salt concentrations is not limited to the archaeal domain.

Nature has devised more than one strategy for microorganisms to cope with high salt concentrations in their environment. While all microbes maintain a cytoplasm that is osmotically at least equivalent to the saline medium outside the cell, the ways osmotic equilibrium is achieved differ greatly among the diverse groups of halophiles. Some (the *Halobacteriales*, *Salinibacter*, members of the *Halanaerobiales*) balance the salt outside with high ionic concentrations, mainly KCl) within the cytoplasm. Such organisms have adapted their intracellular enzymatic machinery to function in high salt. Their proteins are functional at high salt, and generally require salt for activity and stability. Others (*Dunaliella* and other eukaryotes, most halophilic Bacteria, halophilic methanogenic Archaea) exclude salt from their cytoplasm to a large extent, and instead accumulate organic osmotic solutes.

The Deep Sea and Barophiles/Piezophiles

The hydrostatic pressure in the sea increases by about 1 atmosphere (0.1 MPa) every 10 meters. The mean depth of the oceans is about 4 km, equivalent to a pressure of 400 atmospheres or 40 MPa, and the deepest parts of the oceans are more than 10 km deep. Microorganisms living in such environments have to withstand pressures of over a thousand atmospheres.

Some microbial isolates from the deep sea (obligate barophiles or piezophiles) cannot live at normal atmospheric pressure, and these require a pressurized environment for growth. Bacterial isolates related to *Shewanella* and *Moritella* (γ-Proteobacteria), recovered from a depth of up to 11 km in the Mariana Trench, the Japan Trench and the Philippine Trench, require pressures of 70–80 MPa for optimal growth and do not grow below 50 MPa (Kato and Bartlett, 1997; Kato et al., 1998). Strain MT41, isolated by Yayanos et al. (1981) from the Mariana Trench, is probably the most barophilic of all isolates characterized to date: it does not grow at pressures lower than 50 MPa, has its optimum at 70 MPa, and tolerates pressures of at least 100 MPa (see also Yayanos, 2000).

As the deep sea is also a cold environment, barophiles generally exhibit psychrophilic properties as well. Thermophilic barophilic or barotolerant bacteria are associated with deep-sea hot vents. *Thermococcus barophilus*, a hyperthermophile isolated from a Mid-Atlantic Ridge hydrothermal vent, requires high pressure when grown at the highest temperatures, and can then grow up to 100°C and higher (Marteinsson et al., 1999).

Due to methodological problems related to the cultivation of barophiles in the laboratory, our understanding of the molecular mechanisms underlying the barotolerant and/or barophilic behavior of microorganisms lags behind our understanding of most other forms of stress to which the different types of extremophiles are exposed.

Radiation-Resistant Microorganisms

Microorganisms found on terrestrial surfaces, air-borne microorganisms, and microbes living in the upper layers of the sea and other aquatic environments are exposed to direct sunlight, including a significant amount of potentially harmful ultraviolet radiation. Such organisms have to protect themselves against radiation damage. Protection mechanisms include repair mechanisms for damaged DNA, but also screening of the radiation by means of pigments, preventing the harmful radiation from reaching the cell's DNA and other radiation-sensitive targets. Microorganisms exposed to high levels of light are generally pigmented by carotenoids and other pigments absorbing in the 400–500 nm wavelength range (Rothschild, 1999). UV-absorbing pigments are also found. Among the cyanobacteria, we find species that accumulate the UV-absorbing compound scytonemin within their extracellular sheath, effectively screening the radiation. Another type of UV-protecting compound widespread among the cyanobacteria is the mycosporine-like amino acids, UV-absorbing substances that in most cases are found dissolved in the cell's cytoplasm (Castenholz and Garcia-Pichel, 2000).

Some microorganisms can tolerate surprisingly high levels of gamma ionizing radiation. *Deinococcus radiodurans*, an organism phylogenetically belonging to a deep branch in the Bacteria domain, retains viability after exposure to >25 kGy (2.5 Mrad) of gamma radiation. For comparison, exposure to a 5 Gy dose of gamma ionizing radiation is lethal for humans. *D. radiodurans* was initially isolated from canned meat that had been irradiated (Anderson et al., 1956). Subsequently a number of strains of *D. radiodurans* have been reported from diverse sources including sawdust (Ito, 1977) and sediments contaminated with nuclear waste (Fredrickson et al., 2004). The genus *Deinococcus* is by no means the only ionizing radiation resistant organism known but along with the species of the genus *Rubrobacter*, has been shown to survive the highest levels of gamma radiation (Battista, 1997; Carreto et al., 1996; Ferreira et al., 2000; Yoshinaka et al., 1973). Other members of the domain Bacteria that show ionizing radiation resistance but to a lesser degree include some species of the genera *Acinetobacter* (Nishimura *et al.*, 1988), *Chroococcidiopsis* (Billi *et al.*, 2000), *Hymenobacter* (Collins *et al.*, 2000), *Kineococcus* (Phillips *et al.*, 2002), *Kocuria* (Brooks and Murray, 1981), and *Methylobacterium* (Green and Bousfield, 1983; Ito and Iizuka, 1971). Gamma ionizing radiation resistance has also been found in a number of archaeal species including *Pyrococcus furiosus* (DiRuggiero *et al.*, 1997), "Pyrococcus abyssi" (Jolivet *et al.*, 2003a), *Thermococcus gammatolerans* (Jolivet *et al.*, 2003b), *Thermococcus marinus* (Jolivet *et al.*, 2004), and *Thermococcus radiotolerans* (Jolivet *et al.*, 2004).

Why these prokaryotes are resistant to elevated doses of ionizing radiation is unclear, considering that natural sources of radiation on Earth emit at very low levels. Studies on the desiccation resistance of ionizing radiation resistant organisms suggest that the ability of microorganisms to repair their DNA might be a response to DNA damage caused by prolonged desiccation rather than ionizing radiation (Mattimore and Battista, 1996). Environmental evidence would seem to support this idea, considering that higher proportions of ionizing radiation resistant bacteria have been found in arid soils as compared to soils from less arid regions (Rainey *et al.*, 2005). Not only are higher proportions of ionizing radiation resistant bacteria found in these arid soils but the genetic diversity of these organisms is also extensive, as demonstrated by the isolation of nine novel species of the genus *Deinococcus* from a single desert soil sample (Rainey *et al.*, 2005).

The mechanisms that enable *Deinococcus* to survive these extremely high radiation levels are only now becoming clear. A number of ideas to explain the high level of ionizing radiation resistance have been put forward. These include the ability of deinococci to rapidly repair damaged DNA using the normal DNA repair enzymes found in all prokaryotes, the possession of a novel set of DNA repair systems, and use of the multiple chromosomes for homologous recombination (Battista *et al.*, 1999). Recently a special way of packaging of the DNA, that facilitates repair not only of single-strand breaks but of double-strand breaks as well, has been described, and this appears to be one of the basic features in which *Deinococcus* differs from all other cells (Englander *et al.*, 2004). It is interesting to note that this unusual mode of DNA packaging seen in *Deinococcus* is also present in dormant endospores of *Bacillus*, which are extremely desiccation-resistant as well, providing addition support for the link between ionizing radiation and desiccation resistance (Frenkiel-Krispin *et al.*, 2004).

◆◆◆◆◆ THE MECHANISMS OF EXTREMOPHILE BEHAVIOR

An in-depth discussion of the mechanisms that enable the different categories of extremophilic microorganisms to survive and to grow in environments apparently hostile to life is outside the scope of this book; extensive information about this topic can be found in monographs, multi-author review books and review articles cited above. One general aspect, however, deserves to be discussed here, and that is the issue of the nature of the intracellular environment of the cell. In some cases, the intracellular environment is not at all exposed to the stress factor that makes the environment outside the cell so extreme. This is due to the properties of the cell membrane and the activity of energy-dependent processes that allow homeostasis of the intracellular milieu at a level that is far less stressful than the medium outside the cell. In other cases, the cells are unable to exclude the stressful factor from their cytoplasm, and all intracellular components have therefore to be functional at the environmental extremes. Which of these two strategies is used depends on the nature of the stress factor involved and sometimes on the type of organism as well. In the case of temperature (low as well as high) and hydrostatic pressure, no microorganism is able to regulate the intracellular values at a level different from that of their environment. All enzymes of a hyperthermophile should be active at the ambient temperature to enable growth of the cell, and the same is true for a psychrophile that grows in the cold deep sea or in Arctic or Antarctic environment. On the other hand, none of the known acidophiles and alkaliphiles allows the intracellular pH to equilibrate with the low or high extracellular pH, respectively, and intracellular pH is always maintained at a value close to neutrality. A very low permeability of the membrane to protons and absence of uncontrolled proton movement are first prerequisites for any acidophile or alkaliphile. In fact these properties are not at all unusual: a low proton permeability of the membrane is a basic property for any cell, as its bioenergetic processes are based on the generation and exploitation of transmembrane proton gradients. The only components of the cell that thus require special adaptation to the environmental stress are the cytoplasmic membrane, periplasmic proteins, and (if present) exoenzymes excreted from the cell.

The halophiles present us with an interesting case of two fundamentally different strategies that microorganisms have developed to allow life at the extremely high osmotic pressures exerted by salt-saturated brines. No halophilic microorganism can maintain a dilute cytoplasm because of the high permeability of the cytoplasmic membrane to water: a cell with a cytoplasm of high water activity would immediately lose water and dry out. One strategy is based on the accumulation of salts inside the cells at concentrations no less than those in the outside medium, the other on exclusion of salts from the cytoplasm to a large extent and production or accumulation of organic solutes to balance the osmotic pressure of the medium.

The "salt-in" strategy, in which molar concentrations of potassium chloride are accumulated intracellularly, is used by a few specialized groups of halophiles only. We find this mode of life in the extremely halophilic Archaea of the family *Halobacteriaceae* (Lanyi, 1974), but also in a few groups of Bacteria. One recently discovered example is *Salinibacter*, a red extremely halophilic representative of the *Cytophaga/ Flavobacterium/Bacteroides* phylum (Antón et al., 2002; Oren et al., 2002). Other Bacteria that use salt rather than organic solutes to provide osmotic balance are found within the order *Halanaerobiales*, a group of obligatory anaerobic, mostly fermentative halophiles (Oren *et al.*, 1997). The entire intracellular machinery of such organisms has to be functional in the presence of molar concentrations of salt, and indeed, the proteins of *Halobacterium* and other halophiles that use the "salt-in" strategy show unusual properties when compared with their non-halophilic counterparts. They generally show a very high excess of acidic amino acids (Glu, Asp) over basic amino acids (Lys, Arg). In addition, their content of hydrophobic amino acids is relatively low. Such proteins not only are soluble and functional at high salt, they even require molar concentrations of salt for activity and stability. Lack of flexibility and adaptability to changing salt concentrations is therefore the price these organisms have to pay for their ability to grow at the very highest salt concentrations (Dennis and Shimmin, 1997; Lanyi, 1974; Mevarech *et al.*, 2000; Oren, 2000, 2002).

The second strategy that enables microorganisms to grow at high salt concentrations, in some cases up to NaCl saturation, is the use of organic solutes ("osmotic solutes" or "compatible solutes") that are synthesized by the cell or can in some cases be taken up from the medium. This mechanism of osmotic adaptation is used by many groups of microorganisms, and can be found in all three domains of life. The unicellular green alga *Dunaliella*, an inhabitant of salt lakes worldwide, uses glycerol to provide osmotic balance. A wide variety of organic solutes are used for this same purpose by halophilic and halotolerant representatives of the Bacteria. The list includes compounds such as ectoine, glycine betaine, simple sugars, certain amino acids or amino acid derivatives, and others. The "organic solute in" strategy is also found in the archaeal domain: the halophiles among the methanogens accumulate organic solutes rather than salt in their cytoplasm. The use of organic compatible solutes, all highly soluble, uncharged or zwitterionic compounds, allows a high degree of flexibility to the organism: no special adaptation of the intracellular proteins is necessary to function in the presence of high concentrations of these solutes. Therefore, many of these halophilic or halotolerant organisms can grow over a wide range of salt concentrations by regulating their intracellular concentrations of the organic solutes (Oren, 2000; Ventosa *et al.*, 1998). In such cells, only the membrane proteins exposed to the outer medium and any extracellular enzymes excreted into the medium need to be functional at high salt. The highly diverse world of the halophiles thus shows that the problem how to cope with the environmental stress can in some cases be solved in different ways.

A similar case in which resistance to environmental stress can be achieved in multiple ways is ultraviolet radiation. Some cyanobacteria are highly UV radiation-resistant thanks to the presence of the UV-absorbing pigment scytonemin in the sheath that surrounds the cell. Others accumulate UV-absorbing pigments (mycosporine-like amino acids) within their cytoplasm, and thus protect their DNA from radiation damage (Castenholz and Garcia-Pichel, 2000). The prodigious resistance of *Deinococcus radiodurans* to radioactive and other forms of ionizing radiation cannot be explained by such protective compounds, and here the explanation must rather be sought in effective repair mechanisms as well as special ways of packaging of the DNA (Englander *et al.*, 2004). In addition, resistance to desiccation is probably key to the survival of ionizing radiation resistant life forms. Many of these organisms also produce extracellular polysaccharides as a protection against extreme dryness.

The limits of existence of extremophiles are in many cases determined by the properties of their cytoplasmic membrane. A low passive permeability to protons and other ions, as well as a suitable level of fluidity, are prerequisites for the proper functioning of any cell. The comparative studies by van de Vossenberg *et al.* (1998) have convincingly shown that the maximum growth temperature of any species is that temperature at which the membrane becomes too permeable and loses its ability to maintain a proton gradient. The proton permeability of membranes was found to be kept constant at low values, independent of the optimum growth temperature of the organism investigated and of the type of lipids (archaeal or bacterial), over the whole range of temperatures that support life, from psychrophiles to extreme thermophiles. This may be the reason why the most thermophilic microorganisms are all Archaea: the archaeal lipids with their ether bonds are far more stable than the ester bond-based bacterial and eukaryal lipids, and additional stability is achieved in those Archaea that possess a monolayer membrane in which the glycerol moieties at both sides of the membrane are bridged by covalent bonds, this in contrast to the conventional bilayer membranes in which stability is limited by the weak hydrophobic interactions between the two layers. In Bacteria that have a "normal" type of membrane with aliphatic fatty acid chains connected to glycerol via ether bonds, the fluidity of the membrane can be regulated to some extent by the length of the hydrophobic chains, by introduction of double bonds, and by other structural modifications. Thus it was shown that *Polaromonas vacuolata*, one of the best "cold-adapted" microorganisms known, may contain up to 74–79% of 16:1 ω7c, the highest level of this fatty acid reported in any species, with in addition 7–9% of 18:1. Under these conditions, it contains no more than 14–17% of saturated fatty acids (16:0) (Irgens *et al.*, 1996). Polyunsaturated fatty acids are more or less a rarity in the prokaryote world, but they are found in some psychrophilic (as well as barophilic) representatives of the domain Bacteria. Thus, *Psychroflexus*, isolated from Antarctic sea ice, possess fatty acids with four and even five double bonds (Bowman *et al.*, 1998).

◆◆◆◆◆ METHODS FOR STUDYING EXTREMOPHILIC MICROORGANISMS

The quotation from Donn Kushner's 1978 review that opened this chapter points to the fact that work with extremophilic microorganisms often requires special techniques. Standard methods for the handling of non-extremophiles need to be adapted or modified for their application to the study of extremophiles. In many cases, altogether new methods have to be developed to enable the cultivation and the study of thermophiles, barophiles, psychrophiles, and other types of extremophiles. It is the goal of this book to provide the community of microbiologists who already are active in the field, as well as those who are as yet unfamiliar with the specific problems presented by the extremophilic microorganisms, with guidelines and solutions of common problems, including detailed protocols of procedures that have been found useful in the laboratories of the experts in the field extremophile biology.

Most of the chapters in this book deal with the isolation, cultivation and handling of live cultures of the different groups of extremophiles, and relatively few contributions are devoted to methods dealing with genomics, proteomics, and other molecular techniques. The reason for this is obvious: the approaches based on genomics and proteomics indeed contribute an ever increasing amount of useful information on the extremophilic microorganisms and their adaptations to environmental stress, but the techniques used in these studies are in most cases no different from the current standard techniques used for handling genomes and proteomes of non-extremophiles. It is the isolation and cultivation of extremophiles that presents the true challenges. Many of the dominant types of microbes that inhabit extreme environments still defy the microbiologists' attempts to isolate them. However, breakthroughs do occur from time to time. The recent isolation of the flat square extremely halophilic Archaea that dominate the heterotrophic microbial community in saltern crystallizer ponds worldwide, achieved simultaneously by two groups (Bolhuis *et al.*, 2004; Burns *et al.*, 2004), shows that with the proper methods, it is indeed possible to bring ecologically relevant organisms into culture and study them. Genetic manipulation of extremophiles is covered in detail in several chapters in this book. Genetic systems have been developed for a relatively small number of extremophiles, and each of these poses new and unexpected challenges to the scientist. Solutions have been found in many cases, and we hope that the approaches presented in the following chapters will help others in the development of methods to genetically study and manipulate other extremophilic microorganisms.

Why is one of the chapters in this volume devoted to the handling of obligate anaerobes? We, the editors of this book, do not consider anaerobes in general to be extremophiles. The opposite is true to some extent: aerobes all have to cope with the presence of highly toxic derivatives of oxygen: peroxides, superoxides, and hydroxyl radicals. However, the laboratory environment contains oxygen, and many extremophiles, especially many of the hyperthermophilic Archaea, are obligate anaerobes with a high sensitivity to molecular oxygen. Therefore some guidelines on the handling of oxygen-intolerant anaerobes may be useful in combination with the more specific methodological problems presented by the anaerobic extremophiles.

While we have attempted to present as broad as possible, a treatise on methods useful in the study of extremophilic microorganisms, we are aware that this book is by no means the only resource in which information on extremophile methodology can be found. An earlier volume in this series "Methods in Microbiology", devoted to marine microbiology, contained chapters on the handling of marine psychrophiles (Bowman, 2001), marine piezophiles (Yayanos, 2001), and microbes from submarine hydrothermal vents (Reysenbach and Götz, 2001). These chapters in this earlier volume complement the present volume. Other resources that may be useful to the reader of this book are the online handbook "The Prokaryotes" (Dworkin et al., 1999–2005) (soon also to appear in a printed version), as well as the laboratory manuals for the handling of halophilic and thermophilic Archaea published ten years ago (DasSarma and Fleischmann, 1995; Robb et al., 1995). For the handling of halophiles, including cultivation and genetic manipulation studies, the freely available online "Halohandbook" written by Dyall-Smith (2004) is an invaluable resource.

As editors of the present volume we are pleased that we were offered the opportunity to devote an entire volume of "Methods in Microbiology" to the extremophiles. These "exotic" microorganisms challenge the microbiologist with many specific problems, but when the proper methodology has been developed to solve these problems, their study is highly rewarding. We hope that the chapters that follow will prove useful in many laboratories, and that their application will advance our knowledge of the fascinating world of the extremophiles.

References

Anderson, A. W., Nordan, H. C., Cain, R. F., Parrish, G. and Duggan, D. (1956). Studies on a radio-resistant micrococcus. I. Isolation, morphology, cultural characteristics, and resistance to gamma radiation. *Food. Technol.* **10**, 575–577.

Antón, J., Oren, A., Benlloch, S., Rodríguez-Valera, F., Amann, R. and Rosselló-Mora, R. (2002). *Salinibacter ruber* gen. nov., sp. nov., a novel extreme halophilic member of the Bacteria from saltern crystallizer ponds. *Int. J. Syst. Evol. Microbiol.* **52**, 485–491.

Barns, S. M., Delwiche, C. F., Palmer, J. D. and Pace, N. R. (1996). Perspectives on archaeal diversity, thermophily and monophyly from environmental rRNA sequences. *Proc. Natl. Acad. Sci. USA* **93**, 9188–9193.

Battista, J. R. (1997). Against all odds: the survival strategies of *Deinococcus radiodurans*. *Annu. Rev. Microbiol.* **51**, 203–224.

Battista, J. R., Earl, A. M. and Park, M. J. (1999). Why is *Deinococcus radiodurans* so resistant to ionizing radiation? *Trends Microbiol.* **7**, 362–365.

Billi, D., Friedmann, E. I., Hofer, K. G., Caiola, M. G. and Ocampo-Friedmann, R. (2000). Ionizing-radiation resistance in the desiccation-tolerant cyanobacterium *Chroococcidiopsis*. *Appl. Environ. Microbiol.* **66**, 1489–1492.

Blöchl, E., Rachel, R., Burggraf, S., Hafenbradl, D., Jannasch, H. W. and Stetter, K. O. (1997). *Pyrolobus fumarii*, gen. and sp. nov., represents a novel group of archaea, extending the upper temperature limit for life to 113°C. *Extremophiles* **1**, 14–21.

Bolhuis, H., te Poele, E. M. and Rodríguez-Valera, F. (2004). Isolation and cultivation of Walsby's square archaeon. *Environ. Microbiol.* **6**, 1287–1291.

Bowman, J. P. (2001). Methods for psychrophilic bacteria. In *Marine Microbiology. Methods in Microbiology* (J. H. Paul, ed.), vol. 30, pp. 591–614. Academic Press, San Diego.

Bowman, J. P., McCammon, S. A., Lewis, T., Skerratt, J. H., Brown, K. L., Nichols, D. S. and McMeekin, T. A. (1998). *Psychroflexus torquis* gen. nov., sp. nov., a psychrophilic species from Antarctic sea ice, and reclassification of *Flavobacterium gondwanense* (Dobson *et al.* 1993) as *Psychroflexus gondwanense* gen. nov., comb. nov. *Microbiology UK* **144**, 1601–1609.

Brock, T. D. (1978). *Thermophilic Microorganisms and Life at High Temperatures.* Springer-Verlag, New York.

Brock, T. D. and Freeze, H. (1969). *Thermus aquaticus* gen. n. and sp. n., a non-sporulating extreme thermophile. *J. Bacteriol.* **98**, 289–297.

Brooks, B. W. and Murray, R. G. E. (1981). Nomenclature for "*Micrococcus radiodurans*" and other radiation-resistant cocci: *Deinococcaceae* fam. nov. and *Deinococcus* gen. nov., including five species. *Int. J. Syst. Bacteriol.* **31**, 353–360.

Buenger, J. and Driller, H. (2004). Ectoin: An effective natural substance to prevent UVA-induced premature photoaging. *Skin Pharmacol. Physiol.* **17**, 232–237.

Burns, D. G., Camakaris, H. M., Janssen, P. H. and Dyall-Smith, M. L. (2004). Cultivation of Walsby's square haloarchaeon. *FEMS Microbiol. Lett.* **238**, 469–473.

Carreto, L., Moore, E., Nobre, M. F., Wait, R., Riley, P. W., Sharp, R. J. and da Costa, M. S. (1996). *Rubrobacter xylanophilus* sp. nov., a new thermophilic species isolated from a thermally polluted effluent. *Int. J. Syst. Bacteriol.* **46**, 460–465.

Castenholz, R. W. and Garcia-Pichel, F. (2000). Cyanobacterial responses to UV-radiation. In *Ecology of Cyanobacteria: Their Diversity in Time and Space* (B. A. Whitton and M. Potts, eds), pp. 591–611. Kluwer Academic Publishers, Dordrecht.

Cohn, F. (1897). *Die Pflanze. Vorträge aus dem Gebiete der Botanik*, 3rd edn., vol. 2. J. U. Kern's Verlag (Max Müller), Breslau.

Collins, M. D., Hutson, R. A., Grant, I. R. and Patterson, M. F. (2000). Phylogenetic characterisation of a novel radiation resistant bacterium from irradiated pork: description of *Hymenobacter actinosclerus* sp. nov. *Int. J. Syst. Evol. Microbiol.* **50**, 731–734.

Cowan, D. A., Russell, N. J., Mamais, A. and Sheppard, D. M. (2002). Antarctic Dry Valley mineral soils contain unexpectedly high levels of microbial biomass. *Extremophiles* **6**, 431–436.

da Costa, M. S., Duarte, J. C. and Williams, R. A. D. (eds) (1989). *Microbiology of Extreme Environments and its Potential for Biotechnology.* Elsevier Applied Science, London.

Darwin, C. (1839). *Journal of Researches into the Geology and Natural History of the Various Countries Visited by H.M.S. Beagle, under the Command of Captain Fitzroy, R.N. from 1832 to 1836.* Henry Colburn, London.

DasSarma, S. and Fleischmann, E. M. (eds) (1995). *Archaea. A Laboratory Manual. Halophiles*, Cold Spring Harbor Laboratory Press, Cold Spring Harbor.

Dennis, P. P. and Shimmin, L. C. (1997). Evolutionary divergence and salinity-mediated selection in halophilic archaea. *Microbiol. Mol. Biol.* **61**, 90–104.

DiRuggiero, J., Santangelo, N., Nackerdien, Z., Ravel, J. and Robb, F. T. (1997). Repair of extensive ionizing-radiation DNA damage at 95°C in the hyperthermophilic archaeon *Pyrococcus furiosus*. *J. Bacteriol.* **179**, 4643–4645.

Dworkin, M., Falkow, S., Rosenberg, E., Schleifer, K.-H. and Stackebrandt, E. (eds) (1999–2005). *The Prokaryotes: An Evolving Electronic Resource for the Microbiological Community*, 3rd edn., Springer-Verlag, New York, http://link.springer-ny.com/link/service/books/10125/.

Dyall-Smith, M. L. (2004). *The Halohandbook: Protocols for Halobacterial Genetics*. Version 4.9, March 2004. http://www.microbiol.unimelb.edu.au/micro/staff/mds/HaloHandbook/index.html/.

Englander, J., Klein, E., Brumfeld, V., Sharma, A. K., Doherty, A. J. and Minsky, A. (2004). DNA toroids: framework for DNA repair in *Deinococcus radiodurans* and in germinating bacterial spores. *J. Bacteriol.* **186**, 5973–5977.

Ferreira, A. C., Nobre, M. F., Rainey, F. A., Silva, M. T., Waite, R., Burghardt, J., Chung, A. P. and da Costa, M. S. (1997). *Deinococcus geothermalis* sp. nov. and *Deinococcus murrayi* sp. nov., two extremely radiation-resistant and slightly thermophilic species from hot springs. *Int. J. Syst. Bacteriol.* **47**, 939–947.

Ferreira, A. C., Nobre, M. F., Moore, E. D., Rainey, F. A., Battista, J. R. and da Costa, M. S. (2000). Characterization and radiation resistance of new isolates of *Rubrobacter radiotolerans* and *Rubrobacter xylanophilus*. *Extremophiles* **3**, 235–238.

Fredrickson, J. K., Zachara, J. M., Balkwill, D. L., Kennedy, D., Li, S. M., Kostandarithes, H. M., Daly, M. J., Romine, M. F. and Brockman, F. J. (2004). Geomicrobiology of high-level nuclear waste-contaminated vadose sediments at the Hanford site, Washington state. *Appl. Environ. Microbiol.* **70**, 4230–4241.

Frenkiel-Krispin, D., Sack, R., Englander, J., Shimoni, E., Eisenstein, M., Bullitt, E., Horowitz-Scherer, R., Hayes, C. S., Setlow, P., Minsky, A. and Grayer Wolf, S. (2004). Structure of the DNA-SspC complex: Implications for DNA packaging, protection, and repair in bacterial spores. *J. Bacteriol.* **186**, 3525–3530.

Gold, T. (1992). The deep, hot biosphere. *Proc. Natl. Acad. Sci. USA* **89**, 6045–6049.

Gould, G. W. and Corry, J. E. L. (eds) (1980). *Microbial Growth and Survival in Extremes of Environment*, Academic Press, London.

Green, P. and Bousfield, I. J. (1983). Emendation of *Methylobacterium* Patt, Cole, and Hanson 1976; *Methylobacterium rhodinum* (Heumann 1962) comb. nov. corrig.; *Methylobacterium radiotolerans* (Ito and Iizuka 1971) comb. nov., corrig.; and *Methylobacterium mesophilicum* (Austin and Goodfellow 1979) comb. nov. *Int. J. Syst. Bacteriol.* **33**, 875–877.

Gunde-Cimerman, N., Oren, A. and Plemenitaš, A. (eds) (2005). *Adaptation to Life at High Salt Concentrations in Archaea, Bacteria, and Eukarya*, Springer, Dordrecht.

Heinrich, M. R. (ed.) (1976). *Extreme Environments. Mechanisms of Microbial Adaptation*, Academic Press, New York.

Herbert, R. A. and Codd, G. A. (eds) (1986). *Microbes in Extreme Environments*, Academic Press, London.

Horikoshi, K. (ed.) (1999). *Alkaliphiles*, Harwood Academic, Reading.

Horikoshi, K. and Grant, W. D. (eds) (1991). *Superbugs. Microorganisms in Extreme Environments*, Japan Scientific Societies Press, Tokyo – Springer-Verlag, Berlin.

Horikoshi, K. and Grant, W. D. (eds) (1998). *Extremophiles, Microbial Life in Extreme Environments*, Wiley-Liss Publishers, New York.

Hough, D. W. and Danson, M. J. (1999). Extremozymes. *Curr. Opin. Chem. Biol.* **3**, 39–46.

Huber, H., Hohn, M. J., Rachel, R., Fuchs, T., Wimmer, V. C. and Stetter, K. O. (2002). A new phylum of Archaea represented by a nanosized hyperthermophilic symbiont. *Nature* **417**, 63–67.

Hugenholtz, P., Pitulle, C., Hershberger, K. L. and Pace, N. R. (1998). Novel division level bacterial diversity in a Yellowstone hot spring. *J. Bacteriol.* **180**, 366–376.

Irgens, R. L., Gosink, J. J. and Staley, J. T. (1996). *Polaromonas vacuolata* gen. nov., sp. nov., a psychrophilic marine, gas-vacuolate bacterium from Antarctica. *Int. J. Syst. Bacteriol.* **46**, 822–826.

Ito, H. (1977). Isolation of *Micrococcus radiodurans* occurring in radurized sawdust culture media of mushroom. *Agric. Biol. Chem.* **41**, 35–41.

Ito, H. and Iizuka, H. (1971). Taxonomic studies on a radio-resistant *Pseudomonas*. Part XII. Studies on the microorganisms of cereal grain. *Agric. Biol. Chem.* **35**, 1566–1571.

Jolivet, E., Matsunaga, F., Ishino, Y., Forterre, P., Prieur, D. and Myllykallio, H. (2003a). Physiological responses of the hyperthermophilic archaeon "*Pyrococcus abyssi*" to DNA damage caused by ionizing radiation. *J. Bacteriol.* **185**, 3958–3961.

Jolivet, E., L'Haridon, S., Corre, E., Forterre, P. and Prieur, D. (2003b). *Thermococcus gammatolerans* sp. nov., a hyperthermophilic archaeon from a deep-sea hydrothermal vent that resists ionizing radiation. *Int. J. Syst. Evol. Microbiol.* **53**, 847–851.

Jolivet, E., Corre, E., L'Haridon, S., Forterre, P. and Prieur, D. (2004). *Thermococcus marinus* sp. nov., and *Thermococcus radiotolerans* sp. nov., two hyperthermophilic archaea from deep-sea hydrothermal vents that resist ionizing radiation. *Extremophiles* **8**, 219–227.

Karner, M. B., DeLong, E. F. and Karl, D. M. (2001). Archaeal dominance in the mesopelagic zone of the Pacific Ocean. *Nature* **409**, 507–510.

Kashefi, K. and Lovley, D. R. (2003). Extending the upper temperature limit for life. *Science* **301**, 934.

Kato, C. and Bartlett, D. H. (1997). The molecular biology of barophilic bacteria. *Extremophiles* **1**, 111–116.

Kato, C., Li, L., Nogi, Y., Nakamura, Y., Tamaoka, J. and Horikoshi, K. (1998). Extremely barophilic bacteria isolated from the Mariana Trench, Challenger Deep, at a depth of 11,000 meters. *Appl. Environ. Microbiol.* **64**, 1510–1513.

Kelly, D. P. and Wood, A. P. (2000). Reclassification of some species of *Thiobacillus* to the newly designated genera *Acidithiobacillus* gen. nov., *Halothiobacillus* gen. nov. and *Thermithiobacillus* gen. nov. *Int. J. Syst. Evol. Microbiol.* **50**, 511–516.

Kristjansson, J. K. (ed.) (1991). *Thermophilic Bacteria*, CRC Press, Boca Raton, FL.

Kristjansson, J. K. and Hreggvidsson, G. O. (1995). Ecology and habitats of extremophiles. *World J. Microbiol. Biotechnol.* **11**, 17–25.

Kushner, D. J. (ed.) (1978). *Microbial Life in Extreme Environments*, Academic Press, London.

Lanyi, J. K. (1974). Salt-dependent properties of proteins from extremely halophilic bacteria. *Bacteriol. Rev.* **38**, 272–290.

Madigan, M. T. (2000). Bacterial habitats in extreme environments. In *Journey to Diverse Microbial Worlds. Adaptation to Exotic Environments* (J. Seckbach, ed.), pp. 61–72. Kluwer Academic Publishers, Dordrecht.

Madigan, M. T. and Marrs, B. L. (1997). Extremophiles. *Sci. Am.* **276**(4), 66–71.

Madigan, M. T. and Oren, A. (1999). Thermophilic and halophilic extremophiles. *Curr. Opin. Microbiol.* **2**, 265–269.

Marteinsson, V. F., Birrien, J.-L., Reysenbach, A.-L., Vernet, M., Marie, D., Gambacorta, A., Messner, P., Sleytr, U. B. and Prieur, D. (1999). *Thermococcus barophilus* sp. nov., a new barophilic and hyperthermophilic archaeon isolated under high hydrostatic pressure from a deep-sea hydrothermal vent. *Int. J. Syst. Bacteriol.* **49**, 351–359.

Mattimore, V. and Battista, J. R. (1996). Radioresistance of *Deinococcus radiodurans*: functions necessary to survive ionizing radiation are also necessary to survive desiccation. *J. Bacteriol.* **178**, 633–637.

Mevarech, M., Frolow, F. and Gloss, L. M. (2000). Halophilic enzymes: proteins with a grain of salt. *Biophys. Chem.* **86**, 155–164.

Nishimura, Y., Ino, T. and Iizuka, H. (1988). *Acinetobacter radioresistens* sp. nov. isolated from cotton and soil. *Int. J. Syst. Bacteriol.* **38**, 209–211.

Oren, A. (2000). Life at high salt concentrations. In *The Prokaryotes: An Evolving Electronic Resource for the Microbiological Community*, (M. Dworkin, S. Falkow, E. Rosenberg, K.-H. Schleifer and E. Stackebrandt, eds), 3rd edn., Springer-Verlag, New York, release 3.1, http://link.springer-ny.com/link/service/books/10125/.

Oren, A. (2002). *Halophilic Microorganisms and their Environments*. Kluwer Scientific Publishers, Dordrecht.

Oren, A., Heldal, M. and Norland, S. (1997). X-ray microanalysis of intracellular ions in the anaerobic halophilic eubacterium *Haloanaerobium praevalens*. *Can. J. Microbiol.* **43**, 588–592.

Oren, A., Heldal, M., Norland, S. and Galinski, E. A. (2002). Intracellular ion and organic solute concentrations of the extremely halophilic bacterium *Salinibacter ruber*. *Extremophiles* **6**, 491–498.

Phillips, R. W., Wiegel, J., Berry, C. J., Fliermans, C., Peacock, A. D., White, D. C. and Shimkets, L. J. (2002). *Kineococcus radiotolerans* sp. nov., a radiation-resistant, Gram-positive bacterium. *Int. J. Syst. Evol. Microbiol.* **52**, 933–938.

Pick, U. (1999). *Dunaliella acidophila* – a most extreme acidophilic alga. In *Enigmatic Microorganisms and Life in Extreme Environmental Habitats* (J. Seckbach, ed.), pp. 465–478. Kluwer Academic Publishers, Dordrecht.

Rainey, F. A., Ray, K., Ferreira, M., Gatz, B. Z., Nobre, N. F., Bagaley, D., Rash, B. A., Park, M.-J., Earl, A. M., Shank, N. C., Small, A., Henk, M. C., Battista, J. R., Kämpfer, P. and da Costa, M. S. (2005). Extensive diversity of ionizing radiation-resistant bacteria recovered from a Sonoran Desert soil and the description of 9 new species of the genus *Deinococcus* from a single soil sample. *Appl. Environ. Microbiol.* **17**, 5225–5235.

Reysenbach, A.-L. and Götz, D. (2001). Methods for the study of hydrothermal vent microbes. In *Marine Microbiology. Methods in Microbiology* (J. H. Paul, ed.), vol. 30, pp. 639–656. Academic Press, San Diego.

Reysenbach, A.-L., Voytek, M. and Mancinelli, R. (eds) (2001). *Thermophiles: Biodiversity, Ecology, and Evolution*, Kluwer Academic Publishers, Dordrecht.

Rivkina, E. M., Friedmann, E. I., McKay, C. P. and Gilichinsky, D. A. (2000). Metabolic activity of permafrost bacteria below the freezing point. *Appl. Environ. Microbiol.* **66**, 3230–3233.

Robb, F. T., Place, A. R., Sowers, K. R., Schreier, H. J., DasSarma, S. and Fleischmann, E. M. (eds) (1995). *Archaea. A Laboratory Manual. Thermophiles*, Cold Spring Harbor Laboratory Press, Cold Spring Harbor.

Rodriguez-Valera, F. (ed.) (1988). *Halophilic Bacteria*, CRC Press, Boca Raton (2 volumes).

Rodriguez-Valera, F. (ed.) (1991). *General and Applied Aspects of Halophilic Bacteria*, Plenum Publishing Company, New York.

Rothschild, L. J. (1999). Microbes and radiation. In *Enigmatic Microorganisms and Life in Extreme Environmental Habitats* (J. Seckbach, ed.), pp. 549–562. Kluwer Academic Publishers, Dordrecht.

Rothschild, L. J. and Mancinelli, R. L. (2001). Life in extreme environments. *Nature* **409**, 1092–1101.

Schleper, C., Pühler, G., Holz, I., Gambacorta, A., Janovic, D., Santarius, U., Klenk, H. P. and Zillig, W. (1995). Picrophilus gen. nov., fam. nov.: a novel aerobic, heterotrophic, thermoacidophilic genus and family comprising archaea capable of growth around pH 0. *J. Bacteriol.* **177**, 7050–7059.

Seckbach, J. (ed.) (1994). *Evolutionary Pathways and Enigmatic Algae: Cyanidium caldarium (Rhodophyta) and Related Cells*, Kluwer Academic Publishers, Dordrecht.

Seckbach, J. (ed.) (1999). *Enigmatic Microorganisms and Life in Extreme Environmental Habitats*, Kluwer Academic Publishers, Dordrecht.

Seckbach, J. (ed.) (2000). *Journey to Diverse Microbial Worlds. Adaptation to Exotic Environments*, Kluwer Academic Publishers, Dordrecht.

Seckbach, J. and Oren, A. (2004). Introduction to the extremophiles. In *Origins* (J. Seckbach, ed.), pp. 373–393. Kluwer Academic Publishers, Dordrecht.

Shilo, M. (ed.) (1979). *Strategies of Microbial Life in Extreme Environments*, Verlag Chemie, Weinheim.

Stetter, K. O. (1996). Hyperthermophilic procaryotes. *FEMS Microbiol. Rev.* **18**, 149–158.

van de Vossenberg, J. L. C. M., Driessen, A. J. M. and Konings, W. N. (1998). The essence of being extremophilic: the role of the unique archaeal membrane lipids. *Extremophiles* **2**, 163–170.

Ventosa, A. (ed.) (2004). *Halophilic Microorganisms*, Springer-Verlag, Berlin.

Ventosa, A., Nieto, J. J. and Oren, A. (1998). Biology of aerobic moderately halophilic bacteria. *Microbiol. Mol. Biol. Rev.* **62**, 504–544.

Vincent, W. F. (1988). *Microbial Ecosystems of Antarctica*. Cambridge University Press, Cambridge.

Vreeland, R. H. and Hochstein, L. I. (eds) (1993). *The Biology of Halophilic Bacteria*, CRC Press, Boca Raton.

Walker, J. J., Spear, J. R. and Pace, N. R. (2005). Geobiology of a microbial endolithic community in the Yellowstone geothermal environment. *Nature* **434**, 1011–1014.

Waters, E., Hohn, M. J., Ahel, I., Graham, D. E., Adams, M. D., Barnstead, M., Beeson, K. Y., Bibbs, L., Bolanos, R., Keller, M., Kretz, K., Lin, X. Y., Mathur, E., Ni, J. W., Podar, M., Richardson, T., Sutton, G. G., Simon, M., Söll, D., Stetter, K. O., Short, J. M. and Noordewier, M. (2003). The genome of *Nanoarchaeum equitans*: insights into early archaeal evolution and derived parasitism. *Proc. Natl. Acad. Sci.* **100**, 12984–12988.

Wharton, D. A. (2002). *Life at the Limits: Organisms in Extreme Environments*, Cambridge University Press, Cambridge.

Wiegel, J. and Adams, M. W. W. (eds) (1998). *Thermophiles: The Keys to Molecular Evolution and the Origin of Life?* CRC Press, Boca Raton.

Yayanos, A. A. (2000). Deep-sea bacteria. In *Journey to Diverse Microbial Worlds. Adaptation to Exotic Environments* (J. Seckbach, ed.), pp. 161–174. Kluwer Academic Publishers, Dordrecht.

Yayanos, A. A. (2001). Deep-sea piezophilic bacteria. Methods for the study of hydrothermal vent microbes. In *Marine Microbiology. Methods in Microbiology* (J. H. Paul, ed.), vol. 30, pp. 615–637. Academic Press, San Diego.

Yayanos, A. A., Dietz, A. S. and van Boxtel, R. (1981). Obligately barophilic bacterium from the Mariana Trench. *Proc. Natl. Acad. Sci. USA* **78**, 5212–5215.

Yoshinaka, T., Yano, K. and Yamaguchi, H. (1973). Isolation of highly radioresistant bacterium. *Arthrobacter radiotolerans* nov. sp. *Agric. Biol. Chem.* **37**, 2269–2275.

Zavarzin, G. A. and Zhilina, T. N. (2000). Anaerobic chemotrophic alkaliphiles. In *Journey to Diverse Microbial Worlds. Adaptation to Exotic Environments* (J. Seckbach, ed.), pp. 191–208. Kluwer Academic Publishers, Dordrecht.

Zhilina, T. N. and Zavarzin, G. A. (1994). Alkaliphilic anaerobic community at pH 10. *Curr. Microbiol.* **29**, 109–112.

Thermophiles

◆◆◆

2   *In Situ* Activity Studies in Thermal Environments
3   The Isolation of Thermophiles from Deep-sea Hydrothermal Environments
4   Growth of Hyperthermophilic Microorganisms for Physiological and Nutritional Studies
5   Toward the Large Scale Cultivation of Hyperthermophiles at High-Temperature and High-Pressure
6   Analysis of Lipids from Extremophilic Bacteria
7   Membranes of Thermophiles and Other Extremophiles
8   Characterization and Quantification of Compatible Solutes in (Hyper)thermophilic Microorganisms
9   Functional Genomics of the Thermo-Acidophilic Archaeon *Sulfolobus solfataricus*
10 Heat Shock Proteins in Hyperthermophiles
11 Deep-sea Thermococcales and their Genetic Elements: Plasmids and Viruses
12 Genetic Systems for *Thermus*
13 Gene Transfer Systems for Obligately Anaerobic Thermophilic Bacteria
14 Hyperthermophilic Virus–Host Systems: Detection and Isolation
15 Preservation of Thermophilic Microorganisms

2 *In Situ* Activity Studies in Thermal Environments

Nikolay V Pimenov and Elizaveta A Bonch-Osmolovskaya
Winogradsky Institute of Microbiology, Russian Academy of Sciences, Prospekt 60 Let Oktyabrya 7, bldg. 2, 117312 Moscow, Russia

CONTENTS

Introduction
Determining rates of microbiological processes in thermal habitats
Interpretation of *in situ* data for thermal environments
Conclusion

◆◆◆◆◆ INTRODUCTION

Microbial communities of natural thermal environments attract much interest as possible analogues of early Earth ecosystems and sources of new genes valuable for bioengineering purposes. Many novel, metabolically diverse thermophilic prokaryotes representing deep phylogenetic lineages were isolated from terrestrial and submarine thermal habitats during the past two decades (Huber *et al.*, 2000; Miroshnichenko and Bonch-Osmolovskaya, 2005; Stetter, 1996). These ecosystems were also shown to contain a wealth of so far uncultured prokaryotic organisms detectable by their 16S rRNA gene sequences present in environmental clone libraries (Barns *et al.*, 1994; Hugenholz *et al.*, 1998; Nercessian *et al.*, 2003; Reysenbach *et al.*, 2000; Sievert *et al.*, 2000). Even though certain representatives of the new phylogenetic lineages are predominant in microbial communities and, apparently, play an important functional role in thermal ecosystems, neither their metabolism nor their ecological function are currently known. A few organisms originally detected by molecular methods were later isolated and characterized (Kashefi *et al.*, 2002; Miroshnichenko *et al.*, 2002, 2003). However, as long as the metabolism of most of uncultured thermophilic prokaryotes remains unknown, our interpretation of energy and carbon flows, based on the capacities of cultured strains alone may not be adequate. Direct *in situ* investigation of microbial processes in thermal environments can yield information on the activities in thermophilic microbial communities irrespective of our ability to culture their components under laboratory conditions.

There are several independent approaches to *in situ* investigations of microbial activities in different types of environments. The use of collecting chambers, for example, allowed methane flux rates to be determined in shallow aquatic habitats (Conrad and Schütz, 1988). A modification of this method was employed to estimate the rate of methane formation in the thermophilic cyanobacterial mat of Octopus spring, Yellowstone National Park, USA (Ward, 1978). This approach, however, yields data only on the net product formation and is unable to uncover the pathways and the nearest precursors. The use of specific inhibitors, involving suppression of the targeted reaction and comparison with noninhibited controls, was very popular in the 1980s but, to our knowledge, was never applied to microbial communities of thermal environments. The success of both these approaches greatly depends on the sensitivity of the analytical methods employed. Microbial processes in a sample isolated from the surrounding environment are likely to be affected if the incubation time is not kept to a minimum. Therefore, small changes in product concentrations need to be measured and this greatly depends on the use of sensitive and reliable analytical methods.

Experiments with radioactively labelled substrates allow their microbiological transformations to be traced over very short exposition periods. Rate estimation of microbial processes in different ecosystems by radioisotopic methods has been widely used since the early 1950s (Sorokin, 1999). Radioactive substrates (containing ^{14}C, ^{35}S, ^{3}H, etc.) are added to isolated natural samples that are to be incubated for a short period of time (ranging from several hours to one day, on rare occasions to several days) at the sampling site or under laboratory conditions closest to those *in situ*. During incubation, microorganisms present in the sample perform certain microbiological processes involving transformation of the radioactive substrate and its nonradioactive natural analogue. The radiolabelled atoms are incorporated into both microbial cells and soluble or gaseous metabolic products. The microbial processes are terminated by the addition of antimicrobial agents. The radioactive components present in the incubated sample are then separated into different fractions by physicochemical methods and the radioactivity of each fraction is determined by a liquid scintillation counter. The rate of substrate transformation to a given product can be readily calculated from the radioactivity levels of the substrate and the product and the concentration of this substrate in the natural samples.

A huge body of quantitative evidence has accumulated on photosynthetic and chemosynthetic carbon assimilation, sulfate reduction, and methane production and oxidation in oceans, freshwater reservoirs, and soils of different types. On this basis, fluxes of major biogenic elements on our planet were estimated. Because the radioactive method is so sensitive, it can detect microbial processes that fail to be documented by other methods. At the same time, by employing close replication of *in situ* conditions, this method makes possible detection of activity of microorganisms not available in laboratory cultures. In this chapter, we discuss the essentials of this method and its advantages and limitations in application to natural thermal environments – terrestrial hot springs and hydrothermal sediments, shallow-water and deep-sea submarine hot vents, and deep-subsurface thermal habitats.

◆◆◆◆◆ DETERMINING RATES OF MICROBIOLOGICAL PROCESSES IN THERMAL HABITATS

Sampling Site Characterization and Determination of *In Situ* Substrate Concentrations

A preliminary biogeochemical characterization of the sampling site is essential for subsequent radioisotopic assays. The temperature, pH, Eh, and concentrations of dissolved oxygen and other environmentally important compounds (hydrogen sulfide, etc.) have to be known to characterize the conditions under which the microbial community under study is functioning. The natural concentrations of substrates to be added as radiolabelled compounds need to be known both to design the experiment and to calculate the rates of microbial processes *in situ* from the data obtained.

Carbonate and bicarbonate

$NaH^{14}CO_3$ is one of the most frequently used radioisotopic compounds in environmental studies. It is used in rate studies of several key processes. Incorporation of labelled bicarbonate into microbial cells varies directly with the rates of primary organic matter production and inorganic carbon assimilation. Depending on the incubation conditions – in the light or dark – the rate of inorganic carbon assimilation will reflect rates of photosynthetic or chemosynthetic primary production, respectively. The figures obtained under dark incubation will also account for heterotrophic carbon assimilation, its rate normally not exceeding 10% of that of photosynthetic or chemosynthetic processes. In the course of lithotrophic methanogenesis or acetogenesis, $^{14}CH_4$ or ^{14}C-acetate are formed from radioactive bicarbonate, and rates of these processes can also be determined.

The natural concentration of HCO_3^-/CO_2, also known as the total alkalinity of the sample, is determined by titration of water samples with 0.01 N HCl or by using the Aqua Merck analytical test kit (Merck, Germany).

Sulfate

To calculate the rate of sulfate reduction by microorganisms, one has to know its *in situ* concentration. Most often, the sulfate content of water samples (including pore waters) is determined by means of non-suppressed ion chromatography. A DX 500 ion chromatograph (Dionex Corporation), analytical column Ion Pac AS 14, 8 mm × 240 mm, and

1 mM Na_2CO_3 carbonate/1.2 mM $NaHCO_3$ bicarbonate eluent at a flow rate of 1.5 ml min^{-1} is used. In the field, the sulfate content can be determined by nephelometric assays of suspended $BaSO_4$ using Spectroquant Merck analytical test kit (Merck, Germany).

Organic substrates

Acetate is a common fermentation product and, in natural ecosystems, its methyl group can either be reduced to methane in the process of acetoclastic methanogenesis, or oxidized completely to CO_2 in the course of diverse respiratory reactions performed by different groups of microorganisms. Rates of both these processes are determined by using [2-^{14}C]-acetate. In thermal environments, acetate is usually present in significant concentrations as the product of microbial fermentations or thermocatalytic degradation of organic matter in high-temperature zones (Martens, 1990). Acetate in natural samples is usually assayed by gas–liquid chromatography. If the salt concentration of the sample is low (<500 mg l^{-1}), acetate can be determined directly in the sample acidified with concentrated H_3PO_4 (1:1 v/v). With salt concentrations exceeding 500 mg l^{-1} the following protocol can be used:

Procedure

- Acidify a 5-ml portion of the sample to pH 4.0 with 10% H_3PO_4.
- Distil the volatile fatty acids with water vapour to the final volume of 50 ml.
- Neutralize the distillate with 10% $NaHCO_3$ solution and evaporate to dryness at 60°C.
- Dissolve the dry precipitate in 200 µl of 5% H_3PO_4 directly before assays.
- Determine the acetate content on a gas chromatograph equipped with a flame-ionization detector and a glass column filled with Porapak Q at 180°C and argon as a carrier gas. The sensitivity of this method is 10 mg l^{-1} acetate. Acetate can also be determined with the sensitivity of 0.5 mg l^{-1} on a DX 500 ion chromatograph (Dionex Corporation), analytical column Ion Pac AS 14, 8 × 240 mm. The highest sensitivity of acetate assays is shown by HPLC with precolumn derivatization performed with 2-nitrophenylhydrazine (Albert and Martens, 1997).

Other than acetate-labelled organic substrates that were used in studies of thermal habitats are uniformly labelled glucose, cellulose, and casein hydrolysate (Namsaraev et al., 1994; Starynin et al., 1995). In these experiments, the degradation rates of these substrates were determined along with rates of intermediate and end products formation. Reduced sugars do not occur in natural samples in significant amounts, and so only the potential rate of their degradation was determined (Namsaraev et al., 1994). In this study, the concentration of glucose introduced as an isotope solution had to be taken into account in rate calculations. Protein was determined by the method of Lowry et al. (1951) after 3-h hydrolysis in 1 N NaOH at 37°C. It should be noticed that labelled peptides in casein hydrolysate preparation were probably used more readily than non-hydrolysed proteins present in the samples and, therefore, the rate of protein degradation might have been overestimated.

The concentrations of organic substrates are not usually known until after the end of field work and, for that matter, are not taken into consideration when the experiments are actually planned. For this reason, experiments with labelled acetate (as well as other organic substrates) are often performed "blindly" or on the basis of data previously obtained for similar environments. The exact concentration figures for the habitat studied are used only in process rate calculations. In blind experiments, real rates of microbial processes in natural ecosystems will not be attained unless the concentration of the added tracer substrate is significantly lower than the substrate content of the natural sample. If it turns to be similar or higher or if the natural concentration of the substrate is below the sensitivity level of the analytical method employed, the results obtained will characterize only the potential ability of the microbial population to transform the given substrate to the given product. The potential rate obtained will fail to portray correctly the biogeochemical activity of microorganisms in the given habitat but might still be useful as an indicator of their metabolic capacities.

Preparation of Samples

Thermal environments differ widely in their accessibility for sampling, and, sometimes, special equipment will be needed. This relates particularly to deep-sea hydrothermal habitats. Diverse sampling equipment was invented and used for exploration of such areas including different types of box corers, hydraulically operated piston samplers, hermetic titanium bottles, "slurp guns", and a syringe sampler-incubation array for *in situ* incubation experiments (Wirsen et al., 1986). Terrestrial and shallow-water thermal habitats are much more accessible for sampling.

Different sampling procedures were used for *in situ* studies of microbial processes in thermal environments. Samples of cyanobacterial mats developing in hot springs of Yellowstone National Park, USA, and Uzon Caldera, Kamchatka, Russia, were taken with cork borers 8–10 mm in diameter (Bonch-Osmolovskaya et al., 1987; Gorlenko et al., 1987; Ward and Olson, 1980). Depending on the goal of the study, entire mat cores or those sliced into layers were used for incubation. Cores or their parts were put in test vials – Hungate tubes or penicillin bottles – filled with water taken from the same place. In another study, cores measuring 2.8 cm in diameter and 1.4 cm in length were placed in bottles containing spring water and glass beads and vigorously shaken to obtain a homogenous slurry for use in the subsequent experiments (Zeikus et al., 1980).

Sediment samples from hot springs of Yellowstone National Park (Roychoudhury, 2004), shallow-water submarine hot vents (Namsaraev et al., 1994), lakes (Elsgaard et al., 1994; Starynin et al., 1995), deep-sea hydrothermal deposits of Guaymas Basin (Jørgensen et al., 1992; Kallmeyer and Boetius, 2004; Weber and Jørgensen, 2002), were taken as intact cores. In some cases, a slurry was prepared by combining samples with water from the same site, usually in the proportion 1:2, and, to avoid sample oxygenation, the added water was prereduced with dithionite (Elsgaard *et al.*, 1994) or sodium sulfide (Jørgensen, 1978). Slurries were dispensed into sterile test tubes under constant mixing and flow of oxygen-free nitrogen. Samples from Kamchatka hot springs (Bonch-Osmolovskaya *et al.*, 1999) were a mixture of sediment and water from the same place (2:1) dispensed in 15-ml penicillin bottles.

In some studies, sediment cores were used without dilution. Employing a method originally described by Jørgensen (1978), radio-labelled substrates were introduced directly into the cores with a syringe through silicone-filled ports located along the sampling tube at a distance of 1–2 cm (Weber and Jørgensen, 2002). This method preserved the sample structure and kept the interfering effect of the sampling procedure to a minimum. In another study, samples of hydrothermal chimneys in the East Pacific Rise, 13°N, and organic matter produced by invertebrates colonizing these chimneys were divided mechanically into parts of approximately equal volume, dispensed in test vials (15-ml sterile penicillin bottles) under flow of oxygen-free nitrogen, and covered with water obtained from the same site with a titanium bottle (Bonch-Osmolovskaya and Pimenov, in preparation). The whole procedure was carried out immediately after lifting samples aboard ship, using still hot hydrothermal fluid.

Rates of microbial processes in deep subsurface hot environments were until now measured only in formation water samples obtained from a high-temperature oil reservoir (Belyaev *et al.*, 1990; Bonch-Osmolovskaya *et al.*, 2003; Nazina *et al.*, 2000). Samples of formation water from different horizons were separated from oil and dispensed in sterile 0.5–1 serum bottles filled with oxygen-free nitrogen. The bottles were hermetically closed, and water samples were used in tracing experiments either immediately or after transportation to laboratory. Colourless resazurin added to water samples indicated their anoxic state.

Radioactive Substrate Solutions

Radioactive substrates produced by Amersham Biosciences (UK) or Isotope (Russia) are usually used. The volume of radioactive substrate solution added to each sample usually does not exceed 100–200 µl so that its addition would not cause significant changes in Eh, or chemical composition of the sample. Depending on the sampling site, the solutions are prepared in distilled water or with addition of NaCl to obtain the corresponding salinity. The solutions are prepared under a flow of oxygen-free nitrogen or argon and sterilized by filtration or autoclaving at 110°C for 30 min. In some studies, solutions of radioactive substrates are prepared in anoxic natural water from the spring (Ward and Olson, 1980). The radioactivity of $^{14}HCO_3^-$ added to each sample is usually between 50 and 100 µCi (2–4 MBq), when the rate of lithotrophic methanogenesis or acetogenesis is measured, and between 2 and 10 µCi (0.08–0.4 MBq) for rate measurements of CO_2 assimilation. If the radioactivity of several products is to be assayed sequentially in one experimental vial, then the medium amount of the labelled substrate is to be introduced.

The radioactivity of $^{14}CH_3COO^-$ usually ranges from 12.5 to 25 µCi (0.5–1.0 MBq). The radioactivity of $Na_2^{35}SO_4$ added to each sample normally ranges between 20 and 60 µCi (0.8–2.4 MBq).

Incubation

Radioisotopic experiments are usually carried out in triplicates. High sensitivity of the method coupled with heterogeneity of solid samples is likely to lead to significant spread of results across test replicates (as high as 100%). The deviation can be usually decreased by increasing the sample volume. In water samples, in comparison with sediments, the deviation between replicates is usually much lower. In most cases, the volume of test vials varies from 15 to 100 ml. It is also possible to perform *in situ* incubation of sediment samples in 5-ml polypropylene syringes (Roychoudhury, 2004), or perform whole core incubations (Weber and Jørgensen, 2002).

High temperature, being the main specific feature of thermal environments, constitutes a major problem when reproducing natural conditions at the sampling site. Samples have to be dispensed to experimental vials as fast as possible in order not to influence the existing equilibrium between gaseous and soluble components. Mesophilic anaerobic processes are usually studied in vials completely filled with the sample. For samples from high-temperature habitats, however, some headspace has to be left in vials to prevent their opening under internal pressure caused by incubation at elevated temperatures. The headspace should be filled with an inert gas (argon or nitrogen).

Labelled substrates are normally added to samples with sterile syringes upon which the vials are stoppered tightly with screw or crimped aluminium caps and incubated *in situ* or in a laboratory incubator at the same temperature as *in situ*. Terrestrial hot springs make good incubators for isolated samples with radioactive substrates (Figure 2.1). Samples from submarine or deep-subsurface thermal habitats are often incubated in laboratory incubators. Whenever no exact data on the *in situ* temperature are available or the research goal is to determine the effect of temperature on a microbial process, the incubation is carried out at different temperatures in order to compare potential activities of the microbial population (Jørgensen *et al.*, 1992).

Setting up a proper incubation time is an important issue tackled differently by different investigators. It may well be assumed that, after the sample is disturbed, a certain time is needed for the microbial community to restart its metabolic activities. Therefore, insufficient incubation time might result in underestimated rates of the process or even prevent its detection. On the other hand, too long incubation could also produce incorrect process rates because of the depletion of metabolic substrates present in the sample. This particularly applies to dissolved hydrogen of volcanic origin, continually replenished in natural thermal ecosystems but not so in the isolated sample. The limitation factors could be determined in additional experiments. Overall, the most popular

Figure 2.1. Radiotracing experiment in a Uzon Caldera hot spring.

incubation time for tracing experiments is 24 h; however, much shorter (1 h) and much longer (7 days) incubation times have also been used (Jørgensen et al., 1992; Roychoudhury, 2004). The best solution is likely to consist in incubating samples from each site for different periods of time to determine the proper incubation time within the interval of linear product concentration increase.

When thermophilic phototrophic communities are the objects of studies, samples should be exposed to light in a pattern similar to that in nature. Photosynthetic assimilation of inorganic carbon needs to be studied in transparent bottles. At the same time, it is worth remembering that other processes could be indirectly influenced by light. For example, methanogenesis was inhibited significantly in samples of thermophilic cyanobacterial mat exposed to light (transparent bottles) in comparison with control bottles wrapped in aluminium foil (Bonch-Osmolovskaya et al., 1987).

The incubation is terminated with the addition of antimicrobial agents: 40% NaOH (final concentration, 3.5% v/v) when ^{14}C-labelled substrates are used, or zinc acetate (4% v/v), when the experiments are run with ^{35}S-labelled Na-sulfate. At the same time, certain radioactive products are also bound by terminating agents, which prevents their loss. In some experiments, 2% formamide was used as the termination agent. The fixed samples can be preserved at room temperature for 1–2 weeks and, in refrigerator, for as long as 3 months.

High temperature and harsh environmental conditions of thermal habitats make detection of abiotic processes an important issue. Such controls are necessary to exclude possible chemical reactions and radio-isotopic exchange between different chemical species that otherwise would be attributed to microbiological processes. As a rule, the same chemical agents as those used to terminate the process are added to some

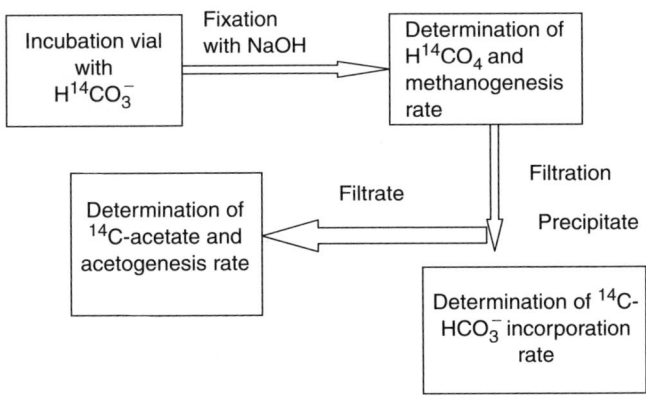

Figure 2.2. Scheme of sequential determination of radioactive products in $H^{14}CO_3^-$ incubation experiments.

of the replicates prior to incubation, for abiotic controls. These are incubated together with test vials and subject to the same procedures. To inhibit microbial processes completely, samples with terminating agents could be preincubated for a day before labelled substrates are added.

Determining the Content of ^{14}C-labelled Products

In experiments with ^{14}C-labelled substrates, the radioactivities of different products formed in the same sample can be measured consecutively. In experiments with $NaH^{14}CO_3$, methane is the first to be flown out, then, after filtration, the radioactivity of volatile fatty acids is measured in filtrate, and incorporation of ^{14}C into cells of prokaryotes is determined in the precipitate (Figure 2.2). The same applies to experiments with $^{14}CH_3COONa$. First, the radioactivity of methane formed in the sample is measured to determine the rate of acetoclastic methanogenesis. Next, the radioactivity of CO_2 formed is determined to find the rate of acetate oxidation in the course of other microbial processes.

Trapping ^{14}C-labelled methane

A method of methane separation to determine its radioactivity was first developed by Laurinavichus and Belyaev (1978) and was only partly modernized since that time. A flowchart of methane combustion and trapping is given in Figure 2.3. Methane is extracted from experimental bottles by air flown at a rate of 40–60 ml min^{-1} during 1 h and combusted at 800°C in an oven in quartz tubes (length, 50 cm; diameter, 0.8 cm) filled with a catalyst – silica gel granules saturated with $CoCl_2$ and acidified with H_3PO_4. Methane can also be combusted in the presence of spiral platinum wire (1 g per tube) as catalyst. $^{14}CO_2$ formed as a product of $^{14}CH_4$ combustion is distilled into a trap containing scintillation liquid Ultima Gold (PerkinElmer) mixed with 2-phenylethylamine and

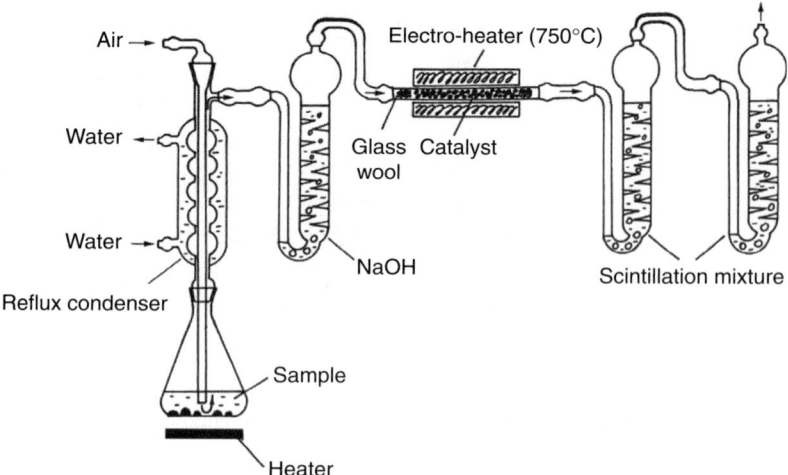

Figure 2.3. Apparatus for $^{14}CH_4$ separation.

ethanol (3:1:1). The radioactivity of the trapping solution is then determined by means of a liquid scintillation counter. Several samples (up to five) can be processed in parallel by heating simultaneously several quartz tubes in a single oven, each tube being connected to a separate trap.

Trapping ^{14}C-labelled CO_2

Separation of CO_2 is performed when all methane is already removed from the vial. The sample is acidified to pH 1.0–3.0 by the addition of 10% H_3PO_4 and then CO_2 is flown out and trapped in a scintillation mixture as previously described.

Determining the incorporation of $H^{14}CO_3^-$ into cells of autotrophic prokaryotes

Samples or their aliquots are filtered on Millipore (pore size 0.2 μm) membrane filters. To remove the remains of dissolved bicarbonate, filters are washed several times with 1% HCl. Then the filters are placed into vials for scintillation counting of radioactive ^{14}C assimilated by microorganisms present in the sample.

The rate of dark CO_2 assimilation in sediments can be determined by the method of persulfate combustion (Rusanov et al., 1998):

Procedure

- Incubate samples incubated with ^{14}C-labelled bicarbonate.
- Fix with 2 M KOH.
- Place the samples in 200-ml glass flasks with 100 ml of water.
- Acidify with 10% H_3PO_4, heat to 60–70°C, and flush the ^{14}C-carbon dioxide in a stream of air for 1 h.
- Add 25 g $K_2S_2O_8$ to each sample, and keep boiling for 1.5 h ("wet combustion"). Trap the evolved $^{14}CO_2$ in scintillation mixture with 2-phenylethylamine, as described in the previous section.

Quantitative assays of ^{14}C-labelled volatile fatty acids

^{14}C-acetate can be formed from $^{14}CO_2$ in the process of acetogenesis or, together with other metabolic products, from labelled organic substrates in the course of their degradation. The total radioactivity of volatile fatty acids can be determined upon removing ^{14}C-bicarbonate from the solution and their distillation with water vapour. The radioactivity of each individual volatile fatty acid can be determined after separation on a DX 500 ion chromatograph (Dionex Corporation), analytical column Ion Pac AS 14, 8 mm × 240 mm, and carbonate/bicarbonate eluents flown at a rate of 1.5 ml min^{-1}).

Determining the Content of ^{35}S-labelled Products

A method to measure the rate of sulfate reduction was initially proposed by Ivanov (1956). It is based on the formation of free hydrogen sulfide after acidification of the medium, which can be readily removed and trapped in an alkaline solution (Figure 2.4). To determine the radioactivity of $H_2^{35}S$ formed in the process of sulfate reduction, 0.5 mM of Na_2S is added to samples as a carrier. The samples are then acidified with 10% H_3PO_4 to pH 3.0 and hydrogen sulfide is distilled into a trap containing scintillation liquid Ultima Gold (PerkinElmer) with 2-phenylethylamine and ethanol (3:1:1). The distillation is continued for about 1 h at a temperature of 50–60°C (Figure 2.4). Finally, the radioactivity of ^{35}S in scintillation mixture is measured in a liquid scintillation counter.

This method can account only for acid-volatile reduced sulfur compounds. However, reduced ^{35}S can also be incorporated into nonacid-volatile sulfur compounds, particularly pyrite. As a consequence, the obtained rate of sulfate reduction might be underestimated. There is a different method, called a single-step chromium reduction, which makes it possible to estimate the radioactivity of the whole pool of reduced sulfur in the sample (Fossing and Jørgensen, 1989). In this method, all reduced sulfur compounds in the sample are converted to H_2S in the presence of Cr(II) and HCl and flushed into a trapping solution (5–10% zinc acetate). There is also a more recent modification of this method (Ulrich et al., 1997).

Calculation of the Process Rates at *In Situ* Conditions

The rate of substrate transformation to a certain product can be readily calculated by the following equation:

$$I = \frac{(r - r_c)C\alpha}{RT}$$

where I is the rate of product formation by microorganisms; r is the radioactivity of the product formed; r_c is the radioactivity of the same product formed in the abiotic control; R is the initial radioactivity of the

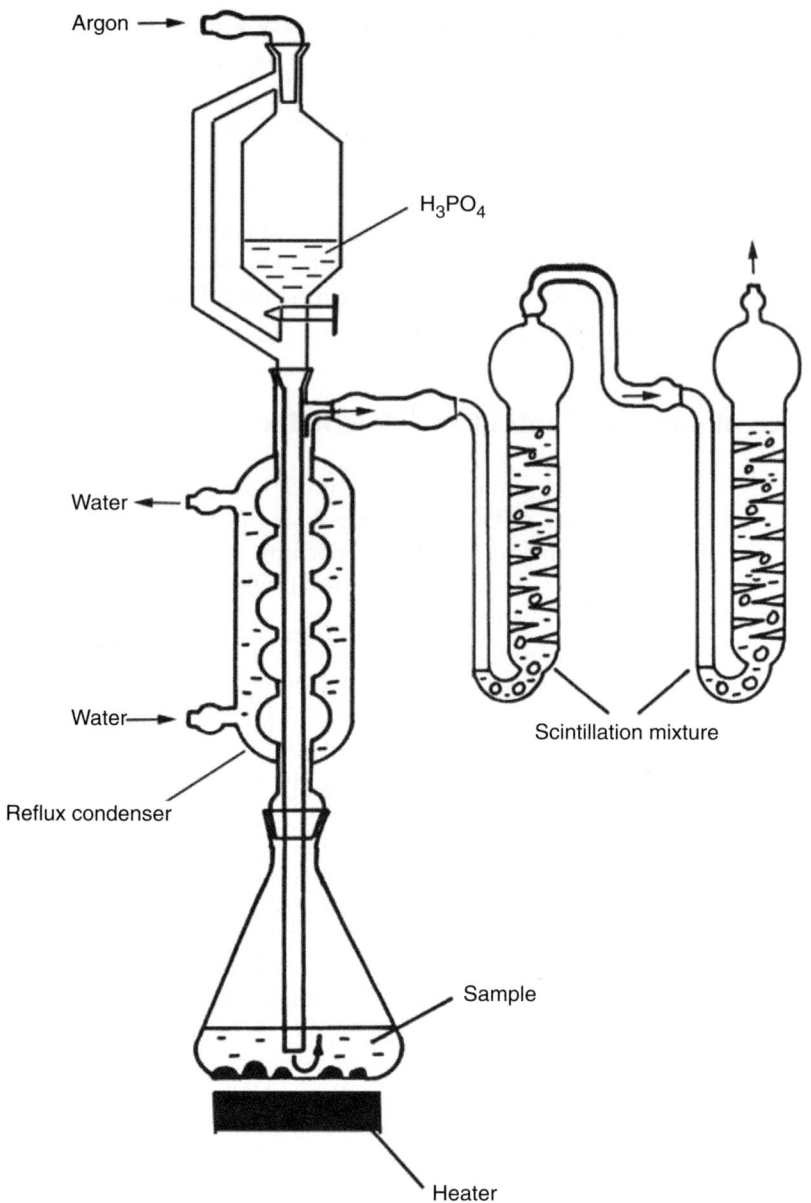

Figure 2.4. Apparatus for $H_2^{35}S$ separation.

labelled substrate added to the sample; C is the natural concentration of this substrate in the sample; α is the correction factor for isotope fractionation (1.06 for ^{14}C, or 1.045 for ^{35}S); and T is the incubation time. The concentration of substrate is usually expressed in millimoles or milligrams of $C_{substrate}$ ($S_{substrate}$, etc.) per litre. The process rate is, therefore, expressed in millimoles (micromoles, nanomoles) of the product or milligrams (micrograms, nanograms) of $C_{product}$ ($S_{product}$, etc.) per litre (or cm^{-3}) per day (or hour). In some articles, the rate of methanogenesis is expressed in microlitres of methane formed per unit volume of water or sediment per unit time. Sometimes, rates of the microbial processes are calculated per unit area, like in the case of phototrophic microbial communities, where the size of illuminated area is an important factor.

◆◆◆◆◆ INTERPRETATION OF *IN SITU* DATA FOR THERMAL ENVIRONMENTS

Methodological Limitations in Thermal Environments

Limitations of *in situ* studies, the most important of which is associated with sample isolation from the surrounding ecosystem, are especially evident in the case of thermal environments. Microbial communities in volcanic habitats are strongly dependent on constant supply of energy substrates and/or electron acceptors of volcanic origin. Therefore, microbial processes in the isolated sample could differ significantly from those taking place in the natural ecosystem.

Another important factor that might influence significantly the outcome of radioisotopic tracing experiments is hydrostatic pressure. Being extremely high in deep-sea and deep-subsurface habitats, it changes the chemistry of water by markedly increasing the solubility of gaseous substrates. Hydrostatic pressure was also shown to affect the growth of some of deep-sea organotrophic thermophilic prokaryotes in laboratory experiments (Alain *et al.*, 2002; Marteinsson *et al.*, 1999). The process of thermophilic sulfate reduction was shown to be stimulated by additional hydrostatic pressure (Kallmeyer and Boetius, 2004). The influence of hydrostatic pressure on mesophilic microbial processes was studied *in situ* in a deep-sea environment (Wirsen *et al.*, 1993). In this study, heterotrophic microbial activity was shown not to be strongly affected by hydrostatic pressure, and process rates yielded by radioisotopic assays in samples incubated at *in situ* temperature and pressure differed at most by 20–30% from those obtained for the same samples incubated aboard ship under atmospheric pressure. This influence, however, could be much more prominent in the case of lithotrophic growth on gaseous energy substrates. For example, adding molecular hydrogen to head space of experimental vials was found to boost the formation of $^{14}CH_4$ from $NaH^{14}CO_3$ in water and chimney samples from the East Pacific Rise (13°N) incubated aboard ship at 60–100°C and atmospheric pressure (Bonch-Osmolovskaya and Pimenov, manuscript in preparation). This fact indicates that the process was limited by hydrogen lost during sample decompression.

Even with these limitations, *in situ* experiments are able to furnish valuable information on thermal ecosystems, both quantitative and qualitative.

The Rates of Microbial Processes in Thermal Environments

Since the beginning of exploration of natural thermal environments in the 1980s, a substantial amount of data on the rates of microbial processes was obtained by means of the radioisotopic approach. Some of these results are presented in Table 2.1.

Terrestrial hot springs are the most readily accessible among thermal environments. They are the source of isolation of many thermophilic prokaryotes and the site of *in situ* investigation of microbial activity. A very specific feature of terrestrial hot springs is the development of a photosynthetic microbial community – cyanobacterial mats that cover large areas and reach several centimeters in thickness (Castenholz, 1984). Radioisotopic methods were applied to measure the rate of inorganic carbon assimilation in the mat of Thermophilny Spring, Uzon Caldera, Kamchatka (Gorlenko *et al.*, 1987). Inhibition of photosystem II (oxygenic photosynthesis) with 7 µM of 3-(3,4-dichlorophenyl)-1,1-dimethylurea (diuron) allowed the contribution of oxygenic and anoxygenic photosynthesis into primary organic matter production to be determined at different temperatures. The production processes in cyanobacterial mats occur concurrently with the degradation processes. By applying radioisotopic methods, it was shown that, depending on the sulfate content, the major terminal degradation process could be lithotrophic methanogenesis (Ward, 1978), sulfate reduction (Ward and Olson, 1980), or their combination (Bonch-Osmolovskaya *et al.*, 1987). Acetoclastic methanogenesis was found to be much slower than the lithotrophic one (Bonch-Osmolovskaya *et al.*, 1987) or failed to be detected at all (Ward, 1978). The processes in thermoalkaliphilic cyanobacterial mats were thoroughly studied in hot springs of Lake Baikal area (Namsaraev *et al.*, 2003).

Terrestrial hot springs in volcanic zones differ widely in their temperature and pH. Using *in situ* experiments with ^{14}C-labelled bicarbonate, rates of methanogenesis, acetogenesis, and inorganic carbon assimilation were determined in three hot springs of Uzon Caldera and Geyser Valley, Kamchatka, with pH 3.5, 7.0 and 8.5 at 55, 70 and 85°C (Bonch-Osmolovskaya *et al.*, 1999). Rates of sulfate reduction at different temperature and pH were determined in similar experiments with ^{35}S-labelled sulfate conducted in hot springs of Yellowstone (Roychoudhury, 2004).

Several microbial processes were assayed by radioisotopic methods in samples of hot sediments and water from shore and bottom hot springs of volcanic Green Lake (Raoul Island, Kermadek Archipelago) with temperatures ranging from 60 to 80°C (Starynin *et al.*, 1995). Active degradation of cellulose was found to occur in subsurface layers of bottom deposits at 60°C, indicating efficient degradation of allochthonous organic matter in this habitat.

Microbial communities inhabiting shallow-water hot vents of Western Pacific were thoroughly studied using radioisotopic methods by Namsaraev *et al.* (1994). Located in the shelf zone, these environments are rich in organic matter, actively degraded at moderate and high temperatures. The fermentation products are actively mineralized *via*

Table 2.1 Rates of microbial processes in thermal environments determined by radioisotope tracing methods

Location	Type of sample	Temperature (°C)	Rate	Reference
Photosynthetic carbon assimilation				
Kamchatka, Uzon Caldera, Thermophilny Spring	Cyanobacterial mat	61–56	1.62 µg C cm^{-2} h^{-1}	Gorlenko et al., 1987
Baikal Rift Zone, Bolsherechensky Spring	Cyanobacterial mat	62–51	0.105 g C m^{-2} d^{-1}	Namsaraev et al., 2003
Dark carbon assimilation				
New Guinea, Matupi Harbour, depth, 27 m	Sediment core	70	6.2 µg C l^{-1} d^{-1}	Namsaraev et al., 1994
Kamchatka, Uzon Caldera, Pulsating Spring	Sediment slurry	85	212.2 µg C l^{-1} d^{-1}	Bonch-Osmolovskaya et al., 1999
Lithotrophic methanogenesis				
Kamchatka, Uzon Caldera, Thermophilny Spring	Cyanobacterial mat	55–40	1.77 µg C cm^{-2} h^{-1}	Bonch-Osmolovskaya et al., 1987
New Zealand, Bay of Plenty, depth, 42 m	Sediment core	85	10.9 µl CH$_4$ l^{-1} d^{-1}	Namsaraev et al., 1994
Kermadek Archipelago, Raoul Island, Green Lake, depth, 6 m	Sediment core at depth 6–9 cm	60	49.9 µl CH$_4$ l^{-1} d^{-1}	Starynin et al., 1995
Mykhpayskoe oil field, Ob region, Russia	Formation water	60	5.44 µl CH$_4$ l^{-1} d^{-1}	Belyaev et al., 1990
Western Siberia, Samotlor oil reservoir, depth, 2299 m	Formation water	84	89 nmol CH$_4$ l^{-1} d^{-1}	Bonch-Osmolovskaya et al., 2003
Acetoclastic methanogenesis				
Kermadek Archipelago, Raoul Island, Green Lake, depth, 6 m	Sediment core at depth 6–9 cm	60	110.8 µl CH$_4$ l^{-1} d^{-1}	Starynin et al., 1995

(*continued*)

Table 2.1 Continued

Location	Type of sample	Temperature (°C)	Rate	Reference
New Guinea, Matupi Harbour, depth, 25 m	Sediment core	75	30.49 µl CH_4 l^{-1} d^{-1}	Namsaraev et al., 1994
Western Siberia, Samotlor oil reservoir, depth, 1850 m	Formation water	60	9.0 nmol CH_4 l^{-1} d^{-1}	Bonch-Osmolovskaya et al., 2003
Sulfate reduction				
Kamchatka, Uzon Caldera, Thermophilny Spring	Cyanobacterial mat	55–40	6.0 µg S cm^{-2} h^{-1*}	Bonch-Osmolovskaya et al., 1987
Yellowstone National Park, Mushroom Spring	Sediment core	59.1	483 nmol cm^{-3} d^{-1}	Roychoudhury et al., 2004
Kermadek Archipelago, Raoul Island, Green Lake, depth, 6 m	Sediment core at depth 1–3 cm	60	1.5 µg S l^{-1} d^{-1*}	Starynin et al., 1995
New Zealand, Bay of Plenty, depth, 42 m	Sediment core	85	448.4 µg S l^{-1} d^{-1*}	Namsaraev et al., 1994
Gulf of California, Guaymas Basin, depth, 2000 m	Sediment core at depth 23 cm	70	3,350 nmol cm^{-3} d^{-1}	Weber and Jørgensen, 2002
Gulf of California, Guaymas Basin, depth, 2013 m	Sediment core	95	6,700 nmol cm^{-3} d^{-1**}	Kallmeyer and Boetius, 2004
Mykhpayskoe oil field, Ob region, Russia	Formation water	60	19.0 µg S cm^{-2} h^{-1*}	Belyaev et al., 1990
Western Siberia, Samotlor oil field, depth, 2150 m	Formation water	84	18 nmol l^{-1} d^{-1*}	Bonch-Osmolovskaya et al., 2003

Notes: *Only the acid-volatile fraction of sulfides was determined.
**pressure 4.5×10^7 Pa.

methanogenesis and sulfate reduction. These processes were found to be most vigorous in shallow-water hot sediments of the Bay of Plenty, New Zealand, and Matupi Harbour, New Guinea.

Deep-sea hydrothermal fields are in many ways the most difficult objects for radioisotopic studies. The access to samples is very complicated and requires expensive equipment and participation of many people. These habitats are characterized by extremely steep gradients of temperature, pH, Eh, and concentrations of most of biogenic elements and are, therefore, very difficult to reconstruct *ex situ*. Different designs of sampling equipment for deep-sea investigations were developed and tested (Karl, 1995; Sagalevitch, 2002). This equipment allows samples as large as several decimeters in length to be taken and brought aboard ship without contact with the surrounding ocean water.

Hydrothermal sediments of the Guaymas Basin were an object of much environmental research, mostly concerned with estimating the rate of sulfate reduction. Experiments with labelled sulfates introduced directly into cores revealed high rates of sulfate reduction in this region (Jørgensen *et al.*, 1992; Weber and Jørgensen, 2002). Another set of experiments performed with slurry prepared from Guaymas hydrothermal sediments also revealed extremely active thermophilic sulfate reduction (Kallmeyer and Boetius, 2004).

The rate of mesophilic assimilation of CO_2 in samples of hydrothermal water from hydrothermal fields of the Mid-Atlantic Ridge was found to vary widely, having a maximum at 1.25 µg C l^{-1} day^{-1} (Pimenov *et al.*, 2000). The rate of carbon assimilation increased with pH, indicating that production processes occur in mixing zones of acidic hydrothermal fluid and neutral and oxygenated water. By contrast, the rates of inorganic carbon assimilation, sulfate reduction, and methanogenesis in the Lost City hydrothermal field of the Mid-Atlantic Ridge were maximal in undiluted hydrothermal fluid with temperature 40–70°C and pH 9.0–9.9 (Rusanov, personal communication). These data indicate the existence of an active sub-seafloor microbial community at that site.

Microbial processes in deep-subsurface thermal environments were studied using samples of formation water from high-temperature oil reservoirs (Belyaev *et al.*, 1990; Nazina *et al.*, 2000; Bonch-Osmolovskaya *et al.*, 2003). The rates of sulfate reduction, and lithotrophic and acetoclastic methanogenesis were found to be much lower than in volcanic habitats.

Potential Microbial Activity in Thermal Environments

Radioisotopic methods are often used to characterize microbial communities in terms of their potential activities under different incubation conditions rather than to determine the actual rates of microbial processes in natural habitats. Sediments of Lake Tanganyika (East Africa) with alkaline reaction (pH 8.5–9.2) and temperatures ranging from 66 to 103°C were used to study the rate of sulfate reduction as a function of pH and temperature (Elsgaard *et al.*, 1994). The sulfate-reducing activity was found to peak at 45 and 65°C and pH 7.0–7.5, with no activity detected at temperatures exceeding 80°C. Experiments with labelled sulfate conducted with samples from Guaymas Basin in a wide range of temperatures showed the existence of several groups of extremely active sulfate reducers having their temperature optima at 38, 80 and 110°C (Jørgensen et al., 1992). However, in another set of experiments performed at the same site, the sulfate reduction activity was detected only in the temperature range 70–100°C (Weber and Jørgensen, 2002).

Addition of possible energy substrates could stimulate the formation of the radioactive product, indicating the ability of microbial population to utilize such substrates as electron donors. Adding molecular hydrogen was found to significantly stimulate sulfate reduction in water and chimney samples from the East Pacific Rise, while the addition of acetate had no effect on this process (Bonch-Osmolovskaya and Pimenov, in preparation). The influence of different electron donors (formate, acetate and lactate) on sulfate reduction was studied *in situ* in hot springs of Yellowstone National Park but no effect was detected (Roychoudhury, 2004). The addition of a mixture of electron donors (yeast extract and peptone or a mixture of organic acids) to hydrothermal samples from Lake Tanganyika was found to increase the rate of this process and change its temperature profile (Elsgaard et al., 1994). Such results could guide future work on enrichment and isolation of new thermophilic microorganisms from the same site.

Detecting New Metabolic Groups of Microorganisms by *In Situ* Activity Assays

Radioisotopic tracing often reveals the activity of metabolic group not known for the given environment, or, occasionally, not known at all (Table 2.2). In some cases, these results, after a while, are substantiated by cultural methods, genomic, or phylogenetic analysis. The activity of moderately thermophilic sulfate-reducing and celullolytic thermophilic prokaryotes in shallow-water hot vents of Pacific Ocean was first shown by radioisotopic methods (Namsaraev et al., 1994). Six year later, representatives of these groups were either isolated or detected by 16S rRNA gene sequencing in studies of Aegean shallow-water hydrothermal systems (Sievert et al., 2000). The detection of endoglucanase genes in the genomes of *Thermococcales* (Ando et al., 2002; Bauer et al., 1999) is in agreement with cellulose-degrading activities detected in shallow-water hot vents at 80°C (Namsaraev et al., 1994).

Some physiological groups detected by radioisotopic methods are still waiting to be enriched and isolated. Radioisotopic studies of sulfate reduction in deep-sea habitats of Guaymas Basin revealed the existence of sulfate-reducing microorganisms active at 100°C, until now not obtained in laboratory cultures (Jørgensen et al., 1992). Another new group of thermophilic prokaryotes – hyperthermophilic

Table 2.2 New microbial activities detected in thermal environments by radioisotopic methods

Location	Type of sample	Temperature (°C)	Rate	Reference
Acetoclastic methanogenesis				
New Zealand, Bay of Plenty, depth, 42 m	Sediment core	85	30.49 µl CH$_4$ l^{-1} d^{-1}	Namsaraev et al., 1994
Western Siberia, Samotlor oil reservoir, depth, 2299 m	Formation water	84	19 nmol CH$_4$ l^{-1} d^{-1}	Bonch-Osmolovskaya et al., 2003
Acetogenesis				
Kamchatka, Uzon Caldera, Pulsating Spring	Sediment slurry	85	1.92 µg C l^{-1} d^{-1}	Bonch-Osmolovskaya et al., 1999
Sulfate reduction				
Gulf of California, Guaymas Basin, depth, 2000 m	Sediment core at 10–20 cm depth	105	19 µmol l^{-1} d^{-1}	Jørgensen et al., 1992
Anaerobic methane oxidation				
Gulf of California, Guaymas Basin, depth, 2013 m	Sediment core	62–85	1.6 nmol cm^{-3} d^{-1}	Kallmeyer and Boetius, 2004

acetogens – was detected by radioisotopic studies of Kamchatka hot springs (Bonch-Osmolovskaya *et al.*, 1999). Hyperthermophilic microorganisms capable of acetoclastic methanogenesis were detected in experiments with labelled acetate both in shallow-water hot vents (Namsaraev *et al.*, 1994) and in deep-sea hydrothermal habitats (Bonch-Osmolovskaya and Pimenov, in preparation) of Pacific Ocean.

Incubation of hydrothermal sediments from Guaymas in the presence of 100% methane gave rise to its anaerobic oxidation at temperatures ranging from 62 to 85°C (Kallmeyer and Boetius, 2004). The rates of the process were significantly higher than in the temperature range 5–33°C, revealing the existence of a thermophilic anaerobic methane-oxidizing consortium.

◆◆◆◆◆ CONCLUSION

Radioisotopic methods are widely used to study large-scale microbial processes. The rates obtained for such ecosystems as oceanic and fresh water sediments, soils, swamps and marshes make it possible to estimate annual production and consumption rates of major biogenic elements. Thermal environments (with a possible exception of deep-subsurface ones) are not very common on the Earth and, thus, fail to have a profound effect on the global ecology. In addition, they are characterized by an extremely wide diversity of physicochemical conditions (pH, temperature, oxygenation, etc.). Therefore, the data obtained for the given sampling site can hardly be extrapolated to the neighbouring one. In particular, this applies to deep-sea environments, where extremely steep gradients make precise measurements of their physicochemical parameters difficult if not impossible. The main limitation of the radioisotopic approach – the need to isolate samples from the surrounding environment – becomes especially significant in the case of thermal environments, where energy substrates and electron acceptors are largely supplied with the hydrothermal fluid.

The conclusions that one can draw from radioisotopic tracing experiments in thermal habitats mostly relate to their qualitative characteristics. By employing this sensitive and reliable method, new microbial processes and, therefore, physiological groups of microorganisms, at least for the habitat studied, were documented.

The radioisotopic approach employed along with cultivation methods and phylogenetic analysis of microbial communities helps us to build an adequate picture of microbial diversity in thermal environments and guide the future isolation work.

Acknowledgements

This work was supported by "Molecular and Cell Biology" Program of the Russian Academy of Sciences.

List of suppliers

Amersham Biosciences UK Limited
Amersham Place, Little Chalfont, Buckinghamshire HP7 9NA, England
Tel: +44-870-606-1921
Fax: +44-1494-544350
http://www1.amershambiosciences.com

Radiochemicals

ISOTOPE
Zagorodnyi prospect 13
Saint Petersburg 191002, Russia
Fax: +7-812-5957000
Tel: +7-812-314-46-86
E-mail: mail@izotop.ru
http://www.izotop.ru

Radiochemicals

Dionex Corporation
1228 Titan Way
Sunnyvale, CA 94085 USA
Tel: +1-408-737-0700
Fax: +1-408- 730-9403

DX 500 IC System, anion column IC-Pack™

Merck KGaA
Frankfurter Str. 250
Darmstadt, D-64293, Germany
Tel.: +49-6151-72-0
Fax: +49-6151-72-2000
E-mail: service@merck.de

Analytical test kits for alkalinity, sulfate

PerkinElmer Life and Analytical Sciences, Inc.
549 Albany Street
Boston, MA 02118-2512, USA
Tel: +1-800-762-4000 (USA)
Outside the USA +1-203-925-4602
Fax: +1-203-944-4904
E-mail: productinfo@perkinelmer.com

Liquid Scintillation Cocktails Ultima Gold™ LLT

References

Alain, K., Marteinsson, V. T., Miroshnichenko, M. L., Bonch-Osmolovskaya, E. A., Prieur, D. and Birrien, J.-L. (2002). *Marinitoga piezophila* sp. nov., a rod-shaped thermo-piezophilic bacterium isolated under high hydrostatic pressure from a deep-sea hydrothermal vent. *Int. J. Syst. Evol. Microbiol.* **52**, 1331–1339.

Albert, D. B. and Martens, C. S. (1997). Determination of low-molecular weight organic acid concentrations in seawater and pore-water samples via HPLC. *Mar. Chem.* **56**, 27–37.

Ando, S., Ishida, H., Kosugi, Y. and Ishikawa, K. (2002). Hyperthermostable endoglucanase from *Pyrococcus horikoshii*. *Appl. Env. Microbiol.* **68**, 430–432.

Barns, S. M., Fundyga, R. E., Jeffries, M. W. and Pace, N. R. (1994). Remarkable archaeal diversity detected in a Yellowstone National Park hot spring environment. *Proc. Natl. Acad. Sci. USA* **91**, 1609–1613.

Bauer, M. W., Driskill, L. E., Callen, W., Snead, M. A., Mathur, E. J. and Kelly, R. M. (1999). An endogluconase, EglA, from the hyperthermophilic archaeon *Pyrococcus furiosus* hydrolyzes β-1,4 bonds in mixed-linkage (1→3), (1→4)-β-D-glucans and cellulose. *J. Bacteriol.* **181**, 284–290.

Belyaev, S. S., Rozanova, E. P., Borzenkov, I. A., Charakhchyan, I. A., Miller, Yu. M., Sokolov, M. Yu. and Ivanov, M. V. (1990). Specificity of the microbiological processes in water-flooded oil field located in the middle Ob region. *Microbiology (Engl. Transl. of Mikrobiologiya)* **59**, 1075–1082.

Bonch-Osmolovskaya, E. A., Gorlenko, V. M., Karpov, G. A. and Starynin, D. A. (1987). Anaerobic destruction of organic matter in microbial mats of the Thermophilny Spring in the Uzon Caldera in Kamchatka. *Microbiology (Engl. Transl. of Mikrobiologiya)* **56**, 812–816.

Bonch-Osmolovskaya, E. A., Miroshnichenko, M. L., Slobodkin, A. I., Sokolova, T. G., Karpov, G. A., Kostrikina, N. A., Zavarzina, D. G., Prokofeva, M. I., Rusanov, I. I. and Pimenov, N. V. (1999). Biodiversity of anaerobic prokaryotes in terrestrial hot springs of Kamchatka. *Microbiology (Engl. Transl. of Mikrobiologiya)* **68**, 343–351.

Bonch-Osmolovskaya, E. A., Miroshnichenko, M. L., Lebedinsky, A. V., Chernyh, N. A., Nazina, T. N., Ivoilov, V. S., Belyaev, S. S., Boulygina, E. S., Lysov, Yu. P., Perov, A. N., Mirzabekov, A. D., Hippe, H., Stackebrandt, E., L'Haridon, S. and Jeanthon, C. (2003). Radioisotopic, culture-based and oligonucleotide microchip analyses of thermophilic microbial communities in a continental high-temperature petroleum reservoir. *Appl. Environ. Microbiol.* **69**, 6143–6151.

Castenholz, R. W. (1984) Composition of hot springs microbial mats: a summary. In *Microbial Mats: Stromatolites. MBL Lectures in Biology* (Y. Cohen, R. W. Castenholz and H. O. Halvorson, eds). vol. 3, pp. 101–120. Alan R. Liss, New York.

Conrad, R. and Schütz, H. (1988) Methods of studying methanogenic bacteria and methanogenic activities in aquatic environments. In *Methods in Aquatic Bacteriology* (B. Austin, ed.), pp. 301–343. J. Wiley & Sons, Chichester.

Elsgaard, L., Prieur, D., Mukwaya, G. M. and Jørgensen, B. B. (1994). Thermophilic sulfate reduction in hydrothermal sediment of Lake Tanganyika, East Africa. *Appl. Environ. Microbiol.* **60**, 1473–1480.

Fossing, H. and Jørgensen, B. B. (1989). Measurement of bacterial sulfate reduction in sediments: evaluation of a single-step chromium reduction method. *Biogeochemistry* **8**, 205–222.

Gorlenko, V. M., Bonch-Osmolovskaya, E. A., Kompantseva, E. I. and Starynin, D. A. (1987). Production of organic matter in microbial mats of the Thermophilny Spring in the Uzon Caldera in Kamchatka. *Microbiology (Engl. Transl. of Mikrobiologiya)* **56**, 692–697.

Huber, R., Huber, H. and Stetter, K. O. (2000). Towards the ecology of hyperthermophiles: biotopes, new isolation strategies and new metabolic propeties. *FEMS Microbiol. Rev.* **24**, 615–623.

Hugenholtz, P., Pitull, C., Hershberger, K. L. and Pace, N. R. (1998). Novel division level bacterial diversity in a Yellowstone hot spring. *J. Bacteriol.* **180**, 366–376.

Ivanov, M. V. (1956). Use of isotopes for study of sulfate reduction in Lake Belovod. *Mikrobiologiya* **25**, 301–313.

Jørgensen, B. B. (1978). A comparison of methods for quantification of bacterial sulfate reduction in coastal marine sediments. *Geomicrobiol. J.* **1**, 11–27.

Jørgensen, B. B., Isaksen, M. F. and Jannasch, H. W. (1992). Bacterial sulfate reduction above 100°C in deep-sea hydrothermal vent sediments. *Science* **258**, 1756–1757.

Kallmeyer, J. and Boetius, A. (2004). Effect of temperature and pressure on sulfate reduction and anaerobic oxidation of methane in hydrothermal sediments of Guaymas Basin. *Appl. Environ. Microbiol.* **70**, 1231–1233.

Karl, D. M. (1995) Ecology of free-living, hydrothermal vent microbial community. In *The Microbiology of Deep-Sea Hydrothermal Vents* (D. M. Karl, ed.), pp. 61–64. CRC Press, Boca Raton.

Kashefi, K., Holmes, D. E., Reysenbach, A.-L. and Lovley, D. R. (2002). Use of Fe(III) as an electron acceptor to recover previously uncultured hyperthermophiles: isolation and characterization of *Geothermobacterium ferrireducens* gen. nov., sp. nov. *Appl. Environ. Microbiol.* **68**, 1735–1742.

Laurinavichus, K. S. and Belyaev, S. S. (1978). The rate of microbial methane production determined by the radioisotopic techniques. *Mikrobiologiya* **47**, 1115–1117.

Lowry, O. H., Rosebrough, N. J., Farr, A. L. and Randall, R. J. (1951). Protein measurement with the Folin phenol reagent. *J. Biol. Chem.* **193**, 265–275.

Martens, C. S. (1990). Generation of short chain organic acid anions in hydrothermally altered sediments of the Guaymas Basin, Gulf of California. *Appl. Geochem.* **5**, 71–76.

Marteinsson, V. T., Birrien, J.-L., Reysenbach, A. L., Vernet, M., Marie, D., Gambacorta, A., Messner, P., Sleytr, U. and Prieur, D. (1999). *Thermococcus barophilus* sp. nov., a new barophilic and hyperthermophilic archaeon isolated under hydrostatic pressure. *Int. J. Syst. Bacteriol.* **49**, 351–359.

Miroshnichenko, M.L. and Bonch-Osmolovskaya, E.A., (2005). Recent developments in the thermophilic microbiology of deep-sea hydrothermal vents. Extremophiles, submitted.

Miroshnichenko, M. L., Kostrikina, N. A., L'Haridon, S., Jeanthon, S., Hippe, S., Stackebrandt, E. and Bonch-Osmolovskaya, E. A. (2002). *Nautilia lithotrophica* gen. nov., sp. nov., a thermophilic sulfur-reducing epsilon proteobacterium isolated from a deep-sea hydrothermal vent. *Int. J. Syst. Evol. Microbiol.* **52**, 1299–1304.

Miroshnichenko, M. L., Kostrikina, N. A., Chernyh, N. A., Pimenov, N. V., Tourova, T. P., Antipov, A. N., Spring, S., Stackebrandt, E. and Bonch-Osmolovskaya, E. A. (2003). *Caldithrix abyssi* gen. nov., sp. nov., a nitrate-reducing, thermophilic, anaerobic bacterium isolated from a Mid-Atlantic

Ridge hydrothermal vent, represents a novel bacterial lineage. *Int. J. Syst. Evol. Microbiol.* **53**, 747–752.

Namsaraev, B. B., Bonch-Osmolovskaya, E. A., Miroshnichenko, M. L., Pikuta, E. V., Kachalkin, V. I., Miller, Yu. M., Propp, L. N. and Tarasov, V. G. (1994). Microbiological processes of carbon cycle in shallow-water hot vents in the Western part of Pacific Ocean. *Microbiology (Engl. Transl. of Mikrobiologiya)* **63**, 59–65.

Namsaraev, Z. B., Gorlenko, V. M., Namsaraev, B. B., Buryuhkaev, S. P. and Yurkov, V. V. (2003). The structure and biogeochemical activity of the phototrophic communities from a Bol'sherechenskii alkaline hot spring. *Mikrobiologiya (Engl. Transl. of Mikrobiologiya)* **72**, 193–202.

Nazina, T. N., Xue, Y.-F., Wang, X.-Y., Belyaev, S. S. and Ivanov, M. V. (2000). Microorganisms of the high-temperature Liaohe oil field of China and their potential for MEOR. *Resour. Environ. Biotechnol.* **3**, 149–160.

Nercessian, O., Reysenbach, A.-L., Prieur, D. and Jeanthon, C. (2003). Archaeal diversity associated with *in situ* samplers deployed on hydrothermal vents on the East Pacific Rise (13°N). *Environ. Microbiol.* **5**, 492–502.

Pimenov, N. V., Lein, A. Yu., Sagalevitch, A. M. and Ivanov, M. V. (2000). Carbon dioxide assimilation and methane oxidation in various zones of the Rainbow Hydrothermal Field. *Microbiology (Engl. Transl. of Mikrobiologiya)* **69**, 623–634.

Reysenbach, A.-L., Longenecker, K. and Kirshtein, J. (2000). Novel bacterial and archaeal lineages from an *in situ* growth chamber deployed at a Mid-Atlantic Ridge hydrothermal vent. *Appl. Environ. Microbiol.* **66**, 3788–3797.

Roychoudhury, A. N. (2004). Sulfate respiration in extreme environments: a kinetic study. *Geomicrobiol. J.* **21**, 33–43.

Rusanov, I. I., Savvichev, S. K., Yusupov, S. K., Pimenov, N. V. and Ivanov, M. V. (1998). Production of exometabolites in the microbial oxidation of methane in marine ecosystems. *Microbiology (Engl. Transl. of Mikrobiologiya)* **67**, 710–717.

Sagalevitch, A. M. (2002) Deep-diving manned submersibles and studies of hydrothermal vents at the P. P. Shirshov Institute of Oceanology. In *Biology of Hydrothermal Systems* (A. V. Gebruk, ed.), pp. 59–71. KMK Press, Moscow.

Sievert, S. M., Kuever, J. M. and Muyzer, G. (2000). Identification of 16S ribosomal DNA-defined bacterial populations at a shallow submarine hydrothermal vent near Milos Island (Greece). *Appl. Environ. Microbiol.* **66**, 3102–3109.

Sorokin, Yu. I. (1999). *Radioisotopic Methods in Hydrobiology.* Springer-Verlag, Berlin.

Starynin, D. A., Namsraev, B. B., Bonch-Osmolovskaya, E. A., Kachalkin, V. I. and Propp, L. N. (1995). Microbial processes in the bottom sediments of Green Lake of Raoul Island (Kermadec Islands the Pacific Ocean). *Microbiology (Engl. Transl. of Mikrobiologiya)* **64**, 219–224.

Stetter, K. O. (1996). Hyperthermophilic prokaryotes. *FEMS Microbiol. Rev.* **18**, 149–158.

Ulrich, G. A., Krumholz, L. R. and Sulfita, J. M. (1997). A rapid and simple method for estimating sulfate reduction activity and quantifying inorganic sulfides. *Appl. Environ. Microbiol.* **63**, 1627–1630.

Ward, D. M. (1978). Thermophilic methanogenesis in a hot spring algal-bacterial mat (71–30°C). *Appl. Environ. Microbiol.* **35**, 1019–1026.

Ward, D. M. and Olson, G. J. (1980). Terminal processes in the anaerobic degradation of an algal-bacterial mat in a high sulfate hot spring. *Appl. Environ. Microbiol.* **40**, 67–74.

Weber, A. and Jørgensen, B. B. (2002). Bacterial sulfate reduction in hydrothermal sediments of the Guaymas Basin, Gulf of California, Mexico. *Deep-Sea Res.* **149**, 827–841.

Wirsen, C. O., Jannasch, H. W. and Molyneaux, S. J. (1993). Chemosynthetic microbial activity at Mid-Atlantic Ridge hydrothermal sites. *J. Geophys. Res.* **98**, 9693–9703.

Wirsen, C. O., Tuttle, J. H and Jannasch, H. W. (1986). Activities of sulfur-oxidizing bacteria at the 21°N East Pacific Rise vent site. *Mar. Biol.* **92**, 449–456.

Zeikus, J. G., Ben-Bassat, A. and Hegge, P. W. (1980). Microbiology of methanogenesis in thermal, volcanic environments. *J. Bacteriol.* **147**, 432–440.

3 The Isolation of Thermophiles from Deep-sea Hydrothermal Environments

Satoshi Nakagawa and Ken Takai

Subground Animalcule Retrieval (SUGAR) Program, Extremobiosphere Research Center, Japan Agency for Marine-Earth Science and Technology (JAMSTEC), 2–15 Natsushima-cho, Yokosuka 237-0061, Japan

◆◆

CONTENTS

Introduction
Sample collection
Sample processing and preservation
Media
Cultivation conditions
Isolation
Concluding remarks and perspectives

◆◆◆◆◆ **INTRODUCTION**

The discovery of deep-sea hydrothermal vents in the Mid-Ocean Ridges (Spiess and RISE group, 1980) revolutionized microbiology in the deep-sea. Before the discovery, the deep-sea had been simply regarded as a cold, high-pressure, "desert-like" environment, but now physicochemically diverse, nutrient-rich, and biologically productive habitats emerged to the microbiologists. Over 60 different thermophilic and hyperthermophilic species with a variety of physiological characteristics have so far been isolated in pure culture from deep-sea hydrothermal environments. In addition, recent studies using culture-independent approaches have provided the growing consensus that these extreme environments still harbor a diversity of yet-uncultivated microorganisms, some of which are definitely involved in the biogeochemical processes *in situ* (Huber *et al.*, 2003; Nakagawa *et al.*, 2005b; Nercessian *et al.*, 2003; Reysenbach *et al.*, 2000; Schrenk *et al.*, 2003; Takai and Horikoshi, 1999; Teske *et al.*, 2002). Consequently, the ecology and physiology of yet-uncultivated microorganisms in the deep-sea hydrothermal environments are of particular interest. The importance of cultivation-dependent studies is increasingly recognized since the cellular and molecular basis of the physiology of isolates provides important clues to the ecophysiological potential and function of the microbial communities (Jeanthon, 2000).

Recent cultivation studies have greatly expanded the diversity of cultivable thermophiles, especially the diversity of moderate to extreme thermophiles that had escaped from the early stage of cultivation explorations mainly targeting hyperthermophiles. This progress is attributed to the increasing and diversifying opportunities for sampling and the technical developments of sampling and cultivation methods, and are also based on our increased understanding of microbial diversity and distribution from culture-independent surveys. In this chapter, we describe the methods for enrichment and isolation of (hyper)thermophiles from deep-sea hydrothermal environments. Excellent reviews of multidisciplinary researches in deep-sea hydrothermal environments were provided in the books edited by Humphris *et al.* (1995), Karl (1995), Van Dover (2000) and Wilcock *et al.* (2004). In the following sections, we employ the term "thermophile" to refer to all microorganisms with optimum growth temperatures above 50°C, including hyperthermophiles that grow optimally above 80°C.

◆◆◆◆◆ SAMPLE COLLECTION

Sample collection has been among the biggest challenges in many fields of deep-sea sciences. For deep-sea water columns and sediments, multiple water samplers, piston corers, grab samplers and dredge samplers have been generally operated from the surface ship. However, for habitat-specific sampling of deep-sea hydrothermal vents and cold seepages, Deep Submergence Vehicles (DSVs) and Remotely Operative Vehicles (ROVs) are required. Typical samples from deep-sea hydrothermal environments consist of fluids (i.e. low- or high-temperature hydrothermal fluids, plumes and ambient seawater) and solids (i.e. sulfide or non-sulfide chimney structures, flange structures, rocks, sediments and animal specimens). Various specialized sampling instruments have been developed for specific research purposes as described in a previous volume in this book series (Reysenbach and Götz, 2001). Briefly, fluid samples are collected by water samplers including Niskin bottles, titanium syringes (Von Damm *et al.*, 1985), and ORI-water samplers (Sakai *et al.*, 1990). Gas-tight water samplers, e.g. the Lupton sampler (Lupton *et al.*, 1985) and WHATS (Tsunogai *et al.*, 2003), have also been developed for geochemical gas measurements. The fluids from these pressure-preserved water samplers conserve the *in situ* chemical conditions to a considerable extent, and are excellent samples for microbiological investigations as well. A variety of solid samples such as chimney structures and animal specimens are taken by using DSV's or ROV's manipulators, and then recovered in an insulated box that is part of the equipment of DSV or ROV. Sediment samples are most often collected by push corers, but pressure-preserved, sterile sediment samplers are also used for this specific purpose (Kato *et al.*, 1997).

It is worth noting that subseafloor-sampling techniques have been developed for various matrices ranging from unconsolidated sediments to hard rocks (Phelps and Fredrickson, 2002). Now it is clear that subseafloor environments provide numerous opportunities for microbial growth and survival (Fredrickson and Fletcher, 2001; Takai et al., 2004c), and this is also applicable to deep-sea hydrothermal fields (Kimura et al., 2003). With the increasing interest in the sub-seafloor biosphere, the sampling opportunities provided by the Integrated Ocean Drilling Program (IODP; formerly ODP) is thus becoming more attractive. Several drilling expeditions to the biosphere beneath deep-sea hydrothermal areas (subvent biosphere) have been conducted within the framework of ODP (Cragg et al., 2000; Kimura et al., 2003; Reysenbach et al., 1998). These ODP attempts, however, did not successfully demonstrate the existence of non-contaminated, indigenous sub-seafloor microbial ecosystems. More interactively, a new scientific deep-sea drilling ship "Chikyu" was recently constructed to achieve up to 7000 m penetration below the seafloor. It is operated by the Japan Agency for Marine-Earth Science and Technology (JAMSTEC) (http://www.jamstec.go.jp). This ship is equipped with riser-drilling mechanics, which enable safe drilling of the seafloor containing gas-enriched fluids such as deep-sea hydrothermal systems and methane-seepage fields.

Hydrothermal habitats harbor thermophiles of both prokaryote domains: Archaea and Bacteria (Figure 3.1). Although Bacteria generally dominate hydrothermal habitats, the ratio of Archaea to Bacteria varies considerably in different sites (Harmsen et al., 1997; Hedrick et al., 1992; Nakagawa et al., 2005b; Schrenk et al., 2003; Takai et al., 2001). The highest temperature habitats such as interior parts of chimney structures appear to be dominated by Archaea (Harmsen et al., 1997; Nakagawa et al., 2005b; Takai et al., 2001, 2004a). For example, Takai et al. (2004a) reported >99% dominance of Archaea in in situ samplers exposed to hydrothermal fluids from the Central Indian Ridge. Several estimations of the abundance of whole microbial communities in various hydrothermal samples have been reported; numbers are in the range of 10^4–10^9 cells per gram of chimney structure (Harmsen et al., 1997; Hedrick et al., 1992; Nakagawa et al., 2005b; Schrenk et al., 2003; Suzuki et al., 2004; Takai and Horikoshi, 1999; Takai et al., 2001), 10^4–10^9 cells per gram of hydrothermal sediment (Kimura et al., 2003; Takai and Horikoshi, 1999), and 10^4–10^8 cells per ml of hydrothermal plumes and ambient seawater (Maruyama et al., 1993; Nakagawa et al., 2005c; Takai et al., 2004f). Superheated vent fluids are characterized by low abundances of microorganisms: numbers reported differ from undetectable to 10^6 cells per ml (Takai and Horikoshi, 1999; Takai et al., 2004a; Nakagawa et al., 2005b). Microorganisms found in the vent fluids have been of particular interest since the microorganisms probably include indigenous inhabitants of the subsurface biosphere (Baross and Deming, 1995). However, samples with less abundant microbial populations are not easily and reliably analyzed by molecular methods (e.g. fluorescent in situ hybridization (FISH)). In addition, even a small amount of mixing with surrounding seawater would easily overwhelm the signals of the indigenous vent fluid population. To overcome

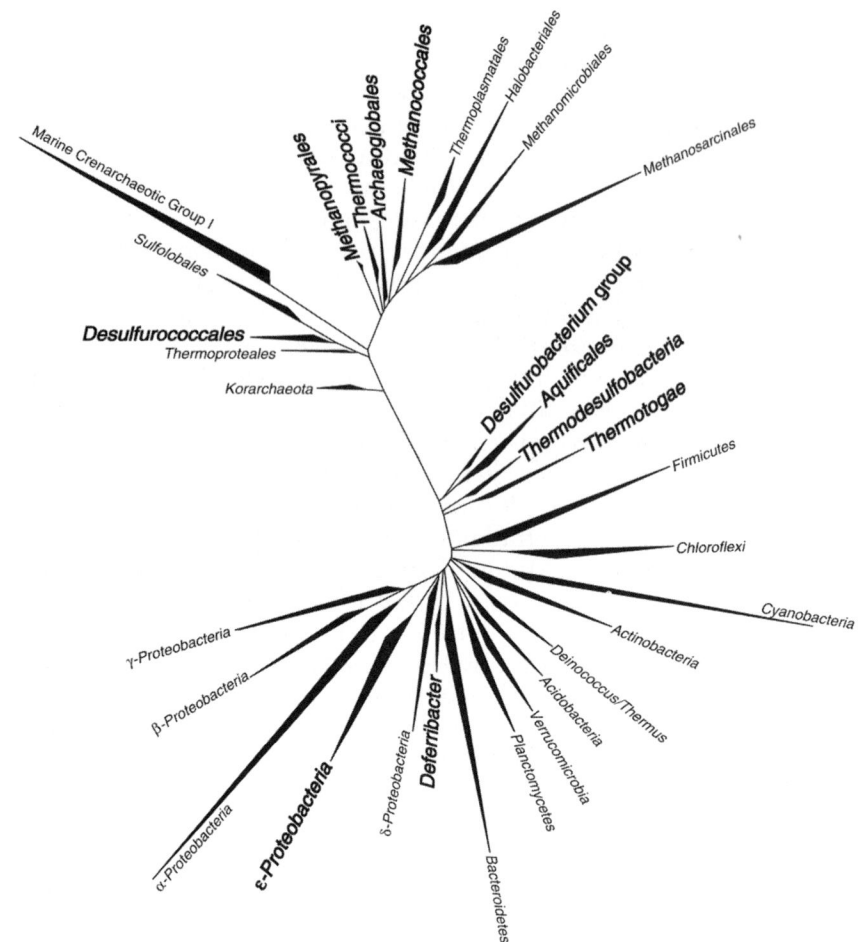

Figure 3.1. 16S rRNA-based tree showing the major groups of Archaea and Bacteria. The tree was constructed, evaluated and optimized using the ARB parsimony tool. Groups including deep-sea thermophiles are indicated by bold face

these problems, various types of *in situ* samplers have been developed. Microbiologists deploy *in situ* samplers, i.e. devices called vent caps (Reysenbach *et al.*, 2000), the *In Situ* Colonization System (ISCS) (Takai *et al.*, 2003a), microcolonizers (López-García *et al.*, 2003b), and the Titanium Ring for *Alvinella* Colonization (TRAC) (Taylor *et al.*, 1999), not only to gather microbial components entrained by vent fluids, but also to investigate the microbial colonization process associated with vent emissions.

We have recently modified the temperature probe and recorder of the ISCS. Similar to the original version described in Takai *et al.* (2003a), the body of ISCS consists of a perforated stainless-steel pipe filled with very porous, natural silica matrices for microbial attachment (Figure 3.2). This device has been inserted into the hydrothermal conduits of chimney structures or deployed adjacent to the vent emissions (Nakagawa *et al.*, 2005b,c; Takai *et al.*, 2004a). During the deployment, the temperature shift of the silica substratum can be recorded by the thermal-resistant (0–400°C) temperature probe inserted into the substratum (Figure 3.2). In all the cases in which the ISCS was inserted into a hydrothermal conduit of a vigorous black or clear smoker, temperatures of the substratum were stable and identical to those of vent fluids.

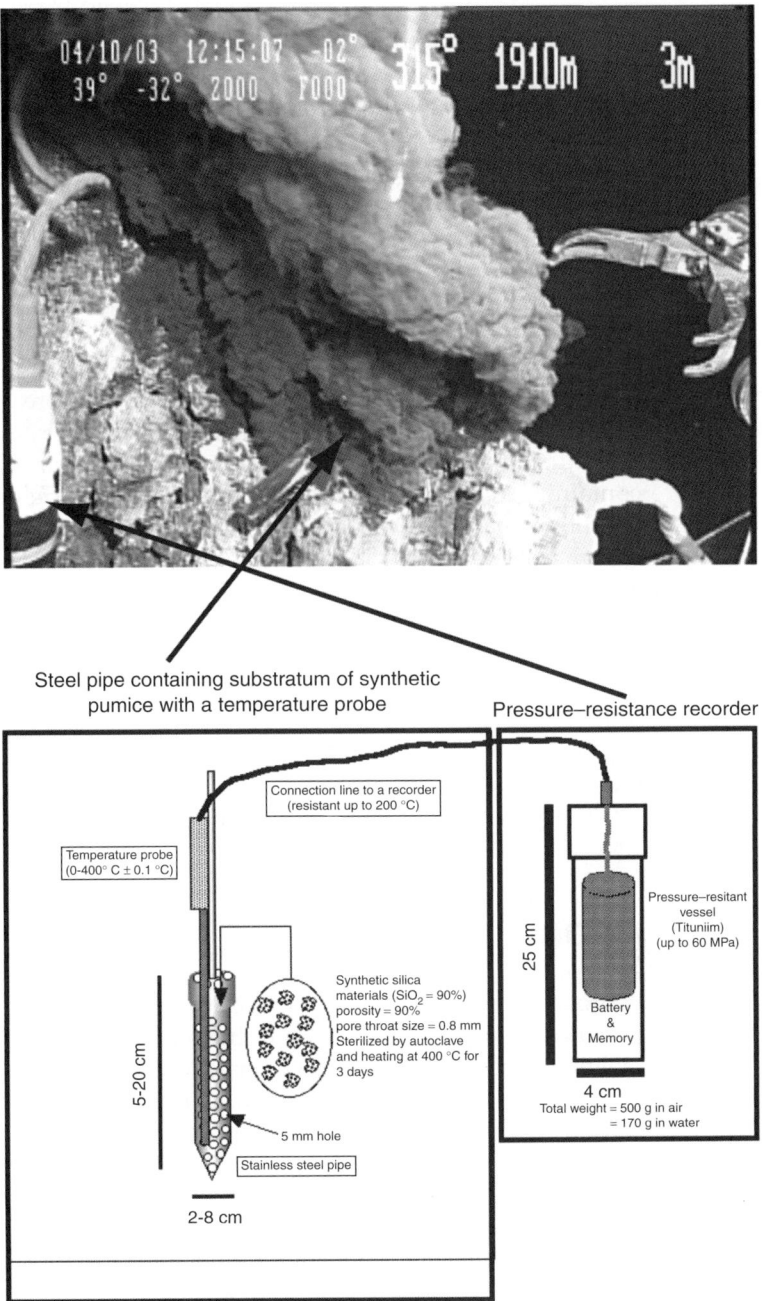

Figure 3.2. *In Situ* Colonization System (ISCS) with a modified temperature probe and data logger. See Plate 3.2 in Colour Plate Section.

◆◆◆◆◆ SAMPLE PROCESSING AND PRESERVATION

Samples should be processed immediately once recovered on board, because most thermophiles are obligate or facultative anaerobes. The previously described exceptions, a few *Thermus* strains (Marteinsson *et al.*, 1995), *Marinithermus hydrothermalis* (Sako *et al.*, 2003), and *Aeropyrum camini* (Nakagawa *et al.*, 2004b), are obligatory aerobic thermophiles capable of growing under air atmosphere.

It is advisable to store samples under anaerobic and reducing conditions to improve the recovery of cultivable thermophiles. We usually store samples in glass bottles (100–250 ml; Schott, Mainz, Germany) tightly sealed with gas-impermeable butyl rubber stoppers (43 mm black lyophilization stoppers; Wheaton Science Products, Millville, USA) under a N_2 atmosphere. Subsampling of solid samples (especially chimney structures) should be immediately performed onboard to assess the distribution and the heterogeneity of the microbial communities across the structures (Harmsen *et al.*, 1997; Hedrick *et al.*, 1992; Nakagawa *et al.*, 2005b; Schrenk *et al.*, 2003; Takai *et al.*, 2001). In order to subsample sediment cores without perturbation of stratigraphy, plastic syringes are used of which the needle end has been cut off. Sectioned solid samples are slurried for inoculating into anaerobic media with sterile seawater containing reducing agents such as sodium sulfide or sodium dithionite (usually used at 0.03–0.05%, w/v). Some researchers store solid samples within anaerobic jars made of sealed polycarbonate containers containing a catalyst to remove oxygen gas. Alternatively, plastic glove bags filled with nitrogen or argon gas are also available. Fluid samples are often stored in glass vials tightly sealed with gas-impermeable butyl rubber stoppers under a N_2 atmosphere. Reducing agents are added to the fluid samples when dissolved oxygen should be completely removed.

The storage temperature may also affect the culturability of microorganisms. Samples processed as described above are usually kept refrigerated in the dark. If not applicable, short-time preservation at room temperature probably does not critically damage thermophiles.

◆◆◆◆◆ MEDIA

A broad array of thermophilic species of both the archaeal and bacterial domains has been isolated and characterized from deep-sea hydrothermal environments (Figure 3.1). Based on their metabolic traits, the deep-sea thermophiles can be classified as follows: fermenters, methanogens, aerobic heterotrophs, and autotrophic or heterotrophic nitrate-, sulfate-, sulfur-, and iron-reducers (Table 3.1). There are some cosmopolitan thermophiles that have been isolated and detected from geographically disparate deep-sea hydrothermal fields (Table 3.2 and Figure 3.3). The most commonly and dominantly found thermophiles are members of the order *Thermococcales*, and the epsilon class of the *Proteobacteria* (ε-*Proteobacteria*). These, together with the thermophiles of the orders *Aquificales*, *Archaeoglobales* and *Methanococcales*, are the major microbial

Table 3.1 Metabolic type of thermophiles previously isolated from deep-sea hydrothermal environments

Metabolic type	Order (Genus)	Substrates or electron donor	Electron acceptor	Carbon source
Fermenters	*Thermococcales* (*Thermococcus, Pyrococcus, Palaeococcus*), *Thermotogales* (*Thermosipho and Marinitoga*), *Clostridiales* (*Tepidibacter, Caminicella, Caloranaerobacter*), and *Desulfurococcales* (*Staphylothermus*)	Organic matter	- (H^+)	Organic matter
	Thermoanaerobacterales (*Caldanaerobacter*)	H_2O, organic matter	CO, -	Organic matter, CO
Methanogens	*Methanococcales* (*Methanocaldococcus, Methanotorris, Methanothermococcus*) and *Methanopyrales* (*Methanopyrus*)	H_2 (formate)	CO_2	CO_2
Aerobic heterotrophs	*Desulfurococcales* (*Aeropyrum*)	H_2	CO_2	CO_2
	Thermales (*Marinithermus*)	Organic matter	O_2	Organic matter
	Thermales (*Vulcanithermus and Oceanithermus*)	Organic matter	O_2	Organic matter
		Organic matter (H_2)	O_2, NO_3^- (NO_2^-, S^0)	Organic matter
Obligate chemolithoautotrophs	*Desulfurococcales* (*Pyrolobus*)	H_2	NO_3^-, $S_2O_3^{2-}$, O_2	CO_2
	Desulfurococcales (*Ignicoccus*), *Desulfurobacterium* group (*Desulfurobacterium, Balnearium, Thermovibrio*), and *Nautiliales* (*Lebetimonas*)	H_2	S^0 (NO_3^-, $S_2O_3^{2-}$, SO_3^{2-})	CO_2

(*continued*)

Table 3.1 Continued

Metabolic type	Order (Genus)	Substrates or electron donor	Electron acceptor	Carbon source
	Thermodesulfobacterales (*Thermodesulfobacterium* and *Thermodesulfatator*)	H_2	SO_4^{2-}	CO_2
	Aquificales (*Persephonella* and *Aquifex*)	H_2 (S^0, $S_2O_3^{2-}$)	O_2, NO_3^- (S^0)	CO_2
	Nautiliales (*Caminibacter*) and Unclassified (*Hydrogenimonas* and *Nitratiruptor*)	H_2	NO_3^-, S^0, O_2	CO_2
Facultative chemolithoautotrophs	*Archaeoglobales* (*Archaeoglobus* and *Geoglobus*)	H_2, organic matter	SO_4^{2-}, $S_2O_3^{2-}$, SO_3^{2-}, Fe(III)	Organic matter, CO_2
	Deferribacterales (*Deferribacter*)	H_2, organic matter	NO_3^-, S^0, Fe(III), arsenate	Organic matter, CO_2
	Nautiliales (*Caminibacter* and *Nautilia*)	H_2, organic matter	NO_3^-, S^0, SO_3^{2-}, colloidal sulfur	Organic matter, CO_2
	Thiotrichales (*Thiomicrospira*)	S^0, $S_2O_3^{2-}$, S^{2-}	O_2	Organic matter, CO_2
Obligately anaerobic heterotrophs	*Desulfurococcales* (*Pyrodictium*)	H_2, organic matter	-, S^0, $S_2O_3^{2-}$	Organic matter
	Desulfuromonadales (*Geothermobacter*)	organic matter	NO_3^-, Fe(III)	Organic matter
	Unclassified (*Caldithrix*)	H_2, organic matter	NO_3^-	Organic matter

components in both culture-dependent and culture-independent analyses (Harmsen *et al.*, 1997; Huber *et al.*, 2002; Nakagawa *et al.*, 2005b,c; Nercessian *et al.*, 2004; Sunamura *et al.*, 2004; Takai *et al.*, 2001, 2004a).

The cultivation media delineate the metabolic traits of microorganisms to be enriched, and they should be designed to reflect the physicochemical conditions of the habitats studied. Despite the low pH of the vent fluids (generally pH < 5) (Von Damm, 1995), apparently thermoacidophiles such as members of the *Sulfolobales* commonly found in terrestrial, acidic hot springs have never been isolated from the deep-sea. The reason may be that thermoacidophiles cannot maintain their intracellular neutral pH at low temperatures, and lyse during sample recovery (Reysenbach and Götz, 2001). A recently described ε-proteobacterium, *Lebetimonas acidiphila*, has the lowest optimum pH (pH 5.2) for growth among the deep-sea thermophiles characterized (Takai *et al.*, 2005).

All deep-sea thermophiles listed in Table 3.2 have a requirement for salt, which is commonly provided in the form of NaCl. For the cultivation of deep-sea thermophiles, various recipes of synthetic seawater containing different trace minerals have been employed by different researchers. The components of synthetic seawater are basically similar to those used for temperate marine microorganisms, although there are significant geochemical differences between vent fluids and seawater (Von Damm, 1995). In addition to the compositions of the basal saline, the choice of electron donors, electron acceptors, nitrogen sources and carbon sources will lead to the enrichment of a specific physiological group of microorganisms.

Growth of most deep-sea thermophiles requires strictly anaerobic techniques. Anaerobic enrichments are usually performed in 15–30 ml Hungate tubes or in 100–250 ml bottles, sealed with gas-impermeable butyl rubber stoppers. Volumes of media should be less than 20% of the bottles or tubes. In most cases, the headspace gas is H_2, H_2/CO_2 (80:20), N_2 or N_2/CO_2 (80:20). Details of the anaerobic technique and experimental setup were described by Balch and Wolfe (1976). The outline of the anaerobic technique is as follows:

- Boil the medium and flush with the desired gas prior to dispensing into tubes. Continue flushing when dispensing the medium.
- Seal the tubes with gas-impermeable butyl rubber stoppers, crimped with aluminum caps, and autoclave.
- After autoclaving, add sterile Na_2S solution (final concentration: 0.03–0.05%, w/v) to remove remaining O_2. Alternatively, especially in cases when sulfide should be avoided, the medium can be reduced with titanium citrate (1.0 mM Ti^{3+} and 2.06 mM sodium citrate) (Zehnder and Wuhrmann, 1976).

Below we list representative recipes for synthetic seawater, trace mineral solution, and vitamin solution, which can be used as the basal medium for the enrichment and isolation of various thermophiles. The following sections on (1) heterotrophic, S^0-utilizing thermophiles, (2) chemolithoautotrophic thermophiles, and (3) hydrogenotrophic methanogens describe specific formulations and brief procedures for the preparation of the media with modified anaerobic techniques.

Table 3.2 Thermophilic species isolated from deep-sea hydrothermal environments

Domain, Phylum	Genus	Species	Source*	Temp. range (°C) (opt.)	Reference
Archaea					
Euryarchaeota	Methanopyrus	kandleri	Guaymas Basin	84–110 (98)	Kurr et al., 1991
	Archaeoglobus	profundus	Guaymas Basin	65–90 (82)	Burggraf et al., 1990
		veneficus	Snake Pit, MAR	65–85 (80)	Huber et al., 1997
	Geoglobus	ahangari	Guaymas Basin	65–90 (88)	Kashefi et al., 2002
	Thermococcus	aggregans	Guaymas Basin	60–94 (88)	Canganella et al., 1998
		guaymasensis	Guaymas Basin	56–90 (88)	Canganella et al., 1998
		barophilus	Snake Pit, MAR	48–95 (85)	Marteinsson et al., 1999
		barossii	Juan de Fuca	60–92 (82.5)	Duffaud et al., 1998
		chitonophagus	Guaymas Basin	60–93 (85)	Huber et al., 1996
		fumicolans	North Fiji Basin	73–103 (85)	Godfroy et al., 1996
		gammatolerans	Guaymas Basin	55–95 (88)	Jolivet et al., 2003
		hydrothermalis	21°N, EPR	55–100 (80–90)	Godfroy et al., 1997
		peptonophilus	South Mariana Trough	60–100 (85)	Gonzalez et al., 1995
		profundus	Iheya Ridge, Okinawa Trough	50–90 (80)	Kobayashi et al., 1994
		siculi	Iheya Ridge, Okinawa Trough	50–93 (85)	Grote et al., 1999
	Pyrococcus	"abyssi"	North Fiji Basin	67–102 (96)	Erauso et al., 1993
		glycovorans	13°N, EPR	75–104 (95)	Barbier et al. 1999
		horikoshii	Iheya Ridge, Okinawa Trough	80–102 (98)	Gonzales et al., 1998
	Palaeococcus	ferrophilus	Myojin, Izu-Bonin Arc	60–88 (83)	Takai et al., 2000
	Methanocaldococcus	jannaschii	20°N, EPR	50–91 (85)	Jones et al., 1983
		fervens	Guaymas Basin	48–92 (85)	Jeanthon et al., 1999
		infernus	Logatchev, MAR	55–91 (85)	Jeanthon et al., 1998
		vulcanius	13°N, EPR	49–89 (80)	Jeanthon et al., 1999

	Genus	species	Location	Temp range (opt)	Reference
Crenarchaeota	Methanotorris	indicus	Kairei, CIR	50–86 (85)	L'Haridon et al., 2003
	Methanothermococcus	formicicus	Kairei, CIR	55–83 (75)	Takai et al., 2004e
		okinawensis	Iheya North, Okinawa Trough	40–75 (60–65)	Takai et al., 2002
	Pyrodictium	"abyssi"	Guaymas Basin	80–110 (97)	Pley et al., 1991
	Pyrolobus	fumarii	TAG, MAR	90–113 (106)	Blöchl et al., 1997
	Ignicoccus	pacificus	9°N, EPR	70–98 (90)	Huber et al., 2000
	Aeropyrum	camini	Suiyo Seamount, Izu-Bonin Arc	70–97 (85)	Nakagawa et al., 2004b
	Staphylothermus	marinus	EPR	65–98 (92)	Fiala et al., 1986
Bacteria					
Aquificae	Persephonella	marina	13°N, EPR	55–80 (73)	Götz et al., 2002
		guaymasensis	Guaymas Basin	55–75 (70)	Götz et al., 2002
		hydrogeniphila	Suiyo Seamount, Izu-Bonin Arc	50–72.5 (70)	Nakagawa et al., 2003
Desulfurobacterium group	Desulfurobacterium	thermolithotrophum	Snake Pit, MAR	40–75 (70)	L'Haridon et al., 1998
	Balnearium	"crinifex"	Axial Volcano, Juan de Fuca	50–70 (60–65)	Alain et al., 2003
		lithotrophicum	Suiyo Seamount, Izu-Bonin Arc	45–80 (70–75)	Takai et al., 2003c
Thermotogae	Thermovibrio	ammonificans	9°N, EPR	60–80 (75)	Vetriani et al., 2004
	Thermosipho	melanesiensis	Lau Basin	50–75 (70)	Antoine et al., 1997
		japonicus	Iheya North, Okinawa Trough	45–80 (72)	Takai and Horikoshi, 2000
		atlanticus	Menez-Gwen, MAR	45–80 (65)	Urios et al., 2004b
	Marinitoga	camini	Menez-Gwen, MAR	25–65 (55)	Wery et al., 2001a
		piezophila	13°N, EPR	45–70 (65)	Alain et al., 2002a

(continued)

Table 3.2 Continued

Domain, Phylum	Genus	Species	Source*	Temp. range (°C) (opt.)	Reference
Thermodesulfobacteria	*Thermodesulfobacterium*	*hydrogeniphilum*	Guaymas Basin	50–80 (75)	Jeanthon et al., 2002
	Thermodesulfatator	*indicus*	Kairei, CIR	55–80 (70)	Moussard et al., 2004
Deferribacteres	*Deferribacter*	*abyssi*	Rainbow, MAR	45–65 (60)	Miroshnichenko et al., 2003d
		desulfuricans	Suiyo Seamount, Izu-Bonin Arc	40–70 (60–65)	Takai et al., 2003b
Deinococcus-Thermus	*Marinithermus*	*hydrothermalis*	Suiyo Seamount, Izu-Bonin Arc	50–72.5 (67.5)	Sako et al., 2003
	Vulcanithermus	*mediatlanticus*	Rainbow, MAR	37–80 (70)	Miroshnichenko et al., 2003c
	Oceanithermus	*profundus*	13°N, EPR	40–68 (60)	Miroshnichenko et al., 2003b
		desulfurans	Suiyo Seamount, Izu-Bonin Arc	30–65 (60)	Mori et al., 2004
Firmicutes	*Tepidibacter*	*thalassicus*	13°N, EPR	33–60 (50)	Slobodkin et al., 2003
		formicigenes	Lucky Strike, MAR	35–55 (45)	Urios et al., 2004a
	Caminicella	*sporogens*	13°N, EPR	45–65 (55–60)	Alain et al., 2002b
	Clostridium	*caminithermale*	Menez-Gwen, MAR	20–58 (45)	Brisbarre et al., 2003
	Caloranaerobacter	*azorensis*	Lucky Strike, MAR	45–65 (65)	Wery et al., 2001b
	Caldanaerobacter (formerly *Carboxydibrachium pacificum*)	*subterraneus*	Iheya Ridge, Okinawa Trough	45–65 (65)	Sokolova et al., 2001
Proteobacteria	*Caminibacter*	*hydrogeniphilus*	13°N, EPR	50–70 (60)	Alain et al., 2002c
		profundus	Rainbow, MAR	45–65 (55)	Miroshnichenko et al., 2004
		mediatlanticus	Rainbow, MAR	45–70 (55)	Voordeckers et al., 2005

Nautilia	lithotrophica	13°N, EPR	37–68 (53)	Miroshnichenko et al., 2002
Lebetimonas	acidiphila	TOTO, Mariana Arc	30–68 (50)	Takai et al., 2005
Hydrogenimonas	thermophila	Kairei, CIR	35–65 (55)	Takai et al., 2004d
Nitratiruptor	tergarcus	Iheya North, Okinawa Trough	40–57 (55)	Nakagawa et al., 2005a
Thiomicrospira	thermophila	TOTO, Mariana Arc	15–55 (35–40)	Takai et al., 2004b
Geothermobacter	ehrlichii	Bag City, Juan de Fuca	35–65 (55)	Kashefi et al., 2003
Unclassified Caldithrix	abyssi	Logatchev, MAR	40–70 (60)	Miroshnichenko et al., 2003a

*MAR, Mid-Atlantic Ridge; EPR, East Pacific Rise; CIR, Central Indian Ridge.

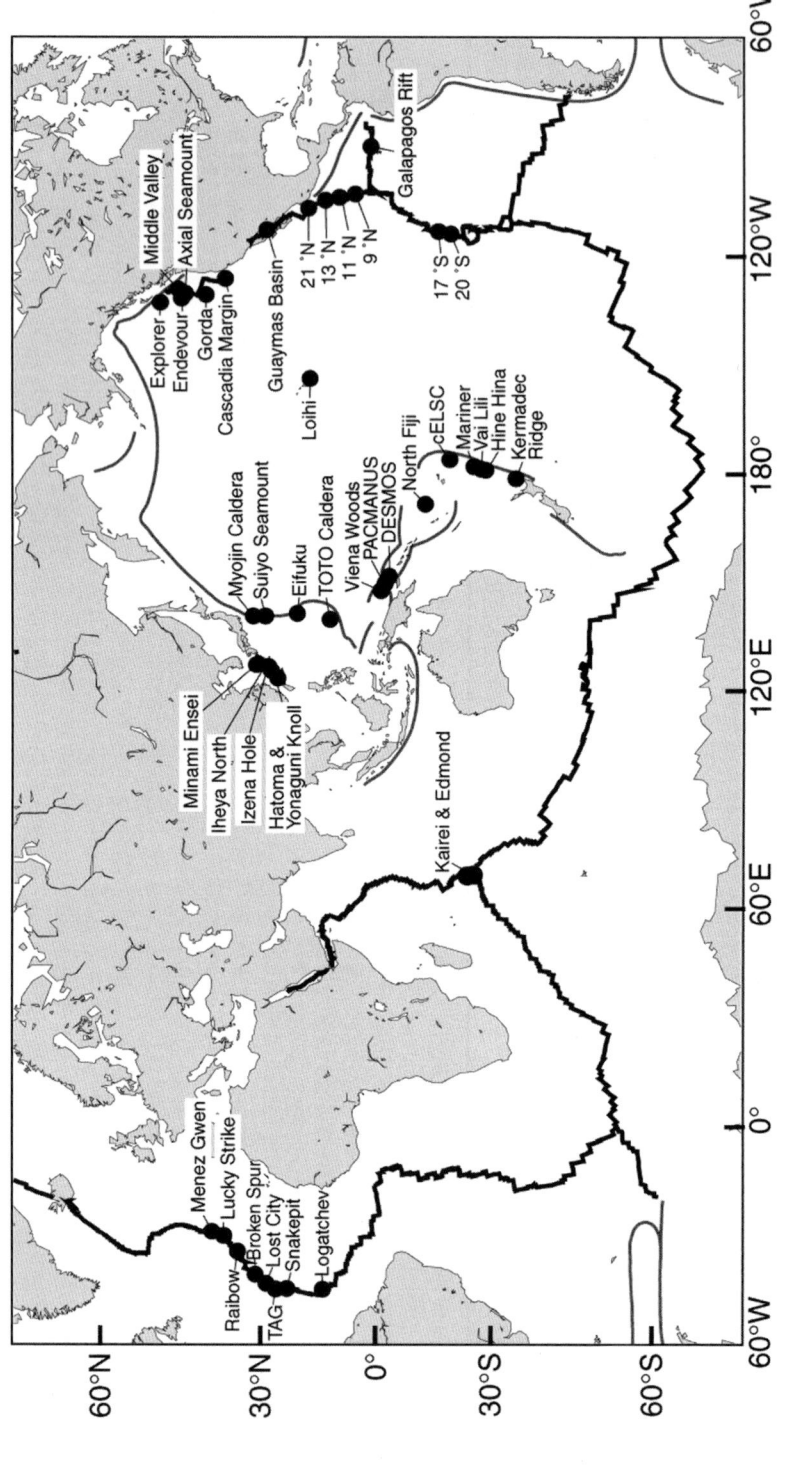

Figure 3.3. Location of deep-sea hydrothermal systems

Basal Saline for Enrichment Media

Synthetic seawater (modified from MJ synthetic seawater of Sako et al., 1996) (per liter of distilled, deionized water (DDW)):

- NaCl   25 g
- $MgCl_2 \cdot 6H_2O$ 4.2 g
- $MgSO_4 \cdot 7H_2O$ 3.4 g
- KCl   0.5 g
- NH_4Cl 0.25 g
- K_2HPO_4 0.14 g
- $CaCl_2 \cdot 2H_2O$ 0.7 g
- $FeSO_4 \cdot 7H_2O$ 0.02 g
- Trace mineral solution (described below)   10 ml

For comparison, the following is the formulation of another useful synthetic seawater (from SME medium described by Stetter et al., 1983) (per liter of DDW):

- NaCl   13.85 g
- $MgCl_2 \cdot 6H_2O$ 2.75 g
- $MgSO_4 \cdot 7H_2O$ 3.5 g
- KCl   0.325 g
- KH_2PO_4 0.5 g
- $CaCl_2 \cdot 2H_2O$ 1.0 g
- NaBr   0.05 g
- H_3BO_3 0.015 g
- $SrCl_2 \cdot 6H_2O$ 0.0075 g
- $(NH_4)_2SO_4$ 0.1 g
- KI   0.025 mg
- $(NH_4)_2Ni(SO_4)_2 \cdot 6H_2O$ 2 mg
- Trace mineral solution (Balch et al., 1979)   10 ml

Trace mineral solution (per liter of DDW; Sako et al., 1996)

- Nitrilotriacetic acid (NTA)   1.5 g
- $MgSO_4 \cdot 7H_2O$ 3 g
- $MnSO_4 \cdot 5H_2O$ 0.5 g
- NaCl   1.0 g
- $FeSO_4 \cdot 7H_2O$ 0.1 g
- $CoSO_4 \cdot 7H_2O$ 0.18 g
- $CaCl_2 \cdot 2H_2O$ 0.1 g
- $ZnSO_4 \cdot 7H_2O$ 0.18 g
- $CuSO_4 \cdot 5H_2O$ 0.01 g
- $KAl(SO_4)_2 \cdot 12H_2O$ 0.02 g
- H_3BO_3 0.01 g
- $Na_2MoO_4 \cdot 2H_2O$ 0.01 g
- $NiCl_2 \cdot 6H_2O$ 0.075 g
- $Na_2SeO_3 \cdot 5H_2O$ 0.05 g

Some researchers have modified the recipe of the trace mineral solution for specific microorganisms. For example, the deep-sea hydrothermal vent (DHV) mineral solution used for the enrichment and isolation of *Thermosipho japonicus* was supplemented with Si, Sr, W, V and Li (Takai and Horikoshi, 2000). However, specific metal requirements of deep-sea thermophiles mostly remain to be determined. Since some thermophiles have a requirement for tungsten (Kletzin and Adams, 1996), we usually add 0.1 g of $Na_2WO_4 \cdot 2H_2O$ per liter to the trace mineral solution (Nakagawa et al., 2003). Addition of tungsten at this concentration appears not to be inhibitory to W-independent microorganisms. Since molybdenum is inhibitory to sulfate-reducing prokaryotes (SRP) (Oremland and Silverman, 1979), it may be better to omit it when targeting SRP. However, at least for deep-sea thermophilic SRP such as *Archaeoglobus* and *Thermodesulfobacterium*, Mo does not have to be removed.

Vitamin solution (100X) (per liter of DDW; Balch et al., 1979)

- Biotin (Vitamin H)   2 mg
- Folic acid (Vitamin Bc)   2 mg
- Pyridoxine-HCl (Vitamin B_6)   10 mg
- Thiamine-HCl $\cdot$ $2H_2O$ (Vitamin B_1)   5 mg
- Riboflavin (Vitamin B_2)   5 mg
- Nicotinic acid (Vitamin B_3)   5 mg
- D-Ca-pantothenate (Vitamin B_5)   5 mg
- Cyanocobalamin (Vitamin B_{12})   1 mg
- p-Aminobenzoic acid (Vitamin Bx)   5 mg
- Lipoic acid   5 mg

Heterotrophic, S^0-utilizing Thermophiles

Heterotrophic, S^0-utilizing thermophiles have been isolated from many different deep-sea hydrothermal fields. The commonly used media for the enrichment of heterotrophic S^0-utilizing thermophiles consist of synthetic seawater, a source of organic carbon such as yeast extract or tryptone (1–5 g l^{-1}) and elemental sulfur (3-30 g l^{-1}). The headspace is generally N_2 or N_2/CO_2 (80:20). By using these media, various species of the orders *Thermococcales* (genera *Palaeococcus*, *Pyrococcus* and *Thermococcus*) and *Thermotogales* (genera *Marinitoga* and *Thermosipho*) have been isolated (Table 3.2). The optimum growth temperatures of members of the genus *Pyrococcus* are approximately 10°C higher than those of *Thermococcus* and *Palaeococcus* species (Table 3.2). The maximum temperatures for growth of the *Thermotogales* are 10–15°C lower than those of the *Thermococcales* (Table 3.2). These organisms gain energy by fermentation of peptides, carbohydrates, or amino acids, mainly producing H_2 and simple organic acids like acetate (Kengen and Stams, 1994). Among these microorganisms, members of the *Thermococcales* have detected in, and isolated from a variety of hot water environments: hydrothermal areas including coastal hydrothermal systems (Zillig *et al.*, 1983), freshwater geothermal pools (Ronimus *et al.*, 1997), and continental oil fields (L'Haridon *et al.*, 1995). They were even detected in non-hydrothermal sub-seafloor sediments (Inagaki *et al.*, 2001; Kormas *et al.*, 2003). In the deep-sea hydrothermal environments, the exterior surface of the chimney structures appears to be the preferred habitat for most of the *Thermococcales* (Harmsen *et al.*, 1997; Nakagawa *et al.*, 2005b; Schrenk *et al.*, 2003; Takai *et al.*, 2001). The energy metabolism of the *Thermococcales* has been of particular interest, since the growth yields of *Thermococcales* are much higher than can be explained when ATP synthesis occurs by substrate-level phosphorylation only (Kengen and Stams, 1994). Interestingly, Sapra *et al.* (2003) demonstrated that the membrane-bound hydrogenase of *Pyrococcus furiosus* is the only protein required for proton reduction with ferredoxin as the electron donor.

Growth of these heterotrophic thermophiles is stimulated in the presence of elemental sulfur (for members of both the *Thermococcales* and the *Thermotogales*) and thiosulfate (for members of the *Thermotogales*). In the presence of sulfur compounds, H_2S is produced instead of H_2 (Fiala and Stetter, 1986). It is still unknown how S^0 reduction is coupled to the electron transport chain. Some *Thermococcales* species utilize Fe(III) as an electron acceptor as well (Slobodkin *et al.*, 1999). Furthermore, although members of the *Thermococcales* had been regarded as obligate heterotrophs, recent research revealed their capability of carboxydotrophic growth (Sokolova *et al.*, 2004). Future elucidation of the metabolic abilities of the *Thermococcales* may provide the explanation for the widespread propagation and predominance of the *Thermococcales* in the global deep-sea hydrothermal systems (Table 3.2 and Figure 3.3).

The following is a useful medium formulation (designated MJYPGS medium, Nakagawa *et al.*, 2005b) for initial isolation of a variety of heterotrophic, S^0-utilizing thermophiles from deep-sea hydrothermal environments (per liter of synthetic seawater (SSW) including trace mineral solution):

MJYPGS medium (per liter of SSW)

- Yeast extract   2 g
- Tryptone   2 g
- Glucose   0.2 g
- Elemental sulfur (S^0)   30 g
- $Na_2S \cdot 9H_2O$ 0.5 g
- Resazurin (redox indicator)   1 mg

Dissolve all chemicals without Na_2S and S^0, adjust the pH to 6.0–7.0, dispense 3 ml portions into 16 ml tubes (Iwaki glass), and autoclave. During gas purging with N_2 or N_2/CO_2 (80:20), add sterile S^0 (sterilized separately by steaming for 1 h on three successive days at 100°C) and 10 % (w/v) Na_2S stock solution (adjusted to pH 7.5 with H_2SO_4 and autoclaved) to each tube. Keep flushing until decoloration of the resazurin indicates a complete reduction of the medium. Cap the tubes with gas-impermeable butyl rubber stoppers and secure with holed screw caps. Apply a headspace pressure of approximately 200 kPa to avoid air intrusion. In addition to *Thermococcales* and *Thermotogales*, many new species of moderately thermophilic *Firmicutes* (genera *Caldanaerobacter*, *Caloranaerobacter*, *Caminicella*, *Clostridium* and *Tepidibacter*) have been recently isolated by using similar heterotrophic media (Table 3.2). Most of these thermophilic heterotrophs can be grown in Marine Broth 2216 (Difco) supplemented with S^0 under a N_2 atmosphere.

It is worth noting that members of the genus *Palaeococcus* have been less frequently isolated among the members of the *Thermococcales*. The genus *Palaeococcus* was comprised of a single species, *P. ferrophilus* (Takai et al., 2000), until the recent isolation of a second species *P. helgesonii* from Vulcano island, Italy (Amend et al., 2003). *P. ferrophilus* was characterized by its requirement for high concentrations of ferrous iron (Takai et al., 2000).

Chemolithoautotrophic Thermophiles

Chemolithoautotrophs are responsible for the primary production in deep-sea vent ecosystems. Molecular hydrogen and sulfide are the main electron donors, and nitrate, oxygen, sulfate and carbon dioxide are the mostly utilized electron acceptors. The commonly used medium formulation for isolation of chemolithoautotrophs other than methanogens consists of synthetic seawater, inorganic carbon sources, vitamin solution, sulfur-compounds and oxidants. The headspace is generally H_2/CO_2 (80:20). Thermophilic chemolithoautotrophs had been represented by hyperthermophilic Archaea such as *Methanocaldococcus*, *Pyrolobus*, and *Ignicoccus* (Blöchl et al., 1997; Huber et al., 2000; Jones et al., 1983). However, a variety of thermophilic, chemolithoautotrophic Bacteria belonging to the order *Aquificales* (genera *Persephonella* and *Aquifex*), the *Desulfurobacterium* group and the ε-*Proteobacteria* have recently been isolated from geographically disparate deep-sea hydrothermal fields (Table 3.2).

Predominant and widespread occurrence of the ε-*Proteobacteria* was first recognized by investigations using culture-independent approaches (Moyer et al., 1995; López-García et al., 2003a; Polz and Cavanaugh, 1995; Reysenbach et al., 2000; Teske et al., 2002). In addition to their prevalence, the ε-*Proteobacteria* are conspicuous by their phylogenetic diversity. At least six phylogenetic subgroups within the ε-*Proteobacteria* were identified by comparative analysis of 16S rRNA gene sequences (Corre et al., 2001; Takai et al., 2003a). Based on the recent investigations, Groups A and D of the ε-*Proteobacteria* consist of moderate thermophiles (species of the genera *Hydrogenimonas*, *Nitratiruptor*, *Caminibacter*, *Nautilia* and *Lebetimonas*) (Table 3.2). Although previous reports have attributed these microorganisms as being microaerophilic sulfur-oxidizers (Taylor et al., 1999; Wirsen et al., 1993), all these moderately thermophilic ε-*Proteobacteria* utilize H_2 as an electron donor. To our knowledge, all Group A members can grow aerobically utilizing oxygen at low partial pressures as an electron acceptor. Members of Group D are obligate anaerobes with the exception of *Caminibacter profundus* (Miroshnichenko et al., 2004).

In spite of the phylogenetic distance between the *Aquificales* and the ε-*Proteobacteria*, these chemolithoautotrophs can be grown in the same media at different temperatures (Nakagawa et al., 2005b,c; Takai et al., 2003a). Our group designed a less selective medium containing a mixture of potential electron donors and acceptors in order to evaluate the abundance of these chemolithoautotrophs (Nakagawa et al., 2005b,c; Takai et al., 2004a). Subsequently, we significantly expanded the diversity of culturable chemolithoautotrophic thermophiles (Nakagawa et al., 2005b,c; Takai et al., 2003a, 2004a). The medium could be easily modified for specific metabolic types of microorganisms (described below). The following is a useful medium formulation (designated MMJHS medium) (per liter of synthetic seawater including trace mineral solution):

MMJHS medium (per liter of SSW)

- $NaNO_3$ 1 g
- $Na_2S_2O_3 \cdot 5H_2O$ 1 g
- $NaHCO_3$ 1 g
- Elemental sulfur (S^0)   30 g
- Vitamin solution   10 ml

Dissolve the chemicals without $NaHCO_3$, vitamin solution and S^0, adjust the pH to 6.0–7.0, dispense 3 ml portions into tubes, and autoclave. Under gas purging with H_2/CO_2 (80:20), add filter-sterilized solutions of $NaHCO_3$ (5%, w/v), vitamin mixture (described above), and S^0 (sterilized as described above) to each tube. Keep flushing vigorously for at least 2–3 min, cap the tubes with butyl rubber stoppers, secure with screw caps, and pressurize the headspace to 350 kPa with H_2/CO_2 (80:20). In our experience, this medium enriches members of Groups A or D of the ε-*Proteobacteria* at 40–60°C, *Persephonella* at 55–80°C and *Aquifex* at 80–90°C (Nakagawa et al., 2005b; Takai et al., 2003a). It should be noted that static cultures of the ε-*Proteobacteria* sometimes show no apparent turbidity, probably indicating that the organisms attach to S^0 particles. Thus, it is advisable to routinely check for growth by microscopy.

The following sections are methods for modifications of the MMJHS medium for specific metabolic traits.

Modification for obligate anaerobes

The addition of Na_2S to the MMJHS medium would allow the growth of anaerobic chemolithoautotrophs preferring more reducing conditions. To prepare the reduced medium, Na_2S (0.05%, w/v) is added during gas purging. Resazurin (0.1 mg per liter) is added as a redox indicator. In our experience, MMJHS medium containing Na_2S has enriched members of the ε-*Proteobacteria* Group D (at 40–60°C) or the *Desulfurobacterium* group (at 65–75°C) (Nakagawa et al., 2005b; Takai et al., 2004a, 2005).

Especially in cases where members of the *Desulfurobacterium* group are targeted, it is advisable to prepare the medium both in the presence or absence of nitrate. Nitrate inhibits the growth of *Balnearium lithotrophicum*, a species within the *Desulfurobacterium* group (Takai *et al.*, 2003c), while some relatives use nitrate as an alternative electron acceptor (Alain *et al.*, 2003; Vetriani *et al.*, 2004).

Modification for sulfate reducers

Hydrogen-oxidizing and sulfate-reducing thermophiles can also be grown in modified MMJHS medium. Members of the genus *Archaeoglobus* and the order *Thermodesulfobacteriales* represent marine thermophilic sulfate reducing prokaryotes capable of utilizing H_2 as an electron donor. Although reports of the cultivation of thermophilic SRP are few, many studies using culture-independent approaches have reported their occurrence in geographically disparate deep-sea hydrothermal fields (Huber *et al.*, 2003; Nakagawa *et al.*, 2005b; Nercessian *et al.*, 2003; Reysenbach *et al.*, 2000; Schrenk *et al.*, 2003; Takai *et al.*, 2004a; Teske *et al.*, 2002). Since organic carbon is stimulatory for growth of some thermophilic SRP (Huber *et al.*, 1997; Jeanthon *et al.*, 2002), we use MMJHS medium supplemented with 0.1% (w/v) yeast extract and 0.025% (w/v) acetate, lactate and pyruvate (Nakagawa *et al.*, 2005b). In addition to the organic carbon, we usually add 0.2% (w/v) sodium sulfate. To prepare the medium, these chemicals are added before autoclaving. It is recommended to remove elemental sulfur (inhibitory), and to add 0.05% (w/v) Na_2S as a reducing agent. It should be noted that recent culture-independent diversity analysis of the dissimilatory sulfite reductase (DSR) gene has suggested a diversity of potential sulfate reducing prokaryotes in deep-sea hydrothermal environments are awaiting to be cultured (Cottrell and Cary, 1999; Dhillon *et al.*, 2003; Nakagawa *et al.*, 2004a; Nercessian *et al.*, 2005).

Modification for iron reducers

Addition of alternative electron acceptors instead of nitrate and sulfur compounds may yield different metabolic types of microorganisms. For example, relatively little is known about use of Fe(III) as an electron acceptor by chemolithoautotrophic thermophiles in deep-sea hydrothermal environments. Iron reducing, chemolithoautotrophic thermophiles are represented by *Geoglobus ahangari* (Kashefi *et al.*, 2002), and *Deferribacter abyssi* (Miroshnichenko *et al.*, 2003d), both of which utilize H_2 or organic substrates as energy sources. Slobodkin *et al.* (2001) suggested the occurrence of Fe(III)-utilizing thermophiles of the ε-*Proteobacteria*, *Deferribacter*, *Clostridiales*, *Thermotogales* and *Thermococcales* in a deep-sea hydrothermal field. In addition, a recently isolated strain of *Pyrodictium* grows *via* hydrogen-oxidation with Fe(III) as an electron acceptor (Kashefi and Lovley, 2003). As an insoluble Fe(III) source in the media, the least crystalline of the Fe(III) oxyhydroxide minerals (prepared as described in Kostka and Nealson, 1998) is most commonly provided at 100 mM (Kashefi *et al.*, 2002). Utilization of other metals (e.g. Mn, As and Se) as electron acceptors by chemolithoautotrophic thermophiles has been poorly explored in deep-sea hydrothermal environments.

Modification for microaerophiles

Low levels of O_2 in the headspace of the MMJHS medium would allow the growth of microaerophilic chemolithoautotrophs. An easy procedure for creating such microaerophilic conditions is to add a defined volume of filter-sterilized O_2 by syringe into culture tubes (Nakagawa et al., 2003). From the energy point of view, oxygen is the most efficient electron acceptor for the microbial hydrogen- and sulfur-oxidations. However, reports on the isolation of obligately microaerobic, thermophilic chemolithoautotrophs are quite scarce. All of the known thermophiles capable of growth under microaerobic conditions are facultative anaerobes (Table 3.2), with the exception of an obligately microaerobic, mixotrophic sulfur-oxidizer, *Thiomicrospira thermophila* (Takai et al., 2004b). The oxygen range for the growth of *T. thermophila* is from 0.3 to 7.0% (v/v). It should be noted that cultivation with agitation affects the growth yield of microaerobes because of increase of dissolved oxygen.

Hydrogenotrophic methanogens

Thermophilic methanogens thus far described from deep-sea hydrothermal environments belong to the orders *Methanopyrales* and *Methanococcales* (Table 3.1). The order *Methanopyrales* consists of a single species, *Methanopyrus kandleri*, which is a hyperthermophilic methanogen originally isolated from Guaymas Basin (Figure 3.3) (Kurr et al., 1991). Additional strains of this species were isolated from the Trans-Atlantic Geotraverse (TAG) and the Snake pit sites in the Mid-Atlantic Ridge (Figure 3.3) (Huber and Stetter, 2001). The order *Methanococcales* consists of 4 genera, *Methanocaldococcus* (hyperthermophiles), *Methanotorris* (extreme thermophiles), *Methanothermococcus* (moderate thermophiles) and *Methanococcus* (mesophiles). Compared to members of the *Methanopyrales*, members of the *Methanococcales* appear to be widely distributed as shown by using culture-dependent (Table 3.2 and Figure 3.3), and culture-independent surveys (Huber et al., 2002; Nakagawa et al., 2005b; Reysenbach et al., 2000; Schrenk et al., 2003; Takai and Horikoshi, 1999; Takai et al., 2001, 2004a; Teske et al., 2002). Members of the genus *Methanopyrus* and order *Methanococcales* both grow on H_2 and CO_2, although some strains of the *Methanococcales* are able to grow on formate instead of H_2 (Jeanthon et al., 1998, 1999; Takai et al., 2002, 2004e). Thus far no acetoclastic, thermophilic methanogens have been isolated from deep-sea hydrothermal environments. Recently, thermophilic methanogens have become of particular interest as they are viewed as the organisms pointing to the existence of a hot sub-seafloor biosphere (Huber et al., 2002). Takai et al. (2004a) suggested the occurrence of hydrogen-driven subseafloor microbial ecosystems consisting mainly of thermophilic methanogens (*Methanocaldococcus*) and fermenters (*Thermococcales*).

All methanogens are obligate anaerobes requiring highly reducing growth conditions. Reduction of media is commonly achieved by the combined use of following agents: sodium sulfide, cysteine-HCl, titanium citrate, dithiothreitol, and thioglycolate (generally used at final concentrations of 0.025–0.05%, w/v). A recommended medium (designated MMJ medium, Takai et al., 2002) for initial isolation of marine thermophilic hydrogenotrophic methanogens includes (per liter of synthetic seawater):

MMJ medium (per liter SSW)

- $NaHCO_3$ 1 g
- Vitamin solution   10 ml
- $Na_2S \cdot 9H_2O$ 0.5 g
- Cysteine · HCl   0.5 g
- Resazurin (redox indicator)   1 mg

Dissolve the chemicals without $NaHCO_3$, Na_2S, Cysteine · HCl and vitamin solution, adjust pH to 6.0–7.0, dispense into tubes and autoclave. During gas purging with H_2/CO_2 (80:20), add filter-sterilized solutions of $NaHCO_3$ (5%, w/v) and vitamin mixture (described above), Na_2S stock solution (prepared as described above), and 10% (w/v) cysteine · HCl stock solution (adjusted to pH 7.5 with NaOH and autoclaved) to each tube. Keep flushing until decoloration of the resazurin occurs, cap the tubes with gas-impermeable stoppers, and pressurize the headspace to 350 kPa with H_2/CO_2 (80:20).

For some methanogens, selenium and magnetite are stimulatory (Jones et al., 1983; Takai et al., 2002). Some researchers have added organic carbon such as yeast extract to the medium for initial enrichment of methanogens, since organic substrates are stimulatory to some thermophilic methanogens (Jeanthon et al., 1998, 1999; Jones et al., 1989). However, complex organic carbon in the first enrichment potentially results in the growth of heterotrophs. In our experience, heterotrophs, especially *Thermococcus* and *Tepidibacter* species, are often co-cultured with methanogens in media supplemented with complex organic carbon. Thus, for enrichments in which organic carbon was added, it is advisable to routinely check the morphology of growing microorganisms. Autofluorescence of the methanogens (containing F_{420}) under UV illumination in the fluorescence microscope would help differentiate methanogens from fermenters. However, it should be noted that hyperthermophilic sulfate reducing prokaryotes of the genus *Archaeoglobus* also exhibit similar blue-green fluorescence (Burggraf et al., 1990). For additional media formulations and procedures for isolation and maintenance of methanogens, see Balch et al. (1979).

Although the cultivation conditions described above cover a variety of metabolic types of numerically abundant and widely distributed thermophiles, it is recommended to consult descriptions of each species for specific nutrient requirements. The general characteristics of most thermophiles are summarized in the 2nd edition of *Bergey's Manual of Systematic Bacteriology* (Boone *et al.*, 2001). Useful databases of media formulations for each species are available in the websites of the Deutsche Sammlung von Mikroorganismen und Zellkulturen GmbH (DSMZ) (http://www.dsmz.de/) and the Japan Collection of Microorganisms (JCM) (http://www.jcm.riken.jp/).

◆◆◆◆◆ CULTIVATION CONDITIONS

In general, no special equipment is necessary for the cultivation of thermophiles. Test tubes and glass bottles can be incubated in oil baths up to temperatures of 130°C and pressures of 400 kPa. Air incubators can also be used, but the temperature should be routinely measured inside control flasks along with the cultures. Some species grow extremely slowly and reach cell densities not exceeding 1×10^7 cells ml^{-1}. It is therefore advisable to check growth in enrichment cultures by microscopy even when there is no visible turbidity. Progress has been made towards the growth of thermophiles in continuous culture to achieve high cell yields for biochemical and molecular biology studies (Krahe *et al.*, 1996). Recently, the continuous culture technique was adopted for primary enrichment of deep-sea thermophiles (Postec *et al.*, 2005). Although some thermophiles inhabiting the deep-sea may require high pressure for growth, no obligately piezophilic thermophiles have been isolated so far (see the chapter by Kato *et al.* in this volume for the isolation and cultivation of piezophiles).

Some strains of thermophiles lyse rapidly when kept at their growth temperatures beyond the stationary growth phase. It is therefore recommended to transfer the cultures to fresh media at the early-stationary growth phase, or to keep the cultures at low temperatures well below the minimum temperature for growth. Most thermophiles remain culturable for several months at refrigerated temperatures under anaerobic conditions. Methods for long-time preservation of thermophilic species were described by Boone *et al.* (2001).

Yet-uncharacterized thermophiles with no cultivable relatives may have requirements for different combinations of energy, carbon, nitrogen and metal sources, at different concentrations. When such yet-unknown microorganisms dominate a microbial habitat, determining the detailed environmental conditions may provide insights into the energy and carbon sources available to the growth. However, this is nearly impossible in deep-sea hydrothermal environments where physical and chemical parameters are always variable. Here, gradient culture techniques may reveal a novel culturable diversity. Previous studies tested gradients of iron/oxygen (Emerson and Moyer, 1997), and sulfide/oxygen (Taylor and Wirsen, 1997), and successfully recovered previously uncultured microorganisms. However, so far there has been no example of thermophiles isolated by using this technique.

It should be noted that various advanced cultivation techniques have been recently developed. For example, Connon and Giovannoni (2002) developed a high throughput culturing method, which utilizes the concept of extinction culturing (Ammerman et al., 1984) to isolate cultures in small volumes of low-nutrient media. In addition, Bruns et al. (2003) also developed a high throughput cultivation assay, which uses an automatic microdrop dispenser to dilute the inoculum into media in microtiter plates. These methods, based on the concept of the most probable number (MPN) technique, can be used to evaluate the abundance of culturable bacteria. Furthermore, Kaeberlein et al., (2002) designed a new method using a diffusion chamber that reproduces the natural nutritional setting. These new techniques have greatly expanded the cultivable diversity of aerobic, heterotrophic mesophiles. There will be technical difficulties in applying these techniques to anaerobic thermophiles. Nevertheless, these new techniques have a great potential to open a way to obtain the yet-cultivable deep-sea thermophiles in culture.

◆◆◆◆◆ ISOLATION

The most commonly used method to obtain pure cultures of thermophiles is the dilution-to-extinction technique. Before performing dilution-to-extinction, it is necessary to count the microbial cell density in the cultures to be diluted. Epifluorescence microscopic counting of 4′,6-diamidino-2-phenylindole (DAPI)-stained cells is most commonly used (Porter and Feig, 1980). Serial 1:10 dilutions are performed until the cell density is diluted to approximately 10^2–10^3 cells ml^{-1}. Serial 1:2 dilutions are then prepared until the total dilution factor exceeds the original cell density at least by two orders of magnitude. The whole dilution-to-extinction series is repeated 3–5 times using the highest positive dilution tubes. Each dilution in the dilution-to-extinction series may be performed in triplicate. The purity should be double-checked by microscopy and by sequencing the 16S rRNA gene using several PCR primers.

Alternatively, some researchers have used solidified media for culture purification. Thermostable solidifying agents used are polysilicate (Stetter, 2001), GP-700 agar (Shimizu Shokuhin, Shimizu, Japan) (Sako et al., 1996), and Gelrite gellan gum (Deming and Baross, 1986). Other solidifying agents are generally unstable at high temperatures above 70°C. Plates should be incubated inside sealed containers to avoid desiccation. The roll-tube technique is also available, which is probably superior to plating especially when the headspace gas serves as a growth substrate. While these solidified media visibly ensure the purification of the isolate, it should be noted that most thermophiles do not readily form colonies, if at all. Some thermophiles have been physically isolated in pure culture by using the FISH technique followed by a newly developed microscopic technique called "optical tweezers", which uses a strongly focussed infrared laser beam (Huber et al., 1995, 1998).

◆◆◆◆◆ CONCLUDING REMARKS AND PERSPECTIVES

Cultivation studies followed by physiological characterization of the isolates will remain a major part of the microbiological studies on the deep-sea hydrothermal environments. However, the rapid development of culture-independent molecular approaches has shown how wide the gap is between the microbial diversity in the environment and the diversity of organisms cultured from these environments. Thus, efforts in developing innovative cultivation methods should be strongly encouraged. Due to methodological biases in both approaches it is increasingly difficult to integrate the results of culture-dependent and culture-independent surveys to reveal the true diversity within the microbial communities. Further development of standardized and quantitative methods for both culture-dependent and culture-independent surveys will be required in the future environmental microbiology and microbial ecology studies in the deep-sea hydrothermal environments.

There is also an increasing need to develop technology for long-time monitoring of *in situ* microbial activities and community structure. Physiological characteristics of isolates determined in the laboratory have provided important insights into their possible *in situ* ecophysiological function. However, only few investigations have undertaken direct assessments of the long-term functions that the microorganisms perform *in situ*. Techniques for long-time monitoring of *in situ* microbial activities and community structure combined with microsensor technology for measuring physicochemical variations in microscale will be necessary to reveal the dynamics and successional changes of the microbial communities in deep-sea hydrothermal environments.

Acknowledgement

Satoshi Nakagawa was supported by the Research Fellowship of the Japanese Society for the Promotion of Science.

List of suppliers

Iwaki Glass Co., Ltd.
50-1Gyoda 1-chome, Funabasi-shi, Chiba 273-0044, Japan
www.igc.co.jp
Tel.: +81-47-460-3700
Fax: +81-47-460-3795

Glass tubes

Schott
Hattenbergstr. 10, Mainz 55122, Germany
www.schott.com

Tel.: +49-6131-66-0
Fax: +49-6131-66-2000

Glass bottles

Shimizu Shokuhin
Sakae-machi 1-3, Shizuoka 420-0859, Japan
www.ssk-ltd.co.jp
Tel.: +81-54-221-8520
Fax: +81-54-221-8531

GP700 agar

Wheaton Science Products
1501 N, 10th Street, Millville, NJ 08332-2093, USA
www.wheatonsci.com
Tel.: +1-856-825-1100
Fax: +1-856-825-1368

Lyophilization stoppers

References

Alain, K., Marteinsson, V. T., Miroshnichenko, M. L., Bonch-Osmolovskaya, E. A., Prieur, D. and Birrien, J.-L. (2002a). *Marinitoga piezophila* sp. nov., a rod-shaped, thermo-piezophilic bacterium isolated under high hydrostatic pressure from a deep-sea hydrothermal vent. *Int. J. Syst. Evol. Microbiol.* **52**, 1331–1339.

Alain, K., Pignet, P., Zbinden, M., Quillevere, M., Duchiron, F., Donval, J.-P., Lesongeur, F., Raguenes, G., Crassous, P., Querellou, J. and Cambon-Bonavita, M.-A. (2002b). *Caminicella sporogenes* gen. nov., sp. nov., a novel thermophilic spore-forming bacterium isolated from an East-Pacific Rise hydrothermal vent. *Int. J. Syst. Evol. Microbiol.* **52**, 1621–1628.

Alain, K., Querellou, J., Lesongeur, F., Pignet, P., Crassous, P., Raguénès, G., Cueff, V. and Cambon-Bonavita, M.-A. (2002c). *Caminibacter hydrogeniphilus* gen. nov., sp. nov., a novel thermophilic, hydrogen-oxidizing bacterium isolated from an East Pacific Rise hydrothermal vent. *Int. J. Syst. Evol. Microbiol.* **52**, 1317–1323.

Alain, K., Rolland, S., Crassous, P., Lesongeur, F., Zbinden, M., Le Gall, C., Godfroy, A., Page, A., Juniper, S. K., Cambon-Bonavita, M.-A., Duchiron, F. and Querellou, J. (2003). *Desulfurobacterium crinifex* sp. nov., a novel thermophilic, pinkish-streamer forming, chemolithoautotrophic bacterium isolated from a Juan de Fuca Ridge hydrothermal vent and amendment of the genus *Desulfurobacterium*. *Extremophiles* **7**, 361–370.

Amend, J. P., Meyer-Dombard, D. R., Sheth, S. N., Zolotova, N. and Amend, A. C. (2003). *Palaeococcus helgesonii* sp. nov., a facultatively anaerobic, hyperthermophilic archaeon from a geothermal well on Vulcano Island, Italy. *Arch. Microbiol.* **179**, 394–401.

Ammerman, J. W., Fuhrman, J. A., Hagström, Å. and Azam, F. (1984). Bacterioplankton growth in seawater. I. Growth kinetics and cellular characteristics in seawater cultures. *Mar. Ecol. Prog. Ser.* **18**, 31–39.

Antoine, E., Cilia, V., Meunier, J. R., Guezennec, J., Lesongeur, F. and Barbier, G. (1997). *Thermosipho melanesiensis* sp. nov., a new thermophilic anaerobic bacterium belonging to the order *Thermotogales*, isolated from deep-sea hydrothermal vents in the southwestern Pacific Ocean. *Int. J. Syst. Bacteriol.* **47**, 1118–1123.

Balch, W. E. and Wolfe, R. S. (1976). New approach to the cultivation of methanogenic bacteria: 2-mercaptoethane-sulfonic aid (HS-CoM)-dependent growth of *Methanobacterium ruminantium* in a pressurized atmosphere. *Appl. Environ. Microbiol.* **32**, 781–791.

Balch, W. E., Fox, G. E., Magrum, L. J., Woese, C. R. and Wolfe, R. S. (1979). Methanogens: a reevaluation of a unique biological group. *Microbiol. Rev.* **43**, 260–296.

Barbier, G., Godfroy, A., Meunier, J. R., Querellou, J., Cambon-Bonavita, M.-A., Lesongeur, F., Grimont, P. A. D. and Raguenes, G. (1999). *Pyrococcus glycovorans* sp. nov., a hyperthermophilic archaeon isolated from the East Pacific Rise. *Int. J. Syst. Bacteriol.* **49**, 1829–1837.

Baross, J. A. and Deming, J. D. (1995). Growth at high temperatures: Isolation and taxonomy, physiology, and ecology In *The Microbiology of Deep-sea Hydrothermal Vents* (D. M. Karl, ed.), pp. 169–217. CRC Press, Boca Raton.

Blöchl, E., Rachel, R., Burggraf, S., Hafenbradl, D., Jannasch, H. W. and Stetter, K. O. (1997). *Pyrolobus fumarii*, gen. and sp. nov., represents a novel group of archaea, extending the upper temperature limit for life to 113°C. *Extremophiles* **1**, 14–21.

Boone, D. R., Castenholz, R. W. and Garrity, G. M. (eds) (2001). *Bergey's Manual of Systematic Bacteriology*, 2nd edn., vol. 1. Springer, New York.

Brisbarre, N., Fardeau, M.-L., Cueff, V., Cayol, J.-L., Barbier, G., Cilia, V., Ravot, G., Thomas, P., Garcia, J.-L. and Ollivier, B. (2003). *Clostridium caminithermale* sp. nov., a slightly halophilic and moderately thermophilic bacterium isolated from an Atlantic deep-sea hydrothermal chimney. *Int. J. Syst. Evol. Microbiol.* **53**, 1043–1049.

Bruns, A., Hoffelner, H. and Overmann, J. (2003). A novel approach for high throughput cultivation assays and the isolation of planktonic bacteria. *FEMS Microbiol. Ecol.* **45**, 161–171.

Burggraf, S., Jannasch, H. W., Nicolaus, B. and Stetter, K. O. (1990). *Archaeoglobus profundus*, sp. nov., represents a new species within the sulfate-reducing archaebacteria. *Syst. Appl. Microbiol.* **13**, 24–28.

Canganella, F., Jones, W. J., Gambacorta, A. and Antranikian, G. (1998). *Thermococcus guaymasensis* sp. nov. and *Thermococcus aggregans* sp. nov., two novel thermophilic archaea isolated from the Guaymas Basin hydrothermal vent site. *Int. J. Syst. Bacteriol.* **48**, 1181–1185.

Connon, S. A. and Giovannoni, S. J. (2002). High-throughput methods for culturing microorganisms in very-low-nutrient media yield diverse new marine isolates. *Appl. Environ. Microbiol.* **68**, 3878–3885.

Corre, E., Reysenbach, A.-L. and Prieur, D. (2001). ε-Proteobacterial diversity from a deep-sea hydrothermal vent of the Mid-Atlantic Ridge. *FEMS Microbiol. Lett.* **205**, 329–335.

Cottrell, M. T. and Cary, S. C. (1999). Diversity of dissimilatory bisulfite reductase genes of bacteria associated with the deep-sea hydrothermal vent polychaete annelid *Alvinella pompejana*. *Appl. Environ. Microbiol.* **65**, 1127–1132.

Cragg, B. A., Summit, M. and Parkes, R. J. (2000). Bacterial profiles in a sulfide mound (site 1035) and an area of active hydrothermal sediments from Middle Valley (Northwest Pacific). *Proc. ODP Sci. Result.* **169**, 1–18.

Deming, J. W. and Baross, J. A. (1986). Solid medium for culturing black smoker bacteria at temperatures to 120°C. *Appl. Environ. Microbiol.* **51**, 238–243.

Dhillon, A., Teske, A., Dillon, J., Stahl, D. A. and Sogin, M. L. (2003). Molecular characterization of sulfate-reducing bacteria in the Guaymas Basin. *Appl. Environ. Microbiol.* **69**, 2765–2772.

Duffaud, G. D., d'Hennezel, O. B., Peek, A. S., Reysenbach, A.-L. and Kelly, R. M. (1998). Isolation and characterization of *Thermococcus barossii*, sp. nov., a hyperthermophilic archaeon isolated from a hydrothermal vent flange formation. *Syst. Appl. Microbiol.* **21**, 40–49.

Emerson, D. and Moyer, C. L. (1997). Isolation and characterization of novel iron-oxidizing bacteria that grow at circumneutral pH. *Appl. Environ. Microbiol.* **63**, 4784–4792.

Erauso, G., Reysenbach, A.-L., Godfroy, A., Meunier, J. R., Crump, B., Partensky, F., Baross, J. A., Marteinsson, V., Barbier, G., Pace, N. R. and Prieur, D. (1993). *Pyrococcus abyssi* sp. nov., a new hyperthermophilic archaeon isolated from a deep-sea hydrothermal vent. *Arch. Microbiol.* **160**, 338–349.

Fiala, G. and Stetter, K. O. (1986). *Pyrococcus furiosus* sp. nov. represents a novel genus of marine heterotrophic archaebacteria growing optimally at 100°C. *Arch. Microbiol.* **145**, 56–61.

Fiala, G., Stetter, K. O., Jannasch, H. W., Langworthy, T. A. and Madon, J. (1986). *Staphylothermus marinus* sp. nov. represents a novel genus of extremely thermophilic submarine heterotrophic archaebacteria growing up to 98°C. *Syst. Appl. Microbiol.* **8**, 106–113.

Fredrickson, J. K. and Fletcher, M. (eds) (2001). *Subsurface Microbiology and Biogeochemistry*, John Wiley & Sons Inc., New York.

Götz, D. K., Banta, A., Beveridge, T. J., Rushdi, A. I., Simoneit, B. R. T. and Reysenbach, A.-L. (2002). *Persephonella marina* gen. nov., sp. nov. and *Persephonella guaymasensis* sp. nov., two novel thermophilic hydrogen-oxidizing microaerophiles from deep-sea hydrothermal vents. *Int. J. Syst. Evol. Microbiol.* **52**, 1349–1359.

Godfroy, A., Meunier, J. R., Guezennec, J., Lesongeur, F., Raguénès, G., Rimbault, A. and Barbier, G. (1996). *Thermococcus fumicolans* sp. nov., a new hyperthermophilic archaeon isolated from a deep-sea hydrothermal vent in the North Fiji Basin. *Int. J. Syst. Bacteriol.* **46**, 1113–1119.

Godfroy, A., Lesongeur, F., Raguénès, G., Quérellou, J., Antoine, E., Meunier, J. R., Guezennec, J. and Barbier, G. (1997). *Thermococcus hydrothermalis* sp. nov., a new hyperthermophilic archaeon isolated from a deep-sea hydrothermal vent. *Int. J. Syst. Bacteriol.* **47**, 622–626.

Gonzalez, J. M., Kato, C. and Horikoshi, K. (1995). *Thermococcus peptonophilus* sp. nov., a fast-growing, extremely thermophilic archaebacterium isolated from deep-sea hydrothermal vents. *Arch. Microbiol.* **164**, 159–164.

Gonzalez, J. M., Masuchi, Y., Robb, F. T., Ammerman, J. W., Maeder, D. L., Yanagibayashi, M., Tamaoka, J. and Kato, C. (1998). *Pyrococcus horikoshii* sp. nov., a hyperthermophilic archaeon isolated from a hydrothermal vent at the Okinawa Trough. *Extremophiles* **2**, 123–130.

Grote, R., Li, L., Tamaoka, J., Kato, C., Horikoshi, K. and Antranikian, G. (1999). *Thermococcus siculi* sp. nov., a novel hyperthermophilic archaeon isolated from a deep-sea hydrothermal vent at the mid-Okinawa Trough. *Extremophiles* **3**, 55–62.

Harmsen, H. J. M., Prieur, D. and Jeanthon, C. (1997). Distribution of microorganisms in deep-sea hydrothermal vent chimneys investigated by whole-cell hybridization and enrichment culture of themophilic subpopulations. *Appl. Environ. Microbiol.* **63**, 2876–2883.

Hedrick, D. B., Pledger, R. D., White, D. C. and Baross, J. A. (1992). In-situ microbial ecology of hydrothermal vent sediments. *FEMS Microbiol. Ecol.* **101**, 1–10.

Huber, R. and Stetter, K. O. (2001). Order I. *Methanopyrales* ord. nov. In *Bergey's Manual of Systematic Bacteriology* (D. R. Boone, R. W. Castenholz and G. M. Garrity, eds), 2nd edn., vol. 1, pp. 353–355. Springer, New York.

Huber, R., Burggraf, S., Mayer, T., Barns, S. M., Rossnagel, P. and Stetter, K. O. (1995). Isolation of a hyperthermophilic archaeum predicted by *in situ* RNA analysis. *Nature* **376**, 57–58.

Huber, R., Stöhr, J., Hohenhaus, S., Rachel, R., Burggraf, S., Jannasch, H. W. and Stetter, K. O. (1996). *Thermococcus chitonophagus* sp. nov., a novel, chitin-degrading, hyperthermophilic archaeum from a deep-sea hydrothermal vent environment. *Arch. Microbiol.* **164**, 255–264.

Huber, H., Jannasch, H. W., Rachel, R., Fuchs, T. and Stetter, K. O. (1997). *Archaeoglobus veneficus* sp. nov., a novel facultative chemolithoautotrophs hyperthermophilic sulfite reducer, isolated form abyssal black smokers. *Syst. Appl. Microbiol.* **20**, 374–380.

Huber, R., Eder, W., Heldwein, S., Wanner, G., Huber, H., Rachel, R. and Stetter, K. O. (1998). *Thermocrinis ruber* gen. nov., sp. nov., a pink-filament-forming hyperthermophilic bacterium isolated from Yellowstone National Park. *Appl. Environ. Microbiol.* **64**, 3576–3583.

Huber, H., Burggraf, S., Mayer, T., Wyschkony, I., Rachel, R. and Stetter, K. O. (2000). *Ignicoccus* gen. nov., a novel genus of hyperthermophilic, chemolithoautotrophic Archaea, represented by two new species, *Ignicoccus islandicus* sp. nov. and *Ignicoccus pacificus* sp. nov. *Int. J. Syst. Evol. Microbiol.* **50**, 2093–2100.

Huber, J. A., Butterfield, D. A. and Baross, J. A. (2002). Temporal changes in archaeal diversity and chemistry in a mid-ocean ridge subseafloor habitat. *Appl. Environ. Microbiol.* **68**, 1585–1594.

Huber, J. A., Butterfield, D. A. and Baross, J. A. (2003). Bacterial diversity in a subseafloor habitat following a deep-sea volcanic eruption. *FEMS Microbiol. Ecol.* **43**, 393–409.

Humphris, S. E., Zierenberg, R. A., Mullineaux, L. S. and Thomson, R. E. (eds) (1995). *Seafloor Hydrothermal Systems*, American Geophysical Union, Washington, DC.

Inagaki, F., Takai, K., Komatsu, T., Kanamatsu, T., Fujioka, K. and Horikoshi, K. (2001). Archaeology of Archaea: geomicrobiological record of Pleistocene thermal events concealed in a deep-sea subseafloor environment. *Extremophiles* **5**, 385–392.

Jeanthon, C. (2000). Molecular ecology of hydrothermal vent microbial communities. *Antonie van Leeuwenhoek* **77**, 117–133.

Jeanthon, C., L'Haridon, S., Reysenbach, A.-L., Vernet, M., Messner, P., Sleytr, U. B. and Prieur, D. (1998). *Methanococcus infernus* sp. nov., a novel extremely thermophilic lithotrophic methanogen isolated from a deep-sea hydrothermal vent. *Int. J. Syst. Bacteriol.* **48**, 913–919.

Jeanthon, C., L'Haridon, S., Reysenbach, A.-L., Corre, E., Vernet, M., Messner, P., Sleytr, U. B. and Prieur, D. (1999). *Methanococcus vulcanius* sp. nov., a novel hyperthermophilic methanogen isolated from

East Pacific Rise, and identification of *Methanococcus* sp. DSM 4213T as *Methanococcus fervens* sp. nov. *Int. J. Syst. Bacteriol.* **49**, 583–589.

Jeanthon, C., L'Haridon, S., Cueff, V., Banta, A., Reysenbach, A.-L. and Prieur, D. (2002). *Thermodesulfobacterium hydrogeniphilum* sp. nov., a thermophilic, chemolithoautotrophic, sulfate-reducing bacterium isolated from a deep-sea hydrothermal vent at Guaymas Basin, and emendation of the genus *Thermodesulfobacterium*. *Int. J. Syst. Evol. Microbiol.* **52**, 765–772.

Jolivet, E., L'Haridon, S., Corre, E., Forterre, P. and Prieur, D. (2003). *Thermococcus gammatolerans* sp. nov., a hyperthermophilic archaeon from a deep-sea hydrothermal vent that resists ionizing radiation. *Int. J. Syst. Evol. Microbiol.* **53**, 847–851.

Jones, W. J., Leigh, J. A., Mayer, F., Woese, C. R. and Wolfe, R. S. (1983). *Methanococcus jannaschii* sp. nov., an extremely thermophilic methanogen from a submarine hydrothermal vent. *Arch. Microbiol.* **136**, 254–261.

Jones, W. J., Stugard, C. E. and Jannasch, H. W. (1989). Comparison of thermophilic methanogens from submarine hydrothermal vents. *Arch. Microbiol.* **151**, 314–318.

Kaeberlein, T., Lewis, K. and Epstein, S. S. (2002). Isolating "uncultivable" microorganisms in pure culture in a simulated natural environment. *Science* **296**, 1127–1129.

Karl, D. M. (ed.) (1995). *The Microbiology of Deep-sea Hydrothermal Vents*, CRC Press, Inc, Boca Raton.

Kashefi, K. and Lovley, D. R. (2003). Extending the upper temperature limit for life. *Science* **301**, 934.

Kashefi, K., Tor, J. M., Holmes, D. E., Gaw Van Praagh, C. V., Reysenbach, A.-L. and Lovley, D. R. (2002). *Geoglobus ahangari* gen. nov., sp. nov., a novel hyperthermophilic archaeon capable of oxidizing organic acids and growing autotrophically on hydrogen with Fe(III) serving as the sole electron acceptor. *Int. J. Syst. Evol. Microbiol.* **52**, 719–728.

Kashefi, K., Holmes, D. E., Baross, J. A. and Lovley, D. R. (2003). Thermophily in the *Geobacteraceae*: *Geothermobacter ehrlichii* gen. nov., sp. nov., a novel thermophilic member of the Geobacteraceae from the "Bag City" hydrothermal vent. *Appl. Environ. Microbiol.* **69**, 2985–2993.

Kato, C., Li, L., Tamaoka, J. and Horikoshi, K. (1997). Molecular analyses of the sediment of the 11000-m deep Mariana Trench. *Extremophiles* **1**, 117–123.

Kengen, S. W. M. and Stams, A. J. M. (1994). Growth and energy conservation in batch cultures of *Pyrococcus furiosus*. *FEMS Microbiol. Lett.* **117**, 305–309.

Kimura, H., Asada, R., Masta, A. and Naganuma, T. (2003). Distribution of microorganisms in the subsurface of the Manus Basin hydrothermal vent field in Papua New Guinea. *Appl. Environ. Microbiol.* **69**, 644–648.

Kletzin, A. and Adams, M. W. (1996). Tungsten in biological systems. *FEMS Microbiol. Rev.* **18**, 5–63.

Kobayashi, T., Kwak, Y., Akiba, T., Kudo, T. and Horikoshi, K. (1994). *Thermococcus profundus* sp. nov., a new hyperthermophilic archaeon isolated from a deep-sea hydrothermal vent. *Syst. Appl. Microbiol.* **17**, 232–236.

Kormas, K. A., Smith, D. C., Edgcomb, V. and Teske, A. (2003). Molecular analysis of deep subsurface microbial communities in Nankai Trough sediments (ODP Leg 190, Site 1176). *FEMS Microbiol. Ecol.* **45**, 115–125.

Kostka, J. and Nealson, K. H. (1998). Isolation, cultivation and characterization of iron- and manganese-reducing bacteria. In *Techniques in Microbial*

Ecology (R. S. Burlage, R. Atlas, D. Stahl, G. Geessey and G. Sayler, eds), pp. 58–78. Oxford University Press, Inc, New York.

Krahe, M., Antranikian, G. and Märkl, H. (1996). Fermentation of extremophilic microorganisms. *FEMS Microbiol. Rev.* **18**, 271–285.

Kurr, M., Huber, R., König, H., Jannasch, H. W., Fricke, H., Trincone, A., Kristjansson, J. K. and Stetter, K. O. (1991). *Methanopyrus kandleri* gen. and sp. nov. represents a novel group of hyperthermophilic methanogens, growing at 110°C. *Arch. Microbiol.* **156**, 239–247.

L'Haridon, S., Reysenbach, A.-L., Glenat, P., Prieur, D. and Jeanthon, C. (1995). Hot subterranean biosphere in a continental oil reservoir. *Nature* **337**, 223–224.

L'Haridon, S., Cilia, V., Messner, P., Raguenes, G., Gambacorta, A., Sleytr, U. B., Prieur, D. and Jeanthon, C. (1998). *Desulfurobacterium thermolithotrophum* gen. nov., sp. nov., a novel autotrophic, sulphur-reducing bacterium isolated from a deep-sea hydrothermal vent. *Int. J. Syst. Bacteriol.* **48**, 701–711.

L'Haridon, S., Reysenbach, A.-L., Banta, A., Messner, P., Schumann, P., Stackebrandt, E. and Jeanthon, C. (2003). *Methanocaldococcus indicus* sp. nov., a novel hyperthermophilic methanogen isolated from the Central Indian Ridge. *Int. J. Syst. Evol. Microbiol.* **53**, 1931–1935.

López-García, P., Duperron, S., Philippot, P., Foriel, J., Susini, J. and Moreira, D. (2003a). Bacterial diversity in hydrothermal sediment and epsilonproteobacterial dominance in experimental microcolonizers at Mid-Atlantic Ridge. *Environ. Microbiol.* **5**, 961–976.

López-García, P., Philippe, H., Gail, F. and Moreira, D. (2003b). Autochthonous eukaryotic diversity in hydrothermal sediment and experimental microcolonizers at the Mid-Atlantic Ridge. *Proc. Natl. Acad. Sci. USA* **100**, 697–702.

Lupton, J. E., Delaney, J. R., Johnson, H. P. and Tivey, M. K. (1985). Entrainment and vertical transport of deep-ocean water by buoyant hydrothermal plumes. *Nature* **316**, 621–623.

Marteinsson, V. T., Birrien, J.-L., Kristjánsson, J. K. and Prieur, D. (1995). First isolation of thermophilic aerobic non-sporulating heterotrophic bacteria from deep-sea hydrothermal vents. *FEMS Microbiol. Ecol.* **18**, 163–174.

Marteinsson, V. T., Birrien, J. L., Reysenbach, A.-L., Vernet, M., Marie, D., Gambacorta, A., Messner, P., Sleytr, U. B. and Prieur, D. (1999). *Thermococcus barophilus* sp. nov., a new barophilic and hyperthermophilic archaeon isolated under high hydrostatic pressure from a deep-sea hydrothermal vent. *Int. J. Syst. Bacteriol.* **49**, 351–359.

Maruyama, A., Mita, N. and Higashihara, T. (1993). Particulate materials and microbial assemblages around the Izena black smoking vent in the Okinawa Trough. *J. Oceanogr.* **49**, 353–367.

Miroshnichenko, M. L., Kostrikina, N. A., L'Haridon, S., Jeanthon, C., Hippe, H., Stackebrandt, E. and Bonch-Osmolovskaya, E. A. (2002). *Nautilia lithotrophica* gen. nov., sp. nov., a thermophilic sulfur-reducing ε-*proteobacterium* isolated from a deep-sea hydrothermal vent. *Int. J. Syst. Evol. Microbiol.* **52**, 1299–1304.

Miroshnichenko, M. L., Kostrikina, N. A., Chernyh, N. A., Pimenov, N. V., Tourova, T. P., Antipov, A. N., Spring, S., Stackebrandt, E. and Bonch-Osmolovskaya, E. A. (2003a). *Caldithrix abyssi* gen. nov., sp. nov., a nitrate-reducing, thermophilic, anaerobic bacterium isolated form a Mid-Atlantic Ridge hydrothermal vent, represents a novel bacterial lineage. *Int. J. Syst. Evol. Microbiol.* **53**, 323–329.

Miroshnichenko, M. L., L'Haridon, S., Jeanthon, C., Antipov, A. N., Kostrikina, N. A., Tindall, B. J., Schumann, P., Spring, S., Stackebrandt, E. and Bonch-Osmolovskaya, E. A. (2003b). *Oceanithermus profundus* gen. nov., sp. nov., a thermophilic, microaerophilic facultatively chemolithoheterotrophic bacterium from a deep-sea hydrothermal vent. *Int. J. Syst. Evol. Microbiol.* **53**, 747–752.

Miroshnichenko, M. L., L'Haridon, S., Nercessian, O., Antipov, A. N., Kostrikina, N. A., Tindall, B. J., Schumann, P., Spring, S., Stackebrandt, E., Bonch-Osmolovskaya, E. A. and Jeanthon, C. (2003c). *Vulcanithermus mediatlanticus* gen. nov., sp. nov., a novel member of the family Thermaceae from a deep-sea hot vent. *Int. J. Syst. Evol. Microbiol.* **53**, 1143–1148.

Miroshnichenko, M. L., Slobodkin, A. I., Kostrikina, N. A., L'Haridon, S., Nercessian, O., Spring, S., Stackebrandt, E., Bonch-Osmolovskaya, E. A. and Jeanthon, C. (2003d). *Deferribacter abyssi* sp. nov., an anaerobic thermophile from deep-sea hydrothermal vents of the Mid-Atlantic Ridge. *Int. J. Syst. Evol. Microbiol.* **53**, 1637–1641.

Miroshnichenko, M. L., L'Haridon, S., Schumann, P., Spring, S., Bonch-Osmolovskaya, E. A., Jeanthon, C. and Stackebrandt, E. (2004). *Caminibacter profundus* sp. nov., a novel thermophile of Nautiliales ord. nov. within the class "Epsilonproteobacteria", isolated from a deep-sea hydrothermal vent. *Int. J. Syst. Evol. Microbiol.* **54**, 41–45.

Mori, K., Kakegawa, T., Higashi, Y., Nakamura, K., Maruyama, A. and Hanada, S. (2004). *Oceanithermus desulfurans* sp. nov., a novel thermophilic sulfur-reducing bacterium isolated from a sulfide chimney in Suiyo Seamount. *Int. J. Syst. Evol. Microbiol.* **54**, 1561–1566.

Moussard, H., L'Haridon, S., Tindall, B. J., Banta, A., Schumann, P., Stackebrandt, E., Reysenbach, A.-L. and Jeanthon, C. (2004). *Thermodesulfatator indicus* gen. nov., sp. nov., a novel thermophilic chemolithoautotrophic sulfate-reducing bacterium isolated from the Central Indian Ridge. *Int. J. Syst. Evol. Microbiol.* **54**, 227–233.

Moyer, C. L., Dobbs, F. C. and Karl, D. M. (1995). Phylogenetic diversity of the bacterial community from a microbial mat at an active, hydrothermal vent system, Loihi Seamount, Hawaii. *Appl. Environ. Microbiol.* **61**, 1555–1562.

Nakagawa, S., Takai, K., Horikoshi, K. and Sako, Y. (2003). *Persephonella hydrogeniphila* sp. nov., a novel thermophilic, hydrogen-oxidizing bacterium from a deep-sea hydrothermal vent chimney. *Int. J. Syst. Evol. Microbiol.* **53**, 863–869.

Nakagawa, T., Nakagawa, S., Inagaki, F., Takai, K. and Horikoshi, K. (2004a). Phylogenetic diversity of sulfate-reducing prokaryotes in active deep-sea hydrothermal vent chimney structures. *FEMS Microbiol. Lett.* **232**, 145–152.

Nakagawa, S., Takai, K., Horikoshi, K. and Sako, Y. (2004b). *Aeropyrum camini* sp. nov., a strictly aerobic, hyperthermophilic archaeon from a deep-sea hydrothermal vent chimney. *Int. J. Syst. Evol. Microbiol.* **54**, 329–335.

Nakagawa, S., Takai, K., Inagaki, F., Horikoshi, K. and Sako, Y. (2005a). *Nitratiruptor tergarcus* gen. nov., sp. nov. and *Nitratifractor salsuginis* gen. nov., sp. nov., nitrate-reducing chemolithoautotrophs of the ε-Proteobacteria isolated from a deep-sea hydrothermal system in the Mid-Okinawa Trough. *Int. J. Syst. Evol. Microbiol.* **55**, 925–933.

Nakagawa, S., Takai, K., Inagaki, F., Chiba, H., Ishibashi, J., Kataoka, S., Hirayama, H., Nunoura, T., Horikoshi, K. and Sako, Y. (2005b). Variability in microbial community and venting chemistry in a sediment-hosted backarc hydrothermal system: impacts of subseafloor phase-separation. *FEMS Microbiol. Ecol.* **54**, 141–155.

Nakagawa, S., Takai, K., Inagaki, F., Hirayama, H., Nunoura, T., Horikoshi, K., and Sako, Y. (2005c). Distribution, phylogenetic diversity and physiological characteristics of epsilon-Proteobacteria in a deep-sea hydrothermal field. *Environ. Microbiol.* **7**, 1619–1632.

Nercessian, O., Reysenbach, A.-L., Prieur, D. and Jeanthon, C. (2003). Archaeal diversity associated with *in situ* samplers deployed on hydrothermal vents on the East Pacific Rise (13 degrees N). *Environ. Microbiol* **5**, 492–502.

Nercessian, O., Prokofeva, M., Lebedinski, A., L'Haridon, S., Cary, C., Prieur, D. and Jeanthon, C. (2004). Design of 16S rRNA-targeted oligonucleotide probes for detecting cultured and uncultured archaeal lineages in high-temperature environments. *Environ. Microbiol.* **6**, 170–182.

Nercessian, O., Bienvenu, N., Moreira, D., Prieur, D. and Jeanthon, C. (2005). Diversity of functional genes of methanogens, methanotrophs and sulfate reducers in deep-sea hydrothermal environments. *Environ. Microbiol.* **7**, 118–132.

Oremland, R. S. and Silverman, M. P. (1979). Microbial sulfate reduction measured by an automated electrical impedance technique. *Geomicrobiol. J.* **1**, 355–372.

Phelps, T. J. and Fredrickson, J. K. (2002). Drilling, coring, and sampling subsurface environments. In *Manual of Environmental Microbiology* (C. J. Hurst, R. L. Crawford, G. R. Knudsen, M. J. McInerney and L. D. Stetzenbach, eds), 2nd edn., pp. 679–695. ASM Press, Washington, DC.

Pley, U., Schipka, J., Gambacorta, A., Jannasch, H. W., Fricke, H., Rachel, R. and Stetter, K. O. (1991). *Pyrodictium abyssi* sp. nov. represents a novel heterotrophic marine archaeal hyperthermophile growing at 110°C. *Syst. Appl. Microbiol* **14**, 245–253.

Polz, M. F. and Cavanaugh, C. M. (1995). Dominance of one bacterial phylotype at a Mid-Atlantic Ridge hydrothermal vent site. *Proc. Natl. Acad. Sci. USA* **92**, 7232–7236.

Porter, K. G. and Feig, Y. S. (1980). The use of DAPI for identifying and counting aquatic microflora. *Limnol. Oceanogr.* **25**, 943–948.

Postec, A., Urios, L., Lesongeur, F., Ollivier, B., Querellou, J. and Godfroy, A. (2005). Continuous enrichment culture and molecular monitoring to investigate the microbial diversity of thermophiles inhabiting deep-sea hydrothermal ecosystems. *Curr. Microbiol.* **50**, 138–144.

Reysenbach, A.-L. and Götz, D. (2001). Methods for the study of hydrothermal vent microbes. In *Methods in Microbiology, Marine Microbiology* (J. H. Paul, ed.), vol. 30, pp. 639–655. Academic Press, San Diego.

Reysenbach, A.-L., Holm, N. G., Hershberger, K., Prieur, D. and Jeanthon, C. (1998). In search of a subsurface biosphere at a slow-spreading ridge. *Proc. ODP Sci. Result* **158**, 355–360.

Reysenbach, A.-L., Longnecker, K. and Kirshtein, J. (2000). Novel bacterial and archaeal lineages from an *in situ* growth chamber deployed at a Mid-Atlantic Ridge hydrothermal vent. *Appl. Environ. Microbiol* **66**, 3798–3806.

Ronimus, R. S., Reysenbach, A.-L., Musgrave, D. R. and Morgan, H. W. (1997). The phylogenetic position of the *Thermococcus* isolate AN1 based on 16S rRNA gene sequence analysis: a proposal that AN1 represents a new species, *Thermococcus zilligii* sp. nov. *Arch. Microbiol.* **168**, 245–248.

Sakai, H., Gamo, T., Kim, E.-S., Shitashima, K., Yanagisawa, F., Tsutsumi, M., Ishibashi, J., Sano, Y., Wakita, H., Tanaka, T., Matsumoto, T., Naganuma, T. and Mitsuzawa, K. (1990). Unique chemistry of the hydrothermal solution in the mid-Okinawa Trough backarc basin. *Geophys. Res. Lett.* **17**, 2133–2136.

Sako, Y., Takai, K., Ishida, Y., Uchida, A. and Katayama, Y. (1996). *Rhodothermus obamensis* sp. nov., a modern lineage of extremely thermophilic marine bacteria. *Int. J. Syst. Bacteriol.* **46**, 1099–1104.

Sako, Y., Nakagawa, S., Takai, K. and Horikoshi, K. (2003). *Marinithermus hydrothermalis* gen. nov., sp. nov., a strictly aerobic, thermophilic bacterium from a deep-sea hydrothermal vent chimney. *Int. J. Syst. Evol. Microbiol.* **53**, 59–65.

Sapra, R., Bagramyan, K. and Adams, M. W. (2003). A simple energy-conserving system: proton reduction coupled to proton translocation. *Proc. Natl. Acad. Sci. USA* **100**, 7545–7550.

Schrenk, M. O., Kelley, D. S., Delaney, J. R. and Baross, J. A. (2003). Incidence and diversity of microorganisms within the walls of an active deep-sea sulfide chimney. *Appl. Environ. Microbiol.* **69**, 3580–3592.

Slobodkin, A. I., Jeanthon, C., L'Haridon, S., Nazina, T., Miroshnichenko, M. and Bonch-Osmolovskaya, E. (1999). Dissimilatory reduction of Fe(III) by thermophilic bacteria and archaea in deep subsurface petroleum reservoirs of western Siberia. *Curr. Microbiol.* **39**, 99–102.

Slobodkin, A., Campbell, B., Cary, S. C., Bonch-Osmolovskaya, E. and Jeanthon, C. (2001). Evidence for the presence of thermophilic Fe(III)-reducing microorganisms in deep-sea hydrothermal vents at 13 degrees N (East Pacific Rise). *FEMS Microbiol Ecol.* **36**, 235–243.

Slobodkin, A. I., Tourova, T. P., Kostrikina, N. A., Chernyh, N. A., Bonch-Osmolovskaya, E. A., Jeanthon, C. and Jones, B. E. (2003). *Tepidibacter thalassicus* gen. nov., sp. nov., a novel moderately thermophilic, anaerobic, fermentative bacterium from a deep-sea hydrothermal vent. *Int. J. Syst. Evol. Microbiol.* **53**, 1131–1134.

Sokolova, T. G., Gonzáles, J. M., Kostrikina, N. A., Chernyh, N. A., Tourova, T. P., Kato, C., Bonch-Osmolovskaya, E. A. and Robb, F. T. (2001). *Carboxydobrachium pacificum* gen. nov., sp. nov., a new anaerobic, thermophilic, CO-utilizing marine bacterium from Okinawa Trough. *Int. J. Syst. Evol. Microbiol.* **51**, 141–149.

Sokolova, T. G., Jeanthon, C., Kostrikina, N. A., Chernyh, N. A., Lebedinsky, A. V., Stackebrandt, E. and Bonch-Osmolovskaya, E. A. (2004). The first evidence of anaerobic CO oxidation coupled with H_2 production by a hyperthermophilic archaeon isolated from a deep-sea hydrothermal vent. *Extremophiles* **8**, 317–323.

Spiess, F. N. and RISE Group. (1980). East Pacific Rise; hot springs and geophysical experiments. *Science* **297**, 1421–1433.

Stetter, K. O. (2001). Family II. *Methanothermaceae*. In *Bergey's Manual of Systematic Bacteriology* (D. R. Boone, R. W. Castenholz and G. M. Garrity, eds), 2nd edn., vol. 1, pp. 233–235. Springer, New York.

Stetter, K. O., König, H. and Stackebrandt, E. (1983). *Pyrodictium* gen. nov., a new genus of submarine disc-shaped sulfur reducing archaebacteria growing optimally at 105°C. *Syst. Appl. Microbiol.* **4**, 535–551.

Sunamura, M., Higashi, Y., Miyako, C., Ishibashi, J. and Maruyama, A. (2004). Two bacteria phylotypes are predominant in the Suiyo seamount hydrothermal plume. *Appl. Environ. Microbiol.* **70**, 1190–1198.

Suzuki, Y., Inagaki, F., Takai, K., Nealson, K. H. and Horikoshi, K. (2004). Microbial diversity in inactive chimney structures from deep-sea hydrothermal system. *Microb. Ecol.* **47**, 186–196.

Takai, K. and Horikoshi, K. (1999). Genetic diversity of archaea in deep-sea hydrothermal vent environments. *Genetics* **152**, 1285–1297.

Takai, K. and Horikoshi, K. (2000). *Thermosipho japonicus* sp. nov., an extremely thermophilic bacterium isolated from a deep-sea hydrothermal vent in Japan. *Extremophiles* **4**, 9–17.

Takai, K., Sugai, A., Itoh, T. and Horikoshi, K. (2000). *Palaeococcus ferrophilus* gen. nov., sp. nov., a barophilic, hyperthermophilic archaeon from a deep-sea hydrothermal vent chimney. *Int. J. Syst. Evol. Microbiol.* **50**, 489–500.

Takai, K., Komatsu, T., Inagaki, F. and Horikoshi, K. (2001). Distribution of archaea in a black smoker chimney structure. *Appl. Environ. Microbiol.* **67**, 3618–3629.

Takai, K., Inoue, A. and Horikoshi, K. (2002). *Methanothermococcus okinawensis* sp. nov., a thermophilic, methane-producing archaeon isolated from a Western Pacific deep-sea hydrothermal system. *Int. J. Syst. Evol. Microbiol.* **52**, 1089–1095.

Takai, K., Inagaki, F., Nakagawa, S., Hirayama, H., Nunoura, T., Sako, Y., Nealson, K. H. and Horikoshi, K. (2003a). Isolation and phylogenetic diversity of members of previously uncultivated ε-Proteobacteria in deep-sea hydrothermal fields. *FEMS Microbiol. Lett.* **218**, 167–174.

Takai, K., Kobayashi, H., Nealson, K. H., and Horikoshi, K. (2003b). *Deferribacter desulfuricans* sp. nov., a novel sulfur-, nitrate- and arsenate-reducing thermophile isolated from a deep-sea hydrothermal vent. *Int. J. Syst. Evol. Microbiol.* **53**, 839–846.

Takai, K., Nakagawa, S., Sako, Y. and Horikoshi, K. (2003c). *Balnearium lithotrophicum* gen. nov., sp. nov., a novel thermophilic, strictly anaerobic, hydrogen-oxidizing chemolithoautotroph isolated from a black smoker chimney in the Suiyo Seamount hydrothermal system. *Int. J. Syst. Evol. Microbiol.* **53**, 1947–1954.

Takai, K., Gamo, T., Tsunogai, U., Nakayama, N., Hirayama, H., Nealson, K. H. and Horikoshi, K. (2004a). Geochemical and microbiological evidence for a hydrogen-based, hyperthermophilic subsurface lithoautotrophic microbial ecosystem (HyperSLiME) beneath an active deep-sea hydrothermal field. *Extremophiles* **8**, 269–282.

Takai, K., Hirayama, H., Nakagawa, T., Suzuki, Y., Nealson, K. H. and Horikoshi, K. (2004b). *Thiomicrospira thermophila* sp. nov., a novel microaerobic, thermotolerant, sulfur-oxidizing chemolithomixotroph isolated from a deep-sea hydrothermal fumarole in the TOTO caldera, Mariana Arc, Western Pacific. *Int. J. Syst. Evol. Microbiol.* **54**, 2325–2333.

Takai, K., Inagaki, F. and Horikoshi, K. (2004c). Distribution of unusual archaea in subsurface biosphere In *The Subseafloor Biosphere at Mid-Oceanic Ridges* (W. S. D. Wilcock, E. F. DeLong, D. S. Kelley, J. A. Baross and S. C. Cary, eds), pp. 369–381. American Geographic Union, Washington, DC.

Takai, K., Nealson, K. H. and Horikoshi, K. (2004d). *Hydrogenimonas thermophila* gen. nov., sp. nov., a novel thermophilic, hydrogen-oxidizing chemolithoautotroph within the ε-*Proteobacteria*, isolated from a black smoker in a Central Indian Ridge hydrothermal field. *Int. J. Syst. Evol. Microbiol.* **54**, 25–32.

Takai, K., Nealson, K. H. and Horikoshi, K. (2004e). *Methanotorris formicicus* sp. nov., a novel extremely thermophilic, methane-producing archaeon isolated from a black smoker chimney in the Central Indian Ridge. *Int. J. Syst. Evol. Microbiol.* **54**, 1095–1100.

Takai, K., Oida, H., Suzuki, Y., Hirayama, H., Nakagawa, S., Nunoura, T., Inagaki, F., Nealson, K. H. and Horikoshi, K. (2004f). Spatial distribution of Marine Crenarchaeota Group I (MGI) in the vicinity of deep-sea hydrothermal systems. *Appl. Environ. Microbiol.* **70**, 2404–2413.

Takai, K., Hirayama, H., Nakagawa, T., Suzuki, Y., Nealson, K. H. and Horikoshi, K. (2005). *Lebetimonas acidiphila* gen. nov., sp. nov., a novel thermophilic, acidophilic, hydrogen-oxidizing chemolithoautotroph within the *"Epsilonproteobacteria"*, isolated from a deep-sea hydrothermal fumarole in the Mariana Arc. *Int. J. Syst. Evol. Microbiol.* **55**, 183–189.

Taylor, C. D. and Wirsen, C. O. (1997). Microbiology and ecology of filamentous sulfur formation. *Science* **277**, 1483–1485.

Taylor, C. D., Wirsen, C. O. and Gaill, F. (1999). Rapid microbial production of filamentous sulfur mats at hydrothermal vents. *Appl. Environ. Microbiol.* **65**, 2253–2255.

Teske, A., Hinrichs, K.-U., Edgcomb, V., de Versa Gomez, A., Kysela, D., Sylva, S. P., Sogin, M. L. and Jannasch, H. W. (2002). Microbial diversity of hydrothermal sediments in the Guaymas Basin: evidence for anaerobic methanotrophic communities. *Appl. Environ. Microbiol.* **68**, 1994–2007.

Tsunogai, U., Toki, T., Nakayama, N., Gamo, T., Kato, H. and Kaneko, S. (2003). WHATS: a new multi-bottle gas-tight sampler for sea-floor vent fluids. *Chikyukagaku* (*Geochemistry*) 37, 101–109 (In Japanese with English abstract).

Urios, L., Cueff, V., Pignet, P. and Barbier, G. (2004a). *Tepidibacter formicigenes* sp. nov., a novel spore-forming bacterium isolated from a Mid-Atlantic Ridge hydrothermal vent. *Int. J. Syst. Evol. Microbiol.* **54**, 439–443.

Urios, L., Cueff-Gauchard, V., Pignet, P., Postec, A., Fardeau, M-L., Ollivier, B. and Barbier, G. (2004b). *Thermosipho atlanticus* sp. nov., a novel member of the *Thermotogales* isolated from a Mid-Atlantic Ridge hydrothermal vent. *Int. J. Syst. Evol. Microbiol.* **54**, 1953–1957.

Van Dover, C. L. (ed.) (2000). *The Ecology of Deep-Sea Hydrothermal Vents*, Princeton University Press, New Jersey.

Vetriani, C., Speck, M. D., Ellor, S. V., Lutz, R. A. and Starovoytov, V. (2004). *Thermovibrio ammonificans* sp. nov., a thermophilic, chemolithotrophic, nitrate ammonifying bacterium from deep-sea hydrothermal vents. *Int. J. Syst. Evol. Microbiol.* **54**, 175–181.

Von Damm, K. L. (1995). Controls on the chemistry and temporal variability of seafloor hydrothermal fluids. In *Seafloor Hydrothermal Systems* (S. E. Humphris, R. A. Zierenberg, L. S. Mullineaux and R. E. Thomson, eds), pp. 222–247. American Geophysical Union, Washington, DC.

Von Damm, K. L., Edmond, J. M., Grant, B. C., Measures, C. I., Malden, B. and Weiss, R. F. (1985). Chemistry of submarine hydrothermal solutions at 21°N, East Pacific Rise. *Geochim. Cosmochim. Acta* **49**, 2197–2220.

Voordeckers, J. W., Starovoytov, V. and Vetriani, C. (2005). *Caminibacter mediatlanticus* sp. nov., a thermophilic, chemolithoautotrophic, nitrate-ammonifying bacterium isolated from a deep-sea hydrothermal vent on the Mid-Atlantic Ridge. *Int. J. Syst. Evol. Microbiol.* **55**, 773–779.

Wery, N., Lesongeur, F., Pignet, P., Derennes, V., Cambon-Bonavita, M.-A., Godfroy, A. and Barbier, G. (2001a). *Marinitoga camini* gen. nov., sp. nov., a rod-shaped bacterium belonging to the order *Thermotogales*, isolated from a deep-sea hydrothermal vent. *Int. J. Syst. Evol. Microbiol.* **51**, 495–504.

Wery, N., Moricet, J. M., Cueff, V., Jean, J., Pignet, P., Lesongeur, F., Cambon-Bonavita, M.-A. and Barbier, G. (2001b). *Caloranaerobacter azorensis* gen. nov., sp. nov., an anaerobic thermophilic bacterium isolated from a deep-sea hydrothermal vent. *Int. J. Syst. Evol. Microbiol.* **51**, 1789–1796.

Wilcock, C. L., DeLong, E. F., Kelley, D. S., Baross, J. A. and Cary, S. C. (eds) (2004). *The Subseafloor Biosphere at Mid-Ocean Ridge*, American Geophysical Union, Washington, DC.

Wirsen, C. O., Jannasch, H. W. and Molyneaux, S. J. (1993). Chemosynthetic microbial activity at Mid-Atlantic Ridge hydrothermal vent sites. *J. Geophys. Res.* **98**, 9693–9703.

Zehnder, A. J. B. and Wuhrmann, K. (1976). Titanium (III) citrate as a nontoxic oxidation-reduction buffering system for the culture of obligate anaerobes. *Science* **194**, 1165–1166.

Zillig, W., Holz, I., Janekovic, D., Schäfer, W. and Reiter, W. D. (1983). The archaebacterium *Thermococcus celer*, represents a novel genus within the thermophilic branch of the archaebacteria. *Syst. Appl. Microbiol.* **4**, 88–94.

4 Growth of Hyperthermophilic Microorganisms for Physiological and Nutritional Studies

Anne Godfroy[1], Anne Postec[1] and Neil Raven[2]

[1] Laboratoire de Microbiologie des Environnements Extrêmes, UMR 6197, Ifremer, BP 70 29280, Plouzané, France; [2] Health Protection Agency, Porton Down, Salisbury SP4 0JG, UK

CONTENTS

Introduction
Cultivation of sulphur-reducing hyperthermophiles in bioreactors
Continuous culture of *Thermococcales* species in a glass gas-lift bioreactor
Future prospects: cultivation of thermophilic and hyperthermophilic microbial communities

◆◆◆◆◆ INTRODUCTION

Hyperthermophilic microorganisms grow at temperatures of 80°C and above. Numerous species belonging to both the Archaea and Bacteria have now been isolated from both continental and volcanic areas. Initially following major isolation efforts, heterotrophic species were recovered, but this has now extended to species involved in all geo-biochemical cycles and employing a wide range of electron donors and acceptors.

In addition to their ecological and evolutionary interest, these organisms were rapidly identified as potential sources of stable biomolecules having potential application in industrial processes. However, an understanding of their growth characteristics and metabolism is necessary to realize their full biotechnological potential. A major limitation to their study has frequently been the small amount of biomass that can be generated.

Within the archaeal hyperthermophiles, some species have been more extensively studied: (1) thermoacidophilic species belonging to the order *Sulfolobales* (2) methanogens and (3) a number of species belonging to the order *Thermococcales*.

A key factor in the study of hyperthermophilic organisms has been the development of cultivation strategies allowing exploration of their metabolic behaviour and the optimization of their growth parameters. At the same time, this has allowed the production of biomass in sufficient quantities for further studies.

Initially, concerns about the damaging effect of high levels of reduced sulphur species by anaerobic hyperthermophiles, or sulphuric acid by oxidizing species, led to the development of multi-hundred litre capacity fermentors with ceramic liners (Kelly and Deming, 1988). Large scale cultivation of hyperthermophilic species was reported by Karl Stetter's group for biomass production of various hyperthermophilic species using an enamelled 300-litre fermentor (Blöchl et al., 1997; Huber et al., 2000; Stetter et al., 1983). A range of cultivation systems have also been reported for the culture of acidophilic hyperthermophiles (Schiaraldi et al., 1999; Worthington et al., 2003) and methanogens (Mukhopadhyay et al., 1999).

♦♦♦♦♦ CULTIVATION OF SULPHUR-REDUCING HYPERTHERMOPHILES IN BIOREACTORS

The cultivation of microorganisms at temperatures up to 110°C, in media that may contain high salt concentrations, and that actively produce hydrogen sulphide represents a major challenge to conventional laboratory fermentation equipment.

Thermococcales have been isolated from both deep and shallow marine ecosystems (with the exception of *Thermococcus sibiricus* (Miroshnichenko et al., 2001) isolated from a high temperature oil reservoir and *Thermococcus zilligii* (Ronimus et al., 1997) and *Thermococcus waiotapuensis* (Gonzalez et al., 1999) isolated from a continental hot spring). The order *Thermococcales* includes 3 genera: *Thermococcus*, *Pyrococcus* and *Palaeococcus*. These species are all anaerobic; they utilize peptides as carbon and energy sources and most of them exhibit better growth in the presence of elemental sulphur producing large amounts of hydrogen sulphide. Some are able to grow in the absence of elemental sulphur, but in serum vial batch culture, the production of molecular hydrogen as a metabolic end product leads to early inhibition of growth.

Within the *Thermococcales*, the shallow marine species *Pyrococcus furiosus* has been the most extensively studied in terms of its physiology, genetics and enzymology, and is generally considered as the model species for heterotrophic hyperthermophiles. The pathways of both peptide and carbohydrate metabolism of this species have been well studied (Adams et al., 2001).

Pressurized bioreactors have been developed in consideration that high pressure may be important for the growth of deep sea sulphur-reducing hyperthermophiles (Kelly and Deming, 1988; Miller et al., 1988; Nelson et al., 1992). However, most hyperthermophiles have been isolated at atmospheric pressure, although pressure effects have been reported for some of them (Canganella *et al.*, 2000; Erauso *et al.*, 1993; Marteinsson *et al.*, 1999a, 1999b).

Simpler and less expensive approaches for the cultivation of hyperthermophiles have also been reported. These have been used successfully both to generate large quantities of biomass and to explore their physiology.

As they are relatively easy to grow, many systems have been developed for the cultivation of *Thermococcales* species, with particular emphasis on *P. furiosus*. Batch fermentation of numerous sulphur-reducing hyperthermophiles has been reported using conventional stainless-steel bioreactors, with proper attention to cleaning and maintenance operations, and numerous enzymes have been purified from the biomass produced (Adams, 1993; Gantelet and Duchiron, 1998; Raffin *et al.*, 2000). In the presence of sulphur, hydrogen sulphide is produced instead leading to moderate cell densities in closed culture (Fiala and Stetter, 1986; Godfroy *et al.*, 2000). Since hydrogen sulphide is reasonably soluble in water, this can lead to growth inhibition, the absence of a real stationary phase and cell lysis (personal observation). Moreover, reaction with many metals occurs to form extremely insoluble sulphide precipitates. However, high cell densities of *Thermococcales* can be obtained by sparging sulphur-containing cultures with anaerobic gases such as nitrogen, in both closed vials or in conventional glass fermentors (Brown and Kelly, 1989) or a modified one as described for the cultivation of "Pyrococcus abyssi" (Godfroy *et al.*, 2000).

The reactor described by Brown and Kelly (Brown and Kelly, 1989) comprised a 2-litre glass round bottomed flask maintained at high temperature by a heating mantle controlled by a proportional controller. The reactor was operated at a volume of 1 litre. The vessel was fitted with a gas inlet sparger and the exit gas was passed through sodium hydroxide to remove hydrogen sulphide. The medium was supplied from a polycarbonate reservoir using a peristaltic pump and anaerobicity maintained by sparging with nitrogen. After each change of dilution rate, sulphur was added to ensure it remained in excess. A constant volume was maintained *via* a dip tube connected in parallel to the medium feed pump.

When grown in the absence of elemental sulphur, species belonging to the order *Thermococcales* were shown to produce hydrogen as an end product of their metabolism and this appeared to limit growth severely unless removed. The glass continuous culture vessel described by Brown and Kelly (1989) has been used in many studies (Rinker and Kelly, 1996, 2000) for cultivation of *P. furiosus*, *T. litoralis* and the bacterium *Marinitoga maritima*.

Fermentation of *P. furiosus* in a conventional stirred bioreactor was described by Krahe *et al.* (1996) generating a high cell density, but it was shown that stirrer speeds above 1800 rpm had a negative effect on growth. Krahe also reported the utilization of dialysis membrane bioreactor system for the cultivation of *P. furiosus* (Krahe *et al.*, 1996). Cultivation at high temperature was carried out using a bioengineering (Wald, Switzerland) membrane bioreactor consisting of two compartments (Holst *et al.*, 1997). The outer compartment comprises a cylindrical high strength polyamide foil fixed to a top and bottom stainless steel plate. The inner compartment is separated from the outer one by a tubular dialysis membrane. Gas sparging can be applied to either the inner or outer compartment, or both if necessary. The reactor is also equipped with two stirring units. Application of a high medium dilution rate in the outer compartment was shown to permit high cell densities to be reached compared to cultivation in closed glass bottle, primarily by the removal of a range of metabolic end products. This system was also used for cultivation of the thermoacidophilic species *Sulfolobus shibatae* (Krahe *et al.*, 1996).

A 2-litre bioreactor has also been developed for the cultivation of *P. furiosus* in continuous culture to high cell concentrations and biomass yields (Raven *et al.*, 1992). The intense sparging of sulphur-free, hydrogen-evolving cultures has significant advantages, both in the quality of biomass produced and in the safety of the procedure. Since the risk of contamination of thermophilic and hyperthermophilic continuous cultures is relatively small, long-term steady state experiments both for biomass production and for the detailed investigation of physiological responses to culture perturbation are made possible.

◆◆◆◆◆ CONTINUOUS CULTURE OF *THERMOCOCCALES* SPECIES IN A GLASS GAS-LIFT BIOREACTOR

This experimental protocol, initially developed for optimization of biomass production of *P. furiosus*, was then used for the development of a sulphur-free minimal medium for *P. furiosus* (Raven and Sharp, 1997), physiological study of "*P. abyssi*" strain ST549 (Godfroy *et al.*, 2000), optimization of culture conditions for the growth of *Thermococcus hydrothermalis* and the development of defined and minimal media (Postec *et al.*, 2005b). This protocol is now routinely used for biomass production of the "*P. abyssi*" type strain GE5.

Description of the Gas-Lift Bioreactor and Equipment

Glass vessel

The glass vessel is a concentric tube reactor (Figure 4.1B) constructed by Radleys (UK) and comprising a central column (Figure 4.1A a) with a rounded base and an external glass jacket (Figure 4.1A b). Two glass ports with Duran-type threads are fitted on opposite side (Figure 4.1A c), one at the bottom and one at the top. The base has a 45-mm diameter port with a glass Duran–type screw thread (Figure 4.1A d) for sparger insertion.

A glass draft tube (Figure 4.1A e) is placed within the central column section. It has two sets of 3 glass lugs near the top and

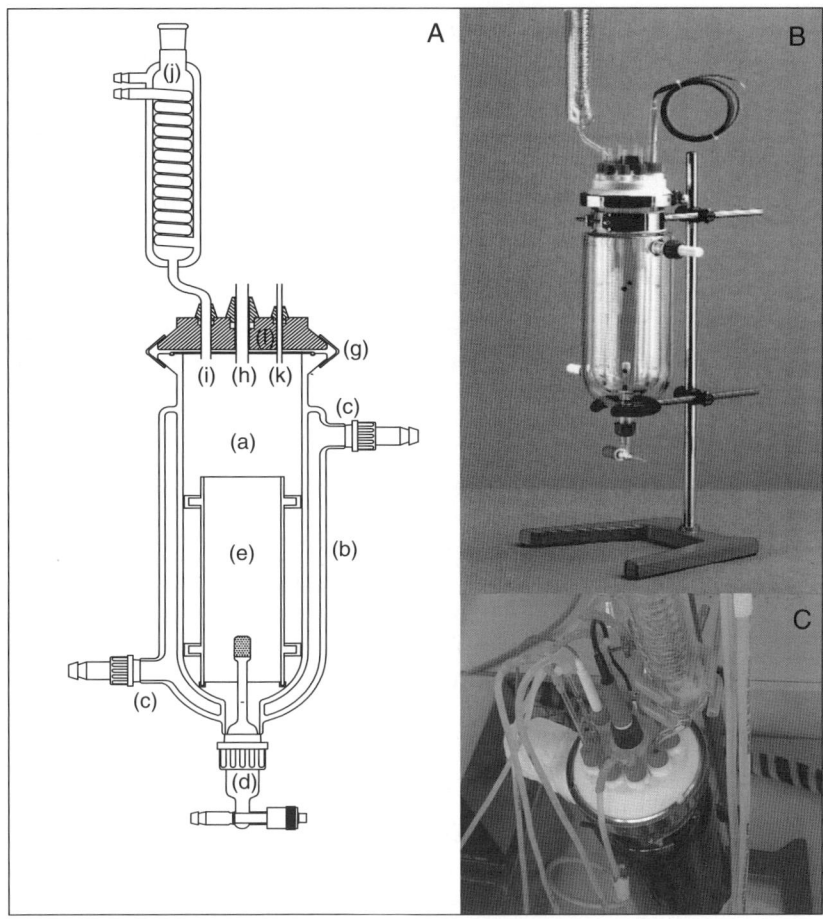

Figure 4.1. The gas-lift bioreactor (A) drawing, (B) picture and (C) top plate.

bottom respectively. Silicone tubing is stretched over each lug and used to position the draft tube within the vessel.

Top plate

A PTFE top plate (Figure 4.1C, A f) is clamped on top of the glass vessel using a stainless steel clamping ring (Figure 4.1A g) with a silicone gasket between.

The top plate is perforated by 13 ports:

- a single central port of 12 mm in diameter (Figure 4.1A h) used to fit a Mettler Toledo gel pH electrode.
- 2 × 8 mm ports (Figure 4.1A i) used to fit a PTFE-covered PT100 temperature probe and a glass condenser (Figure 4.1A i), through which cold water can be passed,
- 8 × 5 mm ports (Figure 4.1A k) used for acid and alkali addition; medium feed and culture draw-off; inoculation and sampling system using glass needles or 1 ml glass pipettes. Water and gas tightness is ensured by compression of 5 ring gaskets. When not used, ports are sealed with small glass rods.

97

Medium/product and alkali/acid bottles

Twenty-litre Nalgene propylene bottles and 750 ml bottles were used for medium and product or acid and alkali respectively. Screw caps have 3 ports: for delivery, drawing-off and a gas vent fitted with a 0.2 μm PTFE filter. Empty bottles are sterilized by autoclaving and filled with either medium or 1 M HCl or 1 M NaOH solutions, sterilized beforehand by filtration through 0.2 μm Sartobran autoclavable filters (Sartorius). All sterile connections are made using autoclavable quick coupling systems (Cole-Parmer).

Medium and product bottles can be kept under anaerobic conditions if necessary by sparging at a low rate with oxygen-free nitrogen.

Pumps and tubing

Masterflex Tygon silicone and Pharmed tubing are used for liquids and gases respectively.

Peristaltic pumps are used for both medium delivery feeding and product draw-off and for alkali or acid feeding. For acid and alkali additions, peristaltic pumps are connected to the pH regulation system. The feeding pump is a programmable pump, while the product draw-off pump is a standard one.

Controls

pH and temperature are controlled by a 4–20 mA Controller and AFS Biocommand system from New Brunswick (Nijmegen, the Netherlands) associated with a pH transmitter (Broadley James) and a temperature-controlled heated circulated bath (Huber). For temperatures up to 95°C, water can be used for circulation, while mineral oil should be used for temperatures above this. Additionally, the gas lift vessel can be connected to any other standard fermentation control system.

Fermentation Protocol

Medium

The basal SME medium (Stetter *et al.*, 1983) was modified according to Sharp and Raven (Sharp and Raven, 1997; Raven *et al.*, 1992) (Table 4.1) and can be used as it is, or can be modified further for nutritional studies (Raven and Sharp, 1997; Postec *et al.*, 2005b). In this medium, L-cysteine is used as reductant.

Bioreactor set-up

Temperature probes, pH probes (calibrated prior to use) and condensers are positioned in their appropriate ports. All other ports are closed with

Table 4.1 Composition of SME medium (Stetter *et al.*, 1983) modified according to Sharp and Raven (Raven *et al.*, 1992; Sharp and Raven, 1997)

Medium	per litre
NaCl	28 g
Yeast extract	1 g
Peptone	2 g
L-cysteine	0.5 g
Resazurin	0.5 mg
Magnesium salt solution	10 ml
Solution A	1 ml
Solution B	1 ml
Solution C	1 ml
Solution D	1 ml

Stock Solutions

Magnesium salt solution	
$MgSO_4 \cdot 7H_2O$	180 g
$MgCl_2 \cdot 6H_2O$	140 g

Solution A	
$MnSO_4 \cdot 4H_2O$	9 g
$ZnSO_4 \cdot 7H_2O$	2.5 g
$NiCl_2 \cdot 6H_2O$	2.5 g
$AlK(SO_4)_2 \cdot 12H_2O$	0.3 g
$CoCl_2 \cdot 6H_2O$	0.3 g
$CuSO_4 \cdot 5H_2O$	0.15 g

Solution B	
$CaCl_2 \cdot 2H_2O$	56 g
NaBr	25 g
KCl	16 g
KI	10 g

Solution C	
K_2HPO_4	50 g
H_3BO_3	7.5 g
$Na_2WO_4 \cdot 2H_2O$	3.3 g
$Na_2MoO_4 \cdot 2H_2O$	0.15 g
Na_2SeO_3	0.005 g

Solution D in 1 M HCl	
$FeCl_2 \cdot 4H_2O$	10 g

(*continued*)

Table 4.1 Continued

Medium	per litre
Vitamins solution in ethanol/water (50:50)	
Pyridoxine·HCl	200 mg
Thiamine·HCl	100 mg
Riboflavin	100 mg
Nicotinic acid	100 mg
DL-calcium pantothenate	100 mg
Lipoic acid	100 mg
Biotin	40 mg
Folic acid	40 mg
Cyanocobalamin	2 mg

glass rods. The vessel can be sterilized (with ports closed and covered as appropriate) by placing in an autoclave, but given the fragility of the system, an *in situ* tyndallization procedure can also be used: The vessel is filled with distilled water and heated at 100°C for 3 h. under aerobic conditions (by sparging the reactor with air) then cooled to room temperature; this procedure is repeated at least 4 or 5 times.

The condenser is then connected to a cold water supply (HS40, Huber refrigerated circulating bath filled with water). Sparging with nitrogen is initiated using a gas flow meter (Aalborg, Monsey, New York) with a 0.2 vol vol^{-1} min^{-1} flow rate. Media feed and product recovery bottles are connected aseptically to the bioreactor and water (used for the sterilization step of the bioreactor) replaced by medium (at least 5 volume changes are needed). At the same time, alkali and acid bottles can be connected aseptically to the bioreactor. A sampling device is also aseptically introduced into the corresponding port.

Temperature and pH regulation systems are switched on and the pH value of the medium standardized by measuring the pH of a small sample by comparison with a standard bench pH meter at room temperature. The system for continuous culture using the gas-lift bioreactor is presented in Figure 4.2.

Starting culture

When the temperature and pH have reached their set values, media feed and product draw off pumps are stopped prior to inoculation with a fresh grown vial culture (20 or 40 ml for a 2-l culture) *via* the inoculation port.

Culture monitoring

The cell density of the culture can be monitored by both direct cell counting using a 0.02 mm depth Thoma chamber (Weber Scientific International, UK) under a phase contrast microscope or by optical density at 600 nm.

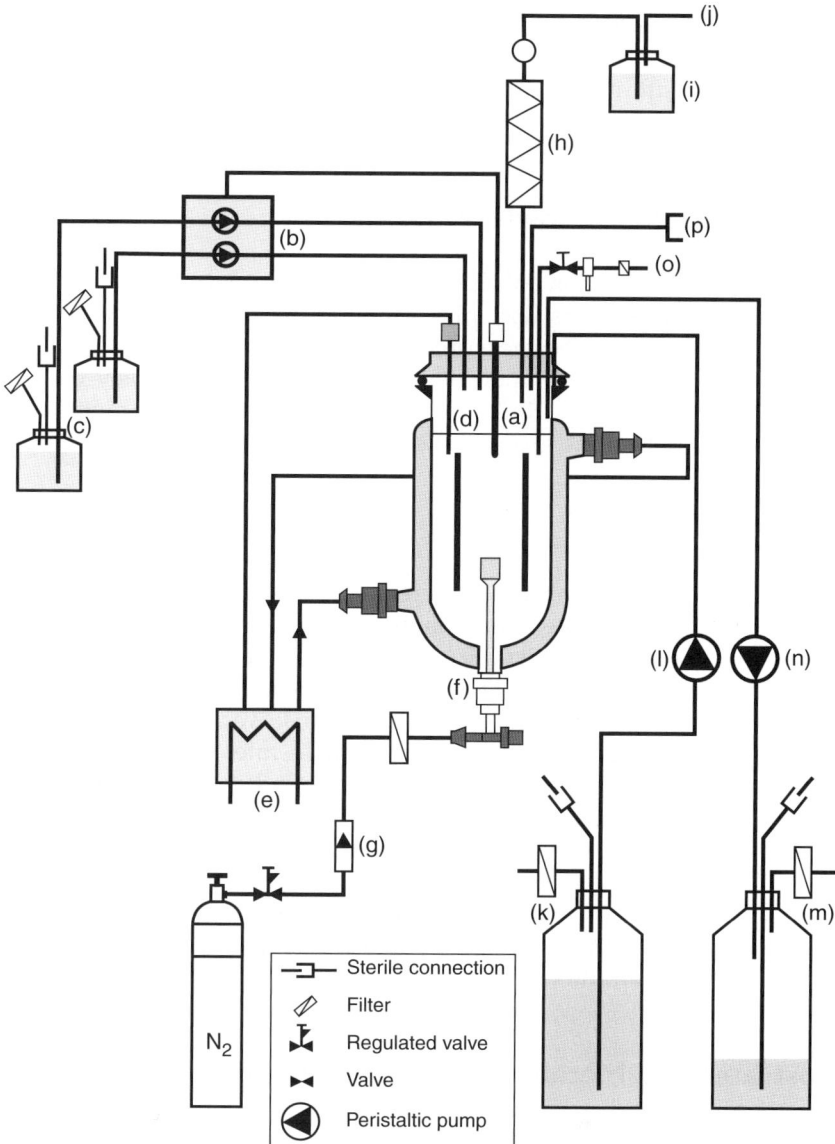

Figure 4.2. Schematic diagram of the experimental system for continuous culture with the gas-lift bioreactor. (a) pH probe; (b) pH controller; (c) acid and alkali bottles; (d) temperature probe; (e) heated circulating bath; (f) sparging assembly; (g) gas flow meter; (h) condenser; (i) condensate trap; (j) gas exhaust; (k) medium bottle; (l) medium pump; (m) product bottle; (n) drawing-off pump; (o) sampling system; and (p) inoculation septum.

When growth has started, the continuous process can be established by starting both feeding and draw-off pumps. For biomass production, a dilution rate of 0.2 h^{-1} or 0.4 h^{-1} is convenient (since from 9.6 to 19.2 l of culture per day are generated). For physiological and nutritional studies, the dilution rate can be lowered in order to obtain at least 3 volume changes in the bioreactor per day. This allows parameters such as pH, temperature and media composition to be altered sequentially each day. with results approximating to equilibrium values being obtained (Postec et al., 2005b; Raven and Sharp, 1997).

Determination of Growth Parameters

In addition to steady state cell concentrations, the influence of various parameters (e.g. pH, temperature, gas flow rate or media composition) can also be measured by determination of maximal growth rate under sequentially modified conditions. As previously described (Rinker and Kelly, 1996; Rinker et al., 1999), wash out experiments can be performed to determine growth rates according to the formula

$$\frac{1}{x}\left(\frac{dx}{dt}\right) = \mu_{max} - D$$

where x = the cell density of the culture; t = time; μ_{max} = the maximal growth rate, and D is the dilution rate (Duarte et al., 1994), but problems with the reliability of medium delivery rate measurements at high dilution rates ($D \geq 2h^{-1}$) have been shown to significantly affect growth rate calculations (personal observation). An alternative method to determine the specific growth rate can be achieved by performing batch culture experiments. A steady state continuous culture under the above culture conditions can be grown at supra-maximal dilution rates in order to lower the cell density significantly. Medium feeding and product draw off are then stopped, and growth under batch conditions followed by regular cell counting (e.g. every 15 min., 3 counts per sample). Growth rate can then be determined by performing a linear regression along the exponential part of the growth curve (Postec et al., 2005b).

Substrates and Metabolic Product Analysis

The gas outflow from the bioreactor can be directly analysed using a MTI M200D micro gas chromatograph equipped with a thermal conductivity detector. A Molecular Sieve column with argon as the carrier gas and a temperature of 30°C was used to detect H_2 and CO_2, while H_2S was detected using a Poraplot U column at 100°C, with helium as the carrier gas.

Substrates (carbohydrates and amino acids) and metabolic end products (organic acids) can be analysed by means of HPLC (Alliance 2690; Waters) as described by Wery et al. (2001). Proteins from the supernatant of the culture are precipitated by a solution of 5-sulfosalicylic acid at 2% (w/v). Amino acids may be analysed after derivatization (with an ethanol/water/triethylamine/phenylisothiocyanate (7:1:1:1 by vol.) solution) according to the conditions given for the Waters Pico Tag method (WAT007360; Waters). Carbohydrates and linear organic acids are separated on an H^+ exclusion column (Polyspher OAKC 1.51270; Merck) at 60°C with an 18 mM H_2SO_4 elution solution (0.35 ml min^{-1}) and detected with a differential refractometer (Refractometer 410, Waters).

◆◆◆◆◆ FUTURE PROSPECTS: CULTIVATION OF THERMOPHILIC AND HYPERTHERMOPHILIC MICROBIAL COMMUNITIES

While the gas-lift bioreactor was originally developed for culture in the absence of sulphur, it has also been shown to be usable for culture in its presence. Particular care regarding gas exhaust elimination must be taken for safety reasons: hydrogen sulphide is trapped in 10 M NaOH and the off gas vented to the atmosphere. Cultures in the presence of sulphur should always be performed in well-ventilated areas and in the presence of a hydrogen sulphide detector. The utilization of the gas-lift bioreactor has also been shown to be a useful tool for the cultivation of microbial communities in addition to pure cultures from deep sea hydrothermal chimneys.

Continuous enrichment cultures have been performed in the gas-lift bioreactor using SME mineral and vitamin solutions added with various organic substrates at 60° and 90°C using a hydrothermal chimney sample as the inoculum. The microbial diversity present was then analysed using molecular tools. Cloning and sequencing of 16S rRNA genes of two independent samples from both 60°C and 90°C cultures was performed. A denaturing gradient gel electrophoresis (DGGE) analysis of the 16S rRNA gene diversity was also performed over the course of the two cultures in order to monitor the dynamics of the microbial population. The results evidenced quite large diversity compared to the diversity previously obtained under the same conditions by classical batch cultures in closed vessels. At 90°C, while archaeal diversity was limited to species belonging to the *Thermococcales*, unexpected bacterial diversity was obtained, including still uncultivated species and already described species growing at temperatures above their previously known optimal growth conditions (Postec *et al.*, 2005c). At 60°C, considerable bacterial diversity was achieved and both heterotrophic and autotrophic species shown to be cultivated, with some successfully purified and formally described (Postec *et al.*, 2005a). These preliminary experiments indicate that such continuous enrichment cultures in bioreactors may allow (i) access to as yet uncultivated microorganisms and (ii) the co-cultivation of thermophilic microbial deep sea hydrothermal vent species (involved in various biogeochemical cycles) permitting the study of the interactions between species under various experimental conditions with emphasis on better simulating environmental conditions.

Acknowledgements

We thank Ed Hartley from Radleys (England) for providing drawings and pictures of the gas-lift bioreactor. We thank Sylvie Gros (Ifremer) for further figures.

List of suppliers

Radleys
Shire Hill, Saffron Walden,
Essex CB11 3AZ, UK.
Contact: Ed Hartley, Production Manager
Tel: +44-(0)1799-513320
http://www.radleys.com

Gas-lift bioreactor, temperature probe, condenser, glass needle and sparging assembly

Sigma
P.O. Box 14508
St Louis, MO 63 178, USA
Tel: 1-800-325-3010
http://www.sigma.cial.com

All chemicals

Difco/BD
1 Becton Drive
Franklin Lakes, NJ USA 07417
Tel 1-201-847-6800
http://www.difco.com

Peptone, yeast extract

Fisher Scientific International Inc. / Bioblock Fischer Scientific
Liberty Lane
Hampton, NH 03842
Tel: 1-603- 926-5911
http://www.bioblock.com
http://www.fisherscientific.com

Pumps and tubing (Masterflex®), quick-coupling system (Cole-Parmer®) and Nalgene® bottles, gas flow meter

Sartorius AG
Weender Landstrasse 94-108
D-37075 Göttingen, Germany
Tel: 49-551-308-333
http://www.sartorius.com

Filters

New Brunswick Scientific
PO Box 4005
Edison, NJ USA
Tel: 1-732-287-1200
http://www.nbsc.com

Control system

MTI Analytical Instruments
41762 Christy St.
Fremont, CA 94538, USA
Tel: 1-510-490-0900

Micro Gas Chromatograph

Waters
34 Maple Street
Milford, Massachusetts 01757, USA
Tel: 1-508-478-2000
http://www.waters.com

HPLC

References

Adams, M. W. W. (1993). Enzymes and proteins from organisms that grow near and above 100°C. *Ann. Rev. Microbiol.* **47**, 627–658.

Adams, M. W. W., Holden, J. F., Menon, A. L. Schut, G. J., Grunden, A. M., Hou, C., Hutchins, A. M., Jenney, F. E., Kim, C., Ma, K. S., Pan, G. L., Roy, R., Sapra, R., Story, S. V. and Verhagen, M. (2001). Key role for sulfur in peptide metabolism and in regulation of three hydrogenases in the hyperthermophilic archaeon *Pyrococcus furiosus*. *J. Bacteriol.* **183**, 716–724.

Blöchl, E., Rachel, R., Burggraf, S., Hafenbradl, D., Jannasch, H. W. and Stetter, K. O. (1997). *Pyrolobus fumarii*, gen. nov., sp. nov., represents a novel group of archaea, extending the upper temperature limit for life to 113°C. *Extremophiles* **1**, 14–21.

Brown, S. H. and Kelly, R. M. (1989). Cultivation techniques for hyperthermophilic archaebacteria: Continuous culture of *Pyrococcus furiosus* at temperature near 100°C. *Appl. Environ. Microbiol.* **55**, 2086–2088.

Canganella, F., Gambacorta, A., Kato, C. and Horikoshi, K. (2000). Effects of hydrostatic pressure and temperature on physiological traits of *Thermococcus guaymasensis* and *Thermococcus aggregans* growing on starch. *Microbiol. Res.* **154**, 297–306.

Duarte, L. C., Nobre, A. P., Girio, F. M. and Amaral-Collaço, M. T. (1994). Determination of the kinetic parameters in continuous cultivation by *Debaromyces hansenii* grown on xylose. *Biotechnol. Techn.* **8**, 859–864.

Erauso, G., Reysenbach, A. L., Godfroy, A., Meunier, J. R., Crump, B., Partensky, F., Baross, J. A., Marteinsson, V., Barbier, G., Pace, N. R. and Prieur, D. (1993). *Pyrococcus abyssi* sp. nov., a new hyperthermophilic archaeon isolated from a deep-sea hydrothermal vent. *Arch. Microbiol.* **160**, 338–349.

Fiala, G. and Stetter, K. O. (1986). *Pyrococcus furiosus* sp. nov. represents a novel genus of marine heterotrophic archaebacteria growing optimally at 100°C. *Arch. Microbiol.* **145**, 56–61.

Gantelet, H. and Duchiron, F. (1998). Purification and properties of a thermoactive and thermostable pullulanase from *Thermococcus hydrothermalis*, a hyperthermophilic archaeon isolated from deep-sea hydrothermal vent. *Appl. Microbiol. Biotechnol.* **49**, 770–777.

Godfroy, A., Raven, N. D. H. and Sharp, R. J. (2000). Physiology and continuous culture of the hyperthermophilic deep-sea vent archaeon *Pyrococcus abyssi* ST549. *FEMS Microbiol. Lett.* **186**, 127–132.

Gonzalez, J. M., Sheckells, D., Viebahn, M., Krupatkina, D., Borges, K. M. and Robb, F. T. (1999). *Thermococcus waiotapuensis* sp. nov., an extremely thermophilic archaeon isolated from a freshwater hot spring. *Arch. Microbiol.* **172**, 95–101.

Holst, O., Manelius, A., Krahe, M., Markl, H., Raven, N. and Sharp, R. (1997). Thermophiles and fermentation technology. *Comp. Biochem. Physiol.* **118A**, 415–422.

Huber, H., Burggraf, S., Mayer, T., Wyschkony, I., Rachel, R. and Stetter, K. O. (2000). *Ignicoccus* gen. nov., a novel genus of hyperthermophilic, chemolithoautotrophic Archaea, represented by two new species, *Ignicoccus islandicus* sp. nov. and *Ignicoccus pacificus* sp. nov. *Int. J. Syst. Evol. Microbiol.* **50**, 2093–2100.

Kelly, R. M. and Deming, J. W. (1988). Extremely thermophilic archaebacteria: biological and engineering considerations. *Biotechnol. Prog.* **4**, 47–62.

Krahe, M., Antranikian, G. and Märkl, H. (1996). Fermentation of extremophilic microorganisms. *FEMS Microbiol. Rev.* **18**, 217–285.

Marteinsson, V. T., Birrien, J. L., Reysenbach, A. L., Vernet, M., Marie, D., Gambacorta, A., Messner, P., Sleytr, U. B. and Prieur, D. (1999a). *Thermococcus barophilus* sp. nov., a new barophilic and hyperthermophilic archaeon isolated under high hydrostatic pressure from a deep-sea hydrothermal vent. *Int. J. Syst. Bacteriol.* **49**, 351–359.

Marteinsson, V. T., Reysenbach, A. L., Birrien, J. L. and Prieur, D. (1999b). A stress protein is induced in the deep-sea barophilic hyperthermophile *Thermococcus barophilus* when grown under atmospheric pressure. *Extremophiles* **3**, 277–282.

Miller, J. F., Almond, E. L., Shah, N. N., Ludlow, J. M., Zollweg, J. A., Streett, W. B., Zinder, S. H. and Clark, D. S. (1988). High-pressure-temperature bioreactor for studying pressure-temperature relationships in bacterial growth and productivity. *Biotechnol. Bioeng.* **31**, 407–413.

Miroshnichenko, M. L., Hippe, H., Stackebrandt, E., Kostrikina, A., Chernyh, N. A., Jeanthon, C., Nazina, T. N., Belyaev, S. S. and Bonch-Osmolovskaya, E. A. (2001). Isolation and characterization of *Thermococcus sibericus* sp. nov. from a Western Siberian high-temperature oil reservoir. *Extremophiles* **5**, 85–91.

Mukhopadhyay, B., Johnson, E. F. and Wolfe, R. S. (1999). Reactor-scale cultivation of the hyperthermophilic methanarchaeon *Methanococcus jannaschii* to high cell densities. *Appl. Environ. Microbiol.* **65**, 5059–5065.

Nelson, C. M., Scuppenhauer, M. R. and Clark, D. S. (1992). High-pressure, high-temperature bioreactor for comparing effects of hyperbaric and hydrostatic pressure on bacterial growth. *Appl. Environ. Microbiol.* **58**, 1789–1793.

Postec, A., Le Breton, C., Fardeau, M. L., Lesongeur, F., Pignet, P., Quérellou, J., Ollivier, B. and Godfroy, A. (2005a). *Marinitoga hydrogenitolerans* sp. nov., a novel member of the order *Thermotogales* isolated from a black smoker chimney on the Mid-Atlantic Ridge. *Int. J. Syst. Evol. Microbiol.* **55**, 1217–1221.

Postec, A., Pignet, P., Cueff-Gauchard, V., Schmitt, A., Querellou, J. and Godfroy, A. (2005b). Optimisation of growth conditions for continuous culture of the hyperthermophilic archaeon *Thermococcus hydrothermalis* and development of sulphur-free defined and minimal media. *Res. Microbiol.* **156**, 82–87.

Postec, A., Urios, L., Lesongeur, F., Ollivier, B., Quérellou, J. and Godfroy, A. (2005c). Continuous enrichment culture and molecular monitoring to investigate the microbial diversity of thermophiles inhabiting the deep-sea hydrothermal ecosystems. *Curr. Microbiol.* **50**, 138–144.

Raffin, J. P., Henneke, G. and Dietrich, J. (2000). Purification and characterization of a new DNA polymerase modulator from the hyperthermophilic archaeon *Thermococcus fumicolans*. *Comp. Biochem. Physiol. Biochem. Mol. Biol.* **127**, 299–308.

Raven, N., Ladwa, N. and Sharp, R. (1992). Continuous culture of the hyperthermophilic archaeum *Pyrococcus furiosus*. *Appl. Microbiol. Biotechnol.* **38**, 263–267.

Raven, N. D. H. and Sharp, R. J. (1997). Development of defined and minimal media for the growth of the hyperthermophilic archaeon *Pyrococcus furiosus* Vc1. *FEMS Microbiol. Lett.* **146**, 135–141.

Rinker, K. D. and Kelly, R. M. (1996). Growth physiology of the hyperthermophilic Archaeon *Thermococcus litoralis*: Development of a sulfur-free defined medium, characterization of an exopolysaccharide, and evidence of biofilm formation. *Appl. Environ. Microbiol.* **62**, 4478–4485.

Rinker, K. D. and Kelly, R. M. (2000). Effect of carbon and nitrogen sources on growth dynamics and exopolysaccharide production for the hyperthermophilic Archaeon *Thermococcus litoralis* and Bacterium *Thermotoga maritima*. *Biotechnol. Bioeng.* **69**, 537–547.

Rinker, K. D., Han, C. J. and Kelly, R. M. (1999). Continuous culture as a tool for investigating the growth physiology of heterotrophic hyperthermophiles and extreme thermoacidophiles. *J. Appl. Microbiol.* **85**(Suppl. S), 118S–127S.

Ronimus, R. S., Reysenbach, A.-L., Musbrave, D. R. and Morgan, H. W. (1997). The phylogenetic position of *Thermococcus* isolate AN1 based on 16S rRNA gene sequence analysis: a proposal that AN1 represents a new species, *Thermococcus zilligii*, sp. nov. *Arch. Microbiol.* **168**, 245–248.

Schiaraldi, C., Marulli, F., Di Lernia, I., Martino, A. and De Rosa, M. (1999). A microfiltration bioreactor to achieve high cell density in *Sulfolobus solfataricus* fermentation. *Extremophiles* **3**, 199–204.

Sharp, R. J. and Raven, N. D. H. (1997). Isolation and growth of hyperthermophiles. In *Applied Microbial Physiology: A Practical Approach* (P. M. Rhodes and P. F. Stanbury, eds), pp. 23–51. IRL Press, Oxford University Press, Oxford.

Stetter, K. O., König, H. and Stackebrandt, E. (1983). *Pyrodictium* gen. nov., a new genus of submarine disc-shaped sulfur-reducing archaebacteria growing optimally at 105°C. *System. Appl. Microbiol* **4**, 535–551.

Wery, N., Lesongeur, F., Pignet, P., Derennes, V., Cambon-Bonavita, M.-A., Godfroy, A. and Barbier, G. (2001). *Marinitoga camini*, gen. nov., sp. nov., a rod-shaped bacterium belonging to the order *Thermotogales*, isolated from a deep-sea hydrothermal vent. *Int. J. Syst. Evol. Microbiol.* **51**, 495–504.

Worthington, P., Blum, P., Perez-Pomares, F. and Elthon, T. (2003). Large-scale cultivation of acidophilic hyperthermophiles for recovery of secreted proteins. *Appl. Environ. Microbiol.* **69**, 252–257.

5 Toward the Large Scale Cultivation of Hyperthermophiles at High-Temperature and High-Pressure

Chan Beum Park[1], Boonchai B Boonyaratanakornkit[2] and Douglas S Clark[2]

[1] Department of Chemical and Materials Engineering, Arizona State University, Tempe, AZ 85287-6006, USA; [2] Department of Chemical Engineering, University of California, Berkeley, CA 94720, USA

CONTENTS

Introduction
Development of a high temperature–pressure (high T–P) bioreactor
Cultivation of *Methanocaldococcus jannaschii* in a high T–P bioreactor
Rapid decompression can cause explosive cell lysis during sampling
Gas–liquid mass transfer limitations
Commentary

◆◆◆◆◆ INTRODUCTION

Deep-sea hydrothermal vents, which are characterized by temperatures up to 350°C and pressures up to at least 350 atm, represent one of the most extreme aquatic environments on Earth (Deming, 1998; Stetter, 1999; Yayanos, 1995). The discovery of life in these extreme settings continues to challenge our concept of the limiting conditions for microbial growth. In the last century, the maximum reported upper temperature limit for microbial growth had increased from 55°C in 1903 (Brock, 1978) to 113°C in the case of the hyperthermophilic archaeon *Pyrolobus fumarii* (Blöchl et al., 1997). Recently, Strain 121, isolated from a hydrothermal vent, was found to grow at temperatures as high as 121°C (Cowan, 2004; Kashefi and Lovley, 2003). Indirect evidence supports the likelihood of bacterial survival at even higher temperatures under elevated pressure. For example, in 1983, Baross and Deming reported that methane was produced by a microbial culture incubated at 300°C and 265 atm (Baross and Deming, 1983), though the identity of the microorganisms that may have been responsible for the methane formation remains unknown and controversial. These findings illustrate that isolation and cultivation of microorganisms under extreme conditions can be very difficult.

Since the 1980s, many efforts had been made to develop bioreactor systems that can mimic the deep-sea hydrothermal vent habitat by operating at high temperatures and high pressures (Table 5.1; Kelly and Deming, 1988; Ludlow and Clark, 1991). One of the major difficulties in the development of high temperature–pressure (T–P) bioreactors is structural stability of the reactor vessel, which cannot be afforded by conventional fermentation systems. Early high T–P bioreactors used a modified-syringe system for pressurization (Thom and Marquis, 1984; Yayanos et al., 1982, 1983), which typically consisted of a syringe fitted with a piston inside a pressure vessel. As an alternative to syringes, compressible bags have been placed inside pressure vessels (Kelly and Deming, 1988). The French press is another tool used for the cultivation of microorganisms at elevated pressure and temperature (Vance and Hunt, 1985). Both syringe and French press bioreactors were relatively easy to develop and operate. However, these systems presented limitations with regard to effective mixing, intermittent sampling, and scale-up. Most other high T–P bioreactors developed later had configurations resembling modified autoclaves or employed a customized pressure vessel constructed of a rigid material, for example, synthetic sapphire (Miller et al., 1988a), 316 stainless steel (Bernhardt et al., 1987; Holden and Baross, 1995; Miller et al., 1988a), or 316 stainless steel that was Teflon-lined (Sturm et al., 1987) or glass-lined (Nelson et al., 1991). The pressurized vessel-type bioreactors allowed for scale-up and intermittent sampling and were sometimes equipped with a magnetic stirrer for efficient mixing of culture medium.

High T–P bioreactors can be categorized into two types: hydrostatic and hyperbaric. Hydrostatic bioreactors, in which the reactor contains only a liquid phase, are relatively easy to construct and operate, and enabled the development of continuous-flow systems (Bubela et al., 1987; Jannasch et al., 1996). Considering that many hyperthermophiles are autotrophic microorganisms that require gaseous substrates such as H_2 and CO_2, sufficient supply of substrate gases is critical for cultivation. In this regard, hydrostatic high T–P bioreactors can be inadequate in that growth in these systems is limited by the amount of gaseous substrate initially dissolved in the liquid medium. In contrast to hydrostatic bioreactors, hyperbaric systems allow adequate quantities of gaseous substrates to be available inside the headspace of the reactor. One of the earliest hyperbaric high T–P bioreactors consisted of a 316 stainless steel pressure vessel housed in an oven with a gas compressor permitting the introduction of gases above gas-cylinder pressure (Miller et al., 1988a,b; Nelson et al., 1991). When this hyperbaric system was used to grow *Methanocaldococcus jannaschii*, an extremely thermophilic methanogen originally isolated from a white-smoker deep-sea hydrothermal vent at a depth of 2600 m (Jones et al., 1983), it was found that the microorganism grew nearly 5 times faster at 750 atm than at 7.8 atm, and that the maximum temperature for

Table 5.1 Overview of high temperature-pressure (T–P) bioreactors

Hydrostatic/hyperbaric	Mixing	Mode	Material	Reactor volume (ml)	Max. operating temp. (°C)	Max. operating pressure (bar)	Reference
Hydrostatic							
Pressure vessel	Ball motion	Batch	Titanium syringe	150	69[1]	1035	Yayanos et al., 1983
French press	None	Batch	Glass serum bottle	<35	75	500	Vance and Hunt, 1985
Pressure vessel	Fluid recycle	Continuous	Stainless steel	1000	150	200	Bubela et al., 1987
Pressure vessel	None	Batch	Glass/plastic syringe	6/60	108[1]	220[1]	Holden and Baross, 1995
Pressure vessel	Magnetic stirrer	Continuous	Titanium	500	NA[2]	710	Jannasch et al., 1996
Pressure vessel	None	Batch	Glass syringe	10	300	400[1]	Marteinsson et al., 1999
Hyperbaric							
Autoclave	None	Batch	Nickel tube	10	400	4000	Bernhardt et al., 1987
Autoclave	Magnetic stirrer	Batch	Stainless steel w/Teflon liner	500	250	200	Sturm et al., 1987
Pressure vessel	Gas flow	Batch[3]	Synthetic sapphire	10	260	350	Miller et al., 1988a
Pressure vessel	Gas flow	Batch[3]	Stainless steel	167	260	1000	Miller et al., 1988b
Pressure vessel	Gas flow[4]	Batch[3]	Stainless steel w/glass liner	1150	200	880	Park and Clark, 2002

[1]The upper limit of the system is not reported; included above are the maximal operating conditions used in the experiments.
[2]Not available since the reactor system was tested only at ambient temperature.
[3]Continuous recycle of gaseous substrates.
[4]With optional feature of stirring with a magnetic bar.

methane production was extended from 92°C at 7.8 atm to 98°C at 250 atm (Miller *et al.*, 1988b).

This chapter describes the development of a custom-made, high T–P bioreactor system for the cultivation of hyperthermophiles under hyperbaric conditions. Equipped with a thermocouple and a heating belt for temperature control, the bioreactor has a total volume of 1.15 l, which is one of largest high T–P bioreactors described to date. It can operate up to 880 atm and 200°C, and pressurization from 7.8 atm to 500 atm can be achieved in 10 min. A step-by-step protocol for cultivating *M. jannaschii* in this system is presented in detail. Relevant issues such as cell disruption during decompression and gas–liquid mass transfer limitation are discussed in later sections.

◆◆◆◆◆ DEVELOPMENT OF A HIGH TEMPERATURE–PRESSURE (HIGH T–P) BIOREACTOR

A schematic diagram of the high T–P bioreactor system is shown in Figure 5.1. Listed below are parts used in the development of the high T–P bioreactor system. In 2000, the approximate cost of components for the entire system was $25 000.

Reactor System Parts List

(1) 316 high-pressure vessel (model GC-17; High Pressure Equipment Co.; $5000)
(2) Heating belt (BS0101-080, 1″ × 8′, 418 Watts; Thermolyne Co.; $100)

- high temperature Al tape, 600°F, 1″ × 60″ (TAX37; Thermolyne Co.; $150)

(3) Thermocouple (Type K; Omega Engineering Inc.; $400)
(4) Proportional-integral-derivative controller (CN3251-R; Omega Engineering Inc.; $420)

- solid state relay (SSR240DCl0; Omega Engineering Inc.; $30)
- finned heat sink (FHS-1; Omega Engineering Inc.; $20)

(5) Diaphragm gas compressor (46-14023; Newport Scientific Inc.; $7000)

- (5A) inlet house air regulator (VI8-04-0000/BI8-04-FGG0; Wilkerson Operations; $200)

(6) Air-activated back-pressure regulator (26-1761-24-069A, 10 000 psi max; Tescom Corp.; $1200)
(7) Liquid pump (Prostar 210 SDM; Varian Inc.; $3700) with pressure module (8700 psi titanium; Varian Inc.; $2900)
(8) Micro-control metering value (60-11HF4-V-SGS; High Pressure Equipment Co.; $200)

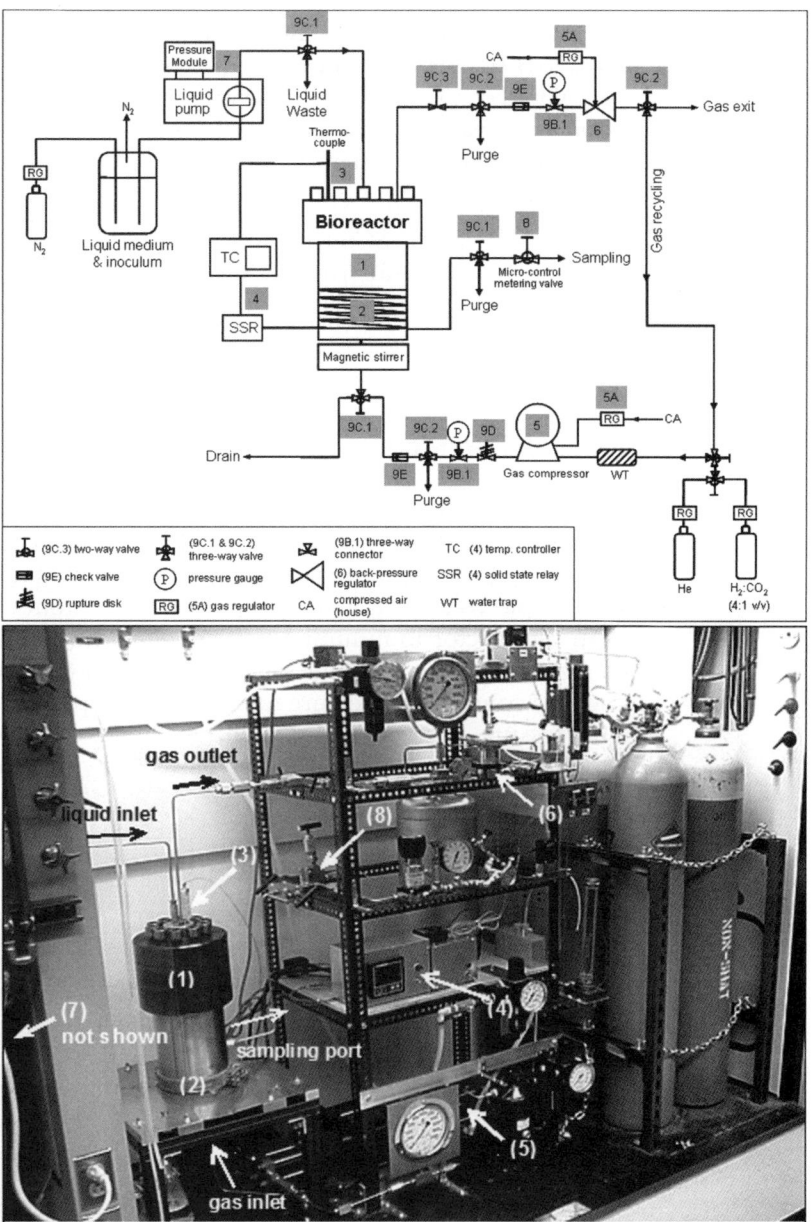

Figure 5.1. Schematic diagram and photograph of the high T–P bioreactor system. (1) 316 high-pressure vessel; (2) heating belt; (3) thermocouple; (4) proportional-integral-derivative controller; (5) gas compressor; (6) back-pressure regulator; (7) liquid pump; (8) micro-control metering value; (9) high-pressure valves and rupture disks (see section entitled Reactor System Parts List for numbering scheme). See Plate 5.1 in Colour Plate Section.

(9) High pressure tubing, valves, rupture disk (High Pressure Equipment Co.; $3000)

- (9A) high-pressure tubing (1/8-inch outer diameter × 0.040-inch inner diameter [9A.1] and 1/4-inch outer diameter × 0.083-inch inner diameter [9A.2], 316 stainless steel)

- (9B) line connectors (60-23HF2-SGS [3-way connector for 1/8-inch line to P gauge; 9B.1], 60-21HF2HF4-SGS [for 1/4-inch line to 1/8-inch line; 9B.2])
- (9C) valves (30-15HF4 [for 1/4-inch line, 3-way; 9C.1], 30-15HF2-SGS [for 1/8-inch line, 3-way; 9C.2], 30-11HF2-SGS [for 1/8-inch line, 2-way; 9C.3])
- (9D) rupture disk assembly
  - 60-63HF2 (in-line connection for 1/8-inch line)
  - 200204-SGS (rupture disk holder)
  - S-5308-01 (rupture disk, rated to 12 500 psig)
- (9E) check valves (30-41HF2-SGS)

Bioreactor, Tubing and Connectors

A conventional 316 stainless steel high-pressure vessel (model GC-17; High Pressure Equipment Co.) was modified in the UC Berkeley College of Chemistry Machine Shop to have multiple inlet and outlet ports and a thermocouple well as shown in Figure 5.2. A custom-made glass liner is used in the reactor to minimize cell contact with the SS316, which may leach materials that are inhibitory to cell growth (Nelson *et al.*, 1991). All tubing and connectors used are rated to endure pressures up to 1360 atm (High Pressure Equipment Co.). The reactor is set up in a fume hood due to the flammability and explosiveness of hydrogen and methane.

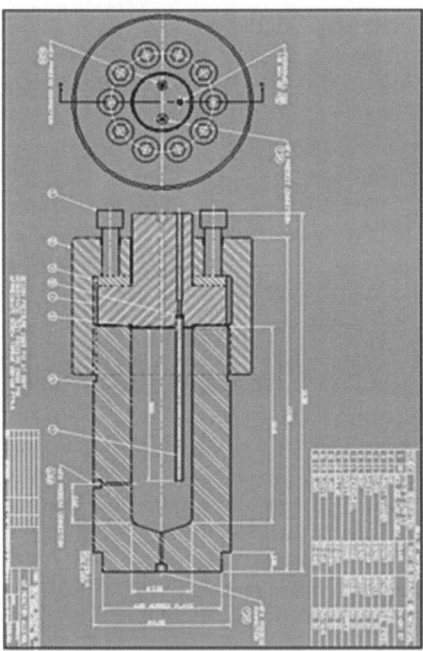

Figure 5.2. Diagram and photograph of the modified pressure vessel. A stainless steel 316 high-pressure vessel (GC-17) from High Pressure Equipment Co. was modified by adding multiple inlet and outlet ports and a thermocouple well. See Plate 5.2 in Colour Plate Section.

Temperature Control

A heating belt (BS0101-080, Thermolyne Co.) is placed around the bioreactor, as shown in Figures 5.1 and 5.2, to heat the culture. A thermocouple (Omega Engineering Inc.) is inserted into the SS316 thermocouple well (extending three-fourths of the way into the bioreactor). The gap between the thermocouple and the top of the well is sealed with a plug made of silicone rubber. The thermocouple provides the signal to a proportional-integral-derivative controller (CN3251-R; Omega Engineering Inc.) for precise control of the temperature through solid state relay to the heating belt.

Pressure Control and Recycle of Gaseous Substrates

A diaphragm compressor (Newport Scientific Inc.) is used to pressurize the bioreactor by supplying He at high pressures. Helium was found to be less inhibitory to bacterial growth than Xe, N_2O, Ar and N_2 (Fenn and Marquis, 1968; Sturm *et al.*, 1987). The bioreactor is typically pressurized with 7.8 atm of $H_2:CO_2$ (4:1 v/v) and He is used for pressurization over 7.8 atm. An oxygen trap (Alltech Inc.) was installed between the diaphragm compressor and the gas cylinder to remove residual oxygen. In order to avoid pressure build-up above safe levels, a rupture disk (High Pressure Equipment Co.) is inserted in the line between the bioreactor and the compressor. A check value is located between the bioreactor and the gas compressor to prevent backflow of culture from the bioreactor during supply of gaseous substrate, which can corrode the compressor. The flow rate of outlet gas is controlled by the pressure of house gas supplied to the gas compressor through a separate line. When gaseous substrates are recycled, the pressure in the bioreactor is controlled with an air-activated back-pressure regulator (Tescom Corp.) placed in the exit line of the gas loop.

Supply of Liquid Medium and Inoculum

Liquid medium and inoculum are supplied anaerobically *via* a liquid pump (Prostar 210 SDM; Varian Inc.) with a pressure module (8700 psi titanium; Varian Inc.). The head of the liquid pump is made of titanium to avoid possible damage by corrosive liquid medium. The liquid medium and inoculum supplied into the bioreactor are sparged with $H_2:CO_2$ (4:1 v/v) three times to eliminate any oxygen present in the liquid. In order to enhance the transfer of gaseous substrate into the culture and keep the culture homogeneous, a 1¼ inch long, egg-shaped magnetic bar is placed in the bioreactor and rotated at speeds up to 1650 rpm *via* a magnetic stir plate mounted below the bioreactor. In certain situations, gas–liquid mass transfer limitations were found to preclude pressure-accelerated growth of *M. jannaschii* (discussed below).

Culture Sampling

Culture samples are taken periodically from the liquid outlet line of the bioreactor with a micro-control metering value (High Pressure Equipment Co.). Cells are harvested after slow de-pressurization of the reactor from 500 atm over a 10-min time span since rapid decompression caused cell lysis (discussed below).

Maintenance of the High T–P bioreactor System

Since most hyperthermophiles grow under strictly anaerobic conditions, liquid medium for cultivating hyperthermophiles typically contains reducing agents such as Na_2S, which is highly corrosive and can damage the SS316 components. It is therefore critical to thoroughly clean the bioreactor system with deionized water after each operation. In addition, even with the check value, liquid medium can flow back into the diaphragm compressor. Thus, it is recommended to disassemble the gas compressor periodically to remove traces of liquid medium deposited inside.

◆◆◆◆◆ CULTIVATION OF *METHANOCALDOCOCCUS JANNASCHII* IN A HIGH T–P BIOREACTOR

This is a step-by-step protocol for cultivating *M. jannaschii* in the high T–P bioreactor. Typically, it takes 2 days for the preparation of the bioreactor and inoculum.

Day 1: Preparing the Reactor

Cleaning

1. Wash the reactor with an anionic detergent (Liquinox, Alconox Inc.) and water (rinse so that no residual detergent remains as this can readily lyse cells).
2. (Optional) Passivate the reactor with Citrisurf 2050 (Stellar Solutions) or 10% w/v nitric acid. Passivation regenerates a thin oxide layer on the surface of the stainless steel making the material less reactive with the culture medium and less inhibitory to *M. jannaschii* growth.

Assembling

1. Place the glass liner and stir bar into the bioreactor and place the lid on the bioreactor.
2. Place reactor screw top over the reactor lid.

a. Screw on the reactor top. Tighten firmly by turning the top with the accompanying metal rod.
b. Screw down the ten knobs on the reactor top firmly with a monkey wrench.
　　i. Tighten knobs in opposite pairs in order to tighten the lid evenly.

3. Attach the two lines on top of the reactor between the reactor lid and the liquid pump/sampling port and the reactor lid and the pressure regulator.

Testing for reactor leaks

1. Turn on the helium tank connected to the reactor.
2. Close all outgoing lines (liquid pump line, pressure regulator line, sampling line, and release line) to pressurize the system.
3. Build up pressure to 1000 psig.
　　a. Supply house air to the diaphragm compressor to pressurize the system.
4. Add Snoop leak detector (Swagelok) around line joints and into the reactor lid to determine if leaks are present.
5. If leaks are detected, carefully tighten the leaking joint and test again. If no leaks are detected, release He slowly from the system.

Medium preparation

1. Prepare 500 ml of medium as described previously (Tsao *et al.*, 1994). This amount is sufficient for two cultivations, each utilizing a 200-ml working volume.
　　a. Mix medium components together and bring to pH 6.8 with 5 M NaOH.
　　b. Sparge 500 ml of medium for 30 min with $H_2:CO_2$ (4:1 v/v).
　　c. Distribute 30 ml of medium into two 125-ml serum bottles (Wheaton) for use as reactor inoculum in each cultivation and seal with a septum and aluminum seal (Bellco Glass).
　　　　i. Sparge sealed serum bottles with $H_2:CO_2$ (4:1 v/v) for 30 seconds to remove residual O_2.
　　d. Distribute the remaining medium evenly into two 1-l bottles and cap the bottles tightly.
2. Prepare 20 ml of cysteine-HCl and Na_2S (0.50 g/20 ml) in 50-ml serum bottles (Wheaton).
　　a. First place 20 ml of H_2O into two 50-ml clear serum bottles.
　　b. Sparge the H_2O with N_2 for 30 min to remove any dissolved oxygen.

c. Then add 0.5 g of Na_2S or cysteine-HCl and seal the 50-ml serum bottles.

3. Autoclave the two 1-l bottles (containing approximately 200 ml of medium in each), the two serum bottles containing medium, and the two serum bottles containing reducing agents.

Inoculum preparation

1. The day before the reactor is started, inoculate *M. jannaschii* into the 125-ml serum bottles containing 30 ml of medium.

    a. Sparge serum bottles of medium for 30 s with H_2-CO_2 (4:1 v/v) before adding 0.6 ml of both the Na_2S and cysteine-HCl (0.5 g/20 ml).
    b. After the disappearance of color (indicating the absence of oxygen), inoculate bottles with 3 ml of inoculum (*M. jannaschii* inoculum is stable when kept at room temperature) and pressurize to 30 psig with H_2:CO_2 (4:1 v/v).

2. Place inoculated medium into a heated, oscillating water bath for growth at 85°C (T_{opt}).

Day 2: Starting the Reactor and Pressurizing

Filling the reactor with medium

1. Carefully remove the cap from the 1-l bottle and insert into the bottle neck a silicone stopper containing a SS316 tube that extends into the medium.
2. Sparge the 1-l bottle with N_2 through the stopper. Adjust the connection so that nitrogen flows into the 1-l bottle and out into a bottle filled with H_2O. Nitrogen flow flushes out O_2 and prevents generation of a vacuum when pumping medium into the bioreactor. The H_2O bottle allows us to confirm N_2 gas flow.
3. Add cysteine-HCl and Na_2S through the stopper (16 ml/800 ml medium).
4. Shake the bottle and wait for the medium to become clear (O_2 is then absent).
5. To prepare the reactor for addition of medium, sparge the reactor with H_2:CO_2 by closing all outgoing lines and filling up the reactor with H_2:CO_2, then release the gas through an outlet port. Repeat 2–3 times.
6. Attach the bottle stopper to the tubing of the liquid pump.
7. Fill up the tubing with medium using a syringe.
8. Begin pumping medium into the reactor.

    a. To fill the line leading into the reactor, first open the valve so that a few ml of medium is pumped into a waste bottle. Close the valve to the waste bottle and open the valve to the reactor to begin filling the reactor.

      b. Activate the liquid pump to begin pumping the medium into the reactor at a rate of 10 ml min^{-1} for 20 min (for 200 ml of medium).
      c. Stop the pump once the reactor has been filled with 200 ml.

9. Turn on the temperature controller and the heating belt to heat the reactor up to 85°C (typically takes about 3 h).

Inoculating the reactor (use 10% v/v inoculum)

1. Pressurize a serum bottle of *M. jannaschii* grown to late exponential phase up to 10 psig (so there is no vacuum in the bottle when pumping inoculum into the reactor).
2. Remove the 1-l bottle and bottle stopper from the liquid-pump tubing.
3. Attach a needle to the pump tubing and poke the needle into the septum of the inoculum bottle.
4. Turn on the liquid pump to begin filling the bioreactor with inoculum.
5. Stop the pump when at least 10% v/v of inoculum has been added to the reactor.
6. Afterwards pump H_2O through the pump to remove deposits and prevent corrosion.

Starting the reactor for batch operation

1. Start the magnetic stirrer.
2. Sparge the reactor with $H_2:CO_2$ to clear out O_2. Repeat 2–3 times.
3. Pressurize the reactor to 100 psig (7.8 atm) with $H_2:CO_2$. Check the liquid pump pressure reading to confirm the reactor pressure.
4. If there is need to pressurize to higher pressures, exchange the $H_2:CO_2$ tank with the He tank and pressurize with the diaphragm compressor.

Starting the reactor for operation with gas recycle

1. Adjust the set-point pressure for the back-pressure regulator (BPR) using the gas regulator that is connected to the BPR.
2. Pressurize the reactor to 100 psig (7.8 atm) with $H_2:CO_2$. Supply He into the reactor with the diaphragm compressor. When the internal pressure of the reactor exceeds the set-point, the BPR will release gas through the exit line until the reactor pressure satisfies the set value.
3. For about 10 min, continue supplying He with the diaphragm compressor and allow the gas to exit out of the BPR since it may contain trace amounts of oxygen that are inhibitory to cell growth.
4. Adjust the three-way valve at the exit of the BPR to redirect the exiting gas to the diaphragm compressor (see the schematic diagram in Figure 5.1). At the same time, turn off the gas supply from the gas cylinders by adjusting the three-way valve found right after the cylinders. At this point, gas should be recycled directly from the BPR into the diaphragm compressor.

Sampling

1. Record pressure readings before sampling in order to determine pressure changes in the system over time (due to cell utilization of gaseous substrates, leaks and sampling).
2. Flush the sampling line by first taking 3–5 ml of sample from the reactor and discarding.
3. After flushing, sample 1 ml for protein and/or OD_{660} measurements.
   a. Take OD_{660} measurements immediately upon sampling. At high pressure (e.g. 260 atm), cell lysis will occur, as described below, precluding accurate determination of cell concentration based on OD_{660}.
   b. Measure protein content with the Bio-Rad protein assay using 0.0083% Triton to completely lyse cells.
4. To harvest intact cells from a high-pressure cultivation, depressurize the reactor slowly (e.g. from 500 atm over a 10-min time period) and remove cells.

◆◆◆◆◆ RAPID DECOMPRESSION CAN CAUSE EXPLOSIVE CELL LYSIS DURING SAMPLING

A recent study by Park and Clark (2002) showed that rapid decompression (ca. 1 s) from high hyperbaric pressure ruptured the cell envelope of *M. jannaschii*. When samples of *M. jannaschii* were withdrawn from the hyperbaric bioreactor at 260 atm through a micro-control metering valve (High Pressure Equipment Co.), cell growth was not evident based on measurements of OD_{660} (Figure 5.3). However, after the same 260 atm

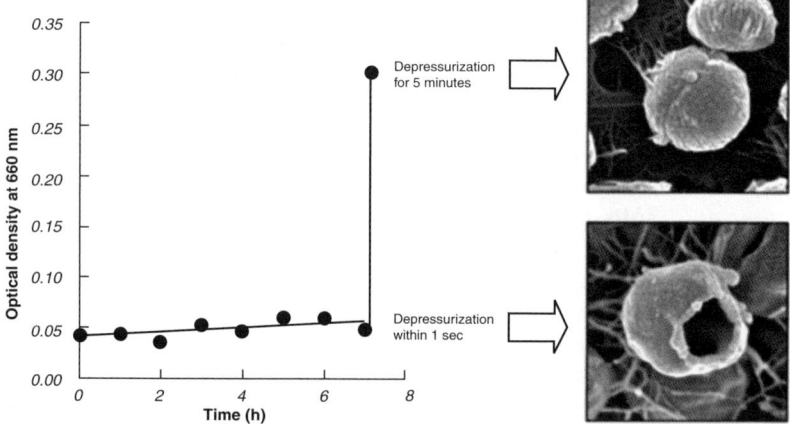

Figure 5.3. Turbidity changes of *M. jannaschii* culture grown at 260 atm and 80°C. Rapid decompression (< 1 s) caused explosive rupturing of the cell membranes as evidenced by the scanning electron micrograph, resulting in negligible turbidity increases for samples removed during cell growth.

Table 5.2 Growth rates at 88°C as a function of pressure and volume of medium in the high pressure reactor. Growth was based on the total protein content in liquid samples determined using the Bio-Rad protein assay reagent

Volume of medium (ml)	Growth rate at 7 atm	Growth rate at 500 atm
200	0.35 ± 0.07 h^{-1}	0.79 ± 0.05 h^{-1}
400	0.032 OD$_{595}$ h^{-1}	0.046 OD$_{595}$ h^{-1}
600	0.017 ± 0.003 OD$_{595}$ h^{-1}	0.014 ± 0.005 OD$_{595}$ h^{-1}

culture was sampled with a longer decompression time (5 min), the OD$_{660}$ increased more than six-fold. Rupture of the cell envelope by rapid decompression was also confirmed by scanning electron micrographs of *M. jannaschii* as shown in Figure 5.3. These results may have important implications for the collection and retrieval of deep-sea microorganisms containing high concentrations of dissolved gases.

◆◆◆◆◆ GAS–LIQUID MASS TRANSFER LIMITATIONS

Gas–liquid mass transfer has been found to limit pressure-accelerated growth of *M. jannaschii* (Boonyaratanakornkit *et al.*, submitted). For example, when *M. jannaschii* was cultivated at working volumes ranging from 200 to 600 ml, growth was linear and no barophilic growth was observed at the larger working volume of 600 ml, and only a slight barophilic effect was observed at 400 ml (Table 5.2). However, when the working volume was reduced to 200 ml thereby increasing the interfacial area per reactor volume and possibly decreasing the thickness of the boundary layer at the gas–liquid interface, growth became exponential and was accelerated by pressure (Table 5.2, Figure 5.4). Apparently, the reduction in working volume shifted cell growth from a mass transfer-limited regime to a biochemically limited regime. The main insight to be gained from these results is that determination of pressure effects on growth should be investigated outside of the mass transfer-limited regime; otherwise, barophilic growth may not be evident.

◆◆◆◆◆ COMMENTARY

The deep-sea 1000 m below sea level accounts for nearly 90% of the ocean, and remains one of the last frontiers on Earth to be explored in the twenty-first century. Among the inhabitants of this exotic biosphere, deep-sea extremophiles are considered to be a potential source of new bioactive compounds such as antiviral molecules, antibiotics, cancer inhibitors, polyunsaturated fatty acids, thermo/osmo-protectants, robust enzymes

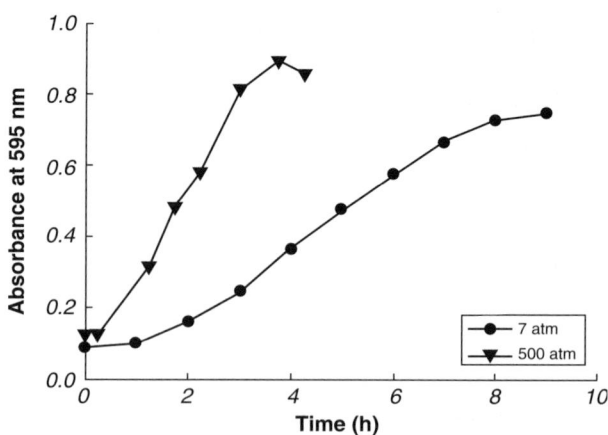

Figure 5.4. Barophilic growth of *M. jannaschii* at 88°C in the high T–P bioreactor utilizing 200 ml of medium. Based on the Bio-Rad protein assay, which was used for protein-content measurements at OD_{595}, the doubling time decreased from 1.8 h at 7 atm to 0.90 h at 500 atm.

and others. In the last 20 years, high T–P bioreactors have been developed to simulate the native environments of these organisms. These systems have evolved from rudimentary pressurized syringe designs to hydrostatic and now hyperbaric bioreactors. Hyperbaric bioreactors such as the one described herein allow the cultivation of organisms utilizing gaseous substrates under conditions whereby dissolved substrate concentrations can be varied independently of the total system pressure. However, there are still several technical challenges to be overcome. For instance, larger commercially available pressure vessels would enable further scale-up, and new probes for on-line analysis of cultures at high pressure would allow better control and characterization. Moreover, pressure vessels and the accompanying high-pressure tubing and connectors are typically made of 316 stainless steel and therefore must be coated with inert materials such as titanium to prevent inhibition of microbial growth. With further improvements to high T–P bioreactors, however, it will become easier to study how organisms have adapted to the extreme environments of deep-sea hydrothermal vents, and to other extreme habitats that define the outermost limits of life.

Acknowledgements

We thank Eric Grunland and the UC Berkeley Chemistry Machine Shop for help in the design and construction of the high-pressure, high-temperature bioreactor system, and Michael Chiu, Jodie Martin and Sohan Patel for assistance in cultivation of *M. jannaschii*. This work was supported by the National Science Foundation, including the ERC program under award number EEC-9731725, and the Schlumberger Fellowship of DSC.

List of suppliers

Bellco Glass
340 Edrudo Road
Vineland, New Jersey 08360
Tel: +1-800-257-7043
Fax: +1-856-691-3247
http://www.bellcoglass.com

Septa, aluminum seals

Bio-Rad
2000 Alfred Nobel Dr
Hercules, California 94547
Tel: +1-800-424-6723
Fax: +1-510-741-5800
http://www.bio-rad.com

Protein assay kit

High Pressure Equipment Co.
P.O. Box 8248, 1222 Linden
Erie, Pennsylvania 16505
Tel: +1-800-289-7447
Fax: +1-814-838-6075
http://www.highpressure.com

316 high-pressure vessel, micro-control metering value, high-pressure tubing, valves, rupture disks

Newport Scientific Inc.
8246 E. Sandy Court
Jessup, Maryland 20794
Tel: +1-301-498-6700
Fax: +1-301-490-2313
http://www.newport-scientific.com

Diaphragm gas compressor

Omega Engineering Inc.
One Omega Drive
Stamford, Connecticut 06907
Tel: +1-800-848-4286
Fax: +1-203-359-7700
http://www.omega.com

Thermocouple, proportional-integral-derivative controller, solid-state relay, finned heat sink

Stellar Solutions
4511 Prime Parkway
McHenry, Illinois 60050
Tel: +1-847-854-2800
Fax: +1-847-854-2830
http://www.stellarsolutions.net

Citrisurf 2050

Swagelok
2441 Spring Court – Unit A
Concord, California 94520
Tel: +1-925-676-4100
Fax: +1-925-798-9833
http://www.swagelok.com

Snoop leak detector

Tescom Corp.
12616 Industrial Blvd
Elk River, Minnesota 55330
Tel: +1-800-447-1204
Fax: +1-763-241-3224
http://www.tescom.com

Air-activated back-pressure regulator

Thermolyne/Barnstead Co.
2555 Kerper Blvd
Dubuque, Iowa 52001
Tel: +1-800-553-0039
Fax: +1-319-589-0516
http://www.barnstead.com

Heating belt, high-temperature aluminum tape

Varian Inc.
2700 Mitchell Dr
Walnut Creek, California 94598
Tel: +1-800-926-3000
Fax: +1-925-945-2102
http://www.varianinc.com

Liquid pump with pressure module

Wilkerson Corp.
Pneumatic Division
Richland, Michigan 49083
Tel: +1-269-629-2550

Fax: +1-269-629-2475
http://www.wilkersoncorp.com

Air regulator

References

Baross, J. A. and Deming, J. W. (1983). Growth of "black smoker" bacteria at temperatures of at least 250°C. *Nature* **303**, 423–426.

Bernhardt, G., Jaenicke, R. and Ludemann, H. D. (1987). High-pressure equipment for growing methanogenic microorganisms on gaseous substrates at high-temperature. *Appl. Environ. Microbiol.* **53**, 1876–1879.

Blöchl, E., Rachel, R., Burggraf, S., Hafenbradl, D., Jannasch, H. W. and Stetter, K. O. (1997). *Pyrolobus fumarii*, gen. and sp. nov., represents a novel group of archaea, extending the upper temperature limit for life to 113°C. *Extremophiles* **1**, 14–21.

Boonyaratanakornkit, B., Lopez, J. C., Park, C. B. and Clark, D. S. (2005). Pressure affects transcription profiles of *Methanocaldococcus jannaschii* despite the absence of barophilic growth under mass-transfer limitations, submitted.

Brock, T. D. (1978). *Thermophilic Microorganisms and Life at High Temperatures.* Springer, New York.

Bubela, B., Labone, C. L. and Dawson, C. H. (1987). An apparatus for continuous growth of microorganisms under oil reservoir conditions. *Biotechnol. Bioeng.* **29**, 289–291.

Cowan, D. A. (2004). The upper temperature for life – where do we draw the line? *Trends Microbiol.* **12**, 58–60.

Deming, J. W. (1998). Deep ocean environmental biotechnology. *Curr. Opin. Biotechnol.* **9**, 283–287.

Fenn, W. O. and Marquis, R. E. (1968). Growth of *Streptococcus faecalis* under high hydrostatic pressure and high partial pressures of inert gases. *J. Gen. Physiol.* **52**, 810–824.

Holden, J. F. and Baross, J. A. (1995). Enhanced thermotolerance by hydrostatic pressure in the deep-sea hyperthermophile *Pyrococcus* strain ES4. *FEMS Microbiol. Ecol.* **18**, 27–34.

Jannasch, H. W., Wirsen, C. O. and Doherty, K. W. (1996). A pressurized chemostat for the study of marine barophilic and oligotrophic bacteria. *Appl. Environ. Microbiol.* **62**, 1593–1596.

Jones, W. J., Leigh, J. A., Mayer, F., Woese, C. R. and Wolfe, R. S. (1983). *Methanococcus jannaschii* sp. nov., an extremely thermophilic methanogen from a submarine hydrothermal vent. *Arch. Microbiol.* **136**, 254–261.

Kashefi, K. and Lovley, D. R. (2003). Extending the upper temperature limit for life. *Science* **301**, 934.

Kelly, R. M. and Deming, J. W. (1988). Extremely thermophilic archaebacteria – biological and engineering considerations. *Biotechnol. Progress* **4**, 47–62.

Ludlow, J. M. and Clark, D. S. (1991). Engineering considerations for the application of extremophiles in biotechnology. *Crit. Rev. Biotechnol.* **10**, 321–345.

Marteinsson, V. T., Birrien, J.-L., Reysenbach, A.-L., Vernet, M., Marie, D., Gambacorta, A., Messner, P., Sleytr, U. B. and Prieur, D. (1999). *Thermococcus barophilus* sp. nov., a new barophilic and hyperthermophilic archaeon isolated under high hydrostatic pressure from a deep-sea hydrothermal vent. *Int. J. Syst. Bacteriol.* **49**, 351–359.

Miller, J. F., Almond, E. A., Shah, N. N., Ludlow, J. M., Zollweg, J. A., Streett, W. B., Zinder, S. H. and Clark, D. S. (1988a). High-pressure-temperature bioreactor for studying pressure-temperature relationships in bacterial growth and productivity. *Biotechnol. Bioeng.* **31**, 407–413.

Miller, J. F., Shah, N. N., Nelson, C. M., Ludlow, J. M. and Clark, D. S. (1988b). Pressure and temperature effects on growth and methane production of the extreme thermophile *Methanococcus jannaschii*. *Appl. Environ. Microbiol.* **54**, 3039–3042.

Nelson, C. M., Schuppenhauer, M. R. and Clark, D. S. (1991). Effects of hyperbaric pressure on a deep-sea archaebacterium in stainless steel and glass-lined vessels. *Appl. Environ. Microbiol.* **57**, 3576–3580.

Park, C. B. and Clark, D. S. (2002). Rupture of the cell envelope by decompression of the deep-sea methanogen *Methanococcus jannaschii*. *Appl. Environ. Microbiol.* **68**, 1458–1463.

Stetter, K. O. (1999). Extremophiles and their adaptation to hot environments. *FEBS Lett.* **452**, 22–25.

Sturm, F. J., Hurwitz, S. A., Deming, J. W. and Kelly, R. M. (1987). Growth of the extreme thermophile *Sulfolobus acidocaldarius* in a hyperbaric helium bioreactor. *Biotechnol. Bioeng.* **29**, 1066–1074.

Thom, S. R. and Marquis, R. E. (1984). Microbial growth modification by compressed gases and hydrostatic pressure. *Appl. Environ. Microbiol.* **47**, 780–787.

Tsao, J. H., Kaneshiro, S. M., Yu, S. S. and Clark, D. S. (1994). Continuous-culture of *Methanococcus jannaschii*, an extremely thermophilic methanogen. *Biotechnol. Bioeng.* **43**, 258–261.

Vance, I. and Hunt, R. J. (1985). Modification of a French press for the incubation of anaerobic-bacteria at elevated pressures and temperatures. *J. Appl. Bacteriol.* **58**, 525–527.

Yayanos, A. A. (1995). Microbiology to 10 500 meters in the deep sea. *Ann. Rev. Microbiol.* **49**, 777–805.

Yayanos, A. A., Dietz, A. S. and Boxtel, R. V. (1982). Dependence of reproduction rate on pressure as a hallmark of deep-sea bacteria. *Appl. Environ. Microbiol.* **44**, 1356–1361.

Yayanos, A. A., Boxtel, R. V. and Dietz, A. S. (1983). Reproduction of *Bacillus stearothermophilus* as a function of temperature and pressure. *Appl. Environ. Microbiol.* **46**, 1357–1363.

6 Analysis of Lipids from Extremophilic Bacteria

Milton S da Costa[1], M Fernanda Nobre[1] and Robin Wait[2]

[1] Departamento de Bioquímica and Departamento de Zoologia, Universidade de Coimbra, Largo Marquês de Pombal, 3001-401 Coimbra, Portugal; [2] Kennedy Institute of Rheumatology, Imperial College, London, UK

◆◆◆

CONTENTS

Introduction
Polar lipids
Fatty acids
Isoprenoid quinones
Concluding remarks

◆◆◆◆◆ **INTRODUCTION**

Species of bacteria have colonized almost all conceivable environments on Earth and many live in environments that we consider extreme and where microbial diversity may be limited to a small number of species. Organisms living at extremes of temperature, pH, salt and those resistant to extreme doses of ionizing radiation have been described that belong to almost all phyla within the domain Bacteria. Because extremophilic bacteria belong to so many different phyla, the lipids associated with these organisms also vary wondrously. In many cases the lipids of extremophiles seem more related to the phylogenetic position of the organisms than to the type of extreme environment which they colonize. In some rare instances the lipid composition does not even appear to be related to the phylogenetic position of the organisms. Some of the species of the order *Thermotogales* have optimum growth temperatures in the neighborhood of 85°C, and do not grow at temperatures below about 55°C. These organisms have, in addition to common straight chain fatty acids, very rare dicarboxylic fatty acids, that appear to span the cell membrane, namely 15,16-dimethyltriacontenedioic acid and 13,14-dimethyloctacosanedioic acid, known as diabolic acids (Carballeira *et al.*, 1997; de Rosa *et al.*, 1988, 1989). It has been proposed that the dicarboxylic fatty acids have a role in stabilization of membranes at very high temperatures (Carballeira *et al.*, 1997), but this is unlikely since these acids were initially identified from a bacterium classified as a strain of the low G + C Gram positive Bacteria of genus *Butyrivibrio* which was mesophilic and did not require NaCl for growth like the thermotogas (Klein et al., 1979).

The phylum "*Deinococcus/Thermus*" represents an ancient line of descent within the domain Bacteria that includes the extremely radiation-resistant bacteria of the genus *Deinococcus* (Battista and Rainey, 2001; Rainey et al., 2005), and the slightly thermophilic or thermophilic members of the family *Thermaceae* (da Costa and Rainey, 2001), represented by the species of the genera *Thermus* (Chung et al., 2000; da Costa et al., 2001); *Meiothermus* (Nobre and da Costa, 2001; Pires et al., 2005), *Marinithermus* (Sako et al., 2003), *Oceanithermus* (Miroshnichenko et al., 2003a; Mori et al., 2004), *Vulcanithermus* (Miroshnichenko et al., 2003b) and *Truepera*, which represents a new line of descent within this phylum (Albuquerque et al., 2005). The species of the genus *Deinococcus* have lower growth temperature ranges than those of the members of the family *Thermaceae* and the latter are not known to be radiation resistant (Zimmerman and Battista, 2005). The species of the genus *Deinococcus* and the genera of the family *Thermaceae* share few easily discernible characteristics, except that all species possess menaquinone 8 and those where peptidoglycan has been detected possess ornithine (Hensel et al., 1986). However, the species of this phylum examined, namely those of the genera *Thermus*, *Meiothermus* and *Deinococcus*, share some polar lipids that are not found in any other known group of bacteria, reflecting a statement made in 1997 by R. G. E. Murray who wrote saying that "To me the most distinctive attribute of the deinococci is not the extreme radiation resistance, but the remarkable polar lipids that are so different from all other bacteria" (R. G. Murray, personal communication). He was referring to a series of glycophospholipids, one of which has also been identified in the species of the genus *Meiothermus* and *Thermus*, namely 2'-O-(1,2-diacyl-sn-glycero-3-phospho)-3'-O-(N-acetylglucosaminyl)-N-D-glyceroyl alkylamine (Anderson and Hansen, 1985; Huang and Anderson, 1989; R. Wait and M. S. da Costa, unpublished results). This lipid does not appear to reflect an adaptation to any known stress, but probably represents a lipid that has originated within this phylum.

Some strains of species of the genera *Thermus* and *Meiothermus* possess long chain 1,2-diol glycolipids along with glycerol-based glycolipids, which have never been found in *Deinococcus* (Rainey et al., 2005; Wait et al., 1997). These lipid structures were initially found in the thermophile *Thermomicrobium roseum* (Pond et al., 1986), which is not closely related to the organisms of the phylum *Deinococcus/Thermus*, but which may reflect a distant common ancestor.

Some extremophilic bacteria possess novel lipids that may reflect the environment that they colonize. The extremely halophilic bacterium *Salinibacter ruber* (Antón et al., 2002) possesses a novel sulfonolipid that could indicate a role in the extremely saline environments that this organism inhabits (Corcelli et al., 2004). However, other organisms of the phylum *Bacterioidetes*, particularly gliding bacteria of the genera *Cytophaga* and *Flexibacter* (Godchaux and Leadbetter, 1983), have slightly different sulfonolipids and it remains to be seen if this unusual sulfonolipid is related or not to growth in extremely saline environments.

These examples show that novel and rare lipids are commonly identified in bacteria and that the examination of lipid composition should be part of a polyphasic approach to the classification of organisms. The analysis of the lipid composition need not be time consuming for most purposes and the information gained helps us understand the diversity of bacteria and of the structures that these organisms contain. Phylogenetic analysis based on 16S rRNA gene sequence comparisons has become ubiquitous and rightly so, because it tells us so much about the position and relationships of organisms in the microbial world. However, our view of bacterial diversity, taxonomy and classification of prokaryotes should not be based primarily (or even exclusively) on the analysis of this parameter as is frequently the case. We should during the characterization of new organisms strive to obtain as much information as possible about physiological, biochemical and chemical parameters to learn as much as possible about these organisms.

The analysis of lipids not only contributes to our knowledge of the diversity of the organisms themselves but also contributes to our knowledge of the diversity of the structural components of the organisms. More importantly the analysis of the lipid composition of bacteria can be used to distinguish among closely related taxa. The so-called chemotaxonomic parameters include the analysis of polar lipids, fatty acids, isoprenoid quinones and peptidoglycan, among other macromolecules of the cells. This chapter will describe methods for the analysis of polar lipids, isoprenoid quinones and fatty acids which can be routinely used for the characterization of extremophilic bacteria though they have variable physiological, biochemical and chemical parameters. The methods are simple enough to be used in most laboratories or common enough so that laboratories can obtain results from others.

◆◆◆◆◆ POLAR LIPIDS

The most common polar lipids of bacteria are phospholipids, glycolipids and glycophospholipids, aminolipids and sulfur-containing lipids. The majority of the polar lipids of bacteria have a glycerol backbone to which fatty acids are attached through ester linkages. Alkyl glycerol ethers are also known among the bacteria, but are relatively rare or present in small amounts. Some organisms such as *Aquifex pyrophilus* (Huber *et al.*, 1992; Jahnke *et al.*, 2001), *Thermodesulfobacterium commune* (Langworthy *et al.*, 1983), *Ammonifex degensii* (Huber *et al.*, 1996) and *Thermotoga maritima* (Carballeira *et al.*, 1997) possess glycerol monoethers and glycerol diethers. Many other thermophilic and mesophilic bacteria also generally possess small amounts of glycerol monoethers (Langworthy and Pond, 1986). Other polar lipids with a long chain 1,2-core are present in *Thermomicrobium roseum*, some species of the phylum

Chloroflexi and the glycolipids of *Thermus* and *Meiothermus*, but are not known in other bacteria (Balkwill et al., 2004; Ferreira et al., 1999; Pond and Langworthy, 1987; Pond et al., 1986; Wait et al., 1997).

Growth of Organisms and Extraction of Polar Lipids

Several methods to extract polar lipids have been described, but the method of Bligh and Dyer (1959) is the most common method used and the one that we have found to be suitable for all bacteria.

The organisms should be grown in liquid medium so that cells from the same phase of growth are always recovered to obtain reproducible results from replicate experiments. Individual polar lipid levels may change with the phase of growth and under different growth conditions and it is, therefore, important to maintain the culture conditions as similar as possible. We normally grow the organisms at the optimum growth temperature and stop the growth of the cultures during mid-exponential or late exponential phase by placing the culture flasks on ice. Some organisms, difficult to cultivate in liquid medium can be cultivated on solid medium. The cells recovered from liquid or solid media are washed twice by centrifugation at 4°C using an appropriate buffer; we generally use 10 mM phosphate buffer at pH 7.5 for most organisms. However, appropriate levels of NaCl for halophiles and other buffer systems should be used to take into account the optimum pH for growth of acidophilic and alkaliphilic organisms.

We generally recover about 1.0–4.0 g wet weight of cells for extraction of polar lipids, but lesser amounts can be used for chemotaxonomic analysis of the polar lipid profiles of bacteria by thin layer chromatography alone.

Precautions: The quantities of cells and volumes of solvents given in the following methodology are for large-scale extraction of polar lipids, but these can be decreased as long as the proportions of the solvents for extraction are maintained. All procedures involving the use of chloroform, other solvents and spray reagents to detect lipids separated by thin layer chromatography (TLC) must be performed in a fume hood. All the spray reagents are hazardous and care must be taken not to inhale them. All glassware should be washed in chloroform:methanol (1:1 by vol.).

- Resuspend the cell pellets in 10 ml of the appropriate washing buffer in large glass centrifuge tubes (90 ml). Add 25 ml of methanol to the suspension and stir magnetically for 10 min. Add 12.5 ml chloroform and stir during 15 min. Centrifuge the suspension at $2000 \times g$ for 15 min in a swinging-bucket centrifuge, at room temperature, until the cells sediment completely. Transfer the extract to a 300 ml Erlenmeyer flask fitted with a sintered-glass stopper. The cell pellet is resuspended in 8 ml water, stirred and re-extracted using the same volumes of chloroform and methanol. Repeat the centrifugation and transfer the extract to the Erlenmeyer flask. If phase separation occurs

at this stage, add methanol drop-wise with slow stirring until a one phase mixture is obtained.
- Add 25 ml chloroform and 50 ml of 0.5 M KCl so that final proportion of chloroform:methanol:water is (2:2:1.8 by vol.). Shake the mixture vigorously to form an emulsion. Distribute the emulsion in glass centrifuge tubes and centrifuge at $2000 \times g$ for about 5 min to obtain phase separation.
- For routine TLC separation of polar lipids, the lower phase is removed once and transferred to round-bottom flasks with sintered-glass joints. Care should be taken to avoid removing the interface and upper phase. For quantitative analysis of the polar lipids, the upper phase is transferred to the same Erlenmeyer, the extraction repeated and the lower phase is removed and added to the first extract.
- Evaporate the lipid extract to dryness under low vacuum at 30–35°C with a rotary evaporator.
- Resuspend the lipid extract in a small volume of chloroform:methanol (1:1 by vol.). The extract should be removed from the round-bottom evaporation flask, for quantitative analysis, at least twice with the solvent mixture to recover as much of the extract as possible.
- Store the extract at $-20°C$ under N_2 until analysis.

Preparative Thin-layer Chromatography (TLC)

It is sometimes necessary to remove pigments and other lipid components from the sample before analysis of the polar lipids by TLC because they may interfere with the migration of the polar components on the plates.

- Silica Gel 60 (Merck 5745, 20×20 cm, 2 mm thickness) plates are dried for 30 min at 120°C and cooled in a desiccator containing dry silica gel until used.
- Add acetone to a chromatography tank in which one of the sides has been lined with filter paper to aid saturation of the atmosphere inside the tank and wait about one hour before placing the chromatographic plate. The solvent should be below the band (origin) where the lipid extract will be applied.
- Evaporate the extract to dryness under a stream of N_2. Resuspend the lipid extract in a small volume (about 250 µl) of chloroform:methanol (1:1 by vol.). Apply all of the extract slowly on the silica gel G plate along a line about 2 cm from the lower edge of the plate. A small amount of chloroform:methanol can be added to the tube to further remove the extract completely and then applied to the chromatography plate. More than one sample can be applied to the same plate, but a space is left between the two sample applications. It is important not to apply the sample closer than about 1.5 cm from the vertical edges of the plate.
- Place the silica gel plate in the chromatography tank with the upper edge leaning against the inside of the tank and wait until the solvent front reaches about three quarters of the height of the plate.
- Remove the plate from the tank and allow the solvent to evaporate at room temperature or blow cold air from a hair dryer to hasten drying.
- Scrape the silica gel of the area surrounding the origin (about 1 cm above and below) to recover the polar lipids which do not migrate in this solvent.

Place the silica on smooth non-absorbent paper, cover with another piece of paper and grind the silica to a fine powder by pressing with finger tips. Place the powdered silica into a glass centrifuge tube. Alternatively add the silica to a glass centrifuge tube and grind the silica to a fine powder with a glass rod. Add a clean magnetic stirring bar to the centrifuge tube. Add 15 ml of chloroform:methanol:water (45:45:10 by vol.), stir slowly on a magnetic stirrer for about 10 min and centrifuge for 5 min in a swinging-bucket centrifuge.

- Remove the extract to a 35 ml Teflon-coated screw-capped tube. Repeat the extraction of the silica and transfer to the same tube.
- Add 7.7 ml of 0.5 M KCl to obtain a chloroform:methanol:water mixture (2:2:1.8 by vol.). Shake the tube vigorously to obtain an emulsion. Centrifuge at $2000 \times g$ until phase separation.
- Remove the lower phase, taking care not to remove the interface or the upper phase, and add to a Teflon-coated screw-capped tube. Evaporate to dryness under a stream of N_2. Store at $-20°C$ until needed.

Silica Gel, Solvents and Spray Reagents

Silica Gel 60 plates (Merck 5626, 0.25 mm thickness, 10×20 cm) are most often used. We cut the plates with a diamond knife to obtain plates that are 10×10 cm, but smaller plates can also be used. Other silica gel plates are available but for most purposes silica gel 60 plates are generally used.

There are many solvent systems that can be used to separate polar lipids on TLC and which are thoroughly described ranging from neutral solvent systems to mixtures of solvents with acidic or basic pH (Christie, 1982; Ratledge and Wilkinson, 1989; Zweig and Sherma, 1972). We generally use an acidic mixture that is adequate for one-dimensional TLC, and neutral and acidic solvent systems for two-dimensional TLC.

The detection reagents described here are generally sufficient to obtain a polar lipid profile from many bacteria, but other reagents can be used for a more detailed presumptive identification of polar lipid types (Christie, 1982; Kates, 1993; Ratledge and Wilkinson, 1989). It is important to remember that none of the spray reagents are completely specific for any type of polar lipid. In some cases, we use standard mixtures of commercially available polar lipids or lipid extracts of bacteria with well-defined polar lipid profiles to test the efficiency and specificity of the spray reagents.

One-Dimensional TLC

One-dimensional thin layer chromatography is generally sufficient for the separation of the polar lipids of most organisms without recourse to two-dimensional TLC (Figure 6.1). Better separation and definition of the lipid bands are obtained if the extract is placed along a line at the origin,

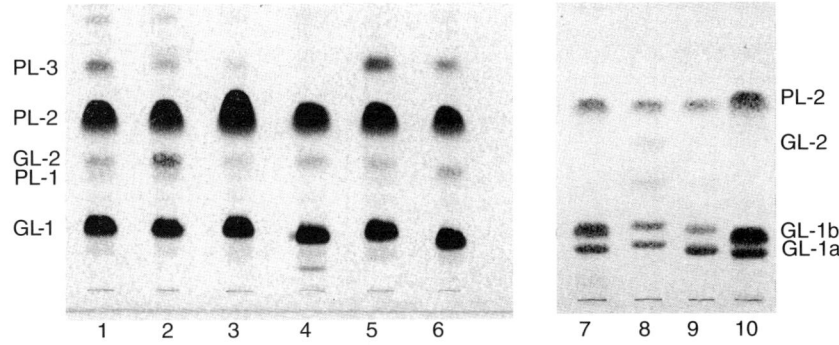

Figure 6.1. One-dimensional thin layer chromatography of the total polar lipids of strains of the genus *Thermus*: 1. *T. filiformis* Rt358; 2. *T. filiformis* Tok4 A2; 3. *T. scotoductus* ITI-252; 4. *T. scotoductus* NH; 5. *T. brockianus* 15038^T; 6. *T. oshimai* SPS-14, and *Meiothermus*: 7. *M. ruber* Loginova 21^T; 8. *M. chliarophilus* ATL-8^T; 9. *M. silvanus* VI-R2^T; 10. *M. cerbereus* GY-1^T. The plates were stained with molybdophosphoric acid followed by heating at 160°C.

instead of a spot. Spotting can lead to overloading of the sample and to vertical smear of the lipids that can mask some of the components.

- Silica gel plates (Merck 5626) are dried for 30 min at 120°C and cooled in a desiccator containing dry silica gel until used.
- Add a solvent mixture composed of chloroform:acetic acid (glacial):methanol: water (80:15:12:4 by vol.) to a chromatography tank with one side lined with filter paper and allow the atmosphere to saturate with the solvent vapors for about one hour before introducing the chromatographic plates.
- Several samples can be applied on the same plate about 1 cm from the lower edge. Apply the extracts on the silica gel plate using a microsyringe or a capillary tube on a line about 1.0 cm in length with about 0.5 cm space between applications (Figure 6.2A). The lines of application can be traced with a pencil before applying the extract.
- Place the silica gel G plate vertically in the chromatography tank with the upper edge leaning against the inner side of the tank, place the lid on the tank and leave until the solvent front reaches the upper edge of the plate.
- Remove the plate from the tank and allow the solvent to evaporate at room temperature or blow cold air from a hair dryer to hasten drying.
- The plate can be chromatographed a second time in the same solvent to improve the separation of the polar lipids.

Two-Dimensional TLC

Two-dimensional TLC may be necessary to resolve complex mixtures of polar lipids or to separate components that migrate close to each other on one-dimensional TLC. Overloading the silica at the application spot is not very important in two-dimensional TLC, because of the individual polar lipids will become separated from each other during the second chromatographic step.

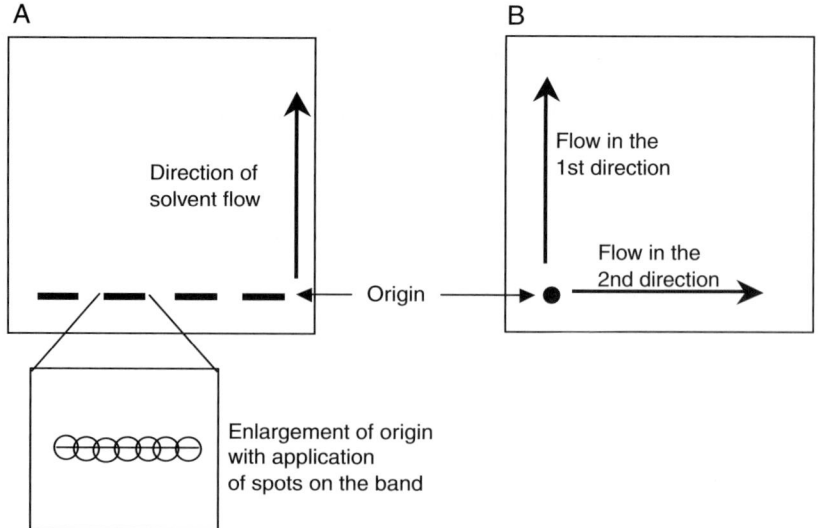

Figure 6.2. TLC plates showing application of samples and direction of flow of solvents for (A) one-dimensional TLC and (B) two-dimensional TLC. The enlargement shows application of small spots along the band for one-dimensional TLC.

- Prepare two chromatography tanks; (1) one tank containing a solvent mixture composed of chloroform:methanol:water (65:25:4 by vol.) and (2) one tank containing a solvent mixture of chloroform:acetic acid (glacial):methanol:water (80:15:12:4 by vol.) and allow the atmosphere of the tank to saturate for about one hour.
- Dry several silica gel G plates (Merck 5626) at 120°C for 30 min and cool in a desiccator containing dry silica gel until used.
- Apply the sample slowly on one spot about 1.5 cm from the lower edge and the left-hand edge of a 10 × 10 cm plate as shown in Figure 6.2B and Figure 6.3.

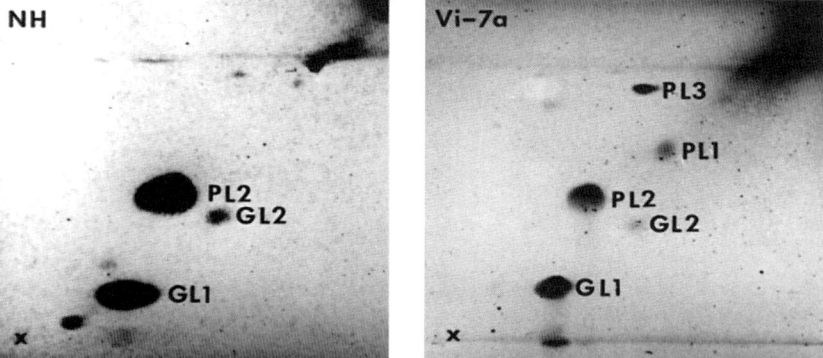

Figure 6.3. Two-dimensional thin layer chromatography (TLC) of the total polar lipids of *Thermus scotoductus* strains NH and VI-7a stained with molybdophosphoric acid followed by heating at 160°C.

- The silica gel G plate is placed in the chromatographic tank (1) until the solvent reaches the upper edge. Remove the plate and allow it to dry with the aid of a cold air stream from a hair dryer.
- Rotate the plate 90° counter-clockwise and place in tank (2). Allow the solvent to ascend to the upper edge of the plate. Remove and dry the plate.

Detection of Polar Lipids on TLC Plates

The reagents for the detection of the polar lipids should be sprayed at a distance of 20–30 cm. Spray the reagents evenly and avoid soaking of the plate until they drip, because detection will not be improved and the silica may fall off the plates.

Phospholipids – molybdenum blue reagent

This reagent is commercially available from Sigma-Aldrich, but we have always made it as follows because of consistently better results:

Solution A

Add 10.0 g molybdenum oxide (MoO_3) to 250 ml of H_2SO_4 (12.5 M). Boil gently in a fume hood until all the MoO_3 dissolves. Allow the solution to cool and store for a few days at room temperature before use.

Solution B

Add 400 mg of molybdenum powder (Mo) to 112.5 ml of solution A. Boil gently in a fume hood for about 15 min to dissolve the Mo. Allow to cool and decant into a glass bottle. Store at room temperature until use.

Spray reagent

Add slowly 20 ml of solution A, 20 ml of solution B and 60–80 ml of water until the reagent becomes gray-green in color. Store the reagent at room temperature in the dark. If the color of this reagent changes to blue, water should be added drop-wise until the reagent becomes gray-green again. A green-gray-colored spray reagent is more effective in staining phospholipids than one with a blue color.

Blue spots appear almost immediately after spraying. The plate can be sprayed a second time to enhance the blue-staining phospholipids. The reagent is very specific for phosphorus containing lipids.

Lipids with free amino groups – ninhydrin

Add 100 mg ninhydrin to 100 ml of 1-butanol:acetic acid (glacial) (95:5 by vol.) Store the reagent at room temperature in the dark.

Spray and place the chromatographic plate in an oven at 120°C for about 5 min. The lipids with free amino groups stain pink. The same plate sprayed with the ninhydrin reagent can be used to stain glycolipids with the α-naphthol reagent, after marking the aminolipids with a pencil.

Glycolipids – α-naphthol-sulfuric acid

- Prepare a 15% solution of α-naphthol in absolute ethanol.
- Add 10.5 ml of the above solution to a mixture containing 6.5 ml of H_2SO_4 (95–97%), 45.5 ml of absolute ethanol and 4 ml water. Store for a few days before using.

Place the plates in an oven heated at 120°C for 5 min after spraying. The glycolipids appear as purple spots.

Glycolipids – diphenylamine

Dilute 20 ml of a 10% solution of diphenylamine in absolute ethanol in 100 ml of HCl (32%) and 80 ml glacial acetic acid. Store the reagent at room temperature in the dark.

Spray lightly; cover the plate with a clean glass plate and heat at 110°C until spots appear (about 40 min). The glycolipids stain blue.

Choline-containing phospholipids – Dragendorff's reagent

Solution A

Dissolve 17 g basic bismuth nitrate ($Bi_5O(OH)_9(NO_3)_4$) in 20% aqueous glacial acetic acid. Store at 4°C.

Solution B

Dissolve 40 g potassium iodide in 100 ml water. Store at 4°C.

Spray reagent

Mix immediately before use 4 parts of solution A and 1 part of solution B with 14 parts of water.

Choline-containing lipids appear orange or red-orange after spraying.

Total lipids – molybdophosphoric acid

The reagent is prepared by adding 5 g of molybdophosphoric acid in 100 ml of absolute ethanol. Store reagent at room temperature in the dark.

Place the chromatographic plate in an oven at 160°C for about 20 min after spraying. The lipids stain grey on a yellow background. The yellow background can be removed by placing the plate in a small covered glass tank containing a few drops of 25% ammonia for a few seconds.

Sulfur containing lipids – Azure A-sulfuric acid

Dissolve Azure A to saturation in H_2SO_4 at room temperature with slow magnetic stirring.

Spray the reagent evenly on the plate. Immerse in a small tank containing 0.04 M H_2SO_4:methanol (3:1 by vol.) to remove excess reagent. The sulfolipids and sulfonolipids stain blue.

Characterization of Polar Lipids by Mass Spectrometry

Mass spectrometry (MS) is one of the most sensitive and powerful techniques available for biomolecular structure determination, and has played a key role in studies of extremophile lipids, both for characterization of intact lipids and for the identification of their constituent fatty acids, long-chain diols, alkylamines and monosaccharides. The requirement for thermal vaporization generally precludes analysis of intact polar lipids by electron ionization (EI) and GC/MS. Such compounds only became routinely amenable to direct MS characterization with the development of so-called soft ionization methods, in which sample molecules are converted to gas phase ions without the need for heating. These methods include fast atom bombardment (FAB), matrix-assisted laser desorption ionization (MALDI) and electrospray ionization (ESI).

We have determined lipid compositions and monosaccharide sequences of a family of glycosylated diacyl glycerols from *Thermus* and *Meiothermus*, using a range of mass spectrometric approaches. Tandem mass spectrometry of acetates and deuteroacetate derivatives established the stereochemistry of the terminal residues, as variously glucopyranose, galactopyranose, galactofuranose or ribopyranose. The presence of hydroxylated fatty acids in several glycolipids was confirmed, and we showed that these are exclusively amide-linked to hexosamine in the glycan head group, and are never ester-linked to glycerol. We also characterized several novel glycolipids, in which similar tri- and tetra-glycosyl head groups were linked to 16-methyl-1,2-heptadecanediol or 15-methyl-1,2-heptadecanediol, rather than to glycerol (Carreto *et al.*, 1996; Ferreira *et al.*, 1999; Wait *et al.*, 1997).

More recently the structure of the major acylglycerol glycolipid from *Thermus oshimai*, NTU-063 was established as β-Glc$p$-(1–6)-β-Glc$p$-(1–6)-β-Glc$p$NAcyl-(1–2)-α-Glc$p$(1–1)glycerol diester, by a combination of NMR spectroscopy, methylation analysis and MS (Lu *et al.*, 2004), which supports our original proposal and supplies the missing anomeric configuration and linkage information.

The analytical strategy entails molecular mass measurement in positive and negative ion modes of the underivatized TLC-purified glycolipids, which enables determination of their composition in terms of sugar, glycerol/diol and acyl units. The glycolipids are then acetylated and the mass spectrometric analysis is repeated after acetylation and deuteroacetylation of two aliquots of sample. Comparison of the masses of the acetylated and deuteroacetylated derivatives enables counting of the number of hydroxyl groups derivatized, which is helpful for determining the true molecular mass, as it is not always obvious if the observed molecular ions are protonated, or are cationized by alkali metal ions. Acetylated glycolipids undergo simple and predictable fragmentation (Dell, 1990), *via* pathways producing predominantly non-reducing terminal carbenium (B_n-type) ions (Domon and Costello, 1988). For example the FAB spectrum of the peracetylated GL-1 from *Thermus* strain SPS-11 had an abundant B_1 ion, diagnostic of a terminal hexose, at *m/z* 331; further B ions at *m/z* 619 (B_2), 1116 (B_3) and 1404 (B_4)

established the sequence of the polar head group as hexose-hexose-(N-acyl)hexosamine-hexose. Glycosidic cleavage is particularly favored at hexosamine, resulting in prominent B_3 fragments which incorporate the N-acylated hexosamine residue which are diagnostic of the N-acyl substituents. The major B_3 ion at m/z 1116 indicates that the hexosamine is N-acylated by iso17:0, other B_3 fragments at m/z 1088 and 1102 indicate that a sub-population is acylated instead by iso15:0 or iso16:0. The ratio of the abundances of m/z 1102 and m/z 1088 indicated that approximately 54% of the fatty acids amide-linked to glucosamine were C17 fatty acids and about 40% were C15 fatty acids, the balance being C16 fatty acids.

Ions at m/z 523 and 551 originate by cleavage of the sugar–glycerol bond and identify the ester-linked substituents; m/z 523 corresponds to acylation with two C15 fatty acids, whereas m/z 551 indicates substitution with one C15 and one C17 acid. Comparison of the relative intensities of the N-acylated B_3 fragments and the acylglycerol ions enables an approximate estimate of the distributions of fatty acids at these locations. The ratio of the relative intensities of the acylglycerol fragments at m/z 551 and 523 was about 1.0/0.9, suggesting that of the total O-linked fatty acids, slightly more than 25% were C17 fatty acids, whereas about 70% were C15 fatty acids.

The mass spectra of glycolipids that are linked to long chain diols rather than to glycerol exhibit similar fragmentation of the sugar chains, but acylglycerol fragments are absent, and are replaced by additional diagnostic fragments which define the diol moiety. Figure 6.4 shows the

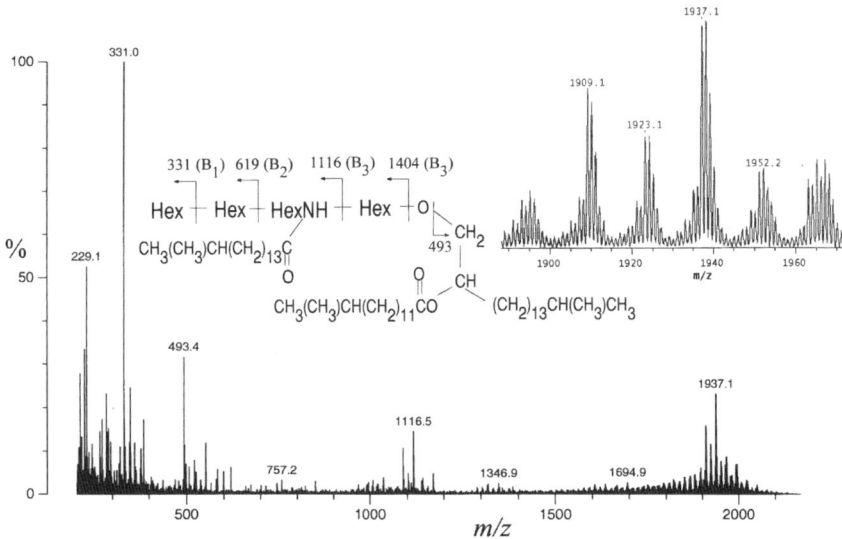

Figure 6.4. Positive ion fast atom bombardment (FAB) mass spectrum of tetraglycosyl lipids from *Thermus filiformis*. The major sodium cationized molecule at m/z 1937.1 originates from a diol-linked glycolipid in which the N-acyl substituent is iso17:0 and the O-acyl group is iso15:0. The sequence of the glycan chain is defined by B-type carbenium ions at m/z 331, 619 and 1116. The fragment at m/z 493 represents loss of the glycan head group with charge retention on the lipid moiety, and is diagnostic of a C18 diol O-acylated with a C15 fatty acid.

mass spectrum of the peracetyl derivative of the major diol-linked glycolipid from *Thermus filiformis*. The fragment at m/z 493 represents loss of the glycan head group with charge retention on the lipid moiety, and is diagnostic of a C18 diol O-acylated with a C15 fatty acid. As before the sugar sequence of the glycan chain is defined by B-type carbenium ions at m/z 331, 619 and 1116.

Acid catalyzed peracetylation of intact polar lipids (Dell, 1990)

Cautiously add 2 volumes of trifluoroacetic anhydride (TFAA) to 1 volume of acetic acid in a screw-capped tube (danger: exothermic reaction; fume hood and eye protection required). Allow to cool before use but do not store; make fresh as required.

- Thoroughly dry the samples (5–50 µg in 5 ml screw-capped tubes) by vacuum centrifugation, then add 100 µl of the TFAA–acetic acid mixture, vortex and leave at room temperature for 10 min.
- Remove the reagent by vacuum centrifugation, dissolve the residue in 1 ml of chloroform and desalt by washing three times with equal volumes of water; discard the water washes.
- Evaporate the chloroform in a vacuum centrifuge, and dissolve the derivatives in 10 µl chloroform:methanol (1:1 v/v) prior to FAB MS.

Deuteroacetates are prepared identically, except that the acetic acid in the reaction mixture is replaced by d_4-acetic acid.

The major phospholipid from *Thermus* species

We have also established the structures of the phospholipids of *Thermus* and *Meiothermus* strains using a combination of GC/MS, FAB MS and NMR spectroscopy. For example, a deprotonated molecule at m/z 1176.0 was observed in the negative ion FAB spectrum of the major phospholipid (PL-2) from *Thermus scotoductus* strain VI-7 (Figure 6.5). On peracetylation and positive ion FAB MS a sodium-cationized molecular ion was observed at m/z 1325.7, suggesting addition of three acetyl groups. Major fragments were observed at m/z 330, the diagnostic mass of a B_1 fragment derived from a terminal hexosamine, and at m/z 551, attributable to a diacyl glycerol fragment esterified with one C15 and one C17 acid (Figure 6.6). GC/MS identified iso15:0 and iso17:0 FAME, and a C17 alkylamine. The mass spectrum of its acetate derivative showed that the latter was also iso branched, because no backbone cleavage fragment was observed at m/z 268 (Figure 6.7). NMR spectroscopy confirmed the presence of terminal alpha-linked glucosamine, and indicated that no other sugars were present. Taken together these data establish the structure as a 2'-O-(1,2-diacyl-sn-glycero-3-phospho)-3'-O-(α-N-acetyl-glucosaminyl)-N-D-glyceroyl alkylamine (Figure 6.6). It is similar to the glucosamine-containing "Lipid 6" from *D. radiodurans* (Huang and Anderson, 1989), differing mainly in that the alkylamine is exclusively saturated and iso-branched, and that the pattern of fatty acid substitution is different.

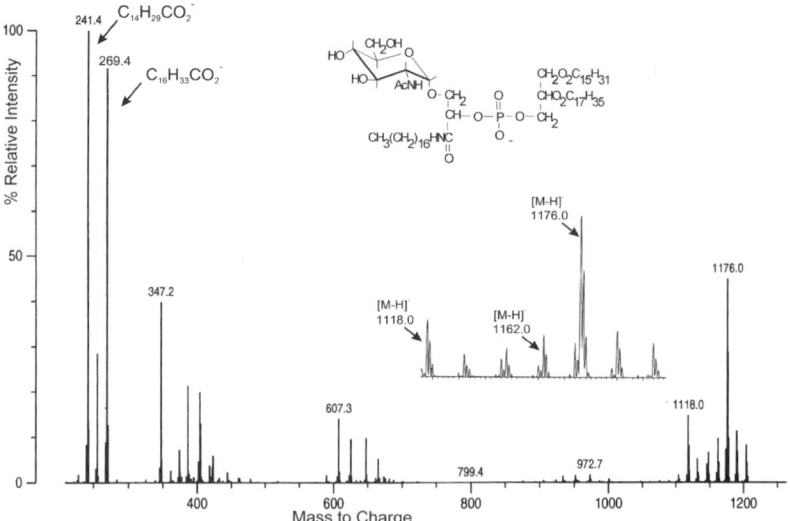

Figure 6.5. Negative ion FAB spectrum of the major phospholipid from *Thermus scotoductus*. The deprotonated molecule at m/z 1176, is consistent with the composition; glycerol, phosphate, C17 alkylamine, glycerate, glycerol, iso15:0, iso17:0 and hexosamine. Carboxylate anions at m/z 269 and 241 show that glycerol is esterified with both 15:0 and 17:0 acids.

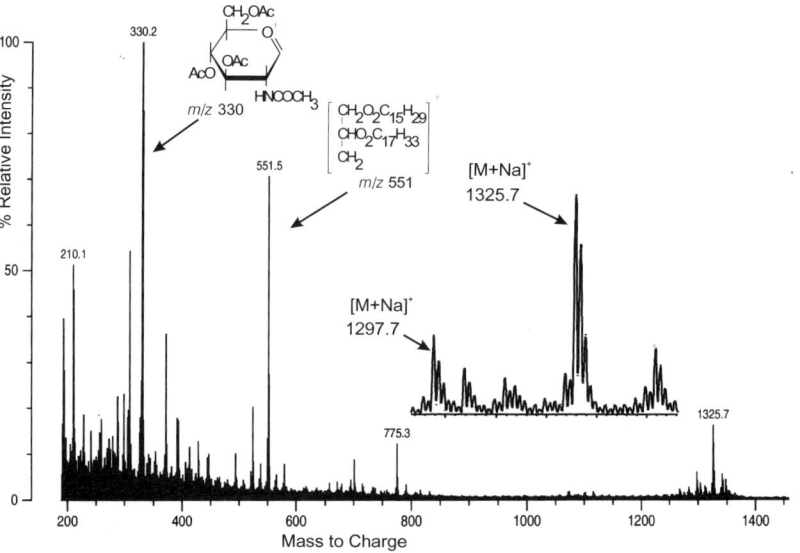

Figure 6.6. Positive ion FAB mass spectrum of peracetylated phospholipid from *T. scotoductus*. The B_1 carbenium ion at m/z 330 is diagnostic of a terminal hexosamine residue. The fragments at m/z 551 and 523 originate by cleavage of the glycerol to phosphate bond, and define the ester-linked substituents (m/z 523 = 15:0/15:0; m/z 551 = 15:0/17:0).

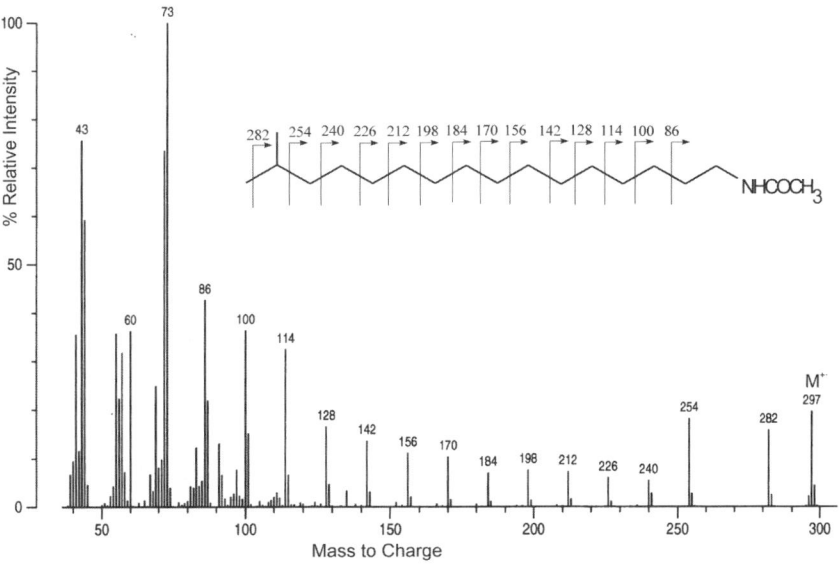

Figure 6.7. EI mass spectrum of 15-methylhexadecylamine, liberated by methanolysis of the major phospholipid from *T. scotoductus*. The presence of an iso branch was inferred from the absence of the alkyl cleavge fragment at *m/z* 268.

◆◆◆◆◆ FATTY ACIDS

Bacteria possess a great variety of fatty acyl chains that include straight chain saturated and monounsaturated fatty acids, terminally branched (iso-branched) and subterminally branched (anteiso-branched), internally-branched fatty acids, 2- and 3-hydroxy fatty acids, cyclopropane fatty acids, ω-cyclic (ω-cyclohexyl- and ω-cycloheptyl-) fatty acids, and dicarboxylic fatty acids, among others. Polyunsaturated fatty acids are rare, but have been encountered most frequently in cyanobacteria. Other fatty acyl chains such as fatty alcohols are also found in generally small amounts.

It is becoming increasingly common for microbiologists interested in classification of organisms to rely on the determination of the fatty acyl composition, because it is relatively easy to perform and can help to resolve or confirm relationships among bacteria and can be used as a chemotaxonomic parameter to distinguish many closely related species. Some species of bacteria have been, in fact, classified on the basis of the fatty acid composition alone, since other diagnostic and commonly used parameters for the characterization of new taxa do not easily distinguish them. It is also sometimes simpler to reliably identify bacteria based on the fatty acid composition than to use other methods that may be more expensive and less reliable. The genus *Legionella*, for example, comprises 49 species that are practically impossible to distinguish from each other on the basis of biochemical and physiological parameters (Adeleke *et al.*, 2001; La Scola *et al.*, 2004). However, 275 strains belonging to the 41 species known at the time could, with the exception of the extremely closely related species *L. erythra* and *L. rubrilucens*, be identified by their fatty acid composition (Diogo et al., 1999). In this study one strain that would later be classified as a new species, was easily distinguished from the other organisms (Lo Presti et al., 1999).

On the other hand, only the strains of two of the nine species of the genus *Thermus* have uniform fatty acid compositions that aid in the identification and classification of the organisms. The fatty acid composition of the strains of the species *T. thermophilus*, isolated from different locations, have very different fatty acid composition (Nobre et al., 1996), that can not be used for the presumptive identification of the strains of this species. The strains of the species *T. filiformis*, which to date, have only been isolated from hot springs in New Zealand, also have extremely variable fatty acid compositions. Anteiso-fatty acids, for example, are the major fatty acids, under all conditions examined, of the type strain of this species while iso-fatty acids are the predominant fatty acids of the other strains and one strain has 3-OH fatty acids (Ferraz et al., 1994).

It is commonly possible to predict the types of fatty acids present in closely related bacterial species, although the relative proportions of the individual fatty acids can not be predicted. However, these predictions sometimes fail. The strains of species of the genus *Meiothermus* can, until now, be easily distinguished from those of the genus *Thermus* by the presence of 2-OH fatty acids which are absent in the species of the latter genus (Nobre et al., 1996). The very rare ω-cyclohexyl- and ω-cycloheptyl-fatty acids are the hallmark fatty acids of the thermo-acidophilic bacteria of the genus *Alicyclobacillus* (Albuquerque et al., 2000; Goto et al., 2003) and were thought to stabilize membranes at high temperatures and acidic conditions (Moore et al., 1997), although they are also found in a few other bacteria, namely the mesophilic and neutrophilic species *Curtobacterium pusillum* (Suzuki et al., 1981) and *Propionibacterium cyclohexanicum* (Kusano et al., 1997). However, one new species of the genus *Alicyclobacillus*, named *A. pomorum*, does not possess ω-cyclic-fatty acids (Goto et al., 2003). In this organism ω-cyclic-fatty acids are replaced by branched chain fatty acids which are also found, in lower relative proportions, in the strains of the other species of this genus. However, this species was described based on one strain, it being possible that other strains of this species have ω-cyclic-fatty acids. Moreover, other species of the genus *Curtobacterium* appear not to possess ω-cyclic-fatty acids (Behrendt et al., 2002).

The majority of the eighteen species of the genus *Deinococcus* whose names were validly described possess large relative proportions of iso- and anteiso-branched fatty acids along in lower levels of straight chain fatty acids; however, *D. radiodurans*, *D. proteolyticus*, *D. radiophilus* and *D. hopiensis* possess saturated and monounsaturated straight chain fatty acids exclusively (Rainey et al., 2005).

Since the proportions of the acyl chains vary with the growth conditions such as the growth temperature, the phase of growth and the medium composition, it is very important to maintain the same culture conditions when examining the fatty acid composition of related organisms. We have grown organisms with different growth temperature ranges at one common temperature to ensure that the fatty acids can be used to discriminate among different species (Rainey et al., 2003).

Fatty acid composition is assessed by gas chromatography (GC) and by comparison with the retention times of known standards. Mass spectrometry is then used to identify those fatty acids that cannot be identified by GC alone.

The Sherlock® Microbial Identification Systems (http://www.midi-inc.com) has facilitated the identification of a large variety of fatty acids from microorganisms by providing a highly standardized system that is capable of identifying a large assortment of fatty acids and other acyl compounds from bacteria and yeast. Identification of acyl chains is based on the equivalent chain length (ECL) which refers to a linear extrapolation of each peak's retention time between two straight chain saturated fatty acid methyl esters reference peaks. The MIS software compares the ECL of each peak in the sample with the expected ECL of fatty acyl compounds in the database. Some peaks may not be identified. Other peaks may be labeled "Summed Feature" because the ECL value corresponds to one fatty acid that cannot be separated from another fatty acid under the chromatographic conditions. The relative concentration of possibly two or more fatty acids is given as one value. It is sometimes possible to tentatively identify a particular fatty acid within a "Summed Feature" by comparing these fatty acids with other fatty acids identified by the database from the sample. For example, in *Meiothermus* spp. one peak labeled as "Summed Feature" may contain two fatty acids, namely 15 iso15:0 2-OH and 16:1 ω7c. As the organism contains other 2-OH-fatty acids, but does not have other monounsaturated straight chain fatty acids, it is very likely that the "Summed Feature" corresponds to 15 iso15:0 2-OH fatty acid alone.

This system was developed, as might be expected, to identify medically important microorganisms and their relatives, but has been expanded to include the identification of a large variety of microorganisms by their fatty acid composition under standard growth conditions. One of the most frequently used methodologies of Microbial Identification System (MIS) takes into account the cultivation of aerobic bacteria at 28°C on Tryptic Soy Broth (BBL 11768), BBL 11849 and Agar in one liter of water for 24 ± 1 h. Under these growth conditions many aerobic bacteria can be identified using the database provided by MIS. These growth conditions can be modified to suit the identification of fatty acids from many bacteria including most, if not all extremophlies. Additional databases can be constructed by the investigator, but without them many strains cannot be identified, because they represent new organisms, they have not been included in the manufacturer's own database or growth conditions have been modified that are not taken into account by the manufacturer. It is not necessary, for comparative purposes, to identify strains; it is merely necessary to identify the majority of the acyl compounds and compare them to closely related organisms grown under the same conditions.

We often use the fatty acid composition to reduce the number of isolates, which may be all very closely related, to a small number that are then subjected to 16S rRNA gene sequence analysis and further characterization. We then compare the fatty acid composition of the new organisms with those of the most closely related organisms for chemotaxonomic purposes. We have now, for example built large data bases for several groups of bacteria which can be used to tentatively identify these organisms. It is our experience that, under controlled conditions, extremely reproducible results can be obtained.

The Sherlock® Microbial Identification System cannot be used with alternate chromatographs or columns than those provided by the manufacturer, because of the interdependency of the system. The manufacturer also sells a standard mixture of fatty acids that is automatically injected to calibrate the process at the onset and after every eleven sample injections. However, some modifications of cultivation, extraction and methylation are possible and very useful in determining the fatty acid composition of bacteria. Moreover, fatty acyl compounds not identified by MIS can be identified by comparison with fatty acid methyl esters from other bacteria where they have been identified. New unknown fatty acids, however, will have to be identified by mass spectroscopy (MS). In some cases, minor unknown fatty acids that are not relevant for characterization of the organisms need not be identified. However, those that are relevant for the classification of the organisms, especially if they are the major components should be identified by MS. We recently identified a new fatty acid by MS that was not found in the MIS database, because it was deemed important for the characterization of a new organism (Albuquerque et al., 2002).

This partially automated system relies on a saponification step followed by acid methanolysis to produce of the methyl esters (Kuykendall et al., 1988), which is reliable for most bacteria. Gas chromatography is performed with Hewlett-Packard chromatograph equipped with an automated injector and a 5% phenyl methyl silicone capillary column (0.2 mm by 25 m) and the fatty acid methyl esters are detected with a flame ionization detector (FID). All reagents should be stored at room temperature in dark glass bottles with Teflon-lined screw caps.

Cultivation and Harvesting of Cells

The bacteria can be grown on solid media, as recommended by the manufacturer or they can be grown in liquid medium. The MIS recommends removing cells from a specific quadrant of the culture plates. Cells, about 40 mg, are scraped from solid media with a new plastic bacteriological loop (10 µl loop) and placed in a 13×100 cm teflon-lined screw-capped tube. Liquid cultures are harvested by centrifugation in the cold and washed once with the appropriate buffer. The cell pellets are transferred to a tube with a bacteriological loop. The cultivation time depends on the organism, the medium used and the temperature, among other growth conditions. As the fatty acid composition can change dramatically with culture conditions, it is very important to always use the same conditions and to harvest at the same growth phase.

Preparation of Fatty Acid Methyl Esters

Saponification

- Mix 45 g NaOH (reagent grade), 150 ml methanol (HPLC grade) and 150 ml ultrapure water until the NaOH pellets are completely dissolved.
- Add 1 ml to the tube containing fresh cell mass, vortex for about 10 s and then heat for 5 min in boiling water bath. Vortex again and incubate in the boiling water bath for 25 min. Cool in water at room temperature.

Methylation

- Add 325 ml of titrated 6.0 N HCl to 225 ml of methanol (HPLC grade).
- Add 2 ml to the tube and vortex for about 10 s and heat in a water bath at 80°C for 10 min. Cool rapidly in water to room temperature.

Extraction

Reagent A

Add 200 methyl-tert-butyl ether (HPLC grade) to 200 hexane (HPLC grade).

Reagent B

Dissolve 10.8 g NaOH (reagent grade) in 900 ml of ultrapure water.

Add 1.25 ml of reagent A to each tube and mix end-over-end with a rocking motion on a laboratory rotator for 10 min. Remove the lower phase with a Pasteur pipette and discard. Keep the upper phase which is washed with 3 ml of reagent B. Mix end-to-end with a rocking motion for 5 min and transfer about two-thirds of the upper phase containing the fatty acid methyl esters to an auto-sampler vial and seal. Continue or store at $-20°C$ under a N_2 atmosphere.

Modifications

We have omitted the saponification step to recover 1,2-long chain diols, which appear not to be recovered or are recovered in low relative proportions following the MIS extraction and methylation protocol and modified the methylation step by incubating the samples at 100°C for 30 min (Carreto *et al.*, 1996; Wait *et al.*, 1997). This modification does not affect the recovery of other fatty acyl methyl esters from bacteria of the genera *Thermus*, *Meiothermus*, *Deinococcus* and *Truepera*. However, cyclopropane and other acyl compounds may be degraded. It is, therefore, important to compare the fatty acid profiles derived from the modified methods with those obtained from the standard method described by the manufacturer.

Mass Spectrometry of Fatty Acid and Diol Derivatives

In a mass spectrometric experiment, sample molecules are ionized and the masses of their gas phase ions are determined. If sufficient excess energy is deposited, or if the ions are additionally excited, e.g. by collision with neutral gas, then a proportion will dissociate in a structure-specific fashion, producing a fragmentation pattern, from which the mass and structure of the original molecule can be deduced. Low molecular mass compounds such as fatty acid methyl esters can be ionized by bombardment with electrons emitted from a heated filament (electron ionization, EI). This is relatively energetic, and considerable fragmentation usually occurs. Since ionization occurs in the vapor phase, EI MS is easily interfaced to gas chromatography, and a combined GC/MS instrument provides detailed structural information on each component of a mixture as it elutes from the chromatograph. Most commercial GC/MS instruments are suitable for fatty acid analysis, whether of magnetic sector, quadrupole, ion trap or time of flight design. The chromatographic columns, injection methods and oven conditions needed are similar to those described above, though some differences in retention behavior and peak shape may be noticed as a result of interfacing to the mass spectrometer.

The mass spectra of methyl esters of saturated fatty acids have easily recognizable molecular ions (typically about 20% of the intensity of the base peak) which define the number of carbon atoms. A series of weaker, regularly spaced fragments 14 m/z units apart of composition $C_nH_{2n-1}O_2$, (at m/z 73, 87, 101 etc), originate from cleavage of the carbon skeleton and, together with the well-known McLafferty rearrangement product at m/z 74 (McLafferty and Tureček, 1993) afford information on the chain structure. Elimination of H_2CH_3 and CH_2CH_2CH from the carboxyl terminus results in fragments at m/z M-29 and M-31. The location of alkyl branches within the chain can usually be inferred from the fragmentation pattern, because cleavage adjacent to the branch will produce a stable secondary carbocation, though information on branch positions is more reliably obtained from the spectra of picolinyl esters or 2-alkenyl-4,4-dimethyloxazoline (DMOX) derivatives. Terminal methyl groups are not easily recognized from the spectra of methyl esters. The spectra of iso-branched and straight chain methyl esters are virtually indistinguishable, but the former can be recognized by their shorter retention times. Anteiso-branched FAME can be identified because the fragment at M-29 is stronger than M-31, whereas the converse is true for iso and normal compounds. However, picolinyl or DMOX derivatives provide a better method for unambiguous branch location.

Unsaturation and the presence of carbocyclic rings decrease the mass of the molecular ion by two m/z units per ring or bond compared to the corresponding saturated acid. The overall appearance of the spectra of ω-cyclohexane ring-containing FAME, such as those present in *Alicyclobacillus* is similar to that of saturated compounds, except for the presence of an ion at m/z 83, which originates from cleavage of the cyclohexane ring. The size and locations of carbocyclic rings are best confirmed from the spectra of picolinyl or DMOX derivatives.

The fragmentation of unsaturated methyl esters is usually more extensive than that of the corresponding saturated compound, but it is not normally possible to deduce the locations of unsaturation because extensive double bond migration occurs under EI conditions. A number of derivatives are available that permit unambiguous localization of double bond positions by MS. The main strategies are either to derivatize the double bonds themselves, so that the resulting spectra contain fragments diagnostic of their positions, or to modify the carboxyl group so as to stabilize the positive charge and prevent bond migration.

bis-Methylthio (dimethyl disulfide) formation represents an example of the former approach (Francis, 1981). These derivatives are prepared from FAME by a simple one-step reaction, have good chromatographic properties and easily interpretable mass spectra. For example, GC/MS analysis of the FAME of a major component with an ECL of 17.83 from *Albidovulum inexpectatum* was consistent with an octadecenoic acid. Synthesis of its corresponding *bis*(methylthio) derivative confirmed this ($M^{\bullet +}$ at m/z 390) and fragment ions at m/z 145 and m/z 245, resulting from cleavage at the original site of unsaturation, enabled localization of the double bond to the $\triangle 11$ position (Albuquerque et al., 2002). A further advantage is that methylthio addition is stereospecific, so chromatographically resolvable erythro and threo products result from E and Z isomers of fatty acids, respectively.

The alternative carboxyl derivatization strategy enables location of features such as alkyl branches and rings, as well as double bonds in the same experiment; however preparation of the derivatives is a little more complex because FAMEs have to be de-esterified first.

Picolinyl esters are easy to prepare (Christie and Stephanov, 1987; Harvey, 1992) and their mass spectra are extremely simple to interpret. However, they elute at substantially higher temperatures than methyl esters and, because of the ring nitrogen, are prone to peak tailing, which degrades chromatographic resolution resulting in peak overlap which may complicate spectral interpretation. For these reasons they have not been extensively used in studies of extremophile lipids.

2-Alkenyl-4,4-dimethyloxazoline (DMOX) derivatives yield informative mass spectra from which the locations of features such as rings, branches and olefinic bonds are readily deduced (Zhang et al., 1988), since the charge stabilization conferred by the heterocyclic ring largely suppresses double bond migration.

These derivatives have been used to establish the structure of a novel FAME from *Albidovulum inexpectatum* which had an ECL 18.08 and a molecular ion at m/z 310. Its DMOX derivative had a molecular ion at m/z 349 and a spectrum characterized by a regularly spaced even mass ion series (m/z 126 + (14)$_n$) originating by cleavage, without rearrangement, at every carbon atom of the alkyl chain. This regular 14 m/z spacing was interrupted between m/z 264 and 224, suggesting that the compound was 11-methyl-11,12-octadecenoic acid (Albuquerque et al., 2002).

Identification of hydroxylated fatty acids

3-Hydroxy acids are present in some *Thermus* species, whereas 2-hydroxy acids are characteristic of *Meiothermus*, though they may be accompanied by 3-hydroxy acids in some cases. Whenever hydroxy acids are present, they N-acylate the hexosamine residue in position two of the sugar chain, and are never glycerol-linked (Carreto *et al.*, 1996; Ferreira *et al.*, 1999; Wait *et al.*, 1997).

The presence of a hydroxyl group provides an additional site for ionization, so the mass spectra of hydroxylated fatty acids usually contain fragments diagnostic of the location of the substituent. Methyl esters of 2-hydroxy acids undergo cleavage between carbons 1 and 2, resulting in an intense ion at m/z M-59. If the hydroxyl group is on carbon 3 a molecular ion is not usually detectable in EI spectra, and the major process observed is cleavage on the alkyl side of the hydroxyl substituted carbon giving an ion at m/z 103, normally the base peak in the spectrum. O-trimethylsilyl (O-TMS) derivatives are useful for the characterization of hydroxylated FAME, since they usually exhibit improved chromatographic behavior, with sharper and more symmetrical peaks. Since the derivatives exhibit increased retention times on non-polar columns, hydroxylated acids can be recognized even without mass spectrometric analysis. Molecular ions are usually weak or absent in the spectra of O-trimethylsilyl ethers, but the molecular mass may usually be deduced from a peak at m/z M-15.

The mass spectrum of the O-TMS ether of the major hydroxylated acid from *Thermus aquaticus*, for example, had a prominent M-15 peak at m/z 357 and an abundant α-cleavage fragment at m/z 175 (the silylated form of the m/z 103 fragment in underivatized hydroxyl FAME) consistent with a 3-hydroxylated 17-carbon compound. It had a retention time shorter than authentic methyl 3-OH heptadecanoate, but identical to 3-hydroxy-15-methylhexadecanoate.

The major hydroxyl acid in *M. ruber* had a molecular ion at m/z 300 (underivatized) which shifted to m/z 372 after trimethylsilylation. The loss of 59 m/z units from the molecular ion (representing a facile cleavage between C-1 and C-2, with charge retention on the substituent-bearing fragment) identified it as a 2-hydroxy 17:0 FAME. Its retention behavior was consistent with an iso-branched alkyl chain, and this was supported by the observation of traces of a later-eluting 2-hydroxy acid in the position expected for the anteiso-branched isomer.

Preparation of *bis*(methylthio) derivatives

- After GC and GC/MS analysis of fatty acid methyl esters, dry down the residual sample in a round bottomed tube with a PTFE-faced screw cap and redissolve in 100 µl of hexane.
- Add 100 µl of dimethydisulfide and two drops of a 6% (w/v) solution of iodine in diethyl ether.
- Allow to react overnight at room temperature.
- Shake with 0.5 ml 5% aqueous sodium thiosulfate to remove iodine and recover the derivatives by extracting (twice) with an equal volume of hexane.

Preparation of 2-alkenyl-4,4-dimethyloxazoline (DMOX) derivatives

- Demethylate the fatty acid methyl esters by overnight treatment with 0.5 ml of 1 M NaOH in 50% aqueous methanol at 60°C.
- Cool, reduce the pH to below 2 by addition of 0.5 ml of 1 M HCl, and recover the free fatty acids by extraction with hexane:chloroform (4:1 by vol.).
- Dry the extracts, redissolve in 50 µl chloroform, mix with 100 µl of 2-methyl-2-amino propanol and heat for 3 h at 285°C in a sealed tube.
- Cool to room temperature, add 2 volumes of chloroform, and wash with 1 ml of distilled water, made alkaline with a few drops of 1 M NaOH solution.
- Remove the water layer and wash the chloroform solution of DMOX derivatives twice more with 1 ml of water.
- Remove the chloroform layer to a clean tube and dry with a vacuum centrifuge.

Preparation of O-trimethylsilyl ethers

- Dry the samples in a suitable reaction tube using a vacuum centrifuge.
- Add 100 µl of bis(trimethylsilyl)trifluoroacetamide and allow to react for 30 min at 60°C (heating block or GC oven).
- Remove the reagent with a vacuum centrifuge and dissolve the derivatives in trimethylpentane or hexane for GC/MS.

Mass spectrometry of long chain diols

Moderate amounts of two late eluting components with equivalent chain lengths (ECL) of 19.060 (major peak) and 19.160 (minor peak), were detected in FAME extracts of some *Thermus* strains (Carreto et al., 1996). On trimethysilylation, the retention times of these components increased to ECLs 19.832 and 19.925, respectively, suggesting the presence of hydroxyl groups.

Abundant protonated molecules were observed at m/z 431 in the isobutane chemical ionization spectra of the TMS derivatives of both compounds. An abundant fragment at m/z 327 was assigned as an α-cleavage ion which would be consistent with the presence of an O-trimethylsilylated alkyl chain of at least 17 carbon atoms (CH_3-$(CH_2)_{15}$-$CHOSi(CH_3)_3 = 327$).

The spectra did not resemble those of hydroxylated fatty acids, but were consistent with octadecanediols, the masses of which would be 286 underivatized, or 430 after trimethylsilylation. Detailed characterization of the structure of the alkyl chains was achieved by mass spectrometry of their acetate derivatives.

Molecular ions are not observed, the highest mass signal observed being m/z 310, attributable to elimination of acetic acid. The structure of the alkyl chain is defined by an important group of ions at odd mass number (m/z 295, 281, 267 and 253) which originate by cleavage, without rearrangement, of the terminal region of the alkyl backbone. The absence of the fragment at m/z 281, which can only be formed by cleavage of two bonds, is diagnostic of an iso branch. Conversely, the fragment at m/z 267 is absent from anteiso-branched octadecanediols.

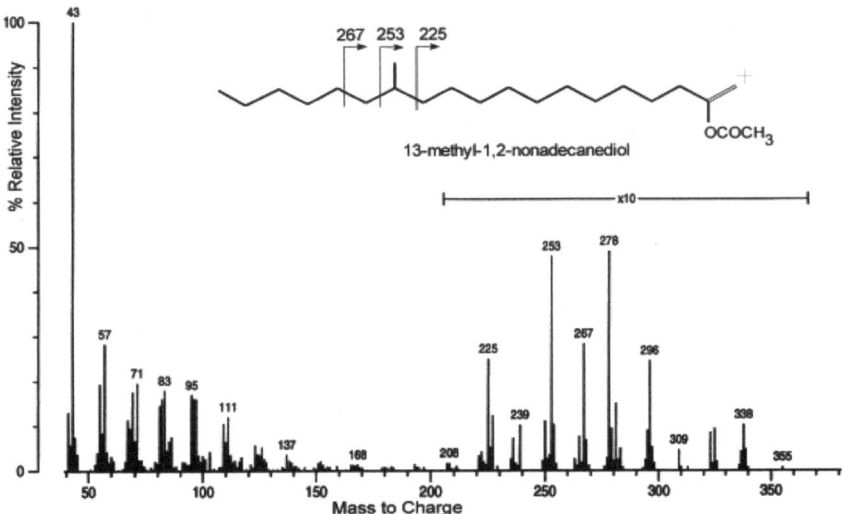

Figure 6.8. Electron ionization mass spectrum of acetylated 13-methyl-1,2-nonadecanediol from *Thermomicrobium roseum*. The fragmentation diagram shows charge-remote fragmentation of the M-60 ion. The position of the methyl branch is indicated by a pair of even-electron fragments at m/z 253 and 267.

Deuteroacetylation provides useful confirmation since the ions at m/z 310, 295, 281, 267 and 253 will be shifted by 3 m/z units, showing they contained a single acetate group. The fragment at m/z 284 will be absent from the spectra of the iso-branched octadecanediols, and that at m/z 270 from the anteiso isomer. Fragments not containing acetate groups, such as those at m/z 268, 250 and 222 will be unaffected by deuteroacetylation.

The long chain diols from *Thermomicrobium roseum* lipids exhibit a greater degree of structural diversity, with both internally branched and unbranched examples. Most of the high mass ions in the spectra of acetate derivatives of the straight chain compounds are odd electron products, attributable to the loss of one or more of the acetate groups (either as acetic acid (M-60) or as ketene (M-42). The main exception is an even electron fragment at M-73, which originates by cleavage between C-1 and C-2. The spectra of the internally branched diacetates are similar, except for the presence of additional even-electron fragment ions, probably originating by a charge-site remote mechanism, which are diagnostic of methyl branch location. In 13-methyl compounds these were observed at m/z 253 and 267 (Figure 6.8). In the acetate derivative of 15-methyl-1,2-henicosanediol by contrast, the analogous ions are shifted to m/z 281 and 295.

Base catalyzed acetylation of long chain diols

- Dry 1 ml of pyridine (Aldrich, HPLC grade) by addition of a spatula full of P_2O_5, vortex and let stand for an hour, then pellet the P_2O_5 with a bench-top centrifuge and aspirate off the pyridine in a fume hood.
- Dissolve the sample (5–50 µg) in 100 µl of the dried pyridine; add 100 µl of acetic anhydride and vortex mix.

- Leave at room temperature for 24 h then, remove the reagents by vacuum centrifugation.
- Dissolve the derivatives in chloroform; and desalt by washing three times with an equal volume of water; discard the washes.

Deuteroacetates are prepared identically except for the use of d_6-acetic anhydride in place of acetic anhydride.

◆◆◆◆◆ ISOPRENOID QUINONES

Isoprenoid quinones are constituents of bacterial cell membranes found in the vast majority of the organisms examined, where they play important functions in electron transport. Ubiquinones (2,3-dimethoxy-5-methyl-1,4-benzoquinone with a polyprenyl side chain of varying length, abbreviated U-n, where n denotes the number of isoprenyl units) are widely distributed in bacteria and eukaryotes, while the chemically related plastoquinones have been identified in plants, algae and cyanobacteria. The menaquinones (2-methyl-3-phytyl-1,4-naphthoquinone, with a polyprenyl side chain of varying length, abbreviated MK-n) are also widely distributed in bacteria. Saturation or hydrogenation of polyprenyl side chain has been reported in corynebacteria, mycobacteria and actinomycetes. Demethylmenaquinones are not uncommon in some bacteria. Other rare quinones have been encountered in some bacteria and care should be taken to account for these (Collins and Jones, 1981; Tindall, 1989).

The respiratory quinones have great taxonomic significance because the type of quinone present in an organism and the isoprenoid chain length reflects the phylogenetic affiliation of the bacteria. For example, the major respiratory quinone of the bacteria of the phylum *Deinococcus/Thermus* is MK-8, although minor amount of MK-7 and MK-9 may also be detected. The vast majority of, if not, all known proteobacteria have ubiquinones with variable isoprene chain lengths. The species of the phylum *Bacterioidetes* (*Cytophaga/Flavobacterium/Bacteroides*) possess menaquinones, generally MK-7 (Collins and Jones, 1981).

Most organisms possess mixtures of either MK or U with different polyprenyl lengths, where one is generally the major respiratory quinone. For example, the species of the genus *Legionella* possess ubiquinones with polyprenyl chains which, depending on the species, range between U-10 and U-14 one of which is the major quinone of the species; however, minor components, namely U-7, U-8, U-9 and U-15 have also been identified in some species (Wait, 1988). There are also cases where two quinones (U or MK) are present in about the same concentration; in *Microcella putealis* MK-12 and MK-13 are present in about the same proportion, there being also very low levels of MK-11 (Tiago et al., 2005).

Isoprenoid quinones are rather delicate cell components that are easily degraded. Therefore, all extraction steps must be performed in subdued light and preferably in brown glass tubes to prevent photo-oxidation; the samples should always be flushed with N_2 before storage at $-20°C$. It is also important to note that all glassware and Teflon-coated screw-capped tubes should be washed thoroughly with hexane to remove hydrophobic impurities before extraction of respiratory lipoquinones.

Growth of Organisms and Extraction of Quinones

- Organisms are generally grown on solid media, when possible. Normally 100 mg of cell dry weight are sufficient for the analysis of lipoquinone composition, but it may be necessary to obtain larger quantities of cells when respiratory quinones are present in very small or vestigial quantities.
- Wet cell paste is resuspended in about 1.0 ml of ultrapure water in a brown glass tube with a Teflon-lined screw cap and freeze-dried. The cells can be stored under N_2 atmosphere at $-20°C$.
- Add 3.0 ml of hexane:methanol (1:2 by vol.) to the lyophilized cells, place a small magnetic bar and flush with N_2. The suspension is stirred with a magnetic stirrer for about 30 min. The samples are placed on ice for 30 min until phase separation takes place.
- Ice cold hexane (3.0 ml) is added to the samples followed by 2.0 ml of 0.3% NaCl under a N_2 atmosphere.
- Centrifuge at $2000 \times g$ for 5 min in a swinging-bucket centrifuge at room temperature for phase separation.
- Remove the upper phase with a Pasteur pipette into a 5 ml brown glass with Teflon-coated screw cap.
- Evaporate until dryness under a N_2 atmosphere and store, if necessary, at $-20°C$.

Separation and Partial Purification of Quinones

- Prepare a solvent composed of hexane:diethyl ether (85:15 by vol.).
- Add the solvent to a chromatography tank with one side lined with filter paper and wait about 60 min for good vapor saturation of the atmosphere. The solvent level in the tank should be lower than the location of application (Origin) of the samples.
- Resuspend the sample in about 0.4 ml diethyl ether. Apply all the extract along a thin line about 1.5 cm from the lower edge of a 10 × 10 cm plastic sheet coated with Silica Gel 60F$_{254}$ (Merck, 5735) under subdued light. Two separate samples can be applied on the same TLC plate. Note that silica gel sheets are generally available in sizes of 20 × 20 cm or 20 × 10 cm and should be cut with scissors to the appropriate dimensions.
- Place the TLC plate in the tank and allow the solvent to ascend to the upper edge of the plate. Remove the silica gel-coated plate and allow the solvent to evaporate. This last step can be speeded up with cold air from a hair dryer.
- Observe the plate under UV light (254 nm) for only a few seconds and mark the contours of the dark bands with a pencil. Protect hands and face from the UV radiation. The menaquinones migrate further from the origin (R_F of about 0.7) while the ubiquinones migrate nearer to the origin (R_F of about 0.4). This step leads to the presumptive identification of the respiratory quinones as ubiquinones or menaquinones and their partial purification.

- Scrape the silica that contains the band with the quinone with a spatula onto non-absorbent paper. Place a piece of paper above the silica and press with finger tips to grind it into a fine powder. The silica powder is poured in a Pasteur pipette that has been previously plugged with hexane-washed glass wool so that this plug retains most of the silica powder.
- Elute the samples twice from the silica with about 0.5 ml hexane:methanol (1:2 by vol.) into a 1.5 ml brown glass Teflon-coated screw cap tube and place it on ice. Add 0.3 ml of cold hexane and four drops of 0.3% NaCl. Mix thoroughly, allow the phases to separate and transfer the upper phase into a new 1.5 ml tube. Care must be taken not to remove the interface or the lower phase of the sample. The purified sample can be stored at $-20°C$ under a N_2 atmosphere.
- The TLC-purified sample is filtered through a hydrophobic membrane filter (0.2 µm pore size, 0.3 mm diameter) to remove silica particles that could block the HPLC column. Evaporate under N_2 until dry.
- Dissolve the sample in about 0.2 ml of methanol:heptane (10:2 by vol.).

High Performance Liquid Chromatography (HPLC)

- The sample is injected into a 20 µl capacity loop using an appropriate syringe. The rest of the quinone extract can be stored at $-20°C$ under a N_2 atmosphere, or it can be evaporated to dryness and dissolved in a small volume to inject a more concentrated extract, if necessary.
- The quinones are separated on a reverse phase ODS 2 (25 cm by 4 mm internal diameter) column equilibrated with degassed methanol:heptane (10:2 by vol.). Degassing of the elutant is achieved by applying vacuum for about 30 min. The samples are eluted with this solvent system at a rate of 2.0 ml min^{-1} at 37°C and detected at 269 nm.

Other solvent systems for elution of lipoquinones are also used. One commonly used elution system is composed of methanol:1-chlorobutane (80:20 by vol.) or (70:30 by vol.), but methanol:heptane (10:2 by vol.) is less corrosive for the HPLC column and tubing and has the same capacity to separate quinones (Wait, 1988).

The affinity of the quinones for the columns is primarily determined by the nature of the isoprene unit at carbon 6, which is non-polar and strongly adsorbed by the packing of the column, so that longer chains tend to be more tightly bound. The isoprene quinones will therefore be eluted from the column in order of the chain length, with shorter chain quinones elution before those with longer chains.

Isoprenoid quinones should be tentatively identified by comparison of their retention times with those of external standards eluted before or after the sample. Some commercial ubiquinones are available, but it is preferable to grow bacteria with known respiratory quinones and use these for comparison with the retention time to identify those from new strains. Short chain quinones are rarer (less than 6 isoprenyl units) than those with longer chains, but some known bacteria have short chain quinones (menaquinones and ubiquinones) that can be used as standards.

We generally store quinone samples for comparison at −20°C under N_2. In some cases, we also mix bacterial samples of known quinone composition with those that are being identified to make sure that the peaks have identical retention times. If the quinones cannot be identified by comparison with known ones then mass spectroscopy must be used for their identification (Wait, 1988). Respiratory quinones may, as stated above, be present in very small amounts and the cell mass used to extract them may have to be increased substantially, as was recently found with a new actinobacterium named *Microcella putealis*, where 300 mg of cell dry weight had to be used to identify vestigial amounts of menaquinones (Tiago *et al.*, 2005).

Identification of Isoprenoid Quinones by Mass Spectrometry

Definitive identification requires the use of physicochemical methods such as nuclear magnetic resonance or mass spectrometry. Mass spectra may be obtained from quinones purified as described above by TLC or HPLC using electron ionization in the positive ion mode and a direct insertion probe. Rapid heating of the probe is necessary to favor evaporation over thermal degradation. Electron ionization mass spectra of quinones usually display abundant molecular ions, and intense low mass ions which are diagnostic of the nature of the quinone nucleus. Ubiquinones, for example, have prominent peaks at *m/z* 197 and 235, derived from their 2,3-dimethoxy-1,4-benzoquinone group, whereas menaquinones have peaks at *m/z* 225 and 187.

Because isoprenoid quinones are thermally labile, the sensitivity obtainable using direct probe EI MS is often poor, and the experiments are technically demanding. It is preferable therefore to employ soft ionization mass spectrometric methods such as fast atom bombardment or electrospray. Fast atom bombardment in the negative mode using 3-nitrobenzyl alcohol as matrix gives rise to abundant semiquinone molecular anion radicals $[M]^{•-}$, revealing the mass of each quinone molecular species present. The relative intensities of these molecular anions reflect the concentration of the individual quinones in the mixture, and generally agree with compositions determined by HPLC (Dennis *et al.*, 1993). More detailed structural information may be obtained by collisional fragmentation and tandem mass spectrometry in the positive ion mode. The fragmentation pathways operating are generally similar to those observed in EI mass spectra.

◆◆◆◆◆ CONCLUDING REMARKS

We hope that this chapter will foster interest in primary lipid analysis using simple methods during the characterization of new organisms or for the identification of bacteria that might not be easy by other methods. The analysis of the polar lipid, fatty acid and quinone composition can provide valuable information on the composition of microorganisms and this information is relatively easy to obtain and so lipids need not be the forgotten components of bacteria.

Acknowledgements

The work in the authors' laboratory has been supported by the European Commission and the Fundação para a Ciência e a Tecnologia, PRODEP, FEDER, POCTI and POCI Programs, Portugal.

References

Adeleke, A. A., Fields, B. S., Benson, R. F., Daneshvar, M. I., Pruckler, J. M., Ratcliff, R. M., Harrison, T. G., Weyant, R. S., Birtles, R. J., Raoult, D. and Halablab, M. A. (2001). *Legionella drozanskii* sp. nov., *Legionella rowbothamii* sp. nov. and *Legionella fallonii* sp. nov.: three unusual new *Legionella* species. *Int. J. Syst. Evol. Microbiol.* **51**, 1151–1160.

Albuquerque, L., Rainey, F. A., Chung, A. P., Sunna, A., Nobre, M. F., Grote, R., Antranikian, G. and da Costa, M. S. (2000). *Alicyclobacillus hesperidum* sp. nov. and a related genomic species from solfataric soils of Sao Miguel in the Azores. *Int. J. Syst. Evol. Microbiol.* **50**, 451–457.

Albuquerque, L., Santos, J., Travassos, P., Nobre, M. F., Rainey, F. A., Wait, R., Empadinhas, N., Silva, M. T. and da Costa, M. S. (2002). *Albidovulum inexpectatum* gen. nov., sp. nov., a nonphotosynthetic and slightly thermophilic bacterium from a marine hot spring that is very closely related to members of the photosynthetic genus *Rhodovulum*. *Appl. Environ. Microbiol.* **68**, 4266–4273.

Albuquerque, L., Simões, C., Nobre, M. F., Pino, N. M., Battista, J. R., Silva, M. T., Rainey, F. A. and da Costa, M. S. (2005). *Truepera radiovictrix* gen. nov., sp. nov., a new radiation resistant species and the proposal of *Trueperaceae* fam. nov. *FEMS Microbiol. Lett.* **247**, 161–169.

Anderson, R. and Hansen, K. (1985). Structure of a novel phosphoglycolipid from *Deinococcus radiodurans*. *J. Biol. Chem.* **260**, 12219–12223.

Antón, J., Oren, A., Benlloch, S., Rodríguez-Valera, F., Amann, R. and Rosselló-Mora, R. (2002). *Salinibacter ruber* gen. nov., sp. nov., a novel, extremely halophilic member of the *Bacteria* from saltern crystallizer ponds. *Int. J. Syst. Evol. Microbiol.* **52**, 485–491.

Balkwill, D. L., Kieft, T. L., Tsukuda, T., Kostandarithes, H. M., Onstott, T. C., Macnaughton, S., Bownas, J. and Fredrickson, J. K. (2004). Identification of iron-reducing *Thermus* strains as *Thermus scotoductus*. *Extremophiles* **8**, 37–44.

Battista, J. R. and Rainey, F. A. (2001). The Genus *Deinococcus*. In *Bergey's Manual of Systematic Bacteriology* (D. R. Boone and R. W. Castenholz, eds), 2nd edn., vol. 1, pp. 396–403. Springer-Verlag, New York.

Behrendt, U., Ulrich, A., Schumann, P., Naumann, D. and Suzuki, K. (2002). Diversity of grass-associated *Microbacteriaceae* isolated from the phyllosphere and litter layer after mulching the sward; polyphasic characterization of *Subtercola pratensis* sp. nov., *Curtobacterium herbarum* sp. nov., and

Plantibacter flavus gen. nov., sp. nov. *Int. J. Syst. Evol. Microbiol.* **52**, 1441–1454.

Bligh, E. G. and Dyer, W. J. (1959). A rapid method of total lipid extraction and purification. *Can. J. Biochm. Physiol.* **37**, 911–917.

Carballeira, N. M., Reyes, M., Sostre, A., Huang, H., Verhagen, M. F. J. M. and Adams, M. W. W. (1997). Unusual fatty acid compositions of the hyperthermophilic archaeon *Pyrococcus furiosus* and the bacterium *Thermotoga maritima*. *J. Bacteriol.* **179**, 2766–2768.

Carreto, L., Wait, R., Nobre, M. F. and da Costa, M. S. (1996). Determination of the structure of a novel glycolipid from *Thermus aquaticus* and demonstration that hydroxy fatty acids are amide linked to glycolipids in *Thermus* spp. *J. Bacteriol.* **178**, 6479–6486.

Christie, W. W. (1982). *Lipid Analysis.* 2nd edn., Pergamon Press, Oxford.

Christie, W. W. and Stephanov, K. (1987). Separation of picolinyl derivatives by high performance liquid chromatography for identification by mass spectrometry. *J. Chromatogr.* **392**, 259–265.

Chung, A. P., Rainey, F. A., Valente, M., Nobre, M. F. and da Costa, M. S. (2000). *Thermus igniterrae* sp. nov. and *Thermus antranikianii* sp. nov., two new species from Iceland. *Int. J. Syst. Evol. Microbiol.* **50**, 209–217.

Collins, M. and Jones, D. (1981). Distribution of isoprenoid quinone structural types in bacteria and their taxonomic implications. *Microbiol. Rev.* **45**, 316–354.

Corcelli, A., Lattanzio, V. M. T., Mascolo, G., Babudri, F., Oren, A. and Kates, M. (2004). Novel sulfonolipid in the extremely halophilic bacterium *Salinibacter ruber*. *Appl. Environ. Microbiol.* **70**, 6678–6685.

da Costa, M. S. and Rainey, F. A. (2001). The Family *Thermaceae*. In *Bergey's Manual of Systematic Bacteriology* (D. R. Boone and R. W. Castenholz, eds), 2nd edn., vol. 1, pp. 403–404. Springer-Verlag, New York.

da Costa, M. S., Nobre, M. F. and Rainey, F. A. (2001). The Genus *Thermus*. In *Bergey's Manual of Systematic Bacteriology* (D. R. Boone and R. W. Castenholz, eds), 2nd edn., vol. 1, pp. 404–414. Springer-Verlag, New York.

Dennis, P. J., Brenner, D. J., Thacker, W. L., Wait, R., Vesey, G., Steigerwalt, A. G. and Benson, R. F. (1993). Five new *Legionella* species isolated from water. *Int. J. Syst. Microbiol.* **43**, 329–337.

de Rosa, M., Gambacorta, A., Huber, R., Lanzotti, V., Nicolaus, B., Stetter, K. O. and Trincone, A. (1988). A new 15,16-dimethyl-30-glyceryl-oxytriacontanoic acid from lipids of *Thermotoga maritima*. *J. Chem. Soc. Chem. Commun.* **19**, 1300–1301.

de Rosa, M., Gambacorta, A., Huber, R., Lanzotti, V., Nicolaus, B., Stetter, K. O. and Trincone, A. (1989). Lipid Structures in *Thermotoga maritima*. In *Microbiology of Extreme Environments and its Potential for Biotechnology* (M. S. da Costa, J. C. Duarte and R. A. Williams, eds), pp. 167–173. Elsevier, London.

Dell, A. (1990). Preparation and desorption mass spectrometry of permethyl and peracetyl derivatives of oligosaccharides. *Meth. Enzymol.* **193**, 647–660.

Diogo, A., Veríssimo, A., Nobre, M. F. and da Costa, M. S. (1999). Usefulness of fatty acid composition for differentiation of *Legionella* species. *J. Clin. Microbiol.* **37**, 2248–2254.

Domon, B. and Costello, C. E. (1988). A systematic nomenclature for carbohydrate fragmentations in FAB MS/MS spectra of glycoconjugates. *Glycoconjugate J.* **5**, 397–409.

Ferraz, A. S., Carreto, L., Tenreiro, S., Nobre, M. F. and da Costa, M. S. (1994). Polar lipids and fatty acid composition of *Thermus* strains from New Zealand. *Antonie van Leeuwenhoek* **66**, 357–363.

Ferreira, A. M., Wait, R., Nobre, M. F. and da Costa, M. S. (1999). Characterization of glycolipids from *Meiothermus* spp. *Microbiology UK* **145**, 1191–1199.

Francis, G. W. (1981). Alkylthiolation for the determination of double bond positions in unsaturated fatty acid esters. *Chem. Phys. Lipids* **29**, 369–374.

Godchaux, W. and Leadbetter, E. R. (1983). Unusual sulfonolipids are characteristic of the *Cytophaga-Flexibacter* group. *J. Bacteriol.* **153**, 1238–1246.

Goto, K., Mochida, K., Asahara, M., Suzuki, M., Kasai, H. and Yokota, A. (2003). *Alicyclobacillus pomorum* sp. nov., a novel thermo-acidophilic, endospore-forming bacterium that does not possess ω-alicyclic fatty acids, and emended description of the genus *Alicyclobacillus*. *Int. J. Syst. Evol. Microbiol.* **53**, 1537–1544.

Harvey, D. (1992). Mass spectrometry of picolinyl and other nitrogen-containing derivatives of lipids. In *Advances in Lipid Methodology - 1* (W. W. Christie, ed.), pp. 19–80. The Oily Press, Ayr.

Hensel, R., Demharter, W., Kandler, O., Kroppenstedt, R. M. and Stackebrandt, E. (1986). Chemotaxonomic and molecular-genetic studies of the genus *Thermus*: evidence for a phylogenetic relationship of *Thermus aquaticus* and *Thermus ruber* to the genus *Deinococcus*. *Int. J. Syst. Bacteriol.* **36**, 444–453.

Huang, Y. and Anderson, R. (1989). Structure of a novel glucosamine-containing phosphoglycolipid from *Deinococcus radiodurans*. *J. Biol. Chem.* **264**, 18667–18672.

Huber, R., Wilharm, T., Huber, D., Trincone, A., Burggraf, S., König, H., Rachel, R., Rockinger, I., Fricke, H. and Stetter, K. O. (1992). *Aquifex pyrophilus* gen. nov., sp. nov., represents a novel group of marine hyperthermophilic hydrogen-oxidizing bacteria. *Syst. Appl. Microbiol.* **15**, 340–351.

Huber, R., Rossnagel, P., Woese, C. R., Rachel, R., Langworthy, T. A. and Stetter, K. O. (1996). Formation of ammonium from nitrate during chemolithoautotrophic growth of the extremely thermophilic bacterium *Ammonifex degensii* gen. nov. sp. nov. *Syst. Appl. Microbiol.* **19**, 40–49.

Jahnke, L. L., Eder, W., Huber, R., Hope, J. M., Hinrichs, K.-U., Hayes, J. M., Des Marais, D. J., Cady, S. L. and Summons, R. E. (2001). Signature lipids and stable carbon isotope analyses of Octopus Spring hyperthermophilic communities compared with those of *Aquificales* representatives. *Appl. Environ. Microbiol.* **67**, 5179–5189.

Kates, M. (1993). Membrane lipids of Archaea. In *The Biochemistry of Archaea (Archaebacteria)* (M. Kates, D. J. Kushner and A. T. Matheson, eds), pp. 261–295. Elsevier, Amsterdam.

Klein, R. A., Hazelwood, G. P., Kemp, P. and Dawson, R. M. C. (1979). A new series of long-chain dicarboxylic acids with vicinal dimethyl branching found as major components of the lipids of *Butyrivibrio* spp. *Biochemistry* **183**, 691–700.

Kusano, K., Yamada, H., Niwa, M. and Yamasato, K. (1997). *Propionibacterium cyclohexanicum* sp. nov., a new acid-tolerant ω-cyclohexyl fatty acid-containing propionibacterium isolated from spoiled orange juice. *Int. J. Syst. Bacteriol.* **47**, 825–831.

Kuykendall, L. D., Roy, M. A., O'Neill, J. J. and Devine, T. E. (1988). Fatty acids, antibiotic resistance, and deoxyribonucleic acid homology groups of *Bradyrhizobium japonicum*. *Int. J. Syst. Bacteriol.* **38**, 358–361.

La Scola, B., Birtles, R. J., Greub, G., Harrison, T. J., Ratcliff, R. M. and Raoult, D. (2004). *Legionella drancourtii* sp. nov., a strictly intracellular amoebal pathogen. *Int. J. Syst. Evol. Microbiol.* **54**, 699–703.

Langworthy, T. A. and Pond, J. L. (1986). Membranes and lipids of thermophiles. In *Thermophiles: General, Molecular and Applied Microbiology* (T. D. Brock, ed.), pp. 107–135. John Wiley and Sons, New York.

Langworthy, T. A., Holzer, G., Zeikus, J. G. and Tornabene, T. G. (1983). Iso- and anteiso-branched glycerol diethers of the thermophilic anaerobe *Thermodesulfobacterium commune*. *Syst. Appl. Microbiol.* **4**, 1–17.

Lo Presti, F., Riffard, S., Meugnier, H., Reyrolle, M., Lasne, Y., Grimont, P. A. D., Grimont, F., Vandenesch, F., Etienne, J., Fleurette, J. and Freney, J. (1999). *Legionella taurinensis* sp. nov., a new species antigenically similar to *Legionella spiritensis*. *Int. J. Syst. Bacteriol.* **49**, 397–403.

Lu, T. L., Chen, C. S., Yang, F. L., Fung, J. M., Chen, M. Y., Tsay, S. S., Li, J., Zou, W. and Wu, S. H. (2004). Structure of a major glycolipid from *Thermus oshimai* NTU-063. *Carbohyd. Res.* **339**, 2593–2598.

McLafferty, F. W. and Tureček, F. (1993). *Interpretation of Mass Spectra*. University Science Books, Mill Valley.

Miroshnichenko, M. L., L'Haridon, S., Jeanthon, C., Antipov, A. N., Kostrikina, N. A., Tindall, B. J., Schumann, P., Spring, S., Stackebrandt, E. and Bonch-Osmolovskaya, E. A. (2003a). *Oceanithermus profundus* gen. nov., sp. nov., a thermophilic, microaerophilic, facultatively chemolithoheterotrophic bacterium from a deep-sea hydrothermal vent. *Int. J. Syst. Evol. Microbiol.* **53**, 747–752.

Miroshnichenko, M. L., L'Haridon, S., Nercessian, O. C., Antipov, A. N., Kostrikina, N. A., Tindall, B. J., Schumann, P., Spring, S., Stackebrandt, E., Bonch-Osmolovskaya, E. A. and Jeanthon, C. (2003b). *Vulcanithermus mediatlanticus* gen. nov., sp. nov., a novel member of the family *Thermaceae* from a deep-sea hot vent. *Int. J. Syst. Evol. Microbiol.* **53**, 1143–1148.

Moore, B. S., Walker, K., Tornus, I., Handa, S., Poralla, K. and Floss, H. G. (1997). Biosynthetic studies of ω-cycloheptyl fatty acids in *Alicyclobacillus cycloheptanicus*. Formation of cycloheptanecarboxylic acid from phenylacetic acid. *J. Org. Chem.* **62**, 2173–2185.

Mori, K., Kakegawa, T., Higashi, Y., Nakamura, K., Maruyama, A. and Hanada, S. (2004). *Oceanithermus desulfurans* sp. nov., a novel thermophilic, sulfur-reducing bacterium isolated from a sulfide chimney in Suiyo Seamount. *Int. J. Syst. Evol. Microbiol.* **54**, 1561–1566.

Nobre, M. F. and da Costa, M. S. (2001). The Genus *Meiothermus*. In *Bergey's Manual of Systematic Bacteriology* (D. R. Boone and R. W. Castenholz, eds), 2nd edn., vol. 1, pp. 414–420. Springer-Verlag, New York.

Nobre, M. F., Carreto, L., Wait, R., Tenreiro, S., Fernandes, O., Sharp, R. J. and da Costa, M. S. (1996). Fatty acid composition of the species of the genera *Thermus* and *Meiothermus*. *Syst. Appl. Microbiol.* **19**, 303–311.

Pires, A. L., Albuquerque, L., Tiago, I., Nobre, M. F., Empadinhas, N., Veríssimo, A. and da Costa, M. S. (2005). *Meiothermus timidus* sp. nov., a new slightly thermophilic yellow-pigmented species. *FEMS Microbiol. Lett.* **245**, 39–45.

Pond, J. P. and Langworthy, T. A. (1987). Effect of growth temperature on the long-chain diols and fatty acids of *Thermomicrobium roseum*. *J. Bacteriol.* **169**, 1328–1330.

Pond, J. P., Langworthy, T. A. and Holzer, G. (1986). Long-chain diols: a new class of membrane lipids from a thermophilic bacterium. *Science* **231**, 1134–1136.

Rainey, F. A., Silva, J., Nobre, M. F., Silva, M. T. and da Costa, M. S. (2003). *Porphyrobacter cryptus* sp. nov., a novel slightly thermophilic, aerobic, bacteriochlorophyll *a*-containing species. *Int. J. Syst. Evol. Microbiol.* **53**, 35–41.

Rainey, F. A., Ray, K., Ferreira, M., Gatz, B. Z., Nobre, M. F., Bagaley, D., Rash, B. A., Park, M.-J., Earl, A. M., Shank, N. C., Small, A. M., Henk, M. C., Battista, J. R., Kämpfer, P. and da Costa, M. S. (2005). Extensive diversity of ionizing radiation-resistant bacteria recovered from a Sonoran Desert soil sample and the description of 9 new species of the genus *Deinococcus* from a single soil sample. *Appl. Environ. Microbiol.* **71**, 5225–5235.

Ratledge, C. and Wilkinson, S. G. (eds), (1989). *Microbial Lipids*, vol. I and II, Academic Press, London.

Sako, Y., Nakagawa, S., Takai, K. and Horikoshi, K. (2003). *Marinithermus hydrothermalis* gen. nov., sp. nov., a strictly aerobic, thermophilic bacterium from a deep-sea hydrothermal vent chimney. *Int. J. Syst. Evol. Microbiol.* **53**, 59–65.

Suzuki, K. I., Saito, K., Kawaguchi, A., Okuda, S. and Komagata, K. (1981). Occurrence of ω-cyclohexyl fatty acids in *Curtobacterium pusillum*. *J. Gen. Appl. Microbiol.* **27**, 261–266.

Tiago, I., Pires, C., Mendes, V., Morais, P. V., da Costa, M., and Veríssimo, A., (2005). *Microcella putealis* gen. nov., sp. nov., a Gram-positive alkaliphilic bacterium isolated from a nonsaline alkaline groundwater. *Syst. Appl. Microbiol.* **28**, 479–487.

Tindall, B. J. (1989). Fully saturated menaquinones in the archaebacterium *Pyrobaculum islandicum*. *FEMS Microbiol. Lett.* **60**, 251–254.

Wait, R. (1988). Confirmation of the identity of *Legionella* by whole cell fatty-acid and isoprenoid quinone profiles. In *A Laboratory Manual for Legionella* (T. G. Harrison and A. G. Taylor, eds), pp. 69–101. John Wiley and Sons, Chichester.

Wait, R., Carreto, L., Nobre, M. F., Ferreira, A. M. and da Costa, M. S. (1997). Characterization of novel long-chain 1,2-diols in *Thermus* species and demonstration that *Thermus* strains contain both glycerol-linked and diol-linked glycolipids. *J. Bacteriol.* **179**, 6154–6162.

Zhang, J. Y., Yu, Q. T., Liu, B. N. and Huang, Z. H. (1988). Chemical modification in mass spectrometry. IV 2-alkenyl-4,4-dimethyloxazolines as derivatives for the double bond location of long-chain olefinic acids. *Biomed. Environ. Mass Spectrom.* **15**, 33–44.

Zimmerman, J. M. and Battista, J. R. (2005). A ring-like nucleoid is necessary for radioresistance in the *Deinococcaceae*. *BMC Microbiol.* **5**, 17.

Zweig, G. and Sherma, J. (eds), (1972). *Handbook of Chromatography*, vol. II, CRC Press, Cleveland.

7 Membranes of Thermophiles and Other Extremophiles

Sonja V Albers, Wil N Konings and Arnold J M Driessen

Molecular Microbiology, Groningen Biomolecular Sciences and Biotechnology Institute, University of Groningen, Haren, The Netherlands

◆◆

CONTENTS

Preface
Isolation of lipids
Liposome preparation and protein permeability measurements

◆◆◆◆◆ PREFACE

In the last decades, an increasing number of microorganisms have been described that are able to grow under extreme conditions such as low or high temperature, acidic or alkaline pH, high salinity or high pressure. These organisms, termed extremophiles, show a multitude of different adaptations to survive in their particular environments. The cytoplasmic membrane plays a crucial role in these adaptations (Albers et al., 2001; Konings et al., 2002). This membrane functions as a barrier between the cytoplasm and the extracellular environment. Membranes are impermeable for ions and most other small molecules, and due to the action of transport proteins, the cytoplasmic membrane controls the ionic composition of the cytoplasm. Moreover, the cytoplasmic membrane has to maintain a proton gradient and an electrical potential across the membrane. The energy stored in the electrochemical gradient of protons can be employed to drive energy requiring processes such as substrate transport, motility and so forth. Extreme acidophilic organisms, for instance, face an extracellular pH of 0.5–2.5 while maintaining an intracellular pH near neutrality. This implies that the membrane has to withstand a huge pH gradient, which is only possible when the endogenous proton permeability of the membrane is very low. Summarizing, the cytoplasmic membrane is important for the maintenance of optimal intracellular conditions and for efficient energy transduction.

The lipid composition of the cytoplasmic membranes of Bacteria and Archaea shows a number of marked differences. Bacterial membranes contain lipids that are composed of two fatty acyl chains that are ester-linked to glycerol. The third hydroxyl group of the glycerol is linked to hydrophilic phospho- or glyco-containing polar headgroups (Figure 7.1A). These lipids are organized in a lipid bilayer such that the polar headgroups are exposed to the water phases while the acyl chains are directed towards the hydrophobic interior of the membrane. In contrast to bacterial lipids, archaeal lipids consist of two phytanyl chains which are linked *via* an ether bond to glycerol or other alcohols like nonitol (Figure 7.1B) (for review: Sprott, 1992). Archaea that grow under moderate conditions contain bilayer-forming lipids with a C_{20} diether lipid core. These lipids form regular phospholipid bilayer membranes similar to those found in Bacteria. In extreme thermophilic and acidophilic Archaea, membrane-spanning ether lipids are found in which the phytanyl chains of two diether lipids are fused to a C_{40-44} core. These so-called *tetraether* or *bipolar* lipids form a monolayer in which the lipids span the entire membrane. Membranes composed of ether lipids are equipped with a higher stability than ester-lipid membranes. This is most likely due to a reduced segmentary motion of tertiary carbon atoms. This restriction in hydrocarbon chain mobility may also result in the reduced ion permeability of archaeal membranes.

To keep the membrane in a liquid crystalline state, thermophilic bacteria increase the chain length of the lipid acyl chain, the ratio of iso/anteiso branching and/or the degree of saturation of the acyl chain (Prado *et al.*, 1988; Reizer *et al.*, 1985; Svobodova and Svoboda, 1988). *Thermotoga maritima* is the only member of the Bacteria known thus far that contains diether lipids to stabilize the fluidity of the membrane at high temperature (Huber *et al.*, 1986). Higher temperatures induce in the archaeon *Methanocaldococcus jannaschii* a change from diether lipids to the more thermostable tetraether lipids (Sprott *et al.*, 1991). In the crenarchaeotes *Sulfolobus* and *Thermoplasma* the degree of cyclization of the C_{40} isopranoid in the tetraether lipids increases at elevated temperatures to decrease the motion of the lipids and therefore restrict the proton permeability of the membrane. Studies have shown that the proton permeability of most bacterial and archaeal membranes at the temperature of growth is maintained within a narrow window (H^+-permeability coefficient near 10^{-9} cm s^{-1}) (van de Vossenberg *et al.*, 1995) (Figure 7.2).

The lipid composition of only a few halo(alkali)philic microorganisms has been studied. Archaea from the genera *Natronobacterium*, *Halorubrum* and *Halobacterium* contain 2,3-diphytanyl-*sn*-glycerol-1-phospho-3'-*sn*-glycerol-1'-methylphosphate as the main phospholipid (Kates *et al.*, 1993). Halobacteria have a high density of negative charges on the polar headgroups on the surface of their membrane (Russell, 1989), which are shielded by the high ionic concentration and thereby prevent a disruption of the lipid bilayer (Kates, 1993). Studies have shown that lipids from more moderate halophilic bacteria could not form liposomes at higher salt concentrations, whereas the liposome-forming ability of lipids from halophilic Archaea was not influenced by salt concentration (Yamauchi *et al.*, 1992). The proton and sodium permeability of liposomes from *Halobacterium salinarum* and *Halorubrum vacuolatum* was found to be low even at high salt concentrations, showing that their membranes

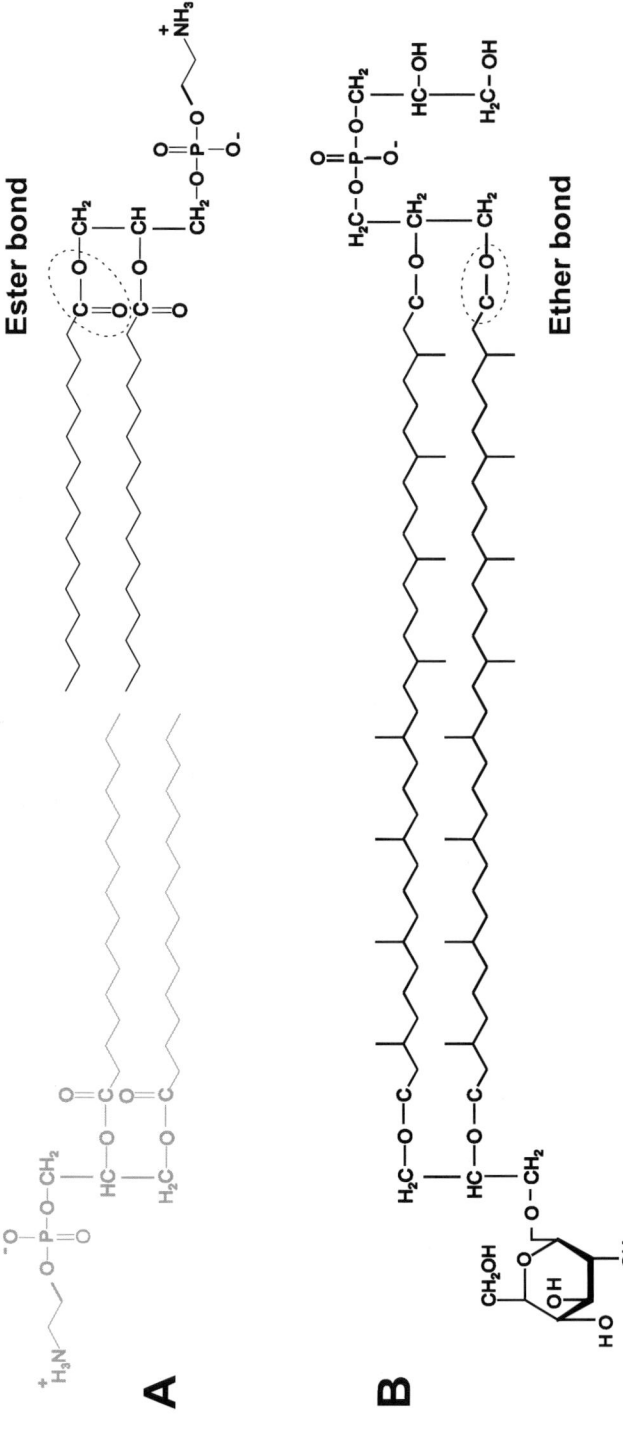

Figure 7.1. Lipids from Archaea and Bacteria. (A) Bilayer-forming lipids in bacteria, phosphatidylethanolamine (PE) from *Escherichia coli*. The acyl chains are connected to the polar headgroup via ester bonds. (B) Monolayer-forming lipids in thermoacidophilic Archaea, main glycophospholipid (MPL) from *Thermoplasma acidophilum*. The acyl chain contains isoprene-like branches and the acyl chains are connected to the polar headgroups *via* ether bonds.

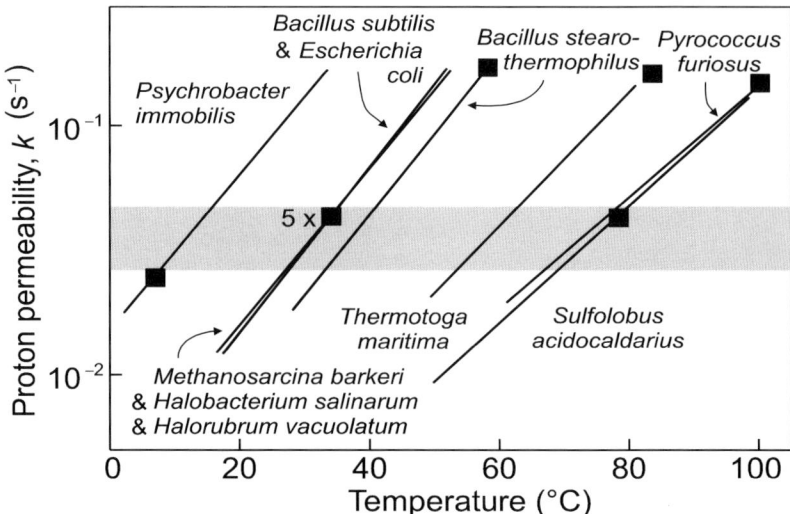

Figure 7.2. Comparison of the proton permeability of liposomes prepared from lipids derived from various Bacteria and Archaea that live at different temperature. At the respective growth temperature, the proton permeability falls within a narrow window (grey bar). From van de Vossenberg et al. (1995) with permission. *Bacillus stearothermophilus* has been renamed as *Geobacillus stearothermophilus*.

are perfectly adapted to their extreme environment (van de Vossenberg et al., 1999).

This chapter aims to provide an overview of different methods for the isolation of lipids from extremophilic organism. Different methods are discussed for the isolation of ether lipids from Archaea and ester lipids from Bacteria. Furthermore, a protocol is provided for the preparation of liposomes from the isolated lipids and the methodology is described to employ these liposomes for proton permeability assays.

◆◆◆◆◆ ISOLATION OF LIPIDS

Principle and Applications

Basically, the method relies on the treatment of freeze-dried cell material with organic solvents that extract lipids from the membranes. The methods for isolation of ether and ester lipids differ. With ether lipids, a more rigorous Soxhlet extraction is employed which involves a lengthy extraction protocol in which the material is heated in organic solvent (Figure 7.3) (Lo and Chang, 1990). This method has been used to isolate the polar lipid fractions of a variety of archaeal species, such as the hyperthermophiles *Sulfolobus acidocaldarius*, *S. solfataricus*, *Pyrococcus furiosus* and also from the mesophile *Methanosarcina barkeri* (van de Vossenberg et al., 1995). The Soxhlet extraction method can also be used for bacteria that contain diether lipids such as *Thermotoga maritima*. For the isolation of bacterial ester lipids mostly a chloroform/methanol extraction is used whereupon the lipids are further purified by

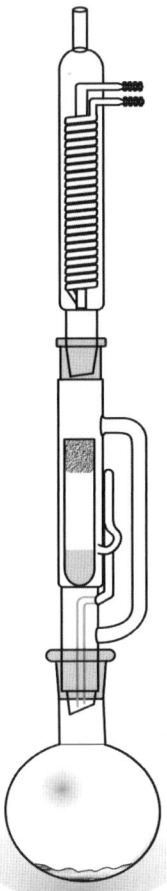

Figure 7.3. Soxhlet lipid extraction apparatus. The main flask contains boiling grains and the organic solvents used for the extraction. The Soxhlet extractor connected to the flask contains a filter with the freeze-dried cells. The setup is closed on top by a refluxer cooled by a constant water stream that condenses the hot organic solvents. Extraction is started by heating the flask whereupon a hot extraction of the lipids from the cells in the filter occurs.

differential acetone–ether wash (Bligh and Dyer, 1959). In acetone, phospholipids are precipitated in order to separate the fatty acids, while the ether step serves to dissolve the phospholipids leaving the (membrane) proteins insoluble (Kagawa and Racker, 1971). This method is generally applicable to bacteria and has for instance been used for the thermophile *Geobacillus stearothermophilus* and the psychrophile *Psychrobacter immobilis* (van de Vossenberg et al., 1995).

Equipment and Reagents

Soxhlet extraction (ether lipids)

- Setup for Soxhlet extraction including 1-l glass Erlenmeyer, a Soxhlet extractor, a cooler connected to water supply and a heater (see Figure 7.3).

- Schleicher & Schuell filters
- Organic solvents (methanol, chloroform)
- Bath sonicator
- Probe sonicator
- Waters Sep Pak Vac™ 20 cc C_{18} columns and a vacuum manifold
- Rotary evaporator

Lipid extraction and purification (Ester lipids)

- Possibility to work under a N_2 atmosphere
- Organic solvents (chloroform, methanol, acetone, diethylether)
- Lysozyme
- Dithiothreitol (DTT)
- 20 mM potassium phosphate at pH 7.0
- Rotary evaporator

Assay

Soxhlet extraction (ether lipids)

- Add 1.5 g freeze-dried cells to the filter, remove excess aqueous solution by filtration and seal the filter with the cells by glass wool. Insert the filter into the Soxhlet extractor, add 400 ml methanol:chloroform 1:1 (v/v) and heat the suspension in the presence of boiling grains for at least for 8 h or overnight. Repeat this procedure three times to increase the yield of lipid recovery.
- Dry the lipids in a pre-weighted 500 ml Erlenmeyer by a rotation evaporator under vacuum. Determine the yield by weighing. Usually 600 mg of "crude lipid" is obtained.
- Add 20 ml H_2O:methanol, 1:1 (v/v) to the dried "crude lipid" and resuspend the material.
- Sonicate the suspension in bath sonicator until the solution is fully dispersed such that no large flocks/aggregates are observed by visual inspection.
- Sonicate the suspension by means of a probe sonicator in aliquots of 3 ml and a big sonication probe at high intensity (10 microns): 10 cycles of 15 s on and 30 s pause.
- Extraction of the lipids over two Waters Sep Pak Vac™ 20 cc C_{18} columns. Remove the filter on top of the column otherwise the lipids will not enter the column.

Per column:
Rinse: 100 ml methanol
Rinse: 100 ml H_2O
Add 30 ml of the crude lipid extract
Elute Fraction 1 with 250 ml methanol:H_2O 1:1 (v/v)
Elute Fraction 2 with 250 ml chloroform:methanol:H_2O 1:2.5:1
Elute Fraction 3 with 100 ml chloroform:methanol:H_2O 65:25:4

- Dry Fraction 2 with rotation evaporator (in 500 ml Erlenmeyer)
- Determine yield (usually slightly more than 100 mg)
- Dissolve the lipids in a sealed test tube in chloroform:methanol:H_2O 65:25:4, at 10–20 mg ml^{-1}. The lipids can be stored at $-20°C$ or for shorter term (1–2 weeks) in the refrigerator at $4°C$.

Lipid extraction (ester lipids)

- Resuspend 50 g cells (wet weight) in 5l ml (final volume) of 20 mM potassium phosphate, pH 7.0
- Add lysozyme (4.2 mg ml^{-1}, final concentration) and incubate 30 min at $37°C$
- All following steps take place in a N_2 atmosphere
- Add 100 ml chloroform and 200 ml methanol and stir for at least 12 h or overnight at $4°C$
- Spin down 10 min at 3000 rpm at $4°C$ and add to the cleared supernatant 95 ml chloroform and 95 ml demineralized water
- Stir for 3 h at room temperature and separate the two phases for 12 h at $4°C$ or room temperature
- Collect the lower phase and dry it in rotation evaporator under vacuum. The typical yield should be around 450 mg of lipid

Lipid purification (ester lipids)

- Dissolve the dried lipids in 2.5 ml chloroform per 0.5 g lipid
- Drip the mixture slowly into 75–100 ml of ice cold acetone containing 1 mM DTT
- Stir 4–5 h at $4°C$ (or overnight in cold room)
- Spin down 15 min at 3000 rpm at $4°C$ and dissolve the pellet in 80 ml of acetone and 1 mM DTT
- Spin down 15 min at 3000 rpm at $4°C$ and dry the pellet thoroughly with N_2 (this step takes about 1 h)
- Dissolve the pellet in 80 ml of diethylether and 1 mM DTT. Stir for 1 h at room temperature
- Spin down 15 min at 3000 rpm
- Take off the supernatant with a pipet and dry the supernatant in a rotation evaporator. The yield should be around 300 mg of lipid
- Dissolve the dried pellet in chloroform and store at $-20°C$ in N_2 atmosphere

◆◆◆◆◆ LIPOSOME PREPARATION AND PROTON PERMEABILITY MEASUREMENTS

Principle and Applications

To analyze the functional properties of the lipids, lipid vesicles are formed by hydration of the lipids into an aqueous buffer. This yields multilamellar vesicles that are subsequently subjected to shear forces

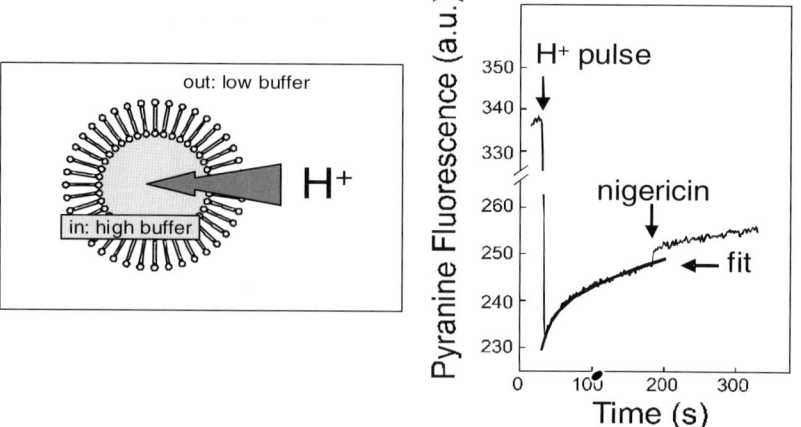

Figure 7.4. Measurements of the proton permeability of liposomes. (A) Experimental setup of the experiment. (B) Actual pH reading in a proton pulse experiment. Liposomes with an internal high buffer capacity are diluted into buffer with low buffering capacity. Pyranine is added to the outside of the liposomes to monitor the changes in external pH. Addition of acid results in a rapid drop in pyranine fluorescence whereupon a recovery phase occurs which signifies protons influx into the liposomes. Finally, nigericin is added to fully equilibrate the protons on the inside and outside. From van de Vossenberg *et al.* (1995) with permission.

(sonication) or preferentially to extrusion through polycarbonate filters in order to yield unilamellar vesicles. These polycarbonate filters contain laser-burned holes with a defined diameter. Extrusion of the multi-lamellar vesicles through these holes results in opening and re-sealing of the membranes, ultimately resulting in unilamellar vesicles with a defined size. For proton permeability measurements, liposomes of the lipids of interest are prepared containing a high buffer capacity inside and are diluted in a buffer with low buffering capacity. The potassium ionophore valinomycin is added to prevent the generation of a counter-acting transmembrane electrical potential due to the electrogenic influx of protons. The external pH is monitored by the fluorescent pH indicator pyranine. When the external pH is lowered by a proton pulse, an immediate drop is observed (Figure 7.4). This is followed by a slow increase in the external pH due to the influx of protons into the lumen of the liposomes, which contain a high buffer capacity. Rapid equilibration of the protons can be obtained after the addition of the ionophore nigericin (Figure 7.4) that mediates the electroneutral exchange of potassium and protons. In the presence of valinomycin, nigericin causes a complete uncoupling of the ion gradients across the membranes. The slow increase of pyranine fluorescence after the proton pulse can be fitted to a first order kinetic rate equation and this is used to obtain the first order rate constant, k_{H+}, of proton influx. This is a measure of the proton permeability of the liposomal membrane. By preparing liposomes with an equal diameter from lipids of various sources, the method allows a comparison of the proton permeability of membrane lipids derived from different organisms.

Equipment and reagents

- Lipofast™ extrusion apparatus and 200 nm polycarbonate filters
- PD 10 columns
- Fluorimeter with heatable magnetically stirred sample compartment
- The ionophores valinomycin and nigericin
- Pyranine

Assay

- Lipids are dried by vacuum rotary evaporation and are hydrated in 50 mM 3-morpholinopropanesulfonic acid (MOPS), pH 7.0, 75 mM KCl, and 25 mM choline to a final concentration of 40 mg ml^{-1}.
- Liposomes are obtained by 5 consecutive freezing in liquid nitrogen and thawing steps, followed by extrusion through 200 nm polycarbonate filters using the Lipofast™ extrusion apparatus. These liposomes are unilamellar with an average size that is close to the pore size of the filter used.
- To exchange the external buffer for one with a lower buffering capacity, 0.5 mM MOPS, pH 7.0, 75 mM KCl, and 75 mM sucrose (buffer A), the liposomes are passed over a Sephadex G-25M PD10 column equilibrated with buffer A and collected.
- Dilute the liposomes to 1.5 mg ml^{-1} in 2 ml buffer A and add these to a 2 ml quartz cuvette. The K$^+$ ionophore valinomycin (1 nmol mg^{-1} lipid) is added from an ethanolic solution in order to prevent the formation of a reversed transmembrane electrical potential ($\Delta\psi$).
- Add pyranine (10 µM) to monitor the changes in the external pH. Use a wavelength of 450 and 508 nm for excitation and emission, respectively.
- After equilibration, add 100 nmol H$^+$ (from a 50 mM H$_2$SO$_4$ stock solution) to lower the external pH. The influx of H$^+$ into the liposomes can be monitored in time as an increase of the pyranine fluorescence.
- Finally add nigericin (1 nmol mg^{-1} lipid), a protonophore, to equilibrate the H$^+$ gradient across the membrane.
- The signal is calibrated after the addition of nigericin by adding small aliquots of base or acid.
- The fluorimetric data are fitted to the first order kinetic equation: in which "a" is the amplitude of the fluorescence signal, k_{H+} is the first order rate constant of proton influx, and "c" the offset. The H$^+$ pulse was imposed at $t=0$, and k_{H+}, is used to compare the proton permeability of different liposomes.

List of suppliers

Amersham Biosciences UK Limited
Amersham Place
Little Chalfont, Buckinghamshire HP7 9NA, England

Tel: +44-0870 606 1921
Fax: +44-01494 544350
http://www4.amershambiosciences.com

PD10 columns

Avestin
2450 Don Reid Drive
Ottawa, Ontario, Canada K1H 1E1
Tel: +1-613-736-0019
Fax: +1-613-736-8086
http://www.avestin.com/

Lipofast™ extrusion apparatus, 200 nm polycarbonate filters

Schleicher & Schuell GmbH
Hahnestraße 3
37586 Dassel, Germany
Phone: +49 5561 791 0
Fax: +49 5561 727 43
www.schleicher-schuell.com

Filters for extraction of lipids

Waters Corporation
34 Maple Street
Milford, MA 01757, USA
Phone: +1-508-478-2000
Fax: +1-508-872-1990
www.waters.com

Columns for purification of lipids

References

Albers, S. V., van de Vossenberg, J. L. C. M., Driessen, A. J. M. and Konings, W. N. (2001). Bioenergetics and solute uptake under extreme conditions. *Extremophiles* **5**, 285–294.

Bligh, E. G. and Dyer, W. J. (1959). A rapid method of total lipid extraction and purification. *Can. J. Biochem. Physiol.* **37**, 911–917.

Huber, R., Langworthy, T. A., König, H., Thomm, M., Woese, C. R., Sleytr, U. B. and Stetter, K. O. (1986). *Thermotoga maritima sp. nov.* represents a new genus of unique extremely thermophilic eubacteria growing up to 90°C. *Arch. Microbiol.* **144**, 324–333.

Kagawa, Y. and Racker, E. (1971). Partial resolution of enzymes catalyzing oxidative phosphorylation. *J. Biol. Chem.* **246**, 5477–5487.

Kates, M. (1993). Membrane lipids of archaea. In *The Biochemistry of Archaea (Archaebacteria)*. (M. Kates, D. J. Kushner and A. T. Matheson, eds), *Comprehensive Biochemistry* (A. Neuberger and L. L. M. van Deenen, series eds), vol. 26, pp. 261–296. Elsevier, Amsterdam.

Kates, M., Moldoveanu, N. and Stewart, L. C. (1993). On the revised structure of the major phospholipids of *Halobacterium salinarium*. *Biochim. Biophys. Acta* **1169**, 46–53.

Konings, W. N., Albers, S. V., Koning, S. M. and Driessen, A. J. M. (2002). The cell membrane plays a crucial role in survival of bacteria and archaea in extreme environments. *Antonie van Leeuwenhoek* **81**, 61–72.

Lo, S. L. and Chang, E. L. (1990). Purification and characterization of a liposomal-forming tetraether lipid fraction. *Biochem. Biophys. Res. Commun.* **167**, 238–243.

Prado, A., Da Costa, M. S. and Madeira, V. M. (1988). Effect of growth temperature on the lipid composition of two strains of *Thermus*. *J. Gen. Microbiol.* **134**, 1653–1660.

Reizer, J., Grossowicz, N. and Barenholz, Y. (1985). The effect of growth temperature on the thermotropic behavior of the membranes of a thermophilic *Bacillus*. Composition-structure-function relationships. *Biochim. Biophys. Acta* **815**, 268–280.

Russell, N. J. (1989). Adaptive modifications in membranes of halotolerant and halophilic microorganisms. *J. Bioenerg. Biomembr.* **21**, 93–113.

Sprott, G. D. (1992). Structures of archaebacterial membrane lipids. *J. Bioenerg. Biomembr.* **24**, 555–566.

Sprott, G. D., Meloche, M. and Richards, J. C. (1991). Proportions of diether, macrocyclic diether, and tetraether lipids in *Methanococcus jannaschii* grown at different temperatures. *J. Bacteriol.* **173**, 3907–3910.

Svobodova, J. and Svoboda, P. (1988). Membrane fluidity in *Bacillus subtilis*. Physical change and biological adaptation. *Folia Microbiol. (Praha)* **33**, 161–169.

van de Vossenberg, J. L. C. M., Ubbink-Kok, T., Elferink, M. G. L., Driessen, A. J. M. and Konings, W. N. (1995). Ion permeability of the cytoplasmic membrane limits the maximum growth temperature of bacteria and archaea. *Mol. Microbiol.* **18**, 925–932.

van de Vossenberg, J. L. C. M., Driessen, A. J. M., Grant, W. D. and Konings, W. N. (1999). Lipid membranes from halophilic and alkali-halophilic Archaea have a low H^+ and Na^+ permeability at high salt concentration. *Extremophiles* **3**, 253–257.

Yamauchi, K., Doi, K., Kinoshita, M., Kii, F. and Fukuda, H. (1992). Archaebacterial lipid models: highly salt-tolerant membranes from 1,2-diphytanylglycero-3-phosphocholine. *Biochim. Biophys. Acta* **1110**, 171–177.

8 Characterization and Quantification of Compatible Solutes in (Hyper)thermophilic Microorganisms

Helena Santos, Pedro Lamosa and Nuno Borges

Instituto de Tecnologia Química e Biológica, Universidade Nova de Lisboa, Rua da Quinta Grande 6, Apartado 127, 2780-156 Oeiras, Portugal

CONTENTS

Introduction
Growth and harvesting of cells
Extraction of compatible solutes
Detection, identification and quantification of compatible solutes by NMR
Biomass quantification
Determination of intracellular volume
Purification of compatible solutes
Identification of unknown compatible solutes

◆◆◆◆◆ INTRODUCTION

The term thermophile is used to designate organisms with optimum growth temperatures between 65 and 80°C while hyperthermophilic organisms are those with optimum growth temperatures above 80°C (Blöchl et al., 1995). Most thermophilic organisms have been isolated from continental geothermal or artificial thermal environments, where the sodium level is generally low, but a few thermophiles such as *Thermus thermophilus* and *Rhodothermus marinus* have been isolated from marine hydrothermal environments (Alfredsson et al., 1988; da Costa et al., 2001). On the other hand, most of the hyperthermophilic organisms isolated to date originate from shallow or abyssal marine geothermal areas (Blöchl et al., 1995). These organisms are only slightly halophilic, being unable to grow at salinities higher than 6.0–8.0% NaCl. Like other halophiles, however, the organisms isolated from hot environments had to develop strategies to preserve cell functionality under saline conditions as well as to cope with fluctuations in the salinity of the external medium. The accumulation of organic compatible solutes appears to be the most common strategy among marine thermophiles and hyperthermophiles (da Costa et al., 1998; Santos and da Costa, 2001, 2002). According to the definition of Brown (1976), compatible solutes are small organic compounds used for osmotic adjustment that do not interfere with cell function. Later studies have revealed that the role of compatible solutes goes beyond protection against osmotic stress to the protection against high temperature, freezing, desiccation and oxygen radicals (Benaroudj et al., 2001; da Costa et al., 1998; Santos and da Costa, 2002; Welsh, 2000). In many thermophiles and hyperthermophiles examined, the level of compatible solutes increases notably when cells are grown at supraoptimal temperatures, leading to the view that these solutes may play a role in the protection of cell components against the damaging effect of heat (Ciulla et al., 1994a; Gonçalves et al., 2003; Martins and Santos, 1995; Martins et al., 1996; Silva et al., 1999). However, the presence of compatible solutes is by no means an essential requisite for thermophily since organisms isolated from fresh water hot springs (*Pyrobaculum islandicum*, *Thermococcus zilligii*, *Fervidobacterium islandicum*) do not accumulate compatible solutes (Lamosa et al., 1998; Martins et al., 1996, 1997).

Compatible solutes of mesophilic microorganisms are generally neutral or zwitterionic and include amino acids and amino acid derivatives, sugars, sugar derivatives, polyols, betaines and ectoines (da Costa et al., 1998; Galinski, 1995). (Hyper)thermophiles accumulate compatible solutes that are rarely or never encountered in mesophiles, such as di-*myo*-inositol phosphate (DIP), diglycerol phosphate (DGP), mannosylglycerate and derivatives of these three compounds (Figure 8.1). Compatible solutes of (hyper)thermophiles are generally negatively charged, while mesophiles accumulate primarily neutral solutes. Notable exceptions are the hyperthermophilic archaea *Pyrobaculum aerophilum* and *Thermoproteus tenax* that exclusively accumulate trehalose (Martins et al., 1997). Di-*myo*-inositol phosphate and mannosylglycerate are the two most widely distributed solutes among non-methanogenic (hyper)thermophiles. Ectoines have never been encountered in (hyper)-thermophiles and betaines are extremely rare. On the other hand, phosphodiester compounds are common compatible solutes of (hyper)-thermophiles but were never found in mesophiles.

The superior efficacy of negatively charged solutes (phosphodiester compounds or carboxylic acids) to enhance protein stability has been demonstrated in several studies from our and other groups (Borges et al., 2002; Kaushik and Bhat, 1999; Lamosa et al., 2000). Therefore, it is tempting to speculate that the unique charged solutes of (hyper)thermophiles were especially selected to preserve cell components at high temperature and that (hyper)thermophiles have adapted their cellular components through evolution to be able to benefit from the most effective protectors against heat available.

This chapter deals primarily with non-methanogenic (hyper)thermophiles. The topic of osmoadaptation in Archaea has been competently and regularly reviewed (Martin et al., 1999; Roberts, 2000, 2004; Robertson and Roberts, 1991) and protocols for the extraction and detection of

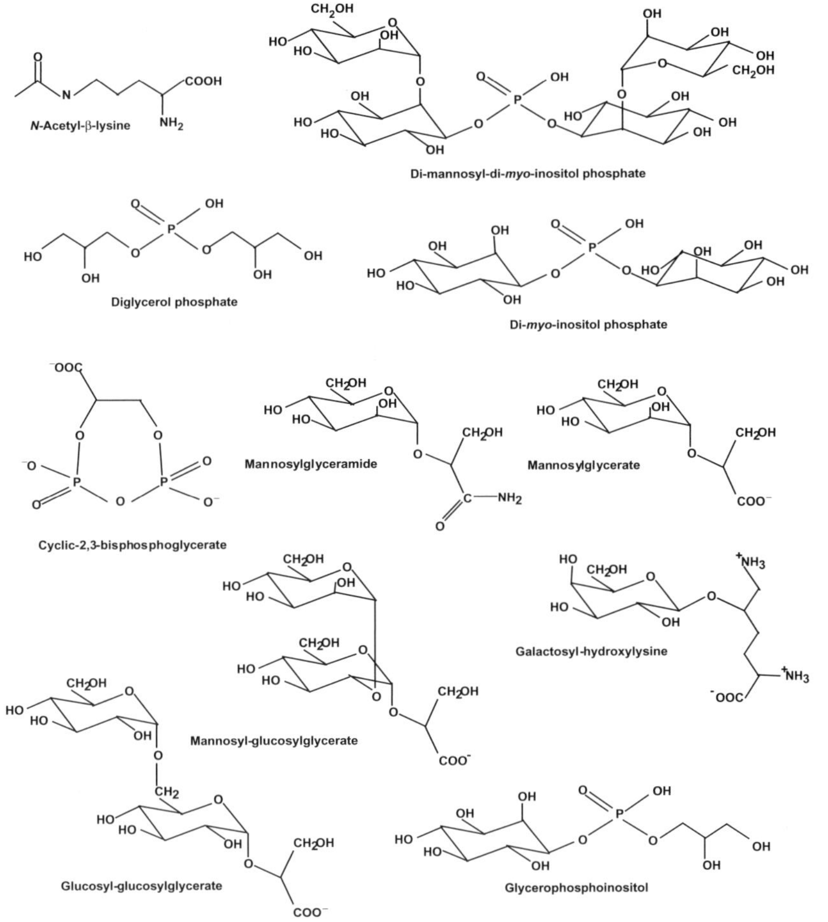

Figure 8.1. Molecular representation of compatible solutes found in thermophiles and hyperthermophiles.

compatible solutes from methanogenic Archaea are also available in the literature (Lai et al., 1995).

◆◆◆◆◆ GROWTH AND HARVESTING OF CELLS

The accumulation of compatible solutes in thermophiles is highly sensitive to the growth conditions, namely the composition of the growth medium, carbon source, gas phase, agitation, growth phase, temperature and salinity.

The first task at hand is the selection of an appropriate growth medium. If *de novo* synthesis of compatible solutes is to be investigated, complex media should be avoided, since some of its components can be taken up and accumulated to high levels repressing *de novo* synthesis. If no defined medium is described for the organism under study, the one recommended by the culture collection supplier can be used as a starting point, replacing the undefined additives (like yeast extract) with trace elements and vitamin solutions and an appropriate carbon source. Several formulations of these solutions are described in the literature and the choice should fall on the ones developed for organisms more closely related to those under study.

Some thermophiles, however, are unable to grow (or grow very poorly) in defined medium. In this case, the undefined additives of the least complex origin should be preferred over more complex formulations (for example casitone should be preferred over peptone) and the medium can be supplemented with other defined carbon sources, if necessary.

Once the composition of the growth medium has been established, cell cultivation should take place under controlled conditions. The fermentation vessel should allow temperature and pH control, choice of stirring conditions, continuous flushing of the gas phase, and sampling under sterile conditions. All the autoclavable components of the medium should be autoclaved for 30 min at 120°C (ensure that all compartments of the fermentor are open during autoclaving to avoid pressure buildup). When the system cools down, open the autoclave and close all openings of the fermentor immediately. Add the non-autoclavable components of the medium (filter sterilized). In the case of anaerobes flush the medium in the fermentor with oxygen-free N_2 (or preferably argon) for at least three hours prior to inoculation. Use a fresh culture as inoculum (10%).

The accumulation of solutes should be studied as a function of the growth temperature and the concentration of NaCl in the medium. To vary the salinity of the growth medium adjust the concentration of NaCl to the desired value, and adapt the cultures to the different extracellular NaCl concentration by making at least two sequential transfers at the new NaCl concentration prior to growth of the final inoculum. The two transfers are not needed when studying the effect of changes in the growth temperature.

Monitor growth by taking samples at selected time intervals and measuring turbidity in a spectrophotometer at 600 nm. In the case of media containing (or developing) solid particles or containing resazurin (oxygen indicator), cell-counting techniques are required (see section "Biomass quantification").

The growth phase at which cells are harvested is critical for reproducible quantitative determination of the level of internal solutes. In some extreme cases, one particular solute can be completely consumed at the onset of the stationary phase. Harvesting the cells should be done at the highest cell density possible (for higher cell mass) but still within the exponential phase of growth, between the mid- and late-exponential phases to obtain consistent results. To compare the accumulation of solutes as a function of temperature or salinity it is important to collect the cells during the same growth phase. Obviously, this does not mean at the same cell density since the growth profiles can vary substantially with the stress imposed. It is advisable to perform three replicates for each set of growth conditions examined. Whenever practicable the profile of solute accumulation should be determined over the entire growth curve.

Harvesting the Cells

1. Centrifuge the cell culture at room temperature (7000 × g, 10 min). In the case of anaerobes, keep the culture under anaerobic conditions by flushing the centrifuge tubes with oxygen-free argon and continuously passing the same gas through the culture prior to closing the tubes.
2. Pour off and discard the supernatant; wash the cells by adding about 200 ml of cold oxygen-free NaCl solution at a concentration identical to that of the growth medium (to prevent solute release). Mix gently, but well and centrifuge again. This washing step is needed since several medium components including organic synthetic buffers, which may be present at high concentrations, are very efficiently extracted in subsequent steps and disturb the analyses of the intracellular content.
3. Add the same NaCl solution to the cell pellet to obtain a final volume of 20 ml of cell suspension. Mix and remove two aliquots (1 ml each) of the cell suspension for biomass quantification (see section "Biomass quantification").
4. Centrifuge the remaining cell suspension (7000 × g, 10 min, 4°C) and discard the supernatant. Store the cell pellet at −20°C until needed for extraction of compatible solutes.

◆◆◆◆◆ EXTRACTION OF COMPATIBLE SOLUTES

The accurate determination of the intracellular content of compatible solutes requires an extraction method that is effective within the concentration range in which these solutes are present.

The use of strong acids (trichloroacetic acid or perchloric acid) is one of the most common procedures for extraction of small metabolites or organic compounds. However, these well-known procedures, which are appropriate for acid-stable compounds, are obviously unsuitable for acid-labile components. The use of an ethanol–chloroform extraction is recommended here. This is a simple procedure for isolation of low molecular mass organic compounds, and leads to little contamination with proteins, lipids, membranes or other macromolecules. The method is easy to implement, highly reliable, and results in clean extracts that can be directly analyzed by nuclear magnetic resonance (NMR), high performance liquid chromatography (HPLC), mass spectrometry and/or thin layer chromatography (TLC).

Ethanol–Chloroform Extraction

1. Harvest and wash the cells as described in the previous section. Add 100 ml of 80% ethanol, mix and transfer the cell suspension to a 250 ml round flask containing a magnetic stir bar.

2. Connect the flask to a coil condenser (do not forget to circulate cold water). Heat the flask in a boiling water bath, with stirring, for 10 min. Centrifuge at 10 000 × g for 20 min at 4°C. Save the supernatant in a clean 250 ml round flask.
3. Add 100 ml of 80% ethanol (v/v) to the pellet, mix well, and transfer the suspension to the round flask mentioned in step 1.
4. Repeat steps 2 and 3 and combine the supernatants. (Note: Usually, three repeats of the extraction procedure are sufficient to recover more than 98% of the organic solutes present in the cell sample.)
5. Remove the ethanol from the pooled supernatants by rotary evaporation at 40°C under reduced pressure.
6. Transfer the residue to a glass centrifuge tube (do not use plastic tubes) and add half the volume of chloroform. Mix vigorously by vortexing. Centrifuge the resulting emulsion (16 000 × g, 15 min, 10°C). Transfer the aqueous phase (upper phase) with a Pasteur pipette to an ice-cold 100 ml round flask. (Note: no need to completely remove the aqueous phase since the extraction procedure will be repeated.)
7. Add one volume of distilled water to the remaining chloroform–water mixture in the glass centrifuge tube, and mix by vortexing. Centrifuge and transfer the aqueous phase to the cold round flask mentioned in step 6.
8. Repeat three times step 7. Combine the aqueous phases. Freeze-dry and store at −20°C until further analysis.

Commentary

If the presence of heat-labile compounds is suspected, the procedure described above can be modified by replacing the three boiling 80% ethanol extractions of 10 min with three extractions of 30 min at room temperature using vigorous stirring.

Perchloric or Trichloroacetic acid Extraction

1. Add 50 ml of 10% trichloroacetic acid (TCA) or 5% perchloric acid (PCA) to the cell pellet (see step 4 of the harvesting protocol) and mix vigorously by vortexing. Keep the suspension on ice for 20 min with stirring. Centrifuge at 30 000 × g for 30 min at 4°C. Transfer the supernatant to an ice-cold round flask. Store the supernatant on ice.
2. Treat the pellet with TCA or PCA, as described in step 1. Collect all supernatant fractions in the same round flask.

For TCA extraction

3. Freeze the extract in liquid nitrogen and freeze-dry to remove TCA. Store the dried residue at −20°C.

For PCA extraction

4. Place the round flask on ice on a magnetic stirrer, and adjust the pH of the extract to 7 with 2 M KOH. If necessary, correct the pH with drops of 5% PCA. A precipitate of potassium perchlorate will develop.

5. Transfer the pH-adjusted PCA extract to a 50 ml centrifuge tube, and wash the round flask with ~5 ml of distilled water. Centrifuge at 30 000 × g for 20 min at 4°C, and transfer the supernatant to a cold round flask. Freeze-dry and store the dried extract at −20°C.

Cautions
(i) Chloroform is toxic and volatile; avoid breathing the vapor and contact with skin, eyes, mucous membranes, respiratory tract. Work in a chemical fume hood with adequate extraction; (ii) Ethanol is volatile and flammable; (iii) Trichloroacetic acid, perchloric acid and potassium hydroxide are highly corrosive to tissue of the mucous membranes and respiratory tract, eyes, and skin; wear gloves and safety glasses and work in a chemical fume hood; (iv) Liquid nitrogen causes severe burns. Wear hand protection and safety glasses when handling liquid nitrogen.

◆◆◆◆◆ DETECTION, IDENTIFICATION, AND QUANTIFICATION OF COMPATIBLE SOLUTES BY NMR

Once a suitable extract is obtained, the compounds present should be identified and quantified. Since *ab initio* we do not know what compounds we may find, we need a technique that: (i) is able to detect all the organic substances present in the extract; (ii) provides structural information, and (iii) can be used for solute quantification. NMR is ideally suited for this task since it meets all three criteria, its only drawback being, its relatively low sensitivity, overcome by the fact that generally compatible solutes accumulate to high levels in the cell. One of the great advantages of NMR is that "it provides an easy way to detect the unexpected". Therefore, the occurrence of novel solutes can be easily noticed without needing to establish specific experimental set-ups. In fact, all the new solutes discovered in thermophiles and hyperthermophiles were identified using NMR, sometimes starting with the direct analysis of whole cells (Ciulla *et al.*, 1994a,b; Lamosa *et al.*, 1998; Martins *et al.*, 1996, 1997; Nunes *et al.*, 1995; Silva *et al.*, 1999).

The usual strategy includes the following steps: (i) Acquisition of ^{1}H, ^{13}C and ^{31}P NMR spectra of cell extracts; (ii) Comparison of the recorded signals with those reported in the literature to identify the compounds; (iii) Quantification of solutes usually by ^{1}H NMR using an internal concentration standard. When the signals cannot be assigned to any known compound, two-dimensional NMR spectroscopy is used to determine the structure of the unknown compounds. This determination should be confirmed by an independent technique. Mass spectrometry for molecular mass determination and chemical synthesis for the production of an authentic sample are excellent complementary strategies.

Procedure

1. For a typical slightly halophilic thermophile, grow a two-liter culture up to an OD_{600} of approximately 0.4 (this will yield a suitable extract for NMR analyses).
2. Dissolve the extract (residue after freeze-drying) in 600 µl of D_2O and transfer it to a 5-mm NMR tube. Add 3-(trimethylsilyl) propane sulfonic acid (TSPSA) for referencing purposes.
3. Acquire a ^{13}C NMR spectrum. If available, a selective ^{13}C probe is optimal, but a broadband detection probe can also be employed. ^{13}C spectra are acquired with ^1H decoupling using a WALTZ sequence during the acquisition time. This has the effect of collapsing the multiple resonances into a single resonance for each carbon atom, which greatly simplifies the analyses. Usually, on a spectrometer operating at 125 MHz, carbon spectra are acquired with a sweep width of 25 kHz, 32K data points, 45° pulse angle, 2.0 s recycling delay, and 10 000–80 000 transients depending on the concentration of the solutes in the sample. The Free Induction Decay (FID) is multiplied by an exponential function (with line broadening of 1–3 Hz) prior to Fourier transform to improve signal-to-noise ratio.
4. Once the ^{13}C NMR spectrum is acquired identification of the signals can be done by resorting to the chemical shift values published in the literature (Table 8.1, Figure 8.1). The chemical shifts are referenced to the signal of TSPSA at 0.0 ppm. With the acquisition conditions mentioned above, the signals arising from the carboxyl groups are poorly visible but the identification of known compounds can usually be accomplished without that information. If the detection of the carboxyl groups is required the relaxation delay has to be increased. The acquisition of ^1H NMR spectra is also helpful in this process and its agreement with the proposed identification should always be confirmed.

Usually the number of different solutes found in a thermophilic or hyperthermophilic organism is small, and therefore the interpretation of the one-dimensional NMR spectra is straightforward. Otherwise, one can resort to the acquisition of carbon-proton correlated spectra with proton detection (HMQC), taking advantage of the great sensitivity improvement provided by this technique. $^1H/^{31}P$ correlation spectra are also acquired to determine the position of the phosphorus atoms when the ^{31}P NMR spectra revealed the presence of this element in the solute structure.

For quantification purposes, fully relaxed ^{13}C spectra have to be acquired, meaning that the relaxation delay has to be increased to at least 60 s. Dioxane (67.4 ppm), added to the sample at a concentration of about 10 mM, is a common concentration standard. With these acquisition conditions the integral of each signal is directly proportional to the number of carbon atoms that contribute to its intensity and the concentration of each species is readily obtained by comparison with the

Table 8.1 Carbon chemical shift values of several organic compounds found in thermophilic or hyperthermophilic organisms

Compound	^{13}C Chemical shift values	Reference
α-alanine	17.3; 51.6; 176.8	Breitmaier and Voelter (1990)
α-aspartate	37.2; 52.9; 175.5; 178.8	Breitmaier and Voelter (1990)
α-glutamate	28.1; 34.5; 55.7; 175.6; 182.3	Breitmaier and Voelter (1990)
β-glutamate	39.2; 47.7; 178.5	Robertson et al. (1990)
α-glutamine	28.3; 33.0; 55.2; 174.4; 179.5	Breitmaier and Voelter (1990)
β-glutamine	37.2; 39.0; 47.3; 175.5; 178.3	Robertson and Roberts (1991)
Proline	24.4; 29.7; 46.5; 61.6; 174.6	Breitmaier and Voelter (1990)
Glycine betaine	54.1; 66.9; 169.7	Breitmaier and Voelter (1990)
Trehalose	61.5; 70.6; 72.0; 73.0; 73.5; 94.0	Breitmaier and Voelter (1990)
α-Mannosylglycerate	61.4; 63.3; 67.2; 70.5; 70.8; 73.5; 77.8; 99.1; 176.6	Silva et al. (1999)
α-Mannosylglyceramide	61.4; 62.8; 67.1; 70.4; 70.7; 73.9; 77.9; 100.1; 175.5	Silva et al. (1999)
α-Glucosylglycerate	61.1; 63.7; 70.0; 72.1; 72.7; 73.7; 79.5; 97.8; 178.0	Robinson and Roberts (1997)
Mannosylglucosylglycerate	63.3; 63.5; 65.7; 69.3; 72.2; 72.5; 72.9; 73.8; 74.6; 75.5; 76.7; 81.0; 96.6; 99.9; 179.2	Our unpublished results
Glucosylglucosylglycerate	60.8; 63.6; 66.0; 69.9; 70.0; 71.1; 71.9; 72.0; 72.3; 73.5; 73.9; 79.7; 98.0; 98.3; 177.3	Our unpublished results
Galactosylhydroxylysine	27.1; 28.4; 43.4; 55.0; 61.4; 68.9; 71.2; 72.8; 75.5; 76.5; 102.2; 176.0	Lamosa et al. (1998)
Diglycerol phosphate	62.5; 66.8; 71.1	Martins et al. (1997)
Di-*myo*-inositol (1,1′) phosphate (DIP)	71.8; 72.5; 72.8; 73.5; 75.5; 77.8	Scholz et al. (1992)
Di-*myo*-inositol (1,3′) phosphate	71.0; 71.6; 71.9; 72.6; 74.2; 77.0	Martins et al. (1996)
Glycerophosphoinositol	62.8; 67.1; 71.4; 71.5; 71.9; 72.1; 72.9; 74.6; 77.0	Our unpublished results
Di-mannosyl-DIP	61.4; 67.2; 70.5; 70.9; 72.0; 73.0; 73.1; 74.3; 76.4; 76.6; 80.0; 101.1	Martins et al. (1996)
Cyclic-2,3-bisphosphoglycerate	69.4; 77.9; 178.8	Kanodia and Roberts (1983)
1,3,4,6-tetracarboxyhexane	29.3; 36.9; 53.6	Ciulla et al. (1994b)
N^ε-acetyl-β-lysine	22.7; 25.1; 30.3; 39.2; 39.6; 50.0; 174.8; 178.8	Sowers et al. (1990)

standard. The need for the long relaxation delay makes this quantification procedure extremely time consuming, so, whenever possible ^{1}H NMR (or an alternative analytical method) is used for quantification of known solutes.

To quantify the solutes using ^{1}H NMR, spectra are typically acquired with a 3 s continuous presaturation of the water signal. The spectrum is usually acquired with 64K data points, 60° flip angle, 30 s relaxation delay, and 16–96 transients depending on the solute concentration. Most thermophiles are halophilic and as a consequence the extracts may contain considerable amounts of salt, which broadens the NMR signals and decreases sensitivity, so dilution of the original extract is advisable. The higher the dilution the lower is the signal-to-noise ratio for the same total time of acquisition, so a compromise has always to be reached. As a rule of thumb, a sample yielding a good ^{13}C spectrum in two hours can be diluted 6 times for ^{1}H analyses. The sample can be prepared using 100 µl of the original extract, 500 µl of D_2O, and 10 µl of a 200 mM standard solution of sodium acetate or formate. Formate has the advantage of resonating in an empty region of the spectrum but the disadvantage of a very long relaxation time meaning that the repetition delay has to be 30 s or longer. Acetate allows shorter repetition delays but occurs in a region that is sometimes crowded with other signals, a simple way to circumvent this problem is to acquire two spectra under the same conditions before and after the addition of the standard, and take the difference in intensity as proportional to the amount of added compound.

Commentary
Once the organic solutes have been detected and identified by NMR, there is obviously no absolute need to quantify them by NMR, and in many cases alternative analytical methods are advantageous. According to the type of solutes in question and equipment availability, amino acids can be very rapidly quantified by HPLC, while for simple sugars and other compounds there are commercially available quantification kits that rely on visible spectroscopy for rapid quantification. However, for odd or new compounds NMR is still the most reliable and straightforward quantification method.

◆◆◆◆◆ BIOMASS QUANTIFICATION

The concentration of compatible solutes is commonly expressed as per milligram of total cell protein; however, other biomass indicators can be used. Biomass can be determined directly by dry weight measurement, or cell counting, or indirectly by quantification of macromolecular components such as proteins or DNA. The major difficulty to quantify total protein or DNA is establishing an optimal cell lysis method. Alkali/heat treatment, French press and sonication are the techniques most widely used to disrupt the cells. Alkali/heat treatment is simple, fast and inexpensive, but reproducibility is poor. French-press disruption is the best method but has several disadvantages: it requires relatively large sample volumes (minimum around 3 ml), and expensive equipment, and is time consuming. Sonication is the most frequently used technique to disrupt cells; however, protein degradation due to heat dissipation can occur. Brief descriptions of the three techniques are presented.

Methods of Cell Lysis

Alkali/heat treatment

Add 450 μl of 1 M NaOH to 50 μl of the cell suspension (aliquot saved in step 3 of the section "Harvesting the cells") and heat the mixture at 100°C for 5 min. Neutralize with 450 μl of 1 M HCl.

French press

Cool the pressure cell in the cold room. Pass the cell suspension through the pressure cell in a French press apparatus applying a pressure of around 140 MPa. Three passages are normally enough to break the cells. Cell disruption can be evaluated with an optical microscope. (Note: To ensure an adequate lysis, the outlet valve should be opened slowly so that the cell suspension flows smoothly preventing a significant pressure drop.)

Sonication

Place a 15 ml elongated plastic tube containing the cell suspension in an ice-water bath with NaCl. Insert the probe in the tube, making sure not to touch its walls. Sonicate the cell suspension for 60 s (550 W cm^{-3}) and then pause for 2 min. Apply two more cycles of sonication. (Important note: Always keep the cell suspension under ice-cold conditions.)

Comments
The length of the sonication period is empirically determined. The following conditions can serve as guiding examples: *Escherichia coli* (3 × 20 s), *Archaeoglobus fulgidus* (3 × 60 s); (ii) Cell disruption can be evaluated with an optical microscope. (iii) Be aware that additional sonication cycles can lead to protein degradation.

Dry Weight Determination

1. Dry cellulose membrane filters inside an open Petri dish with the bottom covered with a large filter paper, in the oven (100°C, 1 h). Cool the filters at room temperature in a desiccator, and then weigh the filters with an analytical balance. Write down the weight of each membrane filter and store them in the desiccator.

2. Using a vacuum unit, filter a known volume of the cell suspension (usually 2 ml). Quickly, wash the cells with 2 ml of ice-cold distilled water to remove contaminating medium and NaCl. Confirm that this treatment will not result in release of cell components by measuring protein content in the filtrate. Place the filter on top of a clean filter paper in a Petri dish. Dry in an oven at 100°C for 12–24 h. At various time periods determine the weight of the membrane filter with the cells until a constant value is reached. Prior to each weight measurement, remove the filter from the oven, and allow the membrane to cool in a desiccator (to prevent water absorption from the air), and then weigh quickly.
3. Calculate the difference in the weight, and express the dry weight in mg ml^{-1}. (Note: Only dry weights greater than 10 mg ml^{-1} give accurate results.)

Cell Counting

1. Place cover glass over the hemacytometer chambers. Carefully and continuously fill both chambers. Leave undisturbed for 1–2 min. (Note: Ensure that there are no air bubbles in the chambers.)
2. Using an optical microscope with a 10X ocular and a 40X objective, count the cells in five (0.2 × 0.2 × 0.1 mm) volumes of the top chamber and repeat the procedure for the lower chamber. The number of cell counts in 5 volumes should be between 50 and 500, otherwise repeat the process after an adequate dilution.
3. Take into account that the total cell counts refer to ($10 \times 0.2 \times 0.2 \times 0.1 \times 10^{-3} = 4 \times 10^{-5}$) ml of cell suspension and calculate the number of cells per ml. Multiply by the dilution factor (if dilution was necessary) to obtain the number of cells per ml in the original cell suspension.

Protein Quantification Methods

The Bradford method, the Lowry method and the bicinchoninic acid (BCA) method, using bovine serum albumin as protein standard, are widely used to measure protein levels. These are quick colorimetric methods that yield reproducible results (Bradford, 1976; Lowry et al., 1951; Smith et al., 1985). The Bradford method is probably the most popular one since it is rapid, has high sensitivity and very few interferences by nonprotein components. However, it has the disadvantage of high protein-to-protein variations. This limitation has been overcome in the procedure involving bicinchoninic acid (Smith et al., 1985). This is the protein assay currently favored in our laboratory. Pierce Chemical Company (Illinois, USA) commercializes a kit with standard solutions for this method, which minimizes errors and improves reproducibility.

DNA Quantification Methods

DNA concentration is often measured by UV absorbance at 260 nm; this method, however, is not very accurate and suffers from interferences from other sample components requiring therefore a high degree of purification of the sample DNA. Fluorimetric methods using fluorescent probes are very sensitive assays with little interference from other cellular components. They are, however, sensitive to chain length. Small fragments of DNA will not fluoresce, meaning that cell lysis has to be effective enough to expose DNA, but mild enough not to break it extensively. One of the most commonly used fluorescent dyes is Hoechst 33258, which is very sensitive and presents a linear dynamic range extending over three orders of magnitude (www.Tecan.de/literatur/Tecan_abstract.asp). However, this dye is sensitive to the GC content of DNA, and accurate measurements require calibration with pure DNA from the organism under study (or with DNA of similar GC content). Another commonly used fluorescent dye is PicoGreen, which is virtually unaffected by the GC content and more sensitive than Hoechst. A protocol for DNA determination using PicoGreen is presented below (http://www.turner biosystems.com/doc/appnotes/998_2630.html).

1. Prepare the PicoGreen fluorescent dye reagent according to the instructions provided by the manufacturer. (Note: the reagent is susceptible to photodegradation, thus protect the working solution from light by covering it with foil. For best results, this solution should be used within a few hours of its preparation.)
2. Prepare a 2 µg ml^{-1} stock solution of Calf thymus dsDNA in 10 mM Tris-HCl, 1 mM EDTA, pH 7.5 (TE) to produce a standard curve.
3. Distribute the 2 µg ml^{-1} DNA stock solution into disposable cuvettes. Add 1.0 ml of the working solution of the dye to each cuvette. Mix well and incubate for 2–5 min at room temperature, protected from light.
4. After incubation, measure the sample fluorescence in a fluorometer using an excitation wavelength of 485 nm and fluorescence measurement at 535 nm. Measure the most fluorescent sample first (1 µg ml^{-1} DNA) to calibrate the instrument sensitivity.
5. Dilute the broken cell suspension in TE (the required dilution will depend greatly on each sample), and filter it through a 0.45 µm cellulose filter to remove particulate material.
6. Add 1.0 ml of the dye solution to each sample. Incubate for 2–5 min at room temperature, protected from light, and measure fluorescence at 535 nm (excitation at 485 nm).
7. Repeat steps 5 and 6 using a higher dilution of the experimental sample to confirm the quantification results.

♦♦♦♦♦ DETERMINATION OF INTRACELLULAR VOLUME

To determine the intracellular concentration of solutes it is essential to measure the intracellular volume. The NaCl concentration of the growth medium can affect the intracellular volume; ideally this parameter should be measured at each NaCl concentration examined. Intracellular volume is usually calculated with the method described by Rottenberg (1979), using the silicone oil technique with tritium-labeled water (probing extracellular and intracellular spaces) and a ^{14}C-labeled compound that does not penetrate the cell, like inulin, dextran or polyethylene glycol (probing extracellular space).

Protocol

1. Incubate 2 ml of a dense cell suspension (equivalent to 10 mg of protein per ml) with ^{3}H$_2$O (10 kBq ml^{-1}) and [^{14}C]polyethylene glycol (30 kBq ml^{-1}) for 10 min at room temperature.
2. Transfer four aliquots of 300 µl to Eppendorf tubes containing 300 µl of silicone oil (d = 1.04) placed over 100 µl of perchloric acid (1 M).
3. Centrifuge at 11 000 × g for 15 min. Under these conditions, cells pass through the silicone oil layer, forming a pellet at the bottom, while the aqueous phase remains on the top of silicone oil.
4. Transfer 250 µl of the aqueous phase above the oil layer to a scintillation vial containing 5 ml of scintillation liquid. Mix the cell pellet with 250 µl of perchloric acid (1 M) and transfer to a scintillation vial containing the scintillation liquid. The scintillation vials containing the two samples are counted with a Beckman liquid scintillation counter using a manually optimized ^{14}C-^{3}H double-labeling program.

The internal volume can then be calculated according to the formula by Rottenberg (1979)

$$V_i = V_s(^3Hp/^3H_s - {^{14}C_p}/{^{14}C_s})$$

where V_i is the internal volume, and the subscripts "s" and "p" refer to supernatant and pellet, respectively. (Note: The calculation of the intracellular volume is based on the assumption that the [^{14}C]polyethylene glycol did not adsorb on to the surface of the cells to a significant extent and tritium reached equilibrium between internal and external spaces).

♦♦♦♦♦ PURIFICATION OF COMPATIBLE SOLUTES

Identification of compatible solutes in a crude extract can be difficult or even impossible in cases where a complex mixture of compatible solutes is accumulated by the organism under study. In these cases, partial purification or even isolation of the new compounds becomes an imperative. Liquid chromatography provides a means to separate molecules with similar chemical and physical properties. Three distinct types of solid matrixes are usually used to isolate solutes: ion-exchange chromatography, gel filtration chromatography and adsorption chromatography. The most useful technique to follow the degree of purification is thin layer chromatography (TLC).

Thin Layer Chromatography

Thin layer chromatography (TLC) and paper chromatography have been widely used to separate and detect compatible solutes. These techniques require low amounts of sample, are fast and inexpensive, and detection is straightforward. An organic solvent or a mixture of organic solvents (mobile phase) migrates through the stationary phase, a thin coating of silica gel, alumina or cellulose. Components will move according to their partition coefficient (concentration of compound in stationary phase relative to concentration of compound in mobile phase). Essentially, the more soluble a compound is in the mobile phase, the further it migrates along the stationary phase. Silica gel plates are commonly used in these analyses. The choice of the solvent system is a very important step in the preparation of a TLC experiment. The common practice consists of searching the scientific literature for a solvent system that separates related compounds. The separation properties can then be improved by empirically changing the percentage of each component in the solvent system. Examples of solvent systems used to separate compatible solutes are presented in Table 8.2.

Once a TLC has been developed, the compounds must be visualized. Unless the compounds are colored or fluorescent, their location on the plate has to be revealed with a chemical reagent to develop color. Several TLC visualization reagents and the type of compounds or structures that are detected are shown in Table 8.3.

Table 8.2 Examples of solvent systems used to separate compatible solutes

Solvent system	Type of compounds separated
Silica gel	
Chloroform/methanol/acetic acid glacial/H_2O (30:50:8:4, v/v/v/v)	Sugars
Butanol/pyridine/water (7:3:1, v/v/v)	Sugar polymers
Butanol/ethanol/water (5:3:2, v/v/v)	Phospho-sugars and sugars
n-Propanol/ammonium 25% (1:1.5, v/v)	Alcohol phosphodiesters
Phenol/water (4:2, v/v)	Betaines
Cellulose paper	
Acetonitrile/0.1 M ammonium acetate (6:4, v/v), pH 4.0	Amino acids and related compounds

Table 8.3 Reagents used in TLC for detection of compatible solutes

Reagents	Compounds detected
Universal reagents	
Phosphomolybdic acid reagent	Most organic compounds
Iodine vapor	Most organic compounds
Specific reagents	
α-naphthol-sulfuric acid reagent	Sugars and related compounds
Ninhydrin reagent	Amino acids and related compounds; N-acetylated diamino acids
Bromocresol green reagent	Betaines

Protocols for the preparation of each TLC visualization reagent, presented in Table 8.3, are given below.

Phosphomolybdic acid reagent: Spray the dried plate with phosphomolybdic acid (5%, w/v, in ethanol), and heat to 120°C with a heat gun.
Iodine vapor: Place the dried TLC plate in a closed vessel containing iodine crystals. Monitor the plate visually as the spots develop a brown color.
α-naphthol-sulfuric acid reagent: Dissolve 1.6 g of α-naphthol in 51 ml of ethanol, and then add 4 ml distilled water, and 6.5 ml of concentrated sulfuric acid. Mix well. Spray the dried TLC plate with α-naphthol-sulfuric acid reagent, and heat at 120°C until spots appear.
Ninhydrin reagent: Spray the dry plate with ninhydrin reagent (0.2%, w/v, in ethanol), and heat at 120°C until spots appear.
Bromocresol green reagent: Dip the dried TLC plate in 0.1 g bromocresol green in 500 ml ethanol and 5 ml 0.1 M NaOH. Dry with a heat gun to develop blue spots on green background.

Cautions
(i) Many organic solvents (such as methanol, chloroform, etc) are highly inflammable and toxic; work in a chemical fume hood with adequate extraction power. (ii) Concentrated acetic acid is highly corrosive to tissue of the mucous membranes and respiratory tract, eyes and skin; wear gloves and safety glasses and work in a chemical fume hood.

Protocol

1. Saturate the chromatographic chamber with the solvent system for 30 min. To ensure that the chamber is completely saturated, place pieces of filter paper in contact with the solvent along two sides of the chamber.
2. Draw a very faint line with a pencil, in the TLC plate (20 × 20 cm), no closer than 2 cm from the bottom or side edges. Divide this line in intervals of 2 cm. Apply a small amount of sample onto the TLC plate with a disposable microcapillary or a micropipette. Dry the sample with a hairdryer, and repeat the spotting process by adding more drops on the original spot. (Note: Never handle the TLC plate with bare fingers.)

3. After cooling-down the spotting area for several minutes at room temperature, place the TLC plate in the chamber (ensure that the spots are above the level of the solvent). Allow the solvent to move nearly to the top of the plate.
4. Remove the plate from the chamber and dry it with a hairdryer in the hood.
5. Spot visualization. In the case of silica gel F245 plates, the compounds can be visualized by UV light. However, the best results are obtained using a specific or universal reagent (see Table 8.3). Always, spray any specific or universal revelator in the hood.
6. Optimization of compound separation. Usually, it is necessary to adjust the polarity of the solvent in order to optimize the separation of the compounds. To alter the polarity of the solvent system take into account the following series (increasing ability to dissolve polar compounds): cyclohexane, heptane, pentane, chlorobenzene, ethylbenzene, toluene, benzene, 2-chloropropane, chloroform, nitrobenzene, ethyl acetate, 2-butanol, ethanol, water, acetone, acetic acid, methanol, pyruvic acid.

Ion-exchange Chromatography

Ion-exchange chromatography separates molecules based on their electrostatic interactions with the fixed charges on a stationary phase. The fixed charges are either negative or positive in the case of cation-exchange or anion-exchange chromatography, respectively.

Protocols

Cation-exchange chromatography

1. Wash 20 g of Dowex 50W-X8 (Bio-Rad) resin by gently stirring it in 100 ml of distilled water for 5 min. Allow the resin to settle, and remove the supernatant by decantation. Repeat one more time.
2. Pack an empty column (20 cm × 1.6 cm) with 10 ml of washed Dowex 50W-X8. Load approximately 100 ml of 1 M HCl to activate the resin. Remove the excess of HCl by abundantly washing the resin with distilled water. Check pH of the flow-through with a pH indicator, and stop washing at around pH 7.0.
3. Dissolve the extract in distilled water and apply onto the top of the column. Elution is performed with a linear gradient of perchloric acid (0–2 M) at a gravity flow.
4. Collect aliquots of 4 ml, and neutralize each sample with KOH (2 M). Analyze the fractions by TLC for the compound(s) of interest. Pool fractions containing compound(s) of interest and adjust the pH to 8, to precipitate the major part of the perchloric acid.
5. The remaining perchloric acid can be removed in an ion retardation column (type AG11-A8, Bio-Rad).

Note
All aqueous solutions must be filtered through a Millipore membrane (0.45 μm), and degassed for 30 minutes.

Caution
All manipulations in the presence of perchloric acid should only be performed in glass vessels.

Anion-exchange chromatography

1. Swell approximately 20 g of QAE-Sephadex A25 (Amersham Biosciences) in 200 ml of distilled water overnight at 4°C. Remove part of the supernatant and agitate the resin slowly. Pack a column (20 cm × 2.6 cm) with 40 ml of QAE-Sephadex A25. Allow the resin to settle, close the column, and connect to the FPLC system. Equilibrate the column with 3 column volumes of 5 mM sodium bicarbonate (pH 9.8) at a flow rate of 4 ml min^{-1}.
2. Dissolve the freeze-dried extract in 2 ml of 5 mM sodium bicarbonate (pH 9.8). Centrifuge at 20 000 × g for 20 min to remove remaining insoluble particles. Load the supernatant onto the column, and wash with 2 volumes of equilibration buffer.
3. After washing the column, initiate the linear gradient from 5 mM sodium bicarbonate to 1 M of sodium bicarbonate (pH 9.8) in 3 volumes, and then 1 volume of 1 M of sodium bicarbonate (pH 9.8) at a flow rate of 4 ml min^{-1}. If an FPLC system is unavailable, the elution can be performed discontinuously in steps of one column volume of 100, 250, 500, 750 and 1000 mM sodium bicarbonate (pH 9.8). The step-elution results in lower resolution.
4. Collect fractions of 4 ml, and look for the compound(s) of interest by TLC. Pool the fractions containing the compound(s) of interest.
5. To remove the sodium bicarbonate, load the sample onto a Dowex 50W-X8 column (for instructions to prepare this column, see above). Elute the column with distilled water.

Note
All aqueous solutions must be filtered with a Millipore membrane (0.45 μm), and degassed under vacuum for 30 min.

Gel Filtration Chromatography

Gel filtration or size exclusion chromatography separates compounds on the basis of molecular size. Compatible solutes that are larger than the pore size of the resin are excluded from the matrix and pass through the resin unimpeded. Smaller compatible solutes that can access the liquid within the pores of the beads are retained longer. Sephadex G-10 (Amersham Biosciences) is usually used to separate small organic compounds.

Protocols

1. A long, narrow column (50 cm × 1.8 cm) is needed to achieve adequate separation. Hydrate the resin in distilled water overnight and degass under vacuum for 1 h to remove air in the resin. Fill the column slowly with the resin. Close the column and wash extensively with de-aerated distilled water in an FPLC system at a flow-rate of 1 ml min^{-1}. Equilibrate the column overnight at 1 ml min^{-1}. (Important note: Do not allow the column to run dry at any stage.)
2. Test the resolution of the column with bromophenol blue (1 ml of 1 mg ml^{-1}). If a sharp peak is observed, the column is ready to be used. If not, leave the column equilibrating with distilled water for one more day, and test it again. Repeat the process until a sharp bromophenol peak is observed.
3. Dissolve the ethanol extract in the minimal amount of distilled water. Centrifuge the sample and load 1 ml onto the column. Since the compatible solutes do not interact with the solid matrix, the resolution of the separation depends highly on the sample volume, therefore, load the smaller feasible volume.
4. Collect fractions of 1 ml in an automatic fraction collector. Analyze small aliquots of each fraction using TLC.

Note

All aqueous solutions must be filtered through a Millipore membrane (0.45 µm), and degassed for 30 min.

Adsorption Chromatography

1. Before preparing the silica gel S (0.032–0.063 mm, Riedel-de-Häen) column, the mobile phase has to be optimized in TLC's to produce adequate separation of the compounds (the protocol is above).
2. To prepare the matrix, gently mix 20 g of silica gel S with 50 ml of the chosen solvent system. Allow the silica to settle, and remove the supernatant by decantation. Repeat the last step. This step removes small particles of silica gel S that can obstruct the column.
3. Pack a column (20 cm × 2 cm) with 20 g of silica gel S slurry. After the column has been filled, tap the sides of the column with a plastic hammer. Equilibrate the column with approximately 3 volumes of the solvent to be used in the separation. These columns are packed and run by gravity flow. (Note: (i) After packing the column, fill it up to the top with solvent to maintain the maximal gravity flow. (ii) It is possible to pressurize the top of the column to enhance the speed of the solvent flow. Note that high flow rates will decrease the resolution of the separation).
4. Dissolve the extract in the solvent system and apply it on top of the column. Allow the sample to be adsorbed on the silica gel. Add the same volume of solvent on top of the column, and allow all the solvent to enter the column. After this washing process, perform elution with

the same solvent system. (Notes: (i) Sometimes the extract is not completely soluble in the organic solvent system; add drops of the most polar component to dissolve the compounds. (ii) Always centrifuge the sample before applying to the column to remove the insoluble fraction and check by TLC or NMR for the presence of the compound of interest in the insoluble fraction.)

5. Collect fractions of 2 ml and analyze them for the presence of the compound(s) by TLC. The fractions containing the compound of interest are pooled and the organic solvent evaporated under vacuum. Some organic compounds, like acetate, are not completely removed by evaporation. In this case, the sample containing acetate can be applied on to a QAE-Sephadex-A25 column (for instructions on the preparation of a QAE-Sephadex A25 column, see above).

◆◆◆◆◆ IDENTIFICATION OF UNKNOWN COMPATIBLE SOLUTES

In the event of an unknown solute being present in the extract, two-dimensional NMR techniques can be used to elucidate its structure. If the sample is simple (low number of compounds) and the new compound presents spectra of low complexity, NMR analyses can proceed without any purification step, however, some degree of purification is always advisable since the presence of other compounds can mask some spectral features of the unknown one. Chromatographic techniques like gel filtration and ion-exchange can be used to achieve some degree of purification. A sample in which the unknown compound is the major component is usually good enough to establish its structure.

Once a suitable sample is obtained NMR analyses can follow. The number and type of spectra that are required vary with the problem at hand. The most commonly used spectra are the following:

COSY – Homonuclear proton–proton correlation of vicinal protons – allows following the several proton signals in a molecule sequentially.

NOESY – Homonuclear proton–proton correlation *via* the nuclear Overhauser effect – allows establishing proximity in space.

TOCSY – Homonuclear proton–proton correlation through sequences of vicinal protons – it allows to find all the signals belonging to the same spin system.

$^1H/^{13}C$ HMQC – Direct heteronuclear carbon–proton correlation – allows the determination of carbon-proton pairs involved in covalent bonds.

$^1H/^{13}C$ HMBC – Heteronuclear multiple bond carbon-proton correlation – allows the establishment of connectivity between proton and carbon atoms at 2–4 bonds distance.

^{31}P spectrum – one-dimensional phosphorus spectrum – allows determining the presence of phosphorus in the molecule, a relatively common feature in solutes from thermophilic sources.

^{1}H/^{31}P HMQC – Heteronuclear phosphorus–proton correlation – allows the determination of the position of the phosphorus atoms in the molecule.

For practical reasons, nitrogen information is not readily available in natural abundance samples, however the presence of amide groups can be investigated acquiring proton spectra in 90% H_2O/10% D_2O (instead of 100% D_2O); signals due to protons linked to the nitrogen atom will appear in the low field region of the spectrum.

Once the structure has been determined by NMR, it should be confirmed by an unrelated technique like Mass Spectrometry, and if possible, an authentic sample should be synthesized and its spectra compared with the natural product to firmly establish its identity.

Example

The procedure followed to identify the structure of galactosyl-5-hydroxylysine will be presented here. This solute was found in extracts of *Thermococcus litoralis* cells grown in peptone and yeast extract as carbon sources (Lamosa et al., 1998). ^{13}C NMR spectra of ethanol extracts of the *T. litoralis* cells showed several sets of resonances that were assigned to di-*myo*-inositol-phosphate, mannosylglycerate, trehalose, β-glutamate, hydroxyproline and aspartate (Figure 8.2). These assignments were obtained by comparison with the chemical shift values found in the literature.

A remaining set of resonances in the ^{13}C NMR spectrum (176.0; 102.5; 76.5; 75.5; 72.8; 71.2; 68.9; 61.4; 55.0; 43.4; 28.4 and 27.1 ppm) could not be initially assigned to a known compound. These were minor signals in a relatively complex sample so the need for purification became obvious. In fact, the identification was significantly facilitated by partial purification of the extract, by ion-exchange chromatography. In this process we found out that the new compound presented a global positive charge at pH 7. The resonances at 176.0 and 102.5 ppm were assigned to a carboxylic group and an anomeric CH group (most likely C_1 of a sugar moiety), respectively, due to their characteristic chemical shift values. The HMQC spectrum (a proton–carbon correlation spectrum through direct covalent bonds) allowed to determine which signals in the proton spectrum were directly bonded to which signals in the carbon spectrum, in particular it showed that the carbon resonance at 102.5 ppm was connected to a proton resonance at 4.49 ppm (Figure 8.3). The same spectrum also showed that resonances at 27.1, 28.4, 43.4 and 61.4 ppm were due to CH_2 groups (two proton signals for a single carbon resonance), while the remaining resonances were due to CH groups. The TOCSY spectrum revealed two spin systems (corresponding to two moieties). The analysis of the COSY spectrum allowed following the signals sequentially within each moiety, and this data in conjunction with the HMQC data showed that the anomeric carbon at 102.5 ppm belonged to a network of a six-carbon monosaccharide (resonances at 102.5; 75.5; 72.8; 71.2; 68.9 and 61.4 ppm). The other moiety (which

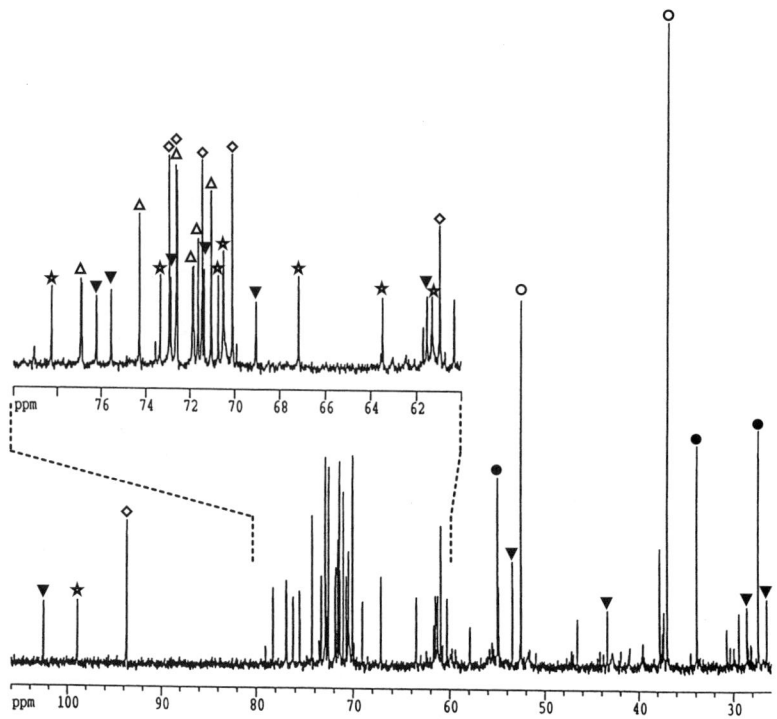

Figure 8.2. Proton decoupled ^{13}C NMR spectrum of an ethanol extract of *Thermococcus litoralis* grown at 85°C in Bacto marine broth containing 4% NaCl. Resonances are due to aspartate (open circles), mannosylglycerate (stars), di-*myo*-inositol-(1,1')-phosphate (open triangles), glutamate (solid circles), β-galactosyl-5-hydroxylysine (solid triangles) and trehalose (open diamonds).

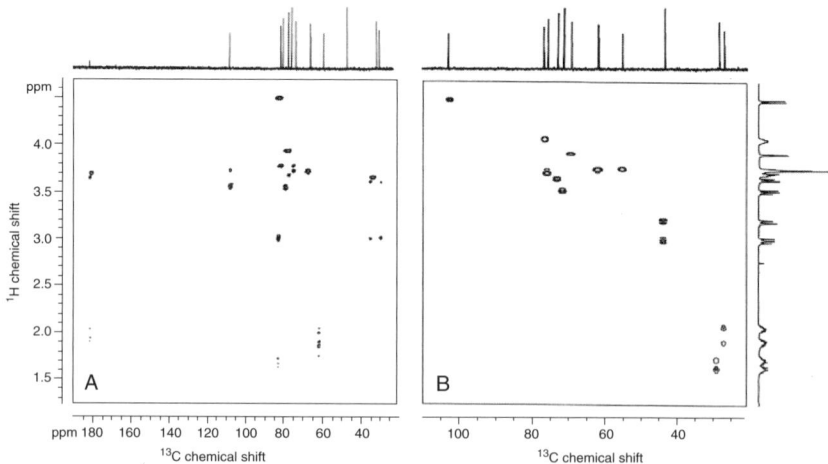

Figure 8.3. ^{1}H/^{13}C correlation spectra of β-galactosylhydroxyproline. (A) heteronuclear multiple bond correlation spectrum, and (B) heteronuclear double quantum correlation spectrum.

had to include the carboxyl group) presented a signal at 55.0 ppm (the region where the α carbons of amino acids commonly appear), and a signal assigned to carbon 5, at 76.5 ppm, a value typical of –CHOH groups. At this stage we suspected the presence of 5-hydroxylysine. The hexose and the amino acid fragments were firmly identified as galactose and 5-hydroxylysine after mild acid hydrolysis of the intact compound and spiking of the hydrolysate with pure galactose and 5-hydroxylysine.

The carbon chemical shift values of the original compound indicated that the galactose moiety was in the pyranose rather than in the furanose configuration (Bock and Pedersen, 1983). The observation of connectivities between H_1 and H_2, H_3 and H_5, but not between H_1 and H_4 in the NOESY spectra, led to the belief that galactose was in the β-pyranosyl configuration. This conclusion was reinforced by the low chemical shift value for galactose H_1 (4.49 ppm) and established by the magnitude of the coupling constants $^3J_{H1,H2} = 7.7$ Hz, $^1J_{C1,H1} = 160$ Hz which unambiguously indicate a β-pyranosyl configuration (Baumann et al., 1992; Bock and Pedersen, 1974; Mehlert et al., 1998). There was also a clear connectivity between H_1 of galactose and H_5 of 5-hydroxylysine in the NOESY spectrum showing the location of the linkage to be between C_1 of galactose and the hydroxyl group of the amino acid. The position of this linkage was confirmed in the HMBC spectrum. On the basis of these results, the unknown compound was firmly identified as β-galactopyranosyl-5-hydroxylysine (inset in Figure 8.4).

◆◆◆◆◆ CONCLUDING REMARKS

Most of the compatible solutes accumulated by thermophiles and hyperthermophiles are negatively charged, which implies the intracellular accumulation of positive counterions, such as potassium. Although well-established analytical methods are available for the determination of potassium (or sodium), the evaluation of the intracellular content of these cations is not easy. NMR coupled with paramagnetic shift reagents provides an attractive means to distinguish between intracellular and extracellular ions in whole cells, but suffers from the drawback of NMR "invisibility", i.e. part of the intracellular potassium (or sodium) may be undetected.

The application of the methods described in this chapter resulted in the identification of several novel compatible solutes synthesized *de novo* by (hyper)thermophiles. The search for novel compatible solutes in organisms from hot environments is far from complete and research in this field is sure to proceed rapidly, fueled by the fundamental interest in the discovery of the physiological role of these solutes in osmo- and thermoadaptation, as well as by their usefulness in an increasing number of biotechnological applications.

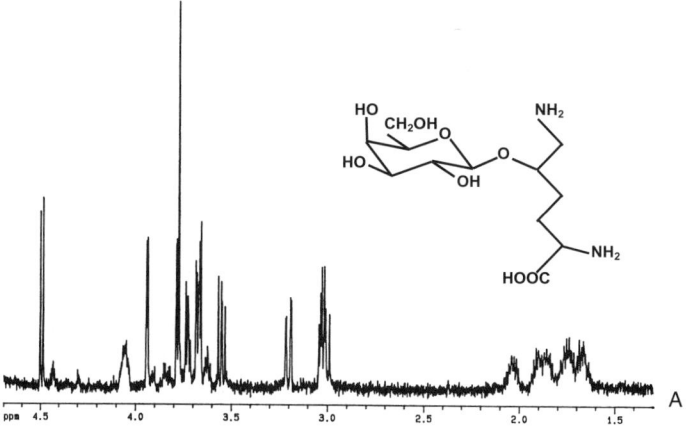

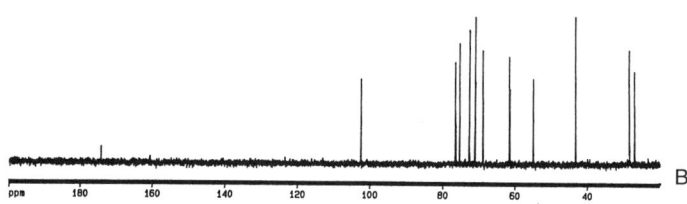

Figure 8.4. ^{1}H NMR (A) and ^{13}C NMR spectra (B) of β-galactopyranosyl-5-hydroxylysine. A structural representation of the compound is shown on the top as an inset.

Acknowledgements

This work was supported by the European Commission, 6th Framework Programme contract COOP-CT-2003-508644 and by Fundação para a Ciência e a Tecnologia, PRODEP, FEDER, and POCTI, Portugal (POCTI/BIA-PRO/57363/2004). Pedro Lamosa and Nuno Borges, acknowledge post-doctoral grants from FCT, Portugal, SFRH/BPD/11511/2002 and SFRH/BPD/14841/2003, respectively.

References

Alfredsson, G. A., Kristjánsson, J. K., Hjörleifsdottir, S. and Stetter, K. O. (1988). *Rhodothermus marinus*, gen. nov., sp. nov., a thermophilic halophilic bacterium from submarine hot springs in Iceland. *J. Gen. Microbiol.* **134**, 299–306.

Baumann, H., Tzianabos, A. O., Brisson, J., Kasper, D. L. and Jennings, H. J. (1992). Structural elucidation of two capsular polysaccharides from one strain of *Bacteroides fragilis* using high-resolution NMR spectroscopy. *Biochemistry* **31**, 4081–4089.

Benaroudj, N., Hee Lee, H. and Goldberg, A. L. (2001). Trehalose accumulation during cellular stress protects cells and cellular proteins from damage by oxygen radicals. *J. Biol. Chem.* **276**, 24261–24267.

Blöchl, E., Burggraf, S., Fiala, G., Lauerer, G., Huber, G., Huber, R., Rachel, R., Segerer, A., Stetter, K. O. and Völkl, P. (1995). Isolation, taxonomy and phylogeny of hyperthermophilic microorganisms. *World J. Microbiol. Biotechnol.* **11**, 9–16.

Bock, K. and Pedersen, C. (1974). A study of ^{13}CH coupling constants in hexopyranoses. *J. Chem. Soc. Perkin II* **3**, 293–297.

Bock, K. and Pedersen, C. (1983). Carbon-13 nuclear magnetic resonance spectroscopy of monosaccharides. *Adv. Carbohydr. Chem. Biochem.* **41**, 27–66.

Borges, N., Ramos, A., Raven, N. D., Sharp, R. J. and Santos, H. (2002). Comparative study of the thermostabilizing properties of mannosylglycerate and other compatible solutes on model enzymes. *Extremophiles* **6**, 209–216.

Bradford, M. M. (1976). A rapid and sensitive method for quantification of microgram quantities of protein utilizing the principle of protein dye-binding. *Anal. Biochem.* **72**, 248–254.

Breitmaier, E. and Voelter, W. (1990). *Carbon-13 NMR Spectroscopy.* VCH Verlagsgesellschaft mbH, Weinheim.

Brown, A. D. (1976). Microbial water stress. *Bacteriol. Rev.* **40**, 803–846.

Ciulla, R. A., Burggraf, S., Stetter, K. O. and Roberts, M. F. (1994a). Occurrence and role of di-*myo*-inositol-1,1'-phosphate in *Methanococcus igneus*. *Appl. Environ. Microbiol.* **60**, 3660–3664.

Ciulla, R. A., Clougherty, C., Belay, N., Krishnan, S., Zhou, C., Byrd, D. and Roberts, M. F. (1994b). Halotolerance of *Methanobacterium thermoautotrophicum* ΔH and Marburg. *J. Bacteriol.* **176**, 3177–3187.

da Costa, M. S., Santos, H. and Galinski, E. A. (1998). An overview of the role and diversity of compatible solutes in *Bacteria and Archaea*. *Adv. Biochem. Eng/Biotechnol.* **61**, 118–153.

da Costa, M. S., Nobre, M. F. and Rainey, F. A. (2001). The genus Thermus In *Bergey's Manual of Systematic Bacteriology* (D. R. Boone and R. W. Castenholtz, eds), 2nd edn., vol. 1, pp. 404–414. Springer-Verlag, New York.

Galinski, E. A. (1995). Osmoadaptation in bacteria. *Adv. Microbiol. Physiol.* **37**, 272–328.

Gonçalves, L. G., Huber, R., da Costa, M. S. and Santos, H. (2003). A variant of the hyperthermophile *Archaeoglobus fulgidus* adapted to grow at high salinity. *FEMS Microbiol. Lett.* **218**, 239–244.

Kanodia, S. and Roberts, M. F. (1983). Methanophosphagen: unique cyclic pyrophosphate isolated from *Methanobacterium thermoautotrophicum*. *Proc. Natl. Acad. Sci. USA* **80**, 5217–5221.

Kaushik, J. K. and Bhat, R. (1999). A mechanistic analysis of the increase in the thermal stability of proteins in aqueous carboxylic acid salt solutions. *Protein Sci.* **8**, 222–233.

Lai, M.-C., Ciulla, R., Roberts, M. F., Sowers, K. R. and Gunsalus, R. P. (1995). Extraction and detection of compatible intracellular solutes. In *Archaea – A Laboratory Manual* (K. R. Sowers and H. J. Schreier, eds), pp. 349–368. Cold Spring Harbor Laboratory Press, Cold Spring Harbor.

Lamosa, P., Martins, L. O., da Costa, M. S. and Santos, H. (1998). Effects of temperature, salinity, and medium composition on compatible solute accumulation by *Thermococcus* spp. *Appl. Environ. Microbiol.* **64**, 3591–3598.

Lamosa, P., Burke, A., Peist, R., Huber, R., Liu, M. Y., Silva, G., Rodrigues-Pousada, C., LeGall, J., Maycock, C. and Santos, H. (2000). Thermostabilization of proteins by diglycerol phosphate, a new compatible solute from the hyperthermophile *Archaeoglobus fulgidus*. *Appl. Environ. Microbiol.* **66**, 1974–1979.

Lowry, O. H., Rosebrough, N. J., Farr, A. J. and Randall, R. J. (1951). Protein measurements with the Folin phenol reagent. *J. Biol. Chem.* **193**, 265–275.

Martin, D. D., Ciulla, R. A. and Roberts, M. F. (1999). Osmoadaption in archaea. *Appl. Environ. Microbiol.* **65**, 1815–1825.

Martins, L. O. and Santos, H. (1995). Accumulation of mannosylglycerate and di-*myo*-inositol-phosphate by *Pyrococcus furiosus* in response to salinity and temperature. *Appl. Environ. Microbiol.* **61**, 3299–3303.

Martins, L. O., Carreto, L. S., da Costa, M. S. and Santos, H. (1996). New compatible solutes related to di-*myo*-inositol-phosphate in members of the order *Thermotogales*. *J. Bacteriol.* **178**, 5644–5651.

Martins, L. O., Huber, R., Huber, H., Stetter, K. O., da Costa, M. S. and Santos, H. (1997). Organic solutes in hyperthermophilic *Archaea*. *Appl. Environ. Microbiol.* **63**, 896–902.

Mehlert, A., Richardson, J. M. and Ferguson, M. A. (1998). Structure of the glycosylphosphatidylinositol membrane anchor glycan of a class-2 variant surface glycoprotein from *Trypanosoma brucei*. *J. Mol. Biol.* **277**, 379–392.

Nunes, O. C., Manaia, C. M., da Costa, M. S. and Santos, H. (1995). Compatible solutes in the thermophilic bacteria *Rhodothermus marinus* and "*Thermus thermophilus*". *Appl. Environ. Microbiol.* **61**, 2351–2357.

Roberts, M. F. (2000). Osmoadaptation and osmoregulation in archaea. *Front. Biosci.* **5**, 796–812.

Roberts, M. F. (2004). Osmoadaptation and osmoregulation in archaea. *Front. Biosci.* **9**, 1999–2019.

Robertson, D. E. and Roberts, M. F. (1991). Organic osmolytes in methanogenic archaebacteria. *Biofactors*. **3**, 1–9.

Robinson, P. M. and Roberts, M. F. (1997). Effects of osmolyte precursors on the distribution of compatible solutes in *Methanohalophilus portucalensis*. *Appl. Environ. Microbiol.* **63**, 4032–4038.

Robertson, D. E., Roberts, M. F., Belay, N., Stetter, K. O. and Boone, D. R. (1990). Occurrence of β-glutamate, a novel osmolyte, in marine methanogenic bacteria. *Appl. Environ. Microbiol.* **56**, 1504–1508.

Rottenberg, H. (1979). The measurement of membrane potential and ΔpH in cells, organelles and vesicles. *Meth. Enzymol.* **55**, 547–569.

Santos, H. and da Costa, M. S. (2001). Organic solutes from thermophiles and hyperthermophiles. *Meth. Enzymol.* **334**, 302–315.

Santos, H. and da Costa, M. S. (2002). Compatible solutes of organisms that live in hot saline environments. *Environ. Microbiol.* **4**, 501–509.

Scholz, S., Sonnenbichler, J., Schäfer, W. and Hensel, R. (1992). Di-*myo*-inositol-1,1'-phosphate: a new inositol phosphate isolated from *Pyrococcus woesei*. *FEBS Lett.* **306**, 239–242.

Silva, Z., Borges, N., Martins, L. O., Wait, R., da Costa, M. S. and Santos, H. (1999). Combined effect of the growth temperature and salinity of the medium on the accumulation of compatible solutes by *Rhodothermus marinus* and *Rhodothermus obamensis*. *Extremophiles* **3**, 163–172.

Smith, P. K., Krohn, R. I., Hermanson, G. T., Mallia, A. K., Gartner, F. H., Provenzano, M. D., Fujimoto, E. K., Goeke, N. M., Olson, B. J. and Klenk, D. C. (1985). Measurement of protein using bicinchoninic acid. *Anal. Biochem.* **150**, 76–85.

Sowers, K. R., Robertson, D. E., Noll, D., Gunsalus, R. P. and Roberts, M. F. (1990). N^ε-acetyl-β-lysine: an osmolyte synthesized by methanogenic archaebacteria. *Proc. Natl. Acad. Sci. USA* **87**, 9083–9087.

Welsh, D. T. (2000). Ecological significance of compatible solute accumulation by micro-organisms: from single cells to global climate. *FEMS Microbiol. Rev.* **24**, 263–290.

9 Functional Genomics of the Thermo-Acidophilic Archaeon *Sulfolobus solfataricus*

John van der Oost[1], Jasper Walther[1], Stan J J Brouns[1], Harmen J G van de Werken[1], Ambrosius P L Snijders[2], Phillip C Wright[2], Anders Andersson[3], Rolf Bernander[4] and Willem M de Vos[1]

[1] Laboratory of Microbiology, Wageningen University, The Netherlands; [2] Biological and Environmental Systems Group, Department of Chemical and Process Engineering, Sheffield University, UK; [3] Department of Biotechnology, KTH – Royal Institute of Technology, Stockholm, Sweden; [4] Department of Molecular Evolution, Evolutionary Biology Center, Uppsala University, Sweden

CONTENTS

Introduction
Thermophiles
Thermophile genomics
Transcriptome analysis
Proteome analysis
Concluding remarks

◆◆◆◆◆ INTRODUCTION

Archaea and bacteria that optimally grow at temperatures above 60°C and 80°C are referred to as thermophiles and hyperthermophiles, respectively (Stetter, 1996). Since their discovery in the late 1960s (Brock and Freeze, 1969), attempts were made to reveal the secrets of the thermal resistance of these microorganisms, initially by physiological, biochemical and genetic analysis (Allers and Mevarech, 2005). In addition, the sequencing of the genomes of many thermophiles during the last decade has allowed for a series of genome-based research lines. Comparative genomics is the *in silico* analysis of genome data that aims to predict the metabolic potential of an organism, including the interconversions of metabolites and the regulation thereof (Ettema *et al.*, 2005; Makarova and Koonin, 2003). Functional genomics is the experimental analysis at the level of RNA (transcriptomics), protein (proteomics), as well as metabolites (metabolomics). In general, such holistic studies aim at addressing the phenotypic response of an organism either to different cultivation conditions, or to genotypic variations. Here we review recent developments of functional genomics of thermophiles in general, and of the thermo-acidophilic archaeon *Sulfolobus solfataricus* in particular.

◆◆◆◆◆ THERMOPHILES

Forty years ago it was generally accepted that life was not possible at temperatures higher than 60°C. In 1969, however, Thomas Brock and co-workers extended the upper temperature limit to 75°C when a microorganism was isolated from thermal springs in Yellowstone National Park (Brock and Freeze, 1969). This organism was characterized as a Gram-negative bacterium that was named *Thermus aquaticus*. Subsequently, Brock's group isolated a remarkable thermo-acidophilic microorganism that they named *Sulfolobus acidocaldarius* from a number of both thermal acid soils and acid hot springs in Yellowstone National Park where there is an abundant source of sulfur, low pH (less than 3.0), and high temperature (65 to 90°C) (Brock et al., 1972). Based on morphological analysis, *Sulfolobus* was initially classified as a thermo-acidophilic bacterium with some remarkable properties (Shivers and Brock, 1973). However, the molecular classification introduced by Carl Woese and colleagues indicated that *Sulfolobus* does not belong to the Bacteria, but rather to a newly-discovered prokaryotic domain: the Archaea (for review see Woese, 1987; Pace, 1997).

The pioneering work of Brock set the stage for further exploration of a wide range of thermal ecosystems worldwide. This has resulted in an ever-growing collection of thermophiles (both Archaea and Bacteria; optimal growth 60–80°C) and hyperthermophiles (mainly Archaea and some Bacteria; optimal growth >80°C) that have been isolated from terrestrial solfataric fields (e.g. the crenarchaea *Sulfolobus solfataricus*, *Sulfolobus acidocaldarius*, *Sulfolobus tokodaii*, isolated from Italy, USA, Iceland, Japan, New Zealand), but also from marine ecosystems ranging from shallow vents (e.g. the bacterium *Thermotoga maritima*, and the euryarchaea *Pyrococcus furiosus*, *Thermococcus kodakaraensis* from Italy and Japan) to deep sea smokers (Pacific, Atlantic Ocean; e.g. the euryarchaea *Methanocaldococcus jannaschii*, Pyrococcus abyssi, *Pyrococcus horikoshii*) (reviewed by Fukui et al., 2005; Huber et al., 2000; Rothschild and Mancinelli, 2001; Stetter, 1996, 1999). The most robust hyperthermophiles described to date are crenarchaea of the family Pyrodictiaceae: *Pyrolobus fumarii* (optimum at 106°C, growth up to 113°C; Blochl et al., 1997) and a related iron-reducing isolate (optimum 106°C, growth between 113 and 121°C; Kashefi and Lovley, 2003). Remarkably, vegetative cultures of *Pyrolobus* and *Pyrodictium* have been demonstrated to survive autoclaving (Kashefi and Lovley, 2003; Stetter, 1999).

◆◆◆◆◆ THERMOPHILE GENOMICS

The last decade several thermophilic Archaea have been selected for complete genomic sequencing (Table 9.1), not only for the expected discovery of robust biocatalysts, but also because of the anticipated insight into: (i) thermophile physiology (unique metabolic enzymes and pathways, especially in Archaea; Ettema *et al.*, 2005); Makarova and Koonin, 2003, (ii) the molecular basis of thermostability of bio-molecules (enhanced numbers of charged residues at surface and subunit/domain interfaces; Cambillau and Claverie, 2000), and (iii) the evolution of the eukaryotic cell (fusion of archaeal and bacterial cell; Rivera and Lake, 2004).

The first complete genome analysis of an archaeon, *Methanocaldococcus jannaschii* (Bult *et al.*, 1996), confirmed the monophyletic position of the Archaea, with respect to the bacteria and the eukaryotes. In addition, Archaea appeared to possess a bacterial-like compact chromosomal organisation with clustering of genes as polycistronic units (operons), and with only few interrupted genes (introns). Moreover, the archaeal systems that drive the flow of genetic information (transcription, translation, replication, DNA repair) correspond to the core of the eukaryal counterparts. These initial observations of bacterial-like "information storage" and eukaryal-like "information processing" have been confirmed by the analyses of subsequently sequenced hyperthermophilic model Archaea: the euryarchaea *Pyrococcus* spp. (*P. furiosus*, P. abyssi, P. horikoshii*) as well as the crenarchaea *Sulfolobus* spp. (*S. solfataricus, S. tokodaii, S. acidocaldarius*) (reviewed by Makarova and Koonin, 2003). The comparative analysis of the genome of the hyperthermophilic bacterium *Thermotoga maritima* with that of *Pyrococcus furiosus* (both isolated from shallow thermal vents at the same beach (Volcano, Italy), led to the conclusion that horizontal (or lateral) gene transfer substantially contributes to the apparent high degree of genome flexibility (Koonin *et al.*, 2001; Nelson *et al.*, 1999). In addition, the comparison of closely related species (*P. furiosus*, P. abyssi, *P. horikoshii*) revealed a high degree of genome plasticity; moreover, it was proposed that that the lateral gain as well as the loss of genes is a modular event (Ettema *et al.*, 2001). Horizontal gene transfer has also been proposed to explain the relatively high degree of homology between genome fragments of the euryarchaeon *Thermoplasma acidophilum* and the crenarchaeon *Sulfolobus solfataricus*, phylogenetically distant Archaea but inhabiting the same environment (65–85°C, pH 2). The *Sulfolobus*-like genes in the *T. acidophilum* genome are clustered into at least five discrete regions, again indicating recombination of larger DNA fragments (Frickey and Lupas, 2004; Ruepp *et al.*, 2000).

After the genome sequence of thermophiles was established, comparative genomics analyses have been performed to assign potential functions for the identified open reading frames. As in all studied genomes, many unique and conserved hypothetical genes (typically half of the total number of genes) were found for which a function could not reliably be predicted. Hence, the main challenge of the

Table 9.1 Thermophilic Archaea and Bacteria that are mentioned in the text

Name	Lifestyle	Genome (Mbp)	Reference
Bacteria			
Thermotoga maritima	Anaerobic chemorganotroph	1.9	Nelson *et al.* (1999)
Archaea – Crenarchaea			
Sulfolobus solfataricus	Aerobic thermo-acidophile, chemorganotroph, sulphur oxidizer	3.0	She *et al.* (2001)
Sulfolobus acidocaldarius	See *S. solfataricus*	2.3	Chen *et al.* (2005)
Sulfolobus tokodaii	See *S. solfataricus*	2.7	Kawarabayasi *et al.* (2001)
Archaea – Euryarchaea			
Methanocaldococcus jannaschii	Anaerobic thermophilic chemo-lithoautotrophic methanogen	1.7	Bult *et al.* (1996)
Pyrococcus furiosus	Anaerobic heterotrophic thermophile, grows on peptides and sugars	1.9	Robb *et al.* (2001)
P. abyssi	Anaerobic heterotrophic thermophile, grows on peptides	1.8	Cohen *et al.* (2003)
Pyrococcus horikoshii	See *P. abyssi*	1.7	Kawarabayasi *et al.* (1998)
Thermoplasma acidophilum	Facultative anaerobic chemorganotrophic thermoacidophile, lacks cell-wall	1.6	Ruepp *et al.* (2000)
Thermoplasma volcanium	See *T. acidophilum*	1.6	Kawashima *et al.* (2000)
Archaea – Nanoarchaea			
Nanoarchaeum equitans	Obligate archaeal parasite, depends on host for metabolites, extremely reduced genome	0.5	Waters *et al.* (2003)

post-genome era is to improve the functional annotation of genes by integrating classical approaches (physiology, biochemistry and molecular genetics) with genomics-based high-throughput approaches (comparative, functional and structural genomics). Obvious targets of comparative and functional analysis of thermophile genomes are the numerous missing links in metabolic pathways as well as the largely unknown regulatory systems (Ettema *et al.*, 2005; Makarova and Koonin, 2003).

The holistic, and often high-throughput analysis of genome structure and function relations is referred to as Functional Genomics. This genomics-based experimental analysis is sub-divided on the basis of the level of analysis: RNA (transcriptomics), protein (proteomics), and metabolites (metabolomics). We here review recent developments of the different types of functional genomics analyses of thermophilic model organisms, with a main focus on the thermo-acidophilic archaeon *Sulfolobus solfataricus*.

◆◆◆◆◆ TRANSCRIPTOME ANALYSIS

A transcriptome is the complete set of RNA transcripts detectable in an organism of interest at a certain moment in time. A typical transcriptome experiment is the analysis either of a single genotype (e.g. wild type) of the selected organism that is grown under distinct cultivation conditions, or of different genotypes (e.g. wild-type and mutant) that are grown under similar conditions. Analysis of the complete set of transcripts may be performed by using a DNA microarray, comprising DNA fragments that correspond to all genes. These fragments can be (i) a genomic fragment (plasmid library), (ii) a PCR-amplified (full-length/partial) copy of the gene, or (iii) one or more oligonucleotides (length either 25-, 50-, 60-, or 70-mer), the sequences of which correspond to gene fragments (Schena, 1996). The DNA fragments are either spotted on a nitrocellulose filter (macroarray), or printed on a glass slide (microarray); alternatively, oligonucleotides (typically 25-mers) may be synthesized on the microarray directly (Affymetrix). When dedicated arrays are available, the experimental procedures are as follows (experimental details are provided below): (i) isolation of total RNA at different time points from a single culture, or simultaneously from two separate cultures, (ii) synthesis of cDNA from each batch of RNA, and incorporation of aminoallyl-dUTP, (iii) labelling of cDNA with distinct variants of a fluorescent dye (e.g. Cy-3 and Cy-5; Amersham Biosciences), (iv) hybridization of a specific DNA microarray with the labelled cDNA, and (v) data acquisition (scanning and image analysis), processing (filtering and normalization) and analysis (averaging, statistical analysis). The resulting ratios of fluorescent signals reflect the relative levels and possibly the fluctuation of expression of all genes included on the DNA microarray.

DNA Microarrays of Thermophiles

DNA microarrays have initially been established as high-throughput functional genomics tools to study eukaryal and bacterial model systems. The first microarray analysis reported on either a hyperthermophile or an archaeon was a pilot study on *Pyrococcus furiosus* that focussed on 271 PCR products of (potential) metabolic genes (Schut et al., 2001; Table 9.2). This initial study was followed by analyses on a complete genome array (Schut et al., 2003; Weinberg et al., 2005; Table 9.2). These studies addressed the adaptation of *Pyrococcus* cells to the availability of sulfur, to different carbon sources, and to cold shock. A similar approach has been taken for the bacterium *Thermotoga maritima*: 269 PCR products of genes predicted to be responsible for the potential of the organism to grow on a wide variety of carbohydrates and (poly)peptides were spotted on glass slides, and hybridized with differentially labelled cDNA derived from RNA from cultures grown with glucose and tryptone (Chhabra et al., 2003; Table 9.2). A complete array of *Thermotoga maritima* has recently been established at TIGR (Nguyen et al., 2004); using chemostat cultures on different sugars catabolite repression has been studied in this thermophilc bacterium.

DNA Microarray Analysis of *Sulfolobus*

DNA microarrays have recently also been developed for *Sulfolobus solfataricus* and *S. acidocaldarius* (Andersson et al., 2005). Based on the genome sequences of *S. solfataricus* (3.0 Mbp) (She et al., 2001) and *S. acidocaldarius* (2.3 Mbp) (R. Garrett, personal communication), internal fragments of genes (100–800 bp) were PCR amplified, checked by agarose gel electrophoresis, and printed on glass slides. Software (DualPrime; http://www.biotech.kth.se/molbio/microarray/) has been developed that allowed for reducing the number of primers required to PCR amplify related (often orthologous) genes or fragments thereof: 2488 genes from *Sulfolobus solfataricus* and 1960 from *Sulfolobus acidocaldarius* (Lundgren et al., 2004). The excellent quality of this *Sulfolobus* microarray has been demonstrated by an experiment in which hybridization with genomic DNA from synchronized cells in different phases of the cell cycle revealed the three origins of replication of the *Sulfolobus* chromosome (Lundgren et al., 2004). A wide range of transcriptomics experiments has been performed with these arrays (Bernander et al., unpublished), including recent experiments in which *S. solfataricus* was grown on sugars (e.g. glucose, arabinose) or on peptides (e.g. tryptone) (Brouns et al., unpublished; Snijders et al., 2006). In both studies, comparative and different functional genomics methods have been integrated. Below, experimental details of a typical transcriptome analysis experiment of *S. solfataricus* are provided.

Table 9.2 Thermophile microarrays – comparison of relevant details; for details (on *Sulfolobus solfataricus*; http://www.biotech.kth.se/molbio/microarray/index.html) see text

#	(1) *Pyrococcus furiosus*	(2) *Pyrococcus furiosus*	(3) *Pyrococcus furiosus*	(4) *Thermotoga maritima*	(5) *Sulfolobus solfataricus*
Reference	Schut et al. (2001)	Schut et al. (2003) Weinberg et al. (2005)	Shockley et al., (2003)	Chhabra et al. (2003), Johnson et al. (2005)	Brouns et al. (unpublished results), Snijders et al. (2006)
Genome ORFs (estimated)	2065	2065	2065	1928	2950
Array ORFs	271	2065	201	269	2488
Array	Aminosilane-coated glass slides	Aminosilane-coated glass slides	CMT-GAPS aminosilane-coated slides (Corning)	As (3)	Aminosilane-coated glass slides
Synth/print	Slide printer (Omnigrid)	As (1)	DNA was attached to the substrate by UV cross-linking in at 250 mJ and baking at 75°C for 2 h	As (3)	
DNA	PCR Full length, or 1000 bp	As (1)	PCR 400–700 bp	PCR Size not indicated	PCR
Cultivation	Batch	Batch	Batch	Batch	Batch

(*continued*)

Table 9.2 Continued

#	(1) *Pyrococcus furiosus*	(2) *Pyrococcus furiosus*	(3) *Pyrococcus furiosus*	(4) *Thermotoga maritima*	(5) *Sulfolobus solfataricus*
Lysis				Cell pellet frozen on dry ice; frozen cells were disrupted by using a mortar and pestle and 12 volumes of lysis/binding solution (Ambion)	Peletted cells dissolved in Lysis Buffer (guanidinium isothiocyanate, citrate, sarcosyl, beta-mercaptoethanol; see text)
Total RNA isolation	Acid phenol extraction	As (1)	RNAqueous kit (Ambion).	As (3)	Acid phenol extraction
RT enzyme	Stratascript RT (Stratagene), purified using a QIAquick PCR purification kit (Qiagen)	As (1)	Stratascript RT (Stratagene), purified using GFX columns (Amersham)	Stratascript RT (Stratagene), purified using GFX columns (Amersham)	Superscript-II RT (Invitrogen), purified using MinElute columns (Qiagen)

RT mix	1 mM dATP, 1 mM dCTP, 1 mM dGTP, 0.3 mM dTTP, 0.5 mM aminoallyl dUTP Random 9-mers (Stratagene)	ARES DNA labelling kit (Molecular Probes)	0.5 mM dATP, 0.5 mM dCTP, 0.5 mM dGTP, 0.3 mM dTTP, 0.2 mM aminoallyl dUTP random 6-mers (Invitrogen)	0.5 mM dATP, 0.5 mM dCTP, 0.5 mM dGTP, 0.3 mM dTTP, 0.2 mM aminoallyl dUTP random 6-mers (Invitrogen)	0.5 mM dATP, 0.5 mM dCTP, 0.5 mM dGTP, mM dTTP, (Amersham Biotech) 0.4 mM aminoallyl dUTP (Sigma) random 6-mers/9-mers (Qiagen)
Dyes	Alexa 488 Alexa 594 (Molecular Probes)	Alexa dyes 488, 546, 594, or 647 (Molecular Probes)	Cy-3, Cy-5 (Amersham Biotech)	Cy-3, Cy-5 (Amersham Biotech)	Cy-3, Cy-5 (Amersham Biotech)
Pre-hybridization	Not indicated	Not indicated	5x SSC, 0.1% SDS, and 1% BSA at 42°C for 45 min	5x SSC, 0.1% SDS, and 1% BSA at 42°C for 45 min	5x SSC, 0.1% SDS, and 1% BSA at 42°C for 40 min
Hybridization probe	Pooled differentially-labelled cDNA	Pooled differentially-labelled cDNA	Pooled differentially-labelled cDNA, with 20 ug competitor DNA (COT1); denatured 3 min at 95°C	Pooled differentially-labelled cDNA, with 40 μg competitor DNA (COT1/polyA); denatured 3 min at 95°C	Pooled differentially-labelled cDNA, with 20 μg competitor DNA (tRNA, hering sperm DNA); denatured 2 min at 95°C

(*continued*)

Table 9.2 Continued

#	(1) Pyrococcus furiosus	(2) Pyrococcus furiosus	(3) Pyrococcus furiosus	(4) Thermotoga maritima	(5) Sulfolobus solfataricus
Hybridization	65°C in a humidity chamber (Arrayit), 10–15 h	65°C (?) in Genetac hybridization station (Genomic Solutions), 10–15 h	Buffer: 50% formamide, 5x SSC, 0.1% SDS; 42°C in hybridization chamber (Corning), 16–20 h	Buffer: 50% formamide, 5x SSC, 0.1% SDS; 42°C in hybridization chamber (Corning), 16–20 h	Buffer: 50% formamide, 5x SSC, 0.1% SDS; 42°C in hybridization chamber, at least 20 h
Wash	Washed twice for 5 min in each of 2x SSC-0.1% SDS (SDS), and 0.2x SSC-0.1% SDS and then rinsed in distilled water and blown dry with compressed air	Washed for 20 in each of 2x SSC-0.1% Tween 20, 0.2x SSC-0.1% Tween 20, 0.2x SSC, and finally rinsed in distilled water and blown dry with compressed air	Washed for 4 min in 1x SSC-0.2% SDS (42°C), for 4 min in 0.1x SSC-0.2% SDS (20°C), and for 4 min in 0.1x SSC (20°C) and allowed to air dry.	Washed for 4 min in 1x SSC-0.2% SDS (42°C), for 4 min in 0.1x SSC-0.2% SDS (20°C), and for 4 min in 0.1x SSC (20°C) and allowed to air dry	Washed for 5 min in 2x SSC-0.1% SDS (42°C), for 10 min in 0.1x SSC-0.1% SDS (20°C), and for 5 min in 0.1x SSC (20°C) and blown dry with compressed air

Scan	Scan Array 5000 spectrometer (Packard, Meriden, Conn.) with the appropriate laser and filter settings and analyzed by using Quantarray (Packard)	Scan Array 5000 slide reader (Perkin-Elmer); identified and quantitated by using the Gleams software package (Nutec, Houston, Tex.).	Scanarray 4000 scanner and Quantarray software, respectively (GSI Lumonics, Billerica, Mass.). Local background intensity was subtracted from each spot signal, and spotting buffer (50% dimethyl sulfoxide (DMSO)) was used to subtract global background. The signal from the Cy-3 channel was normalized to the signal from the Cy-5 channel based on total signal intensity.	Slides were scanned using a Scanarray 4000 scanner (GSI Lumonics and Billerica); Signal intensity data were obtained using Quantarray (GSI Lumonics).	Hybridised slides are scanned using the high-throughput DNA Microarray scanner model G2565BA from Agilent Technologies (see text)
Statistics	Paired t test	Individual t-test; Holm's step-down p value adjustment procedure was performed to give modified p values	Paired t test		Student's t-test (SAM); see text.

Gene amplification and printing of *Sulfolobus* microarray

As mentioned above, different types of DNA microarrays exist: Affymetrix-type with short oligonucleotides (25-mers) synthesized on the array (Schena, 1996), glass slides printed with oligonucleotides (50/75-mers), or glass slides printed with either larger DNA fragments (PCR products were approximately 1000 bp or, in case of smaller coding regions, full length; Table 9.1). The first developed array for *Sulfolobus* consists of PCR-amplified fragments of the majority of the coding regions of two model *Sulfolobus* species. The amplification was done with specific primers (Andersson et al., 2005), using a touchdown PCR protocol, allowing for melting temperature differences between primer pairs. After a quality test of the PCR-products on agarose gel, purification has been performed using multiscreen 384-SEQ filter plates (Millipore). After washing with double demineralized water (ddH$_2$O), the DNA was then dissolved in 30 µl ddH$_2$O, and DMSO was added to a final concentration of 50%. The PCR fragments were spotted in duplicate on amino(propyl)-silane-coated glass slides using a robotic (Qarray; Genetix) slide printer (Lundgren et al., 2004). Apart from this frequently used technology, slides with alternative coatings are available: (i) epoxy-silane, for attaching amine-modified DNA, (ii) aldehyde silane, also for attaching amine-modified DNA, and (iii) poly-L-lysine coating, that forms a dense layer of amine groups (like aminosilane) for initial ionic attachment of the DNA-phosphate groups, which can subsequently be attached covalently to the slide by either baking or by UV irradiation (http://www.eriemicroarray.com/substrates/epoxy.aspx).

Sulfolobus growth and harvest

Sulfolobus solfataricus P2 (DSM1617) was grown on sugars (0.3% (w/v) D-glucose, D-galactose or D-arabinose) or on peptides (0.3% (w/v) tryptone) using the defined medium (Table 9.3), slightly adjusted from Brock's basal salt medium (Brock et al., 1972). To obtain exponentially growing cells, 1000 ml medium is inoculated with appropriate volume of an exponentially growing culture, and grown for 10 generations to an OD$_{600}$ of 0.1–0.3 (approximately 60 h in baffled Erlenmeyer flasks (2 liter) at 80°C in a rotary shaker). Flow cytometry is a convenient method to examine the physiological status of the cells (DNA and cell integrity), and to monitor cell cycle progression (DNA content distribution; relative number of cells in different cell cycle stages) (Bernander and Poplawski, 1997). At the appropriate density, (part of) the cell culture is quickly poured into 50 ml Greiner tubes and cooled on icewater. The cold cell suspension is centrifuged (15 min; 5000 × g; 4°C), after which the supernatant is carefully removed, and tubes are stored at −80°C.

Sulfolobus lysis and RNA isolation

All solutions described in this section are routinely made RNase-free by double autoclaving (for practical suggestions see: http://www.ambion.com/techlib/tb/tb_180.html). In addition, RNase-free water is ddH$_2$O,

Table 9.3 *Sulfolobus* Defined Medium, and Rich Medium (DSM)

Defined medium used for the different carbon sources; adjust pH at room temperature to 4.0 with H_2SO_4; autoclave and store at room temperature.

$(NH_4)_2SO_4$	2.5 g l^{-1}
KH_2PO_4	3.1 g l^{-}
$MgCl_2 \cdot 6H_2O$	203.3 mg l^{-1}
$Ca(NO_3)_2 \cdot 4H_2O$	70.8 mg l^{-1}
$FeSO_4 \cdot 7H_2O$	2.0 mg l^{-1}
$MnCl_2 \cdot 4H_2O$	1.8 mg l^{-1}
$Na_2B_4O_7 \cdot 2H_2O$	4.5 mg l^{-1}
$ZnSO_4 \cdot 7H_2O$	0.22 mg l^{-1}
$CuCl_2 \cdot 2H_2O$	0.06 mg l^{-1}
$Na_2MoO_4 \cdot 2H_2O$	0.03 mg l^{-1}
$VOSO_4 \cdot 2H_2O$	0.03 mg l^{-1}
$CoCl_2 \cdot 6H_2O$	0.01 mg l^{-1}

Carbon source
0.2–0.4% (w/v) monosaccharide, or tryptone

Wolin vitamin stock (2000×); adjust to pH 4.0 with H_2SO_4, filter sterilize and store at RT; add before use of defined medium:

Biotin	2 mg l^{-1}
Folic acid	2 mg l^{-1}
Pyridoxine-HCl	10 mg l^{-1}
Riboflavin	10 mg l^{-1}
Thiamine-HCl	5 mg l^{-1}
Nicotinic acid	5 mg l^{-1}
DL-Ca-Pantothenate	5 mg l^{-1}
Vitamin B12	0.1 mg l^{-1}
p-Aminobenzoic acid	5 mg l^{-1}
Lipoic acid	5 mg l^{-1}

Rich medium (Zillig et al., 1980); Medium for *Sulfolobus solfataricus* strain DSM No. 1617; http://www.dsmz.de/media/med182.htm

Yeast extract (Difco)	2.00 g
KH_2PO_4	3.10 g
$(NH_4)_2SO_4$	2.50 g
$MgSO_4 \cdot 7H_2O$	0.20 g
$CaCl_2 \cdot 2H_2O$	0.25 g
Distilled water	1000 ml

Adjust pH of the medium at room temperature to 3.5 with 10 N H_2SO_4 prior to autoclaving.

and incubated overnight with 0.1% (w/v) diethyl pyrocarbonate (DEPC, commonly used to irreversibly inactivate RNase), after which the solution is autoclaved to remove remaining DEPC. The Lysis Buffer contains 4 M guanidinium isothiocyanate, 25 mM sodium citrate (pH 7.0), 0.5% sarcosyl and 0.07 M β-mercaptoethanol. To minimize handling of the hazardous guanidinium isothiocyanate, it is recommended to prepare a stock solution (stable for at least 3 months when stored in the dark at room temperature): 50 g guanidinium isothiocyanate is dissolved into 58.6 ml ddH$_2$O, after which 3.52 ml 0.75 M sodium citrate (pH 7.0) is added. This solution is heated to 65°C and supplemented with 5.28 ml of sarcosyl (pre-heated to 65°C to liquefy). Preparation of Lysis Buffer is completed by adding 5.2 µl β-mercaptoethanol to 1.0 ml of the latter stock solution. Cell lysis is accomplished as outlined in the box below.

S. solfataricus – Procedure for cell lysis

- Use 1 ml Lysis Buffer to dissolve the pelleted cells (in the 50 ml Greiner tubes), divided over two Eppendorf tubes, and incubated on ice with regular mixing.
- After 15 min add the following *cold* solutions to each tube:
    - 100 µl 2 M Na-acetate (4°C; pH 5.2)
    - 500 µl water-saturated phenol (4°C; pH 4.5).
    - 100 µl 24:1 (v/v) chloroform-isoamyl alcohol mixture (4°C).
- Mix quickly and thoroughly and put on ice for 15 min.
- Next, the samples are centrifuged (15 min; 15 000 × g; 4°C), resulting in distinct phases: RNA is in the top aqueous phase and DNA and proteins are primarily in the phenol or intermediate phase.
- The aqueous phase is carefully transferred to a new Eppendorf tube, and 500 µl cold chloroform-isoamyl alcohol (24:1 (v/v); 4°C) is added and quickly and vigorously mixed.
- After an incubation of 15 min on ice, the solutions are spun down (30 min; 15 000 × g; 4°C). The top (aqueous) phase is transferred to a new tube, and supplemented with an equal volume of cold isopropanol. This is mixed and incubated overnight at −20°C.
- Next, the tubes are centrifuged (15 min; 15 000 × g; 4°C) and the supernatant is removed. The supernatant is dried and resuspended in 174 µl DEPC-treated water. Remaining traces of DNA are subsequently removed by addition of:
    - 5 µl (45 U) RNase-free DNAse I (Ambion),
    - 20 µl 10x DNase buffer (Ambion),
    - 1 µl (40 U) RNAguard RNase inhibitor (Promega).
- This solution is mixed and incubated for 30 min at 37°C, and the RNA is purified by adding the following *cold* solutions:
    - 20 µl 2M Na acetate (4°C; pH 5.2),
    - 100 µl water saturated phenol (4°C; pH 4.5),
    - 20 µl 24:1 (v/v) chloroform-isoamyl alcohol mixture (4°C).

- This is mixed thoroughly and put on ice for 15 min.
- After centrifugation (30 min; 15 000 × g; 4°C), the aqueous phase is transferred to another tube and mixed with an equal amount of *cold* isopropanol, and incubated for at least 2 h, or rather overnight, at −20°C.
- After centrifugation (15 min; 15 000 × g; 4°C), the pellet is washed by adding 300 µl *cold* 75% ethanol.
- The RNA sample is centrifuged again (10 min, 10 000 × g, at 4°C), after which the supernatant is carefully removed. The remaining RNA pellet is briefly dried in the Speedvac (to avoid problems with dissolving the pellets, the pellets should not be dehydrated too much).
- The RNA pellet is dissolved in an appropriate solution (e.g. RNase-free ddH$_2$O) at a concentration of 3–4 µg µl^{-1}, and stored at −80°C.

A typical RNA yield is 10–20 µg per 50 ml cell culture harvested at OD$_{600}$ of 0.1–0.3. The concentration of RNA is routinely determined using a spectrophotometer (e.g. NanoDrop ND-1000; http://www.nanodrop.com). An alternative method of RNA isolation, which has previously been used for transcription analysis of *Sulfolobus* (Brinkman et al., 2002), concerns the straight-forward RNeasy system (Qiagen RNeasy manual; http://www1.qiagen.com/literature/handbooks/PDF/RNAStabilization AndPurification/FromAnimalAndPlantTissuesBacteriaYeastAndFungi/ RNY_Midi_Maxi/1017849HBRNY_MidiMaxi0601WW.pdf). A possible drawback of the latter system is that part of the RNA molecules, especially those smaller then 200 nucleotides, is eluted during the washing steps, resulting in an underestimation of small RNAs (such as 5S rRNA, tRNAs); obviously the home made solutions for the phenol/chloroform extraction are less expensive than the RNeasy kit. On the other hand, the RNeasy method has practical advantages in that it is a quick and easy protocol.

Sulfolobus cDNA synthesis and labelling

The synthesis of cDNA and the subsequent labelling are performed as outlined in the box below.

S. solfataricus – Procedure for cDNA synthesis and labelling

- A purified RNA sample (5–20 µg) is mixed with 5 µg random primer (Qiagen; http://www1.qiagen.com; 2.8 µl 1 mM random hexamer (MW 1791.7 g mol^{-1}), and/or 1.8 µl 1 mM random nonamer (MW is 2718.55 g mol^{-1})) is added, and the total volume is adjusted to 11.6 µl with RNase-free water.
- Incubate for 10 min at 70°C, and then transfer to ice for at least 2 min.
- Add 8.4 µl Reverse Transcriptase (RT)-mix (Table 9.4), mix, and incubate for 2 h at 42°C to allow synthesis of the complementary strand, and incorporation of aminoallyl-dUTP (site for covalent attachment of fluorescent label, see below).

- Add 2 µl 200 mM EDTA and 3 µl I M NaOH, mix, and incubate for 15 min at 70°C (this will stop the reaction, and degrade the RNA).
- Add 3 µl I M HCl to neutralize the mixture.
- cDNA is purified by MinElute columns (Qiagen; alternative DNA purification systems can be used as well), and eventually elute in 10 µl 0.1 M NaHCO$_3$ (pH 9.0).

Table 9.4 *Sulfolobus* cDNA Synthesis – Solutions

Reverse Transcriptase (RT)-mixture:	
5x RT-buffer (Invitrogen)	4.0 µl
50x aadNTP mix (see below)	0.4 µl
DTT (0.1 M)	2.0 µl
SuperscriptII (200 U µl^{-1}; Invitrogen)	2.0 µl
50x aadNTP (4:1 aadUTP:dTTP):	
10 µl each of 100 mM dATP, dGTP, dCTP (Amersham Pharmacia Biotech)	
8 µl 100 mM aminoallyl-dUTP (Sigma)	
2 µl 100 mM dTTP (Amersham Pharmacia Biotech)	
Dissolve all components in 0.1 M KPO$_4$ buffer (pH 8.0)	

The Cy3 and Cy5 dyes (Amersham Biosciences; http://www4.amershambiosciences.com) are each dissolved in 55 µl DMSO and distributed over 10 Eppendorf tubes, dried by Speedvac, and stored at 4°C, preferably in a desiccator. For subsequent labelling with fluorescent Cy3 and Cy5 labels (Amersham Biosciences; alternative labels are available (Table 9.1)), the eluted cDNA solution (10 µl) is transferred to a Cy-dye aliquot tube (Amersham Biosciences), mixed, and incubated for 1.5 h at room temperature, in the dark. The reactions to be co-hybridized are mixed, and purified by a MinElute column (Qiagen; alternative DNA purification systems can be used as well (Table 9.1)). After elution from the spin column, the labelled cDNA is ready for microarray hybridization.

Sulfolobus microarray hybridization

Each microarray is hybridized with two differently labelled cDNA samples: Cy3 or Cy5 (Amersham Biosciences). To enhance the reliability of the analysis, both technical replicates (including "dye swap") as well as biological replicates (using RNA from different cultures) are routinely performed. Replicates obviously increase the accuracy of the microarray analysis (Yang and Speed, 2002). Although proper experimental design to some extent can reduce the number of technical replicates, a minimum of three biological replicates is recommended (Lee et al., 2000).

The slides are transferred to a 50 ml Falcon tube with 50 ml of the preheated pre-hybridization solution (1% (w/v) BSA, 5x SSC (from 20x SSC stock: 3 M NaCl, 300 mM tri-Na citrate, pH 7.0), and 0.1% (w/v) SDS;

the BSA is mixed with the preheated SSC/SDS solution at 42°C until dissolved), and incubated for 40 min at 42°C. Subsequently, slides are washed three times with ddH$_2$O, and once with isopropanol. Slides are dried either by blowing with compressed air or by using a slide centrifuge. Hybridization should be started within one hour; details are described in the box below.

S. solfataricus – Procedure for microarray hybridization

- To 20 µl labelled DNA, the following solutions are added:
    - 1 µl tRNA (10 µg µl^{-1}),
    - 1 µl herring-sperm DNA (10 µg µl^{-1}),
    - 58 µl hybridization mix (final concentration: 50% (v/v) formamide, 5x SSC, 0.1% w/v SDS).

- The probe-solution is incubated for 2 min at 95°C, after which it is cooled on ice for at least 1 min, and spun down.
- A coverslip is applied to the slide and the probe-solution is injected. After transfer to the hybridization chamber (slide container), the array is hybridized at 42°C for at least 20 h.
- Washing consists routinely of the following steps:
    - 5 min in 42°C pre-warmed 2× SSC, 0.1% SDS,
    - 10 min in room temp 0.1× SSC, 0.1% SDS,
    - 5 × 1 min in room temp 0.1× SSC.

- Then slides are dried carefully, either by centrifugation or by blowing with compressed air.

Scanning, data extraction and normalization

After hybridization, the microarrays are scanned, using two different lasers, usually at a final resolution of at least 10% of the spot size using the optimal laser and filter settings (Leung and Cavalieri, 2003). The scanning settings determine the signal-to-noise ratio; optimal settings can be determined by either performing a quick pre-scan while manually adjusting the settings, or by using specific automatic optimization protocols that are linked to some scanners. During scanning, two different images are created that are analyzed by programs such as GenePix Pro (http://www.axon.com/gn_GenePix_File_Formats.html). The availability of open source microarray software has recently been reviewed (Dudoit et al., 2003). Usually, a data filtering is required to exclude low quality spots. Usually, spots excluded from subsequent analysis include (i) spots with very low intensity (i.e. signal below the background plus 2 times the standard deviation (SD)), (ii) spots with unevenly distributed intensity (i.e. the ratio of medians deviates more than 20% from the regression ratio), and (iii) spots with saturated intensity. Before normalization and calculation of ratios, the background can be subtracted from the total intensities of the different spots. Although subtraction of background can improve the detection of some differentially expressed genes, it should be kept in mind that it may result in an increase of the overall variance and thus, in a decrease of the sensitivity of the measurement (Qin and Kerr, 2004). There are many different ways of normalizing data; usually too much normalization of data results in overall reduced ratios, which gives rise to a decreased sensitivity. For that reason, it is suggested to use software that indicates the consequences of distinct normalization steps. If this option is not available, a safe choice would be to restrict to a global lowess (LOcally WEighted Scatterplot Smoothing) normalization for each slide (Cleveland and Devlin, 1988). To determine differentially expressed genes, different statistical tests are available. Differential expression is usually determined via the classical Student's t-test or variations thereof, including SAM (Significance Analysis of Microarrays) and B-test (Lönnstedt and Speed, 2002). The latter variants are especially well suited for analysis of data with relatively few replicates; these methods tend to reduce the number of false positives, and as such are generally more suitable for DNA microarray experiments. The calculated p value is the statistical probability that the difference in gene expression occurs by change, the norm being that a gene is considered differentially expressed at $p < 0.05$ (Cui and Churchill, 2003).

♦♦♦♦♦ PROTEOME ANALYSIS

A proteome is the complete set of proteins present in a given organism under specific conditions at a certain moment. The analysis of proteomes requires (i) separating complex protein mixtures into discrete protein components, (ii) measuring their relative abundances, and (iii) identifying the individual protein components. Two-dimensional gel electrophoresis (2DE) is the classical method to separate proteins on the basis of their charge (isoelectric focusing, IEF) and of their size (sodium dodecyl sulfate polyacrylamide gel electrophoresis, SDS-PAGE). Proteins on a 2D gel can be visualized and subsequently quantified using several methods, the most frequently used are staining by: (i) Coomassie Brilliant Blue (reproducible quantification over linear dynamic range 50–1000 ng; CBB G250 (Sigma)), (ii) acidic silver nitrate (extremely sensitive (range 1–50 ng), although a drawback appears to be inhomogeneous protein staining), (iii) non-covalent fluorescent dyes (range 1–1000 ng; SYPRO Ruby (Molecular Probes) has been reported to be quite reproducible with similar binding characteristics as CBB) (reviewed by Barry et al., 2003).

Separation and identification of complex protein and peptide mixtures are essential steps to understand the function and roles of proteins in the cell. Although new methods using multidimensional liquid chromatography have recently been developed (e.g. Multidimensional Protein Identification Technology (MudPIT) (McDonald and Yates, 2002; Washburn, 2004; Washburn et al., 2002), protein separation by 2DE and subsequent identification by mass spectrometry is still the most frequently used strategy in proteomics. After 2DE, stained or labelled proteins are extracted from individual spots and trypsin digested. Protein identification is routinely performed by either Matrix-Assisted Laser Desorption Ionization Time-of-Flight mass spectrometry (MALDI-TOF MS) or electrospray ionisation tandem mass spectrometry (ESI MS/MS). In the case of MALDI-TOF MS, proteins are usually identified through a peptide mass fingerprint (PMF) type of search. Each digested protein provides a specific fingerprint consisting of tryptic peptide masses. Software algorithms are then used to identify the protein from a database. The popularity of this approach appears to decline, since the explosively increased sequence database and because of the fact that sample impurities can complicate the identification process. ESI MS/MS provides much greater specificity towards database searching and can easily be integrated with LC-based methods for peptide separation. Therefore, MS/MS-based techniques have become the predominant tool for peptide identification. More recently, tandem mass spectrometers with a MALDI interface have become available (MALDI-TOF-TOF) that will help to increase the throughput of MALDI-based identification of 2D gel spots, and offers complementarity in types of peptides identified. In particular, on-line micro capillary reversed phase liquid chromatography interfaced to a tandem mass spectrometer (or to a spotting robot in the case of the MALDI-TOF-TOF) has made the MS/MS analysis more comprehensive, and has allowed for a higher throughput (Lim et al., 2003; Link et al., 1999).

Relative quantification of proteins on 2D gels can be obtained by imaging the intensity of (CBB, silver) stained proteins or fluorescent dye (SYPRO Ruby)-labelled proteins. When two protein samples are labelled with distinct fluorescent probes, the differential analysis can be performed using a single 2D gel. This approach has been termed "Difference in Gel Electrophoresis" (DiGE). More accurate and reproducible methods to measure relative expression of proteins labelled with stable isotopes have recently been established. These methods generate mass-over-charge (m/z) differences for each of the homologous peptides and proteins in the proteome, and expression ratios are measured by comparing peak areas for the protein, or peptide ions, measured in the mass spectrometer. Stable isotopes can be incorporated into proteins by distinct labelling approaches: (1) *chemical*: after cultivation the protein samples are digested with proteases, after which light (hydrogen- or ^{12}C-containing) and heavy (deuterium- or ^{13}C-containing) versions of Isotope-Coded Affinity Tags (ICAT) are covalently linked to cysteinyl residues of peptides, after which the labelled peptides are purified by affinity chromatography (no 2DE required, but 1D gels often help in pre-fractionation) (Gygi et al., 1999; Washburn et al., 2002; Zhou et al., 2002); (2) *enzymatic*: protease-catalyzed hydrolysis of proteins in the presence of normal and heavy water, resulting in peptides with either ^{16}O or ^{18}O at the C-terminus (Oda et al., 1999); and (3) *metabolic labelling*: cultivation with labelled substrates (e.g. ^{12}C and ^{13}C-glucose, or ^{14}N and ^{15}N-ammonium sulfate) and separation of the differently labelled protein samples by 2DE or liquid chromatography (Snijders et al., 2005a).

Proteomics of thermophiles

In a pilot analysis of the proteome of *Methanocaldococcus jannaschii* (Giometti et al., 2001), significant changes in the abundance of a subset of predominant proteins has been observed in response to culture conditions and phase of the growth curve (exponential compared with stationary). Interestingly, several proteins were found to exist in multiple forms with different isoelectric points and molecular weights (see discussion below, *Sulfolobus*); the relative abundance of these protein variants appeared to change with growth conditions. Although variation due to sample treatment cannot be ruled out, these data might reflect post-translational modifications. Although the identity of the modifications remains to be identified, it is tempting to assume that this reflects a means of functional regulation at protein level. In a subsequent study (Giometti et al., 2002), 170 of the most abundant proteins have been identified in total lysates of *M. jannaschii*. To optimize the number of proteins detected, two different protein stains (Coomassie Blue R250 or silver nitrate) and two different first-dimensional separation methods (isoelectric focusing or non-equilibrium pH gradient electrophoresis) were used. Again, evidence of post-translational modification of numerous *M. jannaschii* proteins has been reported, as well as indications of incomplete dissociation of protein–protein complexes.

Lim et al. (2003) have recently performed a comparative analysis of protein identification for a total of 162 protein spots separated by two-dimensional gel electrophoresis from *M. jannaschii* and *Pyrococcus furiosus*, using MALDI-TOF peptide mass mapping and LC–MS/MS. 100% of the gel spots analyzed were successfully matched to the predicted proteins in the two corresponding open reading frame databases by LC-MS/MS while 97% of them were identified by MALDI-TOF mapping. The high success rate from the peptide mass mapping partly correlated with careful sample treatment (desalting/concentrating), but also with optimization of the search parameters, e.g. by incorporating amino acid sequence modifications into database searches. The obtained high sequence coverage in combination with digestion with several proteolytic enzymatic of different specificity is proposed as a method for future analysis of post-translational modifications (Lim et al., 2003).

High-throughput Multidimensional Protein Identification Technology based on microcapillary LC/LC/MS/MS has recently been used to identify 963 proteins of the proteome of *M. jannaschii*, corresponding to as much as 54% of the whole genome (Zhu et al., 2004). Almost half of the identified proteins have an unknown function, being annotated either as "conserved hypothetical" or as "hypothetical" proteins. The majority of the proteins predicted to be involved in distinct metabolic pathways were among the identified proteins. In addition, predicted intein peptides were detected, as well as peptides created by protein splicing. High peptide number, spectrum count, and sequence coverage have been used as indicators of high expression levels (Zhu et al., 2004).

As a means to release all experimental data of proteomics studies, to allow comparison of different proteomics analyses, and to enable integration with other data sets in a systems biology setting, the publicly available GELBANK database has recently been developed for the display of protein profiles generated by two-dimensional gel electrophoresis (http://gelbank.anl.gov). GELBANK is a database of two-dimensional gel electrophoresis (2DE) gel patterns of proteomes from organisms with known genome information, with relevant technical information. It includes the completed, mostly microbial proteomes available from the National Center for Biotechnology Information (Babnigg and Giometti, 2004).

Proteomics of *Sulfolobus*

In two recent studies, the proteome of *Sulfolobus solfataricus* has been analysed (Snijders et al., 2005a, 2006). Proteins corresponding to 349 ORFs were separated and identified using 2DE followed by LC-ESI-MS/MS and database searching (Mass Spectrometry protein sequence DataBase (MSDB); http://csc-fserve.hh.med.ic.ac.uk/msdb.html). Moreover, it was shown that ^{15}N and ^{13}C metabolic labelling for peptide quantification in Archaea has significant advantages compared to traditional gel-based quantification methods (Snijders et al., 2005a, 2005b).

Cultivation and metabolic labelling

S. solfataricus is generally grown aerobically on defined medium in 250-ml flasks. Each flask contains 50 ml basic medium, 25 µl Wolfe's vitamins and glucose as the carbon source with a final concentration of 0.3–0.4% (w/v) (Table 9.3). Cells are routinely inoculated at an optical density of 0.2 (OD_{530}). Cells are harvested at the appropriate phase of the growth curve, and washed twice with basic medium and once with 10 mM Tris/HCl buffer (pH 7.0). The wash steps are necessary to remove salts and contaminants that may interfere with the 2DE protocol, in particular with the iso-electric focussing. After this, cells can be stored at −20°C until further processed. During the process described below, considerable care was taken to ensure that culture-to-culture variation is minimal, and cultures should be prepared in at least triplicate; to further enhance the reproducibility, cultivation in well-controlled fermenters (batch, continuous culture) is currently being developed.

In the case of the ^{15}N or ^{13}C labelling experiments, the same protocol for cell growth may be used, but the nitrogen and carbon sources are replaced by ($^{15}NH_4)_2SO_4$ (Sigma) or ^{13}C-glucose, respectively. The isotope-labelled nutrients are readily incorporated into cell material and have no observable effect on the growth characteristics of *S. solfataricus* (Snijders, unpublished data). In order to ensure full incorporation of the isotope label into all proteins, at least 8 doubling times are required (approximately 48 h). Once full labelling is achieved, the cells can be stored as glycerol stocks until further use.

For a quantitative experiment based on metabolic labelling, two separate growth experiments are set up. In one case, the organism is grown in normal "unlabelled" medium. In the second case, the same organism is grown in the ^{15}N (or ^{13}C)-containing medium. In this way a "light" and a "heavy" proteome are created. Therefore, batch cultures are grown in parallel starting at an OD_{530} of 0.2 with either ^{14}N- or ^{15}N-ammonium sulphate. When the OD_{530} reaches a value of 0.5, the differentially labelled cultures are mixed. To ensure that equal amounts of biomass are mixed, slight corrections in volume should be made to correct for slight deviations with respect to the cultures' optical densities. Subsequently, cell harvest, preparation of cell extracts, 2-DE and protein identification are performed exactly the same way for the labelled/unlabelled cells as for the unlabelled cells (Snijders et al., 2006). In this case, peptide identification and quantification occur at the last (MS) stage of the protocol as described below.

Recently, a useful extension of the stable isotope labelling approach was introduced (Snijders et al., 2005a). In this study, dual labelling with ^{15}N-ammonium sulphate and ^{13}C-glucose was used for both quantification and identification of peptides in S. solfataricus. This variation of the metabolic labelling method is not discussed here further, but clearly offers some strong advantages as three phenotypes can be simultaneously examined.

Cell lysis

The $-20°C$ frozen cells are thawed and immediately resuspended in 1.5 ml of 10 mM Tris-HCl buffer (pH 7.0), and 25 µl of a protease-inhibitor cocktail (Protease Inhibitor Cocktail for use with bacterial cell extracts, Sigma) is added to the cell suspension. Cells are disrupted by sonication for 10 min on ice (Soniprep 150, Sanyo). Insoluble cell material is removed by centrifugation (15 min; $5000 \times g$; 4°C). The protein concentration of the supernatant is then determined using the Bradford Protein Assay (Sigma). At this stage, the supernatant can be stored at $-80°C$.

2-D electrophoresis, protein visualization and image analysis

Below, the details are provided for (i) a recently developed protocol of 2-DE of S. solfataricus cell-free extracts (Snijders et al., 2006), and (ii) a summary of a study in which a systematic optimisation of some critical steps in 2-DE of S. solfataricus lysates has been evaluated (Barry et al., 2003).

S. solfataricus – Procedure for 2-DE

- A sample mix is prepared by mixing the cell-free extract (prepared as described above) with a Rehydration Buffer to yield a final concentration of 50 mM DTT, 8 M Urea, 2% CHAPS (Sigma), 0.2% (w/v) Pharmalyte ampholytes (e.g. pH 3–10) (Fluka) and a trace of Bromophenol Blue.
- Each IPG strip (e.g. pH 3–10; Bio-Rad) is rehydrated overnight with 300 µl (400 µg of protein) of the sample mix.

- Isoelectric focussing (IEF) is performed using a 3-step protocol at 20°C using a Protean IEF cell (Bio-Rad).
- Subsequently, the voltage is linearly ramped to 250 V over 30 min to desalt the strips.
- Next, the voltage is linearly ramped to 1000 V over 2.5 h.
- Finally, the voltage is rapidly ramped to 10 000 V for 40 000 V*h to complete the focussing. At this stage, the strips can be stored overnight at −20°C.
- Focussed strips are incubated for 15 min in a solution containing 6 M urea, 2% SDS, 0.375 M Tris-HCl (pH 8.8), 20% glycerol, and 2% (w/v) DTT.
- The solution is discarded and the strips are incubated in a solution containing 6 M urea, 2% SDS, 0.375 M Tris-HCl (pH 8.8), 20% glycerol, and 4% iodoacetamide.
- After equilibration, proteins are separated in the second dimension using SDS-PAGE performed using a Protean II Multicell (Bio-Rad) apparatus on 10% T (concentration total polymer: % (w/v) acrylamide + % (w/v) N, N'-methylenebisacrylamide), 2.6% C gels (concentration crosslinker: % (w/v) N, N'-methylenebisacrylamide; gel dimensions: 17 cm × 17 cm × 1 mm).
- Electrophoresis is carried out with a constant current of 16 mA/gel for 30 min.
- Subsequently the current is increased to 24 mA/gel for another 7 h.
- Gels are stained using Coomassie Brilliant Blue G250 (Sigma).
- Gels are scanned using a GS-800 densitometer (Bio-Rad) at 100 microns resolution.

All spot detection and quantification is performed with PDQUEST 7.1.0 (Bio-Rad). With this method, Snijders et al. (2006) created a 2D reference map on which approximately 500 spots are visualized. In the case of the metabolic labelling experiments, the gel image was matched to the reference map and protein spots of interest were selected for MS analysis and quantitation.

Using sonicated cell extracts of S. solfataricus cultures, Barry et al. (2003) have performed an extensive comparative study on 2D sample application; it should be noted that in this study the obtained lysates were not subjected to any subsequent purification or fractionation (e.g. removal of membrane fragments and lipids by centrifugation; see below) because the goal was to obtain a single gel system to resolve the complete proteome, including both soluble and membrane proteins. When comparing three different methods for applying the protein samples on the IEF strips (cup-loading, active rehydration and passive rehydration) it was concluded that for basic proteins (using IPG 6-11 strips (Amersham-Pharmacia) on the Protean IEF Cell (Bio-Rad; http://www.bio-rad.com) and its IEF tray) sample application by cup-loading is by far superior over both rehydration methods (greatest number of detectable spots, best gel-to-gel reproducibility, lowest spot quantity variations); in the case of acidic proteins (using IPG 4–7 strips) active rehydration and cup-loading appeared to give the best results (Barry et al., 2003). In this study, gels were stained with SYPRO Ruby (non-covalent fluorescent dye; Molecular Probes; http://probes.invitrogen.com), and analysis was performed with the PDQuest package (Bio-Rad). Experimental details are summarized in Table 9.3.

Protein isolation and identification by MS

Spots of interest are manually excised from the CBB-stained 2D gels, and destained with 200 mM ammonium bicarbonate with 40% acetonitrile at 37°C (twice for 30 min). The gel pieces are incubated overnight in trypsin solution (0.4 µg trypsin (Sigma) in 50 µl of 40 mM ammonium bicarbonate in 9% acetonitrile). Subsequently, peptides are extracted in three subsequent extraction steps using 5 µl of 25 mM ammonium bicarbonate (10 min, room temperature), 30 µl acetonitrile (15 min, 37°C), 50 µl of 5% formic acid (15 min, 37°C) and finally with 30 µl acetonitrile (15 min, 37°C). Using this three-step protocol, the peptide recovery was significantly higher compared to simplified extraction methods. Therefore, the quality of MS spectra was improved and protein sequence coverage was significantly increased (data not shown). All extracts are pooled and dried in a vacuum centrifuge, then stored at −20°C.

The lyophilized peptide mixture is resuspended in 10 µl 0.1% formic acid in 3% acetonitrile. This mixture is separated on a PepMap C-18 RP capillary column (LC Packings), and eluted in a 30-min gradient *via* a LC Packings Ultimate nanoLC directly onto the mass spectrometer. Peptides are analysed using a QStarXL electrospray ionization quadrupole time-of-flight tandem MS (ESI qQ-TOF; Applied Biosystems; http://www.appliedbiosystems.com). The data acquisition on the MS is performed in the positive ion mode using Information Dependent Acquisition (IDA). Peptides with charge states 2 and 3 are selected for tandem mass spectrometry. IDA data were submitted to Mascot for database searching in a sequence query type of search (www.matrixscience.com). The settings are as follows: peptide tolerance 2.0 Da; MS/MS tolerance 0.8 Da; carbamidomethyl modification of cysteine is set as a fixed modification; methionine oxidation is set as a variable modification: maximal 1 missed cleavage site by trypsin was allowed. The search is performed against the Mass Spectrometry protein sequence DataBase (MSDB). MOWSE scores greater than 50 are considered significant (Snijders *et al.*, 2006).

Peptide quantification

Heavy and light versions of a protein or peptide have the same physicochemical properties. Therefore, no distinction can be made between the two versions during cell growth, protein extraction and separation. Only in the MS stage of the protocol, a difference can be observed. Labelled and unlabelled peptides appear as doublets in the MS spectrum. The difference in mass between the unlabelled and labelled peaks corresponds to the number of nitrogen (or carbon) atoms present in the peptide. In the metabolic labelling experiments, peptide identification of the light, ^{14}N version of the peptide is performed as described above. The heavy ^{15}N version of the peptide is identified by changing the isotope abundance of ^{15}N nitrogen to 100% in the Analyst data dictionary. Next, the peak areas of both versions of the same peptide are integrated over time using LC–MS reconstruction tool in the Analyst software (Applied Biosystems). In addition, an extracted ion chromatogram (XIC) is constructed for each peptide. The XIC is an ion chromatogram that shows the intensity values of a single mass (peptide) over a range of scans. This tool is used to check for chromatographic shifts between heavy and light versions of the same peptide. Generally, metabolic labelling with either ^{15}N or ^{13}C does not cause chromatographic shifts.

Protein identification by MS

Using the described proteomics method, 325 unique soluble *S. solfataricus* proteins have been identified in 255 spots from the 2D reference map. The MS technique employed is very sensitive, with reliable MS identification even of the faintest spots visualized by CBB staining. In many cases, multiple proteins per spot were found and preliminary use of pH 4–7 and pH 5–8 zoom gels still usually yielded multiple proteins per Coomassie-stained spot (Snijders, personal communication). Significant MOWSE scores (>51) were found for all 255 spots analysed. Generally, one peptide (intact mass and MS/MS ion spectrum) was sufficient for confident identification of *S. solfataricus* proteins against the MSDB. In most cases, however, multiple peptides of the same protein were found. On an average, the sequence coverage was 30%. Complete sequence coverage was never achieved. There was no relation between the sequence coverage and the protein's size. MOWSE scores greater than 800 were only achieved for proteins larger than 48 kDa.

Apart from looking at abundance levels, proteomics studies are currently ongoing that focus on the attenuation of enzyme activity by protein post-translational modification. Another important observation is that a number of proteins occur in more then 1 spot. Interestingly, this was true for a large number of proteins involved in the TCA cycle (e.g. 2-oxoglutarate oxidoreductase (SSO2815) was found in eight different spots). There are a number of explanations for this, including (1) the protein might exist in multiple forms in the cell, e.g. post-translationally modified *versus* non modified, (2) The protein was modified during protein extraction or during 2-DE (e.g. methionine oxidation), and (3) the protein does not resolve well on the gel and therefore "smears" out over a large pH or mass range. Future proteomics studies are expected to provide additional clues that will reveal the details of quantitative and qualitative modulations of proteins as means of regulating the metabolism of *S. solfataricus*.

◆◆◆◆◆ CONCLUDING REMARKS

The intention of this overview is to describe ongoing developments at the level of functional genomics of archaeal and bacterial thermophiles, with particular emphasis on Archaea of the genus *Sulfolobus*. Obviously, *Sulfolobus* would not have gained the status of "model archaeon" without a long history of classical studies (physiology, biochemistry and molecular genetics) that have been performed over the last three decades in numerous research laboratories. This has resulted in considerable insight in fundamental principles of the archaeal cell, including central metabolic pathways (Snijders *et al.*, 2006, Verhees *et al.*, 2003), replication and cell cycle (Lundgren *et al.*, 2004; Margolin and Bernander, 2004; Robinson *et al.*, 2004), transcription (Bell and Jackson, 2001; Thomm, 1996) and translation (Tumbala *et al.*, 1999). As such, classical research has provided the basis on which modern genomics-related approaches have been built (comparative, functional and structural genomics). At present, important developments in biochemical and molecular genetics analyses are under way, allowing novel approaches for analysis *in vitro* (heterologous expression, directed and random mutagenesis, study of protein/protein and protein/DNA interactions), and *in vivo* (chromosomal knockouts, *in trans* overexpression, phenotype characterization) (current progress reviewed by Allers and Mevarech, 2005; Baliga *et al.*, 2004).

Recent developments of transcriptomics and proteomics tools for thermophiles in general, and *Sulfolobus* in particular have been reviewed. It should be noted, however, that developments in thermophile genomics proceed at an impressive pace: additional hyperthermophile genomes are released on a regular basis, comparative tools become more sophisticated and functional predictions become more reliable, DNA microarrays are available, and recent breakthroughs illustrate the role of mass spectrometry-based proteomics and metabolomics as an indispensable tool for molecular and cellular biology, and for the emerging field of systems biology.

Now the stage is set for yet another challenge in the study of thermophiles, i.e. the integration of classical and modern technologies to acquire knowledge in the functioning of metabolic networks and of the regulatory circuits of these systems. Many breakthroughs are expected in the near future, resulting in a gain of insight into the evolution and functioning of these intriguing thermophile systems.

Acknowledgements

This work was supported by a grant from the European Union in the framework of the SCREEN project (contract QLK3-CT-2000-00649), and by the Netherlands Organisation of Scientific Research (NWO) BioMolecular Informatics Programme, grant 050.50.206. We thank the EPSRC for funding, and Advanced Fellowship support. In addition, we acknowledge The Swedish Research Council as well as The Swedish Foundation for Strategic Research.

References

Allers, T. and Mevarech, M. (2005). Archaeal genetics – the third way. *Nature Rev. Genet.* **6**, 58–73.

Andersson, A., Bernander, R. and Nilsson, P. (2005). Dual-genome primer design for construction of DNA microarrays. *Bioinformatics* **21**, 325–332.

Babnigg, G. and Giometti, C. S. (2004). GELBANK: a database of annotated two-dimensional gel electrophoresis patterns of biological systems with completed genomes. *Nucleic Acids Res.* (Database issue), **32**, D582–D585.

Baliga, N. S., Bjork, S. J., Bonneau, R., Pan, M., Iloanusi, C., Kottemann, M. C., Hood, L. and DiRuggiero, J. (2004). Systems level insights into the stress response to UV radiation in the halophilic archaeon *Halobacterium* NRC-1. *Genome Res.* **14**, 1025–1035.

Barry, R. C., Alsaker, B. L., Robison-Cox, J. F. and Dratz, E. A. (2003). Quantitative evaluation of sample application methods for semipreparative separations of basic proteins by two-dimensional gel electrophoresis. *Electrophoresis* **24**, 3390–3404.

Bell, S. D. and Jackson, S. P. (2001). Mechanism and regulation of transcription in archaea. *Curr. Opin. Microbiol.* **4**, 208–213.

Bernander, R. and Poplawski, A. (1997). Cell cycle characteristics of thermophilic archaea. *J. Bacteriol.* **179**, 4963–4969.

Blöchl, E., Rachel, R., Burggraf, S., Hafenbradl, D., Jannasch, H. W. and Stetter, K. O. (1997). *Pyrolobus fumarii*, gen. and sp. nov., represents a novel group of archaea, extending the upper temperature limit for life to 113°C. *Extremophiles* **1**, 14–21.

Brinkman, A. B., Bell, S. D., Lebbink, R. J., de Vos, W. M. and van der Oost, J. (2002). The *Sulfolobus solfataricus* Lrp-like protein LysM regulates lysine biosynthesis in response to lysine availability. *J. Biol. Chem.* **277**, 29537–29549.

Brock, T. D. and Freeze, H. (1969). *Thermus aquaticus* gen. n. and sp. n., a nonsporulating extreme thermophile. *J. Bacteriol.* **98**, 289–297.

Brock, T. D., Brock, K. M., Belly, R. T. and Weiss, R. L. (1972). *Sulfolobus*: a new genus of sulfur-oxidizing bacteria living at low pH and high temperature. *Arch. Mikrobiol.* **84**, 54–68.

Bult, C. J., White, O., Olsen, G. J., Zhou, L., Fleischmann, R. D., Sutton, G. G., Blake, J. A., FitzGerald, L. M., Clayton, R. A., Gocayne, J. D., Kerlavage, A. R., Dougherty, B. A., Tomb, J. F., Adams, M. D., Reich, C. I., Overbeek, R., Kirkness, E. F., Weinstock, K. G., Merrick, J. M., Glodek, A., Scott, J. L., Geoghagen, N. S. and Venter, J. C. (1996). Complete genome sequence of the methanogenic archaeon, *Methanococcus jannaschii*. *Science* **273**, 1058–1073.

Cambillau, C. and Claverie, J. M. (2000). Structural and genomic correlates of hyperthermostability. *J. Biol. Chem.* **275**, 32383–32386.

Chen, L., Brugger, K., Skovgaard, M., Redder, P., She, Q., Torarinsson, E., Greve, B., Awayez, M., Zibat, A., Klenk, H. P. and Garrett, R. A. (2005). The genome of *Sulfolobus acidocaldarius*, a model organism of the Crenarchaeota. *J. Bacteriol.* **187**, 4992–4999.

Chhabra, S. R., Shockley, K. R., Conners, S. B., Scott, K. L., Wolfinger, R. D. and Kelly, R. M. (2003). Carbohydrate-induced differential gene expression patterns in the hyperthermophilic bacterium *Thermotoga maritima*. *J. Biol. Chem.* **278**, 7540–7552.

Cleveland, W. and Devlin, S. (1988). Locally weighted regression – an approach to regression-analysis by local fitting. *J. Amer. Statist. Assoc.* **83**, 596–610.

Cohen, G. N., Barbe, V., Flament, D., Galperin, M., Heilig, R., Lecompte, O., Poch, O., Prieur, D., Querellou, J., Ripp, R., Thierry, J. C., van der Oost, J.,

Weissenbach, J., Zivanovic, Y. and Forterre, P. (2003). An integrated analysis of the genome of the hyperthermophilic archaeon *Pyrococcus abyssi*. *Mol. Microbiol.* **47**, 1495–1512.

Cui, X. and Churchill, G. A. (2003). Statistical tests for differential expression in cDNA microarray experiments. *Genome Biol.* **4**, 210.

Dudoit, S., Gentleman, R. C. and Quackenbush, J. (2003). Open source software for the analysis of microarray data. *Biotechniques* March 2003, Suppl. 45–51.

Ettema, T., van der Oost, J. and Huynen, M. (2001). Modularity in the gain and loss of genes: applications for function prediction. *Trends Genet.* **17**, 485–487.

Ettema, T., de Vos, W. M. and van der Oost, J. (2005). Discovering new biology by *in silico* archaeology. *Nature Rev. Microbiol.*, **3**, 859–869.

Frickey, T. and Lupas, A. N. (2004). PhyloGenie: automated phylome generation and analysis. *Nucleic Acids Res.* **32**, 5231–5238.

Fukui, T., Atomi, H., Kanai, T., Matsumi, R., Fujiwara, S. and Imanaka, T. (2005). Complete genome sequence of the hyperthermophilic archaeon *Thermococcus kodakaraensis* KOD1 and comparison with *Pyrococcus* genomes. *Genome Res.* **15**, 352–363.

Giometti, C. S., Reich, C. I., Tollaksen, S. L., Babnigg, G., Lim, H., Yates, J. R. 3rd and Olsen, G. J. (2001). Structural modifications of *Methanocaldococcus jannaschii* flagellin proteins revealed by proteome analysis. *Proteomics* **1**, 1033–1042.

Giometti, C. S., Reich, C., Tollaksen, S., Babnigg, G., Lim, H., Zhu, W., Yates, J. and Olsen, G. (2002). Global analysis of a "simple" proteome: *Methanocaldococcus jannaschii*. *J. Chromatogr. B* **782**, 227–243.

Gygi, S. P., Rist, B., Gerber, S. A., Turecek, F., Gelb, M. H. and Aebersold, R. (1999). Quantitative analysis of complex protein mixtures using isotope-coded affinity tags. *Nature Biotechnol.* **17**, 994–999.

Huber, R., Huber, H. and Stetter, K. O. (2000). Towards the ecology of hyperthermophiles: biotopes, new isolation strategies and novel metabolic properties. *FEMS Microbiol. Rev.* **24**, 615–623.

Johnson, M. R., Montero, C. I., Conners, S. B., Shockley, K. R., Bridger, S. L. and Kelly, R. M. (2005). Population density-dependent regulation of exopolysaccharide formation in the hyperthermophilic bacterium *Thermotoga maritima*. *Mol. Microbiol.* **55**, 664–674.

Kashefi, K. and Lovley, D. R. (2003). Extending the upper temperature limit for life. *Science* **301**, 934.

Kawarabayasi, Y., Sawada, M., Horikawa, H., Haikawa, Y., Hino, Y., Yamamoto, S., Sekine, M., Baba, S., Kosugi, H., Hosoyama, A., Nagai, Y., Sakai, M., Ogura, K., Otsuka, R., Nakazawa, H., Takamiya, M., Ohfuku, Y., Funahashi, T., Tanaka, T., Kudoh, Y., Yamazaki, J., Kushida, N., Oguchi, A., Aoki, K. and Kikuchi, H. (1998). Complete sequence and gene organization of the genome of a hyper-thermophilic archaebacterium, *Pyrococcus horikoshii* OT3. *DNA Res.* **5**, 55–76.

Kawashima, T., Amano, N., Koike, H., Makino, S., Higuchi, S., Kawashima-Ohya, Y., Watanabe, K., Yamazaki, M., Kanehori, K., Kawamoto, T., Nunoshiba, T., Yamamoto, Y., Aramaki, H., Makino, K. and Suzuki, M. (2000). Archaeal adaptation to higher temperatures revealed by genomic sequence of *Thermoplasma volcanium*. *Proc. Natl. Acad. Sci. USA* **97**, 14257–14262.

Kawarabayasi, Y., Hino, Y., Horikawa, H., Jin-no, K., Takahashi, M., Sekine, M., Baba, S., Ankai, A., Kosugi, H., Hosoyama, A., Fukui, S.,

Nagai, Y., Nishijima, K., Otsuka, R., Nakazawa, H., Takamiya, M., Kato, Y., Yoshizawa, T., Tanaka, T., Kudoh, Y., Yamazaki, J., Kushida, N., Oguchi, A., Aoki, K., Masuda, S., Yanagii, M., Nishimura, M., Yamagishi, A., Oshima, T. and Kikuchi, H. (2001). Complete genome sequence of an aerobic thermoacidophilic crenarchaeon, *Sulfolobus tokodaii* strain7. *DNA Res.* **8**, 123–140.

Koonin, E. V., Makarova, K. S. and Aravind, L. (2001). Horizontal gene transfer in prokaryotes: quantification and classification. *Annu. Rev. Microbiol.* **55**, 709–742.

Lee, M. L., Kuo, F. C., Whitmore, G. A. and Sklar, J. (2000). Importance of replication in microarray gene expression studies: statistical methods and evidence from repetitive cDNA hybridizations. *Proc. Natl. Acad. Sci. USA* **97**, 9834–9839.

Leung, Y. F. and Cavalieri, D. (2003). Fundamentals of cDNA microarray data analysis. *Trends Genet.* **19**, 649–659.

Lim, H., Eng, J., Yates, J. R. 3rd, Tollaksen, S. L., Giometti, C. S., Holden, J. F., Adams, M. W., Reich, C. I., Olsen, G. J. and Hays, L. G. (2003). Identification of 2D-gel proteins: a comparison of MALDI/TOF peptide mass mapping to mu LC-ESI tandem mass spectrometry. *J. Amer. Soc. Mass Spectrom.* **14**, 957–970.

Link, A. J., Eng, J., Schieltz, D. M., Carmack, E., Mize, G. J., Morris, D. R., Garvik, B. M., Yates and J. R. 3rd. (1999). Direct analysis of protein complexes using mass spectrometry. *Nature Biotechnol.* **17**, 676–682.

Lönnstedt, I. and Speed, T. (2002). Replicated microarray data. *Statistica Sinica* **12**, 31–46.

Lundgren, M., Andersson, A., Chen, L., Nilsson, P. and Bernander, R. (2004). Three replication origins in *Sulfolobus* species: synchronous initiation of chromosome replication and asynchronous termination. *Proc. Natl. Acad. Sci. USA* **101**, 7046–7051.

Makarova, K. S. and Koonin, E. V. (2003). Comparative genomics of archaea: how much have we learned in six years, and what's next? *Genome Biol.* **4**, 115.

Margolin, W. and Bernander, R. (2004). How do prokaryotic cells cycle? *Curr. Biol.* **14**, R768–R770.

McDonald, W. H., Yates and J. R. 3rd. (2002). Shotgun proteomics and biomarker discovery. *Disease Markers* **18**, 99–105.

Nelson, K. E., Clayton, R. A., Gill, S. R., Gwinn, M. L., Dodson, R. J., Haft, D. H., Hickey, E. K., Peterson, J. D., Nelson, W. C., Ketchum, K. A., McDonald, L., Utterback, T. R., Malek, J. A., Linher, K. D., Garrett, M. M., Stewart, A. M., Cotton, M. D., Pratt, M. S., Phillips, C. A., Richardson, D., Heidelberg, J., Sutton, G. G., Fleischmann, R. D., Eisen, J. A., White, O., Salzberg, S. L., Smith, H. O., Venter, J. C. and Fraser, C. M. (1999). Evidence for lateral gene transfer between archaea and bacteria from genome sequence of *Thermotoga maritima*. *Nature* **399**, 323–329.

Nguyen, T. N., Ejaz, A. D., Brancieri, M. A., Mikula, A. M., Nelson, K. E., Gill, S. R. and Noll, K. M. (2004). Whole-genome expression profiling of *Thermotoga maritima* in response to growth on sugars in a chemostat. *J. Bacteriol.* **186**, 4824–4828.

Oda, Y., Huang, K., Cross, F. R., Cowburn, D. and Chait, B. T. (1999). Accurate quantitation of protein expression and site-specific phosphorylation. *Proc. Natl. Acad. Sci. USA* **96**, 6591–6596.

Pace, N. R. (1997). A molecular view of microbial diversity and the biosphere. *Science* **276**, 734–740.

Qin, L. X. and Kerr, K. F. (2004). Empirical evaluation of data transformations and ranking statistics for microarray analysis. *Nucleic Acids Res.* **32**, 5471–5479.

Rivera, M. C. and Lake, J. A. (2004). The ring of life provides evidence for a genome fusion origin of eukaryotes. *Nature* **431**, 152–155.

Robb, F. T., Maeder, D. L., Brown, J. R., DiRuggiero, J., Stump, M. D., Yeh, R. K., Weiss, R. B. and Dunn, D. M. (2001). Genomic sequence of hyperthermophile, *Pyrococcus furiosus*: implications for physiology and enzymology. *Meth. Enzymol.* **330**, 134–157.

Robinson, N. P., Dionne, I., Lundgren, M., Marsh, V. L., Bernander, R. and Bell, S. D. (2004). Identification of two origins of replication in the single chromosome of the archaeon *Sulfolobus solfataricus*. *Cell* **116**, 25–38.

Rothschild, L. J. and Mancinelli, R. L. (2001). Life in extreme environments. *Nature* **409**, 1092–1101.

Ruepp, A., Graml, W., Santos-Martinez, M. L., Koretke, K. K., Volker, C., Mewes, H. W., Frishman, D., Stocker, S., Lupas, A. N. and Baumeister, W. (2000). The genome sequence of the thermoacidophilic scavenger *Thermoplasma acidophilum*. *Nature* **407**, 508–513.

Schena, M. (1996). Genome analysis with gene expression microarrays. *Bioessays* **18**, 427–431.

Schut, G. J., Zhou, J. and Adams, M. W. (2001). DNA microarray analysis of the hyperthermophilic archaeon *Pyrococcus furiosus*: evidence for a new type of sulfur-reducing enzyme complex. *J. Bacteriol.* **183**, 7027–7036.

Schut, G. J., Brehm, S. D., Datta, S. and Adams, M. W. (2003). Whole-genome DNA microarray analysis of a hyperthermophile and an archaeon: *Pyrococcus furiosus* grown on carbohydrates or peptides. *J. Bacteriol.* **185**, 3935–3947.

She, Q., Singh, R. K., Confalonieri, F., Zivanovic, Y., Allard, G., Awayez, M. J., Chan-Weiher, C. C., Clausen, I. G., Curtis, B. A., de Moors, A., Erauso, G., Fletcher, C., Gordon, P. M., Heikamp-de Jong, I., Jeffries, A. C., Kozera, C. J., Medina, N., Peng, X., Thi-Ngoc, H. P., Redder, P., Schenk, M. E., Theriault, C., Tolstrup, N., Charlebois, R. L., Doolittle, W. F., Duguet, M., Gaasterland, T., Garrett, R. A., Ragan, M. A., Sensen, C. W. and van der Oost, J. (2001). The complete genome of the crenarchaeon *Sulfolobus solfataricus* P2. *Proc. Natl. Acad. Sci. USA* **98**, 7835–7840.

Shivvers, D. W. and Brock, T. D. (1973). Oxidation of elemental sulfur by *Sulfolobus acidocaldarius*. *J. Bacteriol.* **114**, 706–710.

Shockley, K. R., Ward, D. E., Chhabra, S. R., Conners, S. B., Montero, C. I. and Kelly, R. M. (2003). Heat shock response by the hyperthermophilic archaeon *Pyrococcus furiosus*. *Appl. Environ. Microbiol.* **69**, 2365–2371.

Snijders, A. P. L., de Vos, M. G. J. and Wright, P. C. (2005a). Novel approach for peptide quantitation and sequencing based on ^{15}N and ^{13}C metabolic labeling. *J. Proteome Res.* **4**, 578–585.

Snijders, A. P. L., de Vos, M. G., de Koning, B. and Wright, P. C. (2005b). A fast method for quantitative proteomics based on a combination between two-dimensional electrophoresis and ^{15}N-metabolic labelling. *Electrophoresis* **26**, 3191–3199.

Snijders, A. P. L., Walther, J., Peter, S., Kinnman, I., de Vos, M. G. J., van de Werken, H. J. G., Brouns, S. J. J., van der Oost, J. and Wright, P.C. (2006). Reconstruction of central carbon metabolism in *Sulfolobus solfataricus* using a two-dimensional gel electrophoresis map, stable isotope labelling and DNA microarray analysis. *Proteomics*, doi: 10.1002/pmic.200402070.

Stetter, K. O. (1996). Hyperthermophilic prokaryotes. *FEMS Microbiol. Rev.* **18**, 149–158.

Stetter, K. O. (1999). Extremophiles and their adaptation to hot environments. *FEBS Lett.* **452**, 22–25.

Thomm, M. (1996). Archaeal transcription factors and their role in transcription initiation. *FEMS Microbiol. Rev.* **18**, 159–171.

Tumbula, D., Vothknecht, U. C., Kim, H. S., Ibba, M., Min, B., Li, T., Pelaschier, J., Stathopoulos, C., Becker, H. and Söll, D. (1999). Archaeal aminoacyl-tRNA synthesis: diversity replaces dogma. *Genetics* **152**, 1269–1276.

Verhees, C. H., Kengen, S. W., Tuininga, J. E., Schut, G. J., Adams, M. W., de Vos, W. M. and van der Oost, J. (2003). The unique features of glycolytic pathways in archaea. *Biochem. J.* **375**, 231–246.

Washburn, M. P. (2004). Utilisation of proteomics datasets generated via multidimensional protein identification technology (MudPIT). *Brief. Funct. Genomic. Proteomic* **3**, 280–286.

Washburn, M. P., Ulaszek, R., Deciu, C., Schieltz, D. M., Yates and J. R. 3rd. (2002). Analysis of quantitative proteomic data generated via multidimensional protein identification technology. *Anal. Chem.* **74**, 1650–1657.

Waters, E., Hohn, M. J., Ahel, I., Graham, D. E., Adams, M. D., Barnstead, M., Beeson, K. Y., Bibbs, L., Bolanos, R., Keller, M., Kretz, K., Lin, X., Mathur, E., Ni, J., Podar, M., Richardson, T., Sutton, G. G., Simon, M., Söll, D., Stetter, K. O., Short, J. M. and Noordewier, M. (2003). The genome of *Nanoarchaeum equitans*: insights into early archaeal evolution and derived parasitism. *Proc. Natl. Acad. Sci. USA* **100**, 12984–12988.

Weinberg, M. P., Schut, G. J., Brehm, S., Datta, S. and Adams, M. W. (2005). Cold shock of a hyperthermophilic archaeon: *Pyrococcus furiosus* exhibits multiple responses to a suboptimal growth temperature with a key role for membrane-bound glycoproteins. *J. Bacteriol.* **187**, 336–348.

Woese, C. R. (1987). Bacterial evolution. *Microbiol. Rev.* **51**, 221–271.

Yang, Y. H. and Speed, T. (2002). Design issues for cDNA microarray experiments. *Nature Rev. Genet.* **3**, 579–588.

Zhou, H., Ranish, J. A., Watts, J. D. and Aebersold, R. (2002). Quantitative proteome analysis by solid-phase isotope tagging and mass spectrometry. *Nature Biotechnol.* **20**, 512–515.

Zhu, W., Reich, C. I., Olsen, G. J., Giometti, C. S., Yates and J. R. 3rd. (2004). Shotgun proteomics of *Methanocaldococcus jannaschii* and insights into methanogenesis. *J. Proteome Res.* **3**, 538–548.

Zillig, W., Stetter, K. O., Wunderl, S., Schulz, W., Priess, H. and Scholz, I. (1980). The *Sulfolobus-Caldariella* group, taxonomy on the basis of the structure of DNA-dependent RNA polymerases. *Arch. Microbiol.* **125**, 259–269.

10 Heat Shock Proteins in Hyperthermophiles

Frank T Robb[1], Hem D Shukla[1] and Douglas S Clark[2]

[1] Center of Marine Biotechnology, University of Maryland Biotechnology Institute, 701 East Pratt Street, Baltimore, MD 21202, USA; [2] Department of Chemical Engineering, University of California, Berkeley, CA 94720, USA

◆◆◆

CONTENTS

Introduction
Gene regulation: transcriptional analysis
Protein characterization
Metabolic engineering
Concluding remarks

◆◆◆◆◆ **INTRODUCTION**

Thermophiles, like almost all cells, are able to survive transient exposure to high temperatures beyond their normal growth ranges (Laksanalamai and Robb, 2004; Laksanalamai *et al.*, 2004; Muchowski and Clark, 1998). Because hyperthermophiles are defined as growing optimally at temperatures above 80°C, methods used to study their heat shock responses are different than for other cells. Since heat shock typically takes place 3–10°C above an organism's optimal growth temperature, it follows that many hyperthermophiles will only enter heat shock at temperatures near to or above 100°C. For example, the experimental heat shock temperatures of *Pyrococcus furiosus* and *Methanocaldococcus jannaschii* are 105°C (Laksanalamai *et al.*, 2001) and 95°C (Boonyaratanakornkit *et al.*, 2005), respectively. The record holders for high growth temperatures are *Pyrolobus fumarii* (Blochl *et al.*, 1997) and Strain 121 (Kashefi and Lovley, 2003), which grow optimally at around 110°C and will probably require autoclave conditions (121°C, 15 psi pressure) for heat shock induction. Indeed, it is reported that Strain 121 continues to grow, albeit very slowly, at 121°C.

During periods of heat shock, hyperthermophiles produce an eclectic set of heat shock proteins, which consist mostly of protein chaperones but also include DNA repair proteins (Ruepp *et al.*, 2001, Shockley *et al.*, 2003, Robb, Laksanalamai, DiRuggiero and Lowe, unpublished).

Following exposure to sublethal heat stress beyond their maximal growth temperatures, *Sulfolobus shibatae* cells display a process called thermoadaptation that results in enhanced survival of the adapted cells when they are exposed to normally lethal temperatures (Trent *et al.*, 1990, 1991). Activity and stability of enzymes at high temperature is an obvious and critically important adaptation for the survival of thermophiles at high temperatures and for their activity and growth at the extremes of their temperature ranges. It has been possible to show that protein-folding pathways from hyperthermophiles may be assembled *in vitro* to stabilize proteins during high temperature catalytic processes (Laksanalamai *et al.*, 2005; Okochi *et al.*, 2002). Many thermophiles also accumulate compatible solutes that stabilize proteins during heat stress, and the enzymes comprising the biosynthetic pathways for these compounds should also be considered to be heat shock proteins (Santos and da Costa, 2002).

This chapter describes a selection of experimental methods used to characterize heat shock proteins, starting with the identification of the heat shock regulons. We have suggested that their small genome size, coupled with the necessity to carry out ongoing protein salvage processes in the cell during normal growth, has resulted in minimal but highly efficient protein folding machinery in hyperthermophiles (Laksanalamai *et al.*, 2004). Some proteins, such as a AAA+ATPase of unknown function, are part of the heat shock response (Neuwald *et al.*, 1999). In addition, a homolog of the Nascent Associated Complex (Wang *et al.*, 1995) has recently been described in the Archaea, although it is not clear how it relates to protein folding. Chaperone functions can be difficult to study in mesophiles due to the multiplicity of chaperone-encoding genes and parallel functions that they carry out. Heat shock regulation and chaperone functions in hyperthermophiles may therefore provide fresh insights into basic protein-folding mechanisms.

◆◆◆◆◆ GENE REGULATION: TRANSCRIPTIONAL ANALYSIS

Transcriptional Analysis of the Heat Shock Response in Hyperthermophiles

Most hyperthermophiles grow to low terminal cell densities, thereby complicating the detection of heat shock proteins. Inducing heat shock in hyperthermophiles may also require cultivation in vessels that are pressurized to prevent boiling. In order to determine the identity of heat shock proteins in hyperthermophiles, an initial approach using global gene regulation is probably the most effective method. Because there are now multiple whole genome analyses of hyperthermophiles, heat shock regulons may be conveniently mapped by microarray analysis. One of the first hyperthermophiles to be analyzed comprehensively in this way was *Methanocaldococcus jannaschii* (Boonyaratarkornkit *et al.*, 2005).

Since this will probably be the method of choice, we will describe typical microarray analyses. *Methanocaldococcus jannaschii* was cultivated in 125-ml serum bottles (Wheaton) containing 30 ml of media as described previously (Miller *et al.*, 1988) with Na_2S as the reducing agent. Bottles were inoculated with 3 ml of inoculum and pressurized to 30 psi with a 4:1 v/v mixture of H_2 and CO_2 substrate, then placed in a reciprocal shaking water bath (Precision Scientific Model 25) at 200 oscillations per min and 85°C. Growth was followed by measuring optical density at 660 nm (Tsao *et al.*, 1994). After reaching mid-exponential phase in ca. 3 h ($OD_{660} \sim 0.20$), the bottles were transferred to another water bath and shaken at 200 oscillations per min at either 65°C (for cold shock) or 95°C (for heat shock). In both cases, bottles reached temperature equilibrium within 3 min. We have also carried out microarray analysis with *Pyrococcus furiosus* DSM 3638 (Laksanalamai *et al.*, in preparation). *P. furiosus* was cultivated in 1-l Wheaton bottles at 95°C and a 20-l New Brunswick Bioflo IV fermentor was inoculated with 2 l of culture. Growth was measured by epifluorescence microscopy following fixation of the cells and staining with acridine orange. When the culture reached mid-exponential growth, the temperature, the fermentor equilibrated to the new temperature within 6 min. Typically, *Pyrococcus furiosus* was cultured at 95°C for 4 h, then shifted to 104°C for 30 min to 2 h. The control was a culture sampled without shifting to 104°C. RNA preparation and microarray analysis from *P. furiosus* was essentially as described above for *M. jannaschii*. Samples (20 ml) were passed through a 0.45 μm nitrocellulose filter (Millipore) to collect cells. Total RNA was then precipitated overnight at –20°C with 3 M sodium acetate and isopropanol. Precipitated RNA was re-suspended and treated with DNase (Invitrogen), and each sample was purified on RNeasy columns (Qiagen).

Array Construction

For a spotted whole genome microarray, primers are designed for all predicted ORFs, typically using available software such as Primer 3 or Oligowiz 2.0 (see http://www.cbs.dtu.dk/services/OligoWiz2/). In our experience, an iterative process of primer selection with the following steps is helpful:

- BLAST all probes against genome sequence.
- Parse BLAST output to extract all probe hits on the genome.
- Filter the BLAST hits to exclude probes that had more than one hit elsewhere in the genome sequence longer than 15 nt.
- Select the best scoring probes for each ORF that survived these filters.

PCR reactions are performed, typically with 1 ng of genomic DNA, 0.8 μM of each forward and reverse primer, and 2.5 U Platinum Taq polymerase (Invitrogen). Resuspended products were analyzed on a 0.8% agarose gel. In our studies, UltraGAPS slides (Corning) were printed using an MD-3 robot (Amersham Pharmacia) with PCR products representing 99% of the ORFs in *M. jannaschii*. Filtered PCR products were rearrayed into 384-well plates with 50% DMSO (Sigma).

Labeling and Hybridization

First-strand cDNA synthesis is usually carried out at 42°C overnight with 2 µg of total RNA per sample, after incubation at 70°C for 10 min with 6 µg of random hexamers (Invitrogen), 400 U of SuperScript II (Invitrogen), 1X first-strand buffer (Invitrogen SuperScript II), 0.01 M DTT (Invitrogen SuperScript II), 12.5 mM dATP (Invitrogen), 12.5 mM dCTP (Invitrogen), 12.5 mM dGTP (Invitrogen), 4.16 mM dTTP (Invitrogen), and 8.33 mM aa-dUTP (Ambion). 400 U of SuperScript II (Invitrogen), 1X first-strand buffer (Invitrogen SuperScript II), 0.01 M DTT (Invitrogen SuperScript II), 12.5 mM dATP (Invitrogen), 12.5 mM dCTP (Invitrogen), 12.5 mM dGTP (Invitrogen), 4.16 mM dTTP (Invitrogen) and 8.33 mM aa-dUTP (Ambion). The reaction is quenched by adding 10 µl 0.5 M EDTA, and the RNA template hydrolyzed with the addition of 10 µl 1 M NaOH followed by incubation at 65°C for 15 min. The labeled RNA is separated from reaction reagents using a Microcon 30 spin column (Millipore) and lyophilized. For NHS-Cy dye coupling, the sample is re-suspended in 4.5 µl 0.1 M sodium carbonate, pH 9.0, and 4.5 µl of either NHS-Cy3 or NHS-Cy5 (Amersham Pharmacia) and allowed to couple to the cDNA for 1 h at room temperature in the dark. The coupling reaction is quenched with 4.5 µl 4 M hydroxylamine and incubated in the dark for 15 min at room temperature. Next, 35 µl 100 mM sodium acetate pH 5.2 is added and unincorporated dye is removed using QIAquick columns (Qiagen). The whole undiluted sample is analyzed using a spectrophotometer to determine dye incorporation and nucleotides per dye; targets with incorporation of over 150 pmol and a nucleotides-per-dye ratio below 50 are used for hybridization. The Cy3 and Cy5 labeled cDNA targets are lyophilized to dryness.

Lyophilized, labeled cDNA targets are re-suspended in 20 µl of hybridization buffer (50% formamide, $5 \times$ SSC, 0.1% SDS, 1 µg µl^{-1} salmon sperm DNA) for 21 min at room temperature, and Cy5- and Cy3-labeled targets are combined. Targets are heated for 10 min at 95°C and cooled on ice for 30 s. Glass lifterslips (Erie Scientific) are placed onto the pre-hybridized slides and target is applied using capillary action. Slides are hybridized at 42°C in the dark for 16 to 20 h. Usually, post-hybridization washes are performed in $1 \times$ SSC, 0.1% SDS at 42°C for 5 min followed by $0.1 \times$ SSC, 0.1% SDS for 5 min, and three washes in $0.1 \times$ SSC for 5 min, all at room temperature. Slides are scanned in the Cy3 and Cy5 channels (532 nm and 635 nm, respectively) and stored as paired TIFF images.

Array data were analyzed using the TIGR TM4 suite (Saeed *et al.*, 2003) and data was normalized using the MIDAS software (TIGR TM4). Using the TMEV package (TIGR TM4), average expression ratios and unadjusted p values are calculated for each gene across twelve total replicates for each temperature jump (i.e. three pairs of flip-dye hybridizations for each of two biological replicates). Genes with expression ratios greater than or equal to a two-fold change with *p* values less than 0.01 are then regarded as differentially expressed.

Real Time RT-PCR

In circumstances where a total microarray analysis is impractical and a draft or complete genomic sequence is available, it may be possible to identify members of the heat shock regulon by means of quantitative RT-PCR. This method is also of practical use in confirming the differential regulation of genes identified as part of a heat shock regulon but having a regulatory range close to the two-fold limit where microarray results are equivocal. In our study of *M. jannaschii* heat shock regulation, RNA was reverse transcribed with M-MLV reverse transcriptase (Promega) according to Promega's protocol. RNA controls and cDNA samples consisting of biological replicates from each condition (95°C temperature shock, 65°C temperature shock, 85°C exponential) were assayed in triplicate. Each sample was assayed in 25 μl reaction volumes with the Quantitect SYBR Green PCR kit (Qiagen) and the iCycler iQ Real-Time PCR Detection System (Bio-Rad) according to Qiagen's protocols. Fold changes were calculated relative to the 85°C samples using this calibration curve.

◆◆◆◆◆ PROTEIN CHARACTERIZATION

Recombinant Expression

Purification of native heat shock proteins is often impractical since they are produced in relatively low amounts in the hyperthermophiles, which grow to very low cell densities under stress. Therefore, bacterial recombinant expression is widely used to obtain heat shock proteins from hyperthermophiles in sufficient quantities to carry out biochemical and biophysical analysis. This approach also allows one to express the chaperones in the bacterial cell, and to test their functional roles within the host. *Escherichia coli* cells are outstanding "factories" for recombinant expression of proteins.

However, overexpression is not always accepted by the metabolic system of the host, and in some cases the response encountered in recombinant systems is the accumulation of target proteins into insoluble aggregates known as inclusion bodies. In some situations, the bacterial stress response is induced in response to the overproduction of high levels of partially or misfolded proteins. The recombinant proteins produced in these conditions are in general misfolded and thus biologically inactive (Villaverde and Carrio, 2003). In recent years expression strategies have been modified to obtain soluble expression. One of the most common protein expression systems in *E. coli* is based on the controlled expression of the gene of interest by the T7 RNA

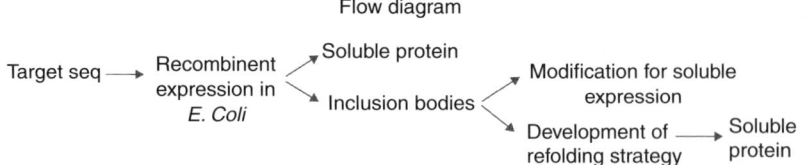

Figure 10.1. Approach for expression of genes from hyperthermophilic Archaea and Bacteria.

polymerase, under the control of the *lac* promoter (Chamberlin and Ring, 1973).

A typical schema for setting up a recombinant expression system for Heat Shock Proteins (HSPs) is shown in Figure 10.1. One of the basic problems associated with recombinant expression of archaeal proteins is the radical difference in the growth temperatures and codon usage of hyperthermophilic Archaea, compared with *E. coli*. Therefore, it is important to consider the approach to recombinant gene expression as follows:

- Selection of appropriate fusion technology for protein solubility.
- Stabilize mRNA by elimination of RNase E sites.
- Overexpress molecular chaperones in the host cell for maximal expression.
- Include tRNA complementing plasmid for recognition and translation of rare codons found in hyperthermophile genomes (alternatively, to synthesize and assemble a gene with *E. coli* codon usage, i.e. codon optimization).

Expression Vectors

The use of various plasmids for recombinant expression is a generic technology with many different methods and host strains. For the sake of brevity, we describe two widely used systems in research laboratories.

pET vectors

The pET Expression System is widely used to express many proteins from thermophiles because of its ability to overexpress proteins relative to the expression of *E. coli* proteins (Studier *et al.*, 1990). The operating principle of the pET Expression System is the tight regulation of the target gene, which is not transcribed unless the T7 RNA polymerase is induced. This leads to selective overexpression of the target protein, often to levels exceeding 20% of the soluble protein in the cell-free extract within a few hours after induction (Studier *et al.*, 1990). The major feature of this system is its ability to maintain target genes transcriptionally silent in the uninduced state, and thus allows recombinant production of "toxic" gene products. The T7 RNA polymerase gene on the host cell chromosome usually has an inducible promoter that is activated by IPTG. The expression plasmids contain several important elements – a *lacI* gene which codes for the *lac* repressor protein, a T7 promoter which is specific to only T7 RNA polymerase (not bacterial RNA polymerase) and also does not occur anywhere in the prokaryotic genome, a *lac* operator which can block transcription, a polylinker, an f1 origin of replication (so that a single-stranded plasmid can be produced when co-infected with M13 helper phage), an ampicillin-resistance gene for selection and an origin of replication. Typically, the host cell used is *E. coli* strain BL(DE3). Control of the pET expression system is accomplished through the *lac* promoter and operator. Although this system is extremely powerful, it is also possible to attenuate expression levels simply by lowering the concentration of inducer. Decreasing the expression level may enhance the soluble yield of some less soluble target proteins.

Applications for proteins expressed in pET vectors vary widely. For example, analytical amounts of a target protein may be needed for activity studies, screening and characterizing mutants, screening for ligand interactions, and antigen preparation. Larger amounts of active protein will typically be required for structural studies. Regardless of scale, it is desirable to express proteins in their soluble, active form. Solubility of a particular target protein is determined by a variety of factors, including the individual protein sequence. The choice of vector and expression host can significantly increase the activity and amount of target protein present in the soluble fraction. However, only one combination of vector, host strain, and culture conditions may work for optimum expression. A vector can enhance solubility and/or folding in one of three ways: (1) provide for fusion to a polypeptide that itself is highly soluble [e.g. glutathione-S-transferase (GST), thioredoxin (Trx), N utilization substance A (NusA)], (2) provide for fusion to an enzyme that catalyzes disulfide bond formation (e.g. thioredoxin, DsbA, DsbC), or (3) provide a signal sequence for translocation into the periplasmic space. When using vectors designed for cytoplasmic expression, folding can be improved in hosts that are permissive for the formation of disulfide bonds in the cytoplasm. These kinds of vectors can also be co-expressed with other vectors. Fusion tags can facilitate detection and purification of the target protein, or may increase the probability of biological activity by increasing the solubility in the cytoplasm or export to the periplasm. If a fusion sequence is tolerated by the application attempted, it is useful to produce fusion proteins carrying the S·Tag^{TM}, T7·Tag, GST·Tag^{TM}, His·Tag.

Duet vectors

There is a need for technologies that can co-express several proteins in controlled ratios of abundance, in an environment that is conducive to protein–complex formation. In order to study protein–protein interaction, protein structure and function, a reliable protein co-expression system for high-throughput protein synthesis is indispensable. Recent studies have suggested that co-expression of one protein can improve the solubility of another which was mainly expressed as inclusion bodies (Yue *et al.*, 2000). Co-expression of multiple proteins in *E. coli* can be achieved by using single plasmid vectors that carry two or more genes, or by using multiple plasmids containing compatible replication origins and drug resistance markers that allow stable maintenance in the same cell (Tan, 2001). Duet vectors (Novagen) are designed for the co-expression of two or more target genes. Three Duet vectors are available in different plasmid backgrounds that carry compatible replicons and antibiotic resistance markers, and may be used together in appropriate host strains to co-express up to eight proteins. The Duet vectors have the following alternative, compatible replication origins and selective markers, as follows: pETDuet-1 carries the ColE1 replicon and *bla* gene (ampicillin resistance), pACYCDuet-1 carries the P15A replicon and *cat* gene (chloramphenicol resistance), and pCDFDuet-1 carries the CloDF13 replicon and *aadA* gene (str), which allows them to coexist in the same cell. Two kanamycin-resistant Duet vectors are available; pRSFDuet-1 carries the RSF1030 replicon, and pCOLADuet-1 carries the ColA replicon. Each vector carries two expression units each controlled by a T7*lac* promoter for high-level protein expression. Each promoter is followed by a optimal ribosome-binding site and multiple cloning sites (MCS) region. A T7 terminator follows the second MCS. The multiple cloning regions have restriction sites that facilitate the cloning of two genes and the transfer from other pET constructs. The potential of Duet vectors to be co-transformed, propagated, and induced for robust target protein co-expression makes them appropriate for the analysis of protein complexes. The user also has the option of expressing proteins on both high- and low-copy number plasmids in order to approximate the stoichiometry needed for intracellular assembly of chaperone complexes with multiple, non-identical subunits.

The Duet vectors provide the option of producing native unfused proteins, or fusions to His-tag sequences for detection and purification of protein complexes. The generic design of MCS regions facilitates the generation of two unfused proteins or one fusion protein with an N-terminal His-tag, and/or one fusion protein with a C-terminal S-tag, as desired for detection, purification, or quantification of protein complexes. Both MCS regions include 8-base pair (bp) rare cutting restriction enzymes. The vectors also carry the *lacI* gene to ensure the expression of sufficient *lac* repressor to control basal expression without titrating *lac* repressor due to the presence of high copy number plasmids.

Co-expression in *E. coli* has important merits over attempting to reconstitute the complex from individually produced components, including enhanced solubility and proper folding of each subunit, resulting in a greatly enhanced yield of active protein complex (Li et al., 1997). In addition, co-expression of subunits greatly facilitates the following types of analyses: characterization of protein–protein interactions by mutagenesis of the subunits, analysis of complex multimeric assemblies by separately characterizing interacting components, analysis of multi-subunit complexes by altering the stoichiometry of their components, and identification and characterization of the interacting subunits in multi-protein complexes through pair-wise co-expression of subunits. This approach has been used to examine the effectiveness of protein folding *in vivo*, using *E. coli* as a test organism for chaperones from *Pyrococcus furiosus* (Laksanalamai et al., 2003, 2005).

Biophysical and Biochemical Methods

Biochemical and biophysical characterization of the recombinant proteins is an important step to confirm their structural and functional integrity. To achieve this, the target protein must be purified, usually to homogeneity by the criteria of overloaded SDS-PAGE analysis. Heat-stable proteins and peptides are usually purified by heating as the first purification step. Ten- to 50-fold purification of the thermophilic proteins in bacterial cell-free extracts can often be accomplished by heating steps that induce specific denaturation and precipitation of the *E. coli* proteins but not of the thermostable heat shock protein(s).

The hyperstable protein structures are stabilized by a combination of physicochemical interactions (e.g. electrostatic and hydrophobic interactions), combined, in at least some cases, with a large degree of flexibility inside the structure of the molecule. If solution conditions change, the molecular structure may be affected, along with a subsequent change in the overall size of the protein. Thus, monitoring the size of a protein molecule is one way of observing its stability and alterations in the structure of the protein.

Molecular interactions and complex interactions can be studied by light scattering, thus providing information about the aggregation state of the protein, for example to judge the multiple subunit assembly processes in the sHSP of *Pyrococcus furiosus* (Laksanalamai et al., 2003, 2004).

Dynamic light scattering (DLS)

This type of light scattering is also known as "photon correlation spectroscopy" (PCS) or "quasi-elastic light scattering" (QELS), and uses the scattered light to measure the rate of diffusion of the protein particles. These motion data are conventionally processed to derive a size distribution for the sample, where the size is given by the "hydrodynamic radius" of the protein particle. This hydrodynamic size depends on both mass and shape (conformation). Dynamic scattering is particularly good at sensing the presence of very small amounts of aggregated protein (<0.01% by weight) and studying samples containing a very large range of masses. It can be quite valuable for comparing stability of different formulations, including real-time monitoring of changes at elevated temperatures. DLS plays a major role in characterizing proteins that function by the formation of multiple subunit complexes. The scattering intensity of a small molecule is proportional to the square of the molecular weight. When a protein denatures, the hydrophobic residues buried within the interior of the folded structure are exposed to the solvent. This entropically unfavorable state leads to aggregation and this non-specific aggregation of denatured proteins is easily monitored with light scattering instrumentation (Georgieva et al., 2004).

In dynamic light scattering, one measures the time dependence of the light scattered from a very small region of solution, over a time range from tenths of a microsecond to milliseconds. These fluctuations in the intensity of the scattered light are related to the rate of diffusion of molecules in and out of the region being studied (Brownian motion), and the data can be analyzed to directly give the diffusion coefficients of the particles doing the scattering. When multiple species are present, a distribution of diffusion coefficients is seen.

One important application for DLS is measuring the hydrodynamic size of molecules in samples containing broad distributions of species of widely differing molecular masses (e.g. a native protein and various sizes of aggregates). Another advantage of dynamic scattering is the ability to study samples directly in their formulation buffers or at high protein concentrations (50 mg ml^{-1} or more). It should be noted, however, that the interpretation of data from high concentration samples (and especially the values of the apparent hydrodynamic radius) can be quite difficult due to solution non-ideality ("molecular crowding") effects. Nonetheless, for a protein in solution at over 50 mg ml^{-1}, the fact that the sample is not homogeneous and contains some large aggregate is unambiguous.

Classical light scattering or "static" scattering

While dynamic scattering is, in principle, capable of distinguishing whether a protein is a monomer or dimer, it is much less accurate for distinguishing small oligomers than is classical light scattering. The latter method provides a direct measure of molecular mass. It is therefore very useful for determining whether the native state of a chaperone is a monomer or oligomer, and for measuring the masses of aggregates or other non-native species. It also can be used for measuring the stoichiometry of complexes between different proteins (e.g. receptor–ligand complexes or antibody–antigen complexes).

Fluorescence

Biological macromolecules are often characterized by their optical properties through the use of spectroscopic measurements. Two types of measurements are most common in characterizing protein molecules by this technique. First is measurement of relative fluorescence intensities and measurement of quantum yield. The most common technique used is relative fluorescence, where the fluorimeter is calibrated to full scale fluorescence intensity under standard conditions. Some perturbations are introduced in the system (change in pH, addition of chemical agent) and the fluorescence intensity is determined relative to the standard conditions. For this, the excitation and emission wavelengths are set for each monochromator. Some protein molecules which have internal amino acids phenyl rings (phenylalanine, tyrosine, tryptophan) are known as intrinsic fluors and that they are fluorescent themselves. This method of measuring intrinsic fluorescence is most often used to study protein conformational changes and to probe the location of active sites and coenzymes in protein molecule. There are also extrinsic fluors that are added into the system. These molecules bind to the rigid hydrophobic environment in the molecule, resulting in a change of fluorescence intensity. Determining the binding of hydrophobic probes and the environments of aromatic amino acid residues is indispensable for characterizing folding and unfolding of protein molecules (Chalfie, 1998; van Borren et al., 2002).

Circular dichroism (CD)

Circular dichroism (CD) spectroscopy is a form of light absorption spectroscopy that measures the difference in absorbance of right- and left-circularly polarized light. CD spectroscopy is very sensitive to the secondary structure of polypeptides, and is particularly powerful for monitoring conformational changes in the secondary structure of a protein (Brahms and Brahms, 1980). It has been shown that CD spectra between 260 and approximately 180 nm can be analyzed for the different secondary structural types: alpha helix, parallel and antiparallel beta sheet, and turns (Johnson, 1992). Protein molecules exhibit absorption in the ultraviolet region of the spectrum by peptide bonds (symmetric chromophores), side chains in proteins, and any prosthetic groups. The lowest energy transition in the peptide chromophore is observed at 210–220 nm with very weak intensity. The commonly used unit in current literature is mean residue ellipticity (degree cm^2 mol^{-1}).

Circular dichroism spectroscopy has also been extensively applied to the structural characterization of peptides. The application of CD for conformational studies in peptides (like proteins) can be largely grouped into: (1) monitoring conformational changes (e.g. monomer-oligomer, substrate binding, denaturation, etc.), and (2) estimation of secondary structural content (e.g. this peptide is 25% helical under these conditions). As already mentioned, CD is particularly well-suited to determine structural changes in both proteins and peptides. Recently, the secondary-structure content of both GTP-free and GTP-inhibited uncoupling protein 1 (UCP1) expressed in yeast has been determined by far-UV CD analysis. Analysis of these results clearly demonstrates that high helical content is the major structural feature of uncoupling protein 1 (Douette et al., 2004). In a typical experiment, circular dichroism scans are taken to determine the secondary structure conformations for each subunit and the reconstituted chaperone complex. Typical scans are taken with the protein in phosphate buffer pH 7.2 and at a concentration of approximately 0.05 mg ml^{-1}. For far-UV CD wavelength scans, measurements are taken from 250–195 nm at 1 nm resolution at 25°C. For the temperature melt, ellipticity is measured at 222 nm at 5°C intervals between 25–95°C with a 45 s equilibrium time.

Differential scanning calorimetry (DSC)

Differential Scanning Calorimetry (DSC) is often used to investigate protein stability, and can be combined with other biophysical methods to link thermodynamics, structure and function (Remmele and Gombotz, 2000; Remmele et al., 1998). DSC measures the heat changes that occur during controlled increase (or decrease) in temperature. It is used to study thermal transitions in biological systems, to determine melting temperatures as well as thermodynamic parameters associated with these changes and is commonly used to characterize the structure and stability of proteins. Heat capacity changes associated with protein unfolding are primarily due to changes in hydration of side chains that were buried in the native state, but become solvent exposed in the denatured state.

A typical DSC thermogram is obtained using a dilute protein solution. Typical protein concentrations of 0.1–3 mg ml^{-1}, 0.7–1.8 ml sample volume, and a temperature range from 5 to 130°C are used, depending on instrument. Protein unfolding is recognized as a sharp endothermic peak centered at a characteristic temperature called the transition midpoint (T_m). The transition midpoint, ΔH (enthalpy) and ΔCp (heat capacity) of the transition are calculated by fitting the data to a two-state transition model using non-linear least squares regression analysis. The higher the T_m, the more stable the protein. DSC has been used to show the binding of HSP70 to exposed hydrophobic residues of citrate synthase following heat shock where HSP70 itself was heat resistant at high temperature and protected the heat-labile protein (Lepock, 2005). It is also possible to obtain information regarding the affinity of binding for a ligand from the difference in stability between the free and bound macromolecule (Freire, 1995), and DSC has been used to examine the specific interaction between regulatory proteins and DNA and is the most direct experimental technique to elucidate the energetics of conformational transitions of biological macromolecules (Kedracka-Krok and Wasylewski, 2003). The oligomeric state of free and inhibitor-bound proline racemase was determined by DSC, and ligand-binding interactions (Straume and Freire, 1992).

Fourier transform infrared (FT-IR)

Like circular dichroism analyses of proteins, FT-IR spectroscopic studies are easily performed and require relatively small amounts of material (~0.1 mg). The infrared spectra of polypeptides exhibit a number of so-called amide bands which represent different vibrational modes of the peptide bond. Of these, the amide I band is most widely used for secondary structure analyses. The amide I band results from the C=O stretching vibration of the amide group coupled to the bending of the N–H bond and the stretching of the C–N bond. These vibrational modes, present as infrared bands approximately in the range of 1600–1700 cm^{-1}, are sensitive to hydrogen bonding and coupling between transition dipole of adjacent peptide bonds and hence are sensitive to secondary structure.

A critical step in the interpretation of IR spectra of proteins is the assignment of the amide I component bands of different types of secondary structure. Amide I bands centered around 1650–1658 cm^{-1} are generally considered to be characteristic of alpha helices in a protein molecule. Random coils and turns also give rise to amide I bands in this region, which complicates analysis. Beta sheets produce highly diagnostic bands in the region 1620–1640 cm^{-1}. Numerous attempts have been made to extract quantitative information on secondary structure of protein from analyses of these amide I bands (Byler and Susi, 1986; Surewicz et al., 1993). Since the potential sources of error in CD and FT-IR analyses of secondary structure content are largely independent, the two methods are highly complementary and can be used in conjunction to increase the accuracy of structural assignments.

N- and C-terminal sequencing of protein/peptide

The amino-terminal sequence analysis of protein and peptide based on the Edman degradation has been widely used for protein structure determination, however it is now widely used to assign protein spots to specific ORF identities in whole proteome studies. In the classical N-terminal sequencing of peptides, phenylisothiocyanate (PITC) reacts with the amino acid residue at the amino terminus under basic conditions to form a phenylthiocarbamyl derivative (PTC–protein). A standard mixture of 19 PTH-amino acids is also injected onto the column for separation (usually as the first cycle of the sequencing run). This chromatogram provides standard retention times of the amino acids for comparison with each Edman degradation cycle chromatogram. To determine the amino acid present at a particular residue number, the chromatogram from the residues of interest is compared with the chromatogram from the previous residue by overlaying one on top of the other. From this, the amino acid for the particular residue can be determined. This process is repeated sequentially to provide the N-terminal sequence of the protein/peptide.

C-terminal sequencing is an orthogonal method of protein sequencing which can provide structure information on proteins with blocked N-termini by means of natural modification. It also provides information on the post-translational processing at the carboxyl terminus of gene products and facilitates the production of more specific probes for gene cloning. Although a number of methods for C-terminal sequencing have been reported, the thiocyanate method that was first described in 1926 by Schlack and Kumpf has been the subject of many studies. In C-terminal sequencing of protein or peptides, acetylisothiocyanate (AITC) is used in acidic conditions as a derivatization reagent to convert the C-terminal amino acid to a thiohydantoin, and an alkaline potassium thiocyanate (KSCN) solution is used to cleave the TH-AAs from the protein. The TH-AAs cleaved from the proteins are analyzed by on-line high-performance liquid chromatography (HPLC) using a graphitized carbon (Hypercarb) column (Anumula and Tang, 1995). The TH-AA standards are used from commercial sources for identification and quantitation. However, TH-AA standards can be prepared easily from the amino acids by treatment with AITC.

SDS-PAGE coupled to laser densitometry

SDS-PAGE analysis can be used to characterize protein molecule in terms of its molecular weight and its monomeric or dimeric forms. The electrophoretic mobility of the SDS–protein complexes is influenced primarily by molecular size: the larger molecules are retarded by the molecular sieving effect of the gel, and smaller molecules have greater mobility. Empirical measurements have shown a linear relationship between the log molecular weight and the electrophoretic mobility. One practical aspect affecting the use of SDS–PAGE with proteins from hyperthermophiles is the extraordinary stability of the proteins. For example, in the case of glutamate dehydrogenase and HSP60, treatment in 2–3% SDS at temperatures up to 105°C is needed in order to achieve denaturation and dissociation of subunits before separation can be observed on the gels (Robb and Laksanalamai, unpublished). This technique is also useful in characterizing different types of subunits in oligomeric proteins. Gel images are acquired with a laser densitometer and imaging system is used to archive the stained gel in digital format. The scanned digital images are processed and protein bands are quantified using a transmittance laser densitometer.

RP-HPLC for protein analysis

Reverse Phase High Performance Liquid Chromatography (RP-HPLC) is a high resolution technique that is able to separate polypeptides of nearly identical sequences, not only small peptides such as those obtained through trypsin digestion, but even for much larger proteins. The separation of small molecules by RP-HPLC involves continuous partitioning of the molecules between the mobile phase and the hydrophobic stationary phase. Polypeptides adsorb to the hydrophobic surfaces after entering the column and remain adsorbed until the concentration of organic modifier reaches the critical concentration necessary to cause desorption. At this point, peptides are desorbed and eluted down the column. Protein conformation is very important in reverse phase separation.

Peptide mapping by LC–MS

Biological mass spectrometry is now an indispensable tool for rapid peptide and protein identification (Henzel et al., 1993). Identification of proteins by mass spectrometry uses peptide masses or the MS–MS fragmentation of a peptide to identify proteins (Yates, 2000). A protein is first digested with trypsin and injected in RP-HPLC coupled with a tandem Mass Spectrometer. LC–MS–MS is a technique that combines the solute separation power of HPLC, with the exquisite detection power of a mass spectrometer. Experiments are run on samples such as in-gel or in-solution digests from purified protein sources. HPLC can separate peptides on the basis of a number of unique or species-specific properties of peptides such as charge, size, hydrophobicity and presence of a specific tag or amino acid(s). HPLC also removes potentially interfering molecules from the sample such as salts, buffers and detergents. Contaminants greatly influence the efficiency of the ionization, and the quality of data generated by the MS is greatly dependent on a clear sample. Subsequently, the fragmentation ion spectra are searched using powerful software like Sequest for proper identification of proteins. The difficulty has been that the HPLC system deals with analytes in the liquid-phase, yet the MS requires a transformation of these ions from the liquid phase, to ions in the gas phase. It is challenging to maintain the required vacuum level in the mass spectrometer because introduction of a liquid at the ion source causes challenges in maintaining a vacuum. For this reason, the gas phase ions are generated before introduction to the MS. Since the introduction of the "thermospray" interface, the LC–MS technique works well with biopolymers such as proteins. The next big improvement was the introduction of the electrospray technique that allows ionization at atmospheric pressure and both are considered to be a method of soft ionization that is a major prerequisite to the analysis of proteins.

Excellent results are obtained by LC–MS–MS for a number of reasons. First, LC–MS–MS will also produce spectra in much the same way as the mass fingerprint generated by the MALDI-ToF, but in addition, MS–MS experiments will generate peptide primary structure (sequence) information from this peptide. This is additional and very specific information that has a greater chance of producing a positive identification from a redundant database. In order to determine whether specific protein–protein interactions occur between protein chaperones and co-chaperones and/or target proteins, protein–protein cross-linking studies can be very informative. For example, the lysine-sensitive cross-linker Bis(sulfosuccinimidyl) suberate (BS3) can be used to determine oligimerization of subunits. Proteins were incubated with 1 mM BS3 for 1 h at 22°C and the reaction quenched with addition of 20 mM Tris for 15 min at 22°C. Molecular weights of the products were determined by SDS-PAGE and Coomassie Blue staining.

Glycosylation/sulfation/phosphorylation site identification

Mass spectrometry has been shown in recent years to be a powerful tool to determine accurate molecular masses and sequences of peptides and proteins and post-translational modifications such as glycosylation, phosphorylation and sulfation. N-glycosylation is normally a co-translational process that occurs during translocation of the nascent protein to the endoplasmic reticulum. Determining whether potential glycosylation sites are actually modified by glycans is important because functions of proteins may be modulated by and/or critically depend upon the presence of glycans at specific sites. Although little is known about post-translational modification of chaperones in hyperthermophiles, phosphorylation is known to be a critical factor in the regulation of eukaryotic small Heat Shock Protein (sHSP) and this aspect of HSP structure and function will be of great interest in future studies.

◆◆◆◆◆ METABOLIC ENGINEERING

Mutations in HSP encoding genes in mesophiles have selectable phenotypes and, in theory, the heat shock responses of extremophiles should also be susceptible to mutation and selection. However, such studies have not been reported yet, and the genetic systems of hyperthermophiles are still in an early stage of development. Another approach that has met with success has been the alteration of the responses of *E. coli* to temperature under the influence of hyperthermophile chaperones that have been cloned and expressed. This has resulted in an upgraded survival and growth responses of *E. coli*, by expression of the small heat shock protein (sHSP) and prefoldin from *P. furiosus* (Laksanalamai et al., 2003, 2004; Laksanalamai and Robb, 2004). This is a convenient method for screening for mutations affecting functions of chaperones, as well as testing the effects of site-directed mutations.

◆◆◆◆◆ CONCLUDING REMARKS

Survival and recovery from heat stress are key adaptations of thermophiles and hyperthermophiles that allow them to survive fluctuating temperature regimes of the geothermal habitats where they usually thrive. The study of heat shock proteins involves decision-making as to which proteins are regulated by heat stress, the mechanisms of heat shock regulation, and the assignment of functions to the heat shock proteins by biophysical and biochemical approaches. The determination of chaperone functions is a very significant challenge, involving recombinant expression and characterization of individual HSPs, and elucidation of the functional roles of HSPs and their cooperative interactions with target proteins. In the future, a systems biology approach may provide the necessary guidelines to facilitate the selection of labor-intensive structure–function studies, and to deduce the pathway order of components comprising the protein folding machinery of hyperthermophiles. These pathways could then be confirmed *in vivo* by genetic analysis, which is now becoming feasible in some hyperthermophilic species.

Acknowledgements

This work was supported by the National Science Foundation (BES-0224733). We are grateful to Boonchai B. Boonyaratanakornkit and Timothy A. Whitehead for assistance in preparing the manuscript.

This is Manuscript #05-121 from the Center of Marine Biotechnology.

List of suppliers

Invitrogen Corporation
1600 Faraday Avenue
PO Box 6482
Carlsbad, California 92008, USA
Tel: 1-760-603-7200
Fax: 1-760-602-6500

Fax: 1-732-457-0557
http://www1.amershambiosciences.com

MD-3 robot

Signa-Aldrich Corp.
St. Louis, MO 63103, USA
Tel: 1-314-771-5765
Fax: 1-314-771-5757
http://www.sigmaaldrich.com

DMSO

Ambion, Inc.
2130 Woodward Street
Austin, TX 78744-1832, USA
Tel: 1-512-651-0200
Fax: 1-512-651-0201
http://www.ambion.com

aa-dUTP

Millipore Corp.
80 Ashby Road
Bedford, MA 01730, USA
Tel: 1-800-645-5476
Fax: 1-800-645-5439
http://www.millipore.com

Microcon 30 spin columns

Bio-Rad Laboratories
1000 Alfred Nobel Drive
Hercules, CA 94547, USA
Tel: 1-510-724-7000
Fax: 1-510-741-5817
http://www.bio-rad.com

iCycler iQ real-time PCR detection system

Novagen
EMD Biosciences, Inc.
10394 Pacific Center Court
San Diego, CA 92121, USA
Tel: 1-858-450-9600
Fax: 1-858-453-3552
http://www.emdbiosciences.com/

Duet vectors

References

Anumula, K. R. and Tang, S. (1995). Novel chemistry for sequencing of proteins from carboxyl terminus yields a simple method. *FASEB J.* **9**, A1477.

Blöchl, E., Rachel, R., Burggraf, S., Hafenbradl, D., Jannasch, H. W. and Stetter, K. O. (1997). *Pyrolobus fumarii*, gen. and sp. nov., represents a novel group of archaea, extending the upper temperature limit for life to 113°C. *Extremophiles* **1**, 14–21.

Boonyaratanakornkit, B. B., Simpson, A. J., Whitehead, T. A., Fraser, C. M., El-Sayed, N. M. and Clark, D. S. (2005). Transcriptional profiling of the hyperthermophilic methanarchaeon *Methanococcus jannaschii* in response to lethal heat and non-lethal cold shock. *Environ. Microbiol.* **7**, 789–797.

Brahms, S. and Brahms, J. (1980). Determination of protein secondary structure in solution by vacuum ultraviolet circular dichroism. *J. Mol. Biol.* **138**, 149–178.

Byler, M. and Susi, H. (1986). Examination of the secondary structure of proteins by deconvolved FTIR spectra. *Biopolymers* **25**, 469–487.

Chalfie, M. (1998). *Green Fluorescent Protein: Properties, Applications, and Protocols*. Wiley-Interscience, New York.

Chamberlin, M. and Ring, J. (1973). Characterization of T7-specific ribonucleic acid polymerase. II. Inhibitors of the enzyme and their application to the study of the enzymatic reaction. *J. Biol. Chem.* **248**, 2245–2250.

Douette, P., Navet, R., Bouillenne, F., Brans, A., Sluse-Goffart, C., Matagne, A. and Sluse, F. E. (2004). Secondary-structure characterization by far-UV CD of highly purified uncoupling protein 1 expressed in yeast. *Biochem. J.* **380**, 139–145.

Freire, E. (1995). Differential scanning calorimetry. *Meth. Mol. Biol.* **40**, 191–218.

Georgieva, D. N., Genov, N., Hristov, K., Dierks, K. and Betzel, C. (2004). Interactions of the neurotoxin vipoxin in solution studied by dynamic light scattering. *Biophys. J.* **86**, 461–466.

Henzel, W. J., Billeci, T. M., Stults, J. T., Wong, S. C., Grimley, C. and Watanabe, C. (1993). Identifying proteins from two-dimensional gels by molecular mass searching of peptide fragments in protein sequence databases. *Proc. Natl. Acad. Sci. USA* **90**, 5011–5015.

Johnson, W. C., Jr. (1992). Analysis of circular dichroism spectra. *Meth. Enzymol.* **210**, 426–447.

Kashefi, K. and Lovley, D. R. (2003). Extending the upper temperature limit for life. *Science* **301**, 934.

Kedracka-Krok, S. and Wasylewski, Z. (2003). A differential scanning calorimetry study of tetracycline repressor. *Eur. J. Biochem.* **270**, 4564–4573.

Laksanalamai, P. and Robb, F. T. (2004). Small heat shock proteins from extremophiles. *Extremophiles* **8**, 1–11.

Laksanalamai, P., Maeder, D. L. and Robb, F. T. (2001). Regulation and mechanism of action of the small heat shock protein from the hyperthermophilic archaeon *Pyrococcus furiosus*. *J. Bacteriol.* **183**, 5198–5202.

Laksanalamai, P., Jiemjit, A., Bu, Z., Maeder, D. L. and Robb, F. T. (2003). Multi-subunit assembly of the *Pyrococcus furiosus* small heat shock protein is essential for cellular protection at high temperature. *Extremophiles* **7**, 79–83.

Laksanalamai, P., Whitehead, T. A. and Robb, F. T. (2004). Minimal protein-folding systems in hyperthermophilic archaea. *Nature Rev. Microbiol.* **2**, 315–324.

Laksanalamai, P., Pavlov, A. R., Slesarev, A. I. and Robb, F. T. (2005). Stabilization of Taq DNA polymerase at high temperature by protein folding pathways from a hyperthermophilic Archaeon, *Pyrococcus furiosus*. *Biotechnol. Bioengin.*, in press.

Lepock, J. R. (2005). Measurement of protein stability and protein denaturation in cells using differential scanning calorimetry. *Methods* **35**, 117–125.

Li, C., Schwabe, J. W., Banayo, E. and Evans, R. M. (1997). Coexpression of nuclear receptor partners increases their solubility and biological activities. *Proc. Natl. Acad. Sci. USA* **94**, 2278–2283.

Miller, J. F., Shah, N. N., Nelson, C. M., Ludlow, J. M. and Clark, D. S. (1988). Pressure and temperature effects on growth and methane production of the extreme thermophile *Methanococcus jannaschii*. *Appl. Environ. Microbiol.* **54**, 3039–3042.

Muchowski, P. and Clark, J. I. (1998). ATP-enhanced molecular chaperone functions of the small heat shock protein human aB-crystallin. *Proc. Natl. Acad. Sci. USA* **95**, 1004–1009.

Neuwald, A. F., Aravind, L., Spouge, J. L. and Koonin, E. V. (1999). AAA(+): A class of chaperone-like ATPases associated with the assembly, operation, and disassembly of protein complexes. *Genome Res.* **9**, 27–43.

Okochi, M., Yoshida, T., Maruyama, T., Kawarabayasi, Y., Kikuchi, H. and Yohda, M. (2002). *Pyrococcus* prefoldin stabilizes protein-folding intermediates and transfers them to chaperonins for correct folding. *Biochem. Biophys. Res. Commun.* **291**, 769–774.

Remmele, R. L., Jr., Nightlinger, N. S., Srinivasen, S. and Gombotz, W. R. (1998). Interleukin-1 receptor (IL-1R) liquid formulation development using differential scanning calorimetry. *Pharm. Res.* **15**, 200–208.

Remmele, R. L., Jr. and Gombotz, W. R. (2000). Differential scanning calorimetry: a practical tool for elucidating stability of liquid pharmaceuticals. *BioPharm* **13**, 36–46.

Ruepp, A., Rockel, B., Gutsche, I., Baumeister, W. and Lupas, A. N. (2001). The chaperones of the archaeon *Thermoplasma acidophilum*. *J. Struct. Biol.* **135**, 126–138.

Saeed, A. I., Sharov, V., White, J., Li, J., Liang, W., Bhagabati, N., Braisted, J., Klapa, M., Currier, T., Thiagarajan, M., Sturn, A., Snuffin, M., Rezantsev, A., Popov, D., Ryltsov, A., Kostukovich, E., Borisovsky, I., Liu, Z., Vinsavich, A., Trush, V. and Quackenbush, J. (2003). TM4: a free, open-source system for microarray data management and analysis. *Biotechniques* **34**, 374–378.

Santos, H. and da Costa, M. (2002). Compatible solutes of organisms that live in hot saline environments. *Environ. Microbiol.* **4**, 501–509.

Schlack, P. and Kumpf, W. (1926). On a new method for determination of the constitution of peptides. *Z. Physiol. Chem.* **164**, 125–170.

Shockley, K. R., Ward, D. E., Chhabra, S. R., Conners, S. B., Montero, C. I. and Kelly, R. M. (2003). Heat shock response by the hyperthermophilic archaeon *Pyrococcus furiosus*. *Appl. Environ. Microbiol.* **69**, 2365–2371.

Straume, M. and Freire, E. (1992). Two-dimensional differential scanning calorimetry: simultaneous resolution of intrinsic protein structural energetics and ligand binding interactions by global linkage analysis. *Anal. Biochem.* **203**, 259–268.

Studier, F. W., Rosenberg, A. H., Dunn, J. J. and Dubendorff, J. W. (1990). Use of T7 RNA polymerase to direct expression of cloned genes. *Meth. Enzymol.* **185**, 60–89.

Surewicz, W. K., Mantsch, H. H. and Chapman, D. (1993). Determination of protein secondary structure by Fourier transform infrared spectroscopy: a critical assessment. *Biochemistry* **32**, 389–394.

Tan, S. (2001). A modular polycistronic expression system for over-expressing protein complexes in *Escherichia coli*. *Protein Expr. Purif.* **21**, 224–234.

Trent, J. D., Osipiuk, J. and Pinkau, T. (1990). Acquired thermotolerance and heat shock in the extremely thermophilic archaebacterium *Sulfolobus* sp. strain B12. *J Bacteriol.* **172**, 1478–1484.

Trent, J. D., Nimmesgern, E., Wall, J. S., Hartl, F. U. and Horwich, A. L. (1991). A molecular chaperone from a thermophilic archaebacterium is related to the eukaryotic protein t-complex polypeptide-1. *Nature* **354**, 434–435.

Tsao, J. H., Kaneshiro, S. M., Yu, S. S. and Clark, D. S. (1994). Continuous-culture of *Methanococcus jannaschii*, an extremely thermophilic methanogen. *Biotechnol. Bioengin.* **43**, 258–261.

van Borren, M., Brandy, N. R., Ravelsloot, J. and Westerhoff, H. V. (2002). Looking into a living cell. In *Fluorescence Spectroscopy, Imaging and Probes: New Tools in Chemical, Physical, and Life Sciences* (R. Kraayenhof, A. J. W. G. Visser and H. C. Gerritsen, eds), pp. 361–372. Springer-Verlag, Berlin.

Villaverde, A. and Carrio, M. M. (2003). Protein aggregation in recombinant bacteria: biological role of inclusion bodies. *Biotechnol. Lett.* **25**, 1385–1395.

Wang, S., Sakai, H. and Wiedmann, M. (1995). NAC covers ribosome-associated nascent chains thereby forming a protective environment for regions of nascent chains just emerging from the peptidyl transferase center. *J. Cell Biol.* **130**, 519–528.

Warters, R. L., III (2001). The nuclear matrix is a thermolabile cellular structure. *Cell Stress Chaperones* **6**, 136–147.

Yates, J. R., III (2000). Mass spectrometry: from genomics to proteomics. *Trends Genet.* **16**, 5–8.

Yue, B.-G., Ajuh, P., Akusjärvi, G., Lamond, A. I. and Kreivi, J. P. (2000). Functional coexpression of serine protein kinase SRPK1 and its substrate ASF/SF2 in *E. coli*. *Nucleic Acids Res.* **28**, E14.

11 Deep-sea Thermococcales and their Genetic Elements: Plasmids and Viruses

Daniel Prieur[1], Gaël Erauso[1], Didier Flament[1], Mélusine Gaillard[1], Claire Geslin[1], Mathieu Gonnet[1], Marc Le Romancer[1], Soizick Lucas[2] and Patrick Forterrre[2]

[1] Laboratoire de Microbiologie des Environnements Extrêmes, UMR 6197 (CNRS, UBO, IFREMER), IUEM and IFREMER, Technopôle Brest-Iroise, 29270 Plouzané, France;
[2] Laboratoire de Biologie Moléculaire du Gène chez les Extremophiles, Département de Microbiologie Fondamentale et Médicale, Institut Pasteur, 25 rue du Dr Roux, 75015 Paris, France

◆◆

CONTENTS

Introduction
Thermococcales
Design of first generation genetic tools
Thermococcales and their plasmids
Virus-like particles
Concluding remarks

◆◆◆◆◆ INTRODUCTION

The discovery of deep-sea hydrothermal vents in 1977, at a depth of 2660 m in the Galapagos Islands area has considerably modified our knowledge of the deep-sea biology, since for the first time an ecosystem apparently independent of solar energy was discovered. But for microbiologists the report in 1979 of deep-sea hot springs (so-called black smokers) with fluid temperatures up to 350°C (fluids remain liquid because of the elevated hydrostatic pressure) was the beginning of intense microbiological surveys dedicated to the search for novel thermophilic and hyperthermophilic microorganisms. Oceanic cruises, using remote-operated vehicles (ROV) and manned submersibles for exploration of hydrothermal fields and collection of samples were organized in the world oceans along oceanic ridges and back-arc basins, at depths ranging from 800 to 3500 m.

A variety of thermophilic and hyperthermophilic Bacteria and Archaea were isolated from fluids, smoker walls and debris, hot sediments, and described as novel genera and species (Prieur, 2002). Most of these are strict anaerobes and show a large variety of energy metabolism, e.g. methanogenesis, fermentation of organic matter, sulfate and sulfur reduction, anaerobic hydrogen oxidation, etc. Among these fascinating microorganisms, *Pyrolobus fumarii* which grows optimally at 110°C (maximum temperature for growth: 113°C) is the most thermophilic organism described so far. Another not yet described isolate, strain 121, whose closest relatives belong to the *Pyrodictium* genus, is remarkable because it was reported to actively grow at 121°C, a temperature used in autoclaves for sterilization (Kashefi and Lovely, 2003).

♦♦♦♦♦ THERMOCOCCALES

The first, and now the most numerous, organisms isolated from deep-sea hydrothermal vents belong to the order Thermococcales, within the Euryarchaeota in the domain Archaea. They are represented by about 40 species (about half of these isolated from deep-sea vents) distributed within three genera: *Pyrococcus, Thermococcus* and *Palaeococcus* (Huber and Stetter, 2001; Takai et al., 2000). They are in general neutrophilic, strictly anaerobic hyperthermophilic peptide and sugar fermenters, that can reduce elemental sulfur to hydrogen sulfide (*Palaeococcus* may use ferric iron as electron acceptor), and they are rather easy to handle and to grow in the laboratory (probably because they are oxygen-resistant, if exposure occurs at room temperature or below). For these reasons they have been frequently isolated, and hundreds of undescribed strains are preserved in many laboratory collections. Some of the published species have been screened for thermostable enzymes, and five of them have been selected for complete genome sequencing and analysis: *Pyrococcus furiosus* (Robb et al., 2001), "*P.* abyssi" (Cohen et al., 2003), *Pyrococcus horikoshii* (Kawarabayasi et al., 1998), *Thermococcus gammatolerans* (data not available) and *Thermococcus kodakarensis* (Fukui et al., 2005).

Among the deep-sea Thermococcales, "*P.* abyssi", isolated from the north-Fiji basin, was earlier selected as a model for the study of archaeal physiology (responses to elevated hydrostatic pressure, starvation or ionizing radiation) (Erauso et al., 1993; Gerard et al., 2001; Jolivet et al., 2003a), as well as enzymology studies and production of thermostable enzymes (alkaline phosphatase, DNA polymerases) (Guegen et al., 2001; Henneke et al., 2005; Zappa et al., 2004). One of the polymerases has been patented for application in high-fidelity PCR and is commercially available. This strain was also used as a model for DNA replication, and its interest considerably increased with the discovery that some proteins involved in DNA replication (PCNA, RF-C, RP-A or CDC6/Orc1) were structurally and functionally conserved between Thermococcales and humans (Henneke et al., 2002; Matsunaga et al., 2001, 2003; Myllykallio et al., 2000).

Thermococcus gammatolerans, although described more recently, should become an interesting model in the future (Jolivet et al., 2003b). This organism is resistant to ionizing radiation in almost the same range as

Deinococcus radiodurans, and for the reasons explained above (conservation of several DNA replication proteins between Thermococcales and humans) represents a promising model for the study of DNA repair mechanisms within Archaea. In the near future, *T. gammatolerans* should be exposed to space conditions (vacuum, radiation) in the framework of a research project recently selected by the European Space Agency (ESA).

◆◆◆◆◆ DESIGN OF FIRST GENERATION GENETIC TOOLS

Among the technical limitations identified for more detailed studies on Thermococcales is the lack of genetic tools such as transformation systems, genetic markers, and transformation and expression vectors.

The discovery of the cryptic plasmid pGT5 in the type strain (GE5) of "P. abyssi" (Erauso *et al.*, 1996), and the availability of a large collection of Thermococcales (several hundreds) from various deep-sea hydrothermal areas led us about 12 years ago to initiate a research program dedicated to mobile genetic elements of Thermococcales and their use for the development of genetic tools (Benbouzid-Rollet *et al.*, 1997; Marteinsson *et al.*, 1995).

The development of genetic tools for Thermococcales requires a series of "building blocks" that were successively designed. The first step consisted in the design of an efficient plating method for organisms growing anaerobically at elevated temperatures (e.g. 85–90°C) (Erauso *et al.*, 1995). The next step was the design of a minimal medium for growth, composed of a mixture of 20 amino acids (Watrin *et al.*, 1995). It was necessary to search for a genetic marker for auxotrophy since usual antibiotics are not efficient against Archaea and consequently Thermococcales, and are not stable at elevated temperatures required by these organisms (Watrin *et al.*, 1996). Also, it was necessary to develop reliable techniques to obtain mutants, and this was done by the use of UV radiation (Watrin and Prieur, 1996). However, this technique was not used to generate "P. abyssi" mutants auxotrophic for uracil. These mutants were obtained spontaneously from "P. abyssi" strain GE9 which does not harbor the plasmid pGT5, by plating about 10^8 cells on minimal medium plates containing 800 μg of 5-fluoroorotic acid (5-FOA) and 50 μg of uracil per ml. Uracil mutants of "P. abyssi" are resistant to 5-FOA, while the wild-type cells are sensitive. Genetic analysis of the mutants obtained revealed that mutations occurred in the *pyrE* and/or *pyrF* genes, encoding key enzymes of the pyrimidine biosynthetic pathway. Two mutants that showed low reversion rates were selected for complementation experiments (Watrin *et al.*, 1999).

To construct a shuttle vector for "P. abyssi" and *Escherichia coli*, we used the plasmid pGT5 from "P. abyssi" strain GE5 and the bacterial vector pLITMUS38. The *pyrE* gene, encoding for the orotate phosphoribosyltransferase of the thermoacidophilic archaeon *Sulfolobus acidocaldarius* (Crenarchaeota) was inserted into the pGT5-based vector, giving the shuttle vector pYS2 (7.2 kb). In fact this construct appeared much bigger

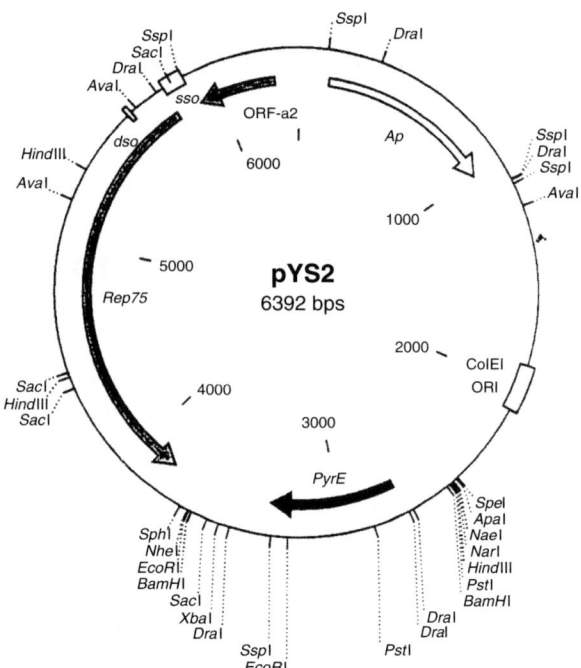

Figure 11.1. Map of the shuttle vector pYS2. Relevant restriction sites are shown. The white box and arrow show the *E. coli* pLITMUS38 plasmid moiety. The gray boxes and arrows correspond to the pyrococcal pGT5 plasmid moiety (*dso, sso*, respectively double-stranded and single-stranded origin of replication; Rep75: replication initiator encoded by the major ORF of pGT5; ORF2: truncated portion of the second main ORF of pGT5). The black arrow indicates the *Sulfolobus acidocaldarius pyrE* gene under the control of its own promoter.

than the original plasmid pGT5 (3.5 kb), and we therefore decided to delete most of ORF2 from pGT5 to obtain a final pYS2 size of 6.4 kb (Figure 11.1) (two ORFS exist in pGT5; ORF1 encodes for a protein involved in the initiation of plasmid replication *via* a rolling circle mechanism; the function of ORF2 is unknown).

"P. abyssi" GE9 *pyrE*⁻ mutants were transformed to prototrophy using a polyethylene glycol/spheroplast method applied to spheroplasts. This method was derived from the protocol previously described for transformation of the extremely halophilic Archaea *Haloferax* spp. and *Halobacterium* spp. (Cline et al., 1989).

Transformation Procedure (Lucas et al., 2002)

Apart from the first centrifugation for collecting cells, all steps are carried out under strictly anaerobic conditions, in an anaerobic chamber as previously described (Erauso et al., 1995).

- Cultivate the mutants at 90°C, 180 rpm, in 100-ml serum bottles containing 50 ml of TAA medium plus uracil (50 µg ml⁻¹). TAA medium consists of a mineral base containing per liter of deionized water: 30 g of sea salts (SIGMA), 3.3 g of PIPES [piperazine-N,N'-bis-(2-ethane sulfonic acid)], 2 ml of a 5% (wt/vol) solution of K_2HPO_4, 1 ml of a 2% (wt/vol) solution of $CaCl_2$, 1 ml of 10 mM Na_2WO_4, 1 ml of 25 mM $FeCl_3$, supplemented with 0.1 g each of the 20 natural amino acids as sole carbon source and 10 ml of the vitamin solution described by Balch et al. (1979). Resazurin (0.1 mg l^{-1}) is used as redox indicator and the pH of the medium is set to 6.8. After replacing the gas phase in the culture bottle by N_2 (100 kPa), anaerobiosis is achieved by injection of 1/100th volume of a 3% (wt/vol) $Na_2S \cdot 9H_2O$ solution.

- Collect the cells when OD_{600} is in the range 0.12–0.15 (5×10^8–10^9 cells ml^{-1} by direct cell counting) by centrifugation at 5000 g for 15 min at 4°C. Keep the cells on ice.
- Wash in 0.2 volumes of ice-cold SM 4x solution (SM 4x solution contains in g l^{-1}: NaCl, 80; KH_2PO_4, 0.8; NaBr, 0.2; $SrCl_2 \cdot 6H_2O$, 0.04; H_3BO_3, 0.08; KCl, 1.0; sodium citrate, 2.0) by centrifugation at 5000 g for 10 min at 4°C.
- Resuspend in 0.01 volumes of spheroplasting buffer (1 M NaCl, 27 mM KCl, 50 mM Tris-HCl (pH 7.5) and 20% sucrose) to reach a final concentration of $\sim 5 \times 10^{10}$ cells ml^{-1}.
- Add EDTA to a final concentration of 10 mM to obtain spheroplasts. Wait 10 min.
- Add transforming DNA (0.5 µg) for 100 µl aliquot of spheroplast suspension and incubate 5 min at 4°C.
- Add an equal volume of a solution containing 50% PEG-600 and 50% spheroplasting buffer, mix gently but thoroughly and keep on ice for 30 min.
- Add 100 µl of dilution solution (SM 4x solution plus 15% (wt/vol) sucrose) and centrifuge at 5000 g for 5 min at 4°C to eliminate PEG. Resuspend the pellet in 100 µl of dilution solution.
- Transfer the cells into 5 ml of reduced YT medium (consisting of the mineral base described above, supplemented with 1 g of yeast extract and 4 g of tryptone) in Hungate tubes and incubate at 85°C for 1 h to allow regeneration of the cell wall.
- Spread aliquots onto plates of TAA medium without uracil using the overlay technique previously described (Erauso et al., 1995), and incubate for 3 days at 85°C.

Colonies of putative transformants were subcultured in liquid TAA medium (without uracil added) to check their uracil prototrophy. They were analyzed by colony hybridization and by isolation of plasmid DNA and back-transformation in E. coli. Transformants obtained grew on the minimal medium as well as the wild-type cells, and showed comparable orotate phosphoribosyltransferase activity (PyrE$^+$). However, although transformation was reproducible, it occurred at a low frequency (10^2–10^3 transformants per µg of pYS2 plasmid DNA and per 10^9 cells of "P. abyssi" auxotrophs). Shuttle vector pYS2 appeared to be very stable and was maintained at high copy number (between 20 and 30 copies per chromosome) in both E. coli and "P. abyssi" (Lucas et al., 2002).

A few months after the publication of our work on the "P. abyssi" shuttle-vector, the first gene-knockout in Thermococcales was described by Sato et al. (2003), using a selection strategy similar to ours. The *trpE* gene of T. kodakarensis was successfully disrupted by inserting a wild-type copy of *pyrF* carried by a suicide plasmid. Resulting transformants had restored uracil prototrophy and became tryptophan auxotrophs. For the transformation procedure, the authors followed the $CaCl_2$ method for *Methanococcus voltae* PS (Bertani and Baresi, 1987) with some modifications.

Transformation Procedure (Sato et al., 2003)

Apart from the first centrifugation for collecting cells, all steps are carried out under strictly anaerobic conditions.

- Cultivate the uracil-auxotrophic mutants (PyrF) at 85°C in 3 ml of YT medium.
- Collect the cells at late exponential phase (4×10^8 cells) by centrifugation at 17 000 g for 5 min.
- Resuspend in 1/15 volume of transformation buffer (80 mM $CaCl_2$ in 0.8X ASW without KH_2PO_4). Keep on ice for 30 min.
- Add transforming DNA (3 µg) dissolved in TE buffer (10 mM Tris-HCl, 1 mM EDTA, pH 8.0). Incubate on ice for 1 h.
- Transfer the cell suspension at 85°C for 45 s and immediately place it on ice for 10 min.
- Cultivate the treated cells in 20 ml of TAA liquid medium without uracil.
- Spread on plates of TAA medium without uracil and incubate for 5–8 days at 85°C. The resulting Pyr^+ strains are analyzed by colony PCR and Southern hybridization.

Recently this gene-knockout approach was successfully applied to study gene function in *T. kodakarensis*. The first case was the disruption of the *rgy* gene encoding the reverse gyrase for which viable mutants could be obtained, demonstrating that in contrast to previous hypotheses, reverse gyrase is not a prerequisite for life at high temperature (Atomi et al., 2004). In the second case, the disruption of the candidate gene *fbp* for fructose-1,6-bisphosphatase (FBPase) allowed its identification as encoding the true FBPase for gluconeogenesis in *T. kodakarensis* (Sato et al., 2004).

◆◆◆◆◆ THERMOCOCCALES AND THEIR PLASMIDS

Despite the existence of a huge collection of Thermococcales strains, very few plasmidic elements have been reported so far in the hyperthermophilic euryarchaeotes.

The first is a small (3.5 kb) cryptic plasmid found in a high copy number in the marine species "P. abyssi" strain GE5 (Erauso et al., 1992). This plasmid, pGT5, is a stable multicopy plasmid (25–30 copies per chromosome) and can be isolated from cultures at different growth phase and in the temperature range from 75 up to 105°C according to the following method of isolation previously described by Charbonnier et al. (1995).

Plasmid Isolation and Purification

Step 1: Harvesting and lysis of the "P. abyssi" strain GE5:

- Harvest the GE5 cells from 1-l of late exponential phase culture by centrifugation at 6000 g for 15 min at room temperature.

- Resuspend cell pellets in a total volume of 8 ml of TNE solution (100 mM Tris-HCl, 100 mM NaCl, 50 mM EDTA, pH 7.5) at room temperature. Homogenize the suspension.
- Transfer the solution into a 30-ml Corex tube and add 1 ml of 10% N-lauryl sarcosine. Invert slowly several times at room temperature.
- Add 1 ml of 10% SDS at room temperature and gently invert the tube.
- Add 0.5 ml of 20 mg ml^{-1} proteinase K and incubate for 3 h at 50°C. After incubation, the lysate must be translucent and viscous.

Step 2: Isolation of the total DNA:

- Transfer the viscous lysate into a 50-ml Teflon tube. Add 11 ml of TE (10 mM Tris-HCl, 0.1 mM EDTA)-saturated phenol (pH 8.0) and stir for 10 min. Spin at 5400 g for 10 min at 20°C. Transfer 90% or more of the aqueous layer into a 50-ml Teflon tube. Repeat twice.
- Add 11 ml of chloroform/isoamyl alcohol (24:1) and stir for 10 min. Spin at 5400 g for 10 min at 20°C. Transfer 90% or more of the aqueous solution into three 15-ml Corex tubes, containing 3 ml in each tube.
- Add 8 ml of 95% ethanol in each tube and incubate for 1 h at −20°C.
- Centrifuge at 15 000 g for 30 min at 4°C (e.g. use a swingout rotor such as a Beckman JS-13 rotor).
- Resuspend pellets in 3 ml of 70% ethanol and transfer all of them into a 15-ml Corex tube. Centrifuge at 12 000 g for 10 min at 20°C. The pellet must be washed again once with 70% ethanol and centrifuged.
- Air dry the pellet for 2 or 3 min.
- Completely resuspend the pellet in 2 ml of TE solution (pH 8.0). Add 3 µl of 10 mg ml^{-1} DNase-free RNase and incubate on a rotator for 1 h at 37°C.

Step 3: Purification of the plasmidic DNA:

- Resuspend cell pellets in a total volume of 4.5 ml of TE solution (pH 7.5).
- Add 4.8 g of cesium chloride and 0.2 ml of a 10 mg ml^{-1} ethidium bromide solution in TE. Mix gently and transfer to a 5.5-ml Beckman Quick-seal ultracentrifuge tube. Spin at 45 000 rpm in a Beckman VTI-65 rotor for 16 h at 17°C.
- Visualize the plasmid band with long wave UV light (365 nm). Remove plasmidic DNA with a syringe with 18-gauge needle, extract ethidium bromide with 5 M NaCl-satured 1-butanol, and dialyze several successive times against 1 l of 10^{-2} mM TE buffer (pH 7.5) at 4°C. Run the DNA on a 1% agarose gel to check for purity.

The purified plasmid pGT5 served as a model to investigate the DNA topology in hyperthermophiles.

DNA Topological Analysis of pGT5 (Charbonnier et al., 1992)

In order to elucidate the DNA topology of pGT5, the carrier strain GE5 was cultured at the optimal growth temperature (95°C) (Erauso et al., 1993).

The extrachromosomal DNA was extracted with the above-described method that involved as few DNA manipulations as possible, in order to recover all the DNA molecules. Then, two-dimensional electrophoresis was performed in the presence of the intercalating drug chloroquine to visualize the topoisomers of the plasmid.

One-dimensional electrophoresis:

- Electrophorese the samples in a 0.7% agarose gel for 16 h at 2 V cm^{-1} in TEB buffer (90 mM Tris-borate, 2 mM EDTA, pH 8.0) at a constant temperature of 25°C.
- After electrophoresis, carefully remove the gel from the apparatus and place it in a Pyrex dish.
- Soak the gel for 8 h at room temperature in TEB buffer containing 2.5 µg ml^{-1} chloroquine.
- Stain with ethidium bromide.
- Rinse the agarose gel with a solution consisting of 1 mM MgSO$_4$ to eliminate the chloroquine.

During the first step of the two-dimensional electrophoresis, pGT5 was shown to be in a relaxed state. Since chloroquine relaxes only negatively supercoiled DNA, this indicated that pGT5 was negatively supercoiled at the temperature of the experiment (25°C). To further elucidate the pGT5 DNA topology status, a second-dimensional electrophoresis was carried out with the following modifications:

Two-dimensional electrophoresis:

- Add a higher concentration of chloroquine (5 µg ml^{-1}) in the TEB buffer in order to separate the topoisomers.
- Perform the electrophoresis in the second dimension at a constant temperature of 25°C for 16 h at 1.3 V cm^{-1}.

The plasmid pGT5 has been positively supercoiled by the addition of higher concentration of chloroquine with a major topoisomer containing 2 ± 0.5 positive superturns. Taking into account the temperature effect on DNA supercoiling, the topological state of pGT5 in physiological conditions (GE5 grows optimally at a temperature of 94°C), was shown to be very close to the relaxed state. In this hyperthermophilic strain of Thermococcales, results of experiments indicate that an increase in the GE5 growth temperature results in a decrease in negative supercoiling of pGT5. This was the first report of a relaxed plasmid at physiological conditions, in contrast with all plasmids known before, which were found negatively supercoiled (Drlica, 1992).

Sequence of Plasmid pGT5: Evidence for Rolling-circle Replication in a Hyperthermophile (Erauso et al., 1996)

pGT5 has been fully sequenced and it was demonstrated for the first time in a hyperthermophilic microorganism that its replication proceeds *via* a rolling-circle mechanism.

pGT5 isolation and sequencing

For this purpose, "P. abyssi" GE5 cells were cultivated under anaerobic conditions at 95°C in a rich medium containing sulfur (Erauso et al., 1993). pGT5 was purified by equilibrium centrifugation in CsCl-ethidium bromide gradients (Sambrook et al., 1989) after sodium dodecyl sulfate lysis of the cells and drastic deproteinization, as described above (Charbonnier et al., 1995).

For sequencing, two Sau3A fragments and two SacI fragments encompassing the Sau3A junctions were cloned into the vectors pAT153 and M13mp18/19, respectively. Both strands were sequenced by primer walking (Strauss et al., 1986) by the dideoxy chain termination method (Sanger et al., 1977). DNA and protein sequence analyses were performed with the Genetics Computer Group (University of Wisconsin Biotechnology, Madison, WI) software package version 8.0 and PC/GENE (Oxford Molecular). Secondary structures were identified with the Genetics Computer Group program RNA-FOLD. The nucleotide and protein sequence similarity searches in GenBank, EMBL, Swissprot and the National Biomedical Research Foundation were done with the programs Fasta and BLAST. The coding probability of the open reading frames (ORFs) was tested by using the program TESTCODE (Genetics Computer Group) together with the programs FRAME (Bibb et al., 1984) and COD-PROK (PC/GENE).

Two main structural features characterize the pGT5sequence: (i) the GC content of 43.4 mol% is similar to that of the "P. abyssi" chromosomal DNA (44.7 mol%); (ii) the presence of numerous direct and inverted repeats, many of them being clustered in four regions (RI to RIV). The putative hairpins produced by inverted repeats in regions RI, RIII and RIV have GC-rich stems, suggesting that they could be maintained *in vivo* at a high temperature. Two major open reading frames (ORF1 and -2) with good coding probability (putative basic proteins of 75 and 46 kDa, respectively) are located on the same strand and covered 85% of the total sequence. The larger open reading frame (ORF1) encodes a putative polypeptide which exhibits sequence similarity with Rep proteins of plasmids using the rolling-circle mechanism for replication. Upstream of this open-reading frame, an 11-bp motif identical to the double-stranded origin (*dso*) of several bacterial plasmids that replicate *via* the rolling-circle mechanism is also detected. Moreover, a putative single-stranded (*sso*) origin exhibits similarities both to bacterial primosome-dependent single-stranded initiation sites and to bacterial primase (*DnaG*) start sites. These data indicate that pGT5 replicates *via* the rolling-circle mechanism and suggest that members of the domain Archaea contain homologs of several bacterial proteins involved in chromosomal DNA replication.

Identification of pGT5 ssDNA

The replication of RC plasmids generates a single-stranded DNA (ssDNA) intermediate that can be detected in crude extracts of host cells. The search for ssDNA of pGT5 was performed according to the method of Strauss et al. (1986).

The lysates were obtained by treating 1-l suspension of rapidly cooled GE5 cells with guanidine thiocyanate according to the following method of Di Ruggiero and Robb (1995):

- Grow GE5 cells in 1-l of a rich medium containing sulfur (Erauso et al., 1993). Harvest the cells during exponential growth by centrifugation in a precooled JA 20 rotor at 4300 g for 20 min at 4°C. The pellet should be 0.5–1.0 g wet weight.
- Wash rapidly in 2 ml of TE buffer (pH 8.0) at 4°C.
- Resuspend the cells in 5 ml of extraction buffer (250 ml of guanidine thiocyanate, 17.6 ml of sodium citrate (0.75 M, pH 7.0), 26.4 ml of sarcosyl (65°C), 293 ml of water (65°C). Filter the buffer through a 0.22-μm polyethersulfone filter (Gelman 09730-26), then add β-mercaptoethanol to a final concentration of 0.75% (v/v)). Mix thoroughly by inversion. Wait 5–10 min at 25°C to achieve complete lysis.
- Treat the lysate with DNA-free RNase (10 mg ml^{-1}).
- Precipitate plasmidic DNA with ethanol as described above.
- Resuspend the pellet in 20 μl of TE buffer (pH 8.0).
- Electrophorese total DNA in 1% agarose gel containing 1 μg ml^{-1} ethidium bromide.

After being transferred to nitrocellulose membranes, DNA, with or without prior denaturation, was hybridized with pGT5 probes labelled using the Genius system (Boehringer). Double-stranded probes are labelled with digoxigenin (DIG)-dUTP incorporated by random priming of the whole pGT5 plasmid.

To prepare sspGT5 RNA probes, pGT5 was subcloned into the Bluescript plasmid (Stratagene), which contains specific promoters around the polylinker. RNA probes were labelled *in vitro* with DIG-UTP using T3 (plus-strand probe) or T7 polymerase (minus-strand probe). To check the functionality of the pGT5 putative ss origin, a 567 bp fragment (position 3108–3231 on the pGT5 sequence) containing region RI was introduced in the ssDNA-overproducing plasmid pHV33Δ*HaeII* (te Riele et al., 1986). This construct was used to transform *E. coli*.

Using this approach, a band that migrated more rapidly than supercoiled pGT5, as is expected for ssDNA, was observed by Southern blotting. Its sensitivity to a S1 nuclease treatment demonstrated that this band was made of ssDNA. The ability of this ssDNA form to hybridize specifically to the minus strand, in addition to the RNA probe corresponding to the same minus strand clearly demonstrated that the pGT5 ssDNA form detected in "P. abyssi" corresponds to the plus strand, as expected for a *bonafide* RC replication intermediate.

pGT5 presents similarities to plasmids from the pC194 family that encode a replication initiator protein, which activates the replication origin by nicking one of the two DNA strands. A functional pGT5 plasmid-encoded replication initiator, the protein Rep75, was then expressed in *E. coli* and its mechanism of action and regulation were investigated (Marsin and Forterre, 1998, 1999; Marsin et al., 2000).

Recently, the complete sequence analysis of a novel small plasmid from the deep-sea *Pyrococcus* sp. strain JT1 was reported (Ward et al., 2002). This plasmid, pRT1 (3373 bp) also replicated *via* the rolling-circle mechanism and its protein Rep was shown to have significant sequence similarity with the Rep75 of pGT5.

Plasmid Diversity

If only two small cryptic plasmids have been described in detail to date, several studies to assess the plasmid diversity in the Thermococcales (and to increase the number of these elements available for laboratory-based genetic studies) have been performed. They revealed the presence of a great number of plasmids in isolates of *Pyrococcus* and *Thermococcus* (Benbouzid-Rollet et al., 1997; Lepage et al., 2004). Recently, a third extensive screening in Thermococcales strains isolated from deep-sea hydrothermal vents located at various sites on the East Pacific Rise, the Southwest Pacific Ocean, the Mid-Atlantic Ridge and the Central Indian Ridge was performed. To isolate new strains of Thermococcales, the plate cultivation technique for strictly anaerobic thermophilic sulfur-metabolizing Archaea previously described by Erauso et al. (1995) was used.

Of over 190 strains purified by plating on Gelrite plates, 40% were found to harbor at least one plasmid.

These plasmids have been classified as a function of their size, origin and restriction fragment length polymorphisms. The plasmid size ranged from 2.8 to more than 35 kb (the most frequent size encountered being between 10 and 15 kb), distributed in 25 distinct restriction fragment length polymorphisms. Each plasmid type, which can be carried by several strains, was unique, and usually corresponded to one specific sample, with few exceptions. To investigate the relationships between the different plasmid types, a representative of each type was used for DNA cross-hybridization on Southern blots using the ECF labelling and detection system (Amersham Biosciences). This analysis showed an unexpectedly high degree of cross-hybridizations between different types of the same hydrothermal site but also between types from distant geographic areas. To elucidate the function of these common sequence regions, a dozen of these elements have been selected for complete sequencing using the following criteria: (i) they were easily and reproducibly extracted using the following standard protocol, (ii) they originated from distant geographic areas or originated from the same site but have distinct restriction patterns, (iii) culture experiments and or electron microscopy observation suggested they may partly represent the genome of a virus. For this purpose, a novel method of plasmid isolation that differs from the Charbonnier's method described previously, was developed. cccDNA was extracted by an alkaline-lysis-based method (Birboim and Doly, 1979).

Plasmid Isolation

- Grow Thermococcales cells at 85°C, 180 rpm, overnight (14–17 h) in YT medium (see above) to late exponential phase (2–4 × 10^8 cells ml^{-1}), transfer the culture (40 ml) into a sterile 50-ml ice-cold polyethylene tube and discard residual sulfur by a short centrifugation (5 min, 1000 g, 4°C).
- Transfer the cell suspension to a new 50-ml tube and harvest cells by centrifugation at 6000 g for 15 min at 4°C.
- Discard the supernatant and resuspend the cell pellet in 600 µl of ice-cold TNE (100 mM Tris-HCl pH 8, 100 mM NaCl, 50 mM EDTA pH 8.0) containing 100 µg ml^{-1} DNase-free RNase A.
- Transfer the cell suspension in a new 2-ml microtube, centrifuge and add 600 µl of freshly prepared lysis solution (0.2 N NaOH, 1% SDS) and mix by rapidly inverting the tube five or six times.
- Add 700 µl of ice-cold neutralization solution (60 ml 5 M potassium acetate, 11.5 ml glacial acetic acid, 28.5 ml H_2O) and homogenize the solution by inverting the tube several times.
- Centrifuge at 15 000 g for 15 min at 4°C; carefully collect the supernatant and transfer to two 2-ml microtubes.
- Add an equal volume of phenol/chloroform/isoamyl alcohol (25:24:1) and mix thoroughly by vigorous shaking for 20–30 s.
- Centrifuge at 15 000 g for 15 min at 4°C, collect the upper (aqueous) phase, transfer it to new microtube, then add an equal volume of chloroform and mix thoroughly by vigorous shaking for 20–30 s (this step is to extract residual phenol).
- Centrifuge at 15 000 g for 5 min at 4°C, collect the upper phase and add 0.7 volumes of isopropanol and allow DNA to precipitate for 10 min at room temperature.
- Centrifuge at 15 000 g for 30 min at 4°C.
- Discard the supernatant and wash the pellet with 500 µl of 70% ethanol. Centrifuge at 15 000 g for 5 min at 4°C.
- Remove carefully the supernatant and dry the pellet by using a speed-vac apparatus (or let it air dry) until all the ethanol is evaporated and add DNase-free RNase (50 µg ml^{-1}).
- Finally redissolve the DNA pellet in 25–50 µl of TE buffer. The sample can be stored for months at −20°C without significant degradation.

Plasmid Cloning and Sequencing

The purified cccDNA was then randomly fragmented by nebulization (Invitrogen). After ends repair, the fragments were sized by agarose-gel electrophoresis and fragments with the desired size (1–3 kb) were extracted from the gel using an extraction kit (QIAEXII, Qiagen). These fragments were then used for the construction of genomic libraries in the pUC18 vector. The clones were sequenced from both ends using the universal M13 primers and assembled using the SeqmanII program of the Lasergene 6.0 package. Preliminary annotation was performed with the help of different public and commercial softwares, mainly the Vector NTi 7.0 package.

At this date, four plasmids have been completely sequenced, originating from the Central Indian Ridge (pCIR10, 13 231 bp), from the Mid-Atlantic Ridge (pIRI48, 12 975 bp and pIRI33, 11 040 bp) and from the East Pacific Rise (pAMT11, 20 533 bp). After preliminary analyses, 17 ORFs were identified in the sequence of pCIR10, 14 ORFs in both pIRI48 and pIRI33, and 35 ORFs in pAMT11.

pIRI33, pIRI48 and pCIR10 shared a common structural organization. Their genome is compact (1.3 ORF per kb), 60% of the putatively protein encoded is homologous. ORF1, which has a significant match with a range of ATP-dependent DNA helicases, is shared by all three plasmids. Comparatively, the fourth plasmid, pAMT11, exhibited a distinct genome organization that did not share any similarity with the former three plasmids. Among all 35 ORFs, ORF3 and ORF12 likely encode a Rep protein and a non-functional integrase, respectively.

◆◆◆◆◆ VIRUS-LIKE PARTICLES

Viruses represent an important reservoir of biodiversity and are the most abundant biological entities on the planet (Breitbart and Rohwer, 2005). Viruses infecting prokaryotes (Bacteria and Archaea) constitute the largest of all viral groups (Ackermann, 2001). However, our knowledge of viruses of Archaea is rather limited with less than 40 archaeal viruses well-described to date (Prangishvili *et al.*, 2001). Among the Thermococcales, the most important order of the phylum Euryarchaeota, only one virus has been isolated and completely characterized to date (Geslin *et al.*, 2003b). Screening of enrichment cultures of Thermococcales has revealed an unexpected diversity among VLPs (Virus-Like Particles) in deep-sea hyperthermophilic environments (Geslin *et al.*, 2003a).

Enrichment Cultures for Virus-Like Particles from Deep-sea Hydrothermal Vents

A systematic search was carried out on samples collected in various geographically distant hydrothermal sites located on the East Pacific Rise (EPR 9°50′N, 104°17′W and 13°N, 104°W; 2500 m depth) and the Mid-Atlantic Ridge (MAR 36°16′N, 33°54′W; 2400 m depth and 37°50′N, 31°50′W; 900 m depth) to investigate the diversity of virus-like particles from deep-sea vents. Pieces of chimney, hydrothermal fluids and sediments were sampled by using manned and/or remote operated deep-sea submersibles.

Sampling

- Transfer small amounts of crude samples to 50-ml sterile glass vials and flood with a sterile solution of 3% (w/v) sea salts (Sigma).

- Close the vials, pressurize with N_2 (100 kPa) and reduce by injection of a $Na_2S \cdot 9H_2O$ solution (final concentration, 0.05% (wt/vol)) and store at 4°C until processing.

Enrichment cultures conditions

Enrichment cultures were established anaerobically in a rich sulfur-containing medium, previously described (Ravot et al., 1995) with the following composition (g l^{-1}): NH_4Cl, 1.0; $MgCl_2 \cdot 6H_2O$, 0.2; $CaCl_2 \cdot 2H_2O$, 0.1; KCl, 0.1; sodium acetate (anhydrous), 0.5; yeast extract, 5.0; bio-Trypcase, 5.0; NaCl, 20; piperazine-N,N'-bis(2-ethanesulfonic acid) (PIPES), 3.45, and resazurin, 0.001.

- Adjust the pH to 7 and sterilize the medium.
- After the medium has cooled down, add the following solutions, separately sterilized by autoclaving: 5 ml of a 6% (wt/vol) KH_2PO_4 solution and 5 ml of 6% (wt/vol) K_2HPO_4 solution.
- Dispense the medium (50 ml) into 100-ml sterile vials and add 1% (wt/vol) elemental sulfur previously sterilized by steaming at 100°C for 1 h on three successive days.
- Obtain anaerobiosis by applying vacuum to the medium and saturate it with N_2.
- Reduce the medium by adding a sterile solution of $Na_2S \cdot 9H_2O$ (final concentration, 0.05% [wt/vol]).
- Inoculate the medium to a final concentration of 3% by adding 1.5 ml of a sample suspension described above.
- Incubate at 85°C with shaking (150–200 rpm) for different times ranging from 15 h (for fast growing cultures) to 14 days.
- Monitor growth by using a phase-contrast microscope.

The culture conditions chosen are routinely used in our laboratory for enrichment of anaerobic, neutrophilic and heterotrophic sulfur-reducing hyperthermophiles from deep-sea hydrothermal vent samples. Such cultures are usually largely dominated by the fast-growing Thermococcales, since the incubation temperature (85°C) allows growth of the most common genera *Thermococcus* and *Pyrococcus* (Harmsen et al., 1997). Examination by a phase-contrast microscope revealed only motile and non-motile cocci in the enrichment cultures, which strengthens the presumption of Thermococcales enrichment.

Detection of virus particles

Enrichment cultures which scored positive are checked for the presence of virus particles by transmission electron microscopy as described by Rice et al. (2001) with some modifications.

- Centrifuge 8 ml of culture (in late exponential phase or early stationary phase) at 4000 g for 15 min.
- Filter the supernatant through Acrodisc PF 0.8/0.2 µm filters (Pall Gelman Laboratory) to remove cells and mineral particles from the original crude sample.
- Ultracentrifuge the filtrate at 100 000 g for 3 h in a Beckman 70.1 Ti rotor.

- Resuspend pelleted particles in 20 µl of TE pH 8 (10 mM Tris-HCl, 1 mM EDTA).
- Spot a droplet of the suspension onto a carbon-coated copper grid for negative staining.
- Allow the specimen to adsorb to the carbon layer for 2 min and remove excess liquid with a piece of filter paper (Whatman).
- Add a droplet of an uranyl acetate solution (2%) to the carbon grid for 40 s and remove excess liquid again.
- After air drying, examine the specimen using a JEOL electron microscope, JEM 100 CX II, operated at 80 keV.

Out of a total of 100 crude samples, ninety enrichment cultures in a medium designed to enrich Thermococcales and cultivated at 85°C were successfully established. Nine distinct virus morphotypes were detected by electron microscopy in 16 different enrichments. Unexpectedly, an important diversity of morphological types was observed: the lemon-shaped type prevailed, but stiff and flexible rods were also observed (Geslin et al., 2003a) (Figure 11.2). Novel morphotypes

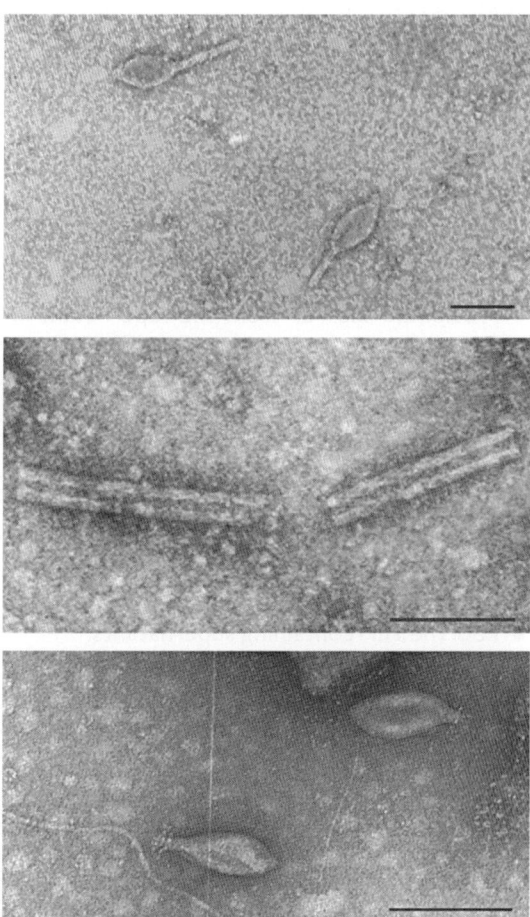

Figure 11.2. Transmission electron micrographs of VLPs after negative staining with 2% uranyl acetate. Scale bar, 100 nm.

such as "spoon-shaped" and spindle particles with bipolar expansions were also discovered which had not been previously reported in prokaryotes but which have recently been reported from terrestrial solfataric fields in Iceland and in the Yellowstone National Park, U.S.A. (Rachel et al., 2002; Rice et al., 2001).

This exploration by electron microscopy of viral morphotypes in samples collected from various oceanic deep-sea vents all around the world suggests that viruses from hot marine environments are as diverse as hyperthermophilic viruses from terrestrial environments (Geslin et al., 2003a).

Isolation and Characterization of a Novel VLP, Designated PAV1, Isolated from "P. abyssi"

The particle named PAV1 (for "P. abyssi" virus 1) is lemon-shaped (120×80 nm) with a short tail terminated by fibers resembling the virus SSV1, the type member of the archaeal virus family Fuselloviridae infecting the hyperthermophilic crenarchaeote *Sulfolobus shibatae* (Geslin et al., 2003b).

To purify PAV1 particles from "P. abyssi", the following procedure can be used:

- Harvest 700 ml of a culture of strain GE23 obtained in Ravot medium (see above) 12 h after inoculation with 10^8 cells.
- Remove cells by low-speed centrifugation (6000 g, 20 min, 4°C).
- Precipitate the VLP from the supernatant in 1 M NaCl with 100 g l^{-1} of PEG 6000 overnight at 4°C with gentle stirring.
- Collect the precipitate by centrifugation (15 200 g for 75 min at 4°C), drain well, and resuspend in TE buffer (10 mM Tris-HCl and 1 mM EDTA (pH 8)).
- Perform a second precipitation with PEG 6000 (10%) and 1 M NaCl for 90 min.
- Collect the precipitate by centrifugation and resuspend in TE as described above.
- Pellet cell debris by centrifugation at 5000 g for 10 min at 4°C.
- Keep the supernatant, and wash the pellet two more times under the same conditions with reduced volumes of TE.
- Pool the VLP-containing supernatants.
- Store overnight at 4°C.
- Remove residual cell debris after centrifugation (5000 g, 15 min, 4°C), concentrate the supernatant by centrifugation at 32 000 g for 105 min, and resuspend the pellet in TE.
- Purify the VLP from this suspension by centrifugation in a CsCl buoyant density gradient (1.298 g ml^{-1}) in a Beckman Optima LE-80 K 70.1 (Beckman Coulter) Ti rotor at 180 000 g for 6 h.
- Detect at 254 nm and collect the fractions containing nucleic acids by using a density gradient fractionator (model 185, Teledyne Isco).
- After dilution in TE buffer, directly concentrate fractions in a Beckman ultracentrifuge at 150 000 g for 3 h and resuspend in TE (20 μl).

VLP Structure

The physical stability of the PAV1 structure was assessed by exposing diluted purified particles to organic solvents, detergents and proteolytic degradation and was checked by TEM observations. The PAV1 viability or infectivity could not be tested because a plaque assay was not available. PAV1 was shown to be sensitive to chloroform (25% wt/vol for 2.5 and 5 min at room temperature with constant agitation), because only spherical particles were observed in treated samples instead of the regular lemon-shaped ones. The VLPs were also partially disrupted by treatment with 0.3% Triton X-100 for 1 and 3 min at room temperature, as well as after incubation for 1 h at 56°C with proteinase K (1 mg ml^{-1}); they were completely destroyed with 0.1% SDS for 3 min at 50°C.

These preliminary results suggested that the envelope of PAV1 was composed of proteins and lipids.

VLP–Host Relationship

PAV1 multiplies and is excreted without damaging the cell. It is maintained in its natural host in a stable "carrier state", i.e. not lytic or lysogenic, a state that seems common in viruses from hyperthermophiles (Zillig et al., 1998). Accordingly, none of the potential stimuli tested, such as UV or γ irradiation, mitomycin C (at 5 or 10 μg ml^{-1}), or hydrostatic pressure shocks (20 and 30 MPa), influenced virion production. Moreover, attempts to infect other potential hosts among our strain collection, either with virus suspension or naked viral DNA, were unsuccessful.

The Genome of PAVI

PAV1 particles contain a double-stranded circular DNA of 18 kb which is also present in high copy number (more than 50 copies per host chromosome) as free plasmid form in the host cells, whereas no integrated prophage DNA was detected in the host chromosome.

The viral genome has been cloned and sequenced (C. Geslin et al., manuscript in preparation).

DNA Manipulation

The host, "P. abyssi" strain GE23, was cultivated in anaerobic conditions at 85°C in rich medium containing sulfur, as described above. Covalently closed circular DNA (cccDNA) was isolated from 300 ml of a culture of exponentially growing cells by the alkaline lysis method. The DNA was then purified by isopycnic centrifugation in a cesium chloride gradient (0.81 g ml^{-1} of CsCl) in the presence of ethidium bromide (0.52 mg ml^{-1}) (Sambrook et al., 1989).

For PAV1 sequencing, two batches of clones were prepared. The purified cccDNA was digested with *Hind*III and *Bam*HI, and the fragments were cloned with a shotgun method into the similarly digested vector pUC28.

Determination of the Nucleotide Sequence

The sequencing was carried out with the "BigDye Terminator" kit (Applied Biosystems), and the reactions of sequences were analyzed with electrophoresis on the ABI PRISM 373XL and ABI PRISM 377 sequencers (Applied Biosystems). Each insert was sequenced starting from the universal M13 forward and reverse primers. For the fragments of more than 1.2 kb the strategy of "gene walking" was adopted with the use of internal primers. The assembly of the sequences in contigs was carried out using program SEQMANII of the Lasergene software (DNASTAR, Madison, WI, USA).

Sequences Analysis and Annotation

GLIMMER (Delcher et al., 1999) and RBS finder (Suzek et al., 2001) were used to find ORFs. Each ORF was submitted to sequence similarity searches (BLASTN, BLASTP, BLASTX) against the NCBI non-redundant protein and nucleic acid databases. Analysis of the complete nucleotide sequence allowed us to identify 24 open-reading frames. BLAST searches revealed that none of these open-reading frames had significant similarities to proteins in the databases. However, searches for structural and functional motifs and domains revealed that two of the largest open-reading frames (676 and 678 amino acids, respectively) possess two sialidase domains, which are common to a large family of glycosyl hydrolases including many viral enzymes (in Bacteria and Eukarya) involved in virus entry or release from the host cell. Four other open-reading frames corresponded to DNA-binding proteins probably involved in replication and maintenance functions, including an open-reading frame homologous to a hypothetical transcription regulator previously identified in several plasmids from hyperthermophilic Archaea (Lipps et al., 2001; Ward et al., 2002).

◆◆◆◆◆ CONCLUDING REMARKS

This shuttle vector and transformation procedure should be improved in the near future, and several projects in this area are currently running in our laboratories. It is also possible that among the tremendous diversity of plasmids and viruses already detected, one particular extrachromosomal element may serve as a novel base for second generation genetic tools.

Concerning the diversity of viruses, there is a remarkable parallel between the richness and novelty of virus and VLPs as obtained from deep-sea hydrothermal vents and that already described from terrestrial solfataric fields. Such extreme environments (high temperatures, low pHs, high level of ionizing radiation, elevated hydrostastic pressures) do not appear to reduce the biological diversity as commonly assumed, but on the contrary seem to be home to a flourishing abundance of novel virus and plasmids that has no equivalent in other "normal" (mesophilic) natural environments. Nevertheless, the methods used to study the archaeal viruses (and plasmids) isolated from extreme environments are not simple to implement. For example, the viability or infectivity of PAV1 could not be tested because a plaque assay was not available. Indeed, it is not easy to obtain a good lawn at high temperature and under anaerobiosis. Such problems exist for almost all the protocols used, and improvements are necessary.

Another important point is the striking similarity between the novel virus morphotypes discovered in hyperthermophiles from the two distinct phyla, the Crenarchaeota and the Euryarchaeota, compared with the head-tailed morphotype (typical of bacteriophages) that seems to prevail in the mesophilic bacterial and archaeal world. This observation raises interesting questions regarding the origin and evolution of viruses and of their hosts. Certainly it emphasizes the importance of getting more of these "exotic" viral forms genome sequenced to enable meaningful studies addressing such fundamental questions. The same novel genetic element may in turn be instrumental to explore the genetics, and the evolution of their host.

List of Suppliers

Amersham Biosciences
800 Centennial Avenue
P.O. Box 1327
Piscataway, NJ 08855-1327, USA
Tel: +1-732-457-8000
Fax: +1-732-457-0557
http://www.amersham.com

ECF labelling system

Applied Biosystems
850 Lincoln Center Drive
Foster City, CA 94404, USA
Tel: +1-800-327-3002; +1-650-638-5800
Fax: +1-650-638-5998
http://www.appliedbiosystems.com

"BigDye Terminator" kit

Beckman Coulter Inc
4300 N. Harbor Boulevard

P.O. Box 3100
Fullerton, CA 92834-3100, USA
Tel: +1-800-742-235
Fax: +1-800-643-4366
http://Beckmancoulter.com

Quick-seal ultracentrifuge tubes

Boehringer Ingelheim GmbH
Binger Str. 173
55216 Ingelheim, Germany
Tel: +49-6132-77-0
Fax: +49-6132-72-0
http://www.boehringer-ingelheim.com

Genius system

DNASTAR Inc.
1228 S. Park St.
Madison, WI 53715, USA
Tel: +1-608-258-7420
Fax: +1-608-258-7439
http://www.dnastar.com

Lasergene software

Invitrogen SARL
BP 96
95613 Cergy Pontoise Cedex France
Tel: 01 34 32 31 00
Fax: 01 30 37 50 07
http://invitrogen.com

Nebulization system

JEOL (Europe) SAS i/o SA
Espace Claude Monet
1 Allée de Giverny
78290 Croissy sur Seine France
Tel: +33 1 30 15 37 37
Fax: +33 1 30 15 37 17
http://www.jeol.fr

Oxford Molecular Group Inc.
2105 S Bascomb Avenue Suite 200
Campbell CA 95008 USA
Tel: 408-879-6300
Fax: 408-879-6302
http://www.oxmol.com

Colour Plate Section

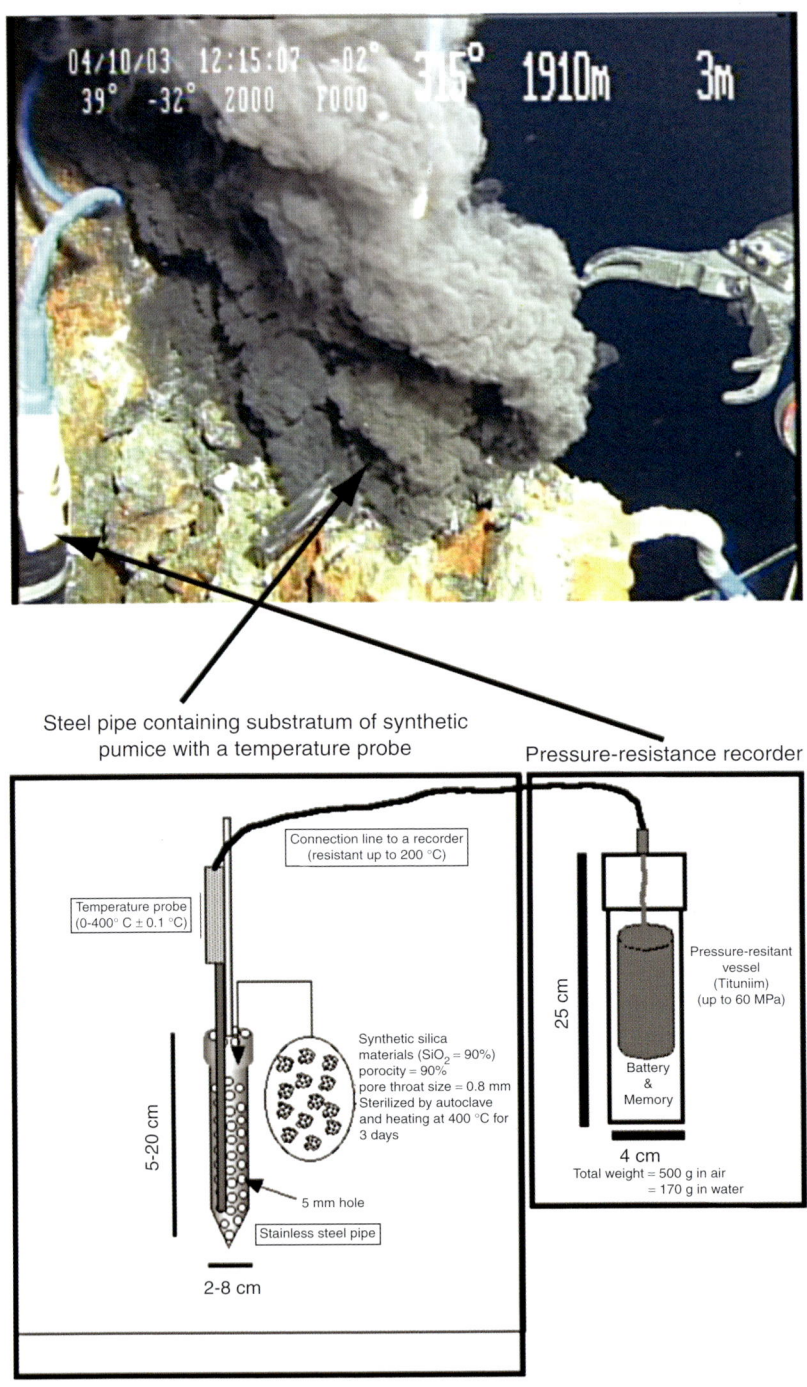

Plate 3.2. *In Situ* colonization system (ISCS) with a modified temperature probe and data logger

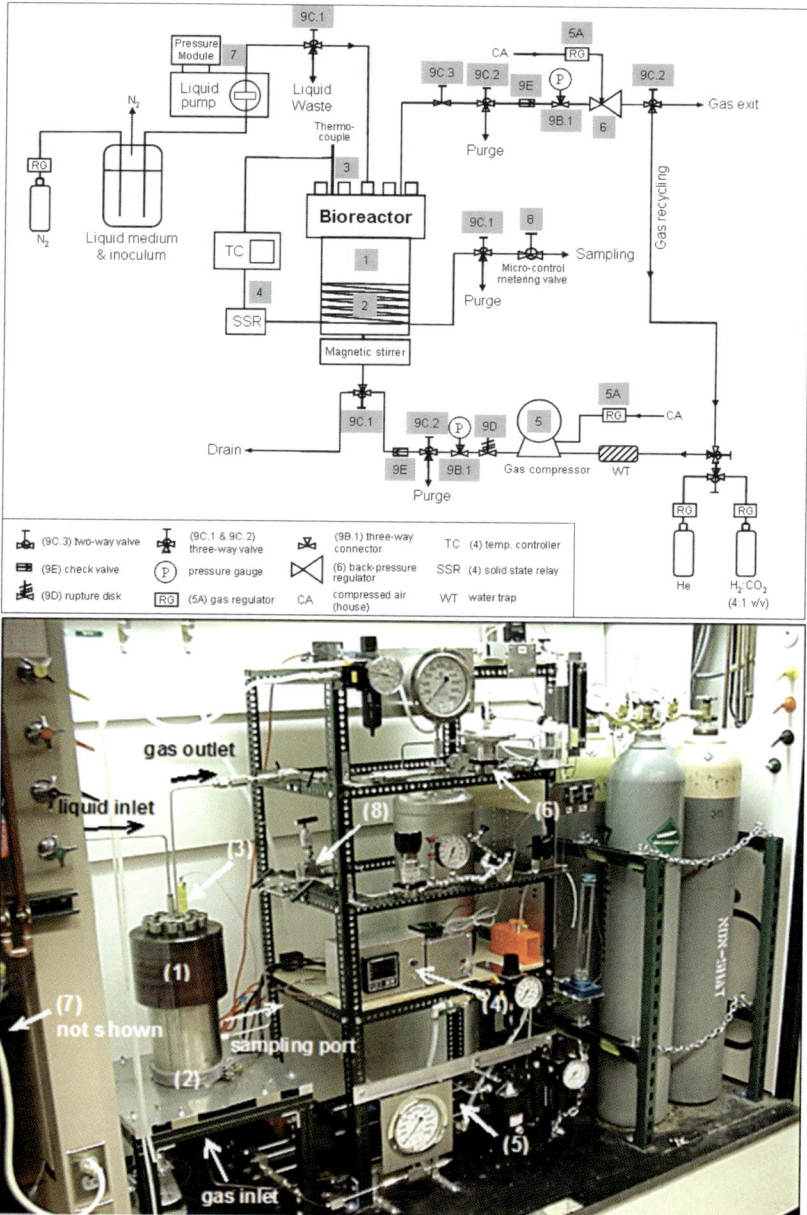

Plate 5.1. Schematic diagram and photograph of the high T–P bioreactor system. (1) 316 high-pressure vessel; (2) heating belt; (3) thermocouple; (4) proportional-integral-derivative controller; (5) gas compressor; (6) back-pressure regulator; (7) liquid pump; (8) micro-control metering value; (9) high-pressure valves and rupture disks (see section entitled Reactor System Parts List for numbering scheme).

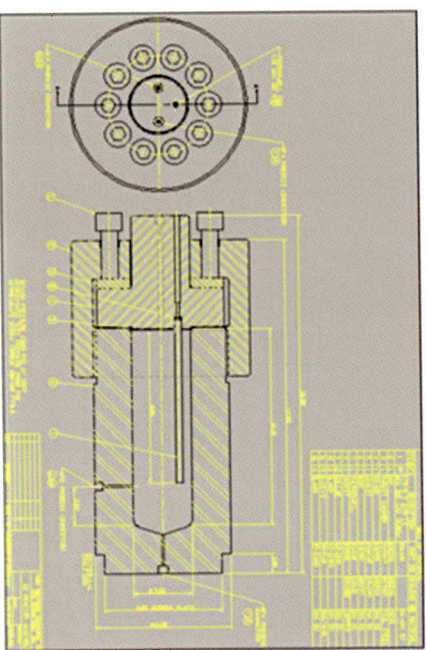

Plate 5.2. Diagram and photograph of the modified pressure vessel. A stainless steel 316 high-pressure vessel (GC-17) from High Pressure Equipment Co. was modified by adding multiple inlet and outlet ports and a thermocouple well.

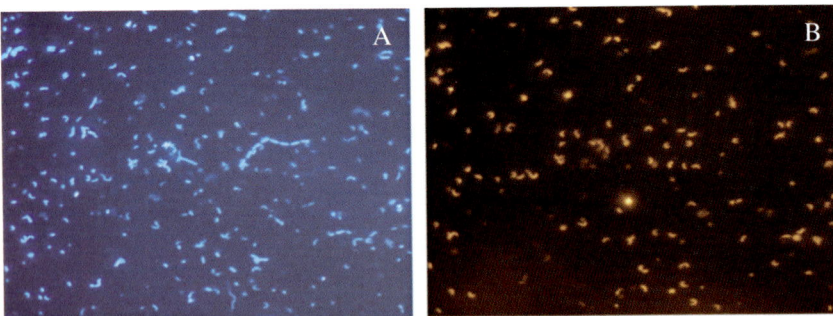

Plate 20.4. Epifluorescence micrographs of bacteria from the Tinto River. Sample obtained from the origin of the river (González-Toril *et al.*, 2003). (A) DAPI-stained cells; (B) same field showing hybridized cells with a specific probe for *L. ferrooxidans*. Probe: LEP636 (Cy3 labeled).

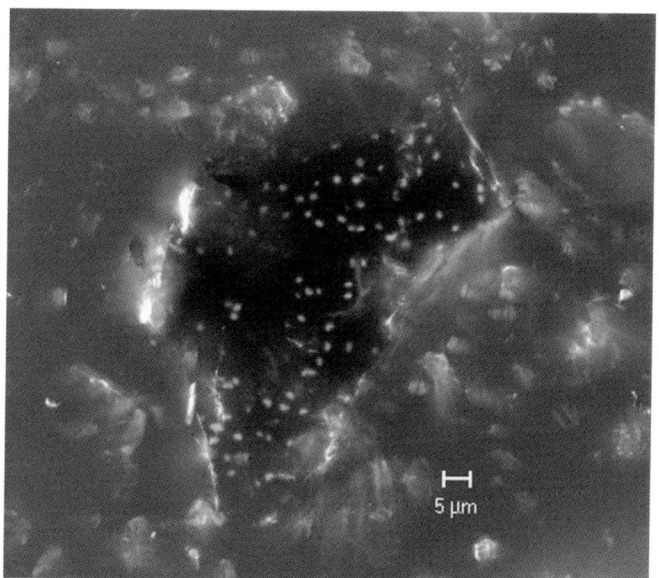

Plate 20.5. Confocal micrographs of bacteria adhered to pyrite. A massive sulfidic mineral sample from the Iberian Pyritic Belt (Tinto River) was exposed to iron-oxidizing bacterial (*L. ferrooxidans* and *At. ferrooxidans*). CARD-FISH using a universal probe for bacterial domain (EUB338 fluorescein labeled) was performed after removal of the solid sample for the incubation media. Attached bacteria, in green, can be observed on the surface of the mineral.

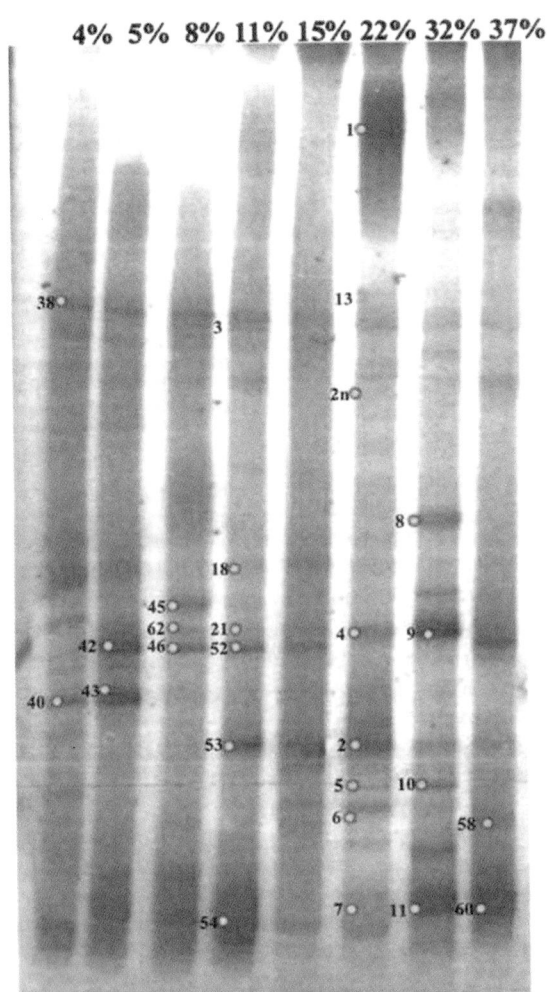

Plate 21.2. Bacterial 16S rRNA gene DGGE profile obtained by the UiB laboratory along the salinity gradient. Bands marked with a numbered dot were excised from the gel, reamplified and sequenced. The closest relatives for these sequences are reported in tables 1–3 in Benlloch et al. (2002) except for bands 18 and 21 (94.9% similarity to marine cyanobacterium Hstpl4), bands 58 and 60 (97.9% similarity to Salinibacter ruber) band 40 (74.8% similarity to Psychroflexus torquis), and band 43 (76.2% similarity to Bacillus halophilus). Reproduced from figure 1, page 350 in Benlloch et al. (2002) with permission of Blackwell Science Ltd.

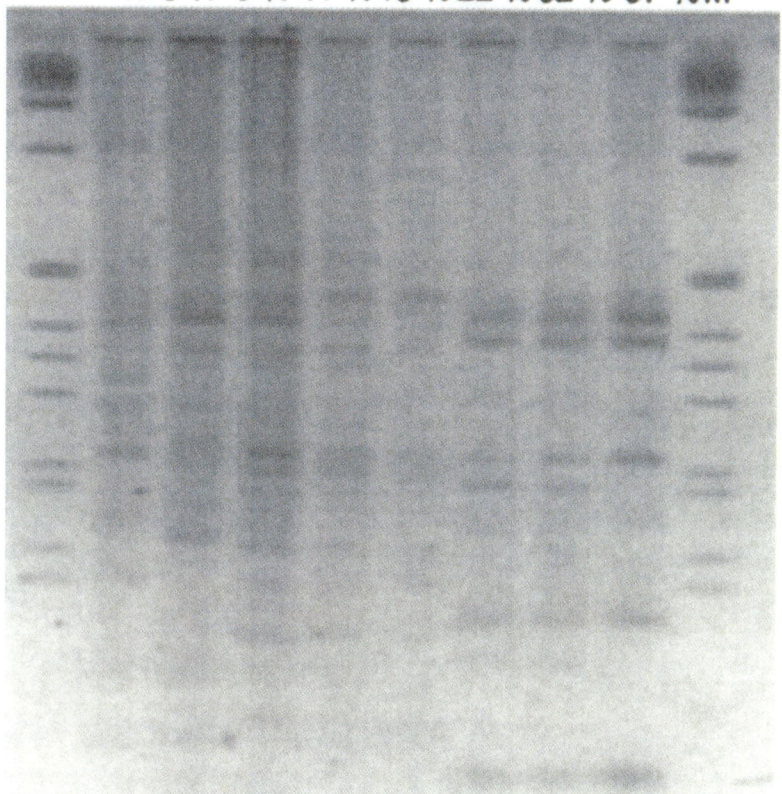

Plate 21.3. Example of a RISA(+RFLP) gel for Archaea. (A) PCR products run in RISA were digested with the restriction enzyme *Hinf*I; M, 1 kb ladder. Reproduced from Figure 6A, page 344 in Casamayor *et al.* (2002) with permission of Blackwell Science Ltd.

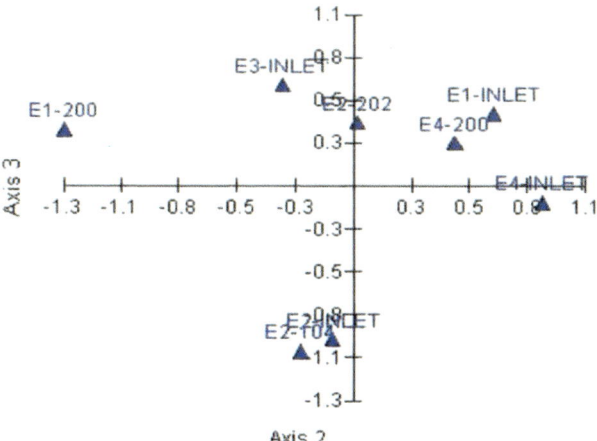

Plate 21.7. Principal component analysis of the BIOLOG data for the samples shown in Figure 15.6.

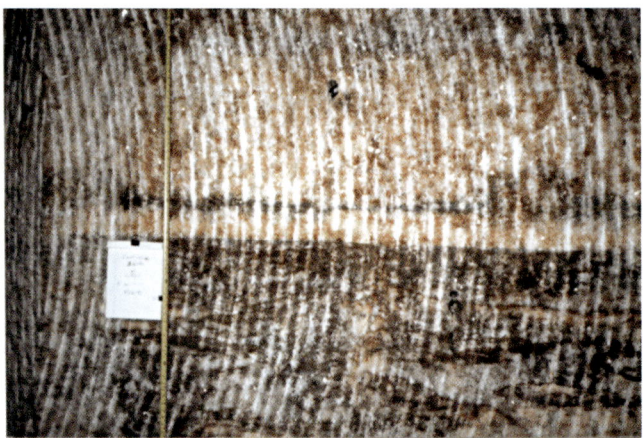

Plate 23.1. Halite layering within a well-preserved Permian aged salt formation. The layers provide a clear example of some of the Steno Principles. Each depositional cycle is clearly illustrated as a horizontal band of similar material. Older layers are toward the bottom of the slide. Several small dissolution pits are present in the middle layers. Note however, that these pits do not extend across multiple layers or out of the image. The ruler marks 2 meters.

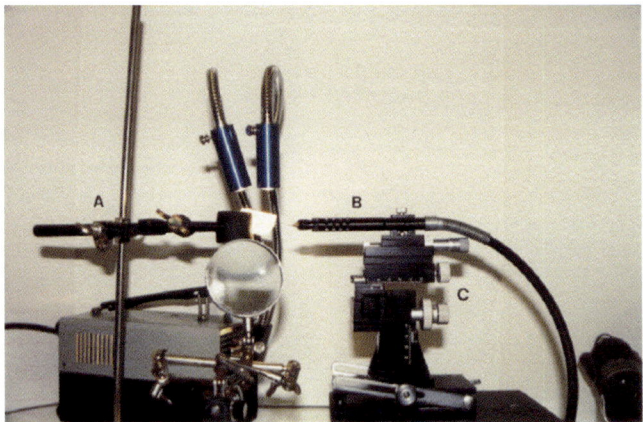

Plate 23.2. Typical homemade crystal drilling system. The drill is secured in the clamp of a micromanipulator which is then used to guide the spinning bit into the crystal. This particular micromanipulator allows alignment of the crystal in two dimensions before beginning the drilling process. Alignment of such a drill system requires being able to see the inclusion from two sides of the crystal. This can be difficult with some materials and behind the glass of a laminar flow hood. Gaining this vision requires smoothing and polishing two surfaces prior to sterilization. Smoothing is best done using a hand held drill (similar to that shown) fitted with a simple grinding bit. Salt crystals can be easily polished by rubbing the sides with a rag wetted with distilled water.

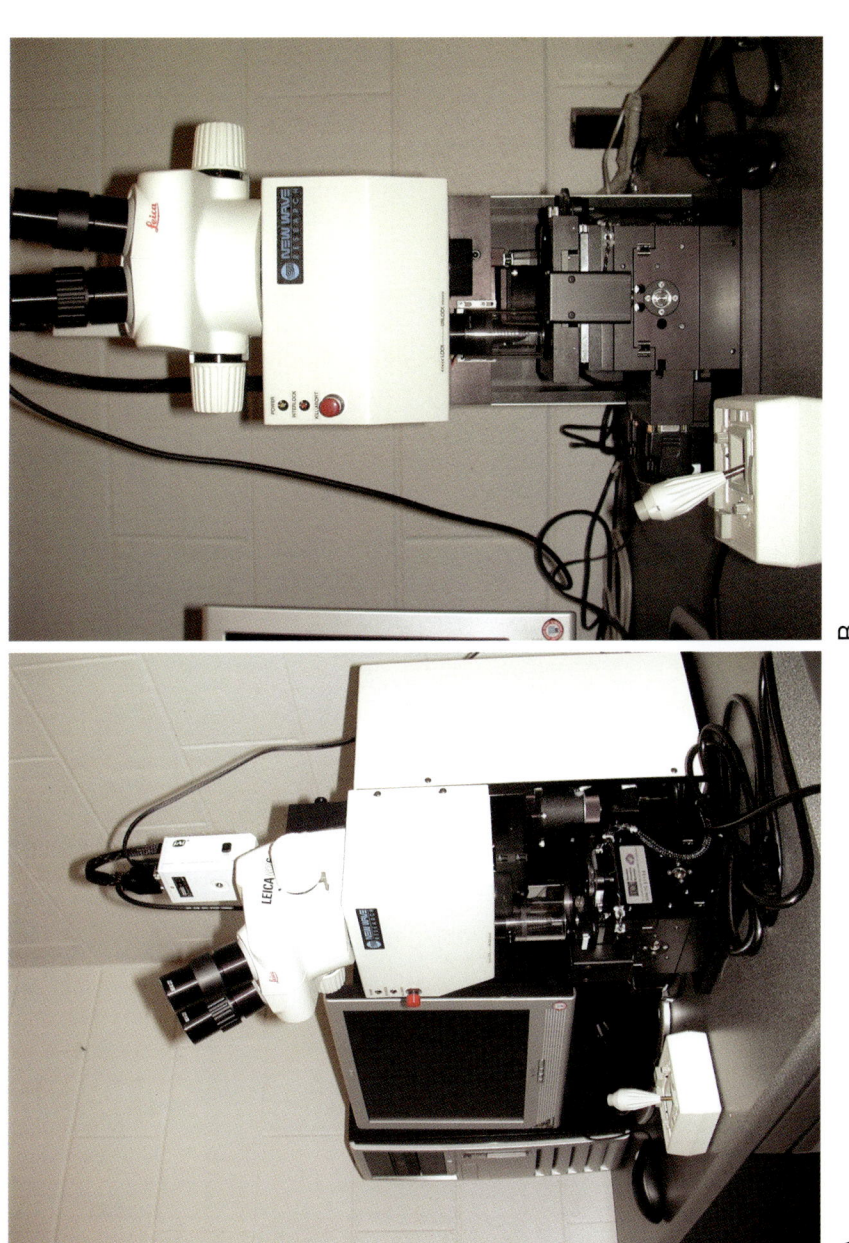

Plate 23.3. Computer controlled Micro-Mill system for sampling crystals and all ancient rock materials. (A) This unit is laser guided and can provide a close approximation of the depth of an inclusion. The unit provides a *video* image of the entire drilling process. A crystal to be sampled is placed beneath the objective lens where it is photographed and measured. Co-ordinates are then laid onto the sample, which are then used to guide the drilling process *via* the computer system seen behind the drill. Alternatively, the drill can be guided by the joystick shown in the foreground. The microscope can be separately placed into a sterile field and computer controlled from outside. (B) A closer view of the microscope and controlling joystick. The objective lens is located on the right with the drill on the left. The precision controls on this instrument allow one to drill a 50-micron hole to a point near the inclusion, then, after a quick change of bits remove a 30-micron core sample from inside the crystal without touching the sides of the bore hole.

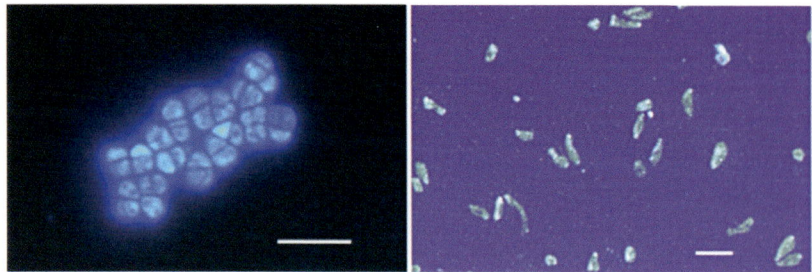

Plate 24.1. Staining of haloarchaea with DAPI. Left: *Halococcus dombrowskii* H4 DSM 14522^T; right: *Halobacterium salinarum* DSM 3754^T (Photograph by S. Rittmann). Bars, 2 µm.

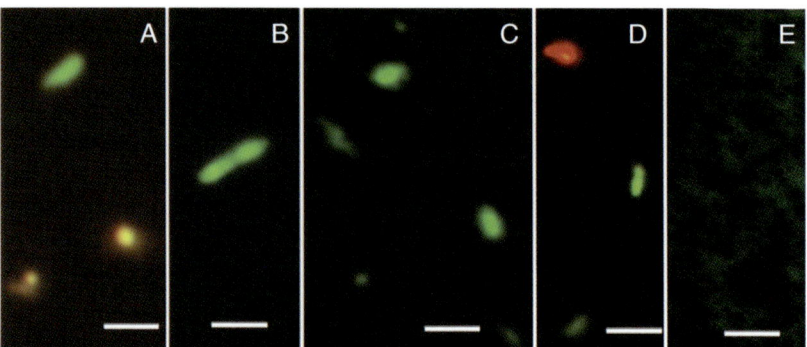

Plate 24.2. Fluorescent particles in environmental samples. Dead Sea water (A, B) and dissolved rock salt (C–E) from the Permian salt deposit at Altaussee, Austria, were treated with the LIVE/DEAD kit without any prior preparation; (E) hazy staining of material in dissolved rock salt. Bars, 2 µm (reprinted with permission from the American Society for Microbiology, Washington D.C.).

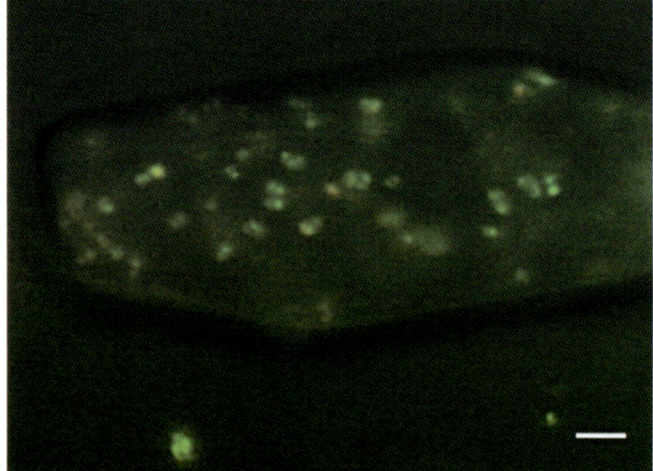

Plate 24.3. Cells of *Halococcus dombrowskii* H4 DSM 14522^T in fluid inclusions. Cells were stained with the LIVE/DEAD kit prior to embedding in crystals of NaCl. Epifluorescence microscopy was performed following entrapment of cells for 3 days. Bar, 5 µm.

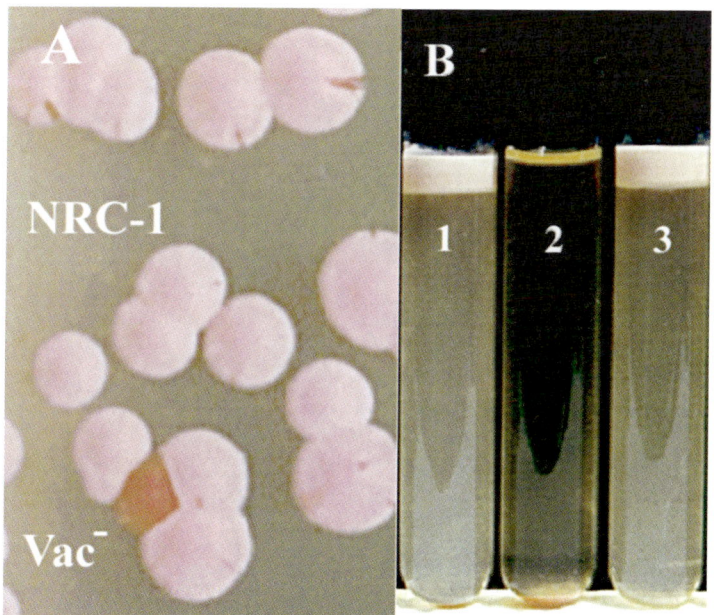

Plate 27.1. *Halobacterium* sp. NRC-1. (A) Colonies of the model haloarchaeal strain *Halobacterium* sp. NRC-1. Among pink wild-type (NRC-1) colonies, an orange gas vesicle-minus (Vac⁻) mutant is visible. (B) Liquid cultures of NRC-1 (tube 1), Vac⁻ mutant SD109 (tube 2) and a SD109 transformant containing the gas vesicle gene cluster on pFL2 (tube 3). Vac⁺ floating cells at the top of tubes 1 and 3 and Vac⁻ non-floating cells at the bottom of the tube 2 are visible. The meniscus is visible at the top of tube 2.

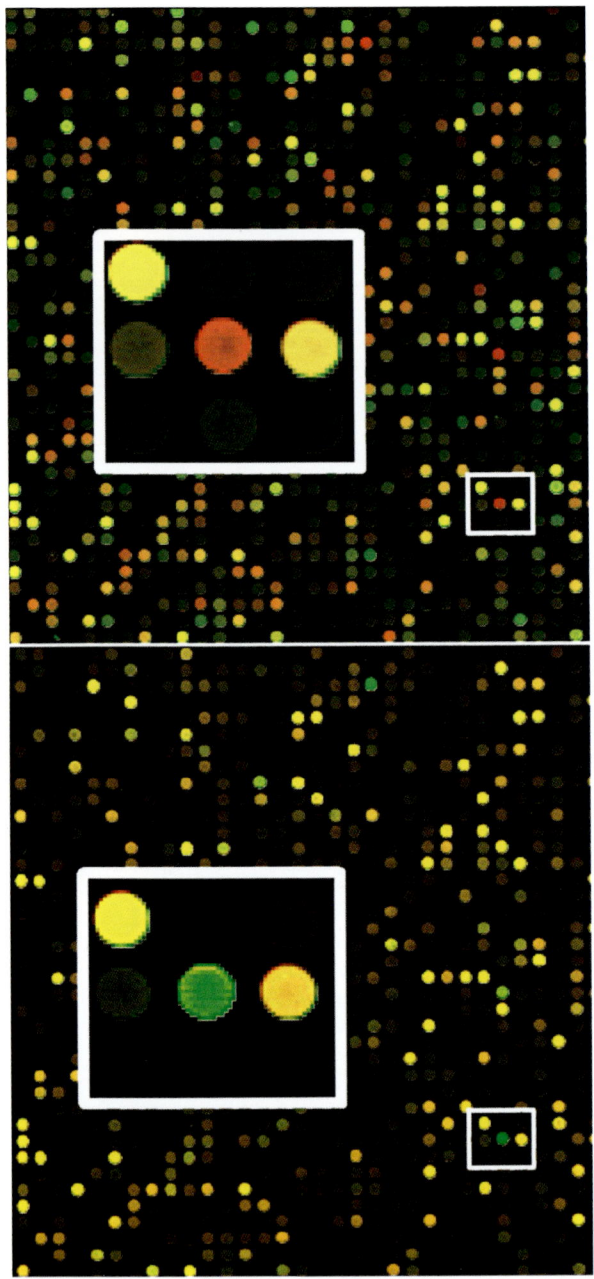

Plate 27.4. Oligonucleotide microarray analysis of the genetically tractable halophilic archaeon *Halobacterium sp.* NRC-1. The upper panel represents an environmental perturbation experiment comparing anaerobic growth *via* TMAO respiration *versus* aerobic growth. The lower panel denotes a genetic perturbation experiment in which a ΔdmsR strain is compared with a dmsR+ strain. The inserts highlight the expression changes of dmsA, coding for the DMSO/TMAO reductase in this extremophile.

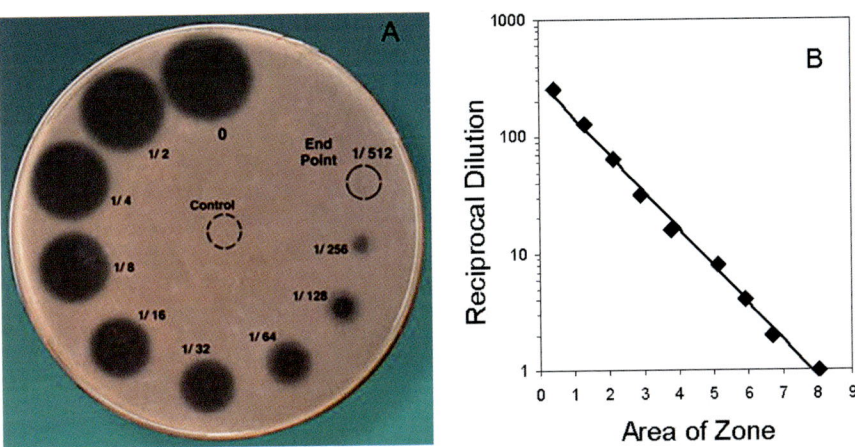

Plate 29.1. (A) Serial two-fold dilutions to extinction of a concentrated solution of halocin R1. The lawn cells are *Halobacterium salinarum* strain NRC817. The negative control is uninoculated medium. The halocin activity of this sample is 512 AU. (B) Correlation between the area of a zone of inhibition (from Figure 1A) and the logarithm of the activity ($R^2 = 0.995$). Note that the "0" dilution was assigned a value of 1 in order to be plotted logarithmically.

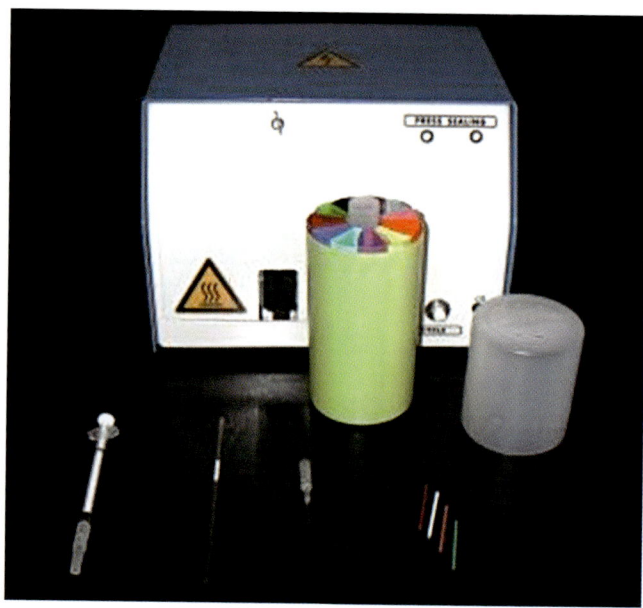

Plate 30.2. Supplies needed to fill cryostraws. From left to right: syringe fitted with nozzle adjustment, cryostraw, sterile filling nozzle and colored labeling rods. The welding machine is seen in the background along with the colored cylinder that holds the straws upright in the liquid nitrogen tank.

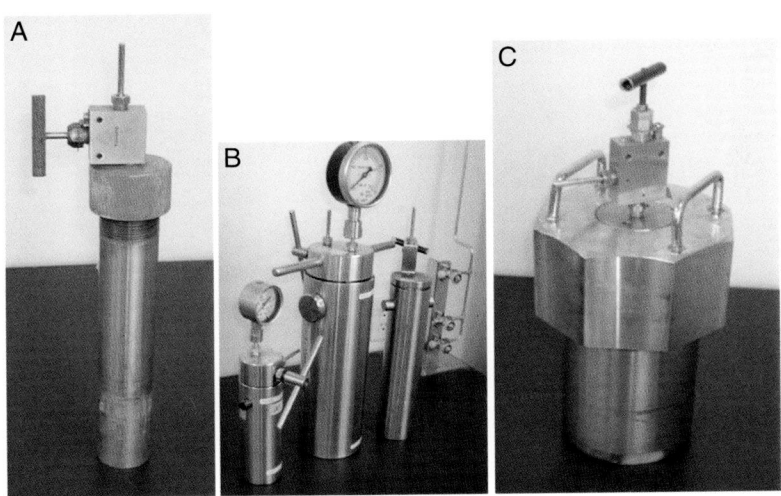

Figure 31.1. Pressure vessels in JAMSTEC projects. (A) Zobell-Morita-type pressure vessel, kindly provided by Prof. Richard Y. Morita. (B) Yayanos-type pressure vessels. Yayanos modified the pressure-retaining system using a pin for easier handling. (C) Large-type (1000 ml) pressure vessel.

Figure 31.2. Sampling sediment from the deep-sea floor using a sterilized sediment sampler with the Shinkai 6500 manipulator. This sampling was performed at a bacterial mat site in the northeast Sea of Japan at a depth of 3134 m on July 1, 2001, during dive 6K#624 (observer, C. Kato).

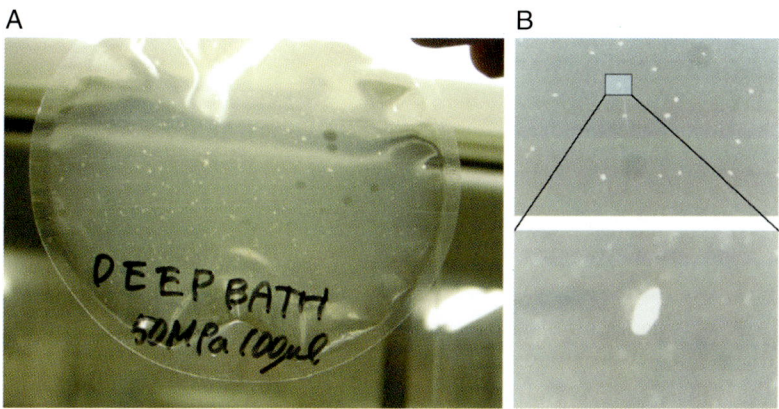

Figure 31.3. Sterilized plastic bag. (A) Plastic bag after high-pressure cultivation. Several microbial colonies formed at a pressure of 50 MPa. (B) Piezophile colonies in the bag. A single colony can be suctioned into a sterilized disposable syringe.

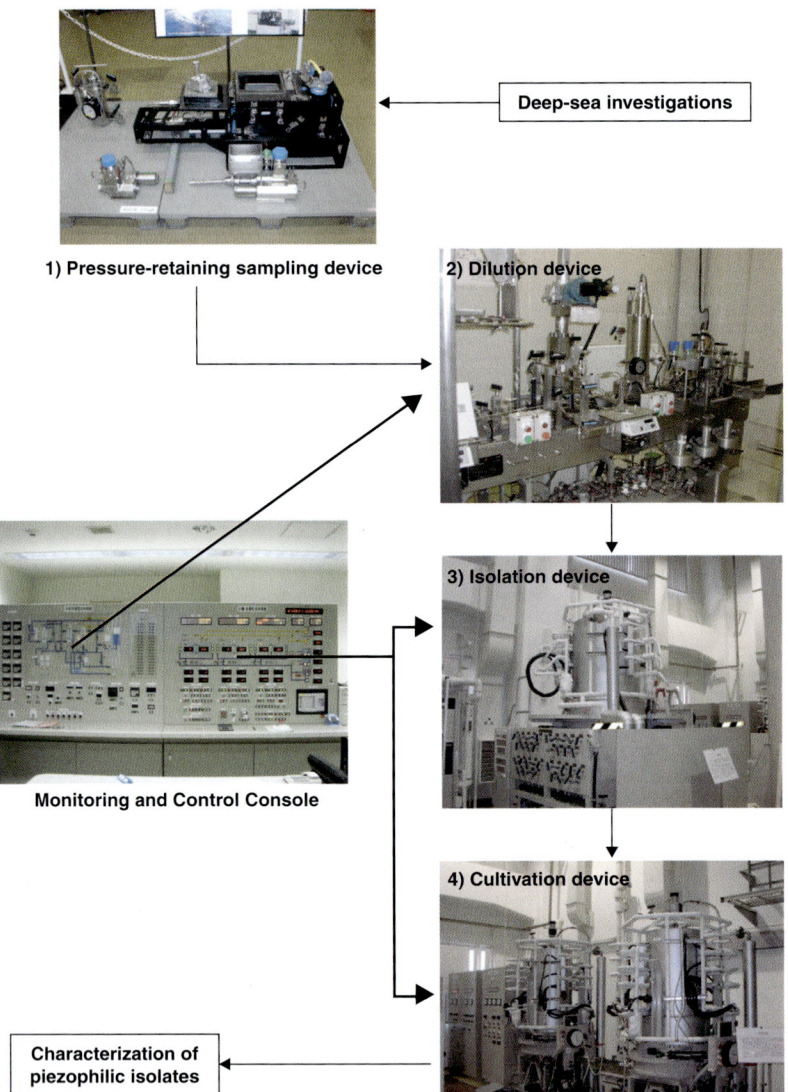

Figure 31.4. The DEEPBATH system. The system is composed of four devices: (1) pressure-retaining sampling device, (2) dilution device under pressure conditions, (3) isolation device and (4) cultivation device. The system is controlled by Monitoring and Control Console.

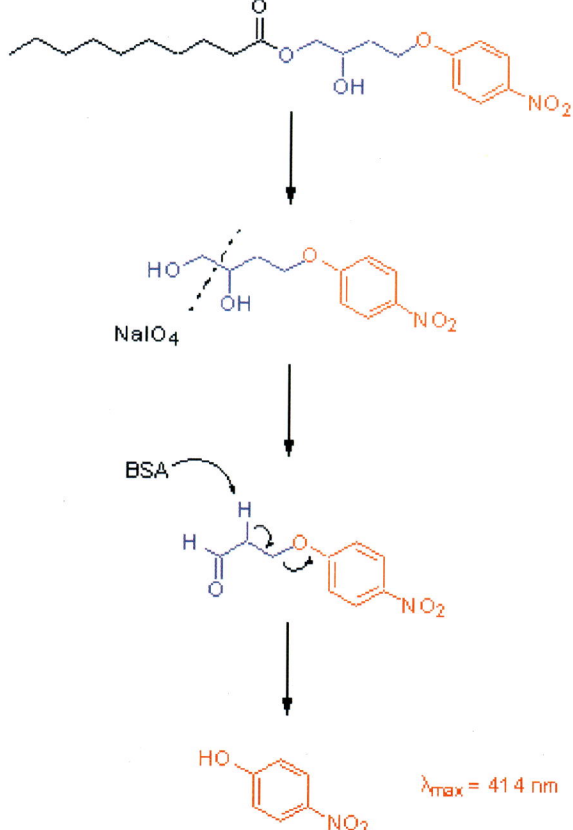

Plate 34.1. Detection of thermostable esterases with a CLIPS-O™ substrate. After incubation with the enzyme, the ester CLIPS-O™ substrate is cleaved into a hydrolyzed intermediate. Both the ester CLIPS-O™ substrate and the hydrolyzed intermediates are extremely stable under a wide range of physico-chemical conditions, including high temperatures and extreme pH. The signal is released solely after a secondary chemical oxidization with sodium periodate, a reaction that only takes place if the intermediate species has been produced.

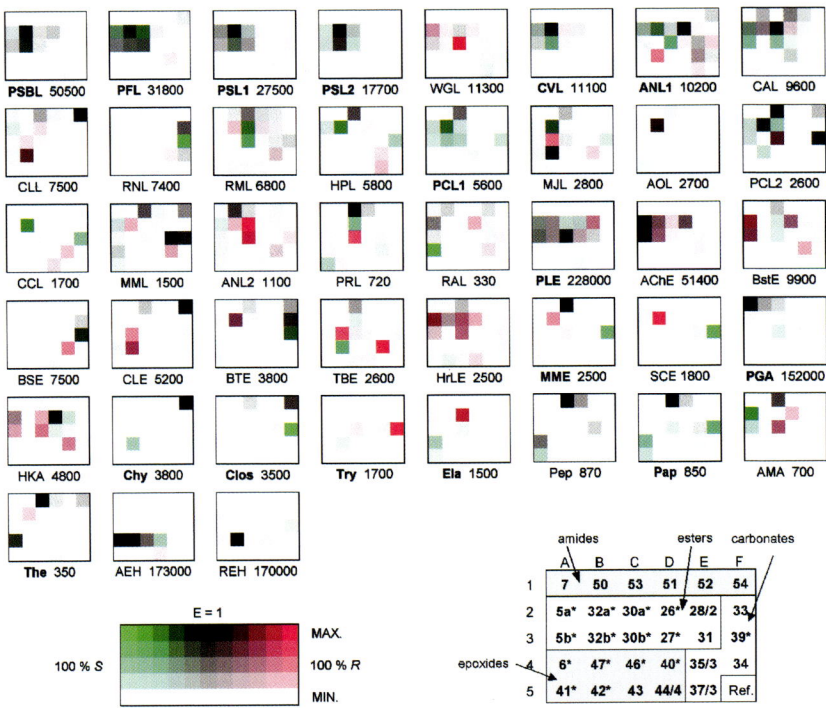

Plate 34.2. Enzymes fingerprinting with a set of CLIPS-O® substrates (modified from Wahler et al., 2002). Substrates marked * in the layout (bottom right) are enantiomeric pairs. Below each array: Enzyme code, and apparent maximum rate in pM s^{-1}, which appears as the darkest (gray scale or color) square in each array. Conditions: 0.1 mg ml^{-1} enzyme, 100 μM substrate, 20 mM borate pH 8.8 with 2.5% v/v DMF, 26°C, 2 mg ml^{-1} BSA, 1 mM NaIO$_4$. Enzymes marked in bold appear homogeneous (>90%) by SDS-PAGE analysis (PSBL – *Pseudomonas* sp. B lipoprotein lipase; PFL – *Pseudomonas fluorescens* lipase; PSL1 – *Pseudomonas* sp. lipoprotein lipase; PSL2 – *Pseudomonas* sp. lipoprotein lipase; WGL – wheat germ lipase; CVL – *Chromobacterium visc.* lipoprotein lipase; ANL1 – *Aspergillus niger* lipase; CAL – *Candida antarctica* lipase; CLL – *Candida lipolytica* lipase; RNL – *Rhizopus niveus* lipase; RML – *Rhizomucor miehei* lipase; HPL – hog pancreatic lipase; PCL1 – *Burkholderia cepacia* (basonym: *Pseudomonas cepacia*) lipase; MJL – *Mucor javanicus* lipase; AOL – *Aspergillus oryzae* lipase; PCL2 – *Burkholderia cepacia* (basonym: *Pseudomonas cepacia*) lipase; CCL – *Candida cylindracea* lipase; MML – *Mucor miehei* lipase; ANL2 – *Aspergillus niger* lipase; PRL – *Penicillium roqueforti* lipase; RAL – *Rhizopus arrhizus* lipase; PLE – pig liver esterase; AChE – *Electrophorus electricus* acetylcholine esterase; BStE – *Geobacillus stearothermophilus* (basonym: *Bacillus stearothermophilus*) esterase; BSE – *Bacillus* sp. esterase; CLE – *Candida lipolytica* esterase; BTE – *Geobacillus thermoglucosidasius* (basonym: *Bacillus thermoglucosidasius*) esterase; TBE – *Thermoanaerobacter brockii* esterase; HrLE – horse liver esterase; MME – *Mucor miehei* esterase; SCE – *Saccharomyces cerevisiae* esterase; PGA – *Escherichia coli* penicillin G acylase; HKA – hog kidney acylase I; Chy – bovine pancreatic chymotrypsin; Clos – *Clostridium histolyticum* clostripain; Try – hog pancreatic trypsin; Ela – porcine pancreatic elastase; Pep – porcine stomach mucosa pepsin; Pap – *Papaya latex* papain; AMA – *Aspergillus melleus* acylase I; The – *Bacillus thermoproteolyticus* thermolysin; AEH – *Aspergillus niger* epoxide hydrolase; REH – *Rhodotorula glutinis* epoxide hydrolase). Fingerprints are presented as arrays of squares with a two-color scale combining stereoselectivity and activity. The patterns observed consistently appear indistinguishable from one measurement to the next.

PC/GENE software

Pall Corporation
2200 Northern Boulevard
East Hills, NY 11548 USA
Tel: 516-484-5400
Fax: 516-484-5228
http://www.pall.com

Acrodisc filters

QIAGEN Inc
27220 Turnberry Lane
Valencia CA 91355 USA
Tel: 800-426-8157
Fax: 800-718-2056
http://www1.qiagen.com

QIAEXII extraction kit

Sigma-Aldrich Corporate Offices
Sigma-Aldrich
3050 Spruce St.
St. Louis, MO 63103 USA
Tel: +1-800-521-8956
http://www.sigmaaldrich.com

Sea salts

Stratagene Corporate
11011 N. Torrey Pines Road
La Jolla, CA 92037 USA
Tel: 858-535-5400
Fax: 512-321-3128
http://www.stratagene.com

Bluescript plasmid

Teledyne Isco Inc
PO Box 82531
Lincoln NE 68501 USA
Phone: 402-464-0231
Fax: 402-465-3064
http://www.isco.com

Density gradient fractionator

Whatman Inc.
200 Park Ave., Ste. 210
Florham Park, NJ 07932 USA
Tel: 973-245-8300

Fax: 973-245-8329
http://www.whatman.com

Filter paper

References

Ackerman, H. W. (2001). Frequency of morphological phage descriptions in the year 2000. *Arch. Virol.* **146**, 843–856.

Atomi, H., Matsumi, R. and Imanaka, T. (2004). Reverse gyrase is not a prerequisite for hyperthermophilic life. *J. Bacteriol.* **186**, 4829–4833.

Balch, W. E., Fox, G. E., Magrum, L. J., Woese, C. R. and Wolfe, R. S. (1979). Methanogens: reevaluation of a unique biological group. *Microbiol. Rev.* **43**, 260–296.

Benbouzid-Rollet, N., López-García, P., Watrin, L., Erauso, G., Prieur, D. and Forterre, P. (1997). Isolation of news plasmids from hyperthermophilic Archaea of the order Thermococcales. *Res. Microbiol.* **148**, 767–775.

Bertani, G. and Baresi, L. (1987). Genetic transformation in the methanogen *Methanococcus voltae* PS. *J. Bacteriol.* **169**, 2730–2738.

Bibb, M. J., Findlay, P. R. and Johnson, M. W. (1984). The relationship between base composition and codon usage in bacterial genes and its use for simple and reliable identification of protein coding sequences. *Gene* **30**, 157–166.

Birboim, H. and Doly, J. (1979). A rapid alkaline extraction procedure for screening recombinant plasmid DNA. *Nucleic Acids Res.* **7**, 1513–1523.

Blöchl, E., Rachel, R., Burggraf, S., Hafenbradl, D., Jannasch, H. W. and Stetter, K. O. (1997). *Pyrolobus fumarii*, gen. and sp. nov., represents a novel group of archaea, extending the upper temperature limit for life to 113°C. *Extremophiles* **1**, 14–21.

Breitbart, M. and Rohwer, F. (2005). Here a virus, there a virus, everywhere the same virus? *Trends Microbiol* **13**, 278–284.

Charbonnier, F., Erauso, G., Barbeyron, T., Prieur, D. and Forterre, P. (1992). Evidence that a plasmid from a hyperthermophilic archaebacterium is relaxed at physiological temperatures. *J. Bacteriol.* **174**, 6103–6108.

Charbonnier, F., Forterre, P., Erauso, G. and Prieur, D. (1995). Purification of plasmids from thermophilic and hyperthermophilic archaebacteria. In *Archaea: A Laboratory Manual* (E. M. Fleischmann, A. R. Place, F. T. Robb and J. J. Schreier, eds), vol. 3, pp. 87–90. Cold Spring Harbor Laboratory Press, Cold Spring Harbor.

Cline, S. W., Lam, W. L., Charlebois, R. L., Schalkwyk, L. C. and Doolittle, W. F. (1989). Transformation methods for halophilic archaebacteria. *Can. J. Microbiol.* **35**, 148–152.

Cohen, G. N., Barbe, V., Flament, D., Galperin, M., Heilig, R., Lecompte, O., Poch, O., Prieur, D., Querellou, J., Ripp, R., Thierry, J. C., van der Oost, J., Weissenbach, J., Zivanovic, Y. and Forterre, P. (2003). An integrated analysis of the genome of the hyperthermophilic archaeon *Pyrococcus abyssi*. *Mol. Microbiol.* **47**, 1495–1512.

Delcher, A. L., Harmon, D., Kasif, S., White, O. and Salzberg, S. L. (1999). Improved microbial gene identification with GLIMMER. *Nucleic Acids Res.* **27**, 4636–4641.

Di Ruggiero, J. and Robb, F. T. (1995). RNA extraction from sulfur-utilizing thermophilic archaea. In *Thermophiles Archaea: a Laboratory Manual* (F. T. Robb and A. R. Place, eds), pp. 97–99. Cold Spring Harbor Laboratory Press, Cold Spring Harbor.

Drlica, K. (1992). Control of bacterial DNA supercoiling. *Mol. Microbiol.* **6**, 425–433.

Erauso, G., Charbonnier, F., Barbeyron, T., Forterre, P. and Prieur, D. (1992). Preliminary characterization of an ultrathermophilic archaebacterium with a plasmid isolated from the North Basin hydrothermal vent. *C.R. Acad. Sci. Paris* **314**, 387–393.

Erauso, G., Reysenbach, A. L., Godfroy, A., Meunier, J. R., Crump, B., Partensky, F., Baross, J. A., Marteinsson, V., Barbier, G., Pace, N. R. and Prieur, D. (1993). *Pyrococcus abyssi*, sp. nov., a new hyperthermophilic archaeon isolated from a deep-sea hydrothermal vent. *Arch. Microbiol.* **160**, 338–349.

Erauso, G., Prieur, D., Godfroy, A. and Raguénes, G. (1995). Plate cultivation technique for strictly anaerobic, thermophilic, sulfur-metabolizing archaea. In *Archaea: A Laboratory Manual* (E. M. Fleischmann, A. R. Place, F. T. Robb and J. J. Schreier, eds), vol. 3, pp. 25–29. Cold Spring Harbor Laboratory Press, Cold Spring Harbor.

Erauso, G., Marzin, S., Benbouzid-Rollet, N., Baucher, M. F., Barbeyron, T., Zivanovic, Y., Prieur, D. and Forterre, P. (1996). Sequence of plasmid pGT5 from the archaeon *Pyrococcus abyssi*: evidence for rolling-circle replication in a hyperthermophile. *J. Bacteriol.* **178**, 3232–3237.

Fukui, T., Atomi, H., Kanai, T., Matsumi, R., Fujiwara, S. and Imanaka, T. (2005). Complete genome sequence of the hyperthermophilic archaeon *Thermococcus kodakaraensis* KOD1 and comparison with *Pyrococcus* genomes. *Genome Res.* **15**, 352–363.

Gerard, E., Jolivet, E., Prieur, D. and Forterre, P. (2001). DNA protection mechanisms are not involved in the radioresistance of the hyperthermophilic archaea *Pyrococcus abyssi* and *P. furiosus*. *Mol. Genet. Genom.* **266**, 72–78.

Geslin, C., Le Romancer, M., Gaillard, M., Erauso, G. and Prieur, D. (2003a). Observation of virus-like particles in high temperature enrichment cultures from deep-sea hydrothermal vents. *Res. Microbiol.* **154**, 303–307.

Geslin, C., Le Romancer, M., Erauso, G., Gaillard, M., Perrot, G. and Prieur, D. (2003b). PAV1, the first virus-like particle isolated from a hyperthermophilic euryarchaeote, "*Pyrococcus abyssi*". *J. Bacteriol.* **185**, 3888–3894.

Gueguen, Y., Rolland, J. L., Lecompte, O., Azam, P., Le Romancer, G., Flament, D., Raffin, J. P. and Dietrich, J. (2001). Characterization of two DNA polymerases from the hyperthermophilic euryarchaeon *Pyrococcus abyssi*. *Eur. J. Biochem.* **268**, 5961–5969.

Harmsen, H. J. M., Prieur, D. and Jeanthon, C. (1997). Distribution of microorganisms in deep-sea hydrothermal vent chimneys investigated by whole-cell hybridization and enrichment culture of thermophilic subpopulations. *Appl. Environ. Microbiol.* **63**, 2876–2883.

Henneke, G., Gueguen, Y., Flament, D., Azam, P., Querellou, J., Dietrich, J., Hubscher, U. and Raffin, J. P. (2002). Replication factor C from the hyperthermophilic archaeon *Pyrococcus abyssi* does not need ATP hydrolysis for clamp-loading and contains a functionally conserved RFC PCNA-binding domain. *J. Mol. Biol.* **323**, 795–810.

Henneke, G., Flament, D., Hubscher, U., Querellou, J. and Raffin, J. P. (2005). The hyperthermophilic euryarchaeota *Pyrococcus abyssi* likely requires the two DNA polymerases D and B for DNA replication. *J. Mol. Biol.* **350**, 53–64.

Huber, R. and Stetter, K. O. (2001). Discovery of hyperthermophilic microorganisms. *Meth. Enzymol.* **330**, 11–24.

Jolivet, E., Matsunaga, F., Ishino, Y., Forterre, P., Prieur, D. and Myllykallio, H. (2003a). Physiological responses of the hyperthermophilic archaeon "*Pyrococcus abyssi*" to DNA damage caused by ionizing radiation. *J. Bacteriol.* **185**, 3958–3961.

Jolivet, E., L'Haridon, S., Corre, E., Forterre, P. and Prieur, D. (2003b). *Thermococcus gammatolerans* sp. nov., a hyperthermophilic archaeon from a deep-sea hydrothermal vent that resists ionizing radiation. *Int. J. Syst. Evol. Microbiol.* **53**, 847–851.

Kashefi, K. and Lovely, D. R. (2003). Extending the upper temperature limit for life. *Science* **301**, 394.

Kawarabayasi, Y., Sawada, M., Horikawa, H., Haikawa, Y., Hino, Y., Yamamoto, S., Sekine, M., Baba, S., Kosugi, H., Hosoyama, A., Nagai, Y., Sakai, M., Ogura, K., Otsuka, R., Nakazawa, H., Takamiya, M., Ohfuku, Y., Funahashi, T., Tanaka, T., Kudoh, Y., Yamazaki, J., Kushida, N., Oguchi, A., Aoki, K. and Kikuchi, H. (1998). Complete sequence and gene organization of the genome of a hyper-thermophilic archaebacterium, *Pyrococcus horikoshii* OT3. *DNA Res.* **5**, 55–76.

Lepage, E., Marguet, E., Geslin, C., Matte-Tailliez, O., Zillig, W., Forterre, P. and Tailliez, P. (2004). Molecular diversity of new Thermococcales isolates from a single area of hydrothermal deep-sea vents as revealed by randomly amplified polymorphic DNA fingerprinting and 16S rRNA gene sequence analysis. *Appl. Environ. Microbiol.* **70**, 1277–1286.

Lipps, G., Ibanez, P., Stroessenreuther, T., Hekimian, K. and Krauss, G. (2001). The protein ORF80 from the acidophilic and thermophilic archaeon *Sulfolobus islandicus* binds highly site-specifically to double-stranded DNA and represents a novel type of basic leucine zipper protein. *Nucleic Acids Res.* **29**, 4973–4982.

Lucas, S., Toffin, L., Zivanovic, Y., Charlier, D., Moussard, H., Forterre, P., Prieur, D. and Erauso, G. (2002). Construction of a shuttle vector for, and spheroplast transformation of, the hyperthermophilic archaeon *Pyrococcus abyssi*. *Appl. Environ. Microbiol.* **68**, 5528–5536.

Marsin, S. and Forterre, P. (1998). A rolling circle replication initiator protein with a nucleotidyl-transferase activity encoded by the plasmid pGT5 from the hyperthermophilic archaeon *Pyrococcus abyssi*. *Mol. Microbiol.* **27**, 1183–1192.

Marsin, S. and Forterre, P. (1999). The active site of the rolling circle replication protein Rep75 is involved in site-specific nuclease, ligase and nucleotidyl transferase activities. *Mol. Microbiol.* **33**, 537–545.

Marsin, S., Marguet, E. and Forterre, P. (2000). Topoisomerase activity of the hyperthermophilic replication initiator protein Rep75. *Nucleic Acids Res.* **28**, 2251–2255.

Marteinsson, V. T., Watrin, L., Prieur, D., Caprais, J. C., Raguenes, G. and Erauso, G. (1995). Phenotypic characterization, DNA similarities, and protein profiles of twenty sulfur-metabolizing hyperthermophilic anaerobic archaea isolated from hydrothermal vents in the southwestern Pacific Ocean. *Int. J. Syst. Bacteriol.* **45**, 623–632.

Matsunaga, F., Forterre, P., Ishino, Y. and Myllykallio, H. (2001). In vivo interactions of archaeal Cdc6/Orc1 and minichromosome maintenance proteins with the replication origin. *Proc. Natl. Acad. Sci. USA* **98**, 11152–11157.

Matsunaga, F., Norais, C., Forterre, P. and Myllykallio, H. (2003). Identification of short 'eukaryotic' Okazaki fragments synthesized from a prokaryotic replication origin. *EMBO Rep.* **4**, 154–158.

Myllykallio, H., Lopez, P., López-García, P., Heilig, R., Saurin, W., Zivanovic, Y., Philippe, H. and Forterre, P. (2000). Bacterial mode of replication with eukaryotic-like machinery in a hyperthermophilic archaeon. *Science* **288**, 2212–2215.

Prangishvili, D., Stedman, K. and Zillig, W. (2001). Viruses of the extremely thermophilic archaeon *Sulfolobus*. *TRENDS Microbiol.* **9**, 39–43.

Prieur, D. (2002). Hydrothermal vents: prokaryotes in deep sea hydrothermal vents. In *Encyclopedia of Environmental Microbiology* (G. Bitton, ed.), pp. 1617–1628. John Wiley & Sons, New York.

Rachel, R., Bettstetter, M., Hedlund, B. P., Häring, M., Kessler, A., Stetter, K. O. and Prangishvili, D. (2002). Remarkable morphological diversity of viruses and virus-like particles in hot terrestrial environments. *Arch. Virol.* **147**, 2419–2429.

Ravot, G., Ollivier, B., Magot, M., Patel, B. K. C., Crolet, J.-L., Fardeau, M.-L. and Garcia, J.-L. (1995). Thiosulfate reduction, an important physiological feature shared by members of the order *Thermotogales*. *Appl. Environ. Microbiol.* **61**, 2053–2055.

Rice, G., Stedman, K., Snyder, J., Wiedenheft, B., Willits, D., Brumfield, S., McDermott, T. and Young, M. J. (2001). Viruses from extreme thermal environments. *Proc. Natl. Acad. Sci. USA* **98**, 13341–13345.

Robb, F. T., Maeder, D. L., Brown, J. R., DiRuggiero, J., Stump, M. D., Yeh, R. K., Weiss, R. B. and Dunn, D. M. (2001). Genomic sequence of hyperthermophile, *Pyrococcus furiosus*: implications for physiology and enzymology. *Meth. Enzymol.* **330**, 134–157.

Sambrook, J., Fritsch, E. F. and Maniatis, T. (1989). *Molecular Cloning: A Laboratory Manual*, 2nd edn., Cold Spring Harbor Laboratory Press, Cold Spring Harbor.

Sanger, F., Nicklen, S. and Coulson, A. R. (1977). DNA sequencing with chain-terminating inhibitors. *Proc. Natl. Acad. Sci. USA* **74**, 5463–5467.

Sato, T., Fukui, T., Atomi, H. and Imanaka, T. (2003). Targeted gene disruption by homologous recombination in the hyperthermophilic archaeon *Thermococcus kodakaraensis* KOD1. *J. Bacteriol.* **185**, 210–220.

Sato, T., Imanaka, H., Rashid, N., Fukui, T., Atomi, H. and Imanaka, T. (2004). Genetic evidence identifying the true gluconeogenic fructose-1,6-bisphosphatase in *Thermococcus kodakaensis* and other hyperthermophiles. *J. Bacteriol.* **186**, 5799–5807.

Strauss, E., Kobori, J., Siu, G. and Hood, L. (1986). Specific primer directed DNA sequencing. *Anal. Biochem.* **154**, 353–360.

Suzek, B. E., Ermolaeva, M. D., Schreiber, M. and Salzberg, S. (2001). A probabilistic method for identifying startcodons in bacterial genomes. *Bioinformatics* **17**, 1123–1130.

Takai, K., Sugai, A., Itoh, T. and Horikoshi, K. (2000). *Palaeococcus ferrophilus* gen. nov., sp. nov., a barophilic, hyperthermophilic archaeon from a deep-sea hydrothermal vent chimney. *Int. J. Syst. Evol. Microbiol.* **50**, 489–500.

te Riele, H., Michel, B. and Ehrlich, S. (1986). Are single-stranded circles intermediates in plasmid DNA replication? *EMBO J* **5**, 631–637.

Ward, D. E., Revet, I. M., Nandakumar, R., Tuttle, J. H., de Vos, W. M., van der Oost, J. and DiRuggiero, J. (2002). Characterization of plasmid pRT1 from *Pyrococcus sp*. strain JT1. *J. Bacteriol.* **184**, 2561–2566.

Watrin, L., Martin Jezequel, V. and Prieur, D. (1995). Minimal amino acid requirements of the hyperthermophilic archaeon *Pyrococcus abyssi*, isolated from deep-sea hydrothermal vents. *Appl. Environ. Microbiol.* **61**, 1138–1140.

Watrin, L. and Prieur, D. (1996). UV and ethyl methanesulfonate effects in hyperthermophilic archaea and isolation of auxotrophic mutants of *Pyrococcus* strains. *Curr. Microbiol.* **33**, 377–382.

Watrin, L., Corre, E. and Prieur, D. (1996). *In vivo* susceptibility of sulfothermophilic archaea to antimicrobial agents. *J. Mar. Biotechnol.* **4**, 215–219.

Watrin, L., Lucas, S., Purcarea, C., Legrain, C. and Prieur, D. (1999). Isolation and characterization of pyrimidine auxotrophs, and molecular cloning of the *pyrE* gene from the hyperthermophilic archaeon *Pyrococcus abyssi*. *Mol. Gen. Genet.* **262**, 378–381.

Zappa, S., Boudrant, J. and Kantrowitz, E. R. (2004). *Pyrococcus abyssi* alkaline phosphatase: the dimer is the active form. *J. Inorg. Biochem.* **98**, 575–581.

Zillig, W., Arnold, H. P., Holz, I., Prangishvili, D., Schweier, A., Stedman, K., She, Q., Phan, H., Garrett, R. and Kristjansson, J. K. (1998). Genetics elements in the extremely thermophilic archaeon *Sulfolobus*. *Extremophiles* **2**, 131–140.

12 Genetic Systems for *Thermus*

Beate Averhoff

Institute of Molecular Biosciences – Molecular Microbiology and Bioenergetics, Johann Wolfgang Goethe University, 60439 Frankfurt am Main, Germany

◆◆

CONTENTS

Introduction
Gene transfer
Genetic tools
Concluding remarks

◆◆◆◆◆ Introduction

Stability of enzymes and activity at high temperatures are important and desirable properties for structural and functional analyses of model enzymes and for industrial applications. Hence, the production of thermostable enzymes has been the major aim of many applied studies. So far, production of nearly all thermostable enzymes is commonly achieved by heterologous expression in a mesophilic host, typically *Escherichia coli* and *Bacillus subtilis*. However, this approach has several limitations, since several thermozymes require either cofactors, appropriate posttranslational processing, active chaperones, high temperatures to fold correctly and/or cannot be overproduced in an active form in a mesophilic host. Likewise thermal adaptation of enzymes from mesophiles would benefit from expression in a thermophilic host. Therefore, the development of genetic tools allowing overexpression of thermostable proteins in a thermophilic context and thermophilic host systems are of great importance.

The extreme thermophile most amenable to genetic manipulation is *Thermus*. Members of the genus *Thermus* are obligate heterotrophs growing on low concentrations of organic material and have been isolated from many natural and artificial thermal environments throughout the world. Strains of *Thermus* spp. are indispensable in research and industrial applications, they have become model organisms in structural biology, and they are important sources of biocatalysts for biotechnological applications (Niehaus *et al.*, 1999; Pantazaki *et al.*, 2002; Silva *et al.*, 2005; Williams and da Costa, 2001; Wimberly *et al.*, 2000).

Thermus aquaticus, the type species of the genus, was the first extremely thermophilic bacterium described and was found to grow at a maximum temperature of 79°C (Brock and Freeze, 1969). Only a few years later *T. thermophilus* strain HB8 and strain HB27 growing at higher maximum temperatures (85°C) and exhibiting high natural transformation frequencies were isolated (Hidaka *et al.*, 1994; Koyama *et al.*, 1986; Oshima and Imahori, 1974). Due to the extraordinary trait of high competence for natural transformation the extreme thermophile *Thermus* is most amenable to genetic manipulation and therefore, searching for and/or establishing of associated genetic tools such as selectable markers and vectors for *Thermus* have been a major undertaking.

◆◆◆◆◆ GENE TRANSFER

Natural Transformation

Natural transformation events in *Thermus* spp., such as *T. thermophilus* HB27, *T. thermophilus* HB8 and other *T. thermophilus* strains and *T. aquaticus* YT1 have been reported for the first time by Koyama *et al.* (1986), who transformed auxotrophic *Thermus* mutants to prototrophy or streptomycin (Str) sensitive *Thermus* strains to Str-resistant transformants. *T. thermophilus* HB27 and *T. flavus* AT62 exhibited highest transformation frequencies of 1 and 0.88%, respectively; lowest transformation frequencies of 0.064% were found with *T. aquaticus* YT1 (Koyama *et al.*, 1986). These findings suggest that among the *Thermus* strains *T. thermophilus* HB27 is the optimal recipient for DNA transfer studies.

Recently, the genome of HB27 was published (Henne *et al.*, 2004) and a whole genome approach was performed to identifiy the first genes of the transformation machinery of *T. thermophilus* HB27 (Friedrich *et al.*, 2001, 2002, 2003). The proteins of the *T. thermophilus* HB27 DNA translocator are the first proteins of a thermostable transformation machinery identified so far. Further studies have led to a model of the DNA translocator and the DNA translocation process in *T. thermophilus* HB27 (Averhoff, 2004; Averhoff and Friedrich, 2003): DNA binds to DNA binding proteins on the cell surface or to proteins associated with a DNA translocator structure comprising of pilin-like proteins. Subsequently, the DNA is either transported along or through a DNA transformation shaft made up of pilin PilA4 and some minor pilin-like proteins such as PilA1, PilA2 and PilA3. A ring-like structure comprising of secretin-like PilQ proteins is suggested to be required for guiding the DNA translocator through the cell envelope. The similarities of the C-terminal part of PilQ to members of the secretin family that are essential for secretin multimerization underline the suggestion that *Thermus* PilQ proteins form a homopolymeric ring structure. Potential retraction of the DNA translocator could depend on the function of PilF, the putative traffic NTPase that may power DNA transport through the ring structures comprising of secretin-like proteins. In the periplasmic space, the DNA might bind to the putative binding protein ComEA which is probably anchored in the inner membrane and delivers the DNA to the inner membrane DNA transport complex. Finally, the DNA is transported through an inner membrane channel generated by ComEC, a polytopic inner membrane protein.

The transformation protocol developed by Koyama et al. (1986) has been applied for transfer of chromosomal and plasmid DNA and is still essentially the standard protocol for DNA transfer in *Thermus*.

Media

Thermus minimal medium (Tanaka et al., 1981)

Solution A (900 ml): sucrose, 20 g; sodium glutamate, 20 g; K_2HPO_4, 0.5 g; KH_2PO_4, 0.25 g; NaCl, 2.0 g and $(NH_4)_2SO_4$, 0.5 g. After pH adjustment to pH 7.0–7.2 the following solutions are added: $NaMoO_2 \cdot 2H_2O$ (1.2 g dissolved in 100 ml of water); $VOSO_4 \cdot 3H_2O$ (0.1 g in 100 ml of water); $MnCl_2 \cdot 4H_2O$ (0.5 g in 100 ml of 0.01 N HCl); $ZnSO_4 \cdot 7H_2O$ and $CuSO_4 \cdot 5H_2O$ (60 mg and 15 mg, respectively, dissolved in 100 ml of water). Biotin (100 μg) and thiamine (1 mg) are added after autoclaving.

Solution B (100 ml): 0.125 g $MgCl_2 \cdot 6H_2O$, 0.025 g of $CaCl_2 \cdot 2H_2O$. The solution is sterilized by autoclaving.

Solution C (100 ml): in 0.01 N H_2SO_4 are dissolved $FeSO_4 \cdot 7H_2O$, 6 g; $CoCl_2 \cdot 6H_2O$, 0.8 g and $NiCl_2 \cdot 6H_2O$, 20 mg. The solution is sterilized by filtration.

For minimal medium 900 ml solution A, 100 ml solution B and 0.1 ml solution C are mixed.

Thermus broth medium (TM) (Koyama et al., 1986)

0.4% Polypeptone (Daigo-Eiyo Chemical Co. Ltd., Osaka Japan) 0.2% yeast extract, 0.1% NaCl and Castenholz basal salt, pH 7.5 (Ramaley and Hixson, 1970). The composition of Castenholz $10 \times$ basal salts stock solution contains (per litre of distilled water): nitrilotriacetic acid, 1 g; $CaSO_4 \cdot 2H_2O$, 0.6 g; $MgSO_4 \cdot 7H_2O$, 1.0 g; NaCl, 0.08 g; KNO_3, 1.03 g; $NaNO_3$, 6.89 g; Na_2HPO_4, 1.10 g; $FeCl_3$ solution (0.28 g l^{-1} of distilled water), 10 ml; and Nitsch's trace element solution, 10 ml, pH 8.2. Nitsch's trace element solution contains (per litre of distilled water): H_2SO_4 (96%), 0.5 ml; $MnSO_4 \cdot H_2O$, 2.2 g; $ZnSO_4 \cdot 7H_2O$, 0.5 g; H_3BO_3, 0.5 g; $CuSO_4 \cdot 5H_2O$, 0.016 g; $Na_2MoO_4 \cdot 2H_2O$; 0.025 g; and $CoCl_2 \cdot 6H_2O$, 0.046 g.

Protocol

1. An auxotrophic mutant or a streptomycin-sensitive wild-type strain can be used as recipients in a natural transformation experiment with chromosomal DNA of a prototroph wild-type strain and a streptomycin-resistant strain, respectively. Auxotrophic mutants of *T. thermophilus* HB27 are generated by exposure of *T. thermophilus* HB27 to *N*-methyl-*N'*-nitro-*N*-nitrosoguanidine (Adelberg et al., 1965). Spontaneous streptomycin-resistant mutants are obtained by plating *Thermus* wild-type strains on TM medium containing 500 μg of streptomycin ml^{-1}.
2. Grow the recipient culture, such as the auxotrophic mutant strain or the streptomycin-sensitive wild-type strain of *T. thermophilus* HB27 overnight at 70°C in TM medium.

3. Dilute the recipient culture into fresh TM broth (1:100) and incubate for 2 h at 70°C. Mix 0.45 ml of the diluted culture with 50 μl DNA of the prototroph parental *Thermus* strain or a streptomycin-resistant *Thermus* strain (final DNA concentration: 10 μg ml^{-1}), incubate for 1 h at 70°C and subsequently cool for 5 min on ice.
4. To terminate the DNA uptake add DNase I (50 μg ml^{-1}). Incubate for 15 min at 37°C.
5. Select prototroph transformants by plating the transformation assay on *Thermus* minimal medium and incubate for 36–48 h at 70°C. Streptomycin-resistant transformants are selected for growth on TM plates containing 50 μg ml^{-1} streptomycin.

The presence of basal salts in the TM medium is essential for high transformation frequencies of *Thermus*: omission of basal salts results in a decrease of transformation frequencies from 9.3 to 2.1% transformants of viable cells (Koyama et al., 1986). The essential components in basal salt required for transformation are Ca^{2+} and Mg^{2+} and highest transformation frequencies of 10.8% are achieved by the supplementation of the TM medium with 0.35 mM Ca^{2+} and 0.4 mM Mg^{2+} instead of basal salts. Furthermore, Koyama et al. (1986) reported that stationary phase cells exhibit a significantly reduced transformation frequency of 1.5% transformants of viable cells.

Conjugation

DNA transfer *via* conjugation has been demonstrated with the self-mobilizable plasmid of strain HB8 which encodes the nitrate reductase operon (*nar*) enabling growth under anaerobic conditions with nitrate as terminal electron acceptor (Ramírez-Arcos et al., 1998). Analogous to the F-plasmid in *E. coli* the conjugative element can integrate into the chromosome and mobilize chromosomal markers. The latter characteristic could be a very useful tool to isolate novel gene loci from different *Thermus* strains.

Media

Thermus broth (TB) medium (Ramirez-Arcos et al., 1998)

Trypticase, 8 g (BBL Microbiology Systems, Cockeysville, MD), yeast extract, 4 g (Oxoid, Hampshire, England), NaCl, 3 g/l of water, pH 7.5. Nitrate TB medium is supplemented with 20 mM KNO_3. To prepare solidified media 1.5% agar is added.

Protocol

1. To allow selection of *T. thermophilus* HB27 exconjugants a spontaneous chloramphenicol-resistant strain of HB27 (HB27Cam^R) is useful. Isolate exconjugants by plating cells (10^8 HB27 cells) on TB medium plates containing 20 μg ml^{-1} chloramphenicol (Ramírez-Arcos et al., 1998).

2. Mix the donor strain HB8 and recipient HB27Cam^R 100:1 in 100 ml nitrate TB medium. After an incubation for 8 h at 70°C under stirring with 100 rpm 500 µl of the mixture has to be transferred into 100 ml nitrate TB medium containing 20 µg ml^{-1} chloramphenicol and subsequently incubated overnight at 70°C to select for exconjugants.
3. To screen for *nar*$^+$ exconjugants inoculate the chloramphenicol-resistant exconjugant cells into 10 ml of nitrate TB medium prepared in capped test tubes. For anaerobic growth, overlay the nitrate TB medium to the top with mineral oil before the tubes are closed and incubated for 24 h at 70°C. Single cultures of donor and recipient strains have to be treated identically as control setups.

Electroporation

The electroporation protocol was developed to transfer the *E. coli*/*Thermus* shuttle vector pMK18 into *T. thermophilus* HB8 and HB27 (de Grado et al., 1998).

Media

Carbonate-rich medium (de Grado et al., 1999)

Dissolve 8 g trypticase, 4 g yeast extract and 3 g NaCl per litre of carbonate-rich mineral water.

Carbonate-rich mineral water: 267 mg of sodium bicarbonate and 25 mg of sulphates (80% calcium sulphates and 20% magnesium sulphates), and minor amounts of chlorides (2 mg), sodium (0.8 mg), and potassium (0.5 mg) per litre. For solidified media 1.5% agar is added.

Protocol

1. Wash exponentially grown HB8 or HB27 cells (OD$_{550}$ 0.8, grown in carbonate-rich medium) twice in 10% glycerol solution at room temperature (5000 g × 5 min).
2. Mix 100 µl aliquots of concentrated cell solutions (2–5 × 10^{10} cells ml^{-1}) with 1 pg of plasmid DNA and incubate for 5 min at 4°C.
3. Electroporate the concentrated cell solutions by applying a maximum of 12.5 V cm^{-1} (pulse length 5 ms).
4. After electroporation incubate the cells for 2 h at 70°C under aeration, transfer to carbonate-rich medium containing 30 µg ml^{-1} kanamycin and subsequentely incubate for 36–48 h at 70°C to select for pMK18-containing transformants.

◆◆◆◆◆ GENETIC TOOLS

Isolation of Insertion and Deletion Mutants

Mutant studies with *Thermus* require the availability of marker genes funtioning at around 70°C. Only a very small number of thermostable antibiotic resistance markers are available so far.

Media

Luria-Bertani (LB) medium (Sambrook et al., 1989)

10 g tryptone; 5 g yeast extract; 10 g NaCl per litre of distilled water.

TT-rich medium (Lasa et al., 1992a)

8 g trypticase, 4 g of yeast extract, 3 g NaCl and 40 mg kanamycin per litre (pH 7.5). 1.5% agar is added to solidify the medium.

Kanamycin resistance marker

Currently, the most widely used thermostable selectable marker in *Thermus* is the kanamycin nucleotidyltransferase gene (*kat*) from *Geobacillus stearothermophilus* (*B. stearothermophilus*), whose thermostability has been increased up to 79°C by in vitro mutagenesis (Liao et al., 1986; Matsumura et al., 1986). Based on the mutated *kat* gene, a resistance cartridge, pKT1 (Figure 12.1) was generated, which can be applied for the disruption of *Thermus* genes *via* homologous recombination between the target gene and a chimeric construct of the target gene inactivated by *in vitro* disruption by the *kat*-resistance marker (Lasa et al., 1992a). The kanamycin-resistance cartridge pKT1 also confers kanamycin resistance

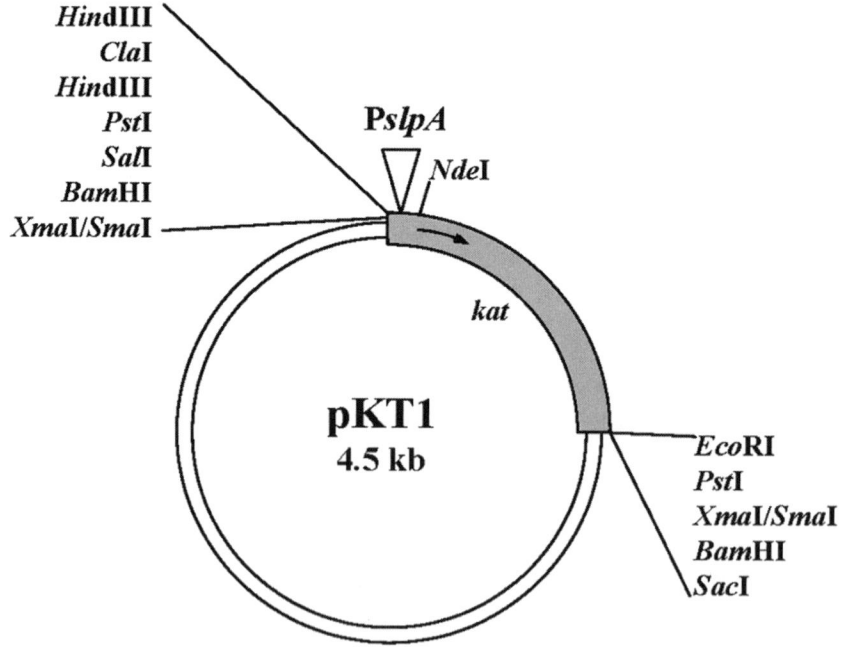

Figure 12.1. The kanamycin resistance cartridge pKT1 for *Thermus* (Lasa et al., 1992a) (reproduced with permission of the publisher). P*slpA*, promoter of the S-layer gene *slpA*; *kat*, thermostable kanamycin resistance gene; the backbone of the cartridge is the *E. coli* vector pBluescriptKS+.

at 37°C due to a bifunctional *E. coli/Thermus* promoter upstream of the *kat* gene. A stable transcription and translation of the *kat* gene in *Thermus* is mediated by transcription and translation signals of the S-layer protein from *Thermus thermophilus* HB8.

Based on the thermostable *kat* resistance marker from *G. stearothermophilus* a second *kat* cartridge, designated pGEM-kat (Figure 12.2) was constructed (Friedrich *et al.*, 2001). A 1.2-kb *Bam*HI fragment carrying the *kat* gene was isolated from the *Thermus/E. coli* shuttle vector pMK18

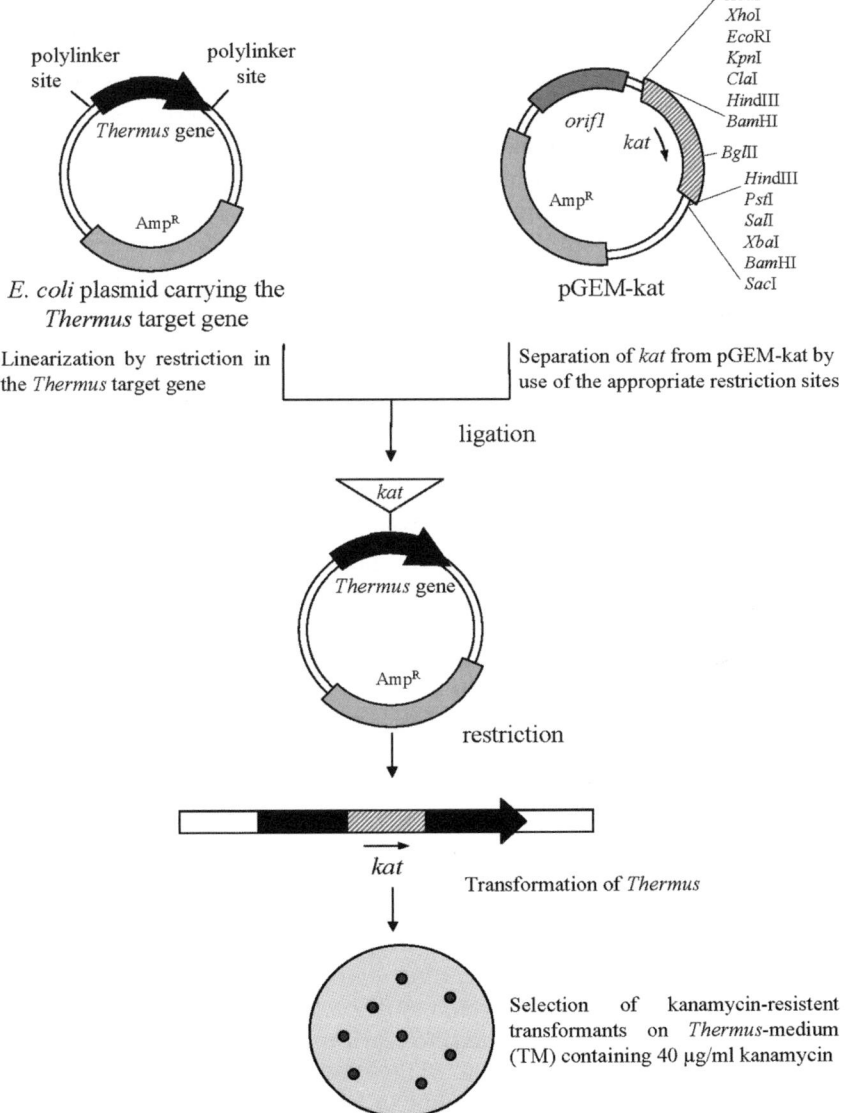

Figure 12.2. Insertional mutagenesis of *Thermus* genes by the use of thermostable *kat* gene in pGEM-kat. *kat*, thermostable kanamycin adenyl transferase gene; Amp^R, ampicillin resistance gene.

(de Grado et al., 1999), which was provided by Berenguer (Centro de Biologia Molecular, Severo Ochoa CSI, Madrid). The *kat* gene was joined with the *E. coli*-cloning vector pGEM (Promega GmbH, Heidelberg, Germany) resulting in the cartridge pGEM-kat (Figure 12.2). The strategy for *Thermus* mutant generation by the use of pGEM-kat is depicted in Figure 12.2 (Friedrich et al., 2001). An analogous strategy using the appropriate restriction sites can be performed with pKT1.

Protocol

1. Isolate the *kat* gene by restriction of pGEM-kat using the unique restriction sites flanking the *kat* gene (Figure 12.2).
2. Insert the *kat* gene into a cloned target gene integrated in an *E. coli* vector, such as pUC18 of pBSK, which confer ampicillin resistance and transform the construct into competent *E. coli* cells (Sambrook et al., 1989). To select for *E. coli* transformants grow the cells at 37°C in Luria Bertani (LB) medium under ampicillin and kanamycin selection (100 µg ampicillin ml^{-1} and 40 µg kanamycin ml^{-1}).
3. Prepare the plasmids of the resulting transformants to allow purification of the inserts which are used for transformation into *T. thermophilus* HB27.
4. For natural transformation into *Thermus* a modified protocol of Koyama et al. (1986) is used. Grow a *T. thermophilus* culture for 12 h at 70°C in TM medium (see transformation protocol) and transfer 200 µl of the culture into a prewarmed 10 ml culture tube, mix the cells with 1–10 µg of chimeric DNA and incubate for 2 h at 65°C by shaking at 150 rpm.
5. Identification of *Thermus* mutants generated by allelic replacement of wild-type DNA by chimeric DNA is performed by kanamycin selection (40 µg kanamycin ml^{-1} TM or TT medium; (Lasa et al., 1992a; Tanaka et al., 1981) for 48 h at 65°C. A minimum of 300 bp flanking the *kat* gene on both sides is sufficient for correct and efficient allelic replacement *via* homologous recombination.

Bleomycin resistance marker

Recently, a thermostable bleomycin resistance marker has been developed by the use of bleomycin-binding protein gene (*shble*) from the mesophilic bacterium *Streptoalloteichus hindustanus* (Brouns et al., 2005). This protein is a small, highly negatively charged, cytoplasmic protein, which forms homodimers that bind two positively charged antibiotic molecules, such as bleomycin. A direct evolution approach using selection in *T. thermophilus* led to various double mutated *shble* genes mediating resistance at 77°C. The bleomycin-resistance cassette was integrated as *NdeI/XbaI*-fragment in pMK18 (see cloning vectors) by replacing the kanamycin nucleotidyltransferase gene. The resulting vector pWUR112 (Figure 12.3) is stably maintained in *E. coli* in the presence of 3 µg ml^{-1} bleomycin.

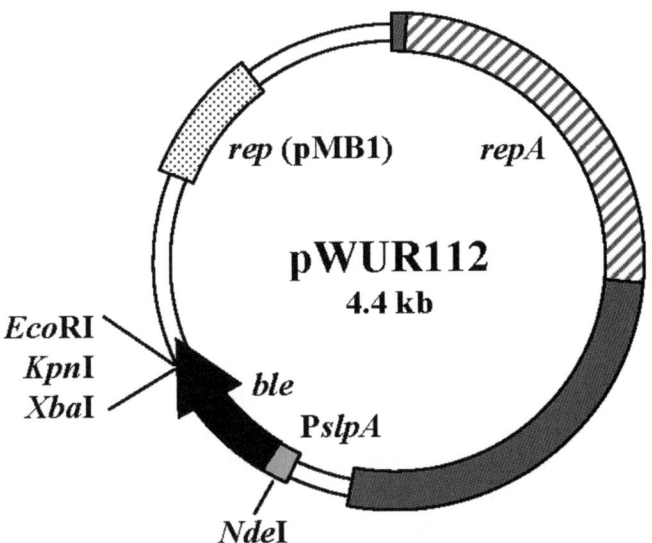

Figure 12.3. The bleomycin resistance cartridge pWUR112 (Brouns et al., 2005). ble, bleomycin resistance gene; rep, pMB1 replicon for replication in E. coli, PslpA, promoter of the S-layer gene slpA; ▨ minimal region required for replication in T. thermophilus HB27.

Protocol

1. Grow the cells at 70°C (at 150 rpm) to an OD_{600} of 0.8 in Ca^{2+} (3.9 mM) and Mg^{2+} (1.9 mM)-rich medium containing 8 g l^{-1} tryptone, 4 g l^{-1} yeast extract and 3 g l^{-1} NaCl dissolved in Evian mineral water (pH 7.7) (Brouns et al., 2005).
2. To transfer pWUR112 in *T. thermophilus* by natural transformation dilute the culture 1:1 in preheated medium and incubate for 1 h before adding pWUR112 DNA (10 μg ml^{-1}) to 0.5 ml of the culture. Further incubate this cell/DNA mixture for 2–3 h at 70°C with occasional shaking before plating onto solid rich medium (3% agar) supplemented with 15 μg ml^{-1} bleomycin (Calbiochem, San Diego, California, USA) for selection.
3. The thermostable bleomycin resistance gene encoded by the NdeI/XbaI-fragment in pWUR112 (Figure 12.3) can be used for multigene knockouts or gene deletion strategies in *Thermus*. Analogous to the kat-gene-based knockout strategies, the shble-based strategies require chimeric constructions of target gene disrupted by the marker gene, transfer into *Thermus* and subsequent homologous recombination of the chimeric constructs and target wild-type genes.

Orotate phosphoribosyltransferase marker

An alternative general system for gene deletion of gene disruption in *T. thermophilus* is based on genes involved in *de novo* pyrimidine biosynthesis (Tamakoshi et al., 1999). The key marker for this gene deletion strategy is the orotate phosphoribosyltransferase gene (*pyrE*) which has been used as a genetic marker in many microorganisms and lower eukaryotes (Begueret et al., 1984; Gruber et al., 1990; Neuhard et al.,

1985). The strategy is based on the finding that a deficiency in *pyrE* results in resistance against the bactericidal compound 5-fluoroorotic acid (5-FOA) while this compound is toxic to wild-type cells. This allows a positive selection of uracil auxotrophic *pyrE* mutants in the presence of 5-FOA. Since the *pyrE* gene of T. thermophilus has been cloned and *pyrE* mutants KT8 and MT111 have been generated this allows the selection of deletion mutants by a general protocol (Tamakoshi et al., 1997, 1999; Yamagishi et al., 1996).

Protocol

1. The strategy to delete a target gene from T. thermophilus consists of two steps: (1) replacement of the target gene of a Δ*pyrE* strain with the *pyrE* gene and (2) deletion of the *pyrE* gene. Therefore the T. thermophilus target gene plus a minimum of 300 bp of flanking *Thermus* DNA upstream and downstream of the target gene has to be cloned in an E. coli vector.
2. The *pyrE* gene cassette is available in plasmid pNIV and can be isolated as a 0.7 kb NdeI-EcoRV fragment (Tamakoshi et al., 1999). This *pyrE* carrying DNA fragment is used to replace the target gene present in an E. coli plasmid by standard cloning protocols (Sambrook et al., 1989). After gene replacement, a minimum of 300 bp of wild-type DNA of the *Thermus* target gene locus has to flank the *pyrE* gene since it is required for homologous recombination of the *pyrE* construct with the wild-type locus.
3. Transform the deletion construct according to the protocol of Koyama et al. (1986), into a T. thermophilus Δ*pyrE* mutant, such as MT111. Screen for MT111 transformants carrying a deletion of the target gene due to gene replacement by the *pyrE* gene (MT111*pyrE*) via homologous recombination between *pyrE* gene flanking DNA and wild-type target locus by growth in minimal medium. The uracil prototroph (*pyrE*$^+$) transformants grow in minimal medium in the absence of uracil (see minimal medium of Tanaka et al. (1981)) in the transformation protocol.
4. To delete the *pyrE* gene from the T. thermophilus MT111*pyrE* genome generate a deletion plasmid devoid of *pyrE* but still carrying the 300 bp flanking the target gene locus. Transfer this construct into T. thermophilus MT111*pyrE* via natural transformation. Deletion of the *pyrE* gene is achieved *via* homologous recombination between the DNA of the deletion construct and the flanking DNA of the *pyrE* locus in *Thermus* MT111*pyrE*. The *pyrE* deletion mutants are 5-FOA-resistant but require uracil, and therefore transformants where the *pyrE* gene is replaced by the deletion derivative are selectable on minimal medium with 20 µg ml^{-1} uracil and 200 µg ml^{-1} 5-FOA.

Transposon Mutagenesis

Transposon mutagenesis is a powerful tool for random mutagenesis of bacterial genomes and insertion of foreign DNA. In 1990, the transfer and transposition of the transmissible transposon Tn916 from *B. subtilis* BS250 containing Tn916 as a chromosomal insert to *T. aquaticus* was reported (Sen and Oriel, 1990).

Protocol

1. Grow the *B. subtilis* BS250 donor strain overnight at 37°C in LB medium with 10 µg tetracycline ml^{-1}. In parallel grow *T. aquaticus* overnight at 75°C in a complex medium containing 8.0 g Bacto peptone, 4.0 g yeast extract and 2.0 g NaCl per litre at pH 7.5.
2. Mix 0.05 ml of the *B. subtilis* donor culture (10^9 CFU ml^{-1}) and 0.5 ml of the *T. aquaticus* recipient culture (10^9 CFU ml^{-1}) in 4.5 ml fresh complex medium and incubate for 4 h at 48°C with 50 µg ml^{-1} DNase I. The mating temperature is very important for the transfer of the conjugative tranposon Tn916, because lower or higher temperatures result in drastically reduced transfer frequencies.
3. To select for *T. aquaticus* transposon mutants exhibiting the Tn916-mediated tetracycline resistance, plate the mating mixture on complex medium and incubate at 75°C in the presence of 10 µg ml^{-1} tetracycline.

E. coli/Thermus Shuttle Vectors

Based on naturally occuring plasmids from *Thermus* species, several vectors that replicate autonomously in *Thermus* have been constructed (Table 12.1). The first generation of vectors, such as pYK109, was based on replicons of *Thermus* cryptic plasmids, the ColE1 replicon for replication in *E. coli* and genes complementing auxotrophic mutants in the synthesis of amino acids or a *Thermus* β-galactosidase gene as selectable markers (Koyama and Furukawa, 1990; Koyama et al., 1990; Raven, 1995). However, a significant background was often observed due to the high recombination frequency of *Thermus* (Hoshino et al., 1993; Koyama et al., 1986).

To circumvent homologous recombination events, a new generation of *Thermus* vectors was developed carrying a chimeric thermostable *kat* resistance cartridge encoding a thermostable kanamycin adenyltransferase controlled by transcriptional signals of the surface layer gene (*slpA*) from *T. thermophilus* HB8 (Matsumura et al., 1986; Lasa et al., 1992a). Based on the *kat* cartridge and the replicon of the cryptic multicopy plasmid pTT8 of *T. thermophilus* HB8 the *Thermus* vector pMKM001 was developed (Mather and Fee, 1992; Matsumura et al., 1986).

The 8.2-kb pMKM001 plasmid was transferred into *T. thermophilus* HB8 *via* natural transformation by a slightly modified protocol of Koyama et al. (1986) that included prior to screening of transformants an incubation at 65°C for 60 min and at 60°C for 90 min. Transformants were then screened at 60°C on *Thermus* medium (TM) with 100 µg kanamycin ml^{-1}. Applicability of this *Thermus* vector is limited due to the small number of suitable restriction sites allowing insertion of DNA and

Table 12.1 E.coli/Thermus shuttle plasmids

Vector	Origin of replication (ORI) for Thermus (host strain)	E.coli replicon	Selection marker in Thermus	Reference
pYK 109	ORI of plasmid pTT8 (T. thermophilus HB8)	ColE1	Tryptophan synthetase gene (trpB)	Koyama et al. (1990)
pLU1 to pLU4	ORI of cryptic plasmids (T. aquaticus)	ColE1	Kanamycin resistance (kat)	Lasa et al. (1992b)
pMY1, pMY2, pMY3	ORI of undescribed, cryptic plasmids (Thermus sp.)	ColE1	Kanamycin resistance (kat)	Lasa et al. (1992b)
pNTsp2	ORI of plasmid pTsp45s (Thermus sp. YS45)	ColE1	Kanamycin resistance (kat)	Wayne and Xu (1997)
pMK18	ORI of a cryptic plasmid (Thermus sp.)	ColE1	Kanamycin resistance (kat)	de Grado et al. (1999)

the lack of an *E. coli* replicon, but this vector is a good starting point for the generation of *E. coli/Thermus* shuttle vectors.

pLUI and pMY series

Based on replicons of naturally occuring *Thermus* plasmids, the *E. coli* ColE1 replicon and the chimeric thermostable *kat* resistance cartridge, several *Thermus/E. coli* shuttle vectors were generated (Table 12.1); pMY1 to pMY3 (Table 12.1) (Lasa et al., 1992b) were transformed into *E. coli* by standard procedures. Transformants were selected on LB medium containing ampicillin (100 µg ml^{-1}) and kanamycin (30 µg ml^{-1}). Transfer into *T. thermophilus* HB8 or HB27 was performed by the method of Koyama et al. (1986), resulting in high transformation efficiencies (1×10^2–1×10^4 cells µg^{-1} DNA). The MIC (minimal inhibitory concentration) of kanamycin for selection of transformants carrying plasmids is >200 µg ml^{-1}. Singular *Pst*I restrictions sites in pLU3, pLU4, pMY1, pMY2 and pMY3 can be used for integration of DNA (Lasa et al., 1992b).

To enable cloning and expression of heterologous DNA in *Thermus* a pMY1 derivative, designated pMY1.1 was generated (Figure 12.4a; Lasa et al., 1992b). Plasmid pMY1.1 carries a *Thermus* promoter and Shine Dalgarno sequence allowing the expression of cloned genes in *Thermus* by a two step strategy. For this strategy a pUC9 derivative, pPS1 (Figure 12.4b), carrying the promoter and the first 99 bp of the *Thermus* S-layer protein gene (*slpA*) is also necessary (Lasa et al., 1992b).

Protocol

1. First clone the target gene downstream of the *slpA* DNA in pPS1 by the use of singular restriction sites *Bam*HI, *Sal*I, *Pst*I or *Hind*III (Figure 12.4b).
2. Then fuse pMY1.1 (Figure 12.4a) to recombinant pPS1 carrying the target gene to be expressed in *Thermus*. To enable the fusion of pMY1.1 and recombinant pPS1 both linearize with *Eco*RI and subsequently

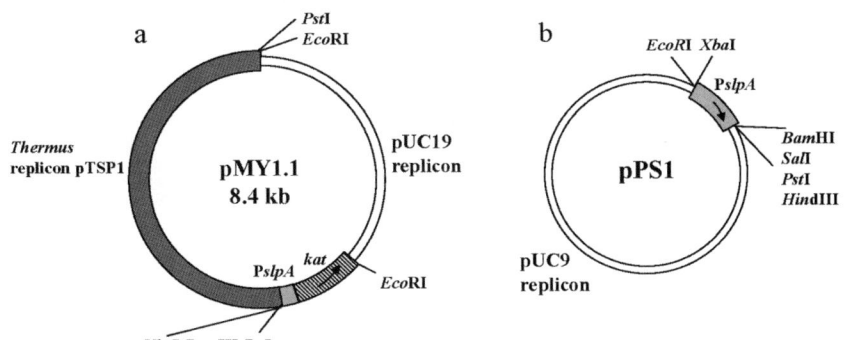

Figure 12.4. The *E. coli/Thermus* shuttle vector pMY1.1 (a) and the promoter vector pPS1 (b) (Lasa et al., 1992b) (reproduced with permission of the publisher). P*slpA*, promoter region of the *Thermus* S-layer gene *slpA*.

ligate the DNA fragments. This step limits the applicability of this vector system for gene expression in *Thermus* since the absence of any *Eco*RI site in the target gene inserted into pPS1 is a prerequisite.

pMK18

Currently, the most widely used vectors for *Thermus* are the high-transformation-efficiency *E. coli/Thermus* shuttle vectors pMK18 and pMK18r (Table 12.1, Figure 12.5; de Grado et al., 1999). pMK18 consists of the minimal replicative origin from a 16-kb plasmid of *Thermus* sp. ATCC 27737, a thermostable *kat*-resistance cartridge conferring kanamycin resistance in *E. coli* (37°C) and in *Thermus* (70°C) and a 1.65-kb *Ava*II-*Nde*I fragment of pUC18 carrying the *E. coli* ColE1 replicon, the multiple cloning site of pUC18 and the 5′-end of the β-galactosidase gene confering α-complementation of the β-galactosidase gene for the *E. coli* host (Figure 12.5; de Grado et al., 1998, 1999).

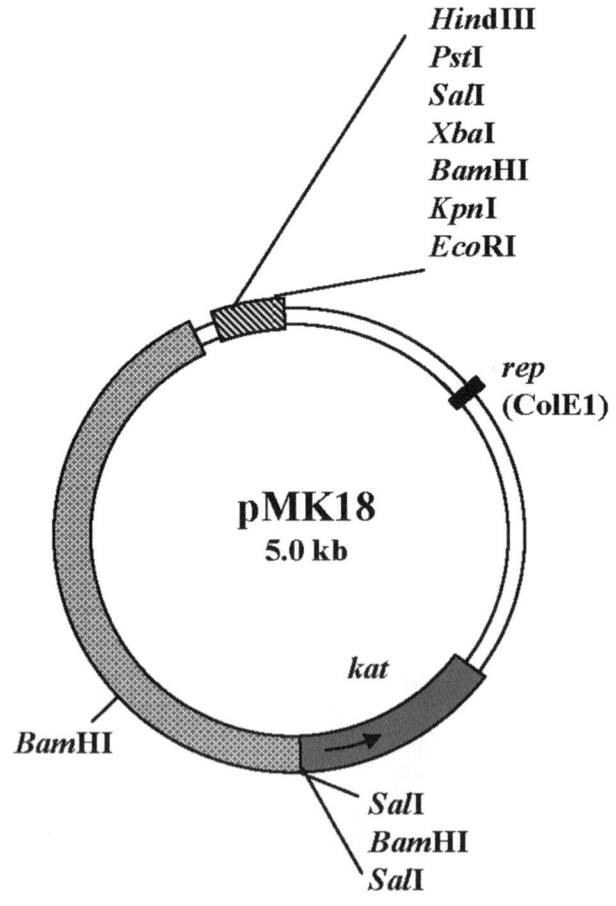

Figure 12.5. The *E. coli/Thermus* shuttle vector pMK18 (de Grado et al., 1999) (reproduced with permission of the publisher). ▨ minimal replicon from an indigenous *Thermus* sp. plasmid, *Nde*I/*Ava*II fragment of pUC18; *rep* (ColE1), ☐ pUC18 sequence.

Protocol

1. Integrate the target gene in pMK18 by the use of a multiple cloning site which retained all pUC18 single restriction sites except for SmaI and BamHI, which are also present in the *Thermus* replicon.
2. To transfer pMK18 into *Thermus*, mix 500 µl of exponentially grown cells (grown on carbonate-rich complex medium, OD_{550} of 0.8) with 500 µl of preheated medium and incubate for 1 h at 70°C. Then add pMK18 DNA and incubate the mixture for 2 h.
3. To select for transformants plate the cells on solid carbonate-rich complex medium (see medium of the electroporation protocol) containing 30 µg ml^{-1} kanamycin and grow for 24–48 h at 70°C.

Highest transformation efficiencies (transformants µg^{-1} of DNA) of 10^8–10^9 were obtained with *T. thermophilus* strain HB27 as recipient and pMK18 DNA obtained from *T. thermophilus* HB27. Twofold reduced transformation efficiencies were obtained with pMK18 isolated from dam^+ dcm^+ *E. coli* cells. With *T. thermophilus* HB8 as recipient, reduced transformation efficiencies of 10^5 were obtained with pMK18 isolated from dam^+ dcm^+ *E. coli* cells, but at least a 100-fold increased transformation efficiency was obtained with pMK18 isolated from strain HB27 (de Grado et al., 1999). From these results it was concluded that strain HB27 is devoid of any restriction modification system, which makes this strain a very suitable host for pMK18. Alternatively, an electroporation protocol can be employed (see electroporation protocol) resulting in pMK18 tranformation efficiencies similar to those obtained *via* natural transformation.

Antisense Expression Vector

The antisense expression vector pMK18r which has been used to study the role of Mn-catalase in *T. thermophilus* is a powerful tool for analyses of the physiological relevance of any gene in *T. thermophilus* (Moreno et al., 2004). Plasmid pKM18r is identical to pMK18 (Figure 12.5) except for the orientation of the *kat* gene and the replicon for *Thermus* which are reversed in pMK18r (de Grado et al., 1999; Moreno et al., 2004). The P*slpA* (promoter of the S-layer gene *slpA*) promoter upstream of the *kat* gene results in strong constitutive expression of the *kat* gene in *T. thermophilus* and moderate expression in *E. coli*. Due to the absence of any transcription terminator, transcription from P*slpA* passes through the multicloning site (MCS) resulting in cotranscription of the *kat* gene and a target gene fused to the *kat* gene in pMK18r. The antisense system is based on the constitutive expression of a "bicistronic" transcript of the *kat* gene mRNA and the target antisense RNA resulting from an inverse orientation of the target gene fused to the *kat* gene.

Protocol

1. Clone the coding region of a target gene in inverse orientation with respect to the *kat* gene into the multicloning site of pMK18r.

2. Transform *T. thermophilus* HB27 with the resulting pMK18r recombinant plamid DNA according to the protocol of Koyama *et al.* (1986) and select transformants on TB medium (see TB medium in the conjugation protocol) containing 30 mg kanamycin l^{-1}.
3. Assay the expression of antisense RNA by RT-PCR with total RNA isolated by standard protocols (Sambrook *et al.*, 1989) from cells grown on TB agar plates.

Promoter Probe Vectors

pPPII

Based on plasmid pYK134 carrying a heat-stable kanamycin resistance (Km^R) gene and the tryptophan synthetase gene (*trpB*) of *Thermus* T2, one of the first *Thermus* promoter probe vectors, designated pPP11 (Figure 12.6), was generated (Hoshino *et al.*, 1993; Maseda and Hoshino, 1995). *Thermus* cells carrying the promoterless kanamycin resistance gene in plasmid pPP11 could not grow in the presence of 20 µg ml^{-1} kanamycin.

Protocol

1. Shear *Thermus* DNA (1.5 µg ml^{-1}) by sonication or restriction with appropriate restriction enzymes. The unique *Eco*RI or *Sal*I restriction sites flanking the 5′-end of the promoterless Km^R gene in pPP11 can be used to integrate restricted DNA.

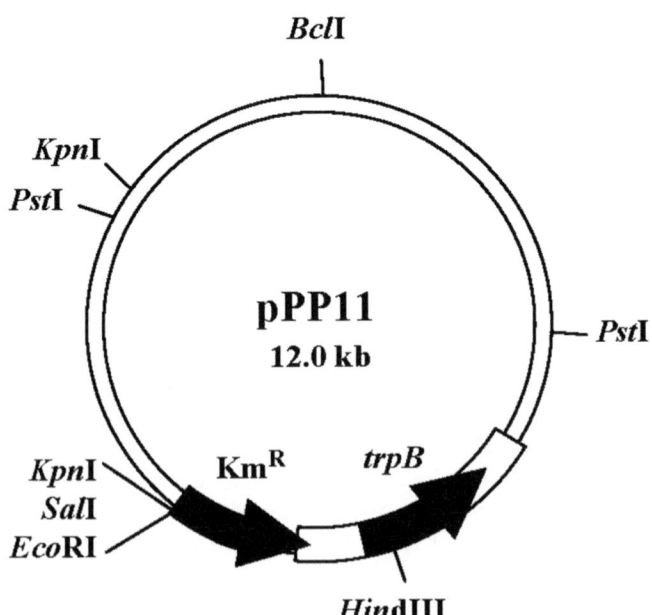

Figure 12.6. The promoter probe vector pPP11 (Maseda and Hoshino, 1995) (reproduced with permission of the publisher). *trpB*, tryptophan synthetase gene.

2. Transfer the ligation mixture into *T. thermophilus* HB27 cells by natural transformation according to the protocol of Koyama *et al.* (1986). Select transformants for growth at 70°C for 36–48 h on TM solid medium (see media of the transformation protocol) containing kanamycin ranging from 40 µg ml^{-1} to 2 000 µg ml^{-1} dependent of the promoter strength.
3. Monitor promoter activities by analyses of the nucleotidyltransferase activity (Sadaie *et al.*, 1980). For the assay, collect *Thermus* cells after 5.5 h of growth in TM medium at 60°C, suspend the cells in N buffer comprising of 20 mM Tris-HCl (pH 7.5), 20 mM MgCl$_2$, 10% glycerol, 0.2 mM dithiothreitol and 50 mM NaCl, and disrupt the cells by sonication.
4. After centrifugation (15 000 × g; 10 min at 4°C) use the cell-free supernatants to perform kanamycin nucleotidyltransferase (KNTase) assays using [^{3}H]-ATP. The assay mixture (20 µl) contains 0.86 mM kanamycin, 0.2 mM (1 µCi) of [^{3}H]-ATP, 4 mM MgCl$_2$, 10 mM mercaptoethanol, 62.5 mM Tris-maleate buffer (pH 6.25). To analyse the KNTase activity, add cell-free supernatant (1 µg of total protein) and incubate the mixture for 30 min at 37°C. Transfer 5 µl of the reaction mixture to phosphocellulose paper, wash with 50 ml distilled water, and count in a scintillation spectrometer. The KTNase activity is expressed as radioactivity incorporated into the kanamycin after 30 min of incubation (Maseda and Hoshino, 1995).

pKANPROII and pTGTII

The next generation of promoter probe vectors, pKANPROII (Figure 12.7a) and pTGT11 (Figure 12.7b), contain replicons for *E. coli* and *Thermus* and allow the detection of promoters by expression of the kanamycin resistance in both *Thermus* and *E. coli*, due to a

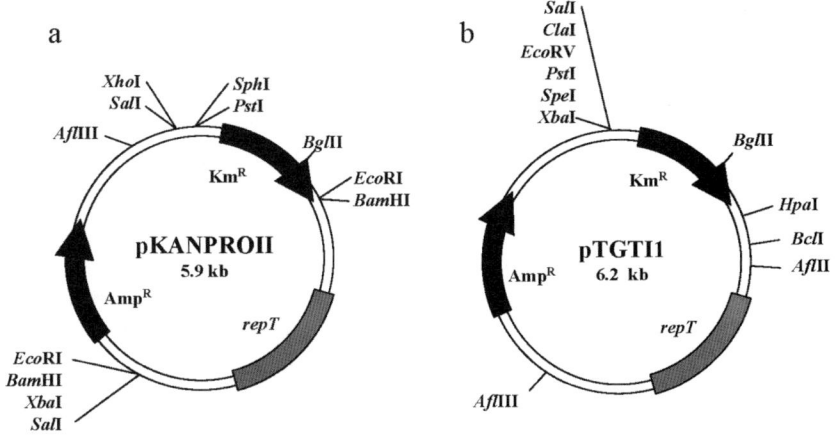

Figure 12.7. The promoter probe vectors pKANPROIII (a) and pTGTI1 (b). (Kayser *et al.*, 2001) (reproduced with permission of the publisher). *repT*, *Thermus* replication origin.

promoterless kanamycin resistance gene functioning in *E. coli* and *Thermus* (Kayser et al., 2001). Both vectors are identical except for more unique restriction sites present in the multicloning site of pTGTI1.

Protocol

1. Ligate *Thermus* chromosomal DNA fragments into the promoter probe vectors by the use of appropriate unique restriction sites and transform the ligation mixture into *T. thermophilus* HB27 according to the protocol of Koyama et al. (1986).
2. Select Km^R transformants on solid TT medium (see medium in the protocol of insertional mutagenesis; Lasa et al., 1992a) at 55°C.
3. Replica plate onto rich medium containing 100 µg ml^{-1}, 500 µg ml^{-1} or 2000 µg ml^{-1} kanamycin to distinguish between weak and strong promoters. Quantify the promoter strength by a KNTase test performed as described in the protocol for the application of the promoter probe vector pPP11 (Maseda and Hoshino, 1995). Independent of the strength of the cloned promoter in pKANPROII and pTGT11 expression of kanamycin nucleotidyltransferase in *E. coli* grown on complex medium at 37°C was always found to be moderate (40 µg ml^{-1} kanamycin).

pTEX1mdh

The vector pTEX1mdh replicates in both *Thermus* spp. and *E. coli*. It contains the malate dehydrogenase (*mdh*) gene from *T. thermophillus* ("T. flavus") as reporter gene (Figure 12.8; Kayser and Kilbane II, 2001). The constitutive J17 promoter isolated from *T. thermophilus* was inserted upstream of the promoterless *mdh* gene and serves as a control for quantification of malate dehydrogenase activities in crude extracts by a spectrophotometric assay. To apply the *mdh*-based promotor probe system, the J17 promoter has to be replaced with target promoters and the construct has to be transferred into a *Thermus mdh* mutant. A *mdh* mutant, designated MM8-5, has been constructed by Kayser and Kilbane II (2001) by replacing the coding region of the malate dehydrogenase (*mdh*) gene in *T. thermophilus* HB27 by a double-crossover homologous event with a thermostable kanamycin resistance cassette (Km^r). The resulting Δ*mdh* Km^r mutant is designated MM8-5.

Protocol

1. Replace the *Eco*RI/*Sph*I DNA fragment upstream of the promotorless *mdh* gene carrying the constitutive *Thermus* J17 promoter in pTEX1mdh with the target promoter region.
2. Transform the resulting construct into the *T. thermophilus* HB27 *mdh* mutant MM8-5 *via* natural transformation.
3. Select transformants at 55°C on TT-rich medium (see TT medium of Lasa et al. (1992a) in the protocol for deletion mutants) with kanamycin (40 µg ml^{-1}). Plate the *mdh* mutant strain as a control.

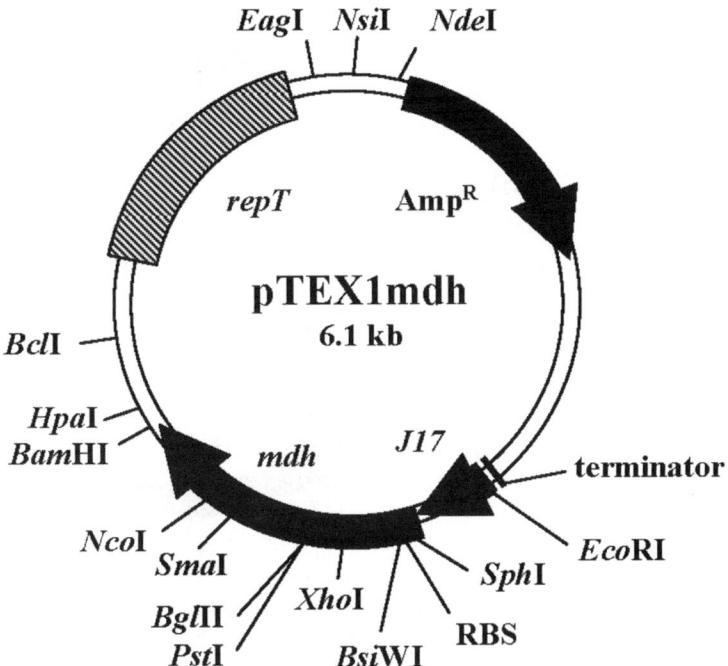

Figure 12.8. The malate dehydrogenase (*mdh*) promoter probe vector pTEX1mdh (Kayser and Kilbane II, 2001) (reproduced with permission of the publisher). *mdh*, promoterless malate dehydrogenase gene from *T. flavus*; J17, a strong *Thermus* promoter also functioning in *E. coli*; *repT*, *Thermus* replicon; Amp^R, ampicillin resistance.

The *mdh* mutant colonies are very small (0.1–0.5 mm) even after several days of incubation due to the *mdh* deletion. In contrast the *mdh*$^+$ transformants form larger colonies in 2–3 days.

4. Quantify the promoter activities by assaying malate dehydrogenase activity of crude extracts. Therefore, grow *Thermus* cultures overnight at 55°C in TT medium with kanamycin and generate cell extracts by sonication and subsequent centrifugation at 10 000 g for 15 min at room temperature to remove cell debris. Determine malate dehydogenase activity by monitoring the decrease of absorbance of NADH at 340 nm due to reduction of oxaloacetate to L-malate.

Integrative promoter probe vector pSJl7mdhA

To examine the promoter activity of genes present as a single copy excluding any copy number or DNA superhelicity effects, an integrative promoter probe vector based on the malate dehydrogenase gene as reporter gene was developed (Kayser and Kilbane II, 2001). The integrative promotor probe vector pSJ17mdhA contains the *E. coli* replicon of pUC19 (Figure 12.9; Kayser and Kilbane II, 2001). pSJ17mdhA does not replicate in *Thermus* spp. but integrates into the *Thermus* chromosome via a double-crossover event by the *scsA* and *gpt* chromosomal region flanking the J17 promotor-*mdh* gene cassette.

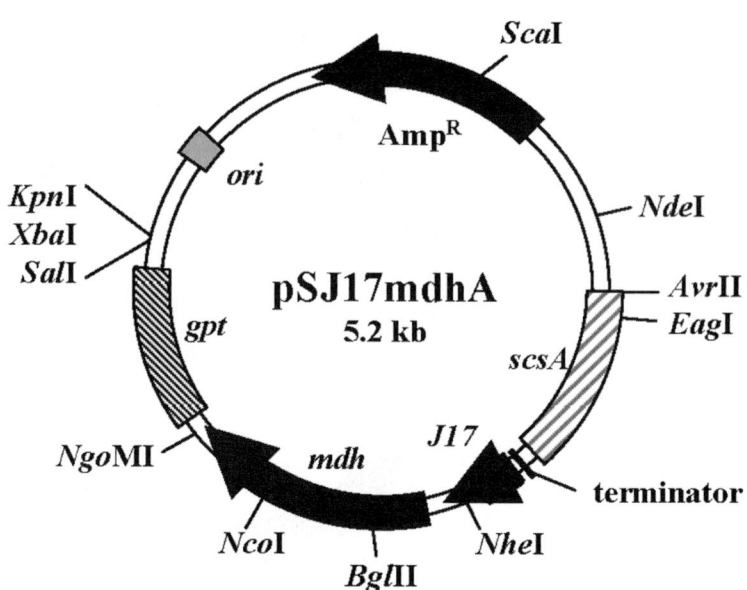

Figure 12.9. The integrative promoter probe vector pSJ17mdhA (Kayser and Kilbane II, 2001) (reproduced with permission of the publisher). *mdh*, promoterless malate dehydrogenase gene from *T. thermophilus* ("T. flavus"); terminator, transcription termination sequence from the *T. flavus* phenylalanyl tRNA synthetase operon; J17, a strong *Thermus* promoter also functioning in *E. coli*; *gtp* and *scsA*, "T. flavus" chromosomal region flanking the "T. flavus" mdh gene.

Integrative promoter probe vector pTPRO1

The integrative promoter probe vector pTPRO1 contains a promoterless kanamycin-resistance gene to examine the expression of genes as a single copy. pTPRO1 contains the *E. coli* replicon of pUC19 but is devoid of a *Thermus* replicon (Figure 12.10; Kayser et al., 2001). A stable single-copy chromosomal integration of pTPRO1 in the genome of *Thermus* is achieved *via* a double-crossover event at the *leuB* (3-isopropylmalate dehydrogenase) locus due to the presence of portions of the "T. flavus" *leuB* gene flanking the kanamycin nucleotidyltransferase gene. Unique restriction sites *NcoI* or *EcoRI* upstream of the promoterless kanamycin resistance gene allow integration of DNA containing promoter sequences. The promoter activities are monitored by analyses of the nucleotidyl-transferase activity (see KNT-activity assay with pPP1; Sadaie et al., 1980).

Expression Vectors

The *Thermus* expression systems available to date are based on four different inducible promoters: (1) the *dnaK* promoter of *T. thermophilus* which is activated by heat shock (Figure 12.11 and 12.12; plasmid pTEX2-dnaK, pTEX7), (2) the arginine-inducible promoter P*arg* (Figure 12.11; pTEX8), (3) the carbon-regulated promoter P*scs-mdh* (Figure 12.11; pTEX9), and (4) the inducible promoter P*nar* of the

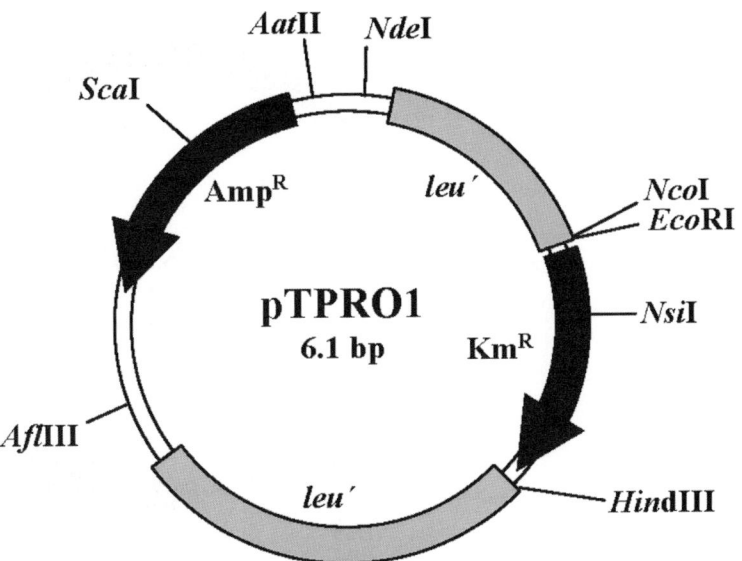

Figure 12.10. The integrative promoter probe vector pTPRO1 (Kayser et al., 2001) (reproduced with permission of the publisher). leu', portions of the 3-isopropylmalate dehydrogenase gene of *T. flavus*, Km^R, promoterless thermostable kanamycin nucleotidyl-transferase gene.

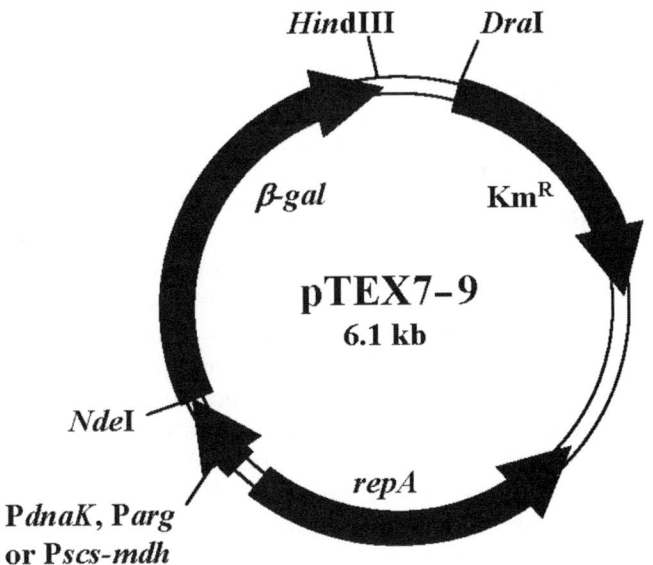

Figure 12.11. The *Thermus* expression vectors pTEX7, pTEX8 and pTEX9 (Park and Kilbane II, 2004) (reproduced with permission of the publisher). pTEX7, pTEX8 and pTEX9 contain promoters P*dnaK*, P*arg* and P*scs-mdh*, respectively. β-gal, β-galactosidase; Km^R, kanamycin resistance; *repA*, *Thermus* replicon.

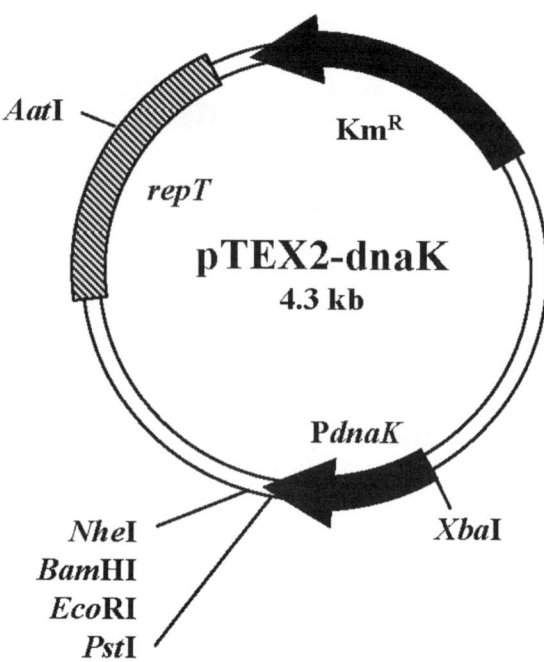

Figure 12.12. The *Thermus* expression vectors pTEX2-dnaK (Kayser *et al.*, 2001) (reproduced with permission of the publisher). Km^R, kanamycin resistance; *repT*, *Thermus* replicon; P*dnaK*, promoter of the *dnaK* gene.

T. thermophilus HB8 nitrate reductase operon (Kayser *et al.*, 2001; Park and Kilbane II, 2004).

pTEX2-dnaK, pTEX7, pTEX8 and pTEX9

The pTEX expression vectors contain a kanamycin nucleotidyltransferase gene expressed by the broad host range promoter vv12 or the *slp* promoter from *Thermus* which results in kanamycin resistance in both *Thermus* and *E. coli* (Kayser *et al.*, 2001; Park and Kilbane II, 2004). Replication in *Thermus* is mediated by the *Thermus* replication origins RepT or RepA whereas replication in *E. coli* is due to the ColE1 replication origin from pUC18/19. The expression vectors contain promoters P*dnaK*, P*arg* and P*scs-mdh*, respectively (Figure 12.11 and Figure 12.12). Furthermore, these vectors contain the *Thermus* sp. A4 β-galactosidase gene as reporter gene which enables selection of transformants by the deep blue colour on X-gal agar plates and quantification of expression level. For selection a β-galactosidase mutant of *T. thermophilus*, designated PPKU (Park and Kilbane II, 2004) has to be used as host strain. The β-galactosidase gene is missing in the expression vector pTEX2-dnaK which carries the inducible promoter P*dnaK* (Figure 12.12).

Protocol

1. Grow *E. coli* cultures containing the expression vectors in LB medium at 37°C with 40 μg ml^{-1} kanamycin. Prepare the expression vectors using standard protocols (Sambrook *et al.*, 1989). Clone the target genes to be expressed in *Thermus* downstream of the inducible promoters by the use of unique restriction sites with *Nde*I in pTEX7, pTEX8 or pTEX9 or *Nhe*I, *Bam*HI, *Eco*RI and *Pst*I in pTEX2-dnaK (Figure 12.11 and 12.12).
2. Transform the recombinant pTEX7, pTEX8 or pTEX9 plasmids into the β-galactosidase mutant strain *T. thermophilus* PPKU whereas for pTEX2-dnaK use the *T. thermophilus* HB27 wild-type strain. Perform the DNA transfer according to the natural transformation protocol of Koyama *et al.* (1986). Select transformants carrying pTEX7 (P*dnaK*) or pTEX9 (P*scs-mdh*) derivatives for growth at 55°C on TT-rich medium (Lasa *et al.*, 1992a) by the deep blue colour due to X-gal conversion mediated by the β-galactosidase and/or kanamycin resistance (40 μg ml^{-1} kanamycin). Select transformants carrying derivatives of pTEX8 (P*arg*) in arginine and uracil-free *Thermus* minimal medium (see media of transformation protocol) with 20 mM pyruvate as carbon source and 10 mM ammonium sulphate as nitrogen source.
3. Induce the heat-inducible promoter P*dnaK* after growth of *T. thermophilus* HB27/pTEX2-dnaK or PPKU/pTEX7 at 55°C by incubation for 10 min at 85°C prior to preparation of cell-free extracts. To maximally induce the arginine-inducible promoter P*arg* grow a HB27/pTEX8 culture at 55°C in arginine and uracil-free medium to an optical density at 600 nm of 0.6 and subsequently incubate for 4 h at 65°C in the presence of 10–30 mM arginine. The P*scs-mdh* promoter in pTEX9 is maximally induced by growth of HB27/pTEX8 at 55°C in *Thermus* minimal medium in the presence of 40 mM malate.

pMKEI, pMKEβgal, pMKEPA, and pMKE2

The pMKE expression vectors contain the inducible nitrate reductase promoter of *T. thermophilus* HB8 and the regulatory sequences of the respiratory nitrate reductase operon of *T. thermophilus* HB8 allowing inducible expression of cloned genes (Figure 12.13 and 12.14; Moreno *et al.*, 2003, 2004). The ColE1 replicon mediates replication in *E. coli*, a minimal replicative origin from a 16-kb plasmid of *Thermus* sp. ATCC 27737 mediates replication in *Thermus* and a thermostable kanamycin resistance is expressed in *E. coli* and *Thermus*. Two pMKE1 derivatives, pMKEβgal and pMKEPA, carrying a thermophilic β-galactosidase and an alkaline phosphatase, respectively, were generated as cytoplasmic or periplasmic reporters. Recently a novel expression vector pMKE2, allowing overexpression of His-tagged enzymes in *T. thermophilus* was developed on the basis of pMKE1 and the His-tag vector pET28b+ (Figure 12.14; Moreno *et al.*, 2005). pMKE2 contains a modified inducible nitrate reductase promoter (P*nar*) resulting in much higher expression of cloned genes in *Thermus* compared to levels of

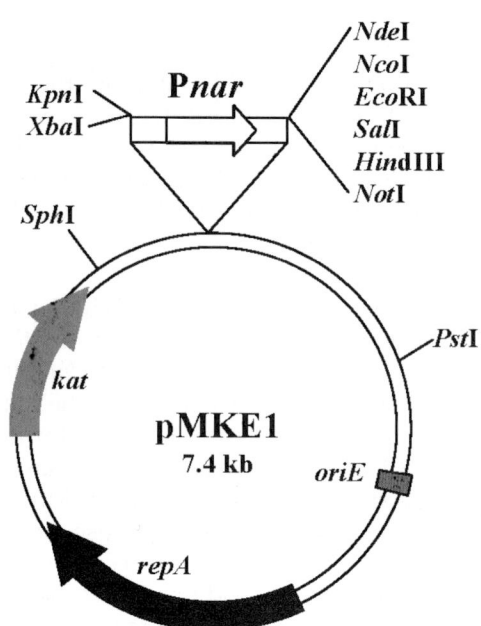

Figure 12.13. The *Thermus* expression vectors pMKE1 (Moreno *et al.*, 2003) (reproduced with permission of the publisher). *kat*, kanamycin resistance; P*nar*, promoter of the nitrate reductase operon; *oriE*, *E. coli* replication origin.

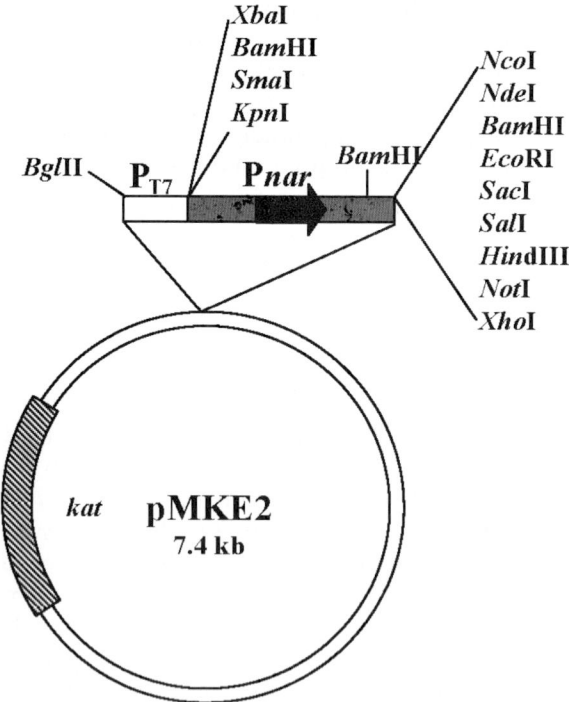

Figure 12.14. The *Thermus* overexpression His-Tag vector pMKE2 (Moreno *et al.*, 2003) (reproduced with permission of the publisher). *kat*, kanamycin resistance; P*nar*, promoter of the nitrate reductase operon; P_{T7}, T7-polymerase promoter.

overexpression in pMKE1. Overproduction using this construct was even higher than overproduction using T7-dependent expression systems in *E. coli* (Moreno *et al.*, 2005).

The expression of P*nar* and modified P*nar* requires both, anoxia and nitrate and is dependent on the presence of regulatory genes associated with the "*nar*" element of the natural host HB8. To allow P*nar*-driven expression of genes in the strictly aerobic *T. thermophilus* HB27 the "*nar*" element was transferred *via* conjugation into HB27 resulting in exconjugants HB27::*nar*. HB27::*nar* was found to exhibit a higher threshold level for P*nar* expression than HB8 and therefore strain HB27 is more suitable for the expression of potentially toxic proteins by the pMKE series in *Thermus* (Moreno *et al.*, 2003).

Protocol

1. Grow *E. coli* cells carrying pMKE vectors in LB medium supplemented with kanamycin (30 mg l^{-1}) at 37°C. Purify plasmids by molecular standard protocols (Sambrook *et al.*, 1989).
2. Insert the target gene in pMKE vectors by the use of unique restriction sites downstream of the P*nar*, transfer recombinant plasmids into *T. thermophilus* HB27::*nar via* natural transformation and select transformants on TB plates with kanamycin (30 mg l^{-1}) (see TB medium of the conjugation protocol).
3. Grow HB27::*nar* transformants aerobically in the absence of nitrate in TB medium at 70°C up to an OD_{550} of 0.2 and induce transcription by the addition of KNO_3 (40 mM), arrest the shaker and further incubate without shaking for up to 20 h.
4. To perform a control by the induction of P*nar*, treat *T. thermophilus* carrying the control plasmids pMKEβgal and pMKEPA, respectively, analogously. Measure the β-galactosidase and phosphatase activities at 70°C with crude HB27::*nar* extracts with *ortho*-nitrophenyl-galactopyranoside (ONPG) in 20 mM Tris-HCl buffer, pH 8, and *para*-nitrophenyl-phosphate in 100 mM Tris-HCl buffer, pH 8, respectively (Miller, 1972).

Thermostabilization of Proteins from Mesophiles in *Thermus*

The improvement of thermal stability of biotechnologically relevant enzymes is one of the major concerns of protein engineering. A suppressor mutation method has been developed and successfully applied for the stabilization of 3-isopropylmalate dehydrogenase (IPMDH) (Kotsuka *et al.*, 1996). This method could be generally applied to optimize proteins heterologously produced in *Thermus*.

Protocol

1. To optimize the IPMDH, a IPMDH mutant (Δ*leuB* mutant) of *T. thermophilus*, MT106, was generated by means of homologous recombination (Tamakoshi *et al.*, 1995).

2. A chimeric *leuB* gene consisting of parts of *T. thermophilus* and *B. subtilis leuB* genes was constructed (Numata et al., 1995). One µg of a 700 bp spanning 3′-region of this gene was dissolved in 500 µl of phosphate buffer (200 mM sodium phosphate (pH 6.0)), and incubated for 30 min at 37°C in the presence of the mutagen N-methyl-N′-nitro-N-nitrosoguanidine (MNNG, 0.2 mg ml^{-1}). MNNG was removed by dialysing twice against 500 ml of 10 mM Tris buffer (pH 7.6) containing 10 mM NaCl and 1 mM EDTA and the mutagenized 3′-region was ligated with the rest of the *leuB* gene prior to amplification in *E. coli*.
3. The randomly mutated *leuB* gene was inserted into the *Thermus* integration vector pIT1 (Tamakoshi et al., 1995) and transferred into *T. thermophilus* by natural transformation, and transformants were incubated at 76°C in 3 ml *Thermus* minimal medium (see media transformation protocol (Tanaka et al., 1981) for 48 h. Instead of pIT1 any of the *E. coli*/*Thermus* vectors developed should be applicable for this suppressor mutant strategy. Plating of the transformants on minimal medium and incubation at 76°C for 2 days should lead to the identification of suppressor mutants with improved thermostability.

◆◆◆◆◆ CONCLUDING REMARKS

The extraordinary trait of high competence of natural transformation and the availability of different gene transfer systems together with methodologies for gene expression in *Thermus* have facilitated the expression of several heterologous genes in *T. thermophilus* (Fridjonsson and Mattes, 2001; Fridjonsson et al., 2002; Koyama et al., 1990; Lasa et al., 1992b; Mather and Fee, 1992; Park et al., 2004; Takagi et al., 1999; Tamakoshi et al., 2001). The genetic systems of *Thermus* have been used for the improvement of thermozymes and overproduction of thermostable and thermoactive enzymes in *Thermus*. High-level overproduction in *Thermus* has already been achieved for several enzymes of biotechnological relevance, such as the *T. thermophilus* Mn-dependent catalase and its DNA polymerase (Hidalgo et al., 2004; Moreno et al., 2005). These applications demonstrate the high potential of *T. thermophilus* as an alternative cell factory for the optimization and overproduction of thermozymes with great biotechnological impact and will encourage the development of genetic tools for *Thermus* also in the future.

References

Adelberg, E. A., Mandel, M. and Chen, G. C. C. (1965). Optimal conditions for mutagenesis by N-methyl-N′-nitroso-N-nitroso-guanidine in *Escherichia coli* K12. *Biochem. Biophys. Res. Commun.* **18**, 788–795.

Averhoff, B. (2004). DNA transport and natural transformation in mesophilic and thermophilic bacteria. *J. Bioenerg. Biomembr.* **36**, 25–33.

Averhoff, B. and Friedrich, A. (2003). Type IV pili-related natural transformation systems: DNA transport in mesophilic and thermophilic bacteria. *Arch. Microbiol.* **180**, 385–393.

Begueret, J., Razanamparany, V., Perrot, M. and Barreau, C. (1984). Cloning gene *ura5* for the orotidylic acid pyrophosphorylase of the filamentous fungus *Podospora anserina*: transformation of protoplasts. *Gene* **32**, 487–492.

Brock, T. D. and Freeze, H. (1969). *Thermus aquaticus* gen. n. and sp. n., a nonsporulating extreme thermophile. *J. Bacteriol.* **98**, 289–297.

Brouns, S. J., Wu, H., Akerboom, J., Turnbull, A. P., de Vos, W. M. and van der Oost, J. (2005). Engineering a selectable marker for hyperthermophiles: Crystal structure of a thermostable bleomycin binding protein. *J. Biol. Chem.* **280**, 11422–11431.

de Grado, M., Lasa, I. and Berenguer, J. (1998). Characterization of a plasmid replicative origin from an extreme thermophile. *FEMS Microbiol. Lett.* **165**, 51–57.

de Grado, M., Castan, P. and Berenguer, J. (1999). A high-transformation-efficiency cloning vector for *Thermus thermophilus*. *Plasmid* **42**, 241–245.

Fridjonsson, O. and Mattes, R. (2001). Production of recombinant α-galactosidases in *Thermus thermophilus*. *Appl. Environ. Microbiol.* **67**, 4192–4198.

Fridjonsson, O., Watzlawick, H. and Mattes, R. (2002). Thermoadaptation of α-galactosidase AgaB1 in *Thermus thermophilus*. *J. Bacteriol.* **184**, 3385–3391.

Friedrich, A., Hartsch, T. and Averhoff, B. (2001). Natural transformation in mesophilic and thermophilic bacteria: identification and characterization of novel, closely related competence genes in *Acinetobacter* sp. strain BD413 and *Thermus thermophilus* HB27. *Appl. Environ. Microbiol.* **67**, 3140–3148.

Friedrich, A., Prust, C., Hartsch, T., Henne, A. and Averhoff, B. (2002). Molecular analyses of the natural transformation machinery and identification of pilus structures in the extremely thermophilic bacterium *Thermus thermophilus* strain HB27. *Appl. Environ. Microbiol.* **68**, 745–755.

Friedrich, A., Rumszauer, J., Henne, A. and Averhoff, B. (2003). Pilin-like proteins in the extremely thermophilic bacterium *Thermus thermophilus* HB27: implication in competence for natural transformation and links to type IV pilus biogenesis. *Appl. Environ. Microbiol.* **69**, 3695–3700.

Gruber, F., Visser, J., Kubicek, C. P. and de Graaff, L. H. (1990). The development of a heterologous transformation system for the cellulolytic fungus *Trichoderma reesei* based on a *pyrG*-negative mutant strain. *Curr. Genet.* **18**, 71–76.

Henne, A., Bruggemann, H., Raasch, C., Wiezer, A., Hartsch, T., Liesegang, H., Johann, A., Lienard, T., Gohl, O., Martinez-Arias, R., Jacobi, C., Starkuviene, V., Schlenczeck, S., Dencker, S., Huber, R., Klenk, H. P., Kramer, W., Merkl, R., Gottschalk, G. and Fritz, H. J. (2004). The genome sequence of the extreme thermophile *Thermus thermophilus*. *Nature Biotechnol.* **22**, 547–553.

Hidaka, Y., Hasegawa, M., Nakahara, T. and Hoshino, T. (1994). The entire population of *Thermus thermophilus* cells is always competent at any growth phase. *Biosci. Biotechnol. Biochem.* **58**, 1338–1339.

Hidalgo, A., Betancor, L., Moreno, R., Zafra, O., Cava, F., Fernandez-Lafuente, R., Guisan, J. M. and Berenguer, J. (2004). *Thermus thermophilus* as a cell factory for the production of a thermophilic Mn-dependent catalase which fails to be synthesized in an active form in *Escherichia coli*. *Appl. Environ. Microbiol.* **70**, 3839–3844.

Hoshino, T., Maseda, H. and Nakahara, T. (1993). Plasmid marker rescue transformation in *Thermus thermophilus*. *J. Ferment. Bioeng.* **76**, 276–279.

Kayser, K. J. and Kilbane, J. J. II (2001). New host-vector system for *Thermus* spp. based on the malate dehydrogenase gene. *J. Bacteriol.* **183**, 1792–1795.

Kayser, K. J., Kwak, J. H., Park, H. S. and Kilbane, J. J. II (2001). Inducible and constitutive expression using new plasmid and integrative expression vectors for *Thermus* sp. *Lett. Appl. Microbiol.* **32**, 412–418.

Kotsuka, T., Akanuma, S., Tomuro, M., Yamagishi, A. and Oshima, T. (1996). Further stabilization of 3-isopropylmalate dehydrogenase of an extreme thermophile, *Thermus thermophilus*, by a suppressor mutation method. *J. Bacteriol.* **178**, 723–727.

Koyama, Y. and Furukawa, K. (1990). Cloning and sequence analysis of tryptophan synthetase genes of an extreme thermophile, *Thermus thermophilus* HB27: plasmid transfer from replica-plated *Escherichia coli* recombinant colonies to competent *T. thermophilus* cells. *J. Bacteriol.* **172**, 3490–3495.

Koyama, Y., Hoshino, T., Tomizuka, N. and Furukawa, K. (1986). Genetic transformation of the extreme thermophile *Thermus thermophilus* and of other *Thermus* spp. *J. Bacteriol.* **166**, 338–340.

Koyama, Y., Okamoto, S. and Furukawa, K. (1990). Cloning of α- and β-galactosidase genes from an extreme thermophile, *Thermus* strain T2, and their expression in *Thermus thermophilus* HB27. *Appl. Environ. Microbiol.* **56**, 2251–2254.

Lasa, I., Caston, J. R., Fernandez-Herrero, L. A., de Pedro, M. A. and Berenguer, J. (1992a). Insertional mutagenesis in the extreme thermophilic eubacteria *Thermus thermophilus* HB8. *Mol. Microbiol.* **6**, 1555–1564.

Lasa, I., de Grado, M., de Pedro, M. A. and Berenguer, J. (1992b). Development of *Thermus-Escherichia* shuttle vectors and their use for expression of the *Clostridium thermocellum celA* gene in *Thermus thermophilus*. *J. Bacteriol.* **174**, 6424–6431.

Liao, H., McKenzie, T. and Hageman, R. (1986). Isolation of a thermostable enzyme variant by cloning and selection in a thermophile. *Proc. Natl. Acad. Sci. USA* **83**, 576–580.

Maseda, H. and Hoshino, T. (1995). Screening and analysis of DNA fragments that show promoter activities in *Thermus thermophilus*. *FEMS Microbiol. Lett.* **128**, 127–134.

Mather, M. W. and Fee, J. A. (1992). Development of plasmid cloning vectors for *Thermus thermophilus* HB8: expression of a heterologous, plasmid-borne kanamycin nucleotidyltransferase gene. *Appl. Environ. Microbiol.* **58**, 421–425.

Matsumura, M., Yasumura, S. and Aiba, S. (1986). Cumulative effect of intragenic amino-acid replacements on the thermostability of a protein. *Nature* **323**, 356–358.

Miller, J. H. (1972). Assay of β-galactosidase. In *Experiments in Molecular Genetics* (T. Platt, B. Miller-Hill and J. H. Miller eds), pp. 319–353. Cold Spring Harbor Laboratory Press, Cold Spring Harbor.

Moreno, R., Zafra, O., Cava, F. and Berenguer, J. (2003). Development of a gene expression vector for *Thermus thermophilus* based on the promoter of the respiratory nitrate reductase. *Plasmid* **49**, 2–8.

Moreno, R., Hidalgo, A., Cava, F., Fernandez-Lafuente, R., Guisan, J. M. and Berenguer, J. (2004). Use of an antisense RNA strategy to investigate the functional significance of Mn-catalase in the extreme thermophile *Thermus thermophilus*. *J. Bacteriol.* **186**, 7804–7806.

Moreno, R., Haro, A., Castellanos, A. and Berenguer, J. (2005). High-level overproduction of His-tagged Tth DNA polymerase in *Thermus thermophilus*. *Appl. Environ. Microbiol.* **71**, 591–593.

Neuhard, J., Stauning, E. and Kelln, R. A. (1985). Cloning and characterization of the *pyrE* gene and of PyrE::Mud1 (Ap *lac*) fusions from *Salmonella typhimurium*. *Eur. J. Biochem.* **146**, 597–603.

Niehaus, F., Bertoldo, C., Kahler, M. and Antranikian, G. (1999). Extremophiles as a source of novel enzymes for industrial application. *Appl. Microbiol. Biotechnol.* **51**, 711–729.

Numata, K., Muro, M., Akutsu, N., Nosoh, Y., Yamagishi, A. and Oshima, T. (1995). Thermal stability of chimeric isopropylmalate dehydrogenase genes constructed from a thermophile and a mesophile. *Protein Eng.* **8**, 39–43.

Oshima, T. and Imahori, K. (1974). Description of *Thermus thermophilus* (Yoshida and Oshima) comb. nov., a nonsporulating thermophilic bacterium from a Japanese thermal spa. *Int. J. Syst. Bacteriol.* **24**, 102–112.

Pantazaki, A. A., Pritsa, A. A. and Kyriakidis, D. A. (2002). Biotechnologically relevant enzymes from *Thermus thermophilus*. *Appl. Microbiol. Biotechnol.* **58**, 1–12.

Park, H. S. and Kilbane, J. J. II (2004). Gene expression studies of *Thermus thermophilus* promoters P*dnaK*, P*arg* and P*scs-mdh*. *Lett. Appl. Microbiol.* **38**, 415–422.

Park, H. S., Kayser, K. J., Kwak, J. H. and Kilbane, J. J. II (2004). Heterologous gene expression in *Thermus thermophilus*: β-galactosidase, dibenzothiophene monooxygenase, PNB carboxy esterase, 2-aminobiphenyl-2,3-diol dioxygenase, and chloramphenicol acetyl transferase. *J. Ind. Microbiol. Biotechnol.* **31**, 189–197.

Ramaley, R. F. and Hixson, J. (1970). Isolation of a nonpigmented, thermophilic bacterium similar to *Thermus aquaticus*. *J. Bacteriol.* **103**, 527–528.

Ramírez-Arcos, S., Fernández-Herrero, L. A., Marín, I. and Berenguer, J. (1998). Anaerobic growth, a property horizontally transferred by an Hfr-like mechanism among extreme thermophiles. *J. Bacteriol.* **180**, 3137–3143.

Raven, N. D. H. (1995). Genetics of Thermus. In *Thermus Species* (R. J. Sharp and R. A. Williams eds), pp. 157–184. Plenum Press, New York.

Sadaie, Y., Burtis, K. C. and Doi, R. H. (1980). Purification and characterization of a kanamycin nucleotidyltransferase from plasmid pUB110-carrying cells of *Bacillus subtilis*. *J. Bacteriol.* **141**, 1178–1182.

Sambrook, J., Fritsch, E. F. and Maniatis, T. (1989). *Molecular Cloning: A Laboratory Manual*. Cold Spring Harbor Laboratory Press, Cold Spring Harbor.

Sen, S. and Oriel, P. (1990). Transfer of transposon Tn916 from *Bacillus subtilis* to *Thermus aquaticus*. *FEMS Microbiol. Lett.* **55**, 131–134.

Silva, Z., Sampaio, M. M., Henne, A., Bohm, A., Gutzat, R., Boos, W., da Costa, M. S. and Santos, H. (2005). The high-affinity maltose/trehalose ABC transporter in the extremely thermophilic bacterium *Thermus thermophilus* HB27 also recognizes sucrose and palatinose. *J. Bacteriol.* **187**, 1210–1218.

Takagi, H., Suzumura, A., Hoshino, T. and Nakamori, S. (1999). Gene expression of *Bacillus subtilis* subtilisin E in *Thermus thermophilus*. *J. Ind. Microbiol. Biotechnol.* **23**, 214–217.

Tamakoshi, M., Yamagishi, A. and Oshima, T. (1995). Screening of stable proteins in an extreme thermophile, *Thermus thermophilus*. *Mol. Microbiol.* **16**, 1031–1036.

Tamakoshi, M., Uchida, M., Tanabe, K., Fukuyama, S., Yamagishi, A. and Oshima, T. (1997). A new *Thermus-Escherichia coli* shuttle integration vector system. *J. Bacteriol.* **179**, 4811–4814.

Tamakoshi, M., Yaoi, T., Oshima, T. and Yamagishi, A. (1999). An efficient gene replacement and deletion system for an extreme thermophile, *Thermus thermophilus*. *FEMS Microbiol. Lett.* **173**, 431–437.

Tamakoshi, M., Nakano, Y., Kakizawa, S., Yamagishi, A. and Oshima, T. (2001). Selection of stabilized 3-isopropylmalate dehydrogenase of *Saccharomyces cerevisiae* using the host-vector system of an extreme thermophile, *Thermus thermophilus*. *Extremophiles* **5**, 17–22.

Tanaka, T., Kawano, N. and Oshima, T. (1981). Cloning of 3-isopropylmalate dehydrogenase gene of an extreme thermophile and partial purification of the gene product. *J. Biochem. (Tokyo)* **89**, 677–682.

Wayne, J. and Xu, S. Y. (1997). Identification of a thermophilic plasmid origin and its cloning within a new *Thermus-E. coli* shuttle vector. *Gene* **195**, 321–328.

Williams, R. A. D. and da Costa, M. S. (2001). The genus *Thermus* and related microorganisms. In *The Prokaryotes: An Evolving Electronic Resource for the Microbiological Community* (Dworkin, M. *et al.*, eds), 3rd, Release 3.7. Springer-Verlag, New York (http://link.springer-ny.com/link/service/books/10125/).

Wimberly, B. T., Brodersen, D. E., Clemons, W. M. Jr., Morgan-Warren, R. J., Carter, A. P., Vonrhein, C., Hartsch, T. and Ramakrishnan, V. (2000). Structure of the 30S ribosomal subunit. *Nature* **407**, 327–339.

Yamagishi, A., Tanimoto, T., Suzuki, T. and Oshima, T. (1996). Pyrimidine biosynthesis genes (*pyrE* and *pyrF*) of an extreme thermophile, *Thermus thermophilus*. *Appl. Environ. Microbiol.* **62**, 2191–2194.

13 Gene Transfer Systems for Obligately Anaerobic Thermophilic Bacteria

Michael V Tyurin[1], Lee R Lynd[1] and Juergen Wiegel[2]

[1] Thayer School of Engineering, Dartmouth College, 8000 Cummings Hall, Hanover, NH 03755, USA;
[2] University of Georgia, 211–215 Biological Sciences Bldg., 1000 Cedar Street, Athens, GA 30602, USA

CONTENTS

Introduction
Availability of sequence information
Exo- and endonuclease activities of thermophilic anaerobes
Replicative shuttle vectors
Electrotransformation protocols
Concluding remarks

♦♦♦♦♦ INTRODUCTION

Among obligately anaerobic thermophilic bacteria, advances relevant to gene transfer methods over the last 10 years have been reported for three Gram-positive species that ferment carbohydrates to ethanol, acetate, lactate, H_2 and CO_2: *Clostridium thermocellum*, *Thermoanaerobacterium saccharolyticum* strain JW/SL-YS485 and *Thermoanaerobacterium thermosaccharolyticum*. Available methods for introducing foreign genes into these organisms are based on electrotransformation (ET). This chapter focuses on methodological aspects of developing gene transfer methods with an emphasis on adapting microbiology protocols for use with these thermophilic anaerobes. In particular, we report gene transfer protocols developed for *C. thermocellum*, *T. saccharolyticum* and *T. thermosaccharolyticum*, and also summarize information relevant to developing such protocols for the hyperthermophilic species *Thermotoga maritima* and *Thermotoga neapolitana*.

C. thermocellum (Cato et al., 1986) is a thermophilic, anaerobic, cellulolytic bacterium that grows on soluble β-glucans, including cellobiose and cellodextrins, as well as on cellulosic substrates, including Avicel, filter paper, solka floc and pretreated mixed hardwood. *C. thermocellum* exhibits one of the highest growth rates on cellulose among described microorganisms (Lynd et al., 2002). The cellulase complex produced by

C. thermocellum, the cellulosome (Bayer *et al.*, 1983), exhibits high activity against crystalline cellulose (Kruus *et al.*, 1995). A non-cellulosomal, highly complex endo-β-1,3-glucanase bound to the outer cell surface of *C. thermocellum* has also been described (Fuchs *et al.*, 2003). *T. saccharolyticum* (Lee *et al.*, 1993) and *T. thermosaccharolyticum* (formerly *Clostridium thermosaccharolyticum*) (Cato *et al.*, 1986) are non-cellulolytic thermophilic, anaerobic bacteria that grow on xylan and a wide range of sugars including glucose. Species of the genus *Thermoanaerobacterium* possess a complex hemicellulolytic system, which allows rapid utilization of sugars derived from hemicellulose in the absence of added saccharolytic enzymes (Liu *et al.*, 1996a,b; Lynd *et al.*, 2001). In high temperature environments where cellulose degradation occurs, *C. thermocellum* is closely associated with *T. saccharolyticum* and *T. thermosaccharolyticum* (Lynd *et al.*, 2002; Mori, 1995; Ng *et al.*, 1981). Co-cultures of *C. thermocellum* with non-cellulolytic, pentose-utilizing organisms such as *T. saccharolyticum*, *T. thermosaccharolyticum*, *Thermoanaerobacter ethanolicus* JW200 and *Moorella thermoacetica* have been suggested for biotechnological processes converting cellulosic biomass into ethanol or other products (Das and Singh, 2004; Freier *et al.*, 1988; Lynd *et al.*, 2002; Saddler and Chan, 1984; Slapack *et al.*, 1987; Wiegel, 1980, 1982a,b; Freier, Mothershed and Wiegel, unpublished data).

The genus *Thermotoga*, currently including 9 species, belongs to the eubacterial domain and represents the hyperthermophilic anaerobic Eubacteria. A major morphological feature of this genus is the presence of an extracellular carbohydrate-rich mechanical barrier, termed a toga, which appears to represent a challenge for genetic manipulation of intact cells. A detailed description of the genus *Thermotoga* may be found elsewhere (http://biology.kenyon.edu/Microbial_Biorealm/bacteria/thermotoga_petrotoga/thermotoga.htm).

◆◆◆◆◆ AVAILABILITY OF SEQUENCE INFORMATION

The *C. thermocellum* genome sequence is available at (http://www.ncbi.nlm.nih.gov/entrez/query.fcgi?db=nucleotide&cmd=search&term=C.+thermocellum). A large number of individual genes associated with the cellulase system of *C. thermocellum* have been cloned, sequenced, and characterized to various degrees as addressed in a recent review (Schwarz *et al.*, 2004). Additional cloned genes include L-lactate dehydrogenase (Ozkan *et al.*, 2004) and acetate kinase/phosphotransacetylase (AF041841: http://www.ncbi.nlm.nih.gov/entrez/viewer.fcgi?db=nucleotide&val=2791837).

Fewer genes have been cloned in *T. saccharolyticum* or *T. thermosaccharolyticum* as compared to *C. thermocellum*, and the genome sequence is not available for either *Thermoanaerobacterium* species. Cloned genes of *T. saccharolyticum* include acetyl xylan esterase, β-xylosidase (Lorenz and Wiegel, 1997), xylose isomerase (Liu *et al.*, 1996a), and endoxylanase

(Liu et al., 1996b), while genes cloned in T. thermosaccharolyticum include glucoamylase (Ducki et al., 1998), and xylose isomerase (Meaden et al., 1994). The genome of Tt. maritima MSB8 is a single circular chromosome consisting of 1 860 725 bp (average G + C content is 46%). About 55% of the 1877 predicted coding regions have functional assignments. Although the core of the Tt. maritima genome may be eubacterial, almost one quarter of the genome is archaeal in nature (http://www.tigr.org/tigr-scripts/CMR2/infxview?db = omnium&asmbl_id = 487&startbp = 1 &stopbp = 100000&width = 1000&height = 300).

◆◆◆◆◆ EXO- AND ENDONUCLEASE ACTIVITIES OF THERMOPHILIC ANAEROBES

Many bacteria produce one or more type II endonucleases that cleave DNA at specified sequences. Methylation of sequences that would otherwise be recognized by restriction systems protects DNA from attack by its own restriction enzymes. Nuclease activity may vary significantly within the same species, depending on the strain. Thus, for example, C. acetobutylicum strain NCIMB 8052 does not exhibit such activity (Oultram et al., 1988), while at least one strong endonuclease system was identified in strain ATCC 824 (Lee et al., 1992; Mermelstein and Papoutsakis, 1993). An MboI-like, 3'-GATC-5'-recognizing restriction activity was identified in C. thermocellum ATCC 27405, as well as some additional isolates identified as C. thermocellum (Klapatch et al., 1996a Ozkan et al., 2001). In C. thermocellum ATCC 27405, isoschizomer activities of BclI-like (TGATCA) and Eco RII-like type (CC(A/T)GG) have also been reported (Young et al., 1989). There are indications that restriction endonucleases CthI and CthII are produced by C. thermocellum ATCC 27405 (http://www.dsmz.de/strains/no001237.htm). In vivo methylation in Dam$^+$ E. coli cells protects DNA from the MboI-like, 3'-GATC-5'-recognizing restriction activity (Klapatch et al., 1996a). Cell extracts of T. thermosaccharolyticum HG-8 (ATCC 31960) did not exhibit restriction enzymatic activity under the conditions examined (Klapatch et al., 1996a). Increased ET efficiency by two orders of magnitude was reported when transforming DNA of plasmid pCTC1 was isolated from cells of T. thermosaccharolyticum compared with the DNA isolated from E. coli cells (Klapatch et al., 1996b), which could be the result of nuclease activity but could also be explained by other factors. There has been no systematic study of exo- or endonuclease activity in T. saccharolyticum JW/SL-YS 485.

◆◆◆◆◆ REPLICATIVE SHUTTLE VECTORS

Several well-known shuttle vectors have been successfully used to transform obligately anaerobic thermophiles (Table 13.1). All these vectors feature antibiotic resistance genes originating from mesophilic Gram-positive bacteria. Vector pIKM1 (Mai et al., 1997), based on pUC19,

Table 13.1 Some replicative vectors used for genetic transformation of obligately anaerobic thermophiles, their antibiotic-resistance markers and recipients/hosts

Plasmid	Recipient	Selective markers in Gram-positive recipient	Reference
pIKM1	C. thermocellum T. saccharolyticum	Em, Lm Km	Desai et al. (2004), Mai et al. (1997), Tyurin et al. (2004)
pCTC1	T. thermosaccharolyticum C. acetobutylicum C. cellulolyticum	Em	Klapatch et al. (1996b), Williams et al. (1990)
pUB110 (57)	Tb. pseudoethanolicus strain 39E (basonym Tb. ethanolicus 39E; C. thermohydrosulfuricum Bacillus subtilis	Neo (Km)	Jennert et al. (2000), Tardif et al. (2001), Soutschek-Bauer et al. (1986), Onyenwoke and Wiegel (2006)
pHV33	Staphylococcus aureus C. thermocellum (protoplasts)	Cm	Sadaie et al. (1980), Tsoi et al. (1987), Zyprian and Matzura (1986)

Em – Erythromycin; Km – Kanamycin; LM – Lincomycin; Neo – Neomycin; Cm – Chloramphenicol.

contains the *mls* gene from pIM13 (Monod *et al.*, 1986) and the thermostable kanamycin-resistance cassette from plasmid pKD102 found in *Streptococcus faecalis* (a ruminant mesophile capable of growth at 45°C) (Trieu-Cuot and Courvalin, 1983). Suicide vectors for gene knockout were developed based on pIKM1. Non-specific integration of the kanamycin-resistance gene was observed. In addition, a vector featuring the kanamycin-resistance gene flanked by parts of cellobiohydrolase gene (*cbhA*) from *C. thermocellum* JW20 was used to obtain integration into the xylanase gene (*xynA*) of the *Thermoanaerobacterium* chromosome *via* homologous recombination (Mai and Wiegel, 1999, 2000). MLS resistance of pIKM1 may be used for selection of transformants of *C. thermocellum* (Tyurin *et al.*, 2004), while resistance to kanamycin is effective for selection of transformants of *T. saccharolyticum* JW/SL-YS 485 (Desai *et al.*, 2004; Mai *et al.*, 1997). When pIKM1 was used for ET of *T. saccharolyticum* JW/SL-YS 485, we were able to select transformants in the presence of kanamycin, but not erythromycin and lincomycin, which suggests absence of *mls* (pIKM1) expression in *T. saccharolyticum* JW/SL-YS 485. We were unable to use kanamycin resistance to select transformants of *C. thermocellum* ATCC 27405 due to a high level of natural resistance of that bacterium to kanamycin (>1 μg ml^{-1}) (Tyurin, unpublished data). Another shuttle vector, pCTC1 (Williams *et al.*, 1990), has an *erm*(AM) gene from pAMβ-1 (Brehm *et al.*, 1987), *oriT* from RP4 to mobilize the vector in a mob-dependent manner in conjugation experiments (Tyurin and Livshits, 1995), with pUC19 as a backbone of the vector. pCTC1 was used for ET of mesophilic clostridia (Tardif *et al.*, 2001; Williams *et al.*, 1990), and *T. thermosaccharolyticum* ATCC 31960 (Klapatch *et al.*, 1996b). pUB110, a rolling-circle replicating plasmid (McKenzie *et al.*, 1987), with a gene encoding kanamycin nucleotidyltransferase (Sadaie *et al.*, 1980), was originally isolated from *Staphylococcus aureus* (Zyprian and Matzura. 1986) and used for genetic transformation of *T. thermohydrosulfuricus* (Soutschek-Bauer *et al.*, 1986). Shuttle vector pHV33 is a hybrid (Dagert *et al.*, 1984), composed of *Staphylococcus aureus* plasmid pC194, replicating *via* a rolling circle mechanism (Horinouchi and Weisblum, 1982), and pBR322. pHV33 was reported successfully transforming protoplasts of *C. thermocellum* (Tsoi *et al.*, 1987). We observe that the plasmids used successfully to transform obligately anaerobic thermophiles utilize both rolling circle and theta replication mechanisms.

◆◆◆◆◆ ELECTROTRANSFORMATION PROTOCOLS

Most gene transfer protocols for Gram-positive bacteria developed in the last decade, and all such protocols for the organisms addressed in this chapter are based on ET. Optimization of ET conditions must generally be undertaken for each particular strain. Such conditions often vary substantially with respect to ET efficiency under a particular set of conditions. Biological variables that can impact ET efficiency include growth medium, conditions used for subculturing prior to pulse application, the temperature at which various manipulations (including

ET and washing) are carried out, and handling (including duration of incubation times) of washed cells prior to, during and after electric pulse application. Electrical variables that can impact ET efficiency include field strength and pulse shape, duration, and number of pulses (Tyurin and Livshits, 1994).

ET protocols are established for *T. saccharolyticum* JW/SL YS485 and *T. thermosaccharolyticum* HG8 (Klapatch et al., 1996b; Mai and Wiegel, 1999; Mai et al., 1997) These protocols use standard electroporation equipment, with most cell manipulations performed under anaerobic conditions but electroporation performed on the bench in the presence of oxygen. An electrotransformation protocol for *C. thermocellum* (strains ATCC 27405 and DSM 1313) has been reported using a custom-built pulse generator and ET cuvettes (Tyurin et al., 2004), which has also been applied successfully to *T. saccharolyticum* JW/SL YS485 (Tyurin et al., 2005). As may be seen in Table 13.2, higher transformation efficiencies have been achieved in recent studies as compared to earlier studies.

Electrotransformation of *T. thermosaccharolyticum*

Cells of *T. thermosaccharolyticum* HG-8 are grown to $OD_{600} \sim 1$, harvested in a cold centrifuge, and washed three times with cold de-ionized water. Then resulting pellet is resuspended in 20% glycerol (ET medium), which may be used immediately or stored at $-80°C$. Under optimal experimental conditions, an exponential electric pulse applied to cell sample in 0.2 cm ET cuvettes using a Bio-Rad Gene Pulser set at 2.0 kV, 800 Ω and 25 µF resulted in a time constant of about 16 ms. ET efficiencies obtained using this protocol were less than 1 transformant $µg^{-1}$ if the transforming DNA was isolated from *E. coli* cells, and 52 transformants $µg^{-1}$ if the DNA was isolated from cells of *T. thermosaccharolyticum* (Klapatch et al., 1996b).

Electrotransformation of *T. saccharolyticum* JW/SL-YS 485

Isonicotinic acid hydrazide (INH, isoniacin), which increased electrocompetence of some *Mycobacterium* cells (Hermans et al., 1990), was used to make cells of *T. saccharolyticum* JW/SL-YS 485 electrocompetent (Mai et al., 1997). It was observed that subculturing in the presence of isoniacin leads to activation of natural autolytic capabilities of this bacterium to the extent that the cells do not undergo spontaneous lysis yet but become highly electrocompetent. In brief, cells of *T. saccharolyticum* are grown in the presence of 4 µg ml^{-1} of isoniacin to OD_{600} 0.6–0.8, then washed twice with cold de-ionized water, concentrated in 0.272 M sucrose (ET medium) and incubated at 48°C until visible signs of autoplasting (that is, protoplasting without addition of muralytic enzymes; Peteranderl et al., 1992) appear. Then 0.1-ml samples mixed with transforming DNA are subjected to single exponential pulses using a Bio-Rad Gene Pulser with settings of 1.25 kV, 400 Ω, 25 µF in pre-chilled

Table 13.2 Electrotransformation efficiencies described for anaerobic thermophiles

Organism	Plasmid	Apparatus/Conditions	ET efficiency ($\mu g\ ml^{-1}$)	Reference
T. saccharolyticum JW/SL YS485	pIKM1	Bio-Rad Gene Pulser, Exponential pulse, 12.5 kV cm^{-1}, 4–8 ms	10^3	Mai and Wiegel (1999), Mai et al. (1997),
T. saccharolyticum JW/SL YS485	pIKM1	Custom-made, Square gate pulse, 25 kV cm^{-1}, 10 ms	7×10^5	Tyurin et al. (2005)
C. thermocellum DSM1313	pIKM1	Custom-made, Square gate pulse, 25 kV cm^{-1}, 10 ms	2×10^5	Tyurin et al. (2004)
ATCC 27405 DSM 4150 & 7072			5×10^4 both ~10^3	Tyurin et al. (2004) Tyurin et al. (2004)
C. thermocellum DSM1313 ATCC 27405	pIKM1	Custom-made, Square gate pulse, 25 kV cm^{-1}, 10 ms	5×10^5 7.5×10^4	Tyurin et al. (2005)
T. thermosaccharolyticum	pCTC1	Bio-Rad Gene Pulser, Exponential pulse, 10 kV cm^{-1}, 16 ms	0.5×10^2	Klapatch et al. (1996b)
T. thermosaccharolyticum	pCTC1	Custom-made, Square gate pulse 25 kV cm^{-1}, 10 ms	10^2	Tyurin et al. (2005)

0.1 cm ET cuvettes. The treated samples are transferred to 5 ml of non-selective broth and incubated for 4 h at 48°C prior to plating on selective 1% agar for subsequent incubation of plates in anaerobic jars. Reported ET efficiencies were 10^1–10^3 transformants μg^{-1} (Desai et al., 2004; Mai et al., 1997). This protocol was used to obtain knockout recombinants of xylanase gene (*xynA*) (Mai and Wiegel, 2000) and L-lactate dehydrogenase (*ldh*) in *T. saccharolyticum* JW/SL-YS485 (Desai et al., 2004).

A New High-efficiency Electrotransformation Protocol with Application to *C. thermocellum* and *T. saccharolyticum*

We developed a protocol to efficiently electrotransform *C. thermocellum* with plasmid pIKM1 (Tyurin et al., 2004). The protocol utilizes a custom-made generator where the cell sample is connected in series with the power capacitor and electronic key (power-modulating tetrode), while cell samples are treated in disposable 2.0-ml polypropylene tubes with custom-made reusable electrodes (2-mm inter-electrode gap) (Tyurin, 1992). Use of tubes as ET cuvettes allows us to conduct ET while the samples are imbedded in an ice block prepared prior to experiments using similar disposable tubes to form shaped wells. ET efficiencies achieved for *C. thermocellum* strain ATCC 27405 were $\sim 7.5 \times 10^4$, and for strain DSM 1313 $\sim 5 \times 10^5$ $\mu g\ ml^{-1}$ (Tyurin et al., 2005). We have successfully applied the same protocol of ET for *T. saccharolyticum* JW/SL-YS485 with pIKM1 ($\sim 7 \times 10^5$ $\mu g\ ml^{-1}$, selection in the presence of 200 $\mu g\ ml^{-1}$ kanamycin) (Tyurin et al., unpublished data) as well as suicide vectors (J. Shaw, personal communication). The same protocol applied to *T. thermosaccharolyticum* resulted in an ET efficiency of about 100 transformants μg^{-1} of pCTC1 isolated from *E. coli* cells. An updated version of the protocol is given below:

- Inoculate frozen ($-80°C$) cells grown in modified DSM 122 liquid medium with cellobiose (http://www.dsmz.de/media/med122.htm) into the same medium and grow overnight. Inoculate 10 ml of DSM 122 broth with 0.1 ml of that culture, incubate for 7–9 h at 58°C, then add 20 ml of DSM 122 cellobiose liquid medium and continue incubation for 8–9 h.
- Pour this culture (or half of it if the OD_{600} is above 0.9–1.0) into a 500 ml bottle with magnetic stirring bar and 250 ml of warm (58°C) DSM 122 broth containing isoniacin (9–10 $\mu g\ ml^{-1}$ for *T. saccharolyticum* JW/SL-YS485, or *T. thermosaccharolyticum* ATCC 31960, and 15–20 $\mu g\ ml^{-1}$ for *C. thermocellum* ATCC 27405 or DSM 1313), and continue incubation at 58°C for 2.5–3 h with the magnetic stirrer set at ~ 150 rpm.
- Pour 14.5 ml of the resulting culture into each of twelve 15-ml disposable polypropylene centrifuge tubes, and chill on ice for 10 min. Centrifuge for 12 min and discard supernatants. Keep tubes with pellets on ice for 3 min. Add 5 ml of ice-cold 0.2 M cellobiose to each tube and re-suspend cells with a Vortex mixer. Combine content of two tubes into one and add 0.2 M cellobiose to final volume 14.5 ml and mix by inverting tubes. Leave on ice for 8 min.

- Harvest cells in a Fisher model 225 centrifuge at a maximal speed for 14 min, and discard supernatant. Resuspend and combine pellets as above. Leave on ice for 4 min and centrifuge again for 15 min. Discard supernatants, and resuspend pellets in 2 ml of 0.2 M cellobiose. Combine suspensions in one tube, leave them on ice for 6 min, and then centrifuge for 15 min. Re-suspend the resulting pellet in 1 ml of 0.2 M cellobiose and keep the suspension on ice until used.
- Distribute 90–110 μl of the suspension into 2.0 ml sterile disposable polypropylene tubes and place prepared samples into ice wells. Add 10–20 μl of DNA (1–5 μg for replicative plasmids, and 5–10 μg for suicidal plasmids), mix by tapping the tubes, insert electrodes and return the assembled sample into an ice well just prior to application of electric pulse.
- Apply single-square-gated 10 ms pulse at 5 kV to the sample, and dilute the sample with 1 ml of ice-cold DSM 122 broth. Place the sample back in the ice well for 10 min.
- Transfer samples to 50-ml polypropylene tubes, and add 5–6 ml (up to 10–12 ml for knockout ET) of cold DSM 122 broth to each sample. Incubate at 58°C for 3–5 h.
- To each tube, add melted warm (58°C) semi-solid DSM 122 agar with respective antibiotic(s), or without antibiotics (for determining cell viability) to "50 ml" mark, and pour into Petri dishes (see Plating section) with dilution as necessary. Incubate for 3–5 (replicative vector) or 5–7 (suicidal vector) days in an anaerobic jar at 58°C (see Plating section).
- Use disposable 200-μl tips and a Pipetman set at 100 μl to aspirate colonies grown inside the selective agar. Transfer aspirated colonies into 10 ml of selective DSM 122 broth in 15 ml disposable polypropylene tubes and incubate 24–48 h at 58°C. Pellet cells for extraction of total DNA after removal of 0.5 ml aliquot for depositing at −80°C.

Notes: Modified DSM 122 is as described elsewhere (http://www.dsmz.de/media/med122.htm) except that the amount of $K_2HPO_4 \cdot 3H_2O$ is reduced to 1.8 g l^{-1}, and glutathione is replaced by 0.3 g l^{-1} of cysteine-HCl. Cleaning of reusable electrodes is performed by washing in water, then wiping using a sterile napkin, polishing the surfaces with jewelry polish, and submerging the whole electrode assembly (the electrodes imbedded into epoxy resin) to dehydrated ethanol for 2–3 min with subsequent drying before reuse.

Plating of Anaerobic Saccharolytic Thermophiles

Plating of bacteria is important to obtain single colonies for evaluation of culture purity, selection of recombinants, or determining viable cell numbers after subjecting cell samples to different manipulations. Use of agar-containing media requires measures to prevent dehydration of agar gels at thermophilic incubation temperatures. Evaporation of water changes the medium composition, which is critical to achieve good plating efficiencies. However, evaporation of water is not the only reason for low plating efficiency. An almost 1000-fold increase in plating efficiency was observed when cells of mesophilic *C. acetobutylicum* ATCC 824 were imbedded into a semi-solid (0.45%) nutrient agar by mixing the cultures with a melted (45°C) medium instead of plating the cell on the surface of the medium (Tyurin *et al.*, 2000). Plating efficiencies for obligately anaerobic thermophiles have not been reported in detail, but our experience suggests that low efficiencies are common. We have had considerable success with a plating protocol featuring saccharolytic thermophiles grown imbedded in agar, rather than on the agar surface. Using this protocol, however, we have observed an approximately 1000-fold decrease in the yield of *C. thermocellum* ATCC 27405 as the agar concentration was varied from 0.5 to 2.0% (Tyurin, unpublished data). Sensitivity of plating efficiency to agar concentration has also been observed for *T. saccharolyticum* JW/SL-YS 485, *T. thermohydrosulfuricus* JW 102, and *Moorella thermoacetica*, for which the use of about 0.7% agar resulted in higher plating efficiencies while at the same time minimized smearing of colonies through the syneresis water (Mai and Wiegel, unpublished data). A plating protocol for *C. thermocellum* is presented below:

- Add warm (58°C) semi-solid (0.55–0.58% agar) nutrient agar to a 50-ml disposable polypropylene tube with a sample or its dilution to "50 ml" mark, tighten the cap and mix by inverting. The agar may include selective agent(s) if selection of recombinants is an objective.
- Pour two 100-mm Petri dishes (25 ml into each), cover the dishes and let the agar solidify for about 30–35 min on a metal plate placed in the anaerobic chamber.
- Put the dishes upside down in anaerobic jar, containing ~10–15 ml of sterile distilled H_2O, tighten the cover and place to incubator set at 58–60°C.

Note: Disposable Petri dishes and polypropylene tubes with loosened caps need to be in the chamber at least 10 h prior to plating. Bottles of hot just autoclaved semi-solid agar need to in the chamber for >4 h at 58–60°C prior to plating. If needed, add antibiotic to warm agar immediately prior to mixing with cells. Keep the cap of hot agar bottle tight when entering the chamber *via* the airlock, and then loosen the cap inside the chamber to assure proper reduction of media. Keep work area, anaerobic jars and rubber gloves clean by periodic wiping with autoclaved napkins soaked in dehydrated 2-propanol.

Selection of Recombinant Anaerobic Thermophiles using Antibiotics

Several antibiotics are widely used for selection of recombinants of Gram-positive bacteria, including aminoglycosides (kanamycin, neomycin), chloramphenicol and its derivative thiamphenicol, erythromycin, lincomycin and tetracycline. We know of no reports using vancomycin, cycloserin, hygromycin, or spectinomycin for applications under thermophilic conditions. Stability, solubility and bio-availability of antibiotics depend on pH, temperature, medium composition, exposure to light, etc. Almost all the above antibiotics, except aminoglycosides and erythromycin, have been found to be unstable after 24–48 h at pH 7.3 in one of the known media for clostridia when the incubation temperature was 50–70°C (Peteranderl et al., 1990; Wu and Welker, 1989). Minimal concentrations of selective antibiotics need to be established for each particular strain in the particular medium in which selection is to be carried out. A protocol for finding the minimal concentration of selective antibiotic follows:

Determination of MIC for selective antibiotic(s)

- Select the medium. Grow selected recipient strain in this liquid medium overnight and dilute the culture to $\sim 10^8$ CFU ml^{-1}.
- Have semi-solid agar (see Plating section) ready for inoculation and mix it with the culture from step 1 in the ratio 1: 500 to reduce the CFU to $\sim 2 \times 10^5$ ml^{-1}.
- Pour 10 ml of inoculated semi-solid agar medium into 15-ml disposable polypropylene tubes and add amounts of a stock solution of the antibiotic to be tested to make a set of tubes with final antibiotic concentrations increasing by 2-fold with each successive tube (e.g. final concentrations of 5, 10, 20 µg ml^{-1}, etc.).
- Let the agar solidify, and incubate the tubes for up to 9 days, recording the day and number of colonies for each antibiotic concentration when colonies are evident.
- Note the minimal concentration of antibiotic where no colonies appear within 9 days.
- Repeat the above procedure stepping back from concentration of the antibiotic identified in the step 5 with 1–5 µg ml^{-1} increment, depending on the highest antibiotic concentration to be found inhibitory in the previous step. Note the new minimal concentration where still no colonies are seen. This is the working concentration for the chosen combination of antibiotic and growth medium for a particular strain. The incubation time should be less than the time frame identified above, if selection of recombinants is performed.

Note: Make stock antibiotic solutions with the recommended solvent and store at the recommended temperature. Do not store your stock solutions in the anaerobic chamber. Use of stock solutions at 100 mg ml^{-1} minimizes addition of oxygen into media when adding the stock solution of antibiotic.

Discrimination between transferred antibiotic resistance and physico-chemical degradation of antibiotics

In some instances, it is important to assess whether an extremophile is resistant to a given antibiotic or whether the growth with antibiotic (especially those with bacteriostatic mode of action) is attributable to the physical and/or chemical instability of the antibiotic. Antibiotic stability can be determined in liquid culture aliquots using a recommended mesophilic test organism. It is beneficial to test aliquots of both sterile medium and inoculated medium, each containing the antibiotic to be evaluated. Tests with inoculated media allow the extent of enzymatic degradation of the antibiotic by the inoculum to be evaluated (Peteranderl et al., 1990). The most comprehensive available study of the stability of antibiotics under thermophilic conditions, by Peteranderl et al. (1990), involved tests at acidic pH values over a temperature range of 50–72°C. As a result, fewer data are available for alkaline conditions and/or temperatures in excess of 72°C. The stability of a few antibiotics has been determined for *Thermotoga* growth conditions at 77°C and incubations times of up to 48 h (Yu et al., 2001). Relatively stable antibiotics were found to be carbenicillin with apparent MIC (given in μg ml^{-1}) increasing after 48 h from 3.9 to 15.6, chloramphenicol from 15.6 to 62.5, hygromycin from 62.5 to 250 and kanamycin from 3.9 to 62.5. Penicillin and thiostrepton were much less stable (Yu et al., 2001). The protocol used by Peteranderl et al. (1990) involved making 2- to 10-fold dilutions (depending on the objective and the desired sensitivity) of culture aliquots (both sterile and inoculated, see above) containing the antibiotic to be tested and appropriate growth medium. The active antibiotic concentration was evaluated using standard method involving mesophilic test organisms. It may be noted that use of sterile 96-well microplates and a multipipettor facilitates analysis of multiple samples. BBL^{TM} Mueller Hinton Broth is the recommended medium to perform such tests. This medium is inoculated with an exponentially growing culture of a certain test-microorganism at a final cell concentration of $1-5 \times 10^5$ ml^{-1}. The recommended test strains are *Escherichia coli* ATCC 25922, *Staphylococcus aureus* ATCC 25923 and *Bacillus megaterium* as described by Thornsberry (1985). After the incubation for 48 h at 35–37°C, growth is determined by observed turbidity using a 96-well microplate reader. When a few microplates are used, there is no need to use a microplate reader. Cells of the test bacterium form visible pellets in the wells, for wells in which growth of the mesophilic test-bacterium occurred, after microplates are kept at (4°C) overnight. A blank triplicate control is required for such tests.

Isolation of Total DNA from Anaerobic Thermophiles

There are several protocols for isolation of total DNA (a combination of genomic and plasmid DNA, if any) for *C. thermocellum* ATCC 27405 or *T. thermosaccharolyticum* ATCC 31960 (HG-8) (Klapatch et al., 1996a). A procedure for isolating plasmid DNA from recombinants of *T. saccharolyticum* JW/SL-YS 485 (Mai et al., 1997) has also been developed.

Total DNA is usually used to validate transformation and/or a chromosomal recombination event. In our experience, use of kits often does not provide adequate quality DNA preparations for *C. thermocellum*, as the residual nuclease activity destroys genomic DNA and can make obtaining reproducible results difficult. We thus recommend a modified protocol for total DNA isolation, which includes enzymatic treatment of the cell wall followed by SDS lysis, RNA digestion (to reduce the total viscosity of the water phase facilitating mixing of phases during extraction), Proteinase K treatment (to digest both native and added proteins – e.g. lysozyme and RNase – decreasing the viscosity of the aqueous phase, and deactivating endo- and exonuclease activities), deproteinization by Tris-equilibrated phenol:chloroform (pH ~8.0) in the presence of 1 M NaCl (to enhance phenol extraction), and precipitation of DNA in the presence of ethanol and sodium acetate. The protocol is a complementary version of a small-scale plasmid isolation procedure developed for lactobacilli, which also produce nucleases (Tyurin, 1990). We have found that this procedure is effective for DNA isolation from *T. saccharolyticum* JW/SL-YS 485 as well as *C. thermocellum*. However, the Qiagen Genomic-tip 100/G kit is equally effective for providing high-quality DNA preparations for *T. saccharolyticum*, and is considerably more convenient and time-efficient to implement as compared to phenol–chloroform extraction. The protocol described below appears, in our hands, to be rapid while yielding high-quality preparations. We do not use mutanolysin, and the concentrations of enzymes for DNA isolation (lysozyme, Proteinase K, RNase), are significantly reduced, while we use only one phenol extraction:

- Repeat step 1 of the ET Protocol, reducing total volume to 10 ml. Add antibiotic(s) to the broth, if necessary. Addition of antibiotic(s) may result in increase of incubation time.
- Collect cells by centrifugation and wash pellet once with 10 ml of cold (0–4°C) solution (0.05 M Tris-HCl, 0.01 M EDTA (pH 8.0) and 0.27 M sucrose). Resuspend pellet in 3 ml of the same solution additionally containing 10 mg ml^{-1} lysozyme and 50 µg ml^{-1} of RNase and incubate 30 min on ice, then 30 min at 37°C.
- Add 0.15 ml of 20% SDS, mix by gentle inverting the tube for 5–6 times (do not shake) (at this point you should have a clear lysate) and incubate at 37°C for 15 min. Then add Proteinase K to 50 µg ml^{-1} and continue the incubation for 20 min.
- Add 5 M NaCl to a final concentration of 1 M and shake gently. Add two volumes phenol:chloroform:isoamyl alcohol (25:24:1 v/v/v) equilibrated with Tris-base to pH 8.0, and mix by gentle shaking for 25 s.
- Centrifuge at 4°C for 10 min and transfer water phase into a clean tube.
- Add two volumes of chloroform and shake gently for 35 s. Centrifuge at 0–4°C for 10 min and transfer water phase into clean tubes. Repeat this step twice if an interphase is still observed after the first chloroform extraction.
- Transfer aqueous phase into a clean tube containing 3 M sodium acetate (pH 5.0) to a final concentration of 0.1 M. Mix well, add 2.5 volumes of cold ethanol, and mix again by shaking. Collect the DNA on a thin sterile glass capillary, made *ex tempore* from a Pasteur pipette, and transfer into a clean tube. Wash the DNA with 1.5 ml of 70% ethanol.
- Centrifuge the sample, discard supernatant and let the DNA dry for 15–25 min at room temperature. Dissolve the DNA in 0.1 ml of 0.01 M Tris, pH 8.0.

Note: This preparations may be stored at −20°C until used and will withstand up to 10–15 freeze–thaw cycles. Broth cultures may be stored at −20°C for periods up to several months prior to DNA isolation.

Detection of Restriction Endonuclease Activity in Anaerobic Thermophiles

It is often useful to characterize newly isolated strains with beneficial or interesting phenotypes for the presence of exo- and endonucleases. Methods for such characterization have been developed for strains of *C. thermocellum* ATCC 27405 and *T. thermosaccharolyticum* ATCC 31960 (Klapatch et al., 1996a).

Preparation of cell extracts

Grow cells to stationary phase then wash once in 0.27 M sucrose, freeze in liquid nitrogen, disrupt by passage through a French press, collect the extract under a nitrogen gas phase, and keep cell extract samples frozen in small aliquots at $-20°C$ until use. Alternatively, cell disruption may be achieved by complete cell autoplasting or protoplasting and freeze-thawing of resulted protoplasts in 0.05 M Tris (pH 8.0) to achieve cell lysis with no detergent added (Burns et al., 1998). Cells may be protoplasted using steps 1 and 2 of the Total DNA isolation protocol given above, except that RNase is omitted. Autoplasting, as described by Peteranderl et al. (1992), or protoplasting needs to be monitored by microscopic examination of small aliquots taken from the reaction volumes at closely spaced time intervals (e.g. 35, 40, 45 and 50 min of incubation). When the rod-shaped cells become "spherical", the sample needs to be frozen at $-20°C$ for 10 h, and then thawed at $37°C$. Additional gentle shaking helps to break the cells, and subsequent centrifugation at $4°C$ for 10 min clears the extract of cell debris. The extract may be stored at $-20°C$ in small aliquots. Alternatively, and in our experience more expediently, protoplasts may be pelleted, re-suspended in 0.01 M Tris (pH 8.0) to form thick suspensions like those described in the ET protocol and subjected to electric pulse at the voltage above optimal for this particular strain. The treated samples are diluted with 0.3 ml of 0.05 M Tris (pH 8.0), centrifuged to pellet the debris, and stored at $-20°C$ in small aliquots.

Examination of Restriction Endonuclease Activity

It is usually desirable to optimize the relative amounts of target DNA and the cell extract for evaluation of nuclease activity. To do this, mix 0.5–1.6 µg of target DNA with 2–4 µl of cell extract, add an appropriate restriction endonuclease buffer, adjust total volume to 20 µl, and incubate for 2–24 h at the regular growth temperature of the thermophile under investigation. Alternatively, incubation can be at $30-37°C$ for 3 h. Digested samples are then examined using gel electrophoresis. It may be necessary to use several "target DNAs" with different GC contents to find out what, if any, sequences are recognized by restriction endonuclease activity in cell extracts. Computer software (e.g. Gene Construction Kit 2.5 or Clone Manager 5) can be used to associate one or more restriction sites with the banding pattern observed in DNA digest gels. This information will be used to identify the right methylase system. When the putative restriction sequence digestion pattern is identified, tests need to be repeated with a known restriction enzyme recognizing that sequence as a control.

Hyperthermophiles

Systems for the hyperthermophilic genus *Thermotoga*

Compared to the (moderate) thermophiles, methods for genetic manipulation of bacterial hyperthermophiles such as *Tt. maritima* and *Tt. neapolitana* are much less developed. To date, only transient transformation has been achieved using antibiotic-resistance genes (Yu et al., 2001), with no other reports of foreign gene expression or targeted knockout to our knowledge. Genetic manipulations of *Thermotoga* species will be fostered as genome sequences become availability and gene transfer systems are developed.

Two potentially useful plasmids have been isolated from hyperthermophilic bacteria and have been tested for use in genetic work: pMC24 from *Tt. maritima* (isolated from Kurile Islands) and pRQ7 (isolated from the Azores) (Akimkina et al., 1999; Yu and Knoll, 1997). These plasmids are very small; in fact one is the smallest plasmid known, and both replicate by a rolling-circle mechanism. Derivatives (pjY1 and 2) of prQ7 have been used in the liposome-mediated DNA uptake and (transient) expression of thermostable chloramphenicol acetyltransferase in both *Thermotoga* species (see Yu et al., 2001 for construction details).

Plasmid preparation (Yu and Noll, 1997)

The cryptic plasmid pRQ7 (846 bp), estimated to be present at 200 copies per cell, can be isolated from an overnight culture using alkaline lysis as described by Lee and Rasheed (1992) or by lysis using hexadecyltrimethyl ammonium bromide, and separated by a standard agarose gel electrophoresis (1% agarose, with 1x Tris Borate-EDTA buffer). The plasmid can be cut using *Eco*R V and *Hin*d III. Modified plasmids have been created by introducing various modifications to obtain plasmid pJY1 (4.81 kb) and pJY2 (4.93 kb), both containing chloramphenicol acetyltransferase and kanamycin nucleotidyltransferase using standard methods, for transforming *Tt. maritima* (pJY2) and *Tt. neapolitana* (pJY1) using the transferred kanamycin resistance (120 µg ml^{-1}) for selection of transformants in liquid culture (Yu et al., 2001).

Spheroplast formation

Transformation of *Tt. maritima* and *Tt. neapolitana* has to date been achieved only with spheroplasts. Spheroplast preparation included harvesting cells from a 100 ml anaerobically grown culture in the early exponential growth phase, resuspending in 1 ml of the spheroplast buffer (20 mM

HEPES, pH 7.4, 4.5 mM NH$_4$Cl, 0.3 mM CaCl$_2$, 0.34 mM K$_2$HPO$_4$, 22 mM KCl, 2 mM MgSO$_4$ and 340 mM NaCl), adding lysozyme to 300 µg ml^{-1}, and incubating for 30 min at 37°C followed by 5 min at 77°C. To maximize the spheroplast formation using other strains of *Thermotoga* or related species, or for modified growth conditions, the progress of spheroplasting needs to be monitored using microscopy. Typically the spheroplasts are prepared *ex tempore*. It has not been investigated yet how long spheroplast suspensions may be stored to maintain capability to be transformed with liposomes.

Liposome-mediated DNA uptake

One ml of spheroplast suspension is mixed with a DNA:liposome mixture and is kept for 1 h at 37°C. Then 0.5 ml is transferred into 10 ml of *Thermotoga* complex medium and incubated for 3–6 h at 77°C to allow cell recovery prior to plating. The DNA–liposome mixture is prepared under anoxic conditions in an anaerobic chamber as follows: 1 µg DNA is suspended in 50 µl of 20 mM HEPES (pH 7.4) to which N-[1-(2,3-dioleoyloxy)-propyl]-N,N,N-trimethylammonium methylsulfate (DOTAP) Liposomal Transfection Reagent is added and mixed. Before addition to the spheroplast suspension, the solution is diluted with 30-µl HEPES-buffer, pH 7.4 and incubated at room temperature for 15 min. The DOTAP:DNA mixture is extremely oxygen sensitive and thus needs to be kept at all the time under N$_2$. The final complete mixture is incubated at 37°C for 1 h.

Plating of transformed spheroplasts using Gelrite soft gel overlays

Spheroplasts are kept in the complex *Thermotoga* medium for 6 h at 77°C before plating using soft Gelrite (Merck) overlay plates. The base plates (50 mm × 9 mm glass Petri dishes) are prepared by pouring 5 ml of 0.7% Gelrite in *Thermotoga* complex medium containing 200 mM HEPES (pH 7.5) (see chapter by Mesbah and Wiegel, this book). These plates are kept in the anaerobic chamber while the other preparations are carried out. Between 10–200 µl of the regenerated spheroplast solution is mixed quickly with 1 ml of anoxic 0.3% (wt/vol) Gelrite containing 0.4% wt/vol glucose, which is then poured onto the base plate. Following addition of the gel overlay, plates are incubated in an upright position for 24 h at 77°C in anaerobic glass jars (Difco), or simple canning jars closed in the anaerobic chamber and then transferred to the corresponding incubator, usually at 77°C. After 24 h, the overlay has firmly attached to the base layer and thus, the plates can then be incubated in the jars for 2–3 days upside down to avoid extensive smearing of growing colonies by the syneresis water.

◆◆◆◆◆ CONCLUDING REMARKS

Since expression of foreign DNA was first demonstrated in *E. coli* 30 years ago, we have seen an explosion in development of new techniques and application of these techniques to gain fundamental insights and develop organisms with desired industrial properties. Notwithstanding the power and significance of these events, the biotechnology "revolution" has to date been limited to a relatively small part of the known diversity of microorganisms. For many microorganisms, important tools of modern biotechnology cannot be applied because of a lack of established gene-transfer protocols. In the case of obligately anaerobic moderately thermophilic bacteria such as *C. thermocellum* and *Thermoanaerobacterium* species, results obtained over the last decade indicate that this barrier is beginning to be overcome. A great deal of work remains to be done, however, before these and similar organisms are amenable to techniques that are routinely applied for "workhorse" microorganisms such as *Escherichia coli, Bacillus subtilis* and *Saccharomyces cerevisiae*. For hypothermophiles of the genus *Thermotoga*, encouraging advances have recently been reported, although the current state of gene transfer capability is even more rudimentary than for the moderate thermophiles discussed in this chapter.

While recognizing that development of new gene-transfer protocols is a significant challenge, the price for not undertaking such development would appear to be high. In particular, failure to develop new gene-transfer protocols will severely limit the breadth of microorganisms available for both fundamental inquiry and industrial applications. In this light, we believe that a larger effort in this field is warranted, and that a systematic, as opposed to largely empirical, approach to transformation systems would be particularly beneficial.

List of Suppliers

Becton, Dickinson and Company
Sparks, MD 21152, USA
Tel.:+1-800-219-7174
http://www.bdbiosciences.com/

BBL™ Mueller Hinton Broth

Bio-Rad Laboratories, Inc.
2000 Alfred Nobel Drive
Hercules, CA 94547, USA
Tel.:+1-800-424-6723
http://www.bio-rad.com/

Bio-Rad Gene Pulser, 0.1 and 0.2 cm disposable cuvettes

Crescent Chemicals Corp.
1324 Motor Parkway
USA-Hauppauge NY 11788, USA
Tel.: +1-631-348-0333
http://www.serva.de/

Gelrite®

QIAGEN Inc. – USA
27220 Turnberry Lane Suite 200
Valencia, CA 91355, USA
Tel.:+1-800-426-8157
http://www1.qiagen.com/

Qiagen Genomic-tip 100/G kit

Roche Diagnostics Corporation
Roche Applied Science
P.O. Box 50414
9115 Hague Road
Indianapolis, IN 46250-0414, USA
Tel.:+1- 800-428-5433
http://www.roche-applied-science.com/

DOTAP Liposomal Transfection Reagent

Sigma-Aldrich
P.O. Box 14508
St. Louis, MO 63178, USA
Tel: +1-800-325-3010
http://www.sigmaaldrich.com/

chemicals, antibiotics

References

Akimkina, T., Ivanov, P., Kostrov, S., Sokolova, T., Bonch-Osmolovskaya, E., Firman, K., Dutta, C. F. and McClellan, J. A. (1999). A highly conserved plasmid from the extreme thermophile *Thermotoga maritima* MC24 is a member of a family of plasmids distributed worldwide. *Plasmid* **42**, 236–240.

Bayer, E. A., Kenig, R. and Lamed, R. L. (1983). Adherence of *Clostridium thermocellum* to cellulose. *J. Bacteriol.* **156**, 818–827.

Brehm, J., Salmond, G. and Minton, N. (1987). *Streptococcus faecalis* plasmid pAM-β1adenine methylase gene. *Nucleic Acids Res.* **15**, 3177.

Burns, B., Mendz, G. and Hazell, S. (1998). Methods for the measurement of a bacterial enzyme activity in cell lysates and extracts. *Biological Procedures Online* **1**, 17–26.

Cato, E. P., George, W. L. and Finegold, S. M. (1986). *Clostridium thermocellum* Vilojen, Fred and Peterson, 7^{AL}. In *Bergey's Manual of Systematic Bacteriology*. (P. H. A. Sneath, N. S. Mair, M. E. Sharpe and J. G. Holt, eds), 2nd edn., vol. 2, pp. 1160–1197. Williams & Wilkins, Baltimore.

Dagert, M., Jones, I., Goze, A., Romac, S., Niaudet, B. and Ehrlich, S. D. (1984). Replication functions of pC194 are necessary for efficient plasmid transduction by M13 phage. *EMBO J.* **3**, 81–86.

Das, H. and Singh, S. K. (2004). Useful byproducts from cellulosic wastes of agriculture and food industry – a critical appraisal. *Crit. Rev. Food Sci. Nutr.* **44**, 77–89.

Desai, S. G., Guerinot, M. L. and Lynd, L. R. (2004). Cloning of L-lactate dehydrogenase and elimination of lactic acid production via gene knockout in *Thermoanaerobacterium saccharolyticum* JW/SL-YS485. *Appl. Microbiol. Biotechnol.* **65**, 600–605.

Ducki, A., Grundmann, O., Konermann, L., Mayer, F. and Hoppert, M. (1998). Glucoamylase from *Thermoanaerobacterium thermosaccharolyticum*: sequence studies and analysis of the macromolecular architecture of the enzyme. *J. Gen. Appl. Microbiol.* **44**, 327–335.

Freier, D., Mothershed, Ch. P. and Wiegel, J. (1988). Characterization of *Clostridium thermocellum* JW20. *Appl. Environ. Microbiol.* **54**, 204–211.

Fuchs, K. P., Zverlov, V. V., Velikodvorskaya, G. A., Lottspeich, F. and Schwarz, W. H. (2003). Lic16A of *Clostridium thermocellum*, a non-cellulosomal, highly complex endo-β-1,3-glucanase bound to the outer cell surface. *Microbiology* **149**, 1021–1031.

Hermans, J., Boschloo, J. O. and de Bont, J. A. M. (1990). Transformation of *Mycobacterium attrum* by electroporation: the use of glycine, lysozyme and isonicotinic acid hydrazide in enhancing transformation efficiency. *FEMS Microbiol. Lett.* **72**, 221–224.

Horinouchi, S. and Weisblum, B. (1982). Nucleotide sequence and functional map of pC194, a plasmid that specifies inducible chloramphenicol resistance. *J. Bacteriol.* **150**, 815–825.

Jennert, K. C. B., Tardif, C., Young, D. I. and Youn, M. (2000). Gene transfer to *Clostridium cellulolyticum* ATCC 35319. *Microbiology* **146**, 3071–3080.

Klapatch, T. R., Demain, A. L. and Lynd, L. R. (1996a). Restriction endonuclease activity in *Clostridium thermocellum* and *Clostridium thermosaccharolyticum*. *Appl. Microbiol. Biotechnol.* **45**, 127–131.

Klapatch, T. R., Guerinot, M. L. and Lynd, L. R. (1996b). Electrotransformation of *Clostridium thermosaccharolyticum*. *J. Indust. Microbiol.* **16**, 342–347.

Kruus, K., Wang, W. K., Ching, J. and Wu, J. H. D. (1995). Exoglucanase activities of the recombinant *Clostridium thermocellum* CelS, a major cellulosome component. *J. Bacteriol.* **177**, 1641–1644.

Lee, S.-Y. and Rasheed, S. (1992). A simple procedure for maximum yield of high-quality plasmid DNA. *BioTechniques* **9**, 676–679.

Lee, S.-Y., Mermelstein, L. D., Bennett, G. N. and Papoutsakis, E. T. (1992). Vector construction, transformation, and gene amplification in *Clostridium acetobutylicum* ATCC 824. *Ann. N.Y. Acad. Sci.* **665**, 39–51.

Lee, Y.-E., Jain, M. K., Lee, C., Lowe, S. E. and Zeikus, J. G. (1993). Taxonomic distinction of saccharolytic thermophilic anaerobes: description of *Thermoanaerobacterium xylanolyticum* gen. nov., sp. nov., and *Thermoanaerobacterium saccharolyticum* gen. nov., sp. nov.; reclassification of *Thermoanaerobium brockii*, *Clostridium thermosulfurogenes*, and *Clostridium thermohydrosulfuricum* E100-69 as *Thermoanaerobacter brockii* comb. nov., *Thermoanaerobacterium thermosulfurigenes* comb. nov., and *Thermoanaerobacter thermohydrosulfuricus* comb. nov., respectively; and transfer of *Clostridium thermohydrosulfuricum* 39E to *Thermoanaerobacter ethanolicus*. *Int. J. Syst. Bacteriol.* **43**, 41–51.

Liu, S. Y., Wiegel, J. and Gherardini, F. C. (1996a). Purification and cloning of a thermostable xylose (glucose) isomerase with an acidic pH optimum from *Thermoanaerobacterium* strain JW/SL-YS 489. *J. Bacteriol.* **178**, 5938–5945.

Liu, S. Y., Gherardini, F. C., Matuschek, M., Bahl, H. and Wiegel, J. (1996b). Cloning, sequencing, and expression of the gene encoding a large

Lorenz, W. W. and Wiegel, J. (1997). Isolation, analysis, and expression of two genes from *Thermoanaerobacterium* sp. strain JW/SL YS485: a β-xylosidase and a novel acetyl xylan esterase with cephalosporin C deacetylase activity. *J. Bacteriol.* **179**, 5436–5441.

Lynd, L. R., Baskaran, S. and Casten, S. (2001). Salt accumulation resulting from base added for pH control, and not ethanol, limits growth of *Thermoanaerobacterium thermosaccharoluticum* HG-8 at elevated feed xylose concentrations in continuous culture. *Biotechnol. Progr.* **17**, 118–125.

Lynd, L. R., Weimer, P. J., van Zyl, W. H. and Pretorius, I. S. (2002). Microbial cellulose utilization: fundamentals and biotechnology. *Microbiol. Mol. Biol. Rev.* **66**, 506–577.

Mai, V. and Wiegel, J. (1999). Recombinant DNA applications in thermophiles. In *Manual of Industrial Microbiology and Biotechnology*. (A. L. Demain and J. E. Davis, eds), 2nd edn., pp. 511–519. ASM Press, Washington D.C.

Mai, V. and Wiegel, J. (2000). Advances in development of a genetic system for *Thermoanaerobacterium* spp.: expression of genes encoding hydrolytic enzymes, development of a second shuttle vector, and integration of genes into the chromosome. *Appl. Environ. Microbiol.* **66**, 4817–4821.

Mai, V., Lorenz, W. W. and Wiegel, J. (1997). Transformation of *Thermoanaerobacterium* sp. strain JW/SL-Y485 with plasmid pIKM1 conferring kanamycin resistance. *FEMS Microbiol. Lett.* **148**, 163–167.

McKenzie, T., Hoshino, T., Tanaka, T. and Sueoka, N. (1987). Correction. A revision of the nucleotide sequence and functional map of pUB110. *Plasmid* **17**, 83–85.

Meaden, P. G., Aduse-Opoku, J., Reizer, J., Reizer, A., Lanceman, Y. A., Martin, M. F. and Mitchell, W. J. (1994). The xylose isomerase-encoding gene (*xylA*) of *Clostridium thermosaccharolyticum*: cloning, sequencing and phylogeny of XylA enzymes. *Gene* **141**, 97–101.

Mermelstein, L. and Papoutsakis, E. T. (1993). In vivo methylation in *Escherichia coli* by the *Bacillus subtilis* phage φ3T1 methyltransferase to protect plasmids from restriction upon transformation of *Clostridium acetobutylicum* ATCC 824. *Appl. Environ. Microbiol.* **59**, 1077–1081.

Monod, M., Denoya, C. and Dubnau, D. (1986). Sequence and properties of pIM13, a macrolide-lincosamide-streptogramin B resistance plasmid from *Bacillus subtilis*. *J. Bacteriol.* **167**, 138–147.

Mori, Y. (1995). Nutritional interdependence between *Thermoanaerobacter thermohydrosulfuricus* and *Clostridium thermocellum*. *Arch. Microbiol.* **164**, 152–154.

Ng, T. K., Ben-Bassat, A. and Zeikus, J. G. (1981). Ethanol production by thermophilic bacteria: fermentation of cellulosic substrates by co-cultures of *Clostridium thermocellum* and *Clostridium thermohydrosulfuricum*. *Appl. Environ. Microbiol.* **41**, 1337–1343.

Onyenwoke, R. and Wiegel, J. (2006). Genus *Thermoanaerobacter*, In *Bergey's Manual of Systematic Bacteriology*. Springer-Verlag, New York, in press.

Oultram, J. D., Loughlin, M., Swinfield, T. J., Brehm, J. K., Thompson, D. E. and Minton, N. P. (1988). Introduction of plasmids into whole cells of *Clostridium acetobutylicum* by electroporation. *FEMS Microbiol. Lett.* **56**, 83–88.

Ozkan, M., Desai, S. G., Zhang, Y., Stevenson, D. M., Beane, J., White, E. A., Guerinot, M. L. and Lynd, L. R. (2001). Characterization of 13 newly isolated strains of anaerobic, cellulolytic, thermophilic bacteria. *J. Indust. Microbiol. Biotechnol.* **27**, 275–280.

Ozkan, M., Yilmaz, E. I., Lynd, L. R. and Ozcengiz, G. (2004). Cloning and expression of the *Clostridium thermocellum* L-lactate dehydrogenase gene in *Escherichia coli* and enzyme characterization. *Can. J. Microbiol.* **50**, 845–851.

Peteranderl, R., Shotts, E. B. Jr. and Wiegel, J. (1990). Stability of antibiotics at growth conditions of thermophilic anaerobes. *Appl. Environ. Microbiol.* **56**, 1981–1983.

Peteranderl, R., Canganella, F., Holzenburg, A. and Wiegel, J. (1992). Induction and regeneration of autoplasts from *Clostridium thermohydrosulfuricum* JW102 and *Thermoanaerobacter ethanolicus* JW 200. *Appl. Environ. Microbiol.* **59**, 3498–3501.

Sadaie, Y., Burtis, K. C. and Doi, R. H. (1980). Purification and characterization of a kanamycin nucleotidyltransferase from plasmid pUB110-carrying cells of *Bacillus subtilis*. *J. Bacteriol.* **141**, 1178–1182.

Saddler, J. N. and Chan, M. K.-H. (1984). Conversion of pretreated lignocellulosic substrates to ethanol by *Clostridium thermocellum* in mono- and co-culture with *Clostridium thermosaccharolyticum*, and *Clostridium thermohydrosulfuricum*. *Can. J. Microbiol.* **30**, 212–220.

Schwarz, W. H., Zverlov, V. V. and Bahl, H. (2004). Extracellular glycosyl hydrolases from clostridia. *Adv. Appl. Microbiol.* **56**, 215–261.

Slapack, G. E., Russell, I. and Stewart, G. G. (1987). *Thermophilic Microbes in Ethanol Production*. CRC Press, Inc., Boca Raton.

Soutschek-Bauer, E., Hartl, L. and Staudenbauer, W. L. (1986). Transformation of *Clostridium thermohydrosulfuricum* DSM 568 with plasmid DNA. *Biotechnol. Lett.* **7**, 705–710.

Tardif, C., Maamar, H., Balfin, M. and Belaich, J. P. (2001). Electrotransformation studies in *Clostridium cellulolyticum*. *J. Indust. Microbiol. Biotechnol.* **27**, 271–274.

Thornsberry, C. (1985). *Methods for Dilution Antimicrobial Susceptibility Tests for Bacteria that Grow Aerobically*. Approved Standard. *Publication M7-A*. National Committee for Clinical Standards, Vilanova, Philadelphia.

Trieu-Cuot, P. and Courvalin, P. (1983). Nucleotide sequence of the *S. faecalis* plasmid gene encoding the 3'5"-aminoglycoside phosphotransferase type III. *Gene* **23**, 331–341.

Tsoi, T. V., Chuvil'skaia, N. A., Atakishieva, Ia. Iu., Dzhavakhishvili, Ts. Ts. and Akimenko, V. K. (1987). [*Clostridium thermocellum* – a new object of genetic studies]. *Molekularnaya Genetika, Mikrobiologiya and Virusologiya* (Russ.) **11**, 18–23.

Tyurin, M. V. (1990). Optimization of a small-scale procedure for plasmid DNA isolation from lactobacilli. *Molekularnaya Genetika, Mikrobiologiya, i Virusologiya* (Russ.) **3**, 29–31.

Tyurin, M. V. (1992). Method for cell suspension treatment with electric current, and an apparatus for the treatment. *Russian patent 2005776*, October.

Tyurin, M. V. and Livshits, V. A. (1994). Some practical aspects of bacterial electrotransformation. *Biologicheskiye Membrany* (Russ.) **11**, 117–139.

Tyurin, M. V. and Livshits, V. A. (1995). Plasmid DNA transfer methods for *Lactobacillus buchneri*: natural competence, conjugative mobilization, and electrotransformation. *Membr. Cell Biol.* **9**, 57–68.

Tyurin, M. V., Padda, R., Huang, K.-X., Wardwell, S., Caprette, D. and Bennett, G. N. (2000). Electrotransformation of *Clostridium acetobutylicum*

ATCC 824 using high-voltage radio frequency modulated square pulses. *J. Appl. Microbiol.* **88**, 220–227.

Tyurin, M. V., Desai, S. G. and Lynd, L. R. (2004). Electrotransformation of *Clostridium thermocellum*. *Appl. Environ. Microbiol.* **70**, 883–890.

Tyurin, M. V., Sullivan, C. R. and Lynd, L. R. (2005). Induced oscillations play an important role in high efficiency transformation of thermophilic anaerobes. *Appl. Environ. Microbiol.* **71**, 8069–8076.

Wiegel, J. (1980). Formation of ethanol by bacteria. A pledge for the use of extreme thermophilic anaerobic bacteria in industrial ethanol fermentation processes. *Experientia* **36**, 1434–1446.

Wiegel, J. (1982a). Ethanol from cellulose. In *New Trends in Research and Utilization of Solar Energy through Biological Systems* (H. Mislin, ed.), pp. 79–83. Birkhäuser Verlag, Basel.

Wiegel, J. (1982b). Ethanol from cellulose. *Experientia* **38**, 151–156.

Williams, D. R., Young, D. I. and Young, M. (1990). Conjugative plasmid transfer from *Escherichia coli* to *Clostridium acetobutylicum*. *J. Gen. Microbiol.* **136**, 819–826.

Wu, L. and Welker, N. E. (1989). Protoplast transformation of *Bacillus stearothermophilus* NUB36 by plasmid DNA. *J. Gen. Microbiol.* **135**, 1315–1324.

Young, M., Minton, N. P. and Staudenbauer, W. L. (1989). Recent advances in the genetics of the clostridia. *FEMS Microbiol. Rev.* **63**, 301–326.

Yu, J.-S. and Noll, K. M. (1997). Plasmid pRQ7 from the hyperthermophilic bacterium *Thermotoga* species strain Q7 replicates by the rolling-circle mechanism. *J. Bacteriol.* **179**, 7161–7164.

Yu, J.-S., Vargas, M., Mityas, C. and Noll, K. M. (2001). Liposome-mediated DNA uptake and transient expression in *Thermotoga*. *Extremophiles* **5**, 53–60.

Zyprian, E. and Matzura, H. (1986). Characterization of signals promoting gene expression on the *Staphylococcus aureus* plasmid pUB110 and development of a gram-positive expression vector system. *DNA* **5**, 219–225.

14 Hyperthermophilic Virus–Host Systems: Detection and Isolation

David Prangishvili
Molecular Biology of the Gene in Extremophiles Unit, Institut Pasteur, rue du Dr. Roux 25, 75724 Paris Cedex 15, France

◆◆◆

CONTENTS

Introduction
Viruses in hot habitats
Isolation of virus–host systems from hot habitats
Viruses isolated from hot habitats

◆◆◆◆◆ INTRODUCTION

Based on results of direct counting of virus particles in water samples by epifluorescent microscopy using highly fluorescent dyes, it has been estimated that their number in a milliliter of sea water reaches $10-100 \times 10^6$ (reviewed by Wommack and Colwell, 2000). Moreover, it is estimated that about 20% of prokaryotic cells are infected with viruses and lysed by them daily (Suttle, 1994). Thus, apparently, viruses are major players in many different ecosystems, strongly affecting their functions such as carbon and nutrient flow, and primary and secondary production. Moreover, by infecting most abundant populations, viruses maintain and regulate diversity in host communities, causing selective killing of winner populations, and *via* generalized and specialized transduction, mediate genetic exchange in the environment (reviewed by Wommack and Colwell, 2000). Despite all these revelations, our knowledge on viral diversity is still rudimentary and viral ecology is still in its infancy. According to available reports, viruses and virus-like particles (VLPs) in freshwater, estuarine, or marine systems with moderate and low temperatures are almost exclusively tailed bacteriophage-like particles, or are tailless polyhedrons (reviewed by Ackermann, 2001; Rachel *et al.*, 2002; Wommack and Colwell, 2000). In addition to these morphotypes, VLPs with star and spindle shapes were found in hypersaline Dead Sea waters and Australian and Spanish salterns (Bath and Dyall-Smith, 1998; Guixa-Boixareu *et al.*, 1996; Oren *et al.*, 1997).

However, the picture of viral diversity is significantly different in geothermally heated hot aquatic habitats. Surprisingly, such environments turned out to be a reservoir of viruses with exceptionally diverse and complex morphotypes, not observed elsewhere in nature. In this chapter, we will summarize our experience with the detection and isolation of viruses from hot terrestrial environments with temperatures above 80°C.

◆◆◆◆◆ VIRUSES IN HOT HABITATS

Enumeration of Viruses in Hot Habitats

Recently Breitbart et al. (2004) reported on the abundance of VLPs, reaching concentrations of a million viruses per milliliter, in hot springs with neutral pH and temperatures around 80°C. They suggested that also in these habitats viruses exert an important influence on microbial community structure and energy flow. The number of viruses in hot environments with acidic pH, below pH 3.5, could be significantly lower. This was reported by Rice et al. (2001), who failed to detect by epifluorescent microscopy any significant numbers of VLPs in hot, acidic spring water samples from Yellowstone National Park, USA. Zillig et al. (1994) had to concentrate VLPs in water samples from hot, acidic springs of Iceland at least 10^4 times in order to be able to observe any VLP by transmission electron microscopy (TEM).

One reason for the relatively low number of VLPs in hot (>80°C), acidic (<pH 3.5) terrestrial springs could lie in the low cell density in such environments. Another plausible reason could be that caused by specific features of interactions of hyperthermophilic viruses with their hosts. With one exception none of the viruses isolated from hot acidic habitats is lytic. They generally persist in the host cells in a carrier state, in a state of certain equilibrium between virus replication and cell multiplication. By favoring such a strategy, virus populations could apparently reduce exposure to harsh environmental conditions.

Sampling and Enrichment Cultures

We generally keep environmental samples after their removal from the source at ambient temperatures (when possible at 4°C) under anaerobic conditions. After settling of most of the mud, samples are injected into previously prepared closed tubes containing 0.2 g of sulfur, 20 µg of resazurin as a redox indicator and a gas phase of 160 kPa of CO_2 and 1 kPa of H_2S to maintain the anaerobic state. If necessary, anaerobic conditions are restored by drop-wise addition of water saturated with H_2S. For acidic samples, prior to their conservation, it was recommended to adjust the pH to about pH 5.5 by the addition of $CaCO_3$ (Zillig et al., 1994). In untreated samples from extreme thermal environments, we usually were unable to observe virus particles by TEM. For this, a significant concentration of particles or an enrichment of samples was required.

Samples from hot springs with acidic pH were enriched at temperatures and pH values close to those of the natural environment, in conditions favorable for growth of members of the hyperthermophilic crenarchaeal order *Sulfolobales*, which are known to dominate in such environments (Huber and Prangishvili, 2004). They may be cultivated heterotrophically on a simple carbon source, autotrophically with S^0 and O_2 or with S^0 and H_2, in either case with CO_2 as the sole carbon source, or mixotrophically. The salt base of the medium contained per liter: 3 g $(NH_4)_2SO_4$, 0.5 g $K_2HPO_4 \times 3H_2O$, 203.3 mg $MgCl_2 \times 6H_2O$, 0.1 g KCl, 70.8 mg $Ca(NO_3)_2 \times 4H_2O$, 0.9 mg $MnCl_2 \times 4H_2O$, 2.24 mg $Na_2B_4O_7 \times 10H_2O$, 0.11 mg $ZnSO_4 \times 7H_2O$, 0.025 mg $CuCl_2 \times 2H_2O$, 0.015 mg $Na_2MoO_4 \times 2H_2O$, 0.015 mg $VOSO_4 \times 5H_2O$, 0.005 mg $CoSO_4 \times 7H_2O$, and 0.005 mg $NiSO_4 \times 6H_2O$. For buffering, medium contained 0.7 g glycine per l and pH was adjusted to pH 3–3.5 with 1:3 diluted sulfuric acid.

In hot environments with neutral or nearly neutral pH, anaerobic microorganisms are known to dominate. We enriched samples from such environments and maintained cells in an atmosphere consisting of N_2 and CO_2 (80/20 v/v) on Allen Medium, pH 6.5 (Allen, 1959) supplemented with 0.001% yeast extract, 0.005% peptone, 0.1 mM $CaSO_4 \cdot 2H_2O$, 0.05 mM Na_2SO_4, 0.1 mM KNO_3 and 3 mM $Na_2S_2O_3 \cdot 5H_2O$. Such conditions enabled cultivation of a large diversity of hyperthermophilic members of the Bacteria and Archaea, as described below in an example of the enrichment from a sample from hot spring with pH 6.5 in Yellowstone National Park, USA (Obsidian Pool).

To the enrichment media, Wolin's vitamin mixture (Wolin, 1963) could be added to a final concentration of 0.5%.

Morphotypes of Viruses and Virus-like Particles in Hot Habitats

In enrichment cultures from samples taken at different terrestrial and marine hydrothermal environments, several groups of researchers observed exceptional morphological diversity of VLPs (Geslin et al., 2003; Häring et al., 2005a; Rachel et al., 2002; Rice et al., 2001; Zillig et al., 1994).

The most comprehensive study was that of virus morphotypes in hot terrestrial environments in Yellowstone National Park, USA, and in Pozzuoli, Italy. In enrichments of samples from two hot springs in Yellowstone National Park, one acidic (pH 1.5–2) and one nearly neutral (pH 6.5), we have observed more than two dozen different morphotypes of VLPs (Figures 14.1 and 14.2). This diversity included rod-shaped particles in three size ranges (Figures 14.1A,E, 14.2D), flexuous filaments, with diverse terminal structures (Figures 14.1B,C,F,G and 14.2E), spindle-shaped particles (Figures 14.1D), typical head-tailed phages with long and short tails (Figure 14.2A–C), spherical particles in two size ranges (Figure 14.2F,G) and unusual particles not previously observed in nature (Figures 14.1H–N, Figure 14.2H–L). Cells in

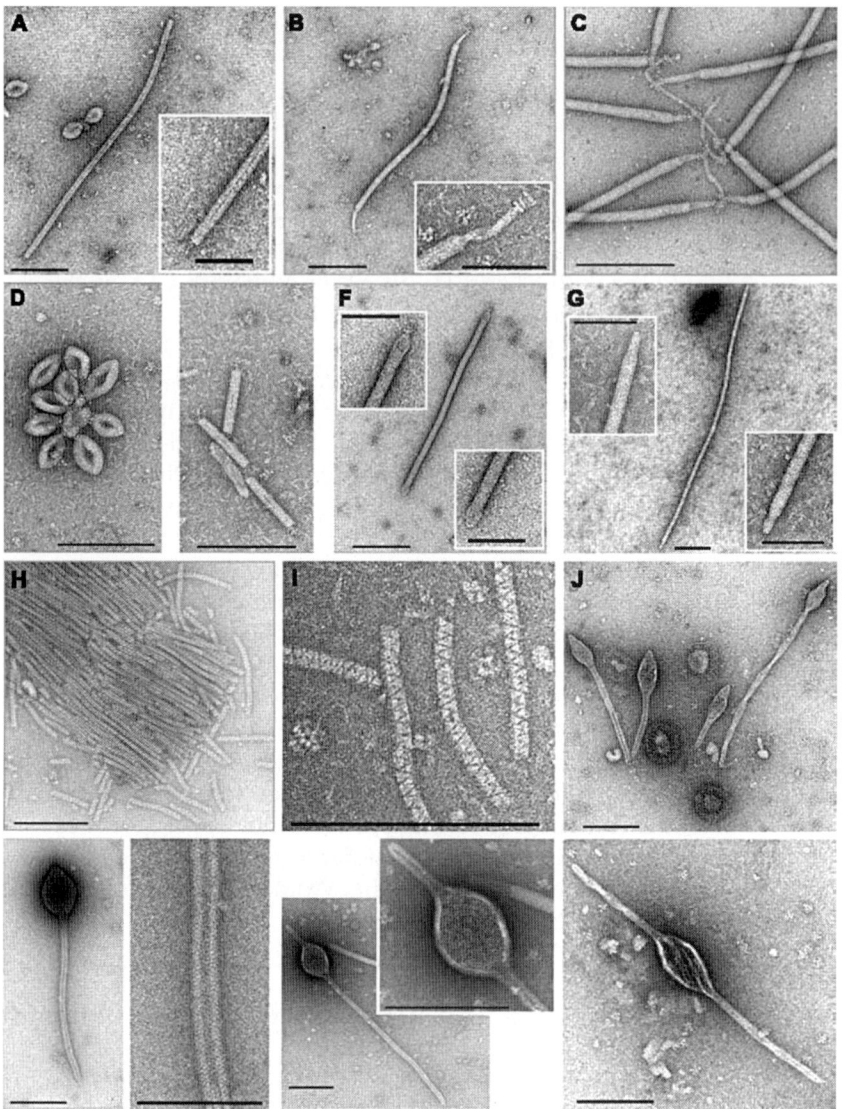

Figure 14.1. Transmission electron micrographs of viruses and VLPs found in an enrichment culture of a sample from a hot acidic spring (85°C, pH 2) in Yellowstone National Park. Samples were negatively stained using 2% uranyl acetate. Bars, 200 nm, 100 nm for insets (reproduced from Rachel et al., 2002).

the enrichment from the acidic hot sample were morphologically homogenous irregular cocci. Only one 16S rRNA gene sequence could be recovered from the enrichment, belonging to a novel species of the hyperthermophilic archaeal genus *Acidianus*, provisionally named "Acidianus hospitalis" (Rachel et al., 2002). By contrast, many different Bacteria and Archaea, which could be regarded as possible hosts, were enriched from the sample from the hot spring with pH 6.5. 16S rRNA gene sequences recovered from the enrichment represented the genera *Thermophilum, Thermoproteus, Pyrobaculum, Thermosphaera* from

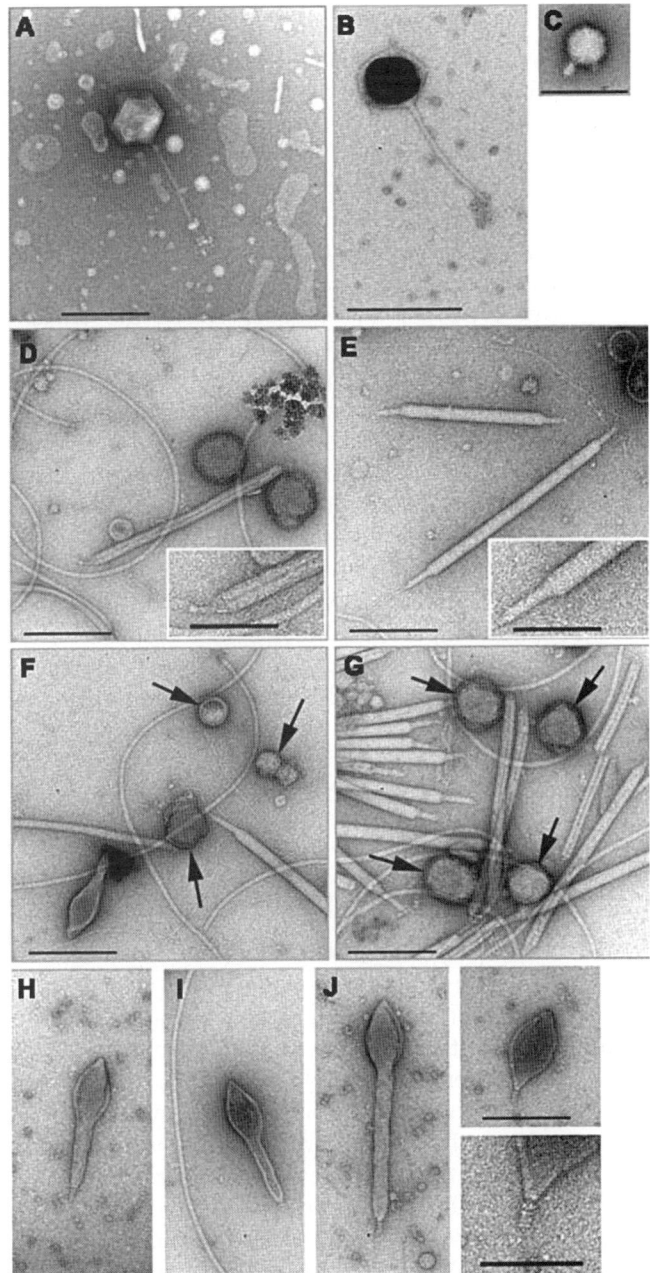

Figure 14.2. Transmission electron micrographs of viruses and VLPs found in an enrichment culture of a sample from a hot spring with nearly-neutral pH (75–93°C, pH 6) in Yellowstone National Park (Obsidian Pool). Samples were negatively stained using 2% uranyl acetate. Bars, 200 nm, 100 nm for insets (reproduced from Rachel *et al.*, 2002).

the archaeal phylum Crenarchaeota, *Archaeoglobus* from the archaeal phylum Euryarchaeota, and *Thermus*, *Geothermobacter*, and *Thermodesulfobacterium* from the domain Bacteria. In addition, sequences from noncultivated members of the crenarchaeota, the "Korarchaeota", the Nanoarchaeota and the bacterial phylum Aquificae have been recovered (Rachel et al., 2002).

In enrichments of a sample from a water reservoir in the crater of the Solfatara volcano at Pozzuoli, Italy, which contained exclusively only species of the genus *Acidianus* (18 different species), we also observed a plethora of VLPs, encompassing rigid rods and flexuous filaments, as well as particles with ellipsoid bodies and two tails of different length, similar to those shown in Figure 14.1 M and N, and unusual bottle-shaped particles (Häring et al., 2005a).

All types of VLPs observed in the enrichments from hot springs of Pozzuoli were shown to be infectious virions of five novel viruses, and they could be isolated and purified (Häring et al., 2005a). In other studies, a much smaller portion of different types of particles observed by TEM could be shown to represent viable virions and be cultured. For example, from two enrichments of samples from the Yellowstone National Park described above, only one filamentous virus (Figures 14.1B) and one spherical virus with a diameter of 100 nm (Figure 14.2G) could be cultured (Bettstetter et al., 2003; Häring et al., 2004). They were infecting strains of the archaeal genera *Acidianus* and *Pyrobaculum*, respectively.

Not all VLPs observed in the enrichments represented viruses. For example, the zipper-like particles shown in Figures 14.1H,I, which were produced by several stains of "A. hospitalis", carried only protein subunits but no nucleic acid, and thus were not of a viral nature (Bettstetter and Prangishvili, unpublished).

◆◆◆◆◆ ISOLATION OF VIRUS–HOST SYSTEMS FROM HOT HABITATS

For the isolation of hyperthermophilic virus–host systems from primary enrichments, basically two approaches were used, which are schematically presented in Table 14.1. One, successfully used in the laboratory of Wolfram Zillig (Table 14.1A), employs isolation from enrichments of single strains and thereafter their screening for virus production. For screening, two criteria were routinely used: (i) an ability of cells to produce an inhibition zone on a lawn of possible host, or/and (ii) presence in cells of extrachromosomal genetic elements. Eventually, virus production has to be confirmed by TEM. The efficiency of this approach depends on the possibility to analyze a large number of autochthonous isolates from each enrichment. In the case when a number of different enrichments itself is very high, the approach is extremely laborious. Another approach, recently successfully used in our group (Table 14.1B), is based on the detection of virus production in enrichments by TEM prior to isolation and analysis of single host strains. In this way, we could avoid the analysis of enrichments which did not contain any detectable virus–host system, and thus increase the efficiency of the screening procedure. However, also in this case, chances for the isolation of host strains are

Table 14.1 Two schemes for the isolation of virus–host systems from hot aquatic samples

A	Enrichment of environmental samples
	↓
	Colony-purification of single isolates
	↓
	Screening single isolates for a production of viruses
	(a) *by infectivity tests*
	(b) *by detection of viral DNA in host cells*
	↓
	Confirmation of virus production by TEM
	↓
	Purification of viruses
	↓
	Conformation of infectivity by propagation in virus-free hosts
B	Enrichment of environmental samples
	↓
	Selection for virus-producing enrichments by TEM
	↓
	Colony-purification of single isolates
	↓
	Screening cultures of single isolates for a presence of viruses by TEM
	↓
	Purification of viruses
	↓
	Conformation of infectivity by propagation in virus-free hosts

dependent on the possibility to analyze a large number of single isolates. Initial analysis of enrichment cultures with TEM has also the significant advantage of revealing a comprehensive picture of viral diversity in enrichments and enables the screening for hosts specifically for each observed virus type.

Detection of Viruses by Transmission Electron Microscopy

Samples for TEM analysis are deposited on carbon-coated copper grids and negatively stained with 2% uranyl acetate, pH 4.5. According to our experience, a practical lower limit for detection of hyperthermophilic viruses by TEM is around 10^5 virions ml^{-1}. Thus, concentration of particles prior to the analysis is required for those cell cultures in which a virus titer is expected to be less than 10^5 ml^{-1}. Removal of cells from growth cultures has to precede virus concentration. Cells can be removed either by low-speed centrifugation (e.g. 4500 rpm for 10 min in the Sorvall GS3

Rotor), and/or by ultra-filtration though a 0.8 µM pore size filter. The most reliable way to concentrate virus particles is by their precipitation by high-speed centrifugation from cell-free culture supernatants (e.g. 38 000 rpm 1 h in the Beckman SW55 rotor), followed by resuspension of precipitated particles in a small volume of growth medium, or distilled water. Although virions of most viruses studied by us were stable after such treatment, it caused partial disruption of a significant population of virions of the spherical virus PSV (Häring et al., 2004) and the bottle-shaped virus ABV (Häring et al., 2005a). Pressure filtration through a 0.2 µm pore size filter is a milder method of concentration of particles; however, some viruses can apparently pass through such a filter.

TEM data alone are not sufficient to consider observed particles as viable virions. For this they have to be isolated and purified, and the infectivity of the purified particles has to be confirmed.

Isolation and Purification of Viruses

After the removal of cells from virus-replicating growth cultures by low speed centrifugation, as described above, virus particles are precipitated from the supernatant by adding NaCl to 1 M and polyethylene glycol 6000 to 10% (w/v) and maintaining at 4°C overnight. The sediment is collected by centrifugation (e.g. 12 000 rpm for 30 min in the Sorvall GS3 Rotor) and suspended in an appropriate buffer. Remaining cells and cell debris can be partially removed from the suspension by low-speed centrifugation (e.g. 25 000 rpm for 5 min in the Heraeus Laborfuge 400R) and the pellet is extracted twice under the same conditions. To the combined supernatants is added CsCl to a final concentration of 0.45 g ml^{-1}, and virus particles are banded by ultracentrifugation for 2 days at 42 000 rpm in the Beckman TI 50 Rotor, or at 48 000 rpm in the Beckman SW55 rotor. The bands are removed with a syringe and dialyzed against an appropriate buffer. All hyperthermophilic viruses which have been isolated so far have a buoyant density in the range of 1.32–1.39 g ml^{-1}. As a buffer in virus purification experiments, we used either 0.02 M MES, containing 1 µM MgCl$_2$ and 0.3 µM Ca(NO$_3$)$_2$, or 20 mM Tris-acetate, pH 6.

Isolation of Host Strains and Analysis of Host Range

For screening for virus hosts among autochthonous stains, single isolates should be colony-purified from the primary enrichment culture on 0.6% Gelrite (Kelco, San Diego) plates in a medium and conditions identical to those used for the enrichment. Generally, plating efficiency of different isolates of hyperthermophilic crenarchaeota is rather high, reaching 90%. An alternative approach, in case difficulties are encountered with growth on plates, involves usage of optical tweezers (Ashkin et al., 1987; Huber et al., 1995).

Natural producers of viruses among isolated strains can be easily identified by TEM analysis. For identification of virus-free stains which can be infected by a purified virus, we usually concentrated cells prior to infection: from a culture grown to mid-exponential phase, cells were precipitated by low-speed centrifugation and suspended in about 1/10 of the original volume. According to our results (Prangishvili, unpublished), infection with hyperthermophilic viruses is a temperature-dependent process and is most efficient at temperatures optimal for growth of the host. After incubation at an appropriate temperature for 1 h, the mixture of concentrated cells and purified virus was used to inoculate fresh growth medium. Virus propagation in a growing culture was analyzed by TEM.

If candidate host strains can grow in lawns, the propagation of virus can be detected by a plaque assay or by a "spot-on-lawn" assay.

Detection of Viruses by Their Infectivity and Plaque Assays

Prerequisites for detection and analysis of viruses by plaque assays are an ability of host cells to form lawns, and an ability of a virus to produce plaques on a lawn of host cells.

Many members of the genus *Sulfolobus* of the hyperthermophilic, acidophilic order *Sulfolobales*, but not of the genus *Acidianus* of the same order, are capable of lawn formation. However, the non-lytic nature of most viruses of acidophilic hyperthermophiles, as discussed above, causes significant problems for plaque formation. Plaques could be developed due to poor growth of infected cells, when viral infection resulted in significant growth retardation at early stages of cell growth. This was a case for members of three different families of non-lytic *Sulfolobus* viruses, *Fuselloviridae*, *Lipothrixviridae*, and *Rudiviridae* (see Table 14.2), and for these, plaque assays could be established. For this, appropriate dilutions of a virus suspension in growth medium (0.1 ml), together with $2-5 \times 10^7$ host cells (0.1 ml) in mid-exponential growth phase, are added to about 2 ml of 0.2–0.3% Gelrite solution in growth medium at 75°C. The mixture is poured over a pre-formed 0.6% Gelrite supporting gel and after solidification the plate is incubated at an appropriate temperature in a jar containing several ml of water to create a humid atmosphere. Plaques on *Sulfolobus* lawns are developed usually between 75 and 80°C for 48 h. It is noteworthy that the plaque morphologies of members of different families of *Sulfolobus* viruses differ from each other (Figure 14.3A–C).

A simplified version of a plaque assay could be a "spot-on-lawn" assay, which aims to screen for virus hosts or to detect virus production in a growth culture. For this, small aliquots (0.5–1 µl) of a virus preparation, or of a growth culture, are applied to freshly poured lawns of a wide diversity of possible host strains, including autochthonous virus-free strains. Lawns are prepared in the same way as described above for a plaque assay. After developing of a lawn at an appropriate temperature, inhibition zones are formed around some spots, indicative of possible

Table 14.2 Viruses of hyperthermophilic Archaea

Virus	Host species	Virus description	DNA: form, size bp	Sequence accession number in GenBank database
Family *Fuselloviridae*				
Genus Fusellovirus				
SSV1	*Sulfolobus*	Martin et al. (1984), Schleper et al. (1992)	ccc, 15465	XO7234
SSV2	*Sulfolobus*	Stedman et al. (2003)	ccc, 14796	AY370762
SS-K1	*Sulfolobus*	Wiedenheft et al. (2004)	ccc, 17385	AY423772
SSVRH	*Sulfolobus*	Wiedenheft et al. (2004)	ccc, 16473	AY388628
Family *Lipothrixviridae*				
Genus Alphalipothrixvirus				
TTV1	*Thermoproteus*	Janekovic et al. (1983)	Linear, 15900	X14855 (85% of DNA)
TTV2		Janekovic et al. (1983)	Linear, 16000	Not sequenced
TTV3		Janekovic et al. (1983)	Linear, 27000	Not sequenced
Genus Betalipothrixvirus				
SIFV	*Sulfolobus*	Arnold et al. (2000b)	Linear, 40852	AF440571 (96% of DNA)
Genus Gammalipothrixvirus				
AFV1	*Acidianus*	Bettstetter et al. (2003)	Linear, 21080	AJ567472

Genus Deltalipotrixvirus					
	AFV2	*Acidianus*	Häring et al. (2005b)	Linear, 31787	AJ854042
Family *Guttaviridae*					
Genus Guttavirus					
	SNDV	*Sulfolobus*	Arnold et al. (2000a)	Linear DNA	Not sequenced
Family *Rudiviridae*					
Genus Rudivirus					
	SIRV1	*Sulfolobus*	Prangishvili et al. (1999)	Linear, 32308	AJ414696
	SIRV2	*Sulfolobus*	Prangishvili et al. (1999)	Linear, 35450	AJ344259
	ARV1	*Acidianus*	Vestergaard et al. (2005)	Linear, 24655	AJ875026
Family *Globuloviridae*					
Genus Globulovirus					
	PSV	*Pyrobaculum, Thermoproteus*	Häring et al. (2004)	Linear, 28337	AJ635162
Family *Ampullaviridae*					
Genus Ampullavirus					
	ABV	*Acidianus*	Häring et al. (2005a)	Linear, 23500	Sequencing under way
Unclassified					
	STIV	*Sulfolobus*	Rice et al. (2004)	Circular, 17663	AY569307

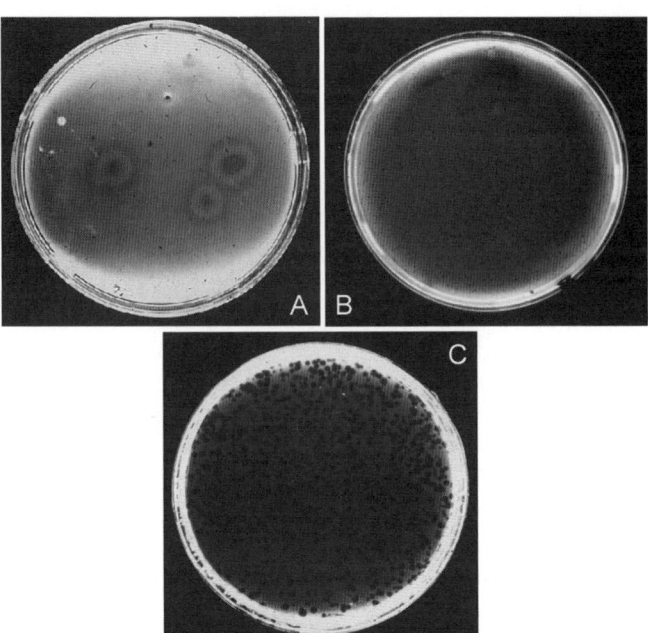

Figure 14.3. Plaques produced by members of three different families of hyperthermophilic viruses: (A) Fusellovirus SSV1 (courtesy of K. Stedman); (B) Lipothrixvirus SIFV; (C) Rudivirus SIRV1, on lawns of S. solfataricus (A), "S. islandicus" HVE10/4 (B), and "S. islandicus" REN2H1 (C) (courtesy of W. Zillig).

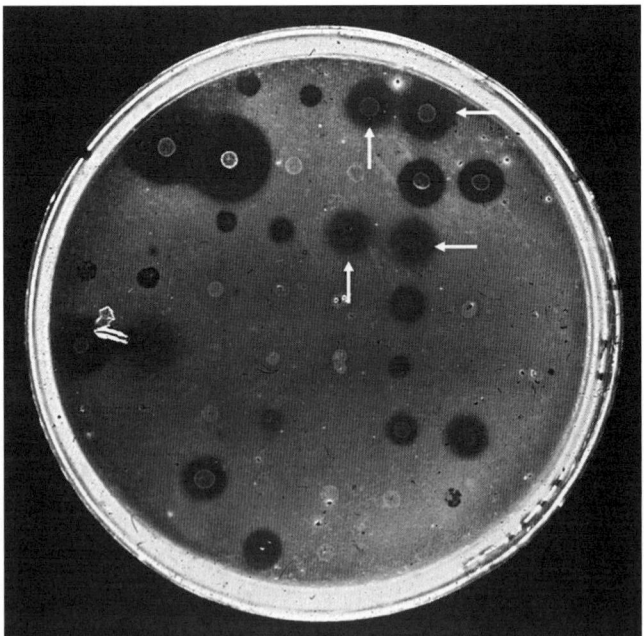

Figure 14.4. A plate with a lawn of S. solfataricus P1 onto which were applied 1 µl drops of growing cultures of 53 different isolates of "S. islandicus". Four arrows indicate inhibition zones around spots with virus-producer strains. The rest of inhibition zones are caused by production of sulfolobicins (Prangishvili, Holz and Zillig, unpublished).

virus replication (Figure 14.4). However, not all inhibition zones are caused by viruses. Any material produced by cells in the "spot" which is toxic for the lawn of cells can cause the appearance of an inhibition zone. As an example, Figure 14.4 shows a plate with spots of 53 single isolates of *Sulfolobus* from enrichments from Icelandic hot springs on a lawn of an autochthonous *Sulfolobus* strain. Only 4 inhibition zones, shown by arrows, were caused by virus production, as confirmed by TEM analysis of growing cultures of corresponding strains; these were different spindle-shaped viruses from the family *Fuselloviridae* (D. Prangishvili and W. Zillig, unpublished). Special studies revealed that the rest of the inhibition zones were caused by proteinaceous toxins, named sulfolobicins (Prangishvili *et al.*, 2000). These toxins were not released in soluble form and were associated with producer cells as well as with cell-derived spherical membrane vesicles 90 to 180 nm in diameter, which erroneously could be considered as virions (Prangishvili *et al.*, 2000). It is noteworthy that among 400 single isolates of *Sulfolobus* analyzed in the whole experiment, 41 were sulfolobicin producers. Enrichment of samples prior to isolation of strains was apparently a reason for the high percentage of the sulfolobicin producers among isolates.

Detection of Viruses by Intracellular Copies of Their Genomes

Replication of viruses could be detected by the presence of their genomes in host cells. Our experience concerns identification of viral genomes represented by an extrachromosomal double-stranded DNA. For its detection, total cellular DNA can be prepared by phenol extraction and subjected to gradient ultracentrifugation in 1 g ml^{-1} CsCl in the presence of 1 mg per ml of ethidium bromide (first at $180\,000 \times g$ at 12°C for three days in the Beckman TI 50 rotor, then at $400\,000 \times g$ overnight in the VTI rotor). In this way, intracellular cccDNA can be separated from the linear chromosomal DNA. However, this method is not suitable for the detection of linear viral genomes, and could be not sensitive enough to detect low copy number elements.

Another and more simple way to detect extrachromosomal double-stranded DNA in a cellular DNA preparation is by an analysis of fragments obtained by digestion with type II restriction endonucleases, in the case when the copy number of viral genomes in a cell is sufficiently high (>10 copies per chromosome). Fragments of viral DNA will appear as prominent stoichiometric bands over the background formed by fragments of chromosomal DNA. This method enables detection of circular as well as linear virus genomes. However, it is an even less sensitive method than the one using gradient centrifugation and even 15 copies of a virus genome per chromosome is easy to overlook. Another disadvantage of the method is that it does not discriminate between plasmid and virus DNA and requires additional analysis to reveal a viral origin of a high copy number extrachromosomal element.

One way to verify a viral origin of extrachromosomal DNA is by transfection of possible host cells with the putative viral DNA.

Until now, there is only one known successful case of transfection of hyperthermophilic crenarchaeaon with a viral DNA. *S. solfataricus* could be transfected by electroporation with circular ds DNA of the virus SSV1 (Schleper *et al.*, 1992). There are a number of ways for detecting successful transfection and virus spreading. The most reliable one is by TEM analysis or by plaque assay. Conclusions about virus spreading based on an analysis of cellular DNA can be misleading. Conjugative plasmids are also capable of self-spreading very efficiently after electroporation. For example, in a course of screening for the production of viruses among 400 novel *Sulfolobus* isolates by analysis of extrachromosomal DNA by transfection and self-spreading, only 4 out of the 15 self-spreading elements detected turned out to be viral genomes (circular genomes of members of the family *Fuselloviridae*) (Zillig *et al.*, 1996); 11 of the self-spreading elements represented conjugative plasmids (Prangishvili *et al.*, 1999b).

Our attempts to transfect *Sulfolobus* cells by electroporation with linear ds DNA of rudiviruses and lipiothrixviruses (see below) failed (Häring and Prangishvili, unpublished).

VIRUSES ISOLATED FROM HOT HABITATS

All viruses isolated so far from hot habitats have double-stranded DNA genomes. However, the result appears not to be caused by the selective nature of procedures used for their detection and isolation. Only one procedure among several used for screening for virus replication, based on detection of intracellular copies of viral genomes, was biased towards the selection of double-stranded DNA viruses.

Due to their unique morphotypes, not observed elsewhere in nature, and unique features of genome organization, about two dozen viruses isolated from hot terrestrial environments have been classified into six virus families, introduced specifically for their classification (reviewed by Häring *et al.*, 2005a; Prangishvili and Garrett, 2004). Above are listed these families and their members, and are given references to accession numbers for their genomes in data bases. Figure 14.5 shows electron micrographs of representatives of different families.

Hosts for all of the viruses which have been isolated from hot habitats are members of the phylum Crenarchaeota of the domain Archaea. Thus, at present it is not clear whether unusual features are associated with hosts or with extreme hot environments. However, the Crenarchaeota represent a significant portion of microbial communities not only in extremely hot but also in cooler aquatic ecosystems, where particles resembling hyperthermophilic viruses have never been observed. Thus, it can be argued that the unusual diversity and uniqueness of the double-stranded DNA viruses described here is a specific feature of a life in hot habitats. Further studies on viral diversity are required to confirm this suggestion and to understand its evolutionary implications.

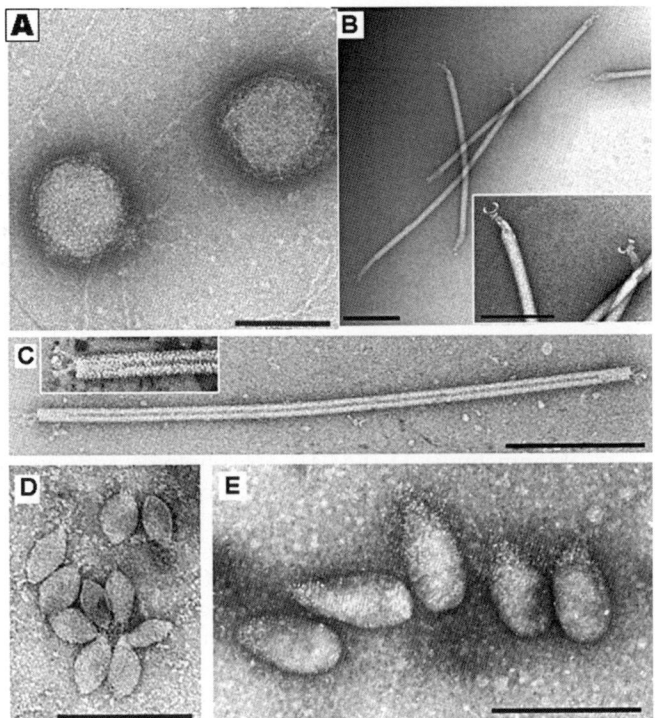

Figure 14.5. Electron micrographs of representatives of five families of viruses of hyperthermophiles: (A) Globulovirus PSV (reproduced from Häring et al., 2004); (B) Lipothrixvirus AFV1 (reproduced from Bettstetter et al., 2003); (C) Rudivirus SIRV1, (D) Fusellovirus SSV1, (E) Guttavirus SNDV (courtesy of W. Zillig), Bars, 200 nm.

Acknowledgements

Thanks to David Musgrave for critical discussions on the manuscript.

References

Ackermann, H.-W. (2001). Frequency of morphological phage descriptions in the year 2000. *Arch. Virol.* **146**, 843–857.

Allen, M. B. (1959). Studies with *Cyanidium caldarium*, an anomalously pigmented chlorophyte. *Arch. Mikrobiol.* **32**, 270–277.

Arnold, H. P., Ziese, U. and Zillig, W. (2000a). SNDV, a novel virus of the extremely thermophilic and acidophilic archaeon *Sulfolobus*. *Virology* **272**, 409–416.

Arnold, H. P., Zillig, W., Ziese, U., Holz, I., Crosby, M., Utterback, T., Weidmann, J. F., Kristjansson, J., Klenk, H. P., Nelson, K. E. and Fraser, C. M. (2000b). A novel lipothrixvirus, SIFV, of the extremely thermophilic crenarchaeon *Sulfolobus*. *Virology* **267**, 252–266.

Ashkin, A., Dziedzic, J. M. and Yamane, T. (1987). Optical trapping and manipulation of single cells using infrared laser beams. *Nature* **330**, 769–771.

Bath, C. and Dyall-Smith, M. L. (1998). His1, an archaeal virus of the *Fuselloviridae* family that infects *Haloarcula hispanica*. *J. Virol.* **72**, 9392–9395.

Bettstetter, M., Peng, X., Rachel, R., Garrett, R. A. and Prangishvili, D. (2003). AFV1, a novel virus of the hyperthermophilic crenarchaeon *Acidianus*. *Virology* **315**, 68–79.

Breitbart, M., Wegley, L., Leeds, S., Schoenfeld, T. and Rower, F. (2004). Phage community dynamics in hot springs. *Appl. Environ. Microbiol.* **70**, 1633–1640.

Geslin, C., Le Romancer, M., Erauso, G., Gaillard, M., Perrot, G. and Prieur, D. (2003). PAV1, the first virus-like particle isolated from a hyperthermophilic euryarchaeote "*Pyrococcus abyssi*". *J. Bacteriol.* **185**, 3888–3894.

Guixa-Boixareu, N., Calderón-Paz, J. I., Heldal, M., Bratbak, G. and Pedrós-Alió, C. (1996). Viral lysis and bacterivory as prokaryotic loss factors along a salinity gradient. *Aquat. Microb. Ecol.* **11**, 215–227.

Häring, M., Peng, X., Brügger, K., Rachel, R., Stetter, K. O., Garrett, R. A. and Prangishvili, D. (2004). Morphology and genome organisation of the virus PSV of the hyperthermophilic archaeal genera *Pyrobaculum* and *Thermoproteus*: a novel virus family, the *Globuloviridae*. *Virology* **323**, 232–242.

Häring, M., Rachel, R., Peng, X., Garrett, R. A. and Prangishvili, D. (2005a). Diversity of viruses in hot springs of Pozzuoli, Italy: a bottle-shaped archaeal virus ABV, representing a novel family of ds DNA viruses, the Ampullaviridae. *J. Virol.*, **147**, 2419–2429.

Häring, M., Vestergaard, G., Brügger, K., Rachel, R., Garrett, R. A. and Prangishvili, D. (2005b). Structure and genome organization of AFV2, a novel lipothrixvirus with unusual terminal and core structures. *J. Bacteriol.*, **187**, 3855–3858.

Huber, H. and Prangishvili, D. (2004). *Sulfolobales*. In *The Prokaryotes: An Evolving Electronic Resource for the Microbiological Community* (M. Dworkin, S. Falkow, E. Rosenberg, K.-H. Schleifer and E. Stackebrandt, eds), 3rd edn., release 3.18. Springer-Verlag, New York, http://link.springer-ny.com/link/service/books/10125/.

Huber, R., Burggraf, S., Meyer, T., Barns, S. M., Rossnagel, P. and Stetter, K. O. (1995). Isolation of a hyperthermophilic archaeum predicted by *in situ* RNA analysis. *Nature* **376**, 57–58.

Janekovic, D., Wunderl, S., Holz, I., Zillig, W., Gierl, A. and Neumann, H. (1983). TTV1, TTV2, and TTV3, a family of viruses of the extremely thermophilic, anaerobic, sulfur-reducing archaebacterium *Thermoproteus tenax*. *Mol. Gen. Genet.* **192**, 39–45.

Martin, A., Yeats, S., Janekovic, D., Reiter, W.-D., Aicher, W. and Zillig, W. (1984). SAV1, a temperate U.V.-inducible DNA virus-like particle from the archaebacterium *Sulfolobus acidocaldarius* isolate B12. *EMBO J.* **3**, 2165–2168.

Oren, A., Bratbak, G. and Heldal, M. (1997). Occurrence of virus-like particles in the Dead Sea. *Extremophiles* **1**, 143–149.

Prangishvili, D. and Garrett, R. A. (2004). Exceptionally diverse morphotypes and genomes of crenarchaeal hyperthermophilic viruses. *Biochem. Soc. Trans.* **32**, 204–208.

Prangishvili, D., Albers, S.-V., Holz, I., Arnold, H. P., Stedman, K. M., Klein, T., Singh, H., Hiort, J., Schweier, A., Kristjansson, J. K. and Zillig, W. (1998). Conjugation in Archaea: frequent occurrence of conjugative plasmids in *Sulfolobus*. *Plasmid* **40**, 190–202.

Prangishvili, D., Arnold, H. P., Götz, D., Ziese, U., Holz, I., Kristjansson, J. K. and Zillig, W. (1999). A novel virus family, the *Rudiviridae*: Structure, virus-host interactions and genome variability of the *Sulfolobus* viruses SIRV1 and SIRV2. *Genetics* **152**, 1387–1396.

Prangishvili, D., Holz, I., Stieger, E., Nickell, S., Kristjansson, J. K. and Zillig, W. (2000). Sulfolobicins, specific proteinaceous toxins produced by strains of the extremely thermophilic archaeal genus Sulfolobus. *J. Bacteriol.* **182**, 2985–2988.

Rachel, R., Bettstetter, M., Hedlund, B., Häring, M., Kessler, A., Stetter, K. O. and Prangishvili, D. (2002). Remarkable diversity of viruses and virus-like particles in hot terrestrial environments. *Arch. Virol.* **147**, 2419–2429.

Rice, G., Stedman, K. M., Snyder, J., Wiedenheft, B., Brumfield, S., McDermott, T. and Young, M. (2001). Novel viruses from extreme thermal environments. *Proc. Natl. Acad. Sci. USA* **98**, 13341–13345.

Rice, G., Tang, L., Stedman, K., Roberto, F., Spuyhlewr, J., Gillitzer, E., Johnson, J. E., Douglas, T. and Young, M. (2004). The structure of a thermophilic archaeal virus shows a double-stranded DNA viral capsid that spans all domains of life. *Proc. Natl. Acad. Sci. USA* **101**, 7716–7720.

Schleper, C., Kubo, K. and Zillig, W. (1992). The particle SSVA from the extremely thermophilic archaeon *Sulfolobus* is a virus: Demonstration of infectivity and of transfection with viral DNA. *Proc. Natl. Acad. Sci. USA* **89**, 7645–7649.

Stedman, K. M., She, Q., Phan, H., Arnold, H. P., Holz, I., Garrett, R. A. and Zillig, W. (2003). Biological and genetic relationships between fuselloviruses infecting the extremely thermophilic archaeon *Sulfolobus*: SSV1 and SSV2. *Res. Microbiol.* **154**, 295–302.

Suttle, C. A. (1994). The significance of viruses to mortality in aquatic microbial communities. *Microb. Ecol.* **28**, 237–243.

Vestergaard, G., Häring, M., Peng, X., Rachel, R., Garrett, R. A. and Prangishvili, D. (2005). A novel rudivirus, ARV1, of the hyperthermophilic archaeal genus *Acidianus*. *Virology* **336**, 83–92.

Wiedenheft, B., Stedman, K. M., Roberto, F., Willits, D., Gleske, A. K., Zoeller, L., Snyder, J., Douglas, T. and Young, M. (2004). Comparative genomic analysis of hyperthermophilic archaeal Fuselloviridae viruses. *J. Virol.* **78**, 1954–1961.

Wolin, E. A., Wolin, M. J. and Wolfe, R. S. (1963). Formation of methane by bacterial extracts. *J. Biol. Chem.* **238**, 2882–2886.

Wommack, K. E. and Colwell, R. R. (2000). Virioplankton. *Microbiol. Mol. Biol. Rev.* **64**, 69–114.

Zillig, W., Kletzin, A., Schleper, C., Holz, I., Janekovic, D., Hain, J., Lanzendörfer, M. and Kristjansson, J. K. (1994). Screening for *Sulfolobales*, their plasmids and their viruses in Icelandic solfataras. *System. Appl. Microbiol.* **16**, 609–628.

Zillig, W., Prangishvili, D., Schleper, C., Elferink, M., Holz, I., Albers, S.-V., Janekovic, D. and Götz, D. (1996). Viruses, plasmids and other genetic elements of thermophilic and hyperthermophilic Archaea. *FEMS Microbiol. Rev.* **18**, 225–236.

15 Preservation of Thermophilic Microorganisms

Stefan Spring
DSMZ-Deutsche Sammlung von Mikroorganismen und Zellkulturen GmbH, Mascheroder Weg Ib, D-38124 Braunschweig, Germany

◆◆

CONTENTS

Introduction
Maintenance by subculturing and preservation
Methods for the long-term preservation of thermophiles
Vaccum drying
Deep freezing

◆◆◆◆◆ INTRODUCTION

Thermophilic and hyperthermophilic microorganisms are highly diverse in respect of their metabolism and phylogeny. Besides high temperatures, most thermophiles have also adapted their lifestyle also to other environmental conditions which can be termed extreme from an anthropocentric point of view, e.g. acidic pH values, highly reduced conditions or hypersalinity. This versatility, among other reasons, may have caused an increasing interest in this group of prokaryotes. The mounting research activity in this field in many laboratories resulted in a need for reliable maintenance methods for thermophilic strains. Especially for comparative genetic or biochemical studies it is often desirable to work with stocks of preserved cultures, or it could be necessary to maintain a large number of mutant strains over longer periods of time. On the other hand, standard methods of preservation often fail in the case of thermophilic microorganisms, especially if they are also strictly anaerobic or acidophilic. Several methods which can be used for the long-term preservation of these fastidious microorganisms are discussed in this chapter. Most of these techniques were originally developed for the reliable preservation of extremophiles at culture collections, but with some expertise, it should be possible to adapt these procedures to the technical facilities available in most research or industrial laboratories.

Resources for material and equipment useful for the described methods are listed at the end of this chapter.

◆◆◆◆◆ MAINTENANCE BY SUBCULTURING AND PRESERVATION

In several laboratories studying thermophilic prokaryotes, important strains are only maintained by frequent subculturing. Subculturing means serial transfer of strains from media with depleted nutrient sources to fresh media. After inoculation into fresh media, cultures are incubated to obtain growth and then eventually stored. To prevent frequent subculturing, the metabolic rate of the organism during storage should be kept at a minimum. Many thermophilic strains stop growing already below 40°C, so that they can be stored easily without refrigeration at room temperature. Refrigeration to 4–8°C can be used to extend the interval of subculturing of some strains, but on the other hand may have a negative effect on the long-term stability of other thermophilic strains. Viability of distinct cultures over time may vary considerably and depends largely on the reached growth phase, storage conditions and quality of the used medium. Hence, time of storage between transfers is normally kept at a minimum to ensure survival of important strains. Subculturing is inexpensive in terms of equipment, applicable to all cultivable strains and avoids problems associated with the resuscitation of preserved stock cultures. However, it can be time-consuming if organisms are handled that require frequent transfers or a whole collection of strains has to be maintained. Besides the time necessary for preparation of media and manipulation of strains, subculturing implicates several risks and disadvantages. Contamination of pure cultures is a permanent threat and the mislabelling or transposition of vials can lead to an interchange of strains. The risk of mishaps increases with the frequency of the manipulation of a strain and can be only minimized by the consequent use of sound microbial techniques. Frequent subculturing can also have a negative effect on the genetic integrity of a strain. Genetic changes or the loss of plasmids may result in the selection of mutant strains and thereby, to a strain drift. A progressive genetic drift in a distinct culture may remain unrecognized if inconspicuous traits of the organism are affected.

Only by using reliable long-term preservation techniques, most of these risks can be efficiently avoided. The investment in laboratory equipment and training of technical staff, which is necessary for the establishment of most preservation techniques, appears to be only a small effort in comparison to the damage caused by the potential loss of important reference strains.

◆◆◆◆◆ METHODS FOR THE LONG-TERM PRESERVATION OF THERMOPHILES

Several well-established methods are available for the long-term preservation of thermophiles. The basic principle common to all of these techniques is the reduction of the amount of freely available water to a value where metabolism is suspended. A decrease in water activity can be achieved either by dehydration or freezing. Numerous procedures have been developed for the gentle cryopreservation or drying of cultures in order to minimize cell damage during preservation. Nevertheless, survival rates of some cultures can be rather low. In general, survival rates depend largely on the sensitivity of a strain against the harmful effects caused by the used preservation method and can vary considerably among closely related strains. Consequently, not all of the techniques described in this chapter are applicable to the whole range of thermophilic microorganisms. The main assets and drawbacks of the most common techniques are discussed below.

Freeze-drying

Freeze-drying or lyophilization is a process which extracts water from a sample by sublimation. Sublimation is the transition of a substance from the solid to the vapour state, without passing through an intermediate liquid phase. Lyophilization involves freezing of the cell pellet so that the water becomes ice, application of vacuum in order to sublimate the ice directly into water vapour and drawing off the water vapour. Cultures preserved by freeze-drying are relatively stable over time and can be stored without further attention. However, due to the low survival rates of susceptible strains, a selection of more resistant subpopulations with different genetic characteristics may take place during freeze-drying. To avoid a genetic drift by the continuous selection of subpopulations, subsequent batches should be always prepared from the same seed stock. Ampoules with lyophilized samples can be transported easily without the risk of damage caused by lyses of sensitive cultures. However, the resuscitation of vacuum-dried samples can be labour-intensive, especially if anaerobic strains were preserved. In addition, most strains show a prolonged lag-phase after resuscitation and need two or three transfers until normal growth takes place.

This preservation method requires an investment in special equipment, which is mainly useful for the purpose of freeze-drying. Some training of technical staff will also be necessary to ensure a safe and smooth flow of lyophilization procedures. Thus this method is only cost-effective for laboratories which have to prepare larger batches of preserved samples from thermophiles or intend to store a large collection of strains without further attention. Thermophiles that are suitable for preservation by lyophilization represent mainly aerobic or aerotolerant, heterotrophic microorganisms, which are relatively robust and show a good growth yield. In contrast, strains that are extremely sensitive to oxygen or reach only very low cell densities in most cases, hardly survive the harsh conditions during freeze-drying and should be better preserved by cryopreservation.

Liquid-drying

Unlike lyophilization, liquid-drying (L-drying; Annear, 1956) involves vacuum-drying of samples from the liquid state without freezing. Liquid-drying has, in general, the same benefits and drawbacks as freeze-drying, but the procedure is less elaborate and may have advantages in the preservation of microorganisms that are particularly sensitive to the initial freezing step involved in lyophilization.

Deep Freezing

The only technique which seems to be applicable to all thermophilic prokaryotes studied so far is the deep freezing of concentrated cell suspensions in fresh media supplemented with a suitable cryoprotectant. The survival rates of cultures preserved by this method are usually quite high and the long-term stability almost unlimited. The resuscitation of cultures preserved by deep freezing is uncomplicated and growth normally takes place without an extended lag phase. However, it is important that the temperature of storage should be kept below $-139°C$, because only then all physical or chemical reactions are suspended (Morris, 1981) and a satisfactory stability of the preserved culture can be achieved. In contrast, the storage of cryopreserved stocks in normal freezers at temperatures above $-70°C$ is not recommended, because free water is still available enabling residual physical or chemical activity, which could damage DNA or other essential cell compounds causing a rapid loss of viability in some strains. Normally, storage in liquid nitrogen ($-196°C$) or the nitrogen vapour phase ($-140°C$) is used to achieve the required low temperatures. With exception of the liquid nitrogen storage tank, the equipment necessary for this method is easy to obtain and inexpensive. A major constraint of this preservation technique may be, however, the continuous need for liquid nitrogen to maintain the required storage temperature. A complete evaporation of liquid nitrogen will cause an inevitable temperature rise in the refrigerator which eventually may kill all sensitive stocks. Hence, care has to be taken to ensure that liquid nitrogen is replenished in regular intervals. Nevertheless, in some geographical regions the adequate supply with liquid nitrogen is problematic. In this case, storage of less sensitive cultures in deep freezers at $-80°C$ may be an alternative. However, it has to be noted that although metabolism is suspended, cells can be damaged by the recrystallization of ice, which still takes place at this temperature.

In the following sections, standard protocols for the preservation of thermophilic prokaryotes are exemplified on representatives of various physiological groups.

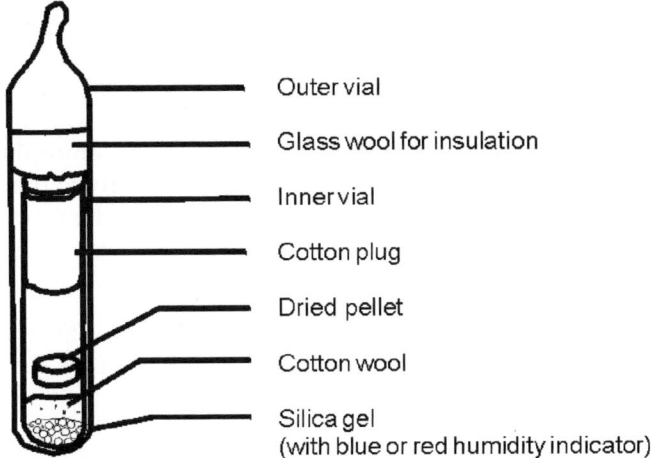

Figure 15.1. Double-vial preparation of a vacuum-dried sample, sealed under vacuum.

♦♦♦♦♦ VACUUM-DRYING

Equipment

Technical equipment necessary for vacuum-drying methods include a vacuum pump, a freeze-dryer equipped with a centrifuge head and manifold or an evacuation jar connected to a moisture or cold trap, constrictors and diverse glass ware. Vacuum-dried samples of cultures can be prepared either in double vial or single vial ampoules. The Deutsche Sammlung von Mikroorganismen und Zellkulturen GmbH (DSMZ) delivers dried cell pellets exclusively as double vial preparations, sealed under vacuum (Figure 15.1). Double-vial ampoules are more elaborate to prepare, but have the advantage that a contamination of the atmosphere by aerosols that can be produced by sudden release of the vacuum in single-vial preparations is efficiently prevented. In addition, the cell pellet is protected from contamination, because inflowing air filters through the sterile cotton plug of the inner vial.

Centrifugal Freeze-drying

Centrifugal freeze-drying is the preferred method for the preservation of fastidious thermophilic strains that are susceptible to the initial freezing in skim milk intrinsic to standard freeze-drying protocols. This method relies on the freezing of the cell suspension by evaporative cooling caused by the loss of water during evacuation of the freeze-drying chamber. To avoid frothing of the suspension due to removal of dissolved gases before freezing is complete, the suspension is centrifuged during the initial stages of drying. In this way, the initial freezing of the cell suspension, which is a critical step of the normal shelf freeze-drying process, can be omitted. At the DSMZ, a special preservation mixture has been developed which is now routinely used for the protection of thermophilic strains during the centrifugal freeze-drying process. The use of this protective solution allows the lyophilization of most aerobic thermophiles, with the exception of strains that grow only to a very low cell density. The supplementation of this solution with amorphous ferrous sulfide also allows the lyophilization of some anaerobic thermophiles without using an oxygen-free gas atmosphere during the processing of samples.

Preparation of vials

1. Plug clean glass vials (44 × 11 mm, flat bottom) loosely with non-absorbent cotton wool plugs (dental rolls size no. 2, 40 mm in length) and sterilize at 121°C for 20 min.
2. Prepare outer glass tubes (135 × 215 mm, soft glass) by placing a few pieces of self-indicating silica gel (e.g. silicagel rubin) into the tubes which are covered with a small amount of cotton wool (see Figure 15.1).

Preparation of the suspending medium

1. *Preservation mixture:* Dissolve 2% (w/v) gelatine, 2% (w/v) yeast extract and 24% (w/v) sucrose in distilled water, bubble with N_2 gas for at least 30 min to remove oxygen and fill in a suitable container, which can be tightly closed, e.g. a serum bottle with butyl rubber stopper and aluminium crimp. Autoclave the protective solution at 121°C for 20 min. The suspending medium is prepared by mixing an equal ratio of freshly prepared sterile cultivation medium and preservation mixture.
2. *Amorphous ferrous sulfide for anaerobic strains:* The ferrous sulfide is prepared according to Brock and O'Dea (1977). Briefly, equimolar amounts of ferrous ammonium sulfate and sodium sulfide react in solution to form amorphous FeS, which precipitates. After settling of the FeS precipitate, the supernatant is discarded. This procedure is repeated several times with an anoxic solution of 0.9% (w/v) NaCl until the supernatant no longer contains detectable amounts of ferrous or sulfide ions. Finally, the purified aqueous suspension of FeS is dispensed under nitrogen gas atmosphere in suitable vessels and autoclaved. The final concentration of amorphous ferrous sulfide in the suspending medium should be approximately 1 mg ml^{-1}.

Preparation of the cell suspension

Depending on the oxygen relationship and acid tolerance of the strain to be preserved, different procedures are necessary for the preparation of the cell suspension. Ideally cells are harvested when they are most active and have reached the mid exponential growth phase. The processing of aerobes is exemplified on representatives of the genera *Aeropyrum* and *Sulfolobus*, whereas the lyophilization of anaerobes is illustrated by a protocol suitable for hyperthermophilic *Pyrococcus* species.

Aerobic heterotrophic thermophiles

Aeropyrum spp. and other oxygen-tolerant hyperthermophiles can be harvested as usual by centrifugation of liquid medium in sterile centrifuge tubes or bottles. The cell pellet is resuspended in approximately 5 ml of suspending medium containing the preservation mixture and aseptically distributed in aliquots of 0.15 ml to each single vial. Several thermoacidophilic species, e.g. *Sulfolobus solfataricus*, are sensitive to low pH values at the end of the growth phase and easily lose viability. Therefore, the growth medium is neutralized prior to harvesting of the cells by adding a small amount of solid, sterilized calcium carbonate. After 10–15 min, the undissolved carbonate forms a deposit which then can be removed from the culture supernatant. In addition, the pH value of the thermoacidophiles growth medium is adjusted to a moderate value (pH 4.0–4.5) prior to mixing with the preservation mixture in order to prevent damage of cells during lyophilization.

Strictly anaerobic thermophiles

Special care has to be taken to avoid exposure to oxygen during harvesting of *Pyrococcus* spp. and other obligate anaerobic thermophiles and hyperthermophiles. Either anaerobically grown cultures have to be transferred under an oxygen-free gas atmosphere to anoxic and gas-tight centrifuge bottles or in a more straightforward approach are cultured in heavy-walled, round-bottomed glass bottles, which can be used for growing of cells and centrifugation. Suitable glass bottles (50–70 ml volume) can be ordered from most glass blowers as custom-made product and should have necks that can be closed with a butyl rubber septum and screw cap as with Hungate-type anaerobe tubes. After centrifugation the centrifuge bottle is opened, a gassing cannula inserted and the supernatant is removed aseptically under a flow of oxygen-free gas. Normally, the gas-mixture used corresponds to the gas atmosphere is used to cultivate the respective strain. However, a mixture of N_2 and CO_2 instead of a H_2 containing gas mixture should be used to avoid risks caused by the generation of flammable gas mixtures in the laboratory. Cell pellets of one or more bottles are collected using suspending medium that has been supplemented with 1 mg ml^{-1} amorphous FeS and transferred to a tube continuously flushed with oxygen-free N_2 gas. Filling of vials is carried out under oxic conditions using for instance an Eppendorf Multipipette with 2.5 ml Combitip. This procedure should be done quickly because the ferrous sulfide provides protection of cells against oxygen only for a limited time.

Drying procedure

Drying is carried out in two stages. Primary drying is achieved by centrifugal freeze-drying, whereas standard shelf freeze-drying is used to obtain double-vial preparations in the second step.

1. *Primary drying*
    1.1. Place vials in the centrifuge head of a freeze-drying machine, switch on the centrifuge and thereafter apply vacuum.
    1.2. Centrifuge at approximately 750 rpm for 2–3 h.
    1.3. Switch off the centrifuge and continue primary drying until the vacuum has dropped to 1–10 mbar.
    1.4. Switch off refrigerator and vacuum pump, allow air to slowly enter the vacuum chamber and remove vials from the centrifuge head.

2. *Secondary drying*
    2.1. Cut off parts of the cotton-wool plugs that are projecting from the vials and place vials in outer glass tubes containing self-indicating silica gel.
    2.2. To protect the cotton-wool plugs from heat during constriction cover the vials with glass wool (Tempstran 475–106) slightly compressed to a layer of 1–2 cm thickness (see Figure 15.1). The outer tubes are constricted just above the glass wool either by hand or by using a semiautomatic ampoule constrictor.
    2.3. After cooling, attach the double-vial preparations to the manifold of a freeze-drying machine for secondary drying for at least 2 h or overnight (see Figure 15.2).
    2.4. At a vacuum of at least 0.1 mbar, the tubes are flame-sealed at the middle of the constriction.

Liquid-drying

The L-drying method has been successfully applied at the DSMZ to several thermophilic, obligate chemolithoautotrophic bacteria, which are difficult to preserve using one of the standard freeze-drying protocols. Examples of successfully preserved thermophiles include *Aquifex pyrophilus*, *Thermocrinis albus*, *Hydrogenobacter thermophilus* and *Sulfurihydrogenibium azorense*. In general, members of these genera and related hydrogen-oxidizing thermophiles are microaerophilic and sensitive to elevated levels of oxygen. Thus, during handling of these cultures care has to be taken to ensure that exposure to atmospheric levels of oxygen is limited to a minimum.

The following outline of an L-drying preservation method suitable for fastidious thermophilic strains is based on the description given by Malik (1990). A setup of the equipment necessary for primary L-drying is shown in Figure 15.3. Alternatively, this method can be carried out without a freeze-drying machine following the simplified procedure described by Malik (1991).

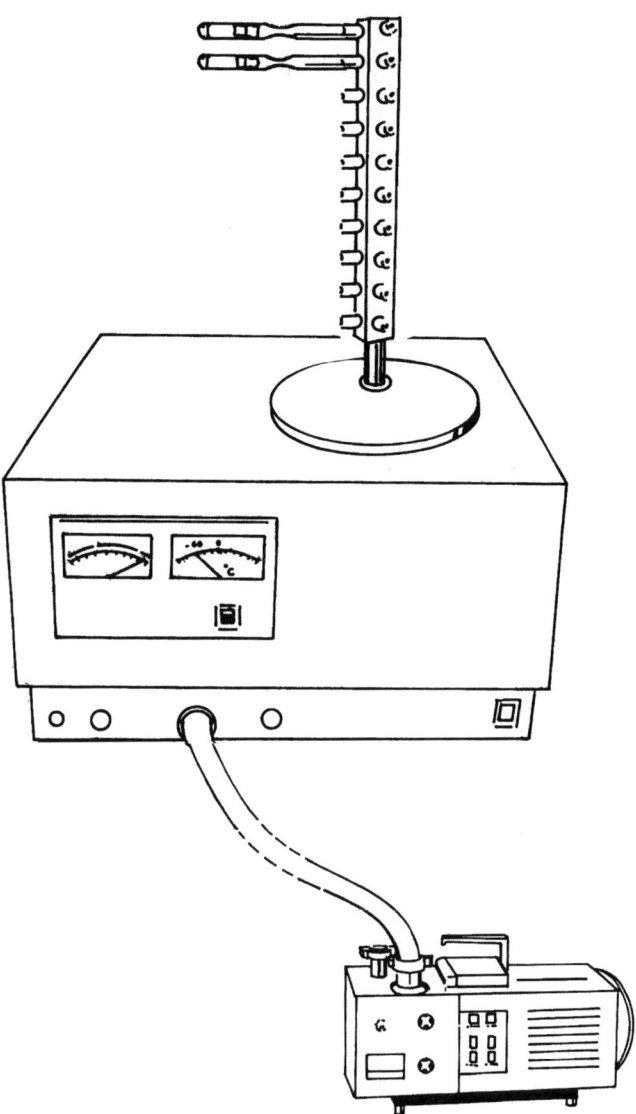

Figure 15.2. Secondary vacuum-drying. Constricted outer tubes containing inner vials are attached to a manifold and mounted on a freeze-drying machine. Adapted from Malik (1990) with permission.

Preparation of the carrier material

A thin disc of carrier material is prepared in order to protect the cell suspension from freezing during the evacuation period.

1. Fill glass vials (44 × 11 mm, flat bottom) with 0.1 ml of 20% (w/v) skim milk containing 1% (w/v) neutral activated charcoal and 5% (w/v) meso-inositol.
2. The vials are loosely plugged with non-absorbent cotton wool and sterilized at 115°C for 13 min.

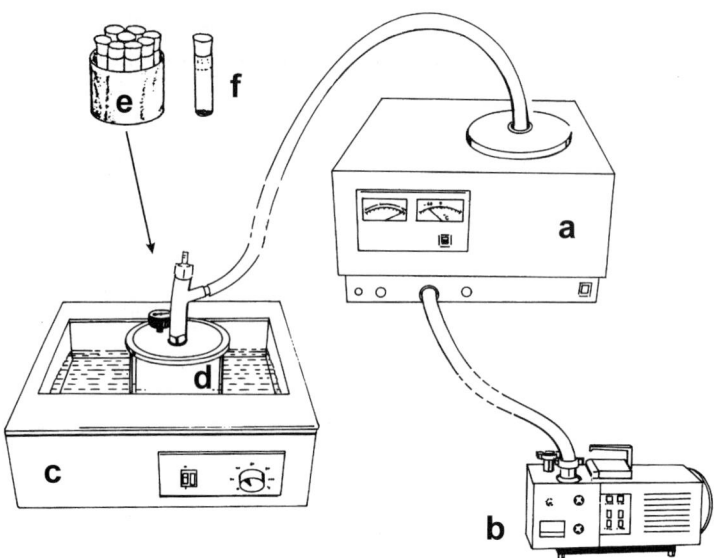

Figure 15.3. L-drying procedure. Setup of the necessary equipment: (a) freeze-drying machine; (b) vacuum pump; (c) water bath kept at a constant temperature of 20°C; (d) metallic evacuation jar equipped with vacuum valve; (e) aluminium Erlenmeyer cap; (f) inner vial with sample and cotton plug. Adapted from Malik (1990) with permission.

3. Freeze the vials at −20°C for several hours and thereafter transfer them to the drying chamber of a freeze-drying machine. Apply vacuum and freeze-dry overnight until vacuum has dropped to 0.1 mbar or less.

Preparation of the cell suspension

Cells are concentrated by harvesting and suspension in a protective medium. A mixture of activated charcoal and meso-inositol has proven most effective in preventing harmful effects on the cells during liquid-drying.

1. Suspend 1% (w/v) activated charcoal and dissolve 3% (w/v) meso-inositol in distilled water (pH adjusted to 7.0), bubble with N_2 gas for at least 30 min to remove oxygen and fill in a suitable container, e.g. a serum bottle with butyl rubber stopper and aluminium crimp, which is tightly closed. Autoclave the protective solution at 115°C for 13 min.
2. Harvest cells as described in the section on centrifugal freeze-drying by aseptic centrifugation and suspend cell pellet in the protective solution, while maintaining anoxic conditions. The obtained cell suspension should have a concentration of about 10^8–10^9 cells ml^{-1}.

Drying procedure

L-drying is carried out in two stages. In the following L-drying procedure, the primary drying is achieved in two steps.

1. To the thin carrier disc of each vial, one drop (25–30 µl) of the cell suspension is transferred aseptically with care so as not to touch the side of the vial.

2. Place ready vials quickly in aluminium Erlenmeyer caps and then transfer into a metallic jar maintained at 20°C in a water bath.
3. After 20–30 min of equilibration, apply a vacuum of 30–80 mbar and dry for 4–6 h.
4. For further drying, adjust vacuum to approx. 20–25 mbar and dry overnight while maintaining the temperature at 20°C.
5. Replace vacuum in the metallic jar with N_2 gas and transfer the vials to soft glass tubes (135 × 15 mm) containing silica gel and cotton plugs. Add glass wool, constrict outer tubes, attach to the manifold of a freeze-drying machine and vacuum-dry for further 2–3 h at 0.01–1 mbar as described in the section on centrifugal freeze-drying.
6. At a vacuum of at least 0.1 mbar, the outer tubes are flame-sealed at the middle of the constriction.

Storage and Recovery

It was shown that the viability of vacuum-dried samples depends on the storage temperature (Banno and Sakane, 1981). As a rule of thumb, it was found that survival values were comparable when ampoules were stored at 5–9°C for a number of years or at 30–37°C for the same number of weeks (Malik, 1999). Thus ampoules should be kept in a dark cold place, or even better, a refrigerator.

Resuscitation

Please observe general safety precautions and wear protective goggles when opening ampoules. To open double-vial ampoules, the pointed part of the outer tube is heated in a Bunsen flame. Place two or three drops of water onto the hot tip to crack the glass. Strike off the glass tip with an appropriate tool (e.g. forceps). The inner vial is taken out and about 0.2–0.5 ml of fresh medium is added to the dried sample in order to dissolve and resuspend it. After about 2–5 min the content of the inner vial is usually rehydrated and can be transferred to a tube with the appropriate cultivation medium. In several cases it was observed that the ingredients of the protective medium can inhibit growth of fastidious strains in the first culture tube. Hence, a serial dilution of at least three or four tubes should be prepared. If the preserved culture can be grown easily on solid media, a few drops of the rehydrated sample should also be tranferred to an agar plate or slant to obtain single colonies in order to check the purity of the strain.

In the case of obligate anaerobic strains, it is important to retain anoxic conditions during all steps after opening of the ampoule. This can be achieved in several ways depending on the used anaerobic technique and available equipment in the laboratory. If the Hungate technique is used, it is recommended to keep the inner vial under a flow of oxygen-free gas by inserting a gassing cannula until the cell pellet is completely resuspended. If the ampoule should be opened in an anaerobic gas chamber, it is necessary to score the ampoule with a sharp file at the middle of its shoulder about one cm from the tip. Transfer the ampoule with the file mark in the anaerobic chamber and strike the ampoule with a file or large forceps to remove the tip. If necessary, wrap the ampoule in tissue paper and enlarge the open end by striking with a file or pencil, then remove the glass wool insulation and the inner vial. Gently raise the cotton plug and sterilize the upper part of the inner vial by wiping it with tissue paper soaked in 70% ethanol. After opening of the ampoule, add approximately 0.5 ml of anoxic medium to resuspend the cell pellet and transfer the suspension to a vial with the recommended cultivation medium (5–10 ml).

Viability testing

It is advisable to determine the survival rate of a preserved culture in order to determine the success of the vacuum-drying procedure. Sometimes, if no living cells can be recovered from lyophilized samples, it has turned out that cultures were incubated too long prior to preservation and thus were already inactive. For determination of the viability before the preservation process, an aliquot of the suspension used to fill the inner vials is inoculated in the appropriate culture medium and serially diluted to extinction in order to determine the approximate number of living cells. The same procedure is then repeated with a vacuum-dried sample. Although the stability of most dried cultures is satisfactory, there are also examples of fastidious strains that lose viability within several years of storage. Therefore, the survival rate of important vacuum-dried stock cultures should be checked in intervals of at least 5 years.

◆◆◆◆◆ DEEP FREEZING

Freezing and storage of cultures in liquid nitrogen has the advantage that the survival rates of the more susceptible microorganisms are usually higher as with methods based on vacuum-drying and that the procedures can be carried out easily under an oxygen-free atmosphere. Therefore, cryopreservation is the most effective method for the maintenance of fastidious thermophiles which grow only to a very low cell density or strains that are extremely sensitive to oxygen, e.g. most representatives of the hyperthermophilic methanogens (*Methanocaldococcus* spp. etc.). Moreover, this procedure is in general applicable to all known prokaryotes and hence has been established as standard preservation method for most seed stocks of prokaryotes held at the DSMZ culture collection.

Equipment

The only major investment which is necessary to establish this method is a cryogenic storage tank for liquid nitrogen or alternatively a mechanical deep freezer, which can cool below −70°C. However, deep-freezers are only second quality and should only be used if a regular supply with liquid nitrogen cannot be guaranteed. The liquid nitrogen container has to be equipped with storage canes and canisters for the storage of glass capillaries in the liquid phase of nitrogen or racks with dividers for the storage of plastic cryotubes in the vapour phase. Further useful equipment includes standard laboratory accessories and diverse glass ware which can be obtained easily at low costs.

Preparation of Cell Suspension

Preparation of suspending medium

Fresh culture medium containing 10% (v/v) glycerol or 5% (v/v) dimethyl sulfoxide (DMSO) is used as a suspending medium. DMSO is often more satisfactory than glycerol, because it requires less time to penetrate the organism. The cryoprotectant is sterilized separately by autoclaving in anoxic test tubes under nitrogen gas atmosphere (DMSO: 10 min, 115°C). Add cryoprotectant aseptically just before use to the appropriate sterile, aerobic or anaerobic cultivation medium.

Preparation of cultures

The cultivation and harvesting of aerobic and anaerobic thermophiles can be done as described in the corresponding paragraphs of the section on centrifugal freeze-drying. Again, the survival rates of aerobic thermoacidophiles like *Metallosphaera* or *Sulfolobus* species can be increased by neutralizing the growth medium prior to harvesting with some solid calcium carbonate.

Distribution in Aliquots and Freezing

Cell suspensions can be distributed for freezing and subsequent storage either in screw cap plastic ampoules (cryotubes) or glass capillaries. Cryotubes have the advantage that the filling with suspension and the recovery of cultures is easy and needs no special training or equipment. However, there is always the risk of an imperfect sealing that could cause seepage of liquid nitrogen into the screw cap vial with subsequent explosion upon thawing. Therefore, it is safer to store cryotubes in the vapour phase, instead of submersing them in liquid nitrogen. In addition, plastic cryotubes can not be used for strictly anaerobic thermophiles. The material of the ampoules is not gas tight so that oxygen can penetrate the vial and irreversibly damage the cells. In contrast, glass capillary tubes can be hermetically sealed and are then absolutely impermeable to liquid nitrogen and gases. They can be stored in the vapour phase as well as in the liquid phase of nitrogen and need very little storage place due to their small size.

It has been found that for the cryopreservation of various filamentous fungi cooling rates in the range of -0.5 to $-200°C$ min^{-1} over the critical period from $+5$ to $-50°C$ were optimal (Smith and Thomas, 1998). However, it seems that controlled cooling rates may not have a significant effect on the survival of the much smaller cells of prokaryotes and hence this technique is not applied in the protocols given below.

Plastic ampoules

Polypropylene screw cap ampoules (1.8 ml vol.) can be obtained sterilized from the manufacturer. Aliquots of the concentrated cell suspension (approx. 0.5–1.0 ml) are distributed aseptically in the cryovials, which are then secured by tightly screwing down the lids. Place the cryotubes in suitable racks and then transfer in the vapour phase of liquid nitrogen for freezing.

Glass capillary tubes

The procedure for the freezing of microorganisms in glass capillary tubes is based on the description given by Hippe (1991) and illustrated in Figure 15.4A.

Preparation of glass capillary tubes

Glass capillary tubes (length 90 mm, outer diameter 1.4 mm, wall thickness 0.26 mm) are rinsed several times in distilled water and then dried. Mark capillaries at approx. 2.5 cm from one end with a permanent marker (water resistant ink), place capillaries in test tubes that are closed with aluminium caps and autoclave at 121°C for 20 min.

Filling and sealing of capillaries

1. Transfer 0.5–1.0 ml of the concentrated cell suspension into a small sterile vial, which is placed in an ice bath. For the preservation of anaerobes, the vial is kept anoxic by gassing with an oxygen-free gas mixture that complies with the gas atmosphere used to cultivate the respective strain. However, CO_2 gas should be replaced with a mixture of 80% N_2 and 20% CO_2, in order to avoid problems related to the high solubility of CO_2 in water, which eventually could lead to the explosion of frozen capillaries upon thawing.
2. One glass capillary is taken from the sterile stock by fitting it to the tip of a micropipettor (e.g. micro-classic pipette controller; Brand GmbH, Wertheim) and enough of the cell suspension is aspirated to fill one-third of the length of the capillary. The volume taken up is approximately 25 µl. The suspension within the capillary is further aspirated until it is about 1 cm from the free end, which is sealed in a fine, hot gas flame. The second seal is made at 2.5 cm from the other end of the capillary (at the mark, which was made prior to sterilizing)

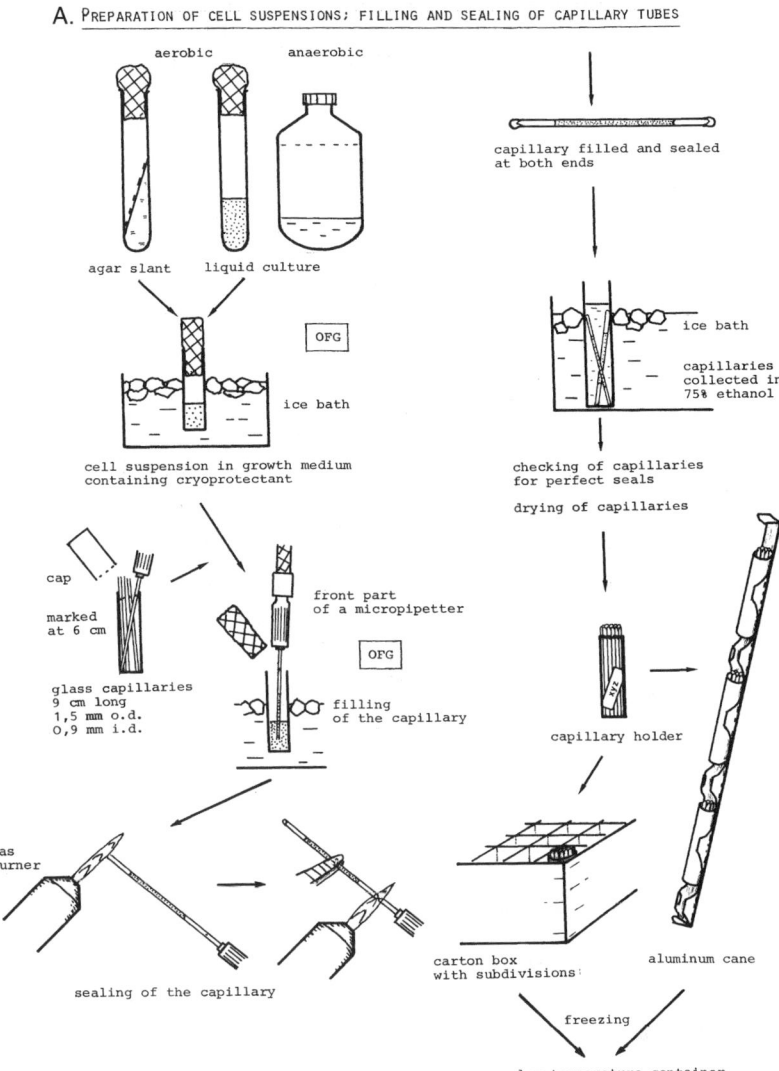

Figure 15.4. Glass capillary tube method for the low-temperature preservation of microorganisms. From Hippe (1991) with permission. (A) Filling, sealing and storage of capillary tubes. (B) Removal of capillaries from freezing storage, opening, and recovery of cell suspension.

by heating and subsequent tearing of the softened glass. The making of the second seal is a critical step of this method and needs some training in order to achieve capillaries that are hermetically sealed.
3. As capillaries are prepared, they are placed in a vial with 75% ethanol for disinfection. The vial is placed in an ice/water bath in order to cool down capillaries immediately after the sealing procedure.
4. All capillaries are examined for correct seals under a stereomicroscope. As with ampoules, improperly sealed glass capillaries will take up nitrogen and will explode on removal from the cold. Nevertheless, a perfect seal is easier to achieve with capillaries. To avoid that moist

B. Removal of Capillaries from Freezing Storage, Opening, and Recovery of the Cell Suspension

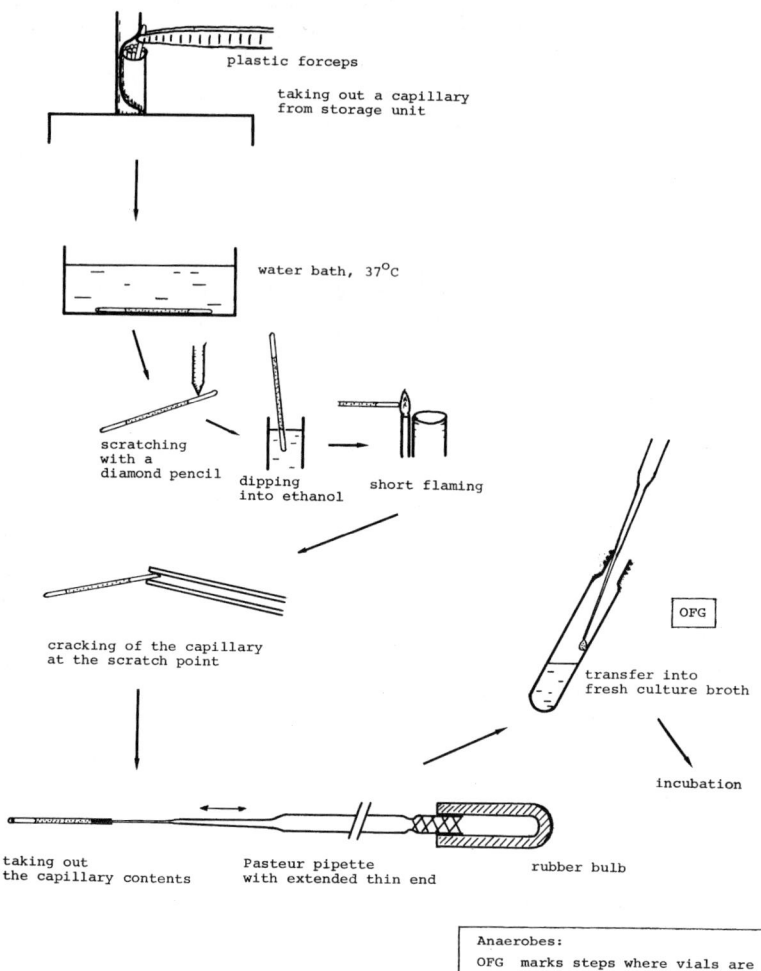

Figure 15.4. *Continued.*

capillaries stick together when frozen, they are dried by placing them between absorbent paper and gently pressing and rolling with the flat hand.

5. The capillaries are stored in a capillary holder (aluminium or polypropylene tube), which is labelled with the strain designation on the outside and then placed in the vapour phase above liquid nitrogen for freezing. The time lapse between preparation of the suspension and freezing should be kept as short as possible. Rapid freezing by immersing capillaries directly in liquid nitrogen should be avoided, because not all capillaries may withstand the developing pressure and could break.

Storage and Recovery

Cryotubes can be easily labelled with special cryo-markers prior to freezing and subsequent storage in racks in the vapour phase above liquid nitrogen.

In contrast, the permanent labelling of capillaries is more difficult, but may be sometimes desirable if samples of different strains should be stored in the same capillary holder. In this case, capillaries can be colour coded by putting them into PVC straws (outer diameter 2 mm) of different colours, which are cut to length and squeezed together at one end. For storage, the capillary holder can be placed into an aluminium cane that is immersed into liquid nitrogen. Usually, several canes are collocated into one canister.

Resuscitation

General safety precautions should be observed for exposure to liquid nitrogen or cryogenic equipment. Wear protective clothing and goggles while handling frozen samples! It has been found that slow warming may cause damage of cryopreserved samples due to the recrystallization of ice, therefore rapid thawing is recommended. To achieve the required fast thawing rates, frozen samples are immersed in warm water (30–37°C) immediately upon removal from the liquid nitrogen container. It has been observed in several cases that, strictly chemolithoautotrophic strains are inhibited by a remaining amount of the cryoprotectant in the first inoculated culture tube. Therefore, it is recommended to prepare a serial dilution of two or three tubes for the resuscitation of sensitive autotrophic strains.

Plastic cryotubes

Thaw the cryotube in a suitable container filled with warm water until the last visible ice has melted. Unscrew the cryotube, remove sample aseptically by using a Pasteur pipette and inoculate a suitable growth medium.

Glass capillary tubes

For the recovery of a strain, one capillary is removed from the liquid nitrogen tank and thawed rapidly in a container with warm water. The capillary is removed from the water bath, dried and opened at one end as shown in Figure 15.4B. In the case of aerobic thermophiles, the small volume of cell suspension is aspirated with a sterile Pasteur pipette that has been drawn out very finely to a length of 4 cm. While aspirating the suspension, the tip of the pipette is slowly moved further into the capillary. The suspension is then transferred to 5 ml of the freshly prepared appropriate medium. For the recovery of anaerobic strains, it is recommended to use a 1 ml disposable (tuberculin) syringe with a 25G hypodermic needle. Prior to use, the syringe is flushed with oxygen-free gas. The capillary is opened at both ends and the contents aspirated into the syringe while avoiding to uptake air bubbles along with the cell suspension. After aspirating the cell suspension from the capillary, the needle is inserted through the rubber closure of an unopened tube with 5–10 ml of anoxic medium. The tube is inverted and the suspension along with some medium is drawn into the syringe. Then, the tube is still in an inverted position, the contents of the syringe is injected into the tube.

Viability testing

The determination of survival rates after cryopreservation can be done according to the description given above in the section on vacuum drying. However, the testing of preserved cultures in regular intervals is not so important than with vacuum-dried samples, because the stability of frozen cultures stored in liquid nitrogen is practically unlimited.

Further detailed information on preservation methods is available from the CABRI consortium, which has developed several guidelines for the maintenance of microorganisms (URL: http://www.cabri.org).

List of suppliers

Air Liquide, Division Matériel Cryogénique
75 Quai d'Orsay
75321 Paris cedex 07, France
http://www.airliquide.com

Cryogenic storage vessels

A. Albrecht GmbH & Co. KG
Hauptstrasse 6-8
D-88326 Aulendorf, Germany

PVC straws

Chart Industries, Inc., MVE Bio-Medical Division
1800 Sandy Plains Industrial Parkway
Marietta, GA 30066, USA
http://www.chartbiomed.com

Cryogenic storage vessels

FGT Feingerätetechnik GbR
Ernst-Thälmann-Str. 27
D-99510 Apolda, Germany
http://www.fgt.de

Ampoule constrictor

Hilgenberg GmbH
Strauchgraben 2

D-34323 Malsfeld, Germany
http://www.hilgenberg-gmbh.de

Glass capillary tubes

Lehmann & Voss & Co.
Alsterufer 19
D-20354 Hamburg, Germany
http://www.lehvoss.de

Glass wool

Martin Christ Gefriertrocknungsanlagen GmbH
P.O. Box 1713
D-37507 Osterode am Harz, Germany
http://www.martinchrist.de

Freeze-drying machines

Nalge Nunc International Corp.
75 Panorama Creek Drive
Rochester, New York 14625-2385, USA
http://www.nuncbrand.com

Cryotubes

Ochs Glasgerätebau GmbH
Pappelweg 26
D-37120 Bovenden/Lenglern, Germany
http://www.labor-ochs.de

Glassware

Paul Hartmann AG
P.O. Box 1420
D-89504 Heidenheim, Germany
http://www.hartmann-online.com

Cotton plugs

Gebr. Rettberg GmbH
Rudolf-Wissell-Strasse 17
D-37079 Göttingen, Germany
http://www.rettberg.biz

Glassware

Sigma-Aldrich Chemie GmbH
Eschenstrasse 5
D-82024 Taufkirchen bei München, Germany
http://www.sigmaaldrich.com

Chemicals, Silicagel rubin

Vacuubrand GmbH & Co. KG
Alfred-Zippe-Str.4
D-97877 Wertheim, Germany
http://www.vacuubrand.de

Vacuum pumps

References

Annear, D. I. (1956). The preservation of bacteria by drying in peptone plugs. *J. Hyg.* **54**, 487–508.

Banno, I. and Sakane, T. (1981). Prediction of prospective viability of L-dried cultures of bacteria after long-term preservation. *Inst. Ferment. Osaka Res. Commun.* **10**, 33–38.

Brock, T. D. and O'Dea, K. (1977). Amorphous ferrous sulfide as a reducing agent for culture of anaerobes. *Appl. Environ. Microbiol.* **33**, 254–256.

Hippe, H. (1991). Maintenance of methanogenic bacteria. In *Maintenance of Microorganisms and Cultured Cells* (B. E. Kirsop and A. Doyle, eds), 2nd edn., pp. 101–113. Academic Press, London.

Malik, K. A. (1990). A simplified liquid-drying method for the preservation of microorganisms sensitive to freezing and freeze-drying. *J. Microbiol. Meth.* **12**, 125–132.

Malik, K. A. (1991). Maintenance of microorganisms by simple methods. In *Maintenance of Microorganisms and Cultured Cells* (B. E. Kirsop and A. Doyle, eds), 2nd edn., pp. 121–132. Academic Press, London.

Malik, K. A. (1999). Preservation of some extremely thermophilic chemolithoautotrophic bacteria by deep-freezing and liquid-drying methods. *J. Microbiol. Meth.* **35**, 177–182.

Morris, G. J. (1981). *Cryopreservation*. Institute of Terrestrial Ecology, Cambridge.

Smith, D. and Thomas, V. E. (1998). Cryogenic light microscopy and the development of cooling protocols for the cryopreservation of filamentous fungi. *World J. Microbiol. Biotechnol.* **14**, 49–57.

Psychrophiles

◆◆

16 Handling of Psychrophilic Microorganisms
17 Proteins from Psychrophiles

16 Handling of Psychrophilic Microorganisms

Nick J Russell[1] and Don A Cowan[2]

[1] Faculty of Life Sciences, Imperial College London, Wye campus, Wye, Ashford, Kent TN25 5AH, UK;
[2] Department of Biotechnology, University of the Western Cape, Bellville 7535, Cape Town, South Africa

◆◆

CONTENTS

Introduction
Sampling from different habitats
Culture media
Culturing in a field laboratory
Isolation of community DNA in a field laboratory
Phylogenetic analysis of psychrophilic communities

◆◆◆◆◆ **INTRODUCTION**

A Cold Planet

In comparison with other planets in our solar system, Earth is regarded as having a generally rather agreeable climate on the basis that it has been colonized by such a diversity of organisms. Nonetheless, from a human perspective, three-quarters of Earth is cold, i.e. more or less permanently below 5°C. It is extreme in the sense that we are comfortable living in the other quarter, often only with the aid of clothing and shelter from the cold (or heat). The major cold regions of Earth are the deep oceans (accounting for nearly 70% of the Earth's area), the poles and high mountains. Within these regions, there is a greater variety of habitats than might first be imagined, differing not only in whether they are permanently cold or undergo thermal fluctuations, but also in parameters such as nutrient status, water activity, salinity and pressure (Table 16.1). Such attributes, rather than temperature *per se*, will have an important bearing on the selection of appropriate media for the isolation of psychrophiles from these habitats. It is also reflected in the wide diversity of types of cold-adapted microorganisms that include aerobes and anaerobes, heterotrophs and autotrophs, photoautotrophs, chemolithotrophs and chemoautotrophs, spore-formers and non-spore-formers, etc.

Table 16.1 Cold habitats on Earth and some of their distinctive features

Category	Habitat	Comments
Deep oceans*	Water column	Generally <5°C and liquid at −2 to −5°C. Pressure increases by 1 bar per 1000 m depth of water.
	Sediment	Average pressure is ~60 bar
Antarctica	Soil	Approx. 1% of the continent exposed as soil during summer. Wide range of soil types from rich ornithogenic to dry gravels. Some specialised microniches (e.g. endolithic and hypolithic).
	Ice/snow	Permanent ice sheets (East and West Antarctic) on land and sea (Ronne, Larsen and Ross). Ice depth at South Pole is ~2.5 km. Seasonal sea ice. Abundant glaciers. Frozen lakes and ponds. Annual snowfall decreases from coast to pole.
	Water	Lakes and ponds vary from freshwater to highly saline. Meltwater streams in summer.
Arctic	Ice/snow	Floating polar ice cap on sea. Seasonal sea ice and land glaciers associated with surrounding land masses.
	Water	Meltponds in ice cap. Rivers, streams and lakes in surrounding land masses.
Mountains	Soil	Wide variety of types.
	Ice/snow	Glaciers. Large seasonal changes in snow cover.
	Water	Lakes and meltwaters, mainly freshwater.
Man-made	Ice etc.	Refrigeration systems and foodstuffs. Temperatures typically between +4°C and −20°C.

*See Chapter 31 for isolation of piezophilic/piezotolerant microbes.

Generally, the more extreme the conditions of an environmental niche, the lower is the diversity of organisms that are capable of growing, and the most extreme environments are dominated by microorganisms. Taking Antarctica and surrounding regions as an example, in the sub-Antarctic islands and peninsula one finds the grass *Deschampsia* and abundant mosses growing; as you move to continental Antarctica, mosses are to be found generally only in coastal areas and lichens are rarer, whilst in the dry valleys where the conditions are particularly harsh, mosses are generally absent (except in highly localized hypolithic habitats) and it is often difficult to find lichens at lower altitudes. In all of these regions, Bacteria, yeast, microalgae and fungi are found, with Bacteria and yeast dominating the most extreme habitats and micro-niches (Cowan and Ah Tow, 2004); only in deep oceanic waters, do the numbers of Archaea match those of Bacteria, but this is based on abundance of molecular/biochemical markers rather than culture-based studies. Archaea, such as methanogens, have been isolated and cultured from a variety of cold habitats, and these microorganisms will require the use of media and (anaerobic) handling methods appropriate to their phenotype. In what follows, no distinction is made between Bacteria and Archaea: the terminology "bacteria" is used, since the general considerations of sampling and handling Bacteria will apply also to Archaea.

Psychrophiles and Psychrotolerants

Despite the fact that so much of our planet is cold, the majority of cold-adapted microorganisms are psychrotolerant (cold-tolerant or psychrotrophic, a terminology favoured by the food industry in particular) rather than psychrophilic (cold-loving). The generally accepted definition of these terms is that given by Morita in a definitive review of psychrophilic bacteria (Morita, 1975). He based his definitions on the cardinal growth temperatures, *viz.* lower limit, optimum and upper limit. Psychrophiles grow at or below zero (0°C) and have an optimum growth temperature ≤15°C and an upper limit of ≤20°C. In contrast, psychrotolerants have optima and upper limits above these temperatures and may well grow at mesophilic temperatures with optima above 30°C, whilst retaining the capacity to grow at or close to zero. In fact, some psychrotolerants grow as fast as psychrophiles at low temperatures, i.e. below 5°C.

Of course, it should be remembered that this classification is artificial. Nature is not bound by such convention, so it is not uncommon to isolate cold-adapted microorganisms that bridge the two definitions. For example, our experience is that Antarctic soil bacterial isolates may have both their optimum and upper limit between 15 and 20°C, or they may have an optimum that is <15°C but an upper limit of >20°C. There is also a problem with defining the lower limit for growth. Russell (1990) argued that this is approximately −10°C, the lower limit for supercooled water when the freezing of bulk water will increase the concentration of salts to levels inimical to growth. The lowest recorded growth temperature that is authenticated by growth curve data is −12°C for "Psychromonas ingrahamii", which has a doubling time of 4 h at this temperature (Breezee et al., 2004), whilst viable bacteria can be found in some natural permafrost environments where the temperatures may be as low as −20°C (Gilichinsky, 2002). Moreover, metabolic activity has been measured in bacteria and lichens at these low temperatures and, despite the constancy of the low temperature, most bacterial isolates from permafrost are psychrotolerant rather than psychrophilic with upper limits around 30°C. This serves to emphasize the fact that the main difference between the two groups of cold-adapted microorganisms is not so much their ability to grow well at low temperatures, but rather the extension of growth to higher temperatures for psychrotolerants, giving them a greater growth temperature range compared with psychrophiles.

It has been stated by a number of authors that psychrophiles are more likely to be found in thermally stable, permanently cold habitats, whilst psychrotolerants dominate those which fluctuate in temperature. However, even in some permanently cold habitats, the dominant populations may still be psychrotolerant. This does not appear to be the result of thermal abuse (warming) during isolation, because investigations in which temperature has been rigorously controlled have come to the same conclusion.

For yeast, fungi and microalgae, these same definitions of psychrophilic and psychrotolerants are usually applied, although some researchers do use slightly different cut-off temperatures. For instance, 25°C is often used as the limiting growth temperature for defining psychrophilic yeasts, most of which have been discovered in Antarctica or Australasia, but not in other locations (Vishniac, 1999), in sharp contrast to other microbial types. The same cut-off temperature of 25°C is also commonly used to define other eukaryotic psychrophilic microorganisms. One of the problems in using the definitions is that different strains belonging to a single yeast species may vary in their upper growth temperature limit by 5–6°C. And the most common ascomycete fungus in Antarctica, *Thelebolus microsporus*, is psychrophilic or psychrotolerant depending on the strain (Onofri, 1999). Notwithstanding these considerations, just as for bacteria, the numbers of psychrophilic yeast, fungal and microalgal species are far outnumbered by psychrotolerants.

Given the fact that most cold-adapted microorganisms can be classified as psychrophiles or psychrotolerants, many researchers have sought to find underlying physiological and biochemical differences for these definitions. Whilst some differences do exist (e.g. in the production of cold-shock proteins in response to a sudden decrease in temperature), no unifying hypothesis has emerged. For the remainder of this review, except when it is important for effective practical manipulations, we shall not distinguish between the two groups of cold-adapted microorganisms and the term "psychrophiles" can be taken to include "psychrotolerants".

◆◆◆◆◆ SAMPLING FROM DIFFERENT HABITATS

Some Thermal Considerations

A *conditio sine qua non* of research on cold-adapted microorganisms, not just their isolation but also identification and subsequent biochemical and molecular investigations, is that the samples, cultures and extracts should be kept cold. Obviously, this applies also to equipment such as scoops, bottles, ice axes and small corers etc. used for collecting, and to the flasks, bottles and other vessels used for transfer. Sterile plastic bags (e.g. Whirl-Pak bags, Nasco) are particularly useful for holding and transferring samples because they can be purchased sterile (or sterilized easily by irradiation or autoclaving), they are light to carry and they do not significantly warm samples even if the bags have not been kept rigorously cold. Good planning is essential not only to ensure that all necessary equipment, reagents etc. are taken into the field, but also to practise sampling protocols in order to minimize the number of transfers, which helps to reduce the chances of contamination.

Psychrotolerant isolates are likely to be more tolerant of thermal abuse (warming) than psychrophiles. Nonetheless, as long as the initial numbers are not too low, it is probably true that populations of even the most extreme psychrophiles can withstand some warming above their upper limit. This point has not been tested systematically and so anyone isolating psychrophiles or psychrotolerants is advised to keep any warming to a practical minimum, because it damages the selective permeability properties of cells, which will lose essential nutrients and other constituents in a time- and temperature-dependent fashion (Russell, 2003). This does not pose a problem for sampling in places where the environment is such that natural "cryoholes" can be dug in ice or permafrost to act as temporary freezers/refrigerators (they may not be cold enough to freeze liquid samples). Small cryogenic storage devices that use liquid nitrogen are also available (Thermolyne Arctic Express, Fisher Scientific), which will keep small volumes of material (therefore better for DNA samples than microbial isolations) frozen for up to a month. In Antarctica, there are facilities on national bases for refrigerating/freezing materials as well as transporting them cooled or frozen to home laboratories by ship or air freight. The Arctic is less well served by cold-storage facilities on bases and for instance, depending on the weather, it can take 2–3 days to return from sampling sites in the Canadian High Arctic, even with the support of modern transport facilities (L. Whyte, personal communication). In other remote cold habitats, such as high mountain ranges with no organized national facilities, it can be very difficult to keep samples permanently cold if the route out to the nearest refrigerator/freezer involves trekking through hot lowland, possibly (sub-)tropical regions. Portable cool-boxes may be of limited benefit in such circumstances and rapid transfer will be the best strategy to prevent thermal abuse. The use of a downloadable electronic data logger placed inside the sample carrier to record temperature is helpful to know the true thermal history of collected samples under these or any circumstances.

Some articles report the isolation of cold-adapted microorganisms using two or even three isolation temperatures, commonly 5 and 25°C, possibly with 15°C, in order to distinguish psychrotolerants and psychrophiles. However, in the light of the comments above on cardinal growth temperatures, such a strategy will fail to yield these two groups. It is much better to incubate plates and liquid cultures for primary isolations of bacteria below 5°C, which will yield both groups of cold-adapted microorganisms. For yeasts, fungi and algae, temperatures as low as 10°C are probably sufficient.

Soils and Sediments

The methodology for sampling soils in cold regions is essentially no different to that for temperate soils, the main concern (apart from keeping everything cool) being to ensure the statistical validity of the sampling regime so that the samples are representative of the particular habitat. This can be a problem if the soil areas are relatively small or rather heterogeneous, such as occurs in many mountainous sites and some regions of Antarctica. There may also be practical problems of taking large samples because of weight restrictions on carrying bulk samples from remote regions. In Antarctica, also for reasons of environmental protection and to comply with the Antarctic Treaty, the extent of sampling may need to be limited (UNOG, 2000).

Permafrost soils present physical difficulties when sampling and problems of the gratuitous introduction of contaminants occur on the tools such as hammers, pick axes and corers used to obtain samples. In such circumstances, it is best to take samples that are as large as possible so the outer layers can be removed aseptically in the laboratory and the inner core can be sub-sampled (see below for a discussion of the treatment of ice cores). Permafrost soils can be cored in the same way as ice and many of the same concerns about contamination with exogenous microorganisms apply. Even at the usual environmental temperatures around $-10°C$, permafrost soils composed of sand, loam or clay (and mixtures) contain at least 1–2% of their water in the liquid state, tightly bound to soil particles. These water films are too thin (5–75 Å) to permit the thermal diffusion of much larger bacteria (0.5–1 µm) from the exterior to the interior of the core. Experiments in which frozen core segments have been seeded directly with *Serratia marcescens* (which is psychrotolerant) for several hours to several months at $-10°C$ have shown that the exogenous bacteria do not penetrate the cores. Thus, the best guard against contamination is the use of clean drilling procedures and maintenance of the core in its frozen state.

Shi et al. (1997) describe such a procedure. Permafrost cores up to 30 cm long are obtained by using a portable gasoline-powered drill that operates without a drilling fluid using a hammer action. The cores are removed from the corer and their surfaces cleaned by shaving with an alcohol-sterilized knife, cut into 5-cm lengths and placed in metal boxes for storage in the field in a cave dug in the permafrost and subsequent transport to the laboratory. At all times, the cores were kept frozen. To test for possible contamination during drilling, an experiment was conducted in which the drilling barrel was seeded with *S. marcescens* for 2 h prior to drilling. The exogenous *S. marcescens* was only ever found on the surface of cores, never in the interior.

In the summer of 2004, two research teams (Canadian and Russian) independently used several new tests, including molecular methods, to check for possible contamination of permafrost and ground-ice core samples during drilling. Their conclusions were the same and the Canadian results have been published by Juck et al. (2005), which gives details of the procedures used. Briefly, fluorescent latex microspheres

(Polyscience Inc., 0.5 µm diameter, to mimic the size of bacteria) were painted on the surfaces of the drilling equipment and allowed to dry, a procedure that ensured better transfer to the core surface during the drilling procedure. The microspheres can be monitored very sensitively by their fluorescence and serve as a microbial surrogate to monitor even low levels of contamination that could be used routinely without running the risk of actually contaminating the ice cores. The microspheres penetrated <1 cm into the core as a result of the drilling process. In order to monitor microbial contamination during subsequent handling and processing of the cores, Juck et al. (2005) used a psychrotolerant strain of Pseudomonas that was genetically modified by insertion of the gene for green fluorescent protein (*gfp*) to measure exogenous bacterial contamination. They used both culturing of the bacterium and PCR amplification of the *gfp* marker gene to show that there was no transfer of bacteria to the core interior.

Soil sampling for phylogenetic analyses

The requirements of sampling for phylogenetic analyses are substantially more stringent than for culture-dependent studies. The requirements for aseptic sample acquisition and handling are as great (a few human skin cells or hairs can introduce a large number of foreign organisms), particularly as the appearance of sequences showing high homology to *Escherichia coli* or *Staphylococcus epidermidis* in 16S rRNA gene environmental libraries is embarrassing evidence of a failure of sterile technique. However, the use of appropriate sampling protocols when undertaking phylogenetic studies of microbial diversity is even more critical. Such protocols must necessarily show evidence of giving sufficiently representative coverage of the habitat, taking the possibility of microheterogeneity into account.

The terrestrial soil microbiologists have been addressing this problem for some years, and well-established protocols have been established and accepted (e.g. Theocharopoulos et al., 2001). Our derivation of these protocols for the sampling of Antarctic mineral soils (Cowan and Cary, unpublished results) is as follows:

a. A transect line is laid across the target area, with intervals (e.g. 50 m for >1 km transects) being located by GPS.
b. At each interval mark a 50 m line, centred on the interval mark, is run out perpendicular to the transect line.
c. Three sampling sites are identified (at the central point and the outer ends of the perpendicular line) and a 1 m square grid (typically coloured string) is laid at each site.
d. Working from the outside of each square, four samples are recovered, one from each quadrant of the grid. We use metal spatulas, which are wiped with ethanol-saturated swabs between samples. Samples, typically 20–100 g are collected in sterile Whirl-Pak bags.
e. A fifth sample is generated by transferring half of each of the four samples to a large sterile plastic bag, mixing vigorously, and resampling.

This sampling protocol, although reasonably tedious, provides the materials both for determination of microheterogeneity across different scales (1 m, 25 m, 50 m or greater, as appropriate) using a technique such as DGGE, as well as for more detailed phylogenetic analysis.

Aquatic Sources

The methodology for sampling cold waters is essentially the same as that for temperate water bodies. It is well established for all types of habitat from small ponds and lakes to surface and deep oceanic collections, and is dealt with in a number of texts (e.g. Sherr et al., 1993). By contrast with other habitats, in deep oceanic waters Archaea match or outnumber Bacteria; this has been demonstrated using surveys of rRNA gene sequences or archaeal lipid signatures, and culture-based methods have yet to be developed to demonstrate this diversity. The special considerations for sampling piezophiles (barophiles) in deep oceanic waters that are under high pressure are dealt with in Chapter 3. The main concerns of sampling in surface oceanic and other relatively shallow waters, such as lakes, ponds, rivers and streams are to sample aseptically and to ensure that sampling is representative. The question arises as to whether samples should be kept refrigerated or frozen, since there is evidence from laboratory studies that some psychrophiles in dilute media analogues of natural waters are easily lysed by freeze–thawing in the absence of cryoprotectants. However, to our knowledge, there has not been a systematic study of whether freezing samples reduces the subsequent biodiversity observed in collected environmental samples. The counterargument is that, since psychrophiles grow readily at refrigeration temperatures, freezing is necessary in order to preserve the natural relative abundance of species for subsequent biodiversity studies of collected samples. For this reason, it is best to make direct platings or to extract DNA from samples in the field in order to determine *in situ* biodiversity. If that is not possible, then one should take duplicate sets of samples, freeze one and store the other at about 5°C.

If possible, aqueous samples should be filtered through membrane filters (Millipore, 0.45 µm for vegetative cells or 0.22 µm if spores are to be collected) using autoclavable or disposable plastic single-operation filtration units. Lightweight, portable hand-driven pumps (Nalgene, Fisher Scientific) are available to facilitate filtration if a power-driven pump cannot be used, for instance in remote field locations. Duplicate sets of samples should be taken so that one set of filters can be stored cool in 50% (v/v) sterile aqueous glycerol and the other, frozen. As above, the dual strategy is necessary because there will be growth of psychrophilic yeasts and fungi in particular on the glycerol (since even filtered psychrophiles can grow as a biofilm on filters kept at refrigeration temperatures), but freeze–thawing will lyse some isolates and the glycerol acts as a cryoprotectant.

Many aquatic psychrophiles will be oligotrophic. Button and coworkers have successfully developed an extinction–dilution procedure in sterile sea water to enhance the recovery of marine oligotrophs (Button et al., 1993), an approach that has been used in combination with a high throughput procedure to grow oligotrophic bacteria (Connon and Giovannoni, 2002).

Ice

It is inadvisable to melt ice in order to obtain primary samples in the field, as this is likely to lyse many members of the microbial community due to the lack of cryoprotectants. If thawing is necessary, if should be done slowly into isotonic medium at <5°C. Great care is needed to maintain aseptic conditions when chipping or cutting solid samples, which should be large enough (up to 10 cm) for the outer layers (a few cm) to be removed and discarded subsequently under controlled conditions in the laboratory. Such a strategy would apply, for example, to sampling glacier ice from vertical surfaces (it is easier to core through horizontal ice layers). Samples should be transferred to sterile bags or containers as quickly after collection as possible and maintained in the frozen state at as low a temperature as is practicable, preferably below −20°C. Plastic containers can be sterilized by soaking overnight in 70% (v/v, aq.) ethanol and UV-irradiating for ≥ 1 h under a microbiocidal lamp. Metal containers can be soaked in ethanol and flamed prior to use. Some researchers prefer to melt away the outer 1–2 cm of collected ice samples by washing with sterile distilled water prior to placing in sterile bags/containers, in order to remove immediate contamination associated with the sampling process. Cores can be cut into convenient lengths using a surface-sterilized saw to make handling and transport easier. Ice temperatures can be measured *in situ* with a sterile thermistor probe, which can be left in place connected to a data logger, if there is concern about thawing during storage in the field and transfer to the laboratory.

The traditional ice sample is a core, and the techniques for taking such cores were developed largely by paleoscientists interested in obtaining ancient records of conditions on Earth long before it was realized that bacteria could remain viable in glacial ice for hundreds of thousands of years and in deep Antarctic ice (or permafrost) for a few millions of years. The cores should be of sufficient diameter (≥ 10 cm) to allow for up to 2–3 cm to be stripped as part of the decontamination process on return to the laboratory (see below). Thus, it is usually necessary to use motorized augurs, which involves the addition of an organic drilling fluid such as kerosene to prevent collapse of the borehole. Besides not being sterile, these fluids also act as growth substrates for a wide range of microorganisms; ice cores are also handled during transport and storage, giving further opportunity for contamination. In the past decade, there have been attempts to obtain ice cores aseptically, but this is not feasible for deep coring. Instead, attention has been focussed on decontaminating the collected cores using a variety of techniques, including different washing regimes, irradiation with ultraviolet light, aseptic subcoring or melting out the interior of the core using a specially

Table 16.2 Decontamination of ice cores

Procedure	Method
Washing	Na hypochlorite (Clorox bleach), HCl, NaOH, ethanol, sterile water, H_2O_2
Mechanical	Scraping with sterile razor blade
Irradiation	Ultraviolet light
Subcoring	Cleaned ice core is aseptically subcored in laboratory
Interior sampling	Device melts core interior, which is collected as liquid water

Summarized from data in Rogers et al. (2004) and Christner et al. (2005).

designed heating/collecting device (Table 16.2). Rogers et al. (2004) have compared the most widely used protocols for decontaminating ice cores. Under controlled laboratory conditions, they generated "sham" ice cores containing either cells or DNA incorporated into the ice at concentrations found in glacier ice; they also painted different species of microorganisms on the outside. It was concluded that only treatment with 5% sodium hypochlorite (diluted commercial bleach solution) eliminated all external contamination by microorganisms and DNA, whilst maintaining the integrity of the ice core.

Christner et al. (2005) provide electron microscopic evidence that the bacteria (mostly bacilli) contaminating the outer layers of ice cores are clearly different from the predominant micrococci of the interior of the cores. Numbers of contaminating bacteria in the outer layers were three orders of magnitude higher than in the interior of the core. These authors deliberately contaminated cores with bacteria (*Serratia marcescens*), a target DNA sequence and a fluorescent dye (rhodamine 6G) in order to monitor decontamination protocols. They showed that the following procedure was sufficient to remove contaminants: initial physical removal of the outermost layer of the core with a sterile microtome blade at $-10°C$ to expose unhandled ice, followed by washing with 95% (w/w, aq.) sterile ethanol and rinsing in sterile deionised water at $4°C$ to disinfect the core. The cleaned core was partially thawed at $+22°C$ and finally the interior was thawed at $+4°C$ [see Christner et al., 2005 for details, including the decontamination procedures for the handling rooms]. When such rigorous protocols are used, it is possible to obtain melted ice-core interiors that lack any externally applied biological or chemical contamination, so it is assumed that recovered viable bacteria are those trapped in the ice during its formation.

The sampling of sea ice can be problematical in that it forms a continuum with the aqueous phase beneath and care must be taken during coring not to disturb the underlying platelet ice and microbial assemblage at the ice–water interface. Notwithstanding this consideration, sea ice can be cored in the usual manner using hand or motorized augers. This is best done outside the summer-melt times, when the ice is more likely to be consistently frozen hard, using the same techniques for subsequent handling etc., as described above for ice in general. If coring has been done carefully so as not to disturb the underlying water, this too can be sampled using a syringe system directly through the ice-core hole (Krembs *et al.*, 1996).

Sea ice, which typically varies in thickness from 1 to 2 m, contains brine-filled channels in which the salt concentration can be as high as 3 M NaCl (equivalent). Therefore, sea ice samples are generally thawed in sterile seawater medium at 0–4°C to prevent hypotonic shocking prior to making isolations on solid media or enrichments in liquid media. Samples can be thawed in successively in salt solutions of progressively lower concentrations. The temperature of sea ice varies from −1.8°C (the ice/sea water eutectic temperature) to −30°C, and the brine volume of the ice can be calculated from the bulk salinity and ice temperature (Frankenstein and Garner, 1967).

Snow

Snow in the natural environment varies widely in its extent of compaction. By analogy with a method used for soil, compacted snow can be sampled by means of a snow pit to obtain historical information about microbial deposition. Freshly fallen snow is usually less compacted and less contaminated, so therefore generally of most interest, e.g. for studying the transfer of psychrophilic microorganisms from the atmosphere to earth. Care should be taken over the sampling site: for example, it should not only be from a "clean" area but also taken upwind of human activity to avoid human contamination. Large volumes (upto 60–80 l) can be collected using aseptic techniques and stored/transported in robust sealable plastic containers that are capable of being sterilized (e.g. ThermoSafe chests) (Carpenter *et al.*, 2000). However, in many field locations that are remote, it will not be possible to transport large volumes, and it will be necessary to melt (particularly uncompacted) snow, which as for ice should be done slowly at about 5°C. The liquid sample can then be passed through a membrane filter (0.45 μm or 0.22 μm Millipore filters, to recover vegetative cells or spores, respectively) as described above to further reduce the weight and volume that has to be carried from a mountain peak, for instance. Again duplicate sets of samples should be taken, one set stored refrigerated and the other frozen.

◆◆◆◆◆ CULTURE MEDIA

Rich *Versus* Dilute Media

Strategies for the recovery of representative microbial populations from low-temperature environments have not been well developed, but a good dictum to follow, especially for ecological studies, is to change as little as possible the growth conditions of the microorganism(s) compared with the natural growth habitat. The frequent use of non-selective rich media such as nutrient broth for psychrophilic isolations makes it seem unlikely that the microorganisms isolated are those most likely to be adapted to the often oligotrophic conditions of many cold habitats, e.g. snow and ice. An exception are ornithogenic soils, e.g. those from penguin colonies, which in comparison with most Antarctic soils contain much more organic carbon and nitrogen in the form of amino acids and urea, so that rich peptone-based media are suitable analogues of the natural habitat. The numbers of permafrost bacteria isolated on rich media are lower than those in dilute media, but the diversity is greater. The use of inappropriate media may fail to recover those species actively growing in the psychrophilic niche, particularly if the salinity of the ecotype is ignored. Cold habitats often contain high concentrations of salts and other cryoprotective compounds. This is particularly true of sea ice and permafrost, but is also relevant to dry soils such as those in the dry valleys of Antarctica, which also have some of the lowest organic carbon contents of any soils. For habitats such as ice and snow, an obvious strategy is to use sterilized *in situ* ice- or snow-melt medium, i.e. without further addition of a carbon/energy source. This can be mixed with agar (1.5% w/v, preferably Noble agar or an equivalent that is low in soluble carbon sources) to provide a solid medium. The same is true of cold aquatic environments. Even so, there will be some isolates that will utilize the agar as carbon/energy source, but our experience is that these are very difficult to sub-culture (even to remove from the agar surface) and so are seldom the subject of further investigations (although they may be significant components of a population where polymeric carbohydrates are the major carbon/energy source).

As a general rule, it is better to use a range of different media (up to 10 different types is not unrealistic) when making primary isolations and always to include some diluted media. The latter can be commercial media (e.g. nutrient broth, tryptone soya agar, Luria-Bertani agar) diluted to one-half or better one-tenth strength, which may be supplemented with low concentrations of yeast extract (0.01–0.001%) to ensure that vitamin and cofactor requirements are met for fastidious organisms. There is also some justification in using even more dilute media (with long incubation times) as there may well be a greater diversity of oligotrophic psychrophiles than hitherto found. Some researchers prefer to use synthetic media, composed to reflect specific attributes of the habitat being sampled. Other parameters that should be taken into account are pH (e.g. ornithogenic soils in Antarctica have relatively higher organic carbon and nitrogen contents, but a very low pH of 3–4, compared with surrounding soils) and redox potential (e.g. the use of specific media for strict anaerobes such as sulphate-reducing bacteria), which should ideally mimic those of the natural habitat. Finally, it may seem obvious, but it is seldom stated, that media for growing cold-adapted microorganisms should be pre-cooled!

Media for Saline Habitats

If the habitat is saline, then of course salt must be included in the media at an appropriate concentration. For isolations of marine psychrophiles the commercially available general medium Marine 2219 agar (Difco Laboratories) is widely used. It can also be prepared from individual constituents dissolved in artificial or natural seawater. If natural seawater is used it should be filtered (0.22 μm Millipore filters) rather than autoclaved. As an alternative, the general dilute medium R2A (Oxoid) or one of a similar composition can be used dissolved in seawater. Many marine microorganisms are attached to detritus and a preliminary screen can be done by filtering seawater through 1 μm filters. This also helps to prevent clogging of smaller pore-size filters used subsequently to collect free-living isolates.

The use of general media in isolations based on simple platings fails to isolate many marine bacteria that are oligotrophic heterotrophs with fastidious growth requirements. Many novel examples of these bacteria have been isolated using the "extinction culturing" technique developed by Button *et al.* (1993), so-called (despite the title of the original article reporting the method) in order to distinguish it from dilution culturing. In the extinction culturing method, natural seawater samples are diluted with unamended sterile seawater to give very low concentrations of bacterial cells of the order of 1–10 cells per tube. Bacterial growth is monitored over a 9-week period using flow cytometry, which is accurate for small numbers. Bacterial doubling times are of the order of 1 day to 1 week. Although this method is rather laborious, it can increase bacterioplankton culturability from 2 to 60%, and has resulted in the isolation of new oligotrophic species. If carbon sources are added to the seawater to concentrations above 5 mg l^{-1}, the culturability is drastically reduced. To improve the efficacy of extinction culturing, Connon and Giovannoni (2002) developed a high-throughput modification using microtitre plates and a detection system that could measure bacterial concentrations as low as 10^3 cells ml^{-1}. They were able to culture up to 14% of the cells observed by direct counting, many of them new species, which is 1 to 3 orders of magnitude better than normally achieved by direct platings of marine waters.

The same general media described above can be used for deep-sea psychrophiles that are also piezophilic or piezotolerant (see Chapter 31 for techniques of handling microorganisms under pressure). A useful addition to media for piezophilic psychrophiles is fluorinert (Sumitoma Chemical), a liquid with high oxygen-binding capacity that can be used to promote oxygen respiration at high pressures during batch culture in sealed vessels, or pure oxygen can be bubbled through media prior to inoculation (Quereshi *et al.*, 1998). However, it should be remembered that many of the deep-sea microorganisms, whether in water or sediment, are not specifically adapted to this habitat, but rather have been transported by down-currents from the surface to the abyss.

Media for Cyanobacteria

Although they are Bacteria, historically and in relation to the development of media for their growth, cyanobacteria were treated distinctly as "blue green algae". A large variety of media have been developed for their cultivation, but there are none that are specific for low temperature growth. Thus the common general media BG-11, MN and ASN-III are also used for psychrophilic cyanobacteria. Medium BG-11 is preferred for freshwater and soil strains, with the nitrate N-source being omitted for isolation of nitrogen-fixing (autotrophic) strains. Marine strains are generally grown in medium ASN-III or more rarely in MN, which has a natural seawater base supplemented with medium BG-11 at half strength (see Pasteur Culture Collection for details, including the preparation of solid media at www.pasteur.fr/recherche/banques/PCC/Media.htm). Glycerol is not very effective in the cryopreservation of cyanobacterial isolates: 2–12% methanol or 4–15% DMSO is better, depending on the strain, added to half-strength medium.

Media for Yeasts and Fungi

The isolation of yeast and fungi requires the judicious use of antibiotics and careful choice of growth substrate concentrations (see below) to suppress competing bacterial growth, which is often faster particularly on rich media. It appears that researchers have paid more attention to this essential problem than to the use of very low temperatures for psychrophilic isolations, which have most often been made at 10–15°C. Bacteria can be suppressed by using penicillin G and streptomycin sulphate, or the use of gentamycin, added to solid media in plates. The isolation of yeasts specifically also requires the suppression of fungal growth. This is achieved by a neat trick in which the soil, water, ice or snow to be sampled is shaken in liquid minimal medium, which inhibits sporulation and causes the fungi to ball up; the enrichment flask is then easily sampled for yeasts by avoiding the balls of fungal mycelium (alternatively, the fungi can be sampled for their selective isolation). This works for both obligately aerobic and fermentative yeasts.

A minimal medium should be used, such as M3C, which contains relatively low concentrations of glutamate (0.2 mM) and glucose (0.2%) as carbon/energy sources, and is supplemented with yeast extract (0.01%), vitamins and trace elements commonly required by yeast. The concentrations of glucose and yeast extract are critical, and the vitamins do not include riboflavin or folic acid that no yeasts have ever been shown to require, thereby giving another selective advantage to this medium. Normal yeast culture media such as glucose–peptone–yeast extract or malt agar should never be used for isolation, as they are too rich and encourage the growth of competing microflora.

Media for Microalgae

As for cyanobacteria, there are no general considerations for growing psychrophilic microalgae. Growth of microalgae is slower than that of heterotrophic bacteria, so avoidance of contamination is a major issue during isolations. Algae prefer alkaline media (pH ~ 8.2) with a low-level light source and inorganic carbon source such as bicarbonate. The choice of medium composition depends on the requirements of the algae; for instance diatoms need a source of silica and many marine microalgae require vitamin supplements. Bold's basal medium (BBM) is a chemically-defined medium useful for many green algae. Many culture collections give details of medium requirements and the needs of individual strains can be found on the UK National Culture Collection strain database (www.ukncc.co.uk/html) and the web site of the Culture Collection of Algae and Protozoa (www.ife.ac.uk/ccap).

Enrichment Strategies

In general terms, many of the problems which limit culturability of organisms from any environment are present, if not accentuated, in the isolation of psychrophiles. In particular, the problem of preferential isolation of fast-growing species (the "microbial weeds") is worse for psychrophiles than for many other groups of organisms purely because of their intrinsically slow growth rates. In this regard, special culturing techniques that reduce the significance of relative growth rates, such as the method of Connon and Giovannoni (2002) described above, are useful. A recently reported method employing the encapsulation and subsequent growth of single cells in microspheres (Zengler et al., 2004), originally developed for the culturing of anchorage-dependent mammalian cells (Lazar et al., 1985) elegantly circumvents the problems of overgrowth of slow-growing organisms by the "weeds". When applied to soil and thermophilic environmental samples, substantial numbers of new species (and even new genera) have been isolated. This technique has not, to our knowledge, been applied to psychrophiles, but would seem to be well suited to the field.

In the constant search for ways to enhance access to the large proportion of "uncultured" organisms in any sample, new culturing strategies based on specific environmental properties have shown considerable promise. Acknowledging that a high proportion of microbial cells in any soil or aquatic environment exist in an attached state, culturing techniques designed to mimic surface adsorbed/biofilm growth conditions have been widely used. The successful coupling of this general method with an enzyme-specific enrichment strategy was employed by Wery et al. (2003). Using a jacketed column packed with 2 mm non-porous glass beads (Potters Europe, UK), a casein-based medium was inoculated with Antarctic soil and cycled through the beaded matrix in order to select for protease-producing species. After 10 days of upward-flow medium cycling and a further 15 days of continuous feed (non-cycling) (both at 10°C), cells were isolated from both attached biofilms and from unattached populations, collected from the outflow. Phylogenetic analysis of isolates showed a wide diversity of genera, including several (e.g. *Chryseobacterium*, *Brevundimonas*) which had hitherto not been isolated from Antarctic environments. It was noted, however, that the morphological diversity of primary colonizers was very much greater than that of the isolates!

Culturing at Low Temperature

Commercial temperature-controlled incubators are available from a variety of manufacturers (e.g. New Brunswick, Gallenkamp) with gyrorotatory platforms, which are convenient for growing batch cultures in flasks up to 2 l in size. The lowest temperature that can be held stably depends on the local ambient temperature, but generally it is difficult to obtain stable temperatures below 5°C unless the external temperature is lowered (we have placed such incubators inside cool/cold rooms in order to achieve temperatures down to and below zero). Circulating water baths connected to chillers can also be used to achieve temperatures close to zero; a modification that achieves better low temperature stability is to use glycerol in the water bath with cultures inside sealed bottles submerged in the liquid (J. Foght, personal communication).

Maintenance and Storage of Cultures

Many psychrophilic microorganisms can be maintained on solid media for up to a year at 0–4°C, with sub-culturing every few months, as long as care is taken to avoid freezing or drying out. Desiccation is best prevented by using slants in screw-cap bottles rather than plates for such medium-term storage. For bacterial isolates, an antifungal agent such as cycloheximide (100 µg ml^{-1}, added from a filter-sterilized concentrated solution in 10% ethanol) and/or nystatin (250 units ml^{-1}, added from a filter-sterilized concentrated solution in methanol) can be included in the media after autoclaving and before pouring slants or plates. The solid media should be surface dry before inoculating, as this helps prevent subsequent contamination.

For long-term storage, it is necessary to use cryopreservation. There are no specific methods for use with psychrophiles and the usual cryoprotectants (e.g. 20% glycerol or 20% DMSO) are suitable. The microbial cell suspensions should be frozen initially at −20°C and then stored at −80°C in small aliquots so that repeated thawing of stored samples is avoided. A simple and convenient method is to mix cell suspensions with about 25 autoclaved small glass beads that have a central hole (these can be obtained from retail haberdashery stores) per small tube prior to freezing. The hole takes up the liquid and when a culture is required, individual beads can be removed aseptically and used to inoculate agar plates directly without having to thaw the complete set of frozen beads.

The vexed question always arises "How long should one incubate plates or liquid cultures?" There is no straightforward answer to this question and opinions vary from 2 weeks to 2 months or longer. On solid media, colonies of psychrophiles are generally visible within 1 month of incubation at 5°C. In very extreme permanently cold environments such as permafrost soils, where biodiversity and competition for nutrients may be low, the growth rates may be extremely low – so low in fact as to be outside the "norms" for routine laboratory experiments (*viz.* doubling times could be years, and even at temperatures above zero, their growth rates might well be longer than anticipated). Gilichinsky and co-workers demonstrated the incorporation of radiolabelled acetate into lipids at temperatures as low as −20°C and, based on the assumption that the rate of acetate incorporation was an indicator of the timescale of growth, they calculated that the bacteria had a doubling time at this low temperature of >160 days (Rivkina *et al.*, 2000). This is compatible with the finding that visible colonies of some permafrost bacteria on solid media may only appear after more than 1 year of incubation at −10°C. Incubation for such extended periods will necessitate taking precautions to prevent agar plates from drying out, by creating a humid atmosphere.

◆◆◆◆◆ CULTURING IN A FIELD LABORATORY

A major problem working in field conditions, often in cramped conditions in a tent, fibre-glass dome or small hut, is to generate a sterile cabinet or designated clean area in which to carry out transfers and platings etc. Material such as plexiglass can be used to construct a small portable glove box, which can be sterilized with 70% (v/v, aq.) ethanol between manipulations. Samples can be stored in, for example, Whirl-pak plastic bags, which are available in a range of sizes, and can be opened and handled aseptically inside the glove box. Ice or permafrost cores can be split in two whilst still in the sterile bag using a steel hammer and anvil (which have been sterilized with alcohol) and sub-samples taken from the interior of the core. Collected snow can also be thawed in such bags, which is best done by raising the temperature in stages to avoid thermal shock and death of microorganisms. The same is true of frozen soil samples, to which appropriate culture media can be added in the bag thereby avoiding unnecessary transfers that give opportunity for contamination. The ever-increasing range of plasticware, which can be purchased already sterilized or can easily be autoclaved, has increased the opportunities for effective field collection and manipulation under aseptic conditions. A domestic pressure cooker heated over a camp stove serves as a convenient autoclave. In general, the best protocols are those that are rigorous but simple with as few transfers as possible. The exercise of ingenuity is usually a desirable attribute: even a modest candle plus strong alcoholic drink can serve in the execution of aseptic technique *in extremis*!

◆◆◆◆◆ ISOLATION OF COMMUNITY DNA IN A FIELD LABORATORY

There are numerous published methods for extraction and purification of community DNA (see review by Roose-Amsaleg *et al.*, 2001).

However, extracting DNA in the field necessarily requires relatively straightforward techniques, preferably using robust portable extraction equipment. We have successfully used a modification of the Miller procedure (Miller et al., 1999) to isolate high quality community DNA from Dry Valley mineral soils. The key items of equipment are a bead-beater (ideally a commercial unit such as a FastPrep® (qBiogene)), although a variable speed vortex mixer can be adapted to hold lysis tubes in a horizontal position), and a microcentrifuge such as the Eppendorf miniSpin. A modern lightweight generator (e.g. the portable Honda 2.0i), preferably equipped with a frequency inverter and both 240 V and 120 V outputs, is invaluable if not essential to successful field microbiology.

For DNA extraction by bead-beating, expensive lysis beads can, with some loss of efficiency, be replaced by sterilized 50–70 mesh quartz sand (Sigma). The key reagents: SDS lysis buffer (100 mM NaCl, 500 mM Tris-HCl, pH 8, 100 g l^{-1} SDS), phosphate buffer (100 mM sodium phosphate, pH 8.0), chloroform:isoamyl alcohol (24:1, v/v), ammonium acetate (7 M aq.), isopropanol, 70% v/v, aq. ethanol and TE buffer (100 mM Tris-HCl, pH 8.0, 1 mM EDTA) can be aliquoted into small volumes (e.g. 2 ml) for safe transport and are stable. Experience has taught us that the accurate manipulation of microliter volumes of solution reagents is greatly facilitated by maintaining the field laboratory at 15°C or even higher.

We have successfully isolated good yields of PCR-quality DNA from Antarctic Dry Valley mineral soils under such conditions (Cary and Cowan, unpublished results). However, some psychrophilic environments, such as the Arctic permafrosts, contain much higher levels of organic polymers (proteins, lignins, humic acids, etc.) and may require further processing by PVPP adsorption or passage through Sephacryl (Pharmacia) before the required level of DNA purity is obtained.

The authors are unaware of any reports of *in situ* recovery of mRNA from psychrophilic environments. However, there seems to be no obvious reason why RNA stabilization techniques used for other systems, such as saturation of samples with RNAlater® (Ambion), should not work equally effectively for psychrophilic samples.

◆◆◆◆◆ PHYLOGENETIC ANALYSIS OF PSYCHROTROPHIC COMMUNITIES

Technically, there is no fundamental difference in undertaking a phylogenetic analysis of a psychrophilic environment than of any other environment. A variety of such studies has been published on a range of habitats, including samples from Antarctic lake sediment (Priscu et al., 1998), Antarctic cryoconite holes (Christner et al., 2003), cryptoendolithic communities (de la Torre et al., 2003), Arctic permafrosts (Shi et al., 1997), deep marine sediments (Ravenschlag et al., 1999), Antarctic Dry Valley mineral soils (Cary and Cowan, unpublished results) and alpine soils

(Lipson and Schmidt, 2004). These investigations have shown that the use of standard ssu rRNA gene primer sets is effective in generating valuable phylogenetic diversity data for all three Domains. In some, but not all psychrophilic environments the task is simplified by a relatively low microbial species diversity compared to more eutrophic temperate environments. The simpler community structure, such as in Dry Valley mineral soils which are essentially dominated by members of the Domain Bacteria and contain relatively few members of the Domains Archaea and Eukarya, also makes the tasks of microbial community analysis and related metagenomic studies relatively easier.

A cautionary note on such phylogenetic studies is required, however. Unlike more temperate environments, the low temperatures and/or water activities of some psychrophilic environments (such as polar terrestrial soils) may contribute to the long-term preservation of dead cells and even naked DNA. Although no comprehensive studies of nucleic stability and longevity have yet been published, a number of diverse reports indicate that some caution must be entertained when undertaking metagenomic (culture-independent) assessments of microbial diversity. For example, the recovery of both DNA and live cells from the estimated 0.25 million-year-old fractions of the Vostok ice core (Priscu *et al.*, 1999) is clear evidence of the long-term cellular and macromolecular stability. It has also been recently reported that naked microbial genomic DNA seeded into Dry Valley mineral soils remains detectable (by PCR amplification) for at least 5 months, and possibly for much longer (Ah Tow and Cowan, 2005).

Acknowledgements

We would like to acknowledge those many colleagues who generously gave their advice and comments, and shared their experiences, including releasing pre-published articles, on handling psychrophiles. In particular, we would like to thank Doug Bartlett, Craig Cary, Julia Foght, David Gilichinsky, John Priscu, Jim Staley, Juergen Wiegel, Helen Vishniac and Lyle Whyte. Of course, responsibility for the facts as presented here rests entirely with the authors.

List of suppliers

Ambion, Inc.
2130 Woodward
Austin, TX 78744-1832
USA
Phone: +1-512-651-0200
Fax: +1-512-651-0201
http://www.ambion.com

RNAlater®

Difco
1 Becton Drive
Franklin Lakes, NJ 07417, USA
Tel: +1-201-847-6800
http://www.difco.com

Growth media

Eppendorf UK Limited
Endurance House
Chivers Way
Histon
Cambridge CB4 9ZR, UK
Tel: +44-1223-200-440
Fax: +44-1223-200-441
http://www.eppendorf.com/uk

Minispin microcentrifuge

qBiogene, Inc.
15 Morgan
Irvine, CA 92618-2005, USA
Tel: +1-440-337-1200
Fax: +1-949-421-2675
http://www.qbiogene.com

FastPrep® Bio101

Fisher Scientific International
Liberty lane
Hampton, NH 03842, USA
Tel: 1-800-766-7000
Fax: 1-800-926-1166
http://www.fisherscientific.com

Chemicals, laboratory equipment incl. shakers (Gallenkamp)

Millipore
290 Concord Road
Billerica, MA 01821, USA
Tel: +1-978-715-4321
http://www.millipore.com

Membrane filters

Nasco Inc
901 Janesville Avenue
Fort Atkinson, WI 53538, USA

Tel: +1-920-563-2448
Fax: +1-920-563-8296
http://www.enasco.com

Whirl-pak plastic bags

New Brunswick Scientific
PO Box 4005
Edison, New Jersey, USA
Tel: +1-732-287-1200
Fax: +1-732-287-4222
http://www.nbsc.com

Temperature controlled shaker incubators

Oxoid Inc
800 Proctor Avenue
Ogdensburg, NY 13669, USA
Tel: +1-613-226-1318
Fax: +1-613-226-3728
http://www.oxoid.com

Growth media

Polyscience Inc
Warrington, PA 18976, USA
Tel: +1-215-343-6848
Fax: 1-800-926-1166
http://www.polyscience.com

Latex microspheres

Sumitomo Chemical
27-1 Shinkawa 2 chome
Chuo-ku, Tokyo 104-8260, Japan
Tel: +81-80-1116-7969
Fax: +81-3-5204-9690
http://www.sumitomo-chem.co.jp

Fluorinert

ThermoSafe
Tel: 1-874 398 0110
Fax: 1-874 398 0653
http://www.polyfoam.com

ThermoSafe sealable plastic containers

References

Ah Tow, L. and Cowan, D. A. (2005). Preservation of bacterial genomes in Antarctic Dry Valley mineral soils. *Extremophiles*, in press.

Breezee, J., Cady, J. and Staley, J. T. (2004). Subfreezing growth of the sea ice bacterium "*Psychromonas ingrahamii*". *Microb. Ecol.* **47**, 300–304.

Button, D. K., Schut, F., Quang, P., Martin, R. and Robertson, B. R. (1993). Viability and isolation of marine bacteria by dilution culture: theory, procedures, and initial results. *Appl. Environ. Microbiol.* **59**, 881–891.

Carpenter, E. J., Lin, S. and Capone, D. G. (2000). Bacterial activity in South Pole snow. *Appl. Environ. Microbiol.* **66**, 4514–4517.

Christner, B. C., Kvitko, B. H. and Reeve, J. N. (2003). Molecular identification of Bacteria and Eukarya inhabiting an Antarctic cryoconite hole. *Extremophiles* **7**, 177–183.

Christner, B. C., Mikucki, J. A., Foreman, C. M., Denson, J. and Priscu, J. C. (2005). Glacial ice cores: a model system for developing extraterrestrial decontamination protocols. *Icarus*, **174**, 572–584.

Connon, S. A. and Giovannoni, S. J. (2002). High-throughput methods for culturing microorganisms in very-low-nutrient media yield diverse new marine isolates. *Appl. Environ. Microbiol.* **68**, 3878–3885.

Cowan, D. A. and Ah Tow, L. (2004). Endangered Antarctic environments. *Annu. Rev. Microbiol.* **58**, 649–690.

de la Torre, J. R., Goebel, B. M., Friedman, E. I. and Pace, N. R. (2003). Microbial diversity of cryptoendolithic communities from the McMurdo Dry Valleys, Antarctica. *Appl. Environ. Microbiol.* **69**, 3858–3867.

Frankenstein, G. and Garner, R. (1967). Equations for determining the brine volume of sea ice from $-5°C$ to $-22.9°C$. *J. Glaciol.* **6**, 943–944.

Gilichinsky, D. (2002). Permafrost as a microbial habitat. In *Encyclopedia of Environmental Microbiology* (G. Bitton, ed.), pp. 932–956. Wiley, New York.

Juck, D. F., Whissell, G., Steven, B., Pollard, W., McKay, C. P., Greer, C. W. and Whyte, L. G. (2005). Utilization of fluorescent microspheres and a green fluorescent protein-marked strain for assessment of microbiological contamination of permafrost and ground ice core samples from the Canadian high Arctic. *Appl. Environ. Microbiol.* **71**, 1035–1041.

Krembs, C., Gradinger, R., Spindler, M. and Goehl, B. (1996). New instruments for current and diffusion measurements at the sea ice-water interface: instrumental design and first results from field measurements in the Arctic Ocean. In *Proceedings of Oceanology International 96. The Global Ocean towards Operational Oceanography*, vol. 3, pp. 95–116. Spearhead Exhibitions, Brighton.

Lazar, A., Silverstein, L., Margel, S. and Mizrahi, A. (1985). Agarose-polyacrolein microsphere beads: a new microcarrier culturing system. *Dev. Biol. Stand.* **60**, 457–465.

Lipson, D. A. and Schmidt, S. K. (2004). Seasonal changes in an alpine soil bacterial community in the Colorado Rocky Mountains. *Appl. Environ. Microbiol.* **70**, 2867–2879.

Miller, D. N., Bryant, J. E., Madsen, E. L. and Ghiorse, W. C. (1999). Evaluation and optimization of DNA extraction and purification procedures for soil and sediment samples. *Appl. Environ. Microbiol.* **65**, 4715–4724.

Morita, R. Y. (1975). Psychrophilic bacteria. *Bacteriol. Rev.* **30**, 144–167.

Onofri, S. (1999). Antarctic microfungi. In *Enigmatic Microorganisms and Life in Extreme Environments* (J. Seckbach, ed.), pp. 325–335. Kluwer Academic Publishers, Dordrecht.

Priscu, J. C., Fritsen, C. H., Adams, E. E., Giovannoni, S. J., Paerl, H. W., McKay, C. P., Doran, P. T., Gordon, D. A., Lanoil, B. D. and Pinckney, J. L. (1998). Perennial Antarctic lake ice: an oasis for life in a polar desert. *Science* **280**, 2095–2098.

Priscu, J. C., Adams, E. E., Lyons, W. B., Voytek, M. A., Mogk, D. W., Brown, R. L., McKay, C. P., Takacs, C. D., Welch, K. A., Wolf, C. F., Kirshtein, J. D. and Avci, R. (1999). Geomicrobiology of subglacial ice above Lake Vostok, Antartica. *Science* **286**, 2141–2144.

Quereshi, M. H., Kato, C. and Horikoshi, K. (1998). Purification of a ccb-type quinol oxidase specifically induced in a deep-sea barophilic bacterium, *Shewanella* sp. strain DB-172F. *Extremophiles* **2**, 93–99.

Ravenschlag, K., Sahm, K., Pernthaler, J. and Amann, R. (1999). High bacterial diversity in permanently cold marine sediments. *Appl. Environ. Microbiol.* **65**, 3982–3989.

Rivkina, E. M., Friedmann, E. I., McKay, C. P. and Gilichinsky, D. A. (2000). Metabolic activity of permafrost bacteria below the freezing point. *Appl. Environ. Microbiol.* **66**, 3230–3233.

Rogers, S. O., Theraisnathan, V., Ma, L. J., Zhao, Y., Zhang, G., Shin, S.-G., Castello, J. D. and Starmer, W. T. (2004). Comparisons of protocols for decontamination of environmental ice samples for biological and molecular examinations. *Appl. Environ. Microbiol.* **70**, 2540–2544.

Roose-Amsaleg, C. L., Garnier-Sillam, E. and Harry, M. (2001). Extraction and purification of microbial DNA from sopil and sediment samples. *Appl. Soil. Microbiol.* **18**, 47–60.

Russell, N. J. (1990). Cold adaptation of microorganisms. *Phil. Trans. Roy. Soc. London B* **326**, 595–611.

Russell, N. J. (2003). Membrane adaptation and solute uptake systems. In *Encyclopedia of Life Support Systems*. EOLSS Publishers Co., Ltd. (Contribution number 6-73-03-02 @ www.eolss.com).

Sherr, E. B., Sherr, B. F. and Cole, J. (eds) (1993). *Current Methods in Aquatic Microbial Ecology*. Lewis publishers, New York.

Shi, T., Reeve, R. H., Gilichinsky, D. A. and Friedmann, E. I. (1997). Characterization of viable bacteria from Siberian permafrost by 16S rDNA sequencing. *Microb. Ecol.* **33**, 169–179.

Theocharopoulos, S. P., Wagner, G., Sprengart, J., Mohr, M. E., Desaules, A., Muntau, H., Christou, M. and Quevauviller, P. (2001). European soil sampling guidelines for soil pollution studies. *Sci. Total Environ.* **264**, 51–62.

UNOG (2000). *The Antarctic Treaty and the Protocol on the Environmental Protection to the Antarctic Treaty* (www.unog.ch/disarm/distreat/antarc.htm).

Vishniac, H. S. (1999). Psychrophilic yeasts. In *Enigmatic Microorganisms and Life in Extreme Environments* (J. Seckbach, ed.), pp. 317–321. Kluwer Academic Publishers, Dordrecht.

Wery, N., Gerike, U., Sharman, A., Chaudhury, J. B., Hough, D. W. and Danson, M. J. (2003). Use of a packed-bed bioreactor for isolation of diverse protease producing bacteria from Antarctic soil. *Appl. Environ. Microbiol.* **69**, 1457–1464.

Zengler, K., Clark, G., Haller, I., Toledo, G., Walcher, M., Rappé, M., Woodnut, G., Short, J. M. and Keller, M. (2004). Accessing microbial diversity by high throughput cultivation. In *Abstracts of ISME 10 – International Symposium on Microbial Ecology*, p. 117. Cancun, Mexico.

17 Proteins from Psychrophiles

Ricardo Cavicchioli[1], Paul MG Curmi[2], Khawar Sohail Siddiqui[1] and Torsten Thomas[1]

[1] School of Biotechnology and Biomolecular Sciences, The University of New South Wales, Sydney, NSW 2052, Australia; [2] School of Physics, The University of New South Wales, Sydney, NSW 2052, Australia and Centre for Immunology, St Vincent's Hospital, Sydney, NSW 2010, Australia

CONTENTS

Introduction
Purification of native and recombinant proteins from psychrophiles
Methods to generate proteins from psychrophiles with modified properties
Methods to study the structure of proteins from psychrophiles
Methods to study the stability of proteins from psychrophiles
Genomics and proteomics of psychrophiles
Biotechnological application of proteins from psychrophiles for use in molecular biology
Conclusion

◆◆◆◆◆ INTRODUCTION

Psychrophiles have adapted to colonize a diverse range of cold environments (Russell, 2005). Similar to other microbial extremophiles (see other chapters throughout this book), psychrophiles must be appropriately adapted to enable not only survival, but microbial growth and proliferation in their niche environments. Throughout evolution, all components of the cell (lipids, intracellular solutes, proteins, etc.) will have been exposed to selection pressures to drive their adaptation. As a result, a successfully competing psychrophile will have inherited a genetic blueprint that describes its cold-adapted cellular components which have arisen by selection over its biological history.

Due to the ecological importance of psychrophiles, their inherent commercial value, and the exciting scientific challenge to understand the molecular basis of cold-adaptation, a large effort has been directed towards understanding mechanisms of psychrophile adaptation (Cavicchioli and Siddiqui, 2004; Cavicchioli *et al.*, 2002; Feller and Gerday, 2003). The cellular components which have received most attention are proteins from psychrophiles (Siddiqui and Cavicchioli, 2006).

Studies on proteins from psychrophiles (also referred to as cold-adapted or cold-active proteins) have advanced rapidly due to the ability to apply methods that have been developed for handling proteins from thermophiles and mesophiles. The standard procedure of purifying, manipulating and storing proteins at low temperatures is inherently suitable for working with proteins from psychrophiles. Modern biochemical and biophysical methods (including crystallography) have proven to be useful for analysing cold-active proteins. No major technical limitations exist for studying proteins from psychrophiles.

The following sections highlight examples of methods which have been applied to proteins from psychrophiles. As described above, many methods are relatively standard and do not require extensive elaboration. We have focused on important applications of methods that have been used to study proteins from psychrophiles, and covered systems developed specifically for proteins from psychrophiles (e.g. a cloning and over-expression system developed specifically in a psychrophilic bacterium) and novel procedures (e.g. chemical modification of proteins).

◆◆◆◆◆ PURIFICATION OF NATIVE AND RECOMBINANT PROTEINS FROM PSYCHROPHILES

Production of Proteins from Psychrophiles

Proteins have been purified from psychrophiles directly from their psychrophilic parent in their native form, and as heterologous proteins following expression in a non-native host (typically *Escherichia coli*). The purpose of the study and the logistics of performing the purification will influence the approach chosen. For example, it may be difficult to obtain sufficient biomass of the psychrophilic host, or the abundance of the protein may be low, thereby limiting the possible yield of a native protein. If large (e.g. protein crystallization, differential scanning calorimetry (DSC) analysis) or small (e.g. enzyme assay) quantities are required, this may influence the strategy chosen. For large quantities a heterologous approach may be beneficial. Alternatively, the gene encoding the protein of interest might not be readily available for recombinant, heterologous expression. This may therefore necessitate an initial purification of the native protein to enable the gene to be cloned by reverse genetics. Moreover, post-translational modification is likely to occur differently in the parent strain compared to a heterologous host. If this is considered important then it will clearly impact on the choice of purification procedure. Purifying both the native and recombinant protein provides opportunities for comparing the biochemical and biophysical properties, and determining the nature of host-specific modifications.

A number of studies have involved screening for a specific enzyme activity from a psychrophilic organism (e.g. amylase activity), followed by traditional protein purification (e.g. chromatography) to enrich or fully purify the enzyme. The purification procedures are well established (e.g. Deutscher, 1990), and not a great deal of specialization is required to handle proteins from psychrophiles. If sufficient protein can be obtained for determining the N- and C-terminal sequence, the gene encoding the protein can be isolated by designing a probe to screen a genomic library of the parent organism, or by directly amplifying the gene from genomic DNA using a PCR-based approach. The gene can then be cloned into a suitable expression system (see below). Heterologous proteins can be purified by traditional protein purification steps. Alternatively, affinity purification can be used, based on affinity-tags that were introduced during recombinant expression (e.g. N- or C-terminal hexa-histidine tags).

Heterologous Expression in *E. coli*

Expression of numerous recombinant proteins in *E. coli* at low temperatures has been reported to increase protein yield (Gottesmann, 1990; Makridies, 1996). This may arise by facilitating proper folding of the polypeptide chain emerging from the ribosome, and by reducing the amount of proteolytic degradation. This might be particularly important for proteins from psychrophiles. The enzyme activity and protein yield of an α-amylase from the Antarctic bacterium *Pseudoalteromonas haloplanktis* (AHA) expressed in *E. coli* was dramatically increased when over-expression was performed at 18°C compared to 37°C (Feller et al., 1998). The protein-folding properties of the native AHA, and the recombinant enzyme purified from *E. coli* cells at 18°C, appear to be identical (Feller et al., 1998). The yield of a recombinant elongation factor 2 (EF-2) protein from a cold-adapted member of the Archaea, *Methanococcoides burtonii*, was also higher in *E. coli* over-expressed at 14°C compared to 23 or 30°C (Thomas, 2001; Thomas and Cavicchioli, 2000).

Recombinant expression of chaperones (Cpn10 and Cpn60) from the psychrophilic bacterium, *Oleispira antarctica* RB-8^T has been shown to allow *E. coli* to grow at relatively high growth rates (0.28 h^{-1}) at 4°C (Ferrer et al., 2004). The low growth temperature has been shown to facilitate efficient expression of a temperature-sensitive esterase in *E. coli*, yielding a 180-fold higher specific enzyme activity compared to the enzyme produced at 37°C. The study found that the low growth temperature *per se* was beneficial for the correct folding of the esterase (Ferrer et al., 2004). Clearly, this system has great potential for the expression of other proteins from psychrophiles. Collectively, the findings for AHA, EF-2 and the esterase indicate that when new proteins from psychrophiles are being over-expressed in *E. coli*, low growth temperature and/or induction conditions should be trialed.

Proteolytic degradation appears to be a major obstacle in achieving high expression levels for recombinant proteins in *E. coli* and a number of studies have shown improvements in yield through the use of protease-negative strains such as *E. coli* BL21 (Gottesmann, 1990;

Makridies, 1996). BL21 is deficient for Lon, the major, intracellular ATP-dependent protease which appears to be responsible for the degradation of a number of naturally unstable proteins and aberrant proteins (Gottesmann, 1990; Makridies, 1996). BL21 has proven useful for the over-expression of archaeal EF-2 (Thomas, 2001; Thomas and Cavicchioli, 2000) and a bacterial, heat-sensitive alkaline phosphatase (Rina et al., 2000).

A cold-adapted Atlantic cod trypsin was shown to be very sensitive to autolytic degradation, thermal inactivation and molecular aggregation even during recombinant expression at low temperatures (18°C) (Jonsdottir et al., 2004). Expression of this protein in E. coli as a fusion partner to the highly soluble protein thioredoxin, and subsequent in vitro cleavage of the fused domain, resulted in an active and highly purified trypsin (Jonsdottir et al., 2004). This example demonstrates the value of using a cleavable protein-fusion for solubilizing and stabilizing inherently unstable proteins from psychrophiles.

Heterologous Expression in Cold-adapted Hosts

In contrast to the focus that has been placed on heterologous expression of proteins from psychrophiles in E. coli, few studies have reported the use of a psychrophilic host for protein over-expression. The most advanced system has been developed in P. haloplanktis. This system utilizes a shuttle plasmid that was constructed from the P. haloplanktis plasmid, pMTBL and the E. coli broad host-range conjugative plasmid, pJB3 (Duilio et al., 2004; Tutino et al., 2001). The sequence of pMTBL was completely determined (4081 bp) and the region responsible for autonomous plasmid replication (ARS) was identified on a segment of approximately 850 bp. The ARS region was introduced into an E. coli vector containing the rolling circle replication origin of transfer (oriT) from pJB3. Transcriptional and translational regulatory elements from the P. haloplanktis aspartate aminotransferase gene were also introduced. The final pFF construct enables cloning in E. coli and conjugative transfer to P. haloplanktis. This shuttle vector is the first (and to date the only) plasmid that enables over-expression in a cold-adapted bacterium (Birolo et al., 2000; Tutino et al., 2002). Proof of principle for the system includes the successful expression and purification of a psychrophilic α-amylase (Tutino et al., 2002).

◆◆◆◆◆ METHODS TO GENERATE PROTEINS FROM PSYCHROPHILES WITH MODIFIED PROPERTIES

Site-directed Mutagenesis

Site-directed mutagenesis has proven to be valuable for analysing the structure–function relationship of proteins from psychrophiles.

The method provides a means of introducing specific nucleotide changes into a gene. As a result, the effects of specific amino acid changes in a protein can be gauged, relative to the protein from the wild-type gene. The resulting structural changes may then be linked to observed changes in function, stability and/or activity.

Site-directed mutagenesis can be used to change, insert or delete single amino acid residues, multiple residues or even entire structural elements (e.g. surface loops). To perform site-directed mutagenesis, the primary DNA sequence of the gene, and therefore inferred sequence of the protein, is required. In addition, a suitable construct in an expression vector (typically in *E. coli*) is required. Tertiary structure information (e.g. protein model or crystal structure) is valuable for guiding the design of the mutagenesis regime. Experiments probing protein structure–function have been successfully conducted based on the identification of conserved residues from multiple sequence alignments of primary sequence data only; however this will limit the nature of the questions which can be addressed.

Many popular site-directed mutagenesis methods use synthetic oligonucleotides to modify the target region. In this approach, the "mutagenic" oligonucleotide contains a portion that is perfectly complementary to the target region with the exception of a specific mismatch to introduce the mutation. The oligonucleotides are hybridized at low-stringency to the target molecule (e.g. the expression clone or a PCR product of the target gene), and DNA synthesis performed using a non-proofreading DNA polymerase (e.g. *Taq* polymerase or exonuclease minus Klenow fragment). The mutated product may then be amplified using PCR, the amplification product ligated to a suitable vector, and the new construct with the mutated gene transformed into a cloning or expression host.

Based on this general principle, a range of protocol variations and licensed procedures have been developed, and a number of commercial mutagenesis kits are available (e.g. GeneEditor™ system from Promega and QuikChange® system from Stratagene).

Specific examples of the application of site-directed mutagenesis related to generating mutant forms of proteins from psychrophiles are summarized below. These provide useful insight into the specific procedures adopted for performing site-directed mutagenesis.

- Transfer of the highly flexible binding site of a cold-active subtilisin to a counterpart from a mesophile resulted in a mutant enzyme with a higher specific activity and reduced thermostability (Tindbaek et al., 2004).
- In a cold-adapted isocitrate lyase, replacement of Q207H and Q217K decreased low-temperature activity (Watanabe and Takada, 2004).
- Introduction of five new amino acid interactions (including disulphide bond formation) resulted in increased thermostability and decreased folding reversibility in an α-amylase (D'Amico et al., 2003a).
- Modifying side-chain flexibility at the active site of a phosphatase from a psychrophile caused low activity at low temperature, and increased the energy of activation of the mutant enzyme (Tsigos et al., 2001).

Directed Evolution

Directed evolution can be a powerful tool for protein engineering, enabling the analysis of protein structure–function relationships through the selection of mutants with particular characteristics (Brakmann, 2001). The general principles of directed evolution are depicted in Figure 17.1.

To perform directed evolution, an expression clone is randomly mutated, and individual clones are subsequently expressed to produce mutant proteins. The protein library is subjected to high-throughput screening for properties of interest (e.g. modified stability, activity, specificity). Selected clones corresponding to proteins with desired properties are subjected to further rounds of mutagenesis and screening in order to further improve the activity. This process resembles an accelerated evolutionary process that is directed by the requirements of the end user (hence "directed evolution").

A large number of random mutagenesis strategies are described in the literature, and these have recently been reviewed (Lutz and Patrick, 2004). Three main methods have been developed: oligonucleotide-directed randomization (ODR), error-prone PCR (epPCR) and *in vitro* recombination (homology-dependent and -independent).

- ODR involves the incorporation of oligonucleotides into an *in vitro* synthesized DNA strand. The method is similar to site-directed mutagenesis (described above), except that degenerate oligonucleotides (i.e. one or more nucleotide in the oligonucleotide is randomized) are used. Using this approach, all 20 standard amino acid residues may be introduced into a specific location within a protein.
- epPCR uses either a low-fidelity polymerase or universal triphosphate nucleotides (e.g. inosine triphosphate) to incorporate mutations into a gene. epPCR tends to produce one or more amino replacements. One form of epPCR, sequence saturation mutagenesis (SeSaM), has proven to be rapid and efficient for producing random mutations (Wong et al., 2004).
- In contrast to ODR and epPCR, *in vitro* recombination exchanges entire regions of DNA between two homologous or non-homologous strands of DNA and is often referred to as "DNA shuffling". As a result, mutant proteins can be generated with rearranged and exchanged structural elements.

Directed evolution tends to produce a limited number of mutants with desired properties, while also generating large numbers of mutants that

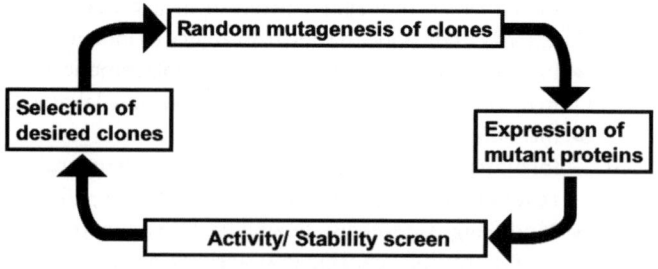

Figure 17.1. General scheme for performing directed evolution.

do not have improved properties. As is typical with a "genetics-based" approach, screening necessitates large numbers to be tested, in contrast to a selection method, which targets desired mutants and suppresses unwanted variants. As a result, simple and efficient protein-expression systems married to high-throughput functional screens are essential for employing a directed evolution approach. High-throughput screens are often performed with simple activity assays (e.g. colorimetric) utilizing a microtitre-plate format with robotics support. To improve chances of success it is desirable to use more than one mutagenesis procedure to improve the properties of a single protein. This will also minimize the potentially negative impact that one procedure may cause. A useful approach may be to employ methods that produce point mutations and domain exchanges. This can be accomplished in a single mutagenesis experiment through the use of *in vitro* recombination in the first cycle of screening, and epPCR in the second round of screening.

Directed evolution can be applied to a gene without knowledge of its tertiary structure, and an understanding of the protein's structure–function is not required. Therefore, unlike a rational approach to mutagenesis which is greatly facilitated by the availability of the protein's tertiary structure, directed evolution can be applied to any gene that has an available primary sequence.

Despite its apparent potential for improving the properties of proteins, only a few reports document its use for proteins from psychrophiles.

- Using directed evolution, variants of a subtilisin with increased thermostability have been reported (Miyazaki and Arnold, 1999). The study identified two amino acids that appeared to play important roles in the modified properties of the mutant proteins.
- In a further study of the same subtilisin, *in vitro* recombination was used to generate variants with greatly improved thermostability without a reduction in specific activity (Miyazaki et al., 2000).
- epPCR and site-directed mutagenesis was used to increase the thermostability of a lipase B from the psychrophilic yeast, *Candida antarctica* (Zhang et al., 2003). Improvements in activity and stability were reported, and the residues responsible for the improvements were identified.

Chemical Modification

Chemical modification of an existing protein can be useful for probing structure, function and stability relationships of proteins. Chemical modification can serve as a replacement or complementary method to recombinant DNA technology for improving the properties of a protein (Table 17.1). This may have particular benefit for proteins targeted for commercial exploitation due to the increasing difficulty in obtaining regulatory approval for genetically modified products. Chemical modification may be applied to native, recombinant, wild-type or mutant proteins. The procedures are relatively inexpensive, may be performed in most laboratory settings, and potentially offer rapid improvements

Table 17.1 Chemical modification methodologies for modifying proteins

Modification	Principle	Reagents	References
Amidation	Carbodiimide activates unionized carboxyl groups to form O-acylisourea intermediates which link to any deprotonated nucleophiles (e.g. amino groups)	Nucleophiles: Glycinamide · HCl; Arginine methylester · 2HCl; Aniline · HCl; Dimethylamine · HCl; Ethylenediamine · 2HCl; Glucosamine · HCl; Oligosaccharide of chitosan	Bokhari et al. (2002), Siddiqui et al. (2000)
Acylation	Amino groups in proteins react with any anhydride to form an amide linkage with carboxyl groups	Acetic anhydride Succinic anhydride	Mozhaev et al. (1988)
Acylation	Amino groups in proteins react with any dianhydride to form an amide linkage with carboxyl group	Pyromellitic dianhydride Benzophenone tetracarboxyl dianhydride	Mozhaev et al. (1988)
Guanidination	Lys is converted to homo-Arg with the transfer of a guanidine group	O-methylisourea; 3,5-dimethylpyrazole-1-carboxamidine nitrate	Cupo et al. (1980), Habib (1972)
Cross-linking	Cross-links formed internally between amino groups of the protein, or between amino groups of protein and chitosan	Glutaraldehyde Glyceraldehyde	Acharya et al. (1988)
Polyethylene glycolyation	Polyethylene glycol (PEG) is activated by p-nitrophenylchloroformate to form PEG-nitrophenylcarbonate, which in turn reacts with NH_2 groups to give a PEG–protein adduct.	Activated PEG	Veronese et al. (1985)

Sugar polymer	Oxidized sugar polymers (dialdehyde polysaccharides, DAP) have vicinal aldehyde groups and react with amino-groups in a protein to form a Schiff's base, which can be reductively aminated.	DAP: dextran, ficoll, chitosan, inulin Reducing agents: $NaBH_4$, $NaCNBH_3$, borane–pyridine complex	Dellacherie et al. (1983), Siddiqui and Cavicchioli (2005)
Carboxy-methylation	S–S bridges are reduced to SH which then forms a bond with the $-CH_2$ group of the reagent to form a carboxymethylated protein.	Iodoacetamide/Iodoacetic acid	Hollecker (1989)
Combined mutagenesis and modification (CMM)	The residue to be modified is mutated to cysteine using site-directed mutagenesis. Cysteine is then thioalkylated using a wide variety of modifiers.	$CH_3-S(O_2)-S-R$, where R = sugar, methyl, ethyl, ethylsulphonate, ethylamine, etc.	DeSantis et al. (1998)
Photoactivated hydrophobic modification	Photo-activation of a carbonyl group to diradical state in a benzophenone derivative, abstracts an electron from a CH_3, CH_2 or CH group in protein to form a radical, and enabling radical–radical coupling.	Benzophenone tetracarboxylic acid	Siddiqui et al. (2004b)
Double modification (carboxyl- and amino-group)	Carboxyl-groups in a protein are aminated, followed by any type of amino-group modification.	Ethylenediamine · 2HCl + carbodiimide and any amino-group modification.	Lopez-Gallego et al. (2005), Siddiqui et al. (1999)

(Bokhari et al., 2002; Rashid and Siddiqui, 1998; Siddiqui et al., 1999, 2000, 2004b).

The reactive groups (amino and carboxyl) are most readily modified (Imoto and Yamada, 1989; Lundblad 1995), although aliphatic groups may also be chemically modified (Siddiqui et al., 2004b). The extent of chemical modification can be monitored in a variety ways, including colorimetric estimation (Fields, 1971), amino acid analysis (Kochhar et al., 2002; Mozhaev et al., 1988), polyacrylamide gel electrophoresis (Hollecker, 1989; Rashid et al., 1997) and mass spectrometry (Siddiqui et al., 2004b, 2005b).

By avoiding modification of the active-site and targeting protein surface groups, chemical modification can lead to improvements in protein stability, activity and efficiency (Eijsink et al., 2004; Mozhaev et al., 1988; Siddiqui et al., 2004b). Moreover, particular amino acids can be targeted to determine the role of side chains in substrate binding and activity (Rashid and Siddiqui, 1998; Siddiqui et al., 2005b). Kinetic and thermal properties of proteins can be changed using chemical modification (Bokhari et al., 2002; Siddiqui et al., 1999, 2000).

Chemical modification has only recently been applied to proteins from psychrophiles. These advances highlight the potential that this technology offers for examining structure–stability–activity relationships, and improving properties of proteins from psychrophiles.

- Improvements in activity and stability were reported by modifying aliphatic groups of shrimp alkaline phosphatase using tetracarboxy-benzophenone derivatives (Siddiqui et al., 2004b).
- Improvements in activity and stability for lipase B from *C. antarctica* were achieved using a range of oxidized polysaccharides covalently attached to the enzyme (Siddiqui and Cavicchioli, 2005).
- The role of disulphide bonds in AHA were probed by chemically modifying cysteine residues with iodoacetic acid and iodoacetamide, and MALDI-TOF-MS was used to characterize the extent of modification (Siddiqui et al., 2005b).

◆◆◆◆◆ METHODS TO STUDY THE STRUCTURE OF PROTEINS FROM PSYCHROPHILES

Structural Biology

Structural biology is having a growing impact on all biological sciences, including microbiology. With the advent of structural genomics programmes, representative structures for nearly all protein folds will be available within the next decade. Already, representative structures exist for all components of many metabolic pathways, allowing a complete molecular understanding of the fundamental molecular processes.

Although these approaches are powerful, there is a need to have structures from particular organisms or classes of organisms in order to address central biological questions. In the case of psychrophilic microorganisms, cold adaptation is a central issue. The existence of representative structures (not from psychrophiles) allows proteins from psychrophiles to be modelled in order to gain insight into questions that address cold adaptation. However, the current methods for homology modelling are still not sufficiently robust to provide accurate pictures of these proteins. What is needed is a near complete set of proteins from psychrophilic organisms.

The methods of structural biology (protein X-ray crystallography and multidimensional nuclear magnetic resonance spectroscopy (NMR)) are quite sophisticated, requiring specialist practitioners. However, the key step to success in structure determination is having a reliable supply of properly folded protein around the 10-mg level. If this has been achieved *via* protein purification from psychrophiles or (more commonly) following heterologous expression, then there is a good opportunity to pursue high-resolution structure determination. An important issue for both NMR and crystallography is whether the protein remains stable and monodisperse in solution at concentrations around 5–10 mg ml^{-1}. Aggregation will thwart crystallization and broaden NMR spectra, preventing structure determination.

Once a suitable quantity of an appropriately "behaving" protein is obtained, the best way to proceed is to establish a collaboration with an appropriate structural biology laboratory. A good collaboration can be achieved when the laboratories are in reasonable proximity and have a common interest in the scientific outcomes. Marrying the respective skills of both laboratories will facilitate success.

To date, there is a dearth of structures for proteins from psychrophiles (Table 17.2). All current structures from psychrophilic microorganisms have been determined by X-ray crystallography, with none yet determined by NMR. Table 17.2 gives some experimental parameters used for the 9 available structures. What is immediately apparent from this table is that there is nothing special about the crystallization of proteins from psychrophiles. They do not require particular stabilizing agents, and they do not require lower crystallization temperatures or particular crystallization screens and conditions. The degree of order in the crystals is also normal. Structures range from moderate resolution (2.7 Å) to ultrahigh resolution (0.93 Å). Thus, there does not appear to be any intrinsic barrier preventing growth of this area of research.

It will be interesting to see if these trends continue (i.e. there are no special tricks to crystallizing proteins from psychrophiles). As more groups work on the structures of proteins from psychrophiles, it may well be that optimization of both expression screens and crystallization screens will enhance success in structure determination. Whether this is the case will only be discovered once more concerted, structural genomics approaches are applied to psychrophiles.

Table 17.2 Crystallographic data for proteins from psychrophilic microorganisms

Protein	Organism	Screens	Protein conditions	Crystallization conditions	Temperature	Special conditions	Resolution	PDB code	Reference
α-amylase	*Pseudoalteromonas haloplanktis*	Std sparse matrix[a] at 4°C and 20°C	15 mg ml^{-1}, 10 mM Tris, pH 8.0, 25 mM NaCl, 1 mM CaCl$_2$, 1 mM NaN$_3$	60–70% MPD, 0.1 M HEPES pH 6.8–7.1	Room temperature	Seeding	1.85 Å	1AQH 1AQM	Aghajari *et al.* (1996, 1998)
Alkaline (metallo) protease	*Pseudomonas aeruginosa*; Antarctic *Pseudomonas* TAC II 18 sp.	Std sparse matrix[a] at 4°C	20 mg ml^{-1}, 10 mM Tris, pH 8.0, 1 mM CaCl$_2$	5–10% PEG 6k either 0.1 M HEPES, pH 7.0 or 0.1 M Tris, pH 8	4°C		2.1 Å	1H71 1G9K	Aghajari *et al.* (2003) Villeret *et al.* (1997)
Triose phosphate isomerase	*Moritella marina*	Fast screen[b]	10 mg ml^{-1}, 10 mM triethanolamine, 25 mM NaCl, 1 mM DTT, 1 mM NaN$_3$	100 mM triethanolamine, pH 7–7.5, either 1.26 M Na citrate + 0.1 M ammonium sulphate or 2.0 M ammonium sulphate + 20 mM 2PG	4°C		2.65 Å	1AW1 1AW2	Alvarez *et al.* (1998)
Citrate synthase	Antarctic bacterial strain DS2-3R		17 mg ml^{-1}, 20 mM Tris pH 8.0, 0.1 M KCl	2.2 M Ammonium sulphate, 20 mM citrate pH 5.6	29°C		2.09 Å	1A59	Russell *et al.* (1998)

Malate dehydrogenase	Aquaspirillium arcticum	10 mg ml^{-1}, 50 mM Tris pH 8.2, 100 mM NaCl	0.1 M Tris pH 8.0, 0.4 M Na-acetate, 35% PEG 4k	18°C		1.9 Å	1B8P 1B8V 1B8U	Kim et al. (1999)
Xylanase	Pseudoalteromonas haloplanktis	Crystal Screen I (Hampton Research) at 4°C	4 mg ml^{-1}, 20 mM MOPS pH 7.5, 50 mM NaCl, 2% trehalose	70% MPD, 0.1 M Na-acetate pH 5.0	4°C	Protein could not be stored above 0.4 mg ml^{-1}	1.3 Å	1H12 1H13 1H14	Van Petegem et al. (2002, 2003)
Sphericase (serine protease)	Antarctic bacilli S41 & S39	Based on related protein	7 mg ml^{-1}, 50 mM Tris pH 7.5, 0.1 M NaCl, 1 mM CaCl$_2$	0.1 M HEPES pH 7.5, 5–25 mM CaCl$_2$, 24% PEG 4k	Room temperature		0.93 Å	1EA7	Almog et al. (2003)
Adenylate kinase	Sporosarcina globispora		20 mg ml^{-1}, 5 mM Ap5A	2.5 mM ammonium sulphate, 1% PEG 1k, 0.1% NaN$_3$, 50 mM HEPES pH 7.0	20°C	Microbatch (4 µl + 6 µl)	2.25 Å	1S3G	Bae and Phillips (2004)
Protein-tyrosine phosphatase	Shewanella sp.	Crystal Screen I (Hampton Research) at 4°C	4.5 mg ml^{-1}	0.1 M Tris pH 8.5, 30% PEG 4k, 0.2 M ammonium sulphate	4°C		1.9 Å	1V73	Tsuruta et al. (2002, 2005)

[a] Jancarik and Kim, 1991.
[b] Zeelen et al., 1994.

◆◆◆◆◆ METHODS TO STUDY THE STABILITY OF PROTEINS FROM PSYCHROPHILES

The stability of a protein is defined as the ability of the protein to resist change in its structure. Two types of stability can be distinguished: conformational stability and kinetic stability (Fagain, 1995).

- Conformational stability refers to the difference in free energy (ΔG) between folded (F) and unfolded (U) protein states, which are in equilibrium (K) with each other (Figure 17.1). Conformational stability is also referred to as, thermodynamic stability. Conformational stability can only be calculated for proteins that undergo reversible unfolding. Proteins with this characteristic tend to be small, single domain proteins. An exception to this is AHA, which is multi-domain (Feller et al., 1999).

$$\text{Folded(F)} \overset{K}{\rightleftharpoons} \text{Unfolded(U)} \overset{k_d}{\to} \text{Denatured(D)}$$

- Kinetic stability is the measure of how rapidly a protein unfolds (U) irreversibly to a denatured state (D) at a given temperature, and is measured by the magnitude of the first order rate constant (k_d) (Figure 17.1). Kinetic stability can be measured for many proteins, even if they aggregate immediately following denaturation and are not capable of reversible unfolding. Large multi-domain proteins are frequently characterized by their kinetic stability.

A number of spectrophotometric (circular dichroism, fluorescence, dynamic fluorescence quenching), calorimetric (differential scanning calorimetry) and electrophoretic (transverse urea gradient gel electrophoresis) methods can be used to study unfolding/folding transitions and measure kinetic and conformational stabilities of proteins (Creighton, 1980; Fagain, 1995; Goldenberg and Creighton, 1984; Shirley, 1995a,b).

Most studies of proteins from psychrophiles using spectrophotometric, calorimetric and electrophoretic techniques have been carried out using multi-domain proteins. These include chitobiase from *Arthrobacter* sp. (Lonhienne et al., 2001), xylanase from *P. haloplanktis* (Collins et al., 2002, 2003), dihydrofolate reductase from *Moritella profunda* (Xu et al., 2003a), ornithine carbamoyltransferase from *Moritella abyssi* (Xu et al., 2003b), phosphoglycerate kinase from *Pseudomonas* sp. (Bentahir et al., 2000), DNA ligase from *P. haloplanktis* (Georlette et al., 2003), β-galactosidase from *P. haloplanktis* (Hoyoux et al., 2001), EF-2 from *M. burtonii* (Siddiqui et al., 2002; Thomas and Cavicchioli, 2000; Thomas et al., 2001) and AHA (D'Amico et al., 2001, 2003b; Feller et al., 1994, 1999; Siddiqui et al., 2005a). AHA has been the subject of extensive studies (D'Amico et al., 2001, 2003b; Feller et al., 1994, 1999; Siddiqui et al., 2005a,b) and is the most well characterized protein from a psychrophile.

Overview of Spectrophotometeric Methods for Studying Protein Conformation

Spectrophotometric methods are suitable for studying protein structure (secondary or tertiary) and stability under a variety of physiochemical

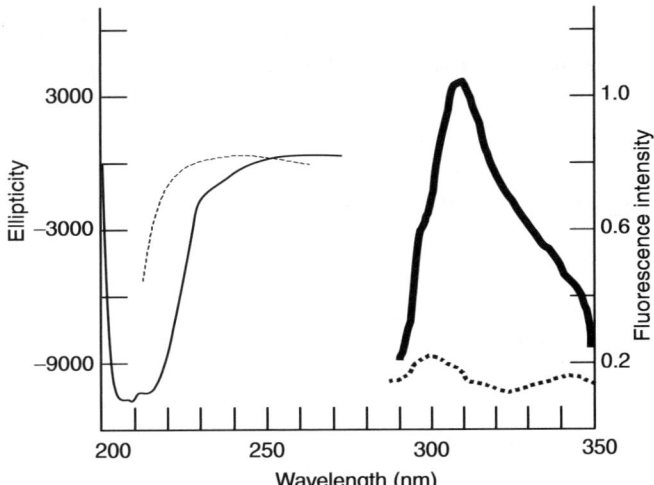

Figure 17.2. Fluorescence and CD spectra for a typical protein. Fluorescence (thick curves) spectra and CD spectra (thin curves) showing folded (continuous line) and fully unfolded states (broken line). CD spectra are in the peptide (far-UV) region. Intrinsic fluorescence spectra shown for an excitation of 278 nm, with a typical protein concentrations of ~0.02 and 1 mg ml^{-1} for fluorescence and CD, respectively.

conditions (e.g. solvent, temperature, denaturants). They may be used to follow structural transitions (unfolding/refolding) of wild-type, mutant or modified proteins. Conformational stability can be determined using fluorescence or circular dichroism (CD) spectroscopy by following protein unfolding as a function of temperature or denaturant (e.g. guanidium-HCl, urea) concentration (Figure 17.2) (D'Amico et al., 2003b; Feller et al., 1999; Pace et al., 1989; Shirley, 1995b).

In order to assess the usefulness of CD or fluorescence spectroscopy for studying a candidate protein, the spectra of the folded and unfolded protein should be determined (Figure 17.2). From the spectra, a number of important features can be assessed.

- The magnitude of the response curve is dependent on the sensitivity of the technique. Fluorescence spectroscopy will generally require an order of magnitude less material in comparison to CD spectroscopy, and therefore may be the method of choice if protein quantity is limiting.
- It is essential to pick a method and wavelength for which the spectra of the folded and unfolded conformations vary by a significant amount (Figure 17.2). The magnitude of the variation between folded and unfolded states will be specific for each protein. If thermal unfolding is being studied it will be better to use CD spectroscopy since fluorescence signals are not very stable at elevated temperatures.
- The signal-to-noise ratio should be maximized to enhance accuracy, and this will need to be empirically determined for each protein sample.
- The specific aim of the analysis (e.g. unfolding of secondary or tertiary structures) will have an important bearing on which method and which wavelength is chosen.

The reader is directed to a number of excellent books that cover theoretical (Shirley, 1995a) and practical (Pace et al., 1989; Schmid, 1989) aspects of protein stability measurement using spectrophotometric methods.

Circular Dichroism (CD) Spectroscopy

CD is observed when there is a difference between the absorption of left- and right-handed circularly polarized light by a protein. The CD spectrum is a function of wavelength and it is dependent on the presence of chiral elements such as protein secondary structure or chiral side-chain environments. The CD spectrum in the far-UV (170–250 nm) provides information about the peptide backbone and can be used to monitor changes in secondary structure (particularly α-helices) during unfolding/folding transitions. CD spectroscopy is of limited value for studying protein β-structures as the CD signal is generally weak. CD signals in the near-UV (250–300 nm) arise from aromatic side chains immobilized in an asymmetric environment of a folded protein. During unfolding, these side chains become exposed to the isotropic solvent and produce a reduced CD signal. Near-UV CD analysis monitors packing interactions or tertiary structure while far-UV CD monitors secondary structure. Both signals give independent measures of the folding/unfolding of a protein. Much higher protein concentrations are required to measure near-UV CD spectra. Most recently, CD spectroscopy has been extended into the vacuum-UV regions (<190 nm) using synchrotron radiation.

Fluorescence Spectroscopy

Proteins naturally absorb light in the UV range due to absorbance by peptide bonds, aromatic amino acids and to a lesser extent, disulphide bonds. In particular, the delocalized π-electrons of aromatic amino acids can absorb in the near-UV. Fluorescence is created when light is generated by an excited electron returning to a ground state. The emission wavelength is always longer than the absorption or excitation wavelength due to the physics of fluorescence. Although all three aromatic residues contribute towards fluorescence emissions, tryptophan has the greatest influence due to its higher absorbance and greater quantum yield. Fluorescence emissions are sensitive to changes in the environment of the fluorescing chromophore, thus, they monitor protein side-chain packing or tertiary structure. Therefore the technique can be useful for studying conformational changes in a protein such as folding/unfolding.

Generation of Unfolding Curves

Under physiological conditions, the equilibrium of most proteins significantly favours the native state, which makes measurements of K difficult (see Figure 17.1). However, in the presence of a denaturing agent or high temperature, the equilibrium constant is shifted towards the unfolded state. Due to the lower stability of proteins from psychrophiles, less drastic conditions (lower denaturant concentrations and temperature) are needed to perturb the equilibrium so as to favour the unfolded state (c.f. proteins from mesophiles or thermophiles). Before experiments can be performed to measure unfolding/folding, it is essential that the unfolding/folding reaction has reached equilibrium. The time taken to reach equilibrium can vary from milliseconds to days depending on the nature of protein, reaction conditions and temperature (Shirley, 1995b).

Using either CD or fluorescence spectroscopy, an unfolding curve can be generated by chemical (0–6 M guanidinium chloride or 0–8 M urea) or thermal denaturation (Figure 17.3). The signal for CD is a measurement of specific optical rotation, and for fluorescence spectroscopy the signal is fluorescence intensity. Unfolding curves have three specific regions (Figure 17.3).

- The pre-transition region reveals how the signal y (fluorescence or CD) of the folded protein (y_F) depends upon the denaturing agent.
- The transition region contains signals from molecules in both the folded and unfolded states as a function of denaturing agent.
- The post-transition region shows how signal y varies for the unfolded protein (y_U).

All these regions should be analysed with at least four points in pre- and post-transition, and 6 points in the transition region (Hoyoux et al., 2001; Pace et al., 1989; Shirley, 1995b). If the transition from the native to unfolded protein is characterized by an abrupt, single-step transition without any plateau or shoulder, then the protein appears to unfold

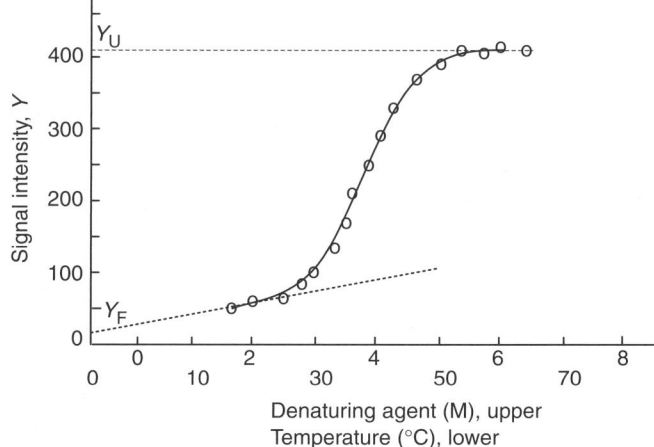

Figure 17.3. Urea and heat-induced unfolding curve for a typical protein from a psychrophile determined using CD or fluorescence spectroscopy. The curve is composed of three regions: the pre-transition region which shows the dependence of y_F on the denaturant; the transition region which shows how the signal intensity (y) varies with the denaturant; the post-transition region which shows the dependence of y_U on the denaturant.

via a two-state process. This should be checked experimentally for reversibility. Thermodynamic analyses can be performed on proteins that fold/unfold reversibly to determine parameters of conformational stability (Figure 17.3).

Analysis of Unfolding Curves and Measurement of Conformational Stability

Chemical denaturation (urea or guanidinium chloride)

From unfolding curves of signal *versus* chemical denaturant (Figure 17.4A), f_U, K and ΔG can be calculated as follows:

$$f_U = (y_F - y)/(y - y_U) \tag{1}$$

where f_U is the fraction of the unfolded protein at any given denaturant concentration, and y_F and y_U are the extrapolated signal intensities of the folded and fully unfolded states calculated for a particular denaturant concentration, respectively.

The equilibrium constant K is given by:

$$K = f_U/(1 - f_U) \tag{2}$$

The Gibbs free energy (ΔG) is calculated from:

$$\Delta G = -RT \ln K \text{ or } \Delta G = -RT \ln[(y_F - y)/(y - y_U)] \tag{3}$$

where R is Universal Gas Constant and T is the absolute temperature.

A straight line (Figure 17.4b) can be fitted to the plot of ΔG *versus* denaturant concentration in the vicinity of the point were ΔG equals zero (i.e. the mid-point of the unfolding reaction). By extrapolating this line to zero denaturant concentration, $\Delta G_{(H_2O)}$ (ΔG in the absence of denaturant) is obtained from the intercept, and m (cal^{-1} mol^{-1} M denaturant) from the slope, where $\Delta G_{(H_2O)} = \Delta G + m$.

Thermal denaturation

For thermal denaturation the approach to calculating thermodynamic values is similar to that used for chemical denaturation. From the unfolding curve of signal *versus* temperature (Figure 17.4c), f_U, K and ΔG can be calculated. ΔG is plotted *versus* temperature using data from the limited region in which ΔG can be measured (Figure 17.4d). The melting temperature (T_m) is the temperature at which half the protein is unfolded (mid-point of thermal unfolding) and ΔG is 0 (Figure 17.4d).

The slope of plot at T_m is equal to ΔS_m, where ΔS_m is the entropy change at T_m.

The enthalpy change ΔH_m at T_m is calculated from:

$$\Delta H_m = T_m \Delta S_m \tag{4}$$

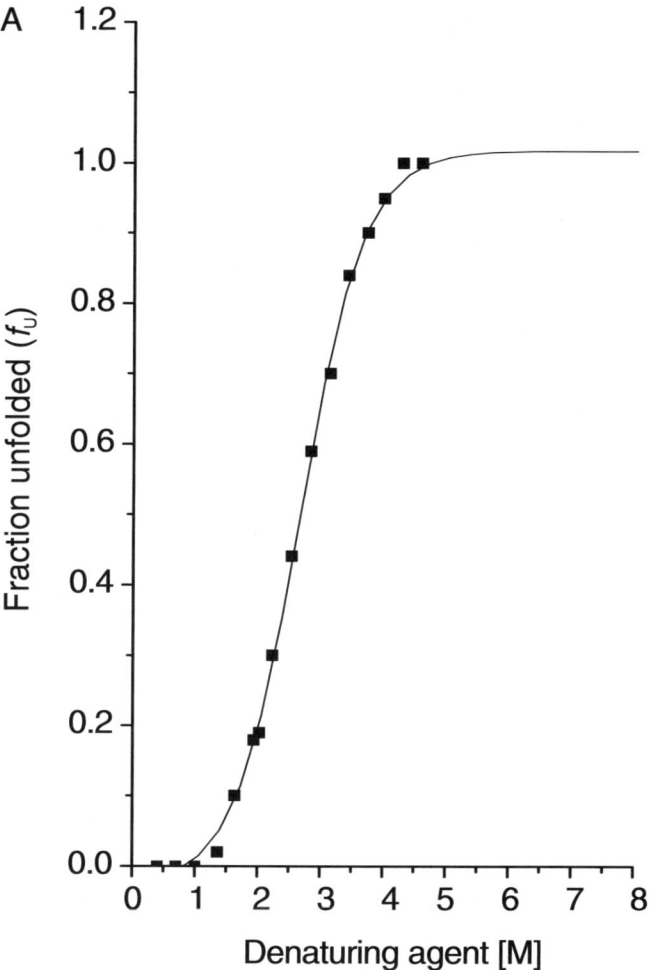

Figure 17.4. Chemical and thermal unfolding curves used for measuring conformational stability. Chemical (A) and thermal (C) induced unfolding curves of a typical protein from a psychrophile. The unfolded fraction (f_U) of the protein is plotted against increasing temperature or increasing concentrations of a denaturing agent. (B) ΔG is plotted *versus* denaturant concentration in the limited region where ΔG can be measured. The plot is then extrapolated to zero denaturant concentration in order to obtain $\Delta G_{(H_2O)}$ in the absence of denaturant. (D) ΔG is plotted *versus* temperature in the linear region in the vicinity of melting temperature (T_m). T_m is shown at the temperature where the protein is half unfolded (midpoint of thermal unfolding) and $\Delta G = 0$.

The conformational free energy of unfolding at any temperature $\Delta G(T)$ can be calculated from the Gibbs–Helmholtz Equation:

$$\Delta G(T) = \Delta H_m(1 - T/T_m) - \Delta C_P[(T_m - T) + T(\ln T/T_m)] \qquad (5)$$

where ΔC_P is the change in heat capacity that can either be determined experimentally by DSC (see below), or approximated by using the

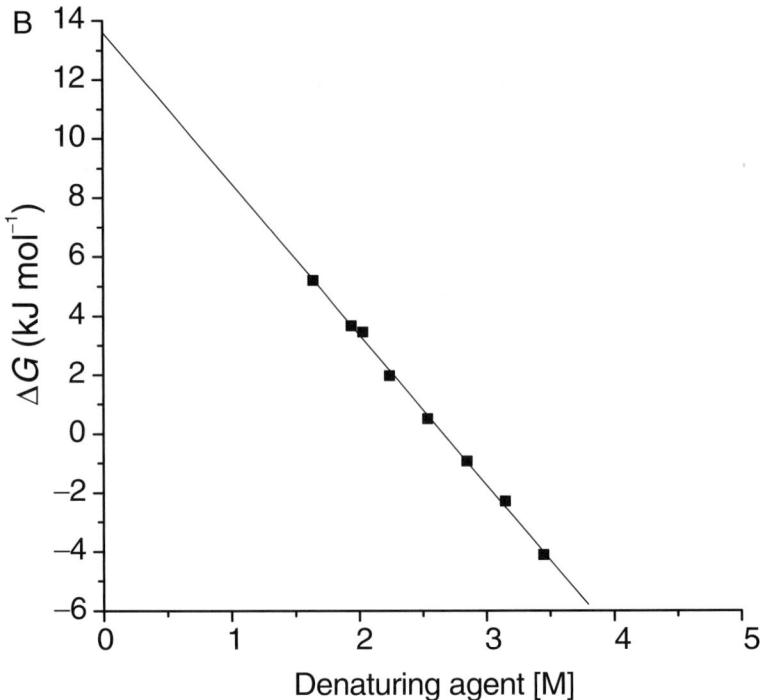

Figure 17.4. *Continued.*

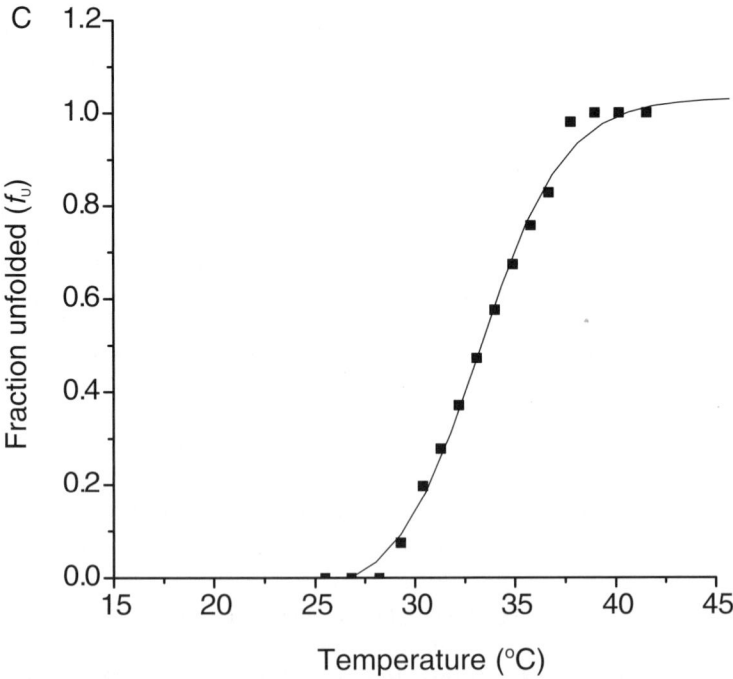

Figure 17.4. *Continued.*

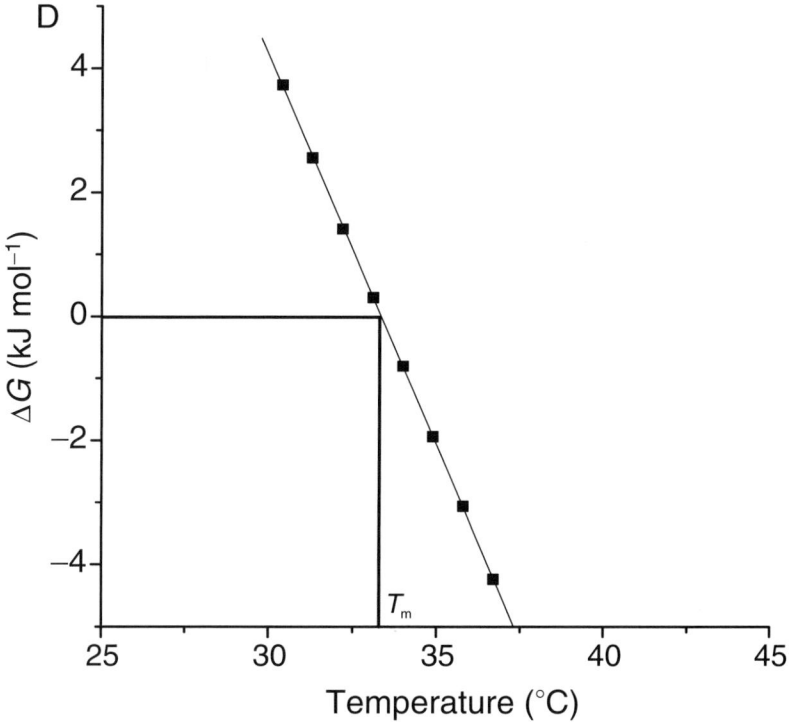

Figure 17.4. *Continued.*

relationship: $\Delta C_p = 12$ cal mol^{-1} K^{-1} multiplied by the number of residues in the protein (D'Amico et al., 2003b; Pace et al., 1989; Schmid, 1989).

Stability curves for thermally adapted proteins can be determined from a plot of $\Delta G(T)$ *versus* temperature (Figure 17.5). The parabola cuts the temperature axis at two points. The right intercept is the melting temperature (where ΔG is 0), and the left intercept is the temperature corresponding to cold denaturation of the protein.

Dynamic Fluorescence Quenching

Dynamic fluorescence quenching is useful for studying the conformational flexibility of proteins from psychrophiles (D'Amico et al., 2003b; Georlette et al., 2003). Dynamic fluorescence quenching employs increasing concentrations of a small quencher molecule (e.g. acrylamide) to explore the accessibility of tryptophan residues inside a protein. The diffusion-driven collisions between acrylamide and tryptophan residues produces decreased fluorescence (i.e. fluorescence quenching). The ability of the acrylamide to quench the fluorescence signal provides an index of protein penetrability by the acrylamide molecules.

Absolute values of the Stern–Volmer quenching constants (K_{sv}) from different proteins cannot be compared, as the location and environment of the tryptophan residues must be identical in each protein. To overcome

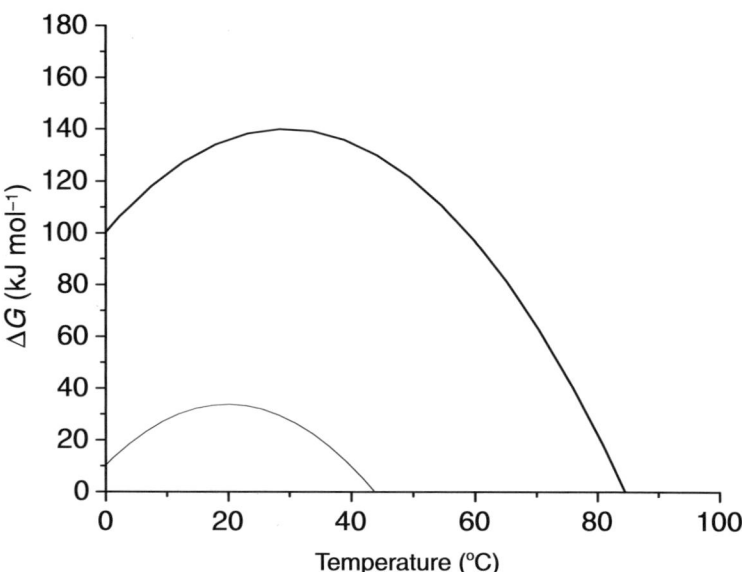

Figure 17.5. Stability curves for typical proteins from a psychrophile and thermophile. Conformational free energy of unfolding (ΔG) is plotted *versus* temperature for proteins from a psychrophile (lower, thin line) and a thermostable homologue (upper, thick line). The curves depict the energy (ΔG) needed to disrupt the protein structure at any temperature. The data are adapted from D'Amico et al. (2003b).

this limitation, it is possible to measure the variation of K_{sv} at different temperatures so that the change in accessibility of the tryptophan residues becomes thermally dependent. Thermally dependent changes in K_{sv} can then be compared.

Acrylamide-dependent quenching of intrinsic tryptophan fluorescence should be monitored at an excitation wavelength of 280 nm and an emission wavelength of 350 nm. Measurements should be taken following consecutive additions of acrylamide. K_{sv} is calculated as follows (Lakowicz, 1983):

$$F/F_0 = 1 + K_{sv}[Q] \qquad (6)$$

where F and F_0 are fluorescence intensity in the presence and absence of molar concentration of acrylamide quencher, Q, respectively (D'Amico et al., 2003b).

A plot of F_0/F (T_2) $-$ (T_1) *versus* acrylamide concentration gives a straight line whose slope equals K_{sv} (Figure 17.6). The intrinsic protein fluorescence "F" is normally corrected for an acrylamide inner filter effect "f" as:

$f = 10^{-\varepsilon[Q]/2}$ where ε is the extinction coefficient for acrylamide (4.3 M^{-1} cm^{-1}) at 280 nm.

Proteins from psychrophiles (e.g. AHA) tend to exhibit higher slopes in Stern–Volmer plots compared to those for proteins from mesophiles or thermophiles (Collins et al., 2003; D'Amico et al., 2003b). This reflects the greater effect of temperature on acrylamide quenching for proteins

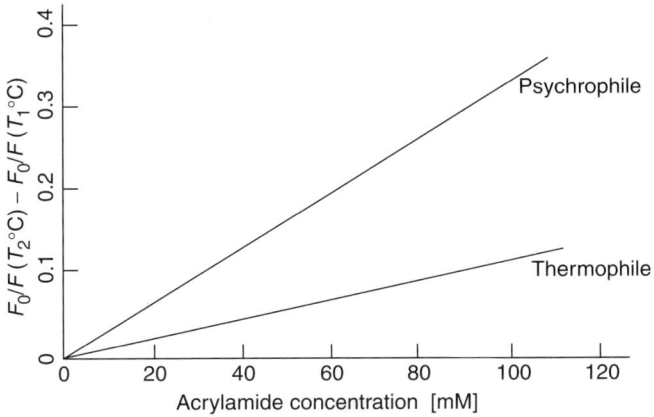

Figure 17.6. Stern–Volmer plots of fluorescence quenching of a typical protein from a psychrophile and a thermophile. A plot of variation of the fluorescence ratios (F_0/F) between two different temperatures (T_1 and T_2) versus acrylamide concentration provides a straight line whose slope equals K_{sv}.

from psychrophiles, and is indicative of a high structural flexibility promoting access of the acrylamide to the protein (Figure 17.6).

Differential Scanning Calorimetry (DSC)

Differential Scanning Calorimetry (DSC) is a calorimetric method that is very useful for studying protein stability. DSC measures the apparent molar heat capacity (C_p) of a protein as a function of temperature (Freire, 1995) (Figure 17.7). The development of DSC methodology has contributed significantly to the understanding of protein unfolding and refolding transitions (Privalov, 1980, 1982; Sturtevant, 1974). It has also provided the means to directly determine kinetic and thermodynamic parameters (D'Amico et al., 2001; Feller et al., 1999; Sanchez-Ruiz, 1988). DSC also provides the ability to determine the stabilization energy (ΔG) of reversibly unfolded proteins (Freire, 1995). This is a useful parameter to calculate as it is possible for two homologous proteins to have similar values for T_m but possess very different stabilities (D'Amico et al., 2001). A good example of this is for AHA, where the ΔG of a double mutant (N150D/V196F) was ~50 kJ mol^{-1} compared with the wild-type enzyme (~32 kJ mol^{-1}), but with little difference in their T_m ($\Delta 2.5°C$) (D'Amico et al., 2001).

Performing DSC

- It is vital that protein concentration and purity are accurately determined. For example, an error of 3% in the specific protein concentration will be translated into 3% error in enthalpy change.
- Approximately 1 mg of protein with a purity exceeding 95% is required for each DSC experiment.

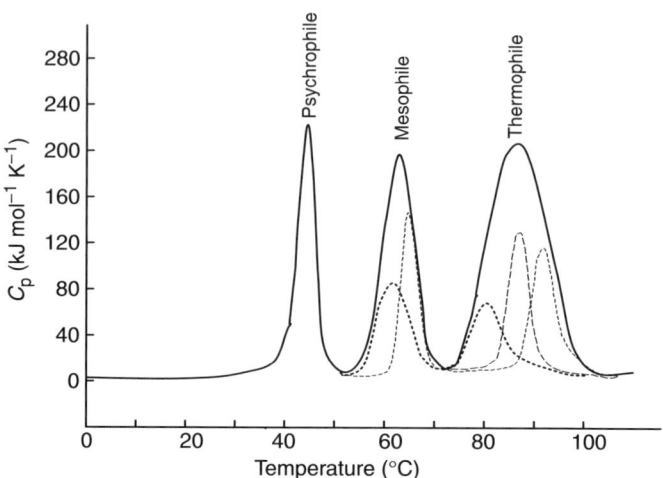

Figure 17.7. DSC thermograms of α-amylases from a pyschrophile, mesophile and thermophile. The data are adapted from D'Amico et al. (2001). The specific heat capacity at a constant pressure C_p (kJ mol^{-1} K^{-1}) (y axis) was plotted against temperature (°C) (x axis). The temperature corresponding to top of the peak is melting temperature (T_m) and the area under the curve is ΔH_{cal}. Deconvolution of the peaks into two and three domains is shown for the protein from the mesophile and thermophile, respectively (dashed lines), and the protein from the psychrophile is shown unfolding as a cooperative unit.

- The protein sample should be extensively dialysed against a suitable buffer. The solubility and pH of the buffer should have minimal temperature-dependency (low dpK_a/dT). For example, TRIS is not a useful buffer, whereas MES, MOPS, HEPES, TAPS, CHES and CAPS and other Good's buffers can be used (Stoll and Blanchard, 1990).
- The sample requires centrifugation or filtering to remove particulate matter, and degassing is required immediately prior to analysis.
- The DSC instrument consists of two cells. One cell should be loaded with the protein sample and the other with the buffer solution. The two cells are heated at a pre-set scan rate, v (K min^{-1}), and the difference in apparent heat capacity (C_p) between the two cells measured over a defined temperature range.
- At least three independent runs should be performed per protein sample (a buffer scan and two protein scans). The two protein scans should assess the unfolding and refolding of the protein, in order to check the reversibility of protein unfolding.

The temperature scan (DSC thermogram) will produce a curve of C_p (kJ mol^{-1} K^{-1}) *versus* temperature (Figure 17.7). For the α-amylases from a psychophile, mesophile and thermophile (Figure 17.7), each thermogram is characterized by a transition from the folded state (left of the peak) to the unfolded/denaturated state (right of the peak). The temperature corresponding to top of the peak is the T_m of the protein, and the area under the curve is the protein's calorimetric enthalpy (ΔH_{cal}). The calorimetric data can also be used to calculate the van 't Hoff enthalpies for the folded and unfolded states and hence the change in van 't Hoff enthalpy (ΔH_{vH}). If the same thermogram is regenerated after an unfolded protein sample is cooled down and reheated again, it implies complete reversibility between folded and unfolded forms. However, if a different thermogram is generated, or no transition is observed after rescanning, then the protein will have undergone an irreversible unfolding to a denatured state. If the ratio between the change in van 't Hoff enthalpy and the change in calorimetric enthalpy is not equal to one ($\Delta H_{vH}/\Delta H_{cal}=1$), then the data are not consistent with a two-state process.

Analysing DSC Thermograms

Thermograms can be analyzed according to two-state (Freire, 1995) or non-two-state models (D'Amico et al., 2001, 2003b; Georlette et al., 2003; Privalov, 1979). Thermodynamic parameters for reversibly unfolding proteins can be directly calculated according to the Gibbs–Helmholtz Equation (Eq. (5)) with ΔH_m and ΔC_p being directly extracted from the thermogram. The ΔC_p value is the difference in molar heat capacity between folded and unfolded states, and ΔH_m is the enthalpy change at T_m. ΔC_p is obtained by subtracting the baseline corresponding to the descending part of the curve (unfolded protein) from that of baseline corresponding to the ascending part of the curve (folded protein).

In many cases, the DSC thermograms reveal that a protein has irreversible unfolding, thereby preventing thermodynamic parameters from being derived. Despite the impediments inherent for proteins showing calorimetric irreversibility, DSC thermograms can be used for interpreting kinetic stability (Sanchez-Ruiz et al., 1988). For the determination of kinetic stability, the rate of irreversible change from the unfolded to the denatured state (k_d) can be determined from the thermogram using the following equation:

$$k_d = v\,\Delta C_p/(\Delta H_{cal} - Q) \tag{7}$$

where Q is the heat absorbed by the protein up to a given temperature, and ΔH_{cal} is the total heat of the process, and is denoted by the area under the curve (Sanchez-Ruiz et al., 1988).

The thermodynamic activation parameters for thermal denaturation can then be determined from the following equations based on Transition State Theory (Lonhienne et al., 2000; Siddiqui et al., 2002):

$$\Delta G^{\#} = -RT\ln(k_d h)/(K_B T) \tag{8}$$

where $\Delta G^{\#}$ is the activation free energy for the irreversible process, T is the absolute temperature, h is Planck Constant (6.63×10^{-34} J s), K_B is Boltzman Constant (1.38×10^{-23} J K^{-1}) and R is the Gas Constant (8.314 J K^{-1} mol^{-1}).

$$\Delta H^{\#} = E_a - RT \tag{9}$$

where $\Delta H^{\#}$ is the activation enthalpy, E_a is activation energy determined from Arrhenius plot of $\ln k_d$ *versus* $1/T$, where E_a is equal to the slope multiplied by R.

$$\Delta S^{\#} = (\Delta H^{\#} - \Delta G^{\#})/T \qquad (10)$$

where $\Delta S^{\#}$ is the activation entropy.

Recently, a modified form of Eq. (8) has been proposed that also takes into account the high viscosity of the solvent that a protein from a psychrophile would encounter at low temperatures (Siddiqui et al., 2004a).

The majority of multi-domain proteins follow an irreversible non-two-state mechanism of unfolding, and as a result their thermograms are scan-rate-dependent. All proteins from psychrophiles reported in the literature to date have been multi-domain proteins.

- Phosphoglycerate kinase displays two distinct denaturation transitions corresponding to a heat-labile and heat-stable domain (Bentahir et al., 2000).
- The EF-2 protein from *M. burtonii* exhibits a complex, irreversible unfolding with at least three, separated unfolding events (Siddiqui et al., 2002; Thomas and Cavicchioli, 2000, Thomas et al., 2001) (Figure 17.8). The Arrhenius plot of K_d *versus* T showed a reduced unfolding activation energy for the cold-adapted EF-2 when compared to its counterpart from the moderate thermophile, *Methanosarcina thermophila* (Figure 17.9). Thermodynamic parameters were calculated from the DSC data, demonstrating that the reduced stability of the protein is due to a reduced enthalpy $\Delta H^{\#}$ value, which is likely to result from fewer or weaker non-covalent bonds breaking during unfolding (Siddiqui et al., 2002).
- The only example of a protein from a psychrophile which displays fully reversible unfolding is AHA (Feller et al., 1999). A N12R mutant of AHA showed ~25% reversibility, as judged by the recovery of the thermogram during a second scan after cooling the sample (D'Amico et al., 2001).

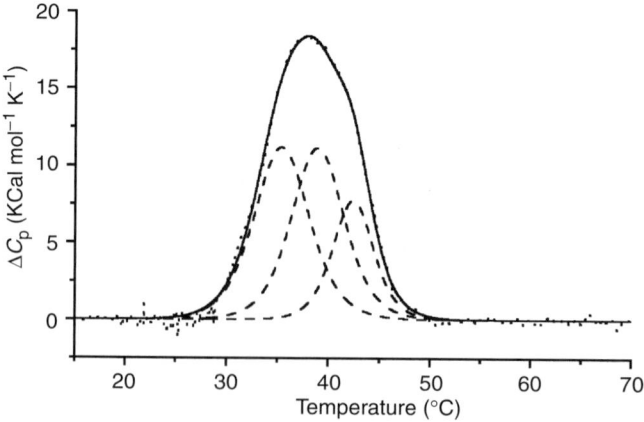

Figure 17.8. Deconvolution of the thermal transition of *M. burtonii* EF-2 at a scan rate of 0.1 K per min. Raw data (dotted line); data fitted to a non-two-state unfolding model with three transitions (full line); approximated unfolding curves for the three unfolding events (dashed line). The data are adapted from Thomas (2001).

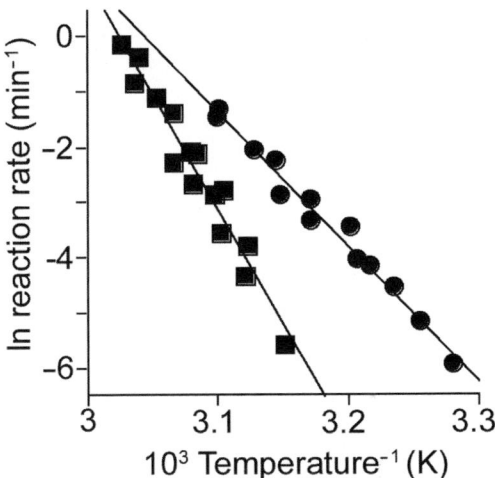

Figure 17.9. Arrhenius plot for the reaction rate of thermal denaturation k_d for EF-2 from *M. burtonii* EF-2 and *M. thermophila*. Values for three different scan rates (0.1, 0.5 and 1.5 K per min) for *M. burtonii* (circles) and *M. thermophila* (squares) proteins. Lines of best fit were used to calculate the activation energy of the reaction. The data are adapted from Thomas (2001).

Transverse Urea Gradient Gel Electrophoresis (TUG-GE)

Transverse urea gradient (TUG) gel electrophoresis (TUG-GE) is a relatively straight forward, inexpensive and powerful method for differentiating protein conformations (Goldenberg 1989; Goldenberg and Creighton, 1984). It also enables thermodynamic and kinetic properties of transitions between conformational states to be analysed (Goldenberg 1989; Goldenberg and Creighton, 1984). TUG-GE has not been used as extensively as spectrophotometric or calorimetric methods for studying protein stability. Recently, it was employed to identify intermediate folded states and to determine which domain unfolds first in the unfolding pathway of the cold-adapted AHA (Siddiqui *et al.*, 2005a).

Unfolding of AHA has previously been shown to proceed *via* cooperative unfolding through a single transition (Feller *et al.*, 1999). These findings were based on biophysical measurements of unfolding/refolding using CD and fluorescence spectroscopy, and DSC (D'Amico *et al.*, 2001; Feller *et al.*, 1999). Combining knowledge from the spectrophotometric, calorimetric and electrophoresis methods has enabled an improved understanding of the structure–stability relationship of this enzyme. It is likely that studies of other proteins from psychrophiles will also benefit by employing a range of technologies, include TUG-GE.

Principle of TUG-GE

TUG-GE analysis of a protein is based on the principle that a fully folded protein is compact and therefore has a low hydrodynamic volume,

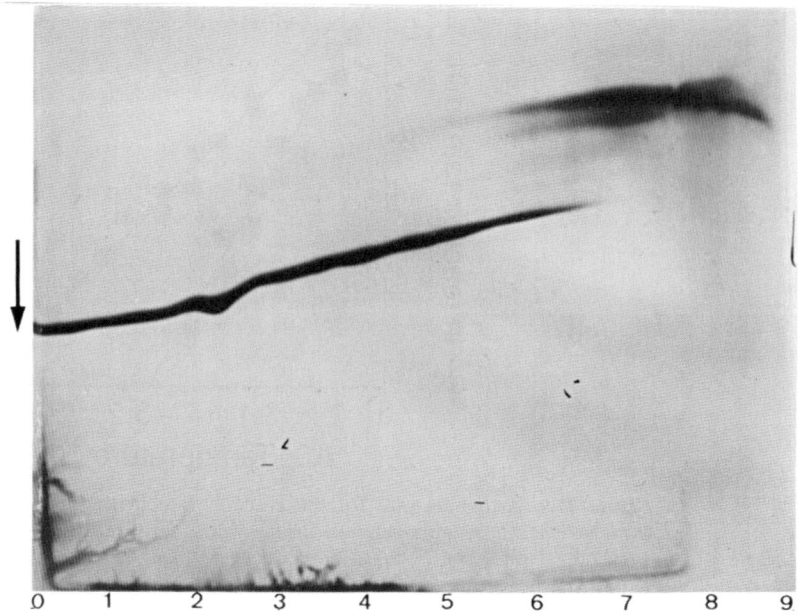

Figure 17.10. TUG-GE of xylose isomerase from a mesophilic *Arthrobacter* sp. The discontinuous curve demonstrates irreversible unfolding. Folded form (lower band); Unfolded form (upper diffuse bands); arrow, direction of electrophoresis; Molar urea concentration (numbers on the x axis). The figure was reproduced from Siddiqui (1990).

whereas a fully unfolded protein has a large hydrodynamic volume. A polyacrylamide gel is cast with a urea gradient and the protein sample electrophoresed perpendicular to the gradient. When exposed to the urea gradient, unfolded forms of the protein will run slower than the folded forms of the protein. As a result, the folded and unfolded forms will separate across the gradient, with partially folded forms lying in between these two extremes (Figure 17.10).

TUG-GE can be used to study proteins that reversibly or irreversibly unfolded. Many large multi-domain proteins do not unfold by cooperative two-state mechanisms and show a discontinuous curve implying irreversible unfolding within the duration of electrophoresis (Figure 17.10). Individual domains may unfold/fold independently at different urea concentrations, producing two or three sequential unfolding transitions.

The technique can reveal irreversible unfolding if conformations inter-convert slowly within the duration of the electrophoretic run (Figure 17.10). This manifests as a diffuse band or smear between folded and unfolded states. At the urea concentrations that correspond to the position of the smear, the rate constant of unfolding/folding is equal to $1/t$, where t is the duration of electrophoresis.

TUG-GE is useful for studying small, single-domain proteins which unfold reversibly by a two-state mechanism and produce a continuous curve across the gradient (Figure 17.11). If the transitions are very rapid compared to the duration of electrophoresis (i.e. the conformations inter-convert many times during the length of the run), then the protein

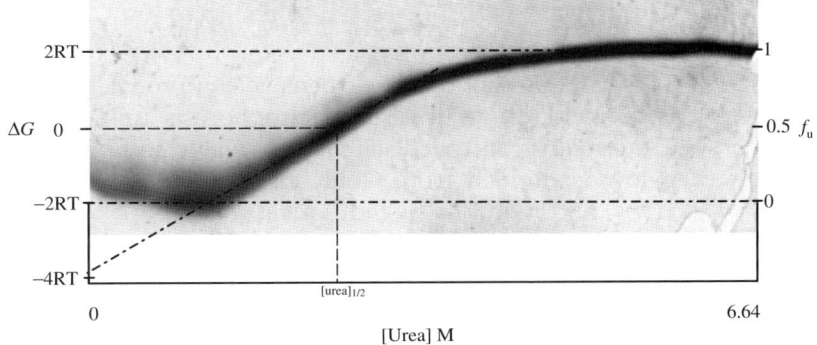

Figure 17.11. TUG-GE of AHA. Characteristic two-state reversible unfolding. The midpoint of urea unfolding, [urea]$_{1/2}$ and ΔG were determined as 2.5 M and 9.6 kJ mol^{-1}, respectively. The direction of electrophoresis was from top to bottom. f_u, fraction of protein unfolded. The figure was reproduced from Siddiqui et al. (2005a).

mobility corresponds to the equilibrium distribution of the conformations. This results in a sharp, continuous band between the two conformations (Figure 17.11). For proteins that show reversible unfolding, thermodynamic parameters, ΔG (stability of folded form relative to the unfolded form) and [urea]$_{1/2}$ (urea concentration corresponding to the mid-point of the curve between the fully folded and fully unfolded forms of the protein) can be determined from the data.

Proteins can be confidently described as following a two-state reversible mechanism of unfolding, if the curve is continuous without any break between the fully folded and fully unfolded parts of the curve (Figure 17.11), and if the pattern of unfolding is identical to that of folding. The latter can be tested by loading a fully folded sample and comparing the profile to that achieved from the electrophoresis of a fully unfolded sample. The patterns should be fully super-imposable. A comprehensive description and interpretation of profiles has been described by Goldenberg and Creighton (1984).

Analysis of TUG-gels

The TUG-gel provides the urea-induced unfolding transition curve (Figure 17.11) from which $K = f_u/(1-f_u)$ can be calculated and extrapolated to zero urea concentration to obtain $\Delta G_{(H_2O)}$ (Goldenberg, 1989; Goldenberg and Creighton, 1984; Siddiqui et al., 2005a). To measure the free energy of unfolding (ΔG) and the mid-point of urea unfolding [urea]$_{1/2}$, the start of the urea gradient and fraction of protein unfolded (f_U) needs to be determined. Near the mid-point of the transition, where $f_U = 0.5$ and $\Delta G = 0$, f_U is the linear function of urea concentration. The net stability in the absence of urea is calculated by extrapolating ΔG from the transition region to 0 M urea concentration (Figure 17.11).

Performing TUG-GE

TUG-GE does not require expensive equipment. The essential equipment is a standard vertical gel-electrophoresis unit, a gradient mixer and a thermally controlled, circulating water bath. The procedure outlined below is derived from Goldenberg (1989), Goldenberg and Creighton (1984) and modified by Siddiqui *et al.* (2005a).

- Urea gradient gels for proteins from psychrophiles may be prepared using a 0–6.64 M urea with an inverse acrylamide gradient of 11–8%.
- A Hoefer Multiple Gel Caster (Amersham Biosciences) (10.5 cm × 10.5 cm with 4 gels) is suitable. One gel sandwich consists of a ceramic-notched plate, a glass plate (10 cm × 8 cm) and a pair of spacers (0.75 mm) held together by two rubber bands. Align gel sandwich pairs in the gel caster with their spacer-secured sides facing the bottom and top of the gel caster.
- Four gel solutions are required: bottom solution (V_1, 14 ml); gradient solution with urea (V_2, 32.5 ml); gradient solution without urea (V_3, 32.5 ml); top solution (V_4, 25 ml). These consist of a buffer solution of appropriate concentration and pH (Table 17.3). These buffer systems are devised primarily for acidic proteins (pI < 7).
- For basic (pI > 7) proteins, the buffer system used is 80 mM β-alanine, 40 mM acetic acid (pH 4.4), and electrophoresis is carried out towards the negative electrode. In low pH buffer systems, it may be beneficial to use riboflavin/TEMED for photo-polymerization of acrylamide.
- TEMED and fresh ammonium persulphate are added immediately before pouring the gels.
- Solution V_1 is carefully poured along the sidewalls in the gel caster so that the final level of the solution is just above the bottom spacer.
- Solution V_2 is poured in the mixing chamber of a gradient mixer with the inter-chamber valve in the closed position and the outlet of the mixing chamber connected to a peristaltic pump through tubing. The gradient maker is placed on a magnetic stirrer with a small stirrer bar in the mixing chamber. The inter-chamber valve is very briefly opened to let a few drops of solution to exit into the reservoir chamber in order to remove air present between two chambers. The solution is pipetted back into the mixing chamber.
- Solution V_3 is poured into the reservoir chamber and the inter-chamber valve is opened. The magnetic stirrer and the pump are started and the urea gradient from 6.64 M to 0 M (bottom to top) is poured into the gel caster. The visual inspection of intensity of the blue colour fading from bottom to the top is a good indicator of the linearity of the gradient.
- After the completion of the gradient, the remaining caster space is filled with solution V_4 using the mixing chamber and the peristaltic pump.
- The gel is polymerized overnight.
- After polymerization the two faces of the gel caster are taken apart and the extra acrylamide outside the gel sandwich is removed.
- Using the gel spacer as a lever, the glass plate is lifted away leaving the gel on the ceramic plate.

Table 17.3 Recipes for 0–6.64 M TUG-gels

[1]Solutions	V$_1$ (bottom)	V$_2$ (6.64 M urea)	V$_3$ (no urea)	V$_4$ (top)
Buffers				
0.86 M Imidazole-HEPES (pH 7.5) (ml)	0.75	1.63	1.63	1.25
1.5 M Tris-HCl (pH 8.8) (ml)	3.75	8.13	8.13	6.25
0.4 M CAPS-NH$_4$OH (pH 10.4) (ml)	0.75	1.63	1.63	1.25
[2]40% Acrylamide (ml) (29:1)	3	6.5	8.9	6.88
[3]Urea (g)	6.4	12.75	–	–
0.4% Bromophenol blue dye (μl)	–	50	–	–
[4]Water	Make up to 15 ml	Make up to 32.5 ml	Make up to 32.5 ml	Make up to 25 ml
Degas the solutions and add polymerization catalysts just before pouring				
10% Ammonium persulphate (μl)	23	42	52	62.5
TEMED (μl)	4.5	9	9	12
Pour in gradient mixer (2 × 100 ml)				
Volume used (ml)	14	32.5	32.5	22–23

[1] All 4 solutions are conveniently made in 50-ml plastic Falcon centrifuge bottles.
[2] Inverse acrylamide gradient is from 11–8%. Bis-acrylamide is 3.3%.
[3] Urea gradient is from 0 to 6.64 M.
[4] Any supplementary solutions, including competitive inhibitors and metals (e.g. MgCl$_2$, CaCl$_2$) can be added in lieu of water.

- Using a long sharp blade, a single well is cut near the notched end across the urea gradient, and the glass plate replaced.
- Two gels are assembled in an electrophoresis apparatus (e.g. Hoefer SE-250) connected to refrigerated water bath (e.g. MultiTemp III, Amersham Biosciences) and pre-run at 40–60 volts for 30–40 min at the required temperature.
- The tank solutions used for various gel buffer systems include: 43 mM imidazole-HEPES (pH 7.5); 0.33% Tris, 1.4% glycine (pH 8.8); 20 mM CAPS-ammonia (pH 10.4). Gels can be run in the presence of metal ions, EDTA or a competitive inhibitor.
- Sample buffer for studying unfolding (folded ⇌ unfolded) and folding (unfolded ⇌ folded) is provided (Table 17.4).
- For unfolded ⇌ folded gels, the protein is incubated in urea containing sample buffer prior to loading on the gel and then subjected to electrophoresis.
- The duration of the electrophoresis will vary depending on buffer and temperature (on the order of 1–3 h may be expected).
- Gels can be stained to visualize the proteins (e.g. Coomassie R250, silver, fluorescent stain) or activity stained (See and Jackowski, 1989).

Table 17.4 Recipes for TUG-gel sample buffers

Solutions	With urea	Without urea
0.4 M Buffer (µl)	3.5	3.5
Glycerol (µl)	–	20
Urea (g)	0.027	–
0.4% Bromophenol blue (µl)	1	1
[1]Protein, 25–30 µg (µl)	5	5
[1]Water (µl)	~30	40
Total volume (µl)	70	70

[1]The amount of protein can be increased and volume of water correspondingly decreased.

◆◆◆◆◆ GENOMICS AND PROTEOMICS OF PSYCHROPHILES

Complete or draft genome sequences of cold-adapted microorganisms are (or will soon be) available for the Archaea, "*Cenarchaeum symbiosum*", *Methanococcoides burtonii* and *Methanogenium frigidum*, and the Bacteria, *Bacillus cereus*, *Colwellia psychrerythraea*, *Desulfotalea psychrophila*, *Photobacterium profundum*, *Pseudoalteromonas haloplanktis*, *Psychrobacter* sp. and *Shewanella frigidimarina*. These provide unique and powerful resources for studying proteins from psychrophiles. Genome sequences represent tailored databases of genes encoding potentially cold-adapted proteins. They can therefore be used to search for classes of enzymes (e.g. hydrolases) or for specific protein targets. Using available clones from gene libraries (e.g. used for genome sequencing), or amplifying directly from genomic DNA, genes of interest can be rapidly cloned and heterologous proteins expressed (see earlier sections in this chapter).

Genome sequences of psychrophiles can also be used for comparative genomic studies for studying the basis of protein adaptation to the cold. Based on predictions generated from these studies, hypotheses can then be experimentally tested. The availability of the first draft genome sequences for cold-adapted Archaea enabled comparisons of archaeal genomes from organisms able to grow from 0 to 110°C (Saunders *et al.*, 2003). Distinguishing features of psychrophiles that emerged from the study included structural and compositional features of proteins, and the identification of novel, nucleic acid binding proteins. An example of this work was the finding from principal components analysis that proteins from the cold-adapted Archaea had a higher content of non-charged polar amino acids, particularly Gln and Thr and a lower content of hydrophobic amino acids, particularly Leu. Using a protein threading approach, 1111 modelled protein structures were generated from 9 methanogen genome sequences. Analysis of the models from the cold-adapted Archaea showed a strong tendency in the solvent accessible area for more Gln, Thr and hydrophobic residues, and fewer charged residues. The approach integrated compositional and structural data, thereby enabling the rationalization of observed trends (i.e. compositional trends for Gln and Thr reflected on the surface structure of the proteins). By involving the analyses of large data sets, conclusions drawn about thermal adaptation are more likely to reflect trends of general importance, compared to findings inferred from a comparative study of single gene sets. This study not only identified a broad range of experimentally testable questions about protein adaptation, but illustrates the type of insight that can be gained from the analysis of raw genome data.

Proteome-based studies provide new avenues for studying proteins from psychrophiles. While the genome sequence highlights the potential genetic capacity of an organism, knowledge of the expressed proteome provides insight into which proteins are produced, and therefore which are required by the cell for growth and survival. Comparing the expressed proteome from two (or more) growth conditions provides insight into which proteins are of particular importance under specific growth conditions. Proteomic (Goodchild *et al.*, 2004a; Saunders *et al.*, 2005) and comparative proteomic (Goodchild *et al.*, 2004b, 2005) studies have been performed on *M. burtonii*. A liquid chromatography-mass spectrometry (LC-MS) approach enabled 528 proteins to be identified from *M. burtonii* growing at 4°C (Goodchild *et al.*, 2004a). Two-dimensional gel electrophoresis (2DE) (Goodchild *et al.*, 2004b) and isotope-coded affinity tag (ICAT) chromatography LC-MS (Goodchild *et al.*, 2005) have also been used for comparative proteomics of cells growing at 4 and 23°C. These developments have not only enabled a greater understanding of the biology of this organism to be derived, but illustrate the capacity to perform high-throughput analyses of proteins from other cold-adapted organisms.

It is clear that the availability of genome sequence data greatly amplifies the number and reliability of protein identifications in proteomics. In addition to being a resource for identifying proteins, the physical organization of genes in the genome sequence provides insight into associated gene partners, and gene regulation (e.g. putative operons). A good illustration of the value of the genome sequence for proteomic studies that goes beyond protein identification was its use for inferring the function of hypothetical proteins (Saunders *et al.*, 2005). From 528 expressed proteins, 135 had no functional classification, being annotated only as hypothetical or conserved hypothetical (Goodchild *et al.*, 2004a). Functional information for 55 of these was inferred by using a comprehensive, integrated analysis of the hypothetical proteins using threading, InterProScan, predicted subcellular localization and visualization of conserved gene context across multiple prokaryotic genomes (Saunders *et al.*, 2005).

Genomics underpins programmes of global functional studies, such as proteomics, and provides unprecedented opportunities for delving into the properties of proteins from psychrophiles.

◆◆◆◆◆ BIOTECHNOLOGICAL APPLICATION OF PROTEINS FROM PSYCHROPHILES FOR USE IN MOLECULAR BIOLOGY

The two most important applied aspects of proteins from psychrophile are their high activity at room temperature and their high thermolability (Cavicchioli et al., 2002). Several molecular biology protocols employ heat inactivation to destroy enzymes for further DNA manipulation. However, prolonged heating at elevated temperatures risks denaturation of double-stranded DNA. Cold-adapted alkaline phosphatases from psychrophilic organisms have been marketed for use in molecular biology due to the ease in which they can be rapidly inactivated without the need for adding harsh chemicals. A heat-labile alkaline phosphatase was sourced from the Antarctic bacterium HK47 for developing this molecular biology application (Kobori et al., 1984). Alkaline phosphatases which are more thermolabile than the enzyme from Antarctic bacterium HK47, include the Red Arctic Shrimp, *Pandalus borealis* enzyme (Olsen et al., 1991) and one from the Antarctic bacterium, strain TAB5 (Rina et al., 2000).

Thermolabile proteases from psychrophiles have been considered for their use in proteolytically digesting molecular biology enzymes, such as DNA polymerases and restriction endonucleases, in order to inactivate them (Moran et al., 2001).

In addition to their thermolability, the high intrinsic activity of proteins from psychrophiles provides avenues for biotechnological exploitation. PCR requires DNA polymerase catalysed reactions to be performed at temperatures typically ranging from 50 to 92°C. For example, DNA may be heated at 92°C to separate two DNA strands followed by primer binding at 50°C and DNA synthesis at 72°C. Recently, DNA helicases have been employed to separate duplex DNA, rather than employing heat to melt apart the DNA strands (Vincent et al., 2004). This has enabled the entire amplification process to be performed at room temperature. Opportunities are clearly available to improve the rate of amplification at room temperature by using DNA helicases and polymerases from psychrophiles.

◆◆◆◆◆ CONCLUSION

From our survey, it is clear that all the techniques of protein science can be applied to proteins from psychrophilic microorganisms. So far, only a small number of such proteins have been studied extensively. In these cases, most methods have been used in the same manner as per proteins of mesophilic and thermophilic origin. What is obvious is that more examples of proteins from psychrophiles require detailed characterization. It may well be that by taking a genomic approach, specific methodologies may arise that increase our knowledge and our ability to handle and study proteins from psychrophiles. This will lead to a better understanding of their properties in general and have impact on basic questions such as cold-adaptation.

Acknowledgements

The authors acknowledge the financial support provided by the Australian Research Council which has contributed significantly to their programmes of research.

List of Suppliers

Promega
2800 Woods Hollow Road
Madison WI 53711, USA
Tel: +1-608-274-4330
Fax: +1-800-356-1970
http://www.promega.com

GeneEditor™

Stratagene
11011 N. Torrey Pines Road
La Jolla, CA 92037, USA
Tel: +1-800-894-1304
Fax: +1-858-535-0034
http://www.stratagene.com

QuikChange®

Amersham Biosciences AB/GE Healthcare
SE-751 84 Uppsala, Sweden
Tel: +46-18-612-0000
Fax: +46-18-612-1200
www.amershambiosciences.com

Gel caster and other electrophoresis equipment

References

Acharya, A. S., Cho, Y. J. and Manjula, B. N. (1988). Cross-linking of proteins by aldotriose: reaction of the carbonyl function of the keto amines generated in situ with amino groups. *Biochemistry* **27**, 4522–4529.

Aghajari, N., Feller, G., Gerday, C. and Haser, R. (1996). Crystallization and preliminary X-ray diffraction studies of α-amylase from the Antarctic psychrophile *Alteromonas haloplanctis* A23. *Protein Sci.* **5**, 2128–2129.

Aghajari, N., Feller, G., Gerday, C. and Haser, R. (1998). Crystal structures of the psychrophilic α-amylase from *Alteromonas haloplanctis* in its native form and complexed with an inhibitor. *Protein Sci.* **7**, 564–572.

Aghajari, N., Van Petegem, F., Villeret, V., Chessa, J. P., Gerday, C., Haser, R. and Van Beeumen, J. (2003). Crystal structures of a psychrophilic metalloprotease reveal new insights into catalysis by cold-adapted proteases. *Proteins* **50**, 636–647.

Almog, O., Gonzalez, A., Klein, D., Greenblatt, H. M., Braun, S. and Shoham, G. (2003). The 0.93 Å crystal structure of sphericase: a calcium-loaded serine protease from *Bacillus sphaericus*. *J. Mol. Biol.* **332**, 1071–1082.

Alvarez, M., Zeelen, J. P., Mainfroid, V., Rentier-Delrue, F., Martial, J. A., Wyns, L., Wierenga, R. K. and Maes, D. (1998). Triose-phosphate isomerase (TIM) of the psychrophilic bacterium *Vibrio marinus*: Kinetic and structural properties. *J. Biol. Chem.* **273**, 2199–2206.

Bae, E. and Phillips, G. N. Jr. (2004). Structures and analysis of highly homologous psychrophilic, mesophilic, and thermophilic adenylate kinases. *J. Biol. Chem.* **279**, 28202–28208.

Bentahir, M., Feller, G., Aittaleb, M., Lamotte-Brasseur, J., Himri, T., Chessa, J.-P. and Gerday, C. (2000). Structural, kinetic and calorimetric characterization of the cold-active phosphoglycerate kinase from the Antarctic *Pseudomonas* sp. TACII18. *J. Biol. Chem.* **275**, 11147–11153.

Birolo, L., Tutino, M. L., Fontanella, B., Gerday, C., Mainolfi, K. P. S., Sannia, G., Vinci, F. and Marino, G. (2000). Aspartate aminotransferase from the Antarctic bacterium *Pseudoalteromonas haloplanktis* TAC 125: Cloning, expression, properties, and molecular modelling. *Eur. J. Biochem.* **267**, 2790–2801.

Bokhari, S. A., Afzal, A. J., Rashid, M. H., Rajoka, M. I. and Siddiqui, K. S. (2002). Coupling of surface carboxyls of carboxymethylcellulase with aniline via chemical modification: extreme thermostabilization in aqueous and water-miscible organic mixtures. *Biotechnol. Progr.* **18**, 276–281.

Brakmann, S. (2001). Discovery of superior enzymes by directed molecular evolution. *Chembiochem* **2**, 865–871.

Cavicchioli, R. and Siddiqui, K. S. (2004). Cold-adapted enzymes. In *Enzyme Technology* (A. Pandey, C. Webb, C. R. Soccol and C. Larroche, eds), pp. 615–638. AsiaTech Publishers, New Delhi.

Cavicchioli, R., Siddiqui, K. S., Andrews, D. and Sowers, K. R. (2002). Low-temperature extremophiles and their applications. *Curr. Opinion Biotechnol.* **13**, 253–261.

Collins, T., Meuwis, M. A., Stals, I., Claeyssens, M., Feller, G. and Gerday, C. (2002). A novel family 8 xylanase, functional and physicochemical characterization. *J. Biol. Chem.* **277**, 35133–35139.

Collins, T., Meuwis, M. A., Gerday, C. and Feller, G. (2003). Activity, stability and flexibility in glycosidases adapted to extreme thermal environments. *J. Mol. Biol.* **328**, 419–428.

Creighton, T. E. (1980). Kinetic study of protein unfolding and refolding using urea gradient electrophoresis. *J. Mol. Biol.* **137**, 61–80.

Cupo, P., El-Deiry, W., Whitney, P. L. and Awad, W. M. (1980). Stabilization of proteins by guanidination. *J. Biol. Chem.* **255**, 10828–10833.

D'Amico, S., Gerday, C. and Feller, G. (2001). Structural determinants of cold adaptation and stability in a large protein. *J. Biol. Chem.* **276**, 25791–25796.

D'Amico, S., Gerday, C. and Feller, G. (2003a). Temperature adaptation of proteins: engineering mesophilic-like activity and stability in a cold-adapted α-amylase. *J. Mol. Biol.* **332**, 981–988.

D'Amico, S., Marx, J. C., Gerday, C. and Feller, G. (2003b). Activity-stability relationship in extremophilic enzymes. *J. Biol. Chem.* **278**, 7891–7896.

Dellacherie, E., Bonneaux, F., Labrude, P. and Vigneron, C. (1983). Modification of human hemoglobin by covalent association with soluble dextran. *Biochim. Biophys. Acta* **749**, 106–114.

DeSantis, G., Bergland, P., Stabile, M. R., Gold, M. and Jones, J. B. (1998). Site-directed mutagenesis combined with chemical modification as a strategy for altering the specificity of the S1 and S1' pockets of subtilisin *Bacillus lentus*. *Biochemistry* **37**, 5968–5973.

Deutscher, M. P. (1990). *Guide to Protein Purification*. Academic Press, San Diego.

Duilio, A., Tutino, M. L. and Marino, G. (2004). Recombinant protein production in Antarctic Gram-negative bacteria. In *Methods in Molecular Biology: Recombinant Gene Expression: Reviews and Protocols* (P. Balbas and A. Lorence, eds), vol. 267, pp. 225–237. Humana Press Inc., Totowa.

Eijsink, V. G. H., Bjork, A., Gaseidnes, S., Sirevag, R., Synstad, B., van den Burg, B. and Vriend, G. (2004). Rational engineering of enzyme stability. *J. Biotechnol.* **113**, 105–120.

Fagain, C. O. (1995). Understanding and increasing protein stability. *Biochim. Biophys. Acta* **1252**, 1–14.

Feller, G. and Gerday, C. (2003). Psychrophilic enzymes: hot topics in cold adaptation. *Nature Rev. Microbiol.* **1**, 200–208.

Feller, G., Payan, F., Theys, F., Qian, M., Hasser, R. and Gerday, C. (1994). Stability and structural analysis of α-amylase from the Antarctic psychrophile *Altermonas haloplanctis* A23. *Eur. J. Biochem.* **222**, 441–447.

Feller, G., Le Bussy, O. and Gerday, C. (1998). Expression of psychrophilic genes in mesophilic hosts: assessment of the folding sate of a recombinant α-amylase. *Appl. Environ. Microbiol.* **64**, 1163–1165.

Feller, G., D'Amico, D. and Gerday, C. (1999). Thermodynamic stability of a cold-active α-amylase from Antarctic bacterium *Alteromonas haloplanctis*. *Biochemistry* **38**, 4613–4619.

Ferrer, M., Chernikova, T. N., Timmis, K. N. and Golyshin, P. N. (2004). Expression of a temperature-sensitive esterase in a novel chaperone-based *Escherichia coli* strain. *Appl. Environ. Microbiol.* **70**, 4499–4504.

Fields, R. (1971). The measurement of amino groups in proteins and peptides. *Biochem. J.* **124**, 581–589.

Freire, E. (1995). Differential scanning calorimetry. In *Protein Stability and Folding* (B. A. Shirley, ed.), pp. 191–218. Humana Press, Totowa.

Georlette, D., Damien, B., Blaise, V., Depiereux, E., Uversky, V. N., Gerday, C. and Feller, G. (2003). Structural and functional adaptations to extreme temperatures in psychrophilic, mesophilic, and thermophilic DNA ligases. *J. Biol. Chem.* **278**, 37015–37023.

Goldenberg, D. P. (1989). Analysis of protein conformation by gel electrophoresis. In *Protein Structure: A Practical Approach* (T. E. Creighton, ed.), pp. 225–250. IRL Press, Oxford.

Goldenberg, D. P. and Creighton, T. E. (1984). Review: gel electrophoresis in studies of protein conformation and folding. *Anal. Biochem.* **138**, 1–18.

Goodchild, A., Raftery, M., Saunders, N. F. W., Guilhaus, M. and Cavicchioli, R. (2004a). The biology of the cold adapted archaeon, *Methanococcoides burtonii* determined by proteomics using liquid chromatography-tandem mass spectrometry. *J. Prot. Res.* **3**, 1164–1176.

Goodchild, A., Saunders, N. F. W., Ertan, H., Raftery, M., Guilhaus, M., Curmi, P. M. G. and Cavicchioli, R. (2004b). A proteomic determination of cold adaptation in the Antarctic archaeon, *Methanococcoides burtonii*. *Mol. Microbiol.* **53**, 309–321.

Goodchild, A., Raftery, M., Saunders, N. F. W., Guilhaus, M. and Cavicchioli, R. (2005). Cold adaptation of the Antarctic archaeon, *Methanococcoides burtonii* assessed by proteomics using ICAT. *J. Prot. Res.* **4**, 473–480.

Gottesmann, S. (1990). Minimizing proteolysis in *Escherichia coli*: a genetic solution. *Method. Enzymol.* **185**, 119–129.

Habib, A. F. S. A. (1972). Guanidination of proteins. *Method. Enzymol.* **25**, 558–566.

Hollecker, M. (1989). Counting integral number of residues by chemical modification In *Protein Structure: A Practical Approach* (T. E. Creighton, ed.), pp. 145–153. IRL Press, Oxford.

Hoyoux, A., Jennes, I., Dubois, P., Genicot, S., Dubail, F., Francois, J. M., Baise, E., Feller, G. and Gerday, C. (2001). Cold-adapted β-galactosidase from the Antarctic psychrophile *Pseudomonas haloplanktis*. *Appl. Environ. Microbiol.* **67**, 1529–1535.

Imoto, T. and Yamada, H. (1989). Chemical modification. In *Protein Structure: A Practical Approach* (T. E. Creighton, ed.), pp. 247–277. IRL Press, Oxford.

Jancarik, J. and Kim, S. H. (1991). Sparse matrix sampling: a screening method for crystallization of proteins. *J. Appl. Cryst.* **24**, 409–411.

Jonsdottir, G., Bjarnason, J. B. and Gudmundsdottir, A. (2004). Recombinat cold-adapted trypsin from Atlantic cod: expression, purification, and identification. *Protein Expres. Purif.* **33**, 110–122.

Kim, S. Y., Hwang, K. Y., Kim, S. H., Sung, H. C., Han, Y. S. and Cho, Y. (1999). Structural basis for cold adaptation: sequence, biochemical properties, and crystal structure of malate dehydrogenase from a psychrophile *Aquaspirillium arcticum*. *J. Biol. Chem.* **274**, 11761–11767.

Kobori, H., Sullivan, C. W. and Shizuya, H. (1984). Heat-labile alkaline phosphatase from Antarctic bacteria: rapid 5' end-labeling of nucleic acids. *Proc. Natl. Acad. Sci. USA* **81**, 6691–6695.

Kochhar, S., Mouratou, B. and Christen, P. (2002). Amino acid analysis by precolumn derivatization with 1-fluoro-2,4-dinitrophenyl-5-l-alanine amide (Marfey's Reagent) In *The Protein Protocols Handbook* (J. M. Walker, ed.), pp. 567–572. Humana Press, Totowa.

Lakowicz, J. (1983). *Principles of Fluorescence Spectroscopy*. Plenum Press, New York.

Lonhienne, T., Gerday, C. and Feller, G. (2000). Psychrophic enzymes: revisiting the thermodynamic parameters of activation may explain local flexibility. *Biochim. Biophys. Acta* **1543**, 1–10.

Lonhienne, T., Zoidakis, J., Vorgias, C. E., Feller, G., Gerday, C. and Bouriotis, V. (2001). Modular structure, local flexibility and cold-activity of a novel chitobiase from a psychrophilic Antarctic bacterium. *J. Mol. Biol.* **310**, 291–297.

Lopez-Gallego, F., Montes, T., Fuentes, M., Alonso, N., Grazu, V., Betancor, L., Guisan, J. M. and Farnandez-Lafuente, R. (2005). Improved stabilization of chemically aminated enzymes *via* multipoint covalent attachment on glyoxyl supports. *J. Biotechnol.* **116**, 1–10.

Lundblad, R. L. (1995). *Techniques in Protein Modification*. CRC Press Inc., Boca Raton.

Lutz, S. and Patrick, W. M. (2004). Novel methods for directed evolution of enzymes: quality, not quantity. *Curr. Opinion Biotechnol.* **15**, 291–297.

Makridies, S. C. (1996). Strategies for achieving high-level expression of genes in *Escherichia coli*. *Microbiol. Mol. Biol. Rev.* **60**, 512–538.

Miyazaki, K. and Arnold, F. H. (1999). Exploring nonnatural evolutionary pathways by saturation mutagenesis: rapid improvement of protein function. *J. Mol. Biol.* **49**, 716–720.

Miyazaki, K., Wintrode, P. L., Grayling, R. A., Rubingh, F. H. and Arnold, F. H. (2000). Directed evolution study of temperature adaptation in a psychrophilic enzyme. *J. Mol. Biol.* **297**, 1015–1026.

Moran, A. J., Hills, M., Gunton, J. and Nano, F. E. (2001). Heat-labile proteases in molecular biology applications. *FEMS Microbiol. Lett.* **197**, 59–63.

Mozhaev, V. V., Siksnis, V. A., Melik-Nubarov, N. S., Galkantaite, N. Z., Denis, G. J., Butkus, E. P., Zaslavsky, B. Y., Mestechkina, N. M. and Martinek, K. (1988). Protein stabilization *via* hydrophilization: covalent modification of trypsin and alpha-chymotrypsin. *Eur. J. Biochem.* **173**, 147–154.

Olsen, R. L., Overbo, K. and Myrnes, B. (1991). Alkaline phosphatase from the hapatopancreas of shrimp (*Pandalus borealis*): a dimeric enzyme with catalytically active subunits. *Comp. Biochem. Physiol. B* **99**, 755–761.

Pace, C. N., Shirley, B. A. and Thomson, J. A. (1989). Measuring the conformational stability of a protein. In *Protein Structure: A Practical Approach* (T. E. Creighton, ed.), pp. 311–330. IRL Press, Oxford.

Privalov, P. L. (1979). Stability of proteins: small globular proteins. *Adv. Protein Chem.* **33**, 167–241.

Privalov, P. L. (1980). Scanning microcalorimeters for studying macromolecules. *Pure Appl. Chem.* **52**, 479–497.

Privalov, P. L. (1982). Stability of proteins: proteins which do not present a single cooperative system. *Adv. Protein Chem.* **35**, 1–104.

Rashid, M. H. and Siddiqui, K. S. (1998). Carboxyl group modification: high temperature activation of charge neutralized and charge reversed β-glucosidase from *Aspergillus niger*. *Biotechnol. Appl. Biochem.* **27**, 231–237.

Rashid, M. H., Najmus, S. A. A., Rajoka, M. I. and Siddiqui, K. S. (1997). Native Enzyme Mobility Shift Assay (NEMSA): A new method for monitoring the carboxyl group modification of carboxymethylcellulase from *Aspergillus niger*. *Biotechnol. Techniques* **11**, 245–247.

Rina, M., Pozidis, C., Mavromatis, K., Tzanodaskalaki, M., Kokkinidis, M. and Bouriotis, V. (2000). Alkaline phosphatase from the Antarctic strain TAB5: properties and psychrophilic adaptations. *Eur. J. Biochem.* **267**, 1230–1238.

Russell, N. J. (2006). Methods for handling of psychrophilic microorganisms. In *Methods in Microbiology: Extremophiles*, (F. A. Rainey and A. Oren, eds), vol. 35, pp. 367–389. Academic Press, London.

Russell, R. J. M., Gerike, U., Danson, M. J., Hough, D. W. and Taylor, G. L. (1998). Structural adaptations of the cold-active citrate synthase from an Antarctic bacterium. *Structure* **6**, 351–361.

Sanchez-Ruiz, J. M., Lopez-Lacomba, J. L., Cortijo, M. and Mateo, P. L. (1988). Differential scanning calorimetry of the irreversible thermal denaturation of thermolysin. *Biochemistry* **27**, 1648–1652.

Saunders, N. F. W., Thomas, T., Curmi, P. M. G., Mattick, J. S., Kuczek, E., Slade, R., Davis, J., Franzmann, P. D., Boone, D., Rusterholtz, K., Feldman, R., Gates, C., Bench, S., Sowers, K., Kadner, K., Aerts, A., Dehal, P., Detter, C., Glavina, T., Lucas, S., Richardson, P., Larimer, F., Hauser, L., Land, M. and Cavicchioli, R. (2003). Mechanisms of thermal adaptation revealed from the genomes of the Antarctic Archaea, *Methanogenium frigidum* and *Methanococcoides burtonii*. *Genome Res.* **13**, 1580–1588.

Saunders, N. F. W., Goodchild, A., Raftery, M., Guilhaus, M., Curmi, P. M. G. and Cavicchioli, R. (2005). Predicted roles for hypothetical proteins in the low-temperature expressed proteome of the Antarctic archaeon *Methanococcoides burtonii*. *J. Prot. Res.* **4**, 464–472.

Schmid, F. X. (1989). Spectral methods of characterizing protein conformation and conformational changes. In *Protein Structure: A Practical Approach* (T. E. Creighton, ed.), pp. 251–285. IRL Press, Oxford.

See, Y. P. and Jackowski, G. (1989). Estimating molecular weights of polypeptides by SDS gel electrophoresis. In *Protein Structure: A Practical Approach* (T. E. Creighton, ed.), pp. 1–21. IRL Press, Oxford.

Shirley, B. A. (1995a). *Methods in Molecular Biology: Protein Stability and Folding: Theory and Practice*, vol. 40. Humana Press, Totowa.

Shirley, B. A. (1995b). Urea and guanidinium hydrochloride denaturation curves. In *Methods in Molecular Biology: Protein Stability and Folding: Theory and Practice* (B. A. Shirley, ed.), vol. 40, pp. 177–190. Humana Press, Totowa.

Siddiqui, K. S. (1990). Proteolytic nicking and carboxyl group modification of *Arthrobacter* D-xylose isomerase. Ph.D. Thesis. London University, UK.

Siddiqui, K. S. and Cavicchioli, R. (2005). Improved thermal stability and activity in the cold-adapted lipase B from *Candida antarctica* following chemical modification with oxidized polysaccharides. *Extremophiles*.

Siddiqui, K. S. and Cavicchioli, R. (2006). Cold adapted enzymes. *Annu. Rev. Biochem.*, in press.

Siddiqui, K. S., Shemsi, A. M., Anwar, M. A., Rashid, M. H. and Rajoka, M. I. (1999). Partial and complete alteration of surface charges of carboxymethylcellulase by chemical modification: thermostabilization in water miscible organic solvent. *Enzyme Microb. Technol.* **24**, 599–608.

Siddiqui, K. S., Najmus, S. A. A., Rashid, M. H. and Rajoka, M. I. (2000). Carboxyl group modification significantly altered the kinetic properties of purified carboxymethylcellulase from *Aspergillus niger*. *Enzyme Microb. Technol.* **27**, 467–474.

Siddiqui, K. S., Cavicchioli, R. and Thomas, T. (2002). Thermodynamic activation properties of elongation factor 2 (EF-2) proteins from psychrotolerant and thermophilic archaea. *Extremophiles* **6**, 143–150.

Siddiqui, K. S., Bokhari, S. A., Afzal, A. J. and Singh, S. (2004a). A novel thermodynamic relationship based on Kramers Theory for studying enzyme kinetics under high viscosity. *IUBMB Life* **56**, 403–407.

Siddiqui, K. S., Poljak, A. and Cavicchioli, R. (2004b). Improved activity and stability of alkaline phosphatases from psychrophilic and mesophilic organisms by chemically modifying aliphatic or amino groups using tetracarboxy-benzophenone derivatives. *Cell. Mol. Biol.* **50**, 657–667.

Siddiqui, K. S., Giaquinto, L., Feller, G., D'Amico, S., Gerday, C. and Cavicchioli, R. (2005a). The active site is the least stable structure in the unfolding pathway of a multi-domain cold-adapted α-amylase. *J. Bacteriol.* **187**, 6197–6205.

Siddiqui, K. S., Poljak, A., Guilhaus, M., Feller, G., D'Amico, S., Gerday, C. and Cavicchioli, R. (2005b). The role of disulfide-bridges in the activity and stability of a cold-active α-amylase. *J. Bacteriol.* **187**, 6206–6212.

Stoll, V. S. and Blanchard, J. S. (1990). Buffer: principles and practice. *Method. Enzymol.* **182**, 24–49.

Sturtevant, J. M. (1974). Some applications of calorimetry in biochemistry and biology. *Ann. Rev. Biophys. Bioeng.* **3**, 35–51.

Thomas, T. (2001). Low-temperature adaptation of elongation factor 2 proteins from Archaea. Ph.D. Thesis. The University of New South Wales, Sydney.

Thomas, T. and Cavicchioli, R. (2000). Effect of temperature on stability and activity of elongation factor 2 proteins from Antarctic and thermophilic methanogens. *J. Bacteriol.* **182**, 1328–1332.

Thomas, T., Kumar, N. and Cavicchioli, R. (2001). Effects of ribosomes and intracellular solutes on activities and stabilities of elongation factor 2 proteins from psychrotolerant and thermophilic methanogens. *J. Bacteriol.* **183**, 1974–1982.

Tindbaek, N., Svendsen, A., Oestergaard, P. R. and Draborg, H. D. (2004). Engineering a substrate-specific cold-adapted subtilisin. *Protein Eng.* **17**, 149–156.

Tsigos, I., Mavromatis, K., Tzanodaskalaki, M., Pozidis, C., Kokkinidis, M. and Bouriotis, V. (2001). Engineering the properties of a cold active enzyme through rational redesign of the active site. *Eur. J. Biochem.* **268**, 5074–5080.

Tsuruta, H., Mikami, B., Yamamoto, C. and Aizono, Y. (2002). Crystallization and preliminary X-ray studies of cold-active protein-tyrosine phosphatase of *Shewanella* sp. *Acta. Crystallogr. D Biol. Crystallogr.* **58**, 1465–1466.

Tsuruta, H., Mikami, B. and Aizono, Y. (2005). Crystal structure of cold-active protein-tyrosine phosphatase from a psychrophile, *Shewanella* sp. *J. Biochem.* (Tokyo) **137**, 69–77.

Tutino, M. L., Duilio, A., Parrilli, R., Remaut, E., Sannia, G. and Marino, G. (2001). A novel replication element from an Antarctic plasmid as a tool for the expression of proteins at low temperature. *Extremophiles* **5**, 257–264.

Tutino, M. L., Parrilli, E., Giaquinto, L., Duilio, A., Sannia, G., Feller, G. and Marino, G. (2002). Secretion of α-amylase from *Pseudoalteromonas haloplanktis* TAB23: two different pathways in different hosts. *J. Bacteriol.* **184**, 5814–5817.

Van Petegem, F., Collins, T., Meuwis, M. A., Gerday, C., Feller, G. and Van Beeumen, J. (2002). Crystallization and preliminary X-ray analysis of a xylanase from the psychrophile *Pseudoalteromonas haloplanktis*. *Acta. Crystallogr. D Biol. Crystallogr.* **58**, 1494–1496.

Van Petegem, F., Collins, T., Meuwis, M. A., Gerday, C., Feller, G. and Van Beeumen, J. (2003). The structure of a cold-adapted family 8 xylanase at 1.3 Å resolution: structural adaptations to cold and investgation of the active site. *J. Biol. Chem.* **278**, 7531–7539.

Veronese, F. M., Largajolli, R., Boccu, E., Benassi, C. A. and Schiavon, O. (1985). Surface modification of proteins: activation of monomethoxypolyethylene glycols by phenylchloroformates and modification of ribonuclease and superoxide dismutase. *Appl. Biochem. Biotechnol.* **11**, 141–152.

Villeret, V., Chessa, J. P., Gerday, C. and Van Beeumen, J. (1997). Preliminary crystal structure determination of the alkaline protease from the Antarctic psychrophile *Pseudomonas aeruginosa*. *Protein Sci.* **6**, 2462–2464.

Vincent, M., Xu, Y. and Kong, H. (2004). Helicase-dependent isothermal DNA amplification. *EMBO Rep.* **5**, 795–800.

Watanabe, S. and Takada, Y. (2004). Amino acid residues involved in cold-adaptation of isocitrate lyase from a pyschrophilic bacterium, *Colwellia maris*. *Microbiology* **150**, 3393–3403.

Wong, T. S., Tee, K. L., Hauer, B. and Schwaneberg, U. (2004). Sequence saturation mutagenesis (SeSaM): a novel method for directed evolution. *Nucleic Acids Res.* **32**, e26.

Xu, Y., Feller, G., Gerday, C. and Glansdorf, N. (2003a). *Moritella* cold-active dihydrofolate reductase: are there natural limits to optimization of catalytic efficiency at low temperature? *J Bacteriol.* **185**, 5519–5526.

Xu, Y., Feller, G., Gerday, C. and Glansdorf, N. (2003b). Metabolic enzymes from psychrophilic bacteria: challenge of adaptation to low temperatures in ornithine carbamoyltransferase from *Moritella abyssi. J. Bacteriol.* **185**, 2161–2168.

Zhang, N., Suen, W. C., Windsor, W., Xiao, L., Madison, V. and Zaks, A. (2003). Improving tolerance of *Candida antarctica* lipase B towards irreversible thermal inactivation through directed evolution. *Protein Eng.* **16**, 599–605.

Zeelen, J. P., Hiltunen, J. K., Ceska, T. A. and Wierenga, R. K. (1994). Crystallization experiments with 2-enoyl-CoA hydratase, using an automated 'fast-screening' crystallization protocol. *Acta. Crystallogr. D Biol. Crystallogr.* **50**, 443–447.

Alkaliphiles

18 Cultivation of Aerobic Alkaliphiles
19 Isolation, Cultivation and Characterization of Alkalithermophiles

18 Cultivation of Aerobic Alkaliphiles

William D Grant

Department of Infection, Immunity and Inflammation, University of Leicester, Leicester, LE1 9HN, UK

◆◆

CONTENTS

Source environments
Types of alkaliphilic microorganisms
Media and culture conditions
Epilogue

◆◆◆◆◆ SOURCE ENVIRONMENTS

The assumptions generally implicit in the use of the term "alkaliphile" are that the microorganisms in question grow optimally or very well at pH values above 9, often with pH optima around 10, showing little or no growth at near neutral pH values. Alkali-tolerant microorganisms, on the other hand, although capable of growth at alkaline pH values, are capable of growing over a wider range of pH values, often including the acid side of the pH range, usually showing pH optima for growth at near neutral pH. These categories by no means cover all the possibilities for microbial growth over the pH range, but suffice as working definitions for most purposes.

Alkaliphilic microorganisms coexist with non-alkaliphilic types in some overall non-alkaline environments such as soils and marine sediments, presumably reflecting the heterogeneous nature of such habitats in terms of microniche environments. Transient alkalinity may arise in microhabitats as a consequence of biological activity such as sulphate reduction or ammonification. Such environments are not particularly diverse in terms of microbial diversity and largely yield spore-forming types related to the diverse *Bacillus* spectrum (Horikoshi, 1999).

There are also man-made alkaline environments such as those produced by cement manufacture, paper manufacture and some food processing (Jones *et al.*, 1994). Again, rather low diversity is to be found in such sites.

A far more diverse population is to be found in stable, naturally occurring high pH environments. There are two main types of such environments, the high Ca^{2+} environments generated by the serpentinization of silicate minerals, exemplified by the hyperalkaline spring waters seen in Jordan (Pedersen et al., 2004) and elsewhere (Tiago et al., 2004). The second main type of alkaline environment is exemplified by the soda lakes and soda deserts. These sodium carbonate (Na_2CO_3)-dominated environments of varying salinity are found in arid and semi-arid areas of the world, including the East African Rift Valley, the plateaux of Mongolia, Inner Mongolia and Tibet, the high desert in the west of the United States and areas around the Kulunda steppe in the former USSR. These particular sites have been subject to microbiological examination, particularly the African lakes, but there are many other largely unexplored examples listed in various reviews (see Grant and Jones, 2000).

High Ca^{2+} Environments

The alkalinity in high-calcium environments is generated by the low temperature weathering of Ca^{2+} and Mg^{2+}-containing silicates olivine ($MgFeSiO_4$) and pyroxene ($MgCaFeSiO_3$) by CO_2-charged surface waters, releasing Mg^{2+}, Ca^{2+} and OH^- into solution. Magnesium is largely immobilized by precipitation as $Mg(OH)_2$, $MgCO_3$ or other carbonates such as $CaMg(CO_3)_2$. Carbonate is further removed by precipitation as $CaCO_3$. However, Ca^{2+} is in vast excess, leaving a $Ca(OH)_2$-dominated brine where solid phase $Ca(OH)_2$ is in equilibrium with soluble $Ca(OH)_2$, maintaining a high pH, very dilute 10 mM [$Ca(OH)_2$] ground water with pH values >12. The weathering of olivine and pyroxene generates a secondary mineral serpentine [$(Mg_3SiO_5(OH)_4)$], hence the term serpentinization for the process. The waters are initially anaerobic due to the concomitant release of Fe^{2+} and the production of H_2 from intermediate Fe hydroxides, but aeration occurs at later stages as the waters emerge to the surface.

Na_2CO_3-dominated Environments

Soda lakes form in shallow depressions where surface evaporation rates exceed the rate of inflow of water, allowing the accumulation of ions. A further important feature of such areas is that the surrounding geology is dominated by high Na^+, low Mg^{2+}/Ca^+ silicates. The East African environment differs from the others in that it is an area of active vulcanism, where hot springs deliver Ca^{2+} and Mg^{2+}-depleted waters into the lake depressions, whereas the other sites are largely charged by rainfall leaching through the surface topography into the catchment basins.

Waters high in Na^+, Cl^- and HCO_3^-/CO_3^{2-} evaporate down and since the concentration of HCO_3^-/CO_3^{2-} greatly exceeds that of any Ca^{2+} and Mg^{2+}, these cations precipitate as insoluble carbonates. Alkalinity develops as a consequence of a shift in the $CO_2/HCO_3^-/CO_3^{2-}$ equilibrium towards CO_3^{2-}, causing the development of a soda

(Na_2CO_3) lake with pH values usually between 10 and 11, occasionally >pH 12. As a consequence of the evaporative concentration, other ions, particularly Cl^- also accumulate, making the lakes somewhat saline and sometimes considerably hypersaline. These lakes are thus high in ionic solutes and essentially devoid of Ca^{2+} and Mg^{2+} compared with the $Ca(OH)_2$-type groundwaters. The detailed mechanisms leading to the formation of such lakes are discussed by Hardie and Eugster (1970). Simplified versions are given in Grant and Jones (2000) and Jones et al. (1994), where ionic compositions of a number of different lakes are also tabulated. East African lakes differ from those at the other sites by having rather uniform temperatures throughout the year and during the diurnal cycle, whereas the other sites show pronounced temperature differences both seasonally and diurnally. These caustic environments support diverse and dense populations of alkaliphiles driven by intense cyanobacterial primary productivity, presumably a reflection of unlimited access to CO_2 via the $CO_2/HCO_3^-/CO_3^{2-}$ equilibrium and high light intensities.

◆◆◆◆◆ TYPES OF ALKALIPHILIC MICROORGANISMS

Aerobic alkaliphiles consist of two main physiological groups: alkaliphiles and haloalkaliphiles. The former have little or no requirement for NaCl in the growth media. Such organisms are to be found in soils, marine sediments and high-Ca^{2+} ground waters. The diversity of alkaliphiles from such sites is usually reported as relatively low, although Tiago et al. (2004) have recently recorded a rather more diverse population in a Ca^{2+}-dominated environment. Dilute soda lakes yield very diverse populations of such alkaliphiles (Grant and Jones, 2000; Jones and Grant, 2000), at least when freshly sampled (stored samples are much less diverse and dominated by *Bacillus* types). Haloalkaliphiles are mostly confined to very hypersaline soda lakes and soda deserts. To date, the majority of isolates from such sites have been haloalkaliphilic Archaea, although some *Bacillus* types are also known (Grant and Jones, 2000; Jones and Grant, 2000).

A survey of those examples of alkaliphiles and haloalkaliphiles brought into culture reveals a remarkable diversity of types, with aerobic alkaliphiles represented in many of the major taxonomic groups of Bacteria but with a rather more restricted range of haloarchaea. The most comprehensive culture survey of alkaline environments was that carried out by Duckworth et al. (1996) who isolated several hundred strains of aerobic, heterotrophic alkaliphiles and haloalkaliphiles from a range of soda lakes in the East African Rift Valley. Phylogenetic analysis revealed many proteobacteria, notably halomonads, pseudomonad-like types, high and low G+C Gram-positives plus a number of alkaliphilic haloarchaea. Some of these have now been published as novel types, including *Halomonas magadiensis* (originally named *H. magadii*) (Duckworth et al., 2000), *Cellulomonas bogoriensis* (Jones et al., 2005),

Alkalimonas delamerensis (Ma *et al.*, 2004), and *Dietzia natronolimnaea* (Duckworth *et al.*, 1998). Other aerobic heterotrophic East African soda lake isolates include *Alcalimnicola halodurans* (Yakimov *et al.*, 2001), *Bacillus bogoriensis* (Vargas *et al.*, 2005), and *Bogoriella caseilytica* (Groth *et al.*, 1997). Soda lakes from elsewhere have yielded *Marinospirillum alkaliphilum* (Zhang *et al.*, 2002a) and "Alkalimonas amylolytica" (Ma *et al.*, 2004), and *Salinicoccus alkaliphilus* (Zhang *et al.*, 2002b). Non-halophilic aerobic heterotrophs from non-soda lake sources include *Alkalibacterium olivapovliticus* from olive wash waters (Ntougias and Russell, 2001), *Marinilactobacillus psychrotolerans* (Ishikawa *et al.*, 2003) from decaying marine organisms, *Oceanobacillus iheyensis* from a deep-sea sediment (Lu *et al.*, 2001), *Exiguobacterium oxidotolerans* from a fish processing plant (Yumoto *et al.*, 2004a), *Alkalibacterium psychrotolerans* from a traditional indigo dye process (Yumoto *et al.*, 2004b) and "Marinobacter alkaliphilus" from serpentine deep-sea floor mud (Takai *et al.*, 2005).

Soda lake environments harbour element cycles comparable to those found in neutral pH environments (Grant, 2003) and lithotrophic alkaliphiles have been described over the last few years. Sulphur-oxidizing *Thioalkalimicrobium*, *Thioalkalivibrio* and *Thioalkalispira* spp. have been described (Banciu *et al.*, 2004; Sorokin *et al.*, 2001a, 2002) from Russian and East African soda lakes and ammonia-oxidizing types have been described in Mongolian soda lakes (Sorokin *et al.*, 2001b). Methylotrophic *Methylomicrobium* (Sorokin *et al.*, 2000) and *Methylophaga* spp. (Doronina *et al.*, 2003a,b) are found in Kenyan and Mongolian soda lakes.

Haloalkaliphilic archaea have been described in a variety of different hypersaline lakes. The haloarchaea form a relatively tight phylogenetic cluster currently comprising 19 genera and around 50 species. Within the family Halobacteriaceae, haloalkaliphilic members constitute a distinct phenotype distributed amongst seven genera with 13 species, all derived form soda lakes. These are *Natronobacterium gregoryi*, *Natronococcus amylolyticus*, *Natronococcus occultus*, *Natronorubrum bangense*, *Natronorubrum tibetense*, *Natronomonas pharaonis*, *Halorubrum tibetense*, *Halorubrum vacuolatum*, *Natrialba magadii*, *Natrialba chahannaoensis*, *Natrialba hulunbeirensis*, *Halobiforma nitratereducens*, *Natronolimnobius baerhuensis* and *Natronolimnobius innermongolicus* (Fan *et al.*, 2004; Hezayen *et al.*, 2002; Itoh *et al.*, 2004; Kamekura *et al.*, 1997; Kanai *et al.*, 1995; Mwatha and Grant, 1993; Soliman and Trüper, 1982; Tindall *et al.*, 1984; Xin *et al.*, 2001; Xue *et al.*, 1999, 2001).

It is only in recent years that dilute high Ca^{2+} alkaline serpentinized ground waters have been subjected to detailed, culture and taxonomic analysis. An early attempt by Bath *et al.* (1987) did not use media appropriate to the chemistry of the environment and relied on non-phylogenetic characterization. Tiago *et al.* (2004) and Pedersen *et al.* (2004) have completed detailed microbiological examination of two distinct sites in Portugal and Jordan. Analysis of cultivated strains revealed mainly alkali tolerance rather than alkaliphily and a dominance of Gram-positive types at the Portuguese site with representatives of *Dietzia*, *Clavibacter*, *Microbacterium*, *Agrococcus*, *Rhodococcus*, *Nesterenkonia*, *Micrococcus*,

Bacillus and *Staphylococcus*. Only five isolates were Gram-negative proteobacteria. The Jordanian site was more or less exclusively analysed by direct molecular methods and although organisms were cultivated, these were not subjected to taxonomic investigation. However, in this case, clone libraries revealed an environment dominated by proteobacteria.

◆◆◆◆◆ MEDIA AND CULTURE CONDITIONS

Regardless of the source and chemistry of sample material, workers have usually attempted to culture organisms on a $NaHCO_3$–Na_2CO_3 buffer medium, usually poising the pH at around 10.5 by the addition of 1% (w/v) Na_2CO_3. It is important to sterilize the Na_2CO_3 separately from the rest of the media components (together with NaCl for highly saline media). It is convenient to dissolve Na_2CO_3 in 500 ml and the other components in 500 ml with agar at 2% (w/v) final concentration for solid media, mixing together just before use. For agar media, the components should be held at 60°C prior to mixing and pouring plates since at lower temperatures, alkaline agar media rapidly solidify. For highly saline media (e.g. 20% (w/v) NaCl) the Na_2CO_3–NaCl mix will not dissolve completely in 500 ml, so here, 700 ml is used, with the organic components dissolved in 300 ml. Incubation of soda lake material is usually at 30°C or 37°C, although other temperatures are used, depending on the nature of the sample material. Enrichments in liquid media with shaking at 200 rpm can be carried out.

There are two basic core media that form the basis of many culture studies aimed at aerobic heterotrophs. The first is designed for relatively low salt sites and is based on the formulation of Horikoshi (1971) originally developed for the isolation of alkaliphiles from soil.

The medium comprises (g l^{-1}):

glucose	10.0
peptone	5.0
yeast extract	5.0
KH_2PO_4	1.0
$MgSO_4 \cdot 7H_2O$	0.2
Na_2CO_3	10.0

Sometimes NaCl is added at 40.0

The second is based on the formulation of Tindall *et al.* (1984) for the culture of haloalkaliphilic Archaea from hypersaline sites.

The medium comprises (g l^{-1}):

Yeast extract	10.0
Casamino acids	7.5
Trisodium citrate	3.0
KCl	2.0

$MgSO_4 \cdot 7H_2O$	1.0
$MnCl_2 \cdot 4H_2O$	0.00036
$FeSO_4 \cdot 7H_2O$	0.05
NaCl	200.0
Na_2CO_3	10.0

Media may show some precipitate and have a pH of 10.5–11 when freshly prepared, but in agar plates the pH drops to 9.5–10.0 after a few days due to the absorption of CO_2. Samples from sites should be cultured as soon as practicable since there is no doubt that diversity is reduced by storage either at room temperature or 4°C. Soda lake samples stored at 4°C for 20 years still yield large numbers of organisms, but the population is dominated by spore-forming Gram-positives, whereas fresh material yields many proteobacteria.

There are many variations of these media that have been used to detect particular enzyme activities or encourage growth of hydrolase-active strains. Duckworth et al. (1996) have listed a large number of such isolates. Thus starch, olive oil, carboxymethyl cellulose or casein can be incorporated into these media sometimes leaving out other carbon sources with the exception of yeast extract at 0.1 or 0.01 (w/v). Such media have little additional buffering capacity and may initially have pH values >11. These media should be reduced in pH < 11 by the addition of acid since, in the author's experience, pH values much above 11 markedly reduce the recovery of organisms.

Variations on these media have been used by many workers, particularly the low salt medium (Ishikawa et al., 2003; Lu et al., 2001; Ma et al., 2004; Vargas et al., 2005; Zhang et al., 2002a).

Tiago et al. (2004) in their analysis of a high-Ca^{2+} alkaline ground water also used yeast extract/peptone/tryptone media with a $NaHCO_3$–Na_2CO_3 buffer system to generate alkaline pH.

These media contained (g l^{-1}):

Medium A	Peptone	5.0
	meat extract	5.0
Medium B	yeast extract	5.0
	tryptone	5.0
	α-ketoglutaric acid (potassium)	1.0

Both contained 100 ml of macronutrient solution (g l^{-1})

$CaSO_4 \cdot 2H_2O$	0.6
$MgSO_4 \cdot 7H_2O$	1.0
NaCl	0.8
KNO_3	1.03
$NaNO_3$	6.89
Na_2HPO_4	1.11

and 10 ml of micronutrient solution (g l^{-1})

$MnSO_4 \cdot H_2O$	0.22
$ZnSO_4 \cdot 7H_2O$	0.05
H_3BO_3	0.05
$CuSO_4 \cdot 5H_2O$	0.0025
$Na_2MoO_4 \cdot 2H_2O$	0.0025
$CoCl_2 \cdot 6H_2O$	0.0046

100 ml of 1 M $NaHCO_3$–Na_2CO_3 buffer, sterilized separately, is used to produce a pH of 9.5 or 1 M Na_2CO_3–KOH to produce a pH of 11. The media were used 10x diluted to reflect the oligotrophic nature of the environment.

Pedersen et al. (2004) actually used groundwater from a similar serpentinized source supplemented with various complex carbon sources to prepare media.

Sorokin et al. (2001a) have devised a mineral medium buffered with Na_2CO_3 for the isolation of chemolithoautrotrophic sulphur bacteria from soda lakes.

The medium contains (g l^{-1}):

Na_2CO_3	21.0
$NaHCO_3$	9.0
NaCl	5.0
K_2HPO_4	0.5
KNO_3	5.0 mM
$MgCl_2 \cdot 6H_2O$	1.0 mM
$Na_2S_2O_3$	40–80 mM

Trace elements (Pfennig and Lippert, 1966) 2 ml

The trace elements comprise (g l^{-1})

EDTA	5.0
$FeSO_4 \cdot 7H_2O$	2.0
$ZnSO_4 \cdot 7H_2O$	0.1
$MnCl_2 \cdot 4H_2O$	0.03
$CoCl_2 \cdot 6H_2O$	0.02
$CuCl_2 \cdot 2H_2O$	0.02
$NiCl_2 \cdot 6H_2O$	0.02
$Na_2MoO_4 \cdot 2H_2O$	0.03

The medium has a final pH of 10–10.1.

A virtually identical medium but with 4 mM NH_4Cl in place of $Na_2S_2O_3$ was used to isolate nitrifying bacteria from the same source (Sorokin et al., 2001b), and similar media replacing the energy source with a head space of CH_4–air used to isolate methane-oxidizers (Sorokin et al., 2000). These media can be sterilized by filtration.

◆◆◆◆◆ EPILOGUE

Alkaline environments harbour large populations of specifically adapted microorganisms that carry out carbon and element transformations similar to that observed at neutral pH. Culture of groups may be attempted using appropriate media made alkaline with Na_2CO_3. There is little point in the addition of high levels of Ca^{2+} and Mg^{2+} since these largely precipitate out as insoluble carbonates. Unfortunately, chemical analysis of the source material is seldom carried out and it is likely that media whose inorganic composition is based on the inoculum source would be more effective than the relatively simple formulations described here. The addition of 10–20% (v/v) filter-sterilized source brine often helps recovery of organisms. It is also worth noting that to date most heterotrophs have been isolated on relatively nutrient-rich media, which in itself must have produced a significant bias in recovery.

References

Banciu, H., Sorokin, D. Y., Galinski, E. A., Muyzer, G., Kleerebezem, R. and Kuenen, J. G. (2004). *Thioalkalivibrio halophilus* sp. nov., a novel obligately chemolithoautotrophic, facultatively alkaliphilic and extremely salt tolerant, sulfur-oxidizing bacterium from a hypersaline alkaline lake. *Extremophiles* **8**, 325–334.

Bath, A. H., Christophi, N., Neal, C., Philip, J. C., Cave, M. R., McKinley, I. G. and Berner, U. (1987). Trace element and microbiological studies of alkaline ground waters in Oman, Arabian Gulf: a natural analogue for cement pore waters. *Rep. Fluid Processes Res. Group. Brit. Geol. Survey* FPLU 87–2.

Doronina, N., Darmaeva, T. R. and Trotsenko, Y. A. (2003a). *Methylophaga natronica* sp. nov., a new alkaliphilic and moderately halophilic restricted facultatively methylotrophic bacterium from soda lake of the southern Transbaikal Region. *Syst. Appl. Microbiol.* **26**, 382–389.

Doronina, N., Darmaeva, T. D. and Trotsenko, Y. A. (2003b). *Methylophaga alcalica* sp. nov., a novel alkaliphilic and moderately halophilic, obligately methylotrophic bacterium from an East Mongolian saline soda lake. *Int. J. Syst. Evol. Microbiol.* **53**, 223–229.

Duckworth, A. W., Grant, W. D., Jones, B. E. and van Steenbergen, R. (1996). Phylogenetic diversity of soda lake alkaliphiles. *FEMS Microbiol. Ecol.* **19**, 181–191.

Duckworth, A. W., Grant, S., Grant, W. D., Jones, B. E. and Meijer, D. (1998). *Dietzia natronolimnaios* sp. nov., a new member of the genus *Dietzia* isolated from an East African soda lake. *Extremophiles* **2**, 359–366.

Duckworth, A. W., Grant, W. D., Jones, B. E., Meijer, D., Marquez, M. C. and Ventosa, A. (2000). *Halomonas magadii* sp. nov., a new member of the genus *Halomonas* isolated from a soda lake of the East African Rift Valley. *Extremophiles* **4**, 53–60.

Fan, H., Xue, Y., Ventosa, A. and Grant, W. D. (2004). *Halorubrum tibetense* sp. nov., a novel haloalkaliphilic archaeon from Lake Zabuye in Tibet, China. *Int. J. Syst. Evol. Microbiol.* **54**, 1213–1216.

Grant, W. D. (2003). Alkaline environments and biodiversity. In *Extremophiles – Basic Concepts* (C. Gerday and N. Glansdorff, eds), 12 pp. Encyclopedia of Life Support Systems (http://www.eolss.net/).

Grant, W. D. and Jones, B. E. (2000). Alkaline environments. In *Encyclopedia of Microbiology* (J. Lederberg, ed.), 2nd edn., pp. 126–133. Academic Press, London.

Groth, I., Schumann, P., Rainey, F. A., Martin, K., Schuetze, B. and Augsten, K. (1997). *Bogoriella caseilytica* gen. nov. sp. nov., a new alkaliphilic actinomycete from a soda lake in Africa. *Int. J. Syst. Bacteriol.* **47**, 788–794.

Hardie, L. A. and Eugster, H. P. (1970). The evolution of closed basin brines. *Mineral. Soc. Ann. Special Publication* **3**, 273–290.

Hezayen, F. F., Tindall, B. J., Steinbüchel, A. and Rehm, B. H. (2002). Characterization of a novel halophilic archaeon *Halobiforma haloterrestris* gen. nov., sp. nov. and transfer of *Natronobacterium nitratireducens* to *Halobiforma nitratireducens* comb. nov. *Int. J. Syst. Evol. Microbiol.* **52**, 2271–2280.

Horikoshi, K. (1971). Production of alkaline enzymes by alkaliphilic microorganisms. I. Alkaline protease produced by *Bacillus* no. 221. *Agric. Biol. Chem.* **36**, 1403–1414.

Horikoshi, K. (1999). Alkaliphiles: some applications of their products for biotechnology. *Microbiol. Mol. Biol. Rev.* **63**, 735–750.

Ishikawa, M., Nakajima, K., Yanagi, M., Yamamoto, Y. and Yamasoto, K. (2003). *Marinilactibacillus psychrotolerans* gen. nov. sp. nov., a halophilic and alkaliphilic marine lactic acid bacterium isolated from marine organisms in temperate and subtropical areas of Japan. *Int. J. Syst. Evol. Microbiol.* **53**, 711–720.

Itoh, T., Yamaguchi, T., Zhan, P. and Takashina, T. (2004). *Natronolimnobius baerhuensis* gen. nov. sp. nov. and *Natronolimnobius innermongolicus* sp. nov., novel haloalkaliphilic archaea isolated from soda lakes in Inner Mongolia, China. *Extremophiles* **9**, 111–116.

Jones, B. E. and Grant, W. D. (2000). Microbial diversity and ecology of soda lakes. In *Journey to Diverse Microbial Worlds* (J. Seckbach, ed.), pp. 177–190. Kluwer Academic Publishers, Dordrecht.

Jones, B. E., Grant, W. D., Collins, N. C. and Mwatha, W. E. (1994). Alkaliphiles: diversity and identification. In *Bacterial Diversity and Systematics* (F. G. Priest, A. Ramos-Cormenzana and B. J. Tindall, eds), pp. 195–230. Plenum Press, New York.

Jones, B. E., Grant, W. D., Duckworth, A. W., Schumann, P., Weiss, N. and Stackebrandt, E. (2005). *Cellulomonas bogoriensis* sp. nov., an alkaliphilic cellulomonad. *Int. J. Syst. Evol. Microbiol.* **55**, 1711–1714.

Kamekura, M., Dyall-Smith, M. L., Upasani, V., Ventosa, A. and Kates, M. (1997). Diversity of alkaliphilic halobacteria: proposals for transfer of *Natronobacterium vacuolatum*, *Natronobacterium magadii* and *Natronobacterium pharaonis* to *Halorubrum*, *Natrialba* and *Natronomonas* gen. nov. respectively as *Halorubrum vacuolatum* comb. nov., *Natrialba magadii* comb. nov. and *Natromonas pharaonis* comb. nov., respectively. *Int. J. Syst. Bacteriol.* **47**, 853–857.

Kanai, H., Kobayashi, T., Aono, R. and Kudo, T. (1995). *Natronococcus amylolyticus* sp. nov., a haloalkaliphilic archaeon. *Int. J. Syst. Bacteriol.* **45**, 762–766.

Lu, J., Nogi, Y. and Takami, H. (2001). *Oceanobacillus iheyensis* gen. nov. sp. nov., a deep sea extremely halotolerant and alkaliphilic species isolated from a depth of 1050 m on the Iheya ridge. *FEMS Microbiol. Lett.* **205**, 291–297.

Ma, Y., Xue, Y., Grant, W. D., Collins, N. C., Duckworth, A. W., van Steenbergen, R. P. and Jones, B. E. (2004). *Alkalimonas amylolytica* gen. nov. sp. nov. and *Alkalimonas delamerensis* gen. nov. sp. nov., novel alkaliphilic bacteria from soda lakes in China and East Africa. *Extremophiles* **8**, 193–200.

Mwatha, W. E. and Grant, W. D. (1993). *Natronobacterium vacuolata* sp. nov., a haloalkaliphilic archaeon isolated from Lake Magadi, Kenya. *Int. J. Syst. Bacteriol.* **43**, 401–404.

Ntougias, S. and Russell, N. J. (2001). *Alkalibacterium olivoapovliticus* gen. nov. sp. nov., a new obligately alkaliphilic bacterium isolated from edible olive wash waters. *Int. J. Syst. Evol. Microbiol.* **51**, 1161–1170.

Pedersen, K., Nilsson, E., Arlinger, J., Hallbeck, L. and O'Neill, A. (2004). Distribution, diversity and activity of microorganisms in the hyperalkaline spring waters of Maqarin. *Extremophiles* **8**, 151–164.

Pfennig, N. and Lippert, K. D. (1966). Über das Vitamin B_{12} Bedürfnis phototropher Schwefelbakterien. *Arch. Mikrobiol.* **55**, 245–255.

Soliman, G. S. H. and Trüper, H. G. (1982). *Halobacterium pharaonis* sp. nov., a new extremely haloalkaliphilic archaebacterium with low magnesium requirement. *Zentralbl. Bakteriol. Hyg. Abt. I Orig.* **C3**, 318–329.

Sorokin, D. Y., Jones, B. E. and Kuenen, J. G. (2000). An obligate methylotrophic, methane-oxidizing *Methylomicrobium* species from a highly alkaline environment. *Extremophiles* **4**, 145–155.

Sorokin, D. Y., Lysenko, A. M., Mityushina, L. T., Tourova, T. P., Jones, B. E., Rainey, F. A., Robertson, L. A. and Kuenen, G. J. (2001a). *Thioalkalimicrobium aerophilum* gen. nov., sp. nov. and *Thioalkalimicrobium sibericum* gen. nov., sp. nov. and *Thioalkalivibrio denitrificans* sp. nov. and *Thioalkalivibrio denitrificans* sp. nov., novel obligately alkaliphilic and obligately chemolithotrophic sulfur-oxidizing bacteria from soda lakes. *Int. J. Syst. Evol. Microbiol.* **51**, 565–580.

Sorokin, D., Tourova, T., Schmid, M. C., Wagner, M., Koops, H. P., Kuenen, G. J. and Jetten, M. (2001b). Isolation and properties of obligately chemolithoautotrophic and extremely alkalitolerant ammonia-oxidizing bacteria from Mongolian soda lakes. *Arch. Microbiol.* **176**, 170–177.

Sorokin, D. Y., Tourova, T. P., Kolganova, T. V., Sjollema, K. A. and Kuenen, J. G. (2002). *Thioalkalispira microaerophila* sp. nov., sp. nov., a novel lithoautotrophic, sulfur-oxidizing bacterium from a soda lake. *Int. J. Syst. Evol. Microbiol.* **52**, 2175–2182.

Takai, K., Moyer, C. L., Miyazaki, M., Nogi, Y., Hirayama, H., Nealson, K. H. and Horikoshi, K. (2005). *Marinobacter alkaliphilus* sp. nov., a novel alkaliphilic bacterium isolated from subseafloor alkaline serpentine mud from Ocean Drilling Programme Site 1200 at South Chamorro Seamount, Marianna Forearc. *Extremophiles* **9**, 17–28.

Tiago, I., Chung, A. P. and Verissimo, A. (2004). Bacterial diversity in a non-saline alkaline environment: heterotrophic aerobic populations. *Appl. Environ. Microbiol.* **70**, 7378–7387.

Tindall, B. J., Ross, H. N. M. and Grant, W. D. (1984). *Natronobacterium* gen. nov. and *Natronococcus* gen. nov., two new genera of haloalkaliphilic archaebacteria. *Syst. Appl. Microbiol.* **5**, 41–57.

Vargas, V. A., Delgado, O. D., Hatti-Kaul, R. and Mattiasson, B. (2005). *Bacillus bogoriensis* sp. nov., a novel alkaliphilic, halotolerant bacterium isolated from a Kenyan soda lake. *Int. J. Syst. Evol. Microbiol.* **55**, 899–902.

Xin, H., Itoh, T., Zhou, P., Suzuki, K. and Nakasi, T. (2001). *Natronobacterium nitratireducens* sp. nov. a halophilic archaeon isolated from a soda lake in China. *Int. J. Syst. Evol. Microbiol.* **51**, 1825–1829.

Xue, Y., Zhou, P. and Tian, X. (1999). Characterization of two novel haloalkaliphilic archaea *Natronorubrum bangense* gen. nov., sp. nov. and *Natronorubrum tibetense* gen. nov., sp. nov. *Int. J. Syst. Bacteriol.* **49**, 261–266.

Xue, Y., Wang, Z., Xue, Y., Zhan, P., Ma, Y., Ventosa, A. and Grant, W. D. (2001). *Natrialba hulunbeirensis* sp. nov. and *Natrialba chahannaoensis* sp. nov., novel haloalkaliphilic archaea from soda lakes in Inner Mongolia Autonomous Region. *Int. J. Syst. Evol. Microbiol.* **51**, 1693–1698.

Yakimov, M. M., Giuliano, L., Chernikova, T. N., Gentile, G., Abraham, W. R., Lunsdorf, H., Timmis, K. N. and Golyshin, P. N. (2001). *Alcalilimnicola halodurans* gen. nov. sp. nov., an alkaliphilic moderately halophilic and extremely halotolerant bacterium, isolated from sediments of soda-depositing Lake Natron, East African Rift Valley. *Int. J. Syst. Evol. Microbiol* **51**, 2133–2143.

Yumoto, I., Hishinuma-Narisawa, M., Hiroto, K., Shingyo, T., Takebe, F., Nodosoka, Y., Matsuyama, H. and Hara, I. (2004a). *Exiguobacterium oxidotolerans* sp. nov., a novel alkaliphile exhibiting high catalase activity. *Int. J. Syst. Evol. Microbiol.* **54**, 2013–2017.

Yumoto, I., Hirota, K., Nokosaka, Y., Yokoto, Y., Hishina, T. and Nakajima, K. (2004b). *Alkalibacterium psychrotolerans* sp. nov., a psychotolerant obligate alkaliphile that reduces an indigo dye. *Int. J. Syst. Evol. Microbiol.* **54**, 2379–2383.

Zhang, W. Z., Xue, Y. F., Grant, W. D., Ventosa, A. and Zhou, P. J. (2002a). *Marinospirillum alkaliphilum* sp. nov., a new alkaliphilic helical bacterium from Haoji soda lake in Inner Mongolia Autonomous Region of China. *Extremophiles* **6**, 33–37.

Zhang, W., Ma, Y., Xue, Y., Zhou, P., Ventosa, A. and Grant, W. D. (2002b). *Salinicoccus alcaliphilus* sp. nov., a new alkaliphilic and moderately halophilic Gram-positive coccus. *Int. J. Syst. Evol. Microbiol.* **52**, 789–793.

19 Isolation, Cultivation and Characterization of Alkalithermophiles

Noha M Mesbah and Juergen Wiegel
Department of Microbiology, The University of Georgia, Athens, GA 30602-2605, USA

◆◆◆

CONTENTS

Introduction
General considerations
Incubation and culture equipment
Enrichment and cultivation methods
Isolation methods
Characterization of alkalithermophiles

◆◆◆◆◆ INTRODUCTION

Alkaline environments can be placed into several broad categories based on the nature of the process generating the alkalinity. All alkaline environments depend on a continuous process, microbial or chemical, to maintain the alkaline pH. Soda lakes and soda deserts are naturally occurring alkaline environments and represent the most stable high-pH environments on Earth, where large amounts of carbonate salts generate pH values of 10–11.5 (Jones *et al.*, 1998; Zavarzin, 1993; Zhilina and Zavarzin, 1994). In general, there are relatively few naturally occurring alkaline environments (in comparison with acidic environments, particularly geothermally heated environments). Examples include the soda lakes of the East African Rift Valley, alkaline hypersaline Mono Lake (California) and the soda lakes of the Wadi An Natrun (Egypt). Several soda lakes also exist in Central Asia, Australia and China (Xin *et al.*, 2001; Xu *et al.*, 2001; Zhang *et al.*, 2002; Zhilina and Zavarzin, 1994).

Few thermophiles are capable of growing at pH values greater than 9.0 (see below for pH measurements at elevated temperatures (Kevbrin *et al.*, 2004; Wiegel, 1998). Other than the recent isolation of anaerobic alkalithermophiles (Chrisostomos *et al.*, 1996; Engle *et al.*, 1995, 1996; Kevbrin *et al.*, 2004; Li *et al.*, 1993, 1994), knowledge of alkalithermophiles is mainly based on characteristics of *Bacillus* species (Cook *et al.*, 2003;

Olsson et al., 2003; Peddie et al., 1999), many of which are not taxonomically classified (Wiegel and Kevbrin, 2004).

In this chapter, methods are described that may be considered as reference procedures for isolation, cultivation and characterization of aerobic and anaerobic alkalithermophiles (Wiegel, 1986; see for detailed description of anaerobic procedures Ljungdahl and Wiegel, 1986). These methods are routinely employed in our laboratory and have proven to give consistent and reproducible results.

◆◆◆◆◆ GENERAL CONSIDERATIONS

With the exception of some basic considerations, methods used for isolation and cultivation of neutrophilic and mesophilic bacteria can also be used for alkalithermophiles. Basic points discussed in this chapter that must be taken into account when working at elevated pH and temperatures include the following:

1. Increased evaporation of medium at elevated temperatures in incompletely sealed culture vessels. This becomes more problematic when aeration is used for aerobic cultures.
2. Strong dependence of medium pH upon temperature; the effect of which increases on increased distance from neutral pH.
3. Caramelization of sugars and complex polysaccharides at high pH values and temperatures. This effect increases under aerobic conditions.
4. Increasing instability of agar at pH values above 8.5 and temperatures above 65°C.
5. Frequent acidification of anaerobic cultures due to formation of acidic fermentation products.

The following sections will focus on the precautions and changes that are necessary for working with alkalithermophiles.

◆◆◆◆◆ INCUBATION AND CULTURE EQUIPMENT

Incubation Equipment

A large variety of incubators are available to grow microorganisms at elevated temperatures. Waterbaths are not convenient for incubations at temperatures greater than 55°C due to rapid evaporation of water. Use of waterbath covers and/or plastic balls floating on the surface of the water provides minimization of the problem.

In the authors' laboratory, hot air incubators equipped with an additional heater and a door switch that shuts off the blower when the door is open are primarily used (Lunaire Environmental Inc.). These incubators can be used for temperatures up to 98°C, and have the advantage of quick recovery of temperature after they are opened (it takes less than 3 min to retain 60°C after opening the door for 10 min). The use of drying ovens is not recommended since the temperature fluctuates too much. If small culture vials are used, then electrically heated aluminum blocks offer a good alternative as the tops of the culture tubes are easily accessible for sampling, pH control and other additions without having to take the tubes out of the incubators. Thus the large temperature fluctuations otherwise encountered with large-scale incubators are avoided. Rotary and reciprocal shakers are available for temperatures up to 70°C for aerobic cultures. There are various brands, and include smaller units to be placed in the above-mentioned incubators.

Tubes and Bottles for Alkalithermophilic Cultures

All culture vessels must be tightly sealed to prevent evaporation of the culture medium. Tubes used for anaerobic techniques, such as the Hungate and Balch tubes are typically used. Hungate tubes are sealed with a butyl rubber septum that is held in place with a plastic screw cap with an opening in the middle to allow inoculation with syringes (Macy et al., 1972). Balch tubes are constructed with flanges at the top and are sealed with butyl rubber stoppers. An aluminum seal is crimped over the stopper and the flange of the tube (Balch et al., 1979). In addition, serum bottles are excellent for use and are available in sizes from 1 to 150 ml, e.g. Wheaton serum bottles and tubes, available from Fisher Scientific. These bottles are sealed with the same butyl rubber stoppers and aluminum seals as the Balch tubes. Special hand crimpers and decapitators are available to handle the aluminum seals. Larger cultivation vessels with 6–20 l capacities are also available. These are constructed with Pyrex glass and can be sealed with butyl rubber stoppers which are held in place either by wiring or metal flanges. Metal flanges can be made at academic instrument shops. Detailed ordering information for anaerobic glassware is provided by Ljungdahl and Wiegel (1986).

For aerobic and gas-consuming cultures, it is important to use culture vessels with a large ratio of surface-to-liquid to avoid the necessity of frequent gassing since solubility of gases is less at higher temperatures. This can be achieved by using a shallow layer of culture fluid. When using rotary shakers, baffled culture flasks are highly recommended. Baffled flasks are commercially available from several sources or, as a cheaper alternative, unbaffled culture vessels can be modified in a glass-blowing shop. To replenish oxygen supply during growth, the culture vessels are flushed through sterile cotton filters with air or O_2–CO_2–N_2 mixtures. The interval of flushing depends on the rate of growth. In addition, an overpressure up to 1 atm is recommended to enhance solubility of gas in liquid medium. In our laboratory, an overpressure of 10 psi is routinely used.

◆◆◆◆◆ ENRICHMENT AND CULTIVATION METHODS

Selection of Media

There are many ways of enriching for alkalithermophiles. Besides adjusting temperature and pH, selective conditions to be used in enrichment will depend on the physiological and nutritional type of the microorganisms to be isolated. In general, the media used for mesophilic microorganisms will work in most instances for alkalithermophiles. Specific exceptions include adjustment of medium pH, preparation and sterilization of media and choice of solidifying agent. The culture medium is made alkaline by addition of sodium carbonate, sodium bicarbonate, sodium hydroxide or an organic buffer system. Buffers with color indicators can be purchased from Fisher Scientific. These types of buffers used in the alkaline range (pH 8–13 at 25°C) are comprised of a mixture of glycine, potassium carbonate, Na^+/K^+-hydrazide and EDTA to keep metals in solution. In addition, most alkalithermophiles have been shown to have a requirement of at least 5 mM sodium chloride in culture medium (Li et al., 1993; Peddie et al., 1999). While a clear relation between sodium and potassium concentrations of alkalithermophiles has not been established yet, addition of 25 mM potassium to culture medium of alkalithermophilic *Clostridium paradoxum* increased the microorganisms' tolerance to sodium threefold (Li et al., 1993). Thus, at least 5 mM of potassium should be included as well (either as part of a phosphate buffer system or as potassium chloride). In our laboratory we use the following standard medium (g l^{-1}): yeast extract, 5.0; tryptone, 4.0; NaCl, 5.0; KCl, 0.5; NH_4Cl, 0.05; Na_2CO_3, 10. Enrichments for oligotrophs can be achieved by reducing concentrations of complex substrates to 0.01 g l^{-1}. For enrichments of halophiles, concentration of NaCl and KCl should be increased accordingly. Na_2CO_3 is added as a buffer and may be replaced by any other buffer indicated in Table 19.1. All carbon sources (with the exception of polymeric substrates) should be filter sterilized separately and added to medium after autoclaving and pH adjustment.

It should be noted, that at pH^{60C} values greater than 8, solubility of Ca^{2+}, Mg^{2+} and phosphate salts decreases and they will precipitate out of the medium. As a result, the concentrations of these cations in the medium can be decreased or they can be kept in solution by addition of EDTA (if growth is not inhibited by EDTA). The use of Tris is not recommended as it has a large temperature coefficient. In general, phosphate-based buffers have the lowest temperature coefficient. Also, it has been noted that, in enrichments for halophilic alkalithermophiles, addition of greater than 15% wt/vol NaCl at pH^{60C} values 9 and above result in turbidity of the medium.

One consideration for isolation of novel microorganisms is also to employ conditions of very low substrate concentrations (around 0.1 mM) to isolate oligotrophic microorganisms, which can play a major role in environments. This requires frequent testing of residual substrates and re-feeding or the employment of a continuous culture approach for

Table 19.1 List of organic buffer solutions employed in cultures of alkalithermophiles and their concentration

Buffer	pKa	pKa (70°C)	Buffering range	Concentration (mM)
MOPS	7.00 (37°C)	6.90	6.5–7.9	10
HEPES	7.31 (37°C)	7.00	6.8–8.2	10
TAPS	8.11 (37°C)	8.01	7.7–9.1	10
Tris · HCl	7.77 (37°C)	ND	7.0–9.0	10
CHES	9.07 (37°C)	9.05	8.6–10	10
CAPS	10.08 (37°C)	9.95	9.7–11.1	20
$NaHCO_3/Na_2CO_3$	10.4 (25°C)	9.88	8.5–11.0	50–100

Abbreviations: MOPS – 4-morpholinopropanesulfonic acid; HEPES – N-(2-Hydroxyethyl) piperazine-N'-(2-ethanesulfonic acid); TAPS – N-[Tris(hydroxymethyl)methyl]-3-amino propanesulfonic acid; Tris – Tris(hydroxymethyl)aminomethane; CHES – 2-(N-Cyclohexylamino) ethanesulfonic acid; CAPS – 3-(Cyclohexylamino)-1-propanesulfonic acid; ND – not determined

enrichments or the use of encapsuled substrates which are then slowly released into the media.

Measurements and Adjustment of Medium pH

There is a strong temperature dependence of the media pH. Measuring pH values at room temperature or at elevated temperature with a pH electrode calibrated at room temperature will lead to an error in the pH value, which can be larger than one pH unit (more acidic) at pH^{25C} values greater than 8.5 (Kevbrin et al., 2004; Wiegel, 1998) (Figure 19.1). Thus correct measurement for pH at elevated temperature should be done by

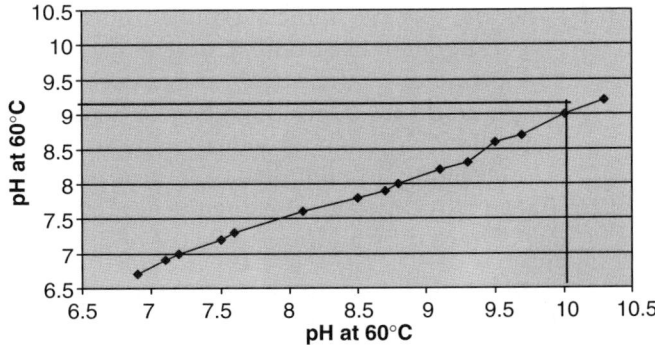

Figure 19.1. Relationship of pH values measured at 25°C to increase in growth temperature (Wiegel, 1998). This was determined in the medium used for *Thermosyntropha lipolytica* (Svetlitshnyi et al., 1996) without the olive oil. The pH was first determined at 25°C using a pH meter calibrated at 25°C then at 60°C with the pH meter calibrated at 60°C. The temperature probe was placed at 60°C and the electrode was allowed to equilibrate at 60°C for 30 min.

using a pH meter calibrated at the respective incubation temperature with a temperature probe attached for compensation of the temperature on the electrode. All standard buffers and solutions used must be preheated to the incubation temperature to include the temperature effect on the pKa-values of the various components. The electrode must be acclimated for at least 20 min. Further, it was requested by Wiegel (1998) that the detailed method of pH measurements for novel taxa be given in the Materials and Methods section of the original descriptions, and that the temperature at which the pH was measured be included as a superscript to the pH value, e.g. pH^{60C} when pH was determined at 60°C with an electrode and pH meter calibrated at 60°C according to the manufacturer's instructions.

The procedure below describes the method used in our laboratory for pH adjustment of medium used for enrichment of alkalithermophiles.

Procedure

1. Prepare medium and adjust pH at room temperature to the approximate desired value. Distribute the medium to individual tubes or bottles and sterilize by autoclaving.
2. Preheat a water bath to the desired incubation temperature. Insert a temperature probe, connected to a pH meter into the water bath and allow the temperature to equilibrate.
3. Preheat all standard buffers to be used for electrode standardization in the water bath. Let the pH electrode equilibrate for at least 30 min in pH buffer 4 (at the elevated temperature, diluted 1:4 in water) prior to standardization. We use a microelectrode (Accumet® combination microelectrode with calomel reference, Cat. No. 13-620-95, Cole-Parmer). This microelectrode has a long 6 in. flexible stem and a small glass bulb that enables pH measurement in small containers such as Eppendorf tubes. Only 100 µl of medium is required for accurate measurement.
4. Calibrate the electrode using preheated pH buffers according to the manufacturer's instructions.
5. The pH of the autoclaved medium can now be measured with the calibrated electrode. Using a syringe needle, aseptically draw 0.1–0.2 ml of medium from the culture vessel and place it in a 0.5-ml Eppendorf tube. Place the tube in a floating tube rack in the waterbath and allow 5–10 min for temperature equilibration. The tube should be closed tightly to minimize CO_2 absorption from the air.
6. Take pH measurement using the electrode. The pH of the medium can be corrected by adding sterile 0.2 or 2 N NaOH and HCl.

A point of caution to be taken when cultures of anaerobic alkalithermophiles are maintained: pronounced pH changes can occur due to formation of organic acids during fermentation. Continuous acidification of the medium can result in retardation or complete arrest of microbial growth. As a result, the pH has to be maintained within the permissive range for growth using the appropriate organic or inorganic buffer systems. Table 19.1 shows a list of buffers commonly used in our laboratory. The pH of the medium should be measured and adjusted regularly, every hour to once a day depending on the growth rate and acid production of the microorganism and buffer capacity of the media. Doubling times as low as 10 min have been observed (e.g. *Thermobrachium celere* growing at 66°C and pH^{66C} 8.2), and adjusted to the correct value using sterile NaOH or HCl. This is essential for anaerobic cultures, where pH decreases of 4 units have been seen over a period of 24 h.

An additional point of consideration is that the pKa of many buffers decreases at elevated temperatures (Table 19.1). This should be taken into account when selecting buffer systems (see discussion below on determination of pH ranges for growth).

Preparation and Sterilization of Media

At temperatures greater than 50°C and pH^{25C} values more alkaline than 8.5, it is better to prepare medium components as separate solutions, sterilized separately then aseptically combined. The separate solutions typically include:

1. growth factors such as yeast extract, tryptone, amino acids and salts
2. carbon source (carbohydrates)
3. buffer solution, such as sodium carbonate, phosphates or organic buffers
4. solidifying agent (discussed in the next section)
5. reducing solution (for anaerobic cultures).

This is especially critical for media containing sugars which caramelize with amino acids or other media ingredients and form inhibitory compounds. The individual solutions are combined together and the pH adjusted as described in the previous section. It is preferable that the carbon source be added after adjusting pH since addition of strong alkaline or acid solutions can locally accelerate degradation of carbohydrates.

Sterilization of medium requires special care. Since spores of thermophiles are known for their increased heat resistance, e.g. D_{10}-times at 121°C of nearly 2 h have been observed for strains of *Morella thermacetica* (Byrer et al., 2000), longer sterilization times, typically a minimum of 45 min are needed even for small volumes such as 10 ml. Owing to this long sterilization time, stock solutions of heat labile substances and carbohydrates should be filter sterilized through 0.22 µM filters.

Anaerobic media can be prepared as previously described by Bryant (1972) and Ljungdahl and Wiegel (1986). Addition of cysteine hydrochloride alone can result in acidification of the medium. Thus a mixture of cysteine hydrochloride and sodium sulfide is used. In our laboratory, a stock solution is prepared by first boiling 0.2 M NaOH solution to remove oxygen. The solution is then flushed with nitrogen gas. After cooling to room temperature, 1.25% wt/vol of cysteine hydrochloride and 1.25% wt/vol of sodium sulfide are added and the pH is titrated to the desired value (usually between 7.5 and 8.0). This solution is then placed in a vial (preflushed with nitrogen), sealed with a butyl rubber stopper, crimped and autoclaved for 30 min at 121°C. Four milliliters of this solution is added per 100 ml of medium just before inoculation.

Preparation of Solid Media

Solid media have the same composition as liquid media with the addition of a solidifying agent. The choice of solidifying agent will depend on the incubation temperature, pH, media composition and the response of the microorganism. Below are comments that must be taken into account when using solidifying agents in cultures of alkalithermophiles.

Agar

Agar is the most commonly used solidifying agent. It is obtained from red marine algae, and is composed of two components, agarose and agaropectin, which are not degraded by most bacterial species except for many marine strains. Molten agar solidifies at 35–45°C, depending on the agar concentration, but temperatures of 90°C or more are needed for remelting. Thus, agar can be used for incubation temperatures up to 70°C (at concentrations of 2–4% wt/vol). However, one should be aware that some microorganisms are inhibited by different levels of elevated agar concentrations (reduced water activity) and do not grow on media with elevated agar concentrations. For alkalithermophiles, it appears also to be important to use high-quality brands of agar to form stable and growth supporting gels (e.g. Difco Noble Agar). Solidified agar (2% wt/vol) is to a limited extent stable at pH^{60C} values up to 9.5. If the pH of the medium is greater than 8.0 (measured at room temperature), then a double strength (2X) agar solution should be autoclaved separately and combined with a 2X medium solution (adjusted to the desired pH) prior to inoculation. The carbon source should be added from a stock solution after the medium components have been combined and final pH adjusted.

Gelrite

Gelrite is an extracellular linear polysaccharide comprised of glucuronic acid, rhamnose and glucose. It is produced by "Pseudomonas elodea" (Kang et al., 1982). Gelrite® is a registered trademark of Merck and Co. Inc., Kelco Division. It forms clear gels with the aid of cations such as Ca^{2+} and Mg^{2+} that can withstand temperatures up to 100°C and pH^{60C} values up to 10.5. A Gelrite medium is prepared by mixing a sterile 2X medium solution with a sterile 2X Gelrite solution. The medium is then inoculated, and polymerizing cations Ca^{2+} or Mg^{2+} are added from filter sterilized 1 M stock solutions to a final concentration of 10 mM. The solidification of Gelrite is pH dependent, and the concentration of the Gelrite solution must be adjusted to the pH of the medium. Table 19.2 shows the amount of Gelrite that is required at different pH values. These amounts were tested at 60°C. We do not have data that indicates whether these

Table 19.2 Dependence of solidification of Gelrite on pH of medium

Medium pH[1]	Gelrite amount $(g\, l^{-1})$[2]
7.0	6.4
8.0	5
9.0	4
10.0	3.4

[1] pH values were measured at 60°C.
[2] solidification of Gelrite was monitored at 60°C.

concentrations will be stable at higher temperatures. Solidification begins quickly after addition of the divalent cation so pouring of plates must be done immediately. In addition, solidification occurs at temperatures below 80°C, so all solutions must be preheated to at least 85°C before combination of the media and inoculation must be carried out at 80°C. If lower concentrations of Gelrite are used for soft overlays, then lower temperatures, typically 75°C, can be used for inoculation.

Gelrite offers the advantage of increased stability at higher temperatures and pH values in comparison with agar. In addition, it forms clear colorless gels that facilitate visualization of growth. However, Gelrite is an organic compound and, although rarely, may be metabolized by some microorganisms. In addition, microorganisms respond differently to the presence of large amounts of Ca^{2+} or Mg^{2+} ions, so the best medium composition for each microorganism must be found by trial and error studies, which can be cumbersome and time consuming.

Silica gel

Silica gel is inorganic and can be successfully used for incubations greater than 65°C and especially suitable for chemolithoautotrophic thermophiles. Silica gel media (Funk and Krulwich, 1964) can be made by dissolving 10 g of powdered silica gel or silicic acid in 100 ml of a 0.7% wt/vol potassium hydroxide solution. This solution is sterilized by autoclaving, and then solidified by addition of a strong mineral acid, usually 2% vol/vol o-phosphoric acid obtaining a pH around 7.0 yielding gelling within 1 min and a firm gel within 15 min. The solidified silica gel then is extensively equilibrated with the growth medium adjusted at the desired pH. The silica gel is overlaid with liquid medium and is allowed to equilibrate for 60 min. This is repeated 3–5 times or until the conductivity or pH of the medium does not change. As is the case with Gelrite, the correct medium composition must be found for each individual microorganism. Silica gel medium has the advantage of being inorganic hence is stable at high temperatures and is not metabolized by microorganisms. These properties make it the medium of choice for isolation of true chemolithoautotrophs. However, its preparation is very laborious and time consuming.

◆◆◆◆◆ ISOLATION METHODS

Isolation of alkalithermophiles is conducted by conventional methods, with the above precautions taken into consideration for media preparation.

Aerobic alkalithermophiles are most easily isolated by the repetitive plating dilution technique. Conventional plastic Petri dishes are only stable up to 65°C (though some brands can withstand up to 75°C), thus glass Petri dishes are used at higher temperatures. In addition, to avoid drying of the medium in Petri dishes which are not tightly closed, it is recommended that they be stacked inside tightly sealed anaerobic jars. The anaerobic jars should contain 5–10 ml of water at the bottom and the jars should be lined with moist filter paper to avoid drying of the medium. Since the jars are tightly closed, they need to be aerated every 24–48 h, depending on growth rate and the extent of loading the jars.

A common problem encountered with incubations of solid medium at higher temperatures is the formation of water of syneresis. Syneresis is the contraction of the solid medium with expulsion of water. To avoid smearing of individual colonies on the surface of the agar plates, the plates should be incubated upside down so that the water accumulates on the lid of the Petri dish. The water of syneresis can then be aseptically withdrawn with a syringe. Again the water of syneresis should be microscopically examined for growth as some microorganisms respond differently to the presence of solidifying agents in the medium.

Soft agar approach

Use of Petri dishes requires at least 2% wt/vol agar in the medium so that it will remain firm at higher temperatures. Since growth of some microorganisms is inhibited by high agar concentrations, growth can be obtained by use of the soft agar overlay technique. About 15–20 ml of medium containing 2% wt/vol solidifying agent is prepared and poured into plates. After this layer solidifies completely, it is coated with another layer (5–10 ml) of inoculated medium containing 1–1.5% wt/vol solidifying agent. Colonies can be picked from medium by the use of loops, applicator sticks or, as we prefer, autoclaved, cotton plugged, long stem Pasteur pipettes where the tip has been drawn out to a capillary. Soft agar plates should be incubated right side up for the first 12–20 h to allow the top layer to adhere tightly to the base layer. The plates may then be inverted for the remaining incubation period in order to minimize smearing of colonies by water of syneresis.

Alternatively, growth can be obtained in shake tubes in media containing 0.5–1.5% wt/vol solidifying agent, yielding a soft gel at elevated temperature. The agar shake roll tube method has been successfully used in our laboratory for isolation of a variety of thermophiles, both aerobic and anaerobic, and alkalithermophiles. This method is useful for adapting microorganisms to grow on solidified media after being enriched in liquid cultures. Some microorganisms require a stepwise adaptation to growth on solid medium, a process that can take several weeks. Soft-agar-containing culture vessels are handled only after they have cooled down to room temperature where they are more viscous and allowed to solidify completely.

The agar (or any other solidifying agent) containing medium is prepared as described in the previous sections and distributed into Balch tubes. The tubes are stoppered with butyl rubber stoppers and crimped. After sterilization and combination of the different medium components, the tubes are inoculated and incubated at 50–60°C (in the case of agar) or 75–80°C (for Gelrite) to prevent solidification of the medium. The medium is then allowed to solidify as a film around the walls of the tube by means of rotating the tubes while they are cooled with tap water or ice. This is best done with a mechanical spinner that evenly rotates the tubes while they are spun. Colonies are sucked into autoclaved cotton-plugged long stem Pasteur pipettes (where the tip has been drawn out to a capillary and bend to a 90° hook of 1–2 mm length). The use of a dissecting microscope aids in controlling the process and used to insure that the colony is indeed sucked into the pipette. In instances where the colonies are of a hard and dry consistency and cannot be drawn into the elongated Pasteur pipettes, the agar plug with the colony is picked with a non-drawn out pipette. The picked colonies are then resuspended into 0.3–0.5 ml of sterile buffer (for anaerobes under a stream of oxygen-free gas and prereduced buffer/media), thoroughly vortexed, then incubated for 20–30 min at the corresponding growth temperature, and vortexed again. Suspensions are then checked using phase contrast microscopy to ensure that it contains the suspension contain mainly single cells. If not, repeat the above process or use a 5 min ultrasonic bath treatment.

As is the case with the Petri dishes, water of syneresis will form. To avoid smearing of individual colonies, the roll tubes should be incubated upside down and in a slightly tilted position (around 30–40°). The water accumulates at the top of the tube below the rubber stopper and can be easily drawn with a syringe. The syneresis water drawn out should be microscopically inspected for growth, as some microorganisms adversely react to solidifying agents. It should be noted that in the case of Gelrite shake roll tubes, the appropriate concentration of Ca^{2+} or Mg^{2+} should be added after inoculation and the tubes rolled immediately to allow even spreading of the Gelrite film around the walls of the tube. We have no experience with using silica gel in roll tubes.

A dilution series can easily be created in shake roll tubes by inoculating the first tube, mixing it gently to prevent introduction of air bubbles in the medium, then using that tube as inoculum for a second tube and so on. By making dilution series, well-separated colonies scattered throughout the solid medium are obtained. According to standard microbiological procedures, colonies should be picked and either streaked out or rolled out for a minimum of three times to ensure purity of the isolate.

Another technique using solid medium in flat bottles has been described by Braun *et al.* (1979). This was originally described for anaerobic bacteria, but it can be used for aerobes as well, provided medium is made aerobically. This technique combines the advantages of shake tubes and Petri dishes. Ten milliliters of liquid, sterile medium prepared with 1–2% wt/vol solidifying agent is introduced into the bottle. The bottles are closed with butyl rubber stoppers and the medium is allowed to solidify with the bottles lying on their flat sides. The bottles are then opened and 0.1 ml of culture is introduced and streaked out. Streaking can be done using a Pasteur pipette with its tip drawn out, as described above. Bottles with well separated colonies can be obtained. A dilution series can be made by inoculating several bottles from a dilution series.

In cases where no growth is obtained on solid medium, isolation can be carried out by repetitive serial dilution in the same manner as with mesophilic microorganisms. This technique requires a large number of dilutions, typically in 1:10 steps followed by 1:2 steps, in liquid medium to ensure purity of the isolate. In all cases, purity of the isolate should be monitored by microscopic examination and verified by 16S rRNA sequence analysis.

◆◆◆◆◆ CHARACTERIZATION OF ALKALITHERMOPHILES

Characterization of alkalithermophiles can be carried out by conventional methods and according to the requirements of the International Journal of Systematic and Evolutionary Microbiology. Of particular interest for alkalithermophiles are their temperature and pH ranges, and optimal temperature and pH values for growth. Since pH, temperature and salt concentration are the main hallmarks of these extremophiles; we recommend that characterization of these parameters be in more detail than required for mesophiles. In other words, focus should be placed on determination of marginal data (i.e. minima, optima and maxima of the pH, temperature and salt ranges) as a function of growth rate in addition to the phylogenetic position.

Determination of Temperature Range

For determination of temperature range, a temperature gradient incubator can be used. The temperature gradient incubator in our laboratory (Scientific Industries, Inc., Bohemia, NY) has a variable temperature gradient with a range from $-5°$ to $110°C$. The incubator has 30 slots which can accommodate Hungate tubes. The temperature within each slot depends on the temperature range being tested, and can be monitored by using temperature probes. Since many thermophiles have large temperature ranges for growth (Wiegel, 1990, 1998), temperature range should be determined over $35–80°C$. Growth of microorganisms in the

Hungate tubes can be monitored either by measurement of optical density at 600 nm or by direct cell counts using a counting chamber.

In cases where a temperature gradient incubator is not available, it can be replaced by a series of water baths set to different temperatures. The temperature range can be determined by first incubating the isolate at three or four temperatures at the edge of the range, for example at 35°, 50°, 65° and 80°C. Then the optimum growth temperature can be determined by adjusting the water bath temperatures between the two temperatures at which the isolate shows the fastest doubling time.

Temperature minima and maxima should be determined by incubating cultures for at least 200 times the shortest doubling time. These data are important from an ecological standpoint where it is desired to know whether the microorganism can grow or not in an environment and not necessarily how fast it grows. For biotechnological considerations, the range of optimal growth, i.e. where the microorganisms grow with the shortest doubling time, is usually of greater importance.

Determination of pH Range

The pH range for growth should be determined over a wide enough range to encounter the most acidic pH and the most alkaline pH allowing growth and, ideally, it should be determined at the optimum temperature for growth. We are typically using intervals of 0.2–0.3 pH units. For pH range measurements, a mixture of buffer solutions should be used to maintain the same components over the desired pH range and minimize the effects of changing the buffer compounds and metabolic by-products on growth characteristics. In our experience, the use of a combination of organic buffers at a final concentration of less than 100 mM is sufficient to maintain constant pH values during the mid-exponential growth phase. Use of a higher concentration of buffers is not recommended as it might inhibit growth. Table 19.2 shows the series of buffers used in our laboratory for determination of pH range. We have successfully used a mixture of 10 mM MOPS + 10 mM TAPS + 50 mM Na_2CO_3 + 10 mM CAPS for determination of pH range. The decrease in buffer pKa at elevated temperatures should be taken into account when designing buffer systems. To minimize inhibition effects by the presence of various buffer components, the exact pH optimum can be established by using the corresponding buffer component alone.

Antibiotic Susceptibility

Alkalithermophiles respond, like their neutrophilic mesophilic counterparts, to antibiotics. However, some antibiotics are not stable at elevated temperatures (Wiegel, 1986). A study by Peteranderl *et al.* (1990) showed variations in potency of antibiotics at different temperatures and over pH range of 5.0–7.3. There are no data on the efficacy of antibiotics at alkaline pH values. However, since antibiotic potency changes at elevated temperature and acidic pH, one must assume that similar effects occur at elevated temperature and alkaline pH as well. Until a similar study is conducted on the effect of temperature and elevated pH on antibiotic stability and activity, results of antibiotic susceptibility tests on alkalithermophiles will remain equivocal.

Determination of Intracellular pH

Whereas the ability of mesophilic aerobic alkaliphiles to maintain intracellular pH homeostasis has been established (Kitada et al., 2000; Krulwich et al., 1998), the adaptations of anaerobic mesophilic alkaliphiles and alkalithermophiles (both aerobic and anaerobic) are not as extensively studied. It has been shown that the anaerobic microorganisms are generally not able to maintain a narrow pH homeostasis and that the internal pH changes with changes in the external pH of the medium (Cook et al., 1996; Speelmans et al., 1993). It is also known that at temperatures greater than 55°C, cytoplasmic membrane permeability to protons increases drastically, further confounding the ability of an alkalithermophile to regulate internal pH (Konings et al., 2002). Thus measuring internal pH of alkalithermophiles is an important parameter in characterization as it will reveal information on the microorganisms' adaptive responses to both alkaline pH and high temperature.

Measurement of intracellular pH on alkalithermophiles can be done by standard methods (Cook, 2000; Cook et al., 1996; Olsson et al., 2003). The increased permeability of the cytoplasmic membrane to protons should be accounted for. This can be done by measuring the proton permeability of liposomes constructed from membranes of alkalithermophiles and the elevated incubation temperature and comparison with that of liposomes prepared from membranes of mesophiles at lower temperature. Preparation of liposomes and permeability measurements can be done as described (van de Vossenberg et al., 1999a,b).

Extraction of DNA

Extraction of DNA from isolates is necessary in order to amplify the 16S rRNA gene, determine mol% G+C content and for DNA hybridization experiments. However, the activity of several of the enzymes typically used during DNA extraction procedures such as Proteinase K and lysozyme decreases at pH values greater than 8. In addition, all buffers and solutions used during DNA extraction procedures have pH 8.0–8.5. Thus, the high pH in the culture medium of alkalithermophiles will interfere with regular DNA extraction procedures.

To successfully extract DNA from an alkalithermophilic culture, the cells should be harvested by centrifugation. The cell pellet (typically 40–50 mg wet weight) should be washed a minimum of 3 times in phosphate-buffered saline or phosphate buffer adjusted to pH 8. The cell pellet is harvested again by centrifugation. DNA extraction can then be performed using traditional protocols. We have not been successful in extraction of DNA from alkalithermophilic isolates using commercial kits. In our laboratory we have successfully used the protocol described by Wilson (1997) for extraction of DNA from pure cultures of bacteria. This method is based on protease/detergent cell lysis, followed by removal of cell wall debris, polysaccharides and remaining proteins by selective precipitation with hexadecyltrimethylammonium bromide (CTAB). After three successive phenol:chloroform:isoamyl alcohol extractions to remove remaining contaminants, high molecular weight DNA is recovered from the remaining aqueous supernatant by precipitation with isopropanol. Cell pellets of the isolates were all washed with 500 µl phosphate-buffered saline before extraction.

List of suppliers

Cole-Parmer Instrument Company
625 East Bunker Court
Vernon Hills, IL 60061-1844, USA
Tel.: 1-847-549-7600
Fax: 1-847-247-2929
http://www.coleparmer.com

Combination microelectrode

Fisher Scientific
2000 Park Lane Drive
Pittsburgh, PA 15275, USA
Tel.: 1-412-490-8300
Fax: 1-412-490-8759
http://www.fishersci.com

Wheaton serum tubes and bottles, Buffers with color indicators, crimpers and decrimpers

Lunaire Environmental Inc.
2121 Reach Road
Williamsport, PA 1770, USA
Tel.: 1-570-326-1770
Fax: 1-570-326-7304
http://www.lunaire.com

Hot air incubators

Merck and Co., Inc., Kelco Division
1500 Rahway Avenue
Avenel, NJ 70012297, USA
Tel.: 1-908-423-1000

Fax: 1-908-594-4662
http://www.merck.com

Gelrite

Scientific Industries, Inc.
70 Orville Drive
Bohemia, NY 11716, USA
Tel.: 1-631-567-4700
Fax: 1-631-567-5896
http://www.scientificindustries.com

Temperature gradient incubator

References

Balch, W. E., Fox, G. E., Magrum, L. J., Woese, C. R. and Wolfe, R. S. (1979). Methanogens: reevaluation of a unique biological group. *Microbiol. Rev.* **43**, 260–296.

Braun, M., Schoberth, S. and Gottschalk, G. (1979). Enumeration of bacteria forming acetate from H_2 and CO_2 in anaerobic habitats. *Arch. Microbiol.* **120**, 201–204.

Bryant, M. P. (1972). Commentary on the Hungate technique for culture of anaerobic bacteria. *Am. J. Clin. Nutr.* **25**, 1324–1328.

Byrer, D. E., Rainey, F. A. and Wiegel, J. (2000). Novel strains of *Moorella thermoacetica* form unusually heat resistant spores. *Arch. Microbiol.* **174**, 334–339.

Chrisostomos, S., Patel, B. K. C., Dwivedi, P. P. and Denman, S. E. (1996). *Caloramator indicus* sp. nov., a new thermophilic anaerobic bacterium isolated from deep-seated nonvolcanically heated waters of an Indian artesian aquifer. *Int. J. Syst. Bacteriol.* **46**, 497–501.

Cook, G. M. (2000). The intracellular pH of the thermophilic bacterium *Thermoanaerobacter wiegelii* during growth and production of fermentation acids. *Extremophiles* **4**, 279–284.

Cook, G. M., Russell, J., Reichert, A. and Wiegel, J. (1996). The intracellular pH of *Clostridium paradoxum*, an anaerobic, alkaliphilic, and thermophilic bacterium. *Appl. Environ. Microbiol.* **62**, 4576–4579.

Cook, G. M., Keis, S., Morgan, H. W., von Ballmoos, C., Matthey, U., Kaim, G. and Dimroth, P. (2003). Purification and biochemical characterization of the F_1F_0-ATP synthase from thermoalkaliphilic *Bacillus* sp. strain TA2.A1. *J. Bacteriol.* **185**, 4442–4449.

Engle, M., Li, Y., Woese, C. R. and Wiegel, J. (1995). Isolation and characterization of a novel alkalitolerant thermophile, *Anaerobranca horikoshii* gen. nov., sp. nov. *Int. J. Syst. Bacteriol.* **45**, 454–461.

Engle, M., Li, Y., Rainey, F., DeBlois, S., Mai, V., Reichert, A., Mayer, F., Messner, P. and Wiegel, J. (1996). *Thermobrachium celere* gen. nov., sp. nov., a rapidly growing thermophilic, alkalitolerant, and proteolytic obligate anaerobe. *Int. J. Syst. Bacteriol.* **46**, 1025–1033.

Funk, H. B. and Krulwich, T. A. (1964). Preparation of clear silica gels that can be streaked. *J. Bacteriol.* **88**, 1200–1201.

Hyun, H. H., Zeikus, J. G., Longin, R., Millet, J. and Ryter, A. (1983). Ultrastructure and extreme heat resistance of spores from thermophilic *Clostridium* species. *J. Bacteriol.* **156**, 1332–1337.

Jones, B. E., Grant, W. D., Duckworth, A. W. and Owenson, G. G. (1998). Microbial diversity of soda lakes. *Extremophiles* **2**, 191–200.

Kang, K. S., Veeder, G. T., Mirrasoul, P. J., Kaneto, T. and Cotterell, I. W. (1982). Agar-like polysaccharide produced by a *Pseudomonas* species: production and basic properties. *Appl. Environ. Microbiol.* **43**, 1086–1091.

Kevbrin, V. V., Romanek, C. S. and Wiegel, J. (2004). Alkalithermophiles: a double challenge from extreme environments. In *Origins. Genesis, Evolution and Diversity of Life* (J. Seckbach, ed.), pp. 395–412. Kluwer Academic Publishers, Dordrecht.

Kitada, M., Kosono, S. and Kudo, T. (2000). The Na^+/H^+ antiporter of alkaliphilic Bacillus sp. *Extremophiles* **4**, 253–258.

Konings, W. N., Albers, S.-V., Koning, S. and Driessen, A. J. M. (2002). The cell membrane plays a crucial role in survival of bacteria and archaea in extreme environments. *Antonie van Leeuwenhoek* **81**, 61–72.

Krulwich, T. A., Ito, M., Hicks, D. B., Gilmour, R. and Guffanti, A. A. (1998). pH homeostasis and ATP synthesis: studies of two processes that necessitate inward proton translocation in extremely alkaliphilic *Bacillus* species. *Extremophiles* **2**, 217–222.

Li, Y., Mandelco, L. and Wiegel, J. (1993). Isolation and characterization of a moderately thermophilic anaerobic alkaliphile, *Clostridium paradoxum* sp. nov. *Int. J. Syst. Bacteriol.* **43**, 450–460.

Li, Y., Engle, M., Weiss, N., Mandelco, L. and Wiegel, J. (1994). *Clostridium thermoalcaliphilum* sp. nov., an anaerobic and thermotolerant facultative alkaliphile. *Int. J. Syst. Bacteriol.* **44**, 111–118.

Ljungdahl, L. G. and Wiegel, J. (1986). Working with anaerobic bacteria. In *Manual of Industrial Microbiology and Biotechnology* (A. L. Demain and N. A. Solomon, eds), pp. 84–96. American Society for Microbiology, Washington DC.

Lundahl, G. (2003). A method of increasing test range and accuracy of bioindicators: *Geobacillus stearothermophilus* spores. *J. Pharm. Sci. Technol.* **57**, 249–262.

Macy, J. M., Snellen, J. E. and Hungate, R. E. (1972). Use of syringe methods for anaerobiosis. *J. Clin. Nutr.* **25**, 1318–1323.

Olsson, K., Keis, S., Morgan, H. W., Dimroth, P. and Cook, G. M. (2003). Bioenergetic properties of the thermoalkaliphilic *Bacillus* sp. strain TA2.A1. *J. Bacteriol.* **185**, 461–465.

Peddie, C. J., Cook, G. M. and Morgan, H. W. (1999). Sodium-dependent glutamate uptake by an alkaliphilic, thermophilic *Bacillus* strain, TA2.A1. *J. Bacteriol.* **181**, 3172–3177.

Peteranderl, R., Shottis, E. B. and Wiegel, J. (1990). Stability of antibiotics and growth conditions for thermophilic anaerobes. *Appl. Environ. Microbiol.* **56**, 1981–1983.

Speelmans, G., Poolman, B., Abee, T. and Konings, W. N. (1993). Energy transduction in the thermophilic anaerobic bacterium *Clostridium fervidus* is exclusively coupled to sodium ions. *Proc. Natl. Acad. Sci. USA* **90**, 7975–7979.

Svetlitshnyi, V., Rainey, F. and Wiegel, J. (1996). *Thermosyntropha lipolytica* gen. nov. sp. nov., a lipolytic anaerobic alkalitolerant thermophilic bacterium utilizing short- and long-chain fatty acids in syntrophic coculture with a methanogenic archaeum. *Int. J. Syst. Bacteriol.* **46**, 1131–1137.

van de Vossenberg, J. L. C. M., Driessen, A. J. M., da Costa, M. S. and Konings, W. N. (1999a). Homeostasis of the membrane proton permeability in *Bacillus subtilis* grown at different temperatures. *Biochim. Biophys. Acta* **1419**, 97–104.

van de Vossenberg, J. L. C. M., Driessen, A. J. M., Grant, D. and Konings, W. N. (1999b). Lipid membranes from halophilic and alkalihalophilic Archaea have a low H^+ and Na^+ permeability at high salt concentration. *Extremophiles* **3**, 253–257.

Wiegel, J. (1986). Methods for isolation and study of thermophiles. In *Thermophiles: General, Molecular and Applied Microbiology* (T. D. Brock, ed.), pp. 17–37. John Wiley & Sons, New York.

Wiegel, J. (1990). Temperature spans for growth: a hypothesis and discussion. *FEMS Microbiol. Rev.* **75**, 155–170.

Wiegel, J. (1998). Anaerobic alkalithermophiles, a novel group of extremophiles. *Extremophiles* **2**, 257–267.

Wiegel, J. and Kevbrin, V. (2004). Diversity of aerobic and anaerobic alkalithermophiles. *Biochem. Soc. Trans.* **32**, 193–198.

Wilson, K. (1997). Preparation of genomic DNA from bacteria. In *Current Protocols in Molecular Biology* (F. M. Ausubel, R. Brent, R. E. Kingston, D. D. Moore, J. G. Seidman, J. A. Smith and K. Struhl, eds), pp. 2.4.1–2.4.5. John Wiley & Sons, New York.

Xin, H., Itoh, T., Zhou, P., Suzuki, K. and Nakase, T. (2001). *Natronobacterium nitratireducens* sp. nov., a haloalkaliphilic archaeon isolated from a soda lake in China. *Int. J. Syst. Evol. Microbiol.* **51**, 1825–1829.

Xu, Y., Wang, Z., Xue, Y., Zhou, P., Ma, Y., Ventosa, A. and Grant, W. D. (2001). *Natrialba hulunbeirensis* sp. nov. and *Natrialba chahannaoensis* sp. nov., novel haloalkaliphilic archaea from soda lakes in Inner Mongolia Autonomous Region, China. *Int. J. Syst. Evol. Microbiol.* **51**, 1693–1698.

Zavarzin, G. (1993). Epicontinental soda lakes as probable relict biotopes of terrestrial biota formation. *Microbiology* (Moscow) **62**, 473–479.

Zhang, W., Xue, Y., Ma, Y., Zhou, P., Ventosa, A. and Grant, W. D. (2002). *Salinicoccus alkaliphilus* sp. nov., a novel alkaliphile and moderate halophile from Baer Soda Lake in Inner Mongolia Autonomous Region, China. *Int. J. Syst. Evol. Microbiol.* **52**, 789–793.

Zhilina, T. N. and Zavarzin, G. (1994). Alkaliphilic anaerobic community at pH 10. *Curr. Microbiol.* **29**, 109–112.

Acidophiles

◆◆

20 The Isolation and Study of Acidophilic Microorganisms

Acidophiles

20 The Isolation and Study of Acidophilic Microorganisms

Elena González-Toril[1], Felipe Gómez[1], Moustafa Malki[2] and Ricardo Amils[1,2]

[1] Centro de Astrobiología (CSIC-INTA), 28850 Torrejón de Ardoz, Spain; [2] Centro de Biología Molecular (UAM-CSIC), Universidad Autónoma de Madrid, Cantoblanco, 28049 Madrid, Spain

◆◆◆

CONTENTS

Introduction
Isolation of acidophilic microorganisms
Growth conditions for the most characteristic acidophilic prokaryotes
Molecular ecological methodology
Acidophilic eukaryotes

◆◆◆◆◆ INTRODUCTION

Organisms that thrive under extreme conditions have recently attracted considerable attention due to their peculiar ecology, physiology as well as their biotechnological potential (Brierley and Brierley, 1999; Margesin and Schinner, 2001; Niehaus et al., 1999; Rawlings, 2002; Russell, 2000). Acidic environments are especially interesting because, in general, the low pH of the habitat is the consequence of microbial metabolism (González-Toril et al., 2001) and not a condition imposed by the system, as is the case for other extreme environments (temperature, ionic strength, high pH, radiation, pressure, etc.).

Acidic, metal-rich environments have two major origins. The first one is associated with volcanic activities. The acidity in these locations may derive from the microbial oxidation of elemental sulfur:

$$S^0 + 1.5O_2 + H_2O \rightarrow SO_4^{2-} + 2H^+$$

which is produced as a result of the condensation reaction between oxidized and reduced sulfur containing gases:

$$2H_2S + SO_2 \rightarrow 3S^0 + 2H_2O$$

In these environments temperature gradients are easily formed. These sites may therefore be colonized by a variety of acidophilic microorganisms, with different optimal temperatures (Johnson, 1999).

Acidic, metal-rich environments can also be found associated with mining activities. Metal and coal mining operations expose sulfidic minerals to the combined action of water and oxygen, which facilitate microbial attack, producing the so-called Acid Mine Drainage, a serious environmental problem. The most abundant sulfidic mineral, pyrite, is of particular interest in this context.

The biological oxidation of pyrite has been studied in detail. The process occurs in several steps, with the overall reaction being:

$$4FeS_2 + 14H_2O + 15O_2 \rightarrow 4Fe(OH)_3 + 8SO_4^{2-} + 16H^+$$

These environments vary greatly in their physico-chemical characteristics and also in their microbial ecology. High temperature may also occur as a consequence of biological activity, facilitating colonization by thermophilic acidophiles.

The mechanisms by which microbes obtain energy by oxidizing sulfidic minerals, a process of biotechnological interest known as bioleaching, have been controversial for many years (Ehrlich, 2002). The recent demonstration that ferric iron, present in the cell envelopes of leaching microorganisms, is responsible for the electron transfer from the sulfidic minerals to the electron transport chain (Gehrke et al., 1995), has clarified the situation. The difference seems to exist at the level of the chemical attack mechanism, which is dependent on the structure of the sulfidic substrates. Three metal sulfides, pyrite, molybdenite and tungstenite, which can only be oxidized by ferric iron, undergo oxidation through the so-called thiosulfate mechanism:

$$FeS_2 + 6Fe^{3+} + 3H_2O \rightarrow S_2O_3^{2-} + 7Fe^{2+} + 6H^+$$
$$S_2O_3^{2-} + 8Fe^{3+} + 5H_2O \rightarrow 2SO_4^{2-} + 8Fe^{2+} + 10H^+$$

with sulfate as the main product (Sand et al., 2001). Most other sulfides (e.g. galena, chalcopyrite, sphalerite, etc.) are susceptible to proton attack as well as to ferric iron oxidation. They are oxidized through a different mechanism, the so-called polysulfide mechanism (Sand et al., 2001):

$$8MS + 8Fe^{3+} + 8H^+ \rightarrow 8M^{2+} + 4H_2S_n + 8Fe^{2+} \ (n \geq 2)$$
$$4H_2S_n + 8Fe^{3+} \rightarrow S_8^0 + 8Fe^{2+} + 8H^+$$

In this case, elemental sulfur is produced which can be further microbially oxidized to sulfuric acid.

The ferrous iron produced in all these reactions is reoxidized by iron-oxidizing microorganisms to ferric iron:

$$2Fe^{2+} + 0.5O_2 + 2H^+ \rightarrow 2Fe^{3+} + H_2O$$

The main role of acidophilic chemolithotrophic microorganisms is to maintain a high concentration of the chemical oxidant, ferric iron, in solution.

The acidophilic strict chemolithotroph, *Acidithiobacillus ferrooxidans* (formerly *Thiobacillus ferrooxidans*), was first isolated from an acidic pond in a coal mine, more than fifty years ago (Colmer et al., 1950). Although *At. ferrooxidans* can obtain energy, oxidizing both sulfide and ferrous iron, much more attention has been paid to the sulfur oxidation reaction due to bioenergetic considerations (Amils et al., 2004; Ehrlich, 2002; Pronk et al., 1992). The discovery that some strict chemolithotrophs, e.g. *Leptospirillum ferrooxidans*, could grow using ferrous iron as sole source of energy, and that this microorganism is mainly responsible for metal bioleaching and acid mine drainage generation, has completely changed this perspective (Edwards et al., 2000; Golyshina et al., 2000; Rawlings, 2002). Furthermore, it is now well established that iron can be oxidized anaerobically, coupled to anoxygenic photosynthesis or to anaerobic respiration using nitrate as an electron acceptor (Benz et al., 1998; Widdel et al., 1993).

Most of the characterized strict acidophilic microorganisms have been isolated from volcanic areas or acid mine drainage from mining activities. The Tinto River (Iberian Pyritic Belt) is an unusual ecosystem due to its acidity (mean pH 2.3, buffered by Fe^{3+} iron), length (100 km), high concentration of heavy metals (Fe, As, Cu, Zn, Cr...) and unexpected level of microbial diversity (Amaral-Zettler et al., 2003; López-Archilla et al., 1993, 2001). Recently, it has been proved that the extreme conditions of the Tinto system are much older than the oldest mining activities known in the area, strongly suggesting that they are natural and not the consequence of industrial contamination (Fernández-Remolar et al., 2003, 2005). Due to its size and easy access, the Tinto River is considered an interesting model system for acidic environments.

In this chapter, we will describe the methodologies, conventional as well as molecular, used to characterize the microbial diversity existing in the extreme acidic environment of the Tinto River. These methodologies have been adapted from different sources. Most of the protocols should be useful for other acidic environments, although the importance of small variables in these systems (pH, temperature, redox potential, metal concentrations, etc.) should be kept in mind and optimization is strongly recommended, especially in the use of molecular ecology tools.

◆◆◆◆◆ ISOLATION OF ACIDOPHILIC MICROORGANISMS

Although molecular ecology methods allow rapid characterization of the diversity of complex ecosystems, isolation of the different constituents is essential to the study of their phenotypic properties in order to evaluate their role in the system and their biotechnological potential. Acidic environments are poorly characterized environments due to the physiological peculiarities of the microorganisms associated with them. Furthermore, strict acidophilic chemolithotrophs are, in general, not easy to grow, especially in solid media (Hallberg and Johnson, 2001).

Prokaryotic microorganisms that are metabolically active in extreme acidic environments are distributed in the domains Bacteria and Archaea, and can be classified according to their energy sources. A variety of chemoautolithotrophic microorganisms capable of oxidizing iron and sulfur-containing minerals have been isolated from acidic environments. The mineral-oxidizing bacteria found in these natural environments are ubiquitous and the most commonly encountered have been characterized as *Acidithiobacillus ferrooxidans*, *Leptospirillum* spp. and *Acidithiobacillus thiooxidans* (Kelly and Harrison, 1989), and more recently *Acidithiobacillus caldus* (Hallberg and Lindström, 1994), *Sulfobacillus* spp. and the members of the archaeal *Ferroplasmaceae* family.

In acidic environments, there is also a variety of acidophilic heterotrophs such as those belonging to the genera *Acidiphilium*, or facultative heterotrophs such as *Acidimicrobium* spp. and "*Ferrimicrobium*" spp. (Johnson and Roberto, 1997). Acidophiles can also be categorized using other physiological criteria, such as temperature (mesophiles, moderate thermophiles and thermophiles), optimal pH, or on the basis of their carbon source (autotrophs, heterotrophs and mixotrophs). In general, the most extremely thermophilic acidophiles correspond to the domain Archaea.

Recently, a variety of acidic/halophilic environments have been described in southwestern Australia (Benison and Laclair, 2003). Although microbial activity has been detected in the system, no isolations have been reported so far.

◆◆◆◆◆ GROWTH CONDITIONS FOR THE MOST CHARACTERISTIC ACIDOPHILIC PROKARYOTES

Domain Bacteria

The majority of the best characterized acidophilic microorganisms belong to the bacterial domain. Acidophiles are widely spread within the bacterial phylogenetic tree. Most are Gram-negative (all the species of the *Acidithiobacillus* genus, *Leptospirillum* spp. and *Acidiphilium* spp.), but recently several Gram-positive microorganisms belonging to different genera have been isolated from acidic environments and biohydrometallurgical operations (*Sulfobacillus* spp., *Acidimicrobium* spp. and "*Ferrimicrobium*" spp.). Most of these microorganisms have been isolated from the Tinto River ecosystem, which as mentioned could be considered a representative model system for this type of habitat. Table 20.1 compares the main characteristics of representative chemolithotrophic acidophilic bacteria.

Table 20.1 Comparison of phenotypic properties of acidophilic iron-oxidizing bacteria

Characteristics	*Leptospirillum ferrooxidans*	*Acidithiobacillus ferrooxidans*	*"Ferrimicrobium acidiphilum"*	*Sulfobacillus acidophilus*	*Acidimicrobium ferrooxidans*
Cell morphology	vibrios 1 μm, spirilla	rods 1–2 μm	rods 1–3 μm, filaments	rods 3–5 μm	rods 1–1.5 μm, filaments
Optimal pH	1.5–2.0	2.5	2.0–2.5	2	2
Mol% G+C	51–56	58–59	52–55	55–57	67–68.5
S⁰ oxidation	−	+	−	+	−
Iron reduction	−	+	+	+	+
Endospores	−	−	−	+	−
Motility	++	+	+	+	++
Utilization of yeast extract	−	−	+	+	+
Growth at 50°C	−	−	−	+	−

Acidithiobacillus ferrooxidans

At. ferrooxidans was the first strictly chemolithotrophic acidophilic microorganism isolated from acidic mining waters. This Gram-negative bacterium is a 0.5 μm × 1.0 μm motile rod, using CO_2 as a carbon source and obtaining its energy from the oxidation of ferrous iron, elemental sulfur, and reduced sulfur compounds (Colmer et al., 1950). *At. ferrooxidans* was believed to be the dominant bacterium responsible for metal sulfide solubilization (Ehrlich, 2002). Different media have been described in the literature for the growth of *At. ferrooxidans* under aerobic conditions (Leathen et al., 1956; Mackintosh, 1978; Silverman and Lundgren, 1959; Temple and Colmer, 1951; Tuovinen and Kelly, 1973). We use Mackintosh media for most of our isolations because we believe that the introduction of trace elements, which is the main difference with the other media, is beneficial for some strains. The medium is made of three solutions that are prepared separately:

- solution A (basal salts): $(NH_4)_2SO_4$, 132 mg; $MgCl_2 \cdot 6H_2O$, 53 mg; KH_2PO_4, 27 mg; $CaCl_2 \cdot 2H_2O$, 147 mg; dissolved in 950 ml of distilled water, pH 1.8 adjusted with 10 N H_2SO_4.
- solution B (energy source): $FeSO_4 \cdot 7H_2O$, 20 g dissolved in 50 ml of 0.25 N H_2SO_4, the pH of this solution should be 1.2.
- solution C (trace elements): $MnCl_2 \cdot 2H_2O$, 62 mg; $ZnCl_2$, 68 mg; $CoCl_2 \cdot 6H_2O$, 64 mg; H_3BO_3, 31 mg; Na_2MoO_4, 10 mg; $CuCl_2 \cdot 2H_2O$, 67 mg; dissolved in 1 l of distilled water, pH 1.8 adjusted with 10 N H_2SO_4.

Solutions are autoclaved separately at 120°C for 30 min. Prior to use, solutions A, B and 1 ml of C are mixed under sterile conditions. The final pH of the medium should be around 2. Media with more acidic pHs can be obtained by adjusting the pH of solution A to more acidic values. The same protocol can be used when elemental sulfur or pyrite are used as energy source. Sulfur can be sterilized at 120°C for 30 min in an autoclave or by tindallization. Pyrite can be sterilized by dry heat.

A different protocol has been described for the growth of *At. ferrooxidans* using thiosulfate as energy source: KH_2PO_4, 3 g; $MgSO_4 \cdot 7H_2O$, 0.5 g; $(NH_4)_2SO_4$, 3 g; $CaCl_2 \cdot 2H_2O$, 0.25 g; $Na_2S_2O_3 \cdot 5H_2O$, 5 g; dissolved in 1 l of distilled water. The medium is prepared without thiosulfate, adjusted to pH 4.4–4.7 and autoclaved at 120°C for 15 min. Thiosulfate is not stable at lower pH values. The thiosulfate solution in water is sterilized by filtration and added to the basal medium.

Different solid media have been developed for *At. ferrooxidans* genetic manipulation studies. Solid medium 2.2 (Peng et al., 1994) contains a mixture of both, ferrous iron and thiosulfate, as energy sources and is prepared in four solutions: Solution A: $Na_2S_2O_3 \cdot 5H_2O$, 2 g dissolved in 10 ml of H_2O. Solution B: $FeSO_4 \cdot 7H_2O$, 2 g dissolved in 10 ml of H_2O, pH adjusted to 2.0 with 2 N H_2SO_4. Solution C: $(NH_4)_2SO_4$, 4.5 g; KCl, 0.15 g and $MgSO_4$, 0.5 g. Solution D: agar, 6.0 g added to 480 ml of H_2O. Solution A and B are filter-sterilized, while solutions C and D are autoclaved at 120°C for 15 min. When solutions C and D have cooled down to 45°C, solutions A and D and solutions B and C are premixed, and the resulting solutions mixed together. The final pH of the medium should be 4.6.

At. ferrooxidans is generally assumed to be an obligate aerobic organism (Kelly and Harrison, 1989). However, under anaerobic conditions, ferric iron can replace oxygen as an electron acceptor for the oxidation of elemental sulfur (Brock and Gustafson, 1976; Corbett and Ingledew, 1987). At pH 2, the free energy of the reaction $S^0 + 6Fe^{3+} + 4H_2O \rightarrow H_2SO_4 + 6Fe^{2+} + 6H^+$ is negative ($\Delta G = -314$ kJ mol^{-1}), so it can be used for energy transduction.

The mineral medium used for anaerobic growth (Muyzer et al., 1987) contains the following components per liter of distilled water: $(NH_4)_2SO_4$, 152 mg; K_2HPO_4, 41 mg; $MgSO_4 \cdot 7H_2O$, 490 mg; $CaCl_2 \cdot 2H_2O$, 9 mg; KCl, 52 mg; $ZnSO_4 \cdot 7H_2O$, 1 mg; $CuSO_4 \cdot 5H_2O$, 2 mg; $MnSO_4 \cdot H_2O$, 1 mg; $Na_2MoO_4 \cdot 2H_2O$, 0.5 mg; $CoCl_2 \cdot 6H_2O$, 0.5 mg; $Na_2SeO_4 \cdot 10H_2O$, 1 mg; $NiCl_2 \cdot 6H_2O$, 1 mg; yeast extract, 5 mg; $Fe_2(SO_4)_3$, 20 g; and powdered elemental sulfur, 10 g. The solution's pH is adjusted to 1.9 with 10 N H_2SO_4. Elemental sulfur is placed in screw-capped tubes or bottles and sterilized at 120°C for 30 min. Before use, sulfur is aseptically layered onto the surface of the sterile basal medium.

Acidithiobacillus thiooxidans

At. thiooxidans is a phylogenetically close relative of *At. ferrooxidans*. Like *At. ferrooxidans* it is also an autotrophic Gram-negative sulfur oxidizer, but it differs from the latter by its inability to oxidize iron. The characteristics of *At. thiooxidans* have been described by Kelly and Harrison (1989) and Kelly and Wood (2000). It grows in liquid medium using reduced sulfur compounds: elemental sulfur, thiosulfate or tetrathionate. The basal salts medium for growing *At. thiooxidans* was described by Harrison (1983). The medium contains the following components (per liter of distilled water): NH_4Cl, 0.10 g; KH_2PO_4, 3 g; $MgCl_2 \cdot 6H_2O$, 0.10 g; $CaCl_2 \cdot 2H_2O$, 0.14 g; powdered elemental sulfur, 10 g. All the ingredients, except the sulfur, are dissolved in distilled water and the pH adjusted to 4.2 and then autoclaved. Elemental sulfur is sterilized as described above.

Acidithiobacillus caldus

At. caldus is a Gram-negative moderately thermophilic acidophile found in mining environments such as coal spoil heaps, where the oxidative dissolution of sulfide-containing minerals occurs (Hallberg and Linström, 1994; Marsh and Norris, 1983). This bacterium obtains its carbon by reductive fixation of atmospheric CO_2. *At. caldus* can oxidize a wide range of reduced sulfur compounds, but, like *At. thiooxidans*, is incapable of oxidizing ferrous iron or pyrite. The main product of the oxidation of reduced sulfur compounds is H_2SO_4, so this bacterium is able to grow at rather low pH values.

The basal medium for growth of *At. caldus* contains in g l^{-1} of distilled water: $(NH_4)_2SO_4$, 3; $Na_2SO_4 \cdot 10H_2O$, 3.2; KCl, 0.1; K_2HPO_4, 0.05;

$MgSO_4 \cdot 7H_2O$, 0.5; $Ca(NO_3)_2$, 0.01; and the following trace elements in mg l^{-1}: $FeCl_3 \cdot 6H_2O$, 11; $CuSO_4 \cdot 5H_2O$, 0.5; H_3BO_3, 2; $MnSO_4 \cdot H_2O$, 2; $Na_2MoO_4 \cdot 2H_2O$, 0.8; $CoCl_2 \cdot 6H_2O$, 0.6; and $ZnSO_4 \cdot 7H_2O$, 0.9. The basal salts solution is adjusted to pH 2.5 with H_2SO_4 and autoclaved before the filter-sterilized trace elements are added. Tetrathionate at a final concentration of 5 mM is used as the energy source. Cells are incubated at 45°C and the medium stirred by bubbling CO_2-enriched air (2%, vol/vol).

Leptospirillum spp.

Leptospirilli are Gram-negative autotrophic acidic bacteria (optimal pH 1.5–1.8) that can only grow with iron as a source of energy (Pivovarova et al., 1981). As a result, leptospirilli have a high affinity for ferrous iron and unlike *At. ferrooxidans* their ability to oxidize ferrous iron is not inhibited by ferric iron. They are aerobic respirers and they are extremely tolerant to low pH. Based on phylogenetic studies three species can be defined. These are referred to as *L. ferrooxidans* (optimal growth temperature from 26 to 30°C), *L. ferriphilum* (optimal temperature between 30 and 40°C) and group III leptospirilli (for which there are no isolates so far) (Bond et al., 2000; González-Toril et al., 2003 Rawlings, 2002).

L. ferrooxidans can be grown in batch culture in the Mackintosh media already described for *At. ferrooxidans*.

Sulfobacillus spp.

Sulfobacilli are low G+C Gram-positive, acidophilic moderate thermophilic, endospore-forming bacteria. They have been isolated from geothermal environments and biohydrometallurgical processes. *Sulfobacillus* spp. grow autotrophically between 40 and 60°C, using ferrous iron, sulfur and sulfide minerals as energy sources (Norris et al., 1996). This genus is also able to grow mixotrophically in media with ferrous iron and yeast extract, or heterotrophically using glucose as carbon and energy source. This bacterium can be grown in media with the following composition:

- Solution A: $(NH_4)_2SO_4$, 3 g; KCl, 0.10 g; K_2HPO_4, 0.50 g; $MgSO_4 \cdot 7H_2O$, 0.50 g; $Ca(NO_3)_2$, 0.01 g; dissolved in 680 ml of distilled water. This solution is adjusted with sulfuric acid to pH 2.0–2.2.
- Solution B: 4.2 g of $FeSO_4 \cdot 7H_2O$ dissolved in 300 ml of distilled water supplemented with 1 ml of 10 N H_2SO_4.
- Solution C: 0.2 g of yeast extract dissolved in 20 ml of distilled water.

After sterilization the three solutions are combined. Final pH should be around 2.

Other protocols have been described in which the basal media is basically the same and different substrates are used for chemolithotrophic, heterotrophic and mixotrophic growth (pyrite 2.5 g l^{-1}; ferrous sulfate 33.4 g l^{-1}; elemental sulfur 0.5 g l^{-1} or glucose 0.5 g l^{-1}) (Suzina et al., 1999).

Actinobacteria

Two species of Gram-positive acidophiles of the class *Actinobacteria* have been isolated from acidic environments, *Acidimicrobium ferrooxidans* and "Ferrimicrobium acidiphilum". Both are iron oxidizers, but they differ in their responses to temperature (Johnson, 1999; Norris et al., 1996).

Acidimicrobium ferrooxidans

Am. ferrooxidans is a moderately thermophilic bacterium that grows autotrophically when oxidizing iron under aerobic conditions. Different strains have been isolated with different carbon requirements (Hallberg and Johnson, 2001; Norris et al., 1996). The medium for growing *Acidimicrobium* spp. is composed of a basal medium to which a second solution with a different composition is added depending on the autotrophic or heterotrophic mode of growth. The basal salt solution is composed by the following reagents, in g l^{-1} of distilled water: $MgSO_4 \cdot 7H_2O$, 0.5; $(NH_4)_2SO_4$, 0.4; K_2HPO_4, 0.2 and KCl, 0.1. The pH is adjusted to 2.0 with H_2SO_4.

For autotrophic growth on ferrous sulfate, 13.9 g l^{-1} of $FeSO_4 \cdot 7H_2O$ are added per liter of basal medium and the pH is adjusted to 1.7 with H_2SO_4 prior to sterilization.

For heterotrophic growth on yeast extract, 10 mg l^{-1} of $FeSO_4 \cdot 7H_2O$ are added to the basal media. After sterilization by autoclaving, yeast extract is added from a sterile stock solution to a final concentration of 0.25 g l^{-1}.

Some *Am. ferrooxidans* strains are able to oxidize reduced carbon compounds using ferric iron as an electron acceptor under anaerobic conditions (Bridge and Johnson, 1998).

"Ferrimicrobium acidiphilum"

This bacterium is closely related to *Am. ferrooxidans* on the basis of 16S rRNA sequence homology. "Ferrimicrobium acidiphilum" is an iron-oxidizing mesophilic obligate heterotroph, unable to form endospores (Hallberg and Johnson, 2001). The medium used to grow "Ferrimicrobium" spp. is identical to the one described above for *Acidimicrobium*. Under limited aeration, it has also the ability to anaerobically respire reduced carbon compounds using ferric iron as an electron acceptor.

Acidiphilium spp.

This group of acidophilic heterotrophic Gram-negative bacteria belonging to the α-Proteobacteria phylum normally appears associated with strictly chemolithoautotrophic bacteria like acidithiobacilli or leptospirilli. These organisms are able to grow aerobically at 30°C in a chemically defined medium (Hiraishi and Kitamura, 1984) composed of a basal medium with the following components per liter of distilled water: $(NH_4)_2SO_4$, 2.0 g; KCl, 0.1 g; K_2HPO_4, 0.25 g and $MgSO_4 \cdot 7H_2O$, 0.25 g, supplemented with

0.1 g of yeast extract and 1 g of glucose as energy and carbon source. The pH is adjusted to 3 with 10 N H_2SO_4 prior sterilization at 120°C for 20 min.

Desulfosporosinus spp.

These Gram-positive bacteria are able to use sulfate as an electron acceptor for anaerobic respiration (sulfate reducing bacteria). The presence of sulfate reducing bacteria in acidic environments has not yet been clearly established. However, the existence of acidophilic sulfate reducing bacteria would be very interesting for bioremediation processes (Hallberg and Johnson, 2001). Some sulfate reducing bacteria activities have been detected in anaerobic sediments of acidic environments, and in most cases the organisms involved were related to this genus (González-Toril *et al.*, 2006; Hallberg and Johnson, 2003). The media described by Postgate (1984) is being used for enrichment cultures of sulfate reducing bacteria in the Tinto ecosystem. Media contain per liter of distilled water: Na-lactate, 12 g; Na_2SO_4, 4.5 g; $CaCl_2 \cdot 2H_2O$, 0.06 g; Na-citrate, 0.3 g; NH_4Cl, 1 g; KH_2PO_4, 0.5 g; $MgSO_4 \cdot 7H_2O$, 2 g; yeast extract, 1 g. To avoid formation of precipitates, dry salts should not be mixed together before dissolution. They should be added successively to the water while stirring. The pH is adjusted to 4 and the solution autoclaved.

Domain Archaea

Acidophilic aerobic or facultatively anaerobic Archaea that colonize biotopes such as acid mine drainage and solfatara fields, where reduced forms of iron and sulfur are present, are represented by two different phylogenetic groups, the orders *Sulfolobales* and *Thermoplasmatales*. These groups differ in their phenotypic properties. Some representatives of the *Sulfolobales*, e.g. *Acidianus brierleyi* and members of the genus *Metallosphaera*, obtain energy by oxidizing sulfur, sulfidic minerals and reduced iron. Other species of the order *Sulfolobales*, e.g. *Sulfolobus acidocaldarius* and *Sulfolobus solfataricus*, utilize sulfur and reduced sulfur compounds. *Sulfolobus metallicus* exploits sulfidic ores and elemental sulfur as energy sources. Although other members of the *Sulfolobales* are able to grow chemolithoautotrophically, *S. metallicus* and *Acidianus ambivalens* are the only obligate chemolithoautotrophs known. In contrast, in the class *Thermoplasmata* only two species of the genus *Ferroplasma*, *F. acidiphilum* and "*F. acidarmanus*", are acidophilic chemolithotrophs capable of oxidizing iron, although they require traces of yeast extract to support autotrophic growth. The other members of the genus, as well as the genera *Thermoplasma* and *Picrophilus*, have a heterotrophic metabolism. In the Tinto ecosystem only members of the *Thermoplasmata* class, *Thermoplasma acidophilum* and *Ferroplasma acidiphilum* have been identified (González-Toril *et al.*, 2003). "*F. acidarmanus*", an extreme acidophile able to grow at negative pH, has been isolated from biofilms of pyritic sediments in the Iron Mountain mine in California (Edwards *et al.*, 2000). *Ferroplasma*-like microorganisms are detected recurrently in different commercial bioleaching plants, indicating that they play an important role in this type of artificial ecosystem.

Ferroplasma spp.

Ferroplasma spp. grow well in a modified medium 9K (Silverman and Lundgren, 1959), containing per liter of distilled water: $MgSO_4 \cdot 7H_2O$, 0.4 g; $(NH_4)_2SO_4$, 0.2 g; KCl, 0.1 g; K_2HPO_4, 0.1 g and $FeSO_4 \cdot 7H_2O$, 25 g. The medium is supplemented with 0.02% of yeast extract and trace elements. The pH of the medium is adjusted to 1.7 by addition of 10 N H_2SO_4.

Acidianus brierleyi

This microorganism was the first thermophilic acidophile ever isolated, although it was described after other extreme thermophiles were reported (Brierley and Brierley, 1973). The medium to grow *A. brierleyi* contains per liter of distilled water: $(NH_4)_2SO_4$, 3 g; $K_2HPO_4 \cdot 3H_2O$, 0.5 g; $MgSO_4 \cdot 7H_2O$, 0.5 g; KCl, 0.1 g; $Ca(NO_3)_2$, 0.01 g; yeast extract, 0.2 g and elemental sulfur, 10 g. The pH is adjusted with 10 N H_2SO_4 to 1.5–2.5. Yeast extract (10% in distilled water) is autoclaved separately. Sulfur can be sterilized by steaming for 3 h on 3 successive days.

Sulfolobus acidocaldarius

S. acidocaldarius is an extremely acidophilic, thermophilic, chemolithotrophic microorganism that has the ability to use ferrous iron oxidation as a source of energy (Brock *et al.*, 1972). It closely resembles *A. brierleyi* morphologically and physiologically, however the two species differ genotypically. The growth medium for *S. acidocaldarius* contains per liter of distilled water: $(NH_4)_2SO_4$, 1.3 g; KH_2PO_4, 0.28 g; $MgSO_4 \cdot 7H_2O$, 0.25 g; $CaCl_2 \cdot 2H_2O$, 0.07 g; $FeCl_3 \cdot 6H_2O$, 0.02 g; $MnCl_2 \cdot 4H_2O$, 1.8 mg; $Na_2B_4O_7 \cdot 10H_2O$, 4.5 mg; $ZnSO_4 \cdot 7H_2O$, 0.22 mg; $CuCl_2 \cdot 2H_2O$, 0.05 mg; $Na_2MoO_4 \cdot 2H_2O$, 0.03 mg; $CoSO_4$, 0.01 mg and yeast extract, 1 g. pH is adjusted to 2 with 10 N H_2SO_4.

◆◆◆◆◆ MOLECULAR ECOLOGICAL METHODOLOGY

As mentioned in the previous section, classical microbial ecology analysis is limited by the unavoidable need for isolation of the microorganisms prior to their characterization (Amann *et al.*, 1990). Although it is obvious that isolation of microorganisms is indispensable for their full characterization, the introduction of molecular biology methods (such as fluorescent *in situ* hybridization, denaturing gradient gel electrophoresis or cloning) has enabled a significant advance in microbial ecology (Amann *et al.*, 1990, 1995), especially in the study of extreme environments such as acidic habitats, in which conventional methods are severely limited, and some may even lead to equivocal conclusions, with occasionally grievous economic consequences.

A good example of this type of problem is the enormous amount of research devoted to *At. ferrooxidans*, mainly to improve commercial bioleaching operations. After its isolation from acid mine drainage in the late 1940s, this microorganism was considered the principal agent in bioleaching, because the available enrichment/isolation methods repeatedly selected it from many different acidic environments (Ehrlich, 2002). However, recent molecular ecology studies have challenged its importance, because instead of *At. ferrooxidans*, most bioleaching operations show high numbers of other iron-oxidizing microorganisms such as *Leptospirillum* spp., *Ferroplasma* spp. and *Sulfobacillus* spp., questioning the observations obtained through enrichment protocols (Edwards et al., 2000; Golyshina et al., 2000; Rawling, 1997; Sand et al., 1992; Schrenk et al., 1998).

In this section, we will describe the molecular ecology methodologies currently applied to the characterization of acidic environments, which have been tested in the Tinto River model system (González-Toril et al., 2003).

Ribosomal RNAs

In recent years the use of ribosomal RNAs (rRNAs) and their genes have produced an authentic revolution in microbial ecology (Akkermans et al., 1994). The sequencing of these genes has allowed a whole range of microorganisms, mainly prokaryotes, to be studied without running into selective enrichment and isolation problems. Most molecular ecology techniques are based on these genes (Wintzingerode et al., 1997).

A Flow Chart for the Study of an Acidic Extreme Environment

Molecular ecology techniques have some biases and limitations, so it is strongly advised to use different complementary techniques to ensure accurate interpretation of the models.

The first technique to be applied to the study of a given sample is Fluorescent *in situ* Hybridization (FISH). This technique can identify and quantify the different microorganisms existing in the sample. As a first approximation, universal probes are recommended to obtain information about the number of Bacteria and Archaea in the sample. At this level more specific bacterial or archaeal probes are not very useful, because, in general, we do not yet know what kind of microorganisms exist and in what numbers. From these data we can make a gross estimate of the biomass present in the sample, and decide whether or not to extract DNA and/or RNA for 16S rRNA gene amplification, resolve the amplified genes and eventually sequence them. These techniques do not provide reliable quantification data, but they are very effective for microbial identification. Denaturing Gradient Gel Electrophoresis (DGGE) allows fast identification of microorganisms and it is very useful for the study of the spatial distribution and temporal evolution of microbial communities.

DGGE has, however, the limitation that only short sequences of around 600 bp can be obtained, which is sufficient for identification but not for phylogenetic studies. A third complementary technique to be used is the cloning of the 16S rRNA genes of the microorganisms present in the sample. Sequencing a representative number of clones, the population of the sample is defined and phylogenetic studies can then be performed. With the 16S rRNA gene sequences obtained from the sample, we can use DGGE or cloning to identify the microorganisms present in the sample. These new sequences are then used to design specific probes for *in situ* hybridization and quantification of the different genera and species present in the sample.

Denaturing Gradient Gel Electrophoresis (DGGE)

DGGE is a fingerprinting technique, whereby DNA fragments of similar length but with different base-pair sequences can be resolved (Muyzer et al., 1996). Separation by DGGE is based on the electrophoretic mobility of a partially melted double stranded DNA molecule in polyacrylamide gels containing a linear gradient of DNA denaturants (urea and formamide), with less mobility than the completely helical form of the molecule. The DNA molecules run through the gel until denaturation occurs as a consequence of exposure to the denaturing gradient. Denaturation of double stranded DNA is dependent on its melting temperature (Tm), and the Tm is a consequence of the sequence, so the retention of DNA molecules in the gel is determined by its sequence. Thus DNA fragments of the same length and with different sequences can be separated (Muyzer et al., 1996). The technique was designed to locate single base mutations in genes, and it has been very useful in molecular pathology (Myers et al., 1987). In 1993, Muyzer et al. applied this technique to the study of an ecosystem for the first time (Muyzer et al., 1993).

Currently, the most useful gene for prokaryotic identification is the 16S rRNA gene. From an environmental sample we can carry out a DNA and/or RNA extraction followed by amplification of the 16S rRNA genes through the Polymerase Chain Reaction (PCR), using universal primers for Bacteria or Archaea domains. DGGE is useful to evaluate the number of 16S rRNA genes corresponding to different microorganisms present in the sample. Spatial or temporal variations of microbial populations can be studied using these fingerprints. Furthermore, each band corresponding to a different class of 16S rRNA genes can be excised and the corresponding DNA sequenced. Sequences can be compared with a sequence database and the closest microorganisms identified. In addition, a preliminary phylogenetic study with the retrieved sequences could be performed (for instance using parsimony), but we must bear in mind that the sequences are too short for a real phylogenetic study. The newly generated sequences can also be used to design specific hybridization probes for the microorganisms present in the sample.

Using this approach, around 50% of the sequences can be detected in DNA fragments of up to 500 bp. This percentage can be increased to nearly 100% by the attachment of GC-rich sequences to the DNA fragment, which will then act as a high temperature melting domain (Sheffield et al., 1989) (Table 20.2).

A standard protocol for DGGE

Molecular ecology techniques that have been applied to the study of extreme acidic environments have the same basis as those applied to the characterization of other habitats, so the protocols are adaptations to the peculiar physico-chemistry of these ecosystems. When appropriate, we will give some tips for the most commonly encountered problems.

Sampling

It is extremely critical to obtain the samples correctly for a molecular ecology study. In the case of DGGE and cloning, the sample must be immediately frozen at $-20°C$ until its use. Freezing ensures the proper conservation of DNA, and especially RNA which is extremely sensitive to RNAses. The most common acidic environments where samples are encountered are:

1. Water. It could be collected directly, but in most cases the number of cells in suspension is very low (especially in extreme conditions), so filtration is a good procedure to increase the sample biomass. In the case of Tinto River water, samples of at least 1 l volume should be collected and filtered through a Millex-GS Millipore filter (0.22 µm, 50 mm diameter). After filtration, which can be done in the field, filters are stored at $-20°C$ until further processing. Depending on the biomass present in the water column, different volumes of water should be filtered.
2. Sediment. Samples are kept at $-20°C$. The size of the samples depends on the biomass present in the sediment.
3. Biofilms. These are kept at $-20°C$ without any additional handling.

Possible problems with extreme acidic samples:

DNA degradation. One possible cause might be due to cellular lysis during freezing and thawing and exposure of the sensitive nucleic acids to hydrolytic conditions due to the low pH of the samples and the high concentration of heavy metals. The best solution to overcome this problem is to extract DNA immediately after sampling. When this is not possible, the samples should be washed extensively with acidic water to remove heavy metals (very common in the case of acid mine drainage or the Tinto River) and later with phosphate saline buffer (PBS: 130 mM NaCl, 10 mM sodium phosphate, pH 7.2). After the samples are neutralized they can be stored at $-20°C$.

DNA extraction

In the case of methodologies based on the amplification of 16S rRNA genes (as DGGE and cloning), the main problems are related to DNA extraction and amplification. When a good PCR product is obtained,

Table 20.2 Primers used for 16S rRNA gene amplification in DGGE

Primer[a]	Target site[b]	Sequence (5′ to 3′)	Specificity	Reference
341F-GC[c]	341–357	CCT ACG GGA GGC AGC AG	Most Bacteria	Muyzer et al. (1996)
907RM	907–926	CCG TCA ATT CMT TTG AGT TT	Most known organisms	Muyzer et al. (1996)
ARC344F-GC	344–363	ACG GGG YGC AGC AGG CGC GA	Most Archaea	Raskin et al. (1994)
ARC915R	915–934	GTG CTC CCC CGC CAA TTC CT	Most Archaea	Stahl and Amann (1991)

[a]F (forward) and R (reverse) indicate the orientation of the primers in relation to the rRNA.
[b]*Escherichia coli* numbering of Brosius et al. (1981).
[c]GC is a 40-nucleotide GC-rich sequence attached to the 5′ end of the primer. The GC sequence is 5′-CGC CCG CCG CGC CCC GCG CCC GTC CCG CCC CCG CCC-3′.

a conventional protocol can be used for the rest of the technique. There are numerous protocols described for DNA extraction. In principle any one could be used for an acidic extreme sample, but there are some considerations to be made before starting the extraction:

1. Sediments and biofilm samples can be treated directly; water samples should be filtered to avoid loss of cells.
2. Heavy metals have to be removed. Samples from acidic environments are rich in heavy metals. Although in principle these are not a problem for the DNA extraction, but their presence should be avoided because metals can generate random breaks in nucleic acids, causing serious problems in the amplification step. Removal of heavy metals can be easily achieved with extensive washes with acidic water (pH 1–2). Iron is a good indicator of the efficiency of metal removal by acidic water. When the brownish-reddish color disappears from the solution one can consider that most heavy metals are gone.
3. After washing, the sample must be neutralized with PBS.

There are different commercial kits for DNA extraction that work quite well after the acidic sample has been neutralized, for example FastDNA Spin Kit for Soil BIO 101 (Q-BIOgene). On the other hand, conventional DNA extraction methods have some advantages as bigger samples can be treated at a lower cost:

1. Each neutralized Millex-GS Millipore filter is cut out and each piece treated with 2 ml of SET buffer (25% sucrose, 50 mM Tris·HCl pH 8, 2 mM EDTA) overnight at $-20°C$ in a 15-ml Falcon tube. Sediments or biofilms can be treated directly with SET buffer.
2. After thawing the sample, 120 µl of SDS (25%) and 2.8 U of pronase (Boehringer) are added, and the mixture incubated at 4°C for 30 min.
3. Protein precipitation with phenol–chloroform–isoamyl alcohol (25:24:1) and 0.5 ml of sodium acetate (2 M, pH 5.2) is repeated until complete elimination of protein in the interface.
4. Nucleic acids are precipitated overnight by addition of 2.5 volumes of ice-cold ethanol (96%) at $-20°C$.
5. Nucleic acids are precipitated by centrifugation, washed with ethanol 70%, vacuum dried and resuspended in 10 mM Tris·HCl pH 8.5.

After DNA extraction it is a good idea to clean the DNA preparation to facilitate the amplification and the conservation of the DNA sample. For DNA cleaning there are several commercial columns available. Also dialysis membranes can be used. Clean DNA can be stored at 4°C during the time that it is routinely used and at $-20°C$ for long-term storage.

Polymerase Chain Reaction (PCR)

Muyzer *et al.* proposed a standard PCR method for DGGE in the Molecular Microbiology Ecology Manual (1996), which has been adapted for DNA extracted from acidic environments. If the DNA preparation obtained is clean and the concentration is sufficient, this reaction normally works well with this type of DNA. The primers used for the amplification depend on the microorganisms that we want to identify. For a preliminary study it is convenient to use universal primers for the bacterial or archaeal domains (Table 20.2). Forward primers have a 40-nucleotide GC-rich sequence attached to the 5′ end of the primer. The GC sequence is 5′-CGC CCG CCG CGC CCC GCG CCC GTC CCG CCG CCC CCG CCC-3′. The annealing temperature is set 10°C above the expected annealing temperature and lowered 1°C every second cycle until 55°C, at which temperature nine additional cycles are carried out. This procedure, called "Touchdown" PCR, reduces the formation of spurious products during the amplification process:

1. Add to a PCR tube:

   - 10x PCR buffer   5 µl
   - dNTP mix (2.5 mM)   5 µl
   - BSA (3 mg ml^{-1})   5 µl
   - 50 pmol forward primer   0.5 µl
   - 50 pmol reverse primer   0.5 µl
   - target DNA   1 µl ($\approx$100 ng)
   - Taq DNA polymerase (5 U µl^{-1})   0.5 µl
   - Milli-Q water   to a final volume of 50 µl

2. PCR reaction:

   - 1 cycle

     95°C   5 min

   - 19 cycles

     95°C   1 min
     64°C   1 min Decrease 1°C every second cycle ("Touchdown")
     72°C   3 min

   - 9 cycles

     95°C   1 min
     55°C   1 min
     72°C   3 min

   - 1 cycle

     95°C   1 min
     55°C   1 min
     72°C   10 min
     4°C   constant

3. Analyze 2 µl of the amplified products by conventional agarose gel electrophoresis to determine the size and concentration of the PCR amplification products.

Denaturing gradient gel electrophoresis

The method described is an adaptation of the standard protocol described by Muyzer *et al.* in the Molecular Microbiology Ecology Manual (1996). First, DGGE equipment is necessary. Several companies, such as CBS Scientific Co., INGENY or Bio-Rad can provide the required equipment. For the gradient gel formation a gradient former and a magnetic stirrer is required. The use of a peristaltic pump is recommended to cast reproducible gradient gels. The gradient interval can be changed depending on the needs. A useful procedure to prepare DGGE plates for a DCode System of Bio-Rad is the following:

1. Clean plates, spacers (1.5-mm thick) and comb with soap and ethanol. Assemble the gel sandwich, and attach the exit tube from the peristaltic pump between the glass plates.
2. Fill the aquarium with 0.5x TAE buffer (20x TAE: 800 mM Tris · HCl pH 7.8, 20 mM EDTA, 400 mm sodium acetate pH 7.4) and turn the heater on. Heat the buffer to 60°C.
3. Add the following reagents to a 50 ml tube.

    - Solution A:

6% Acrylamide/0% Urea-Formamide	16 ml
N,N,N',N'-Tetramethylethylenediamine (TEMED)	10 µl
10% (w/v) ammonium persulfate (APS)	50 µl

    - Solution B:

6% Acrylamide/80% Urea-Formamide	16 ml
TEMED	10 µl
APS	50 µl

    Acrylamide/Urea-Formamide solution:

    6% Acrylamide/0% Urea-Formamide solution:

    - acrylamide   30 g
    - bisacrylamide   0.81 g
    - 20x TAE   12.5 ml
    - Milli-Q water until 500 ml. Store at 4°C in the dark.

    6% Acrylamide/80% Urea-Formamide solution:

    - acrylamide   30 g
    - bisacrylamide   0.81 g
    - 20x TAE   12.5 ml
    - Urea   168 g
    - Formamide deionized 160 ml
      with AG501-X8 mixed
      bed resin (Bio-Rad)
    - Milli-Q water until 500 ml. Store at 4°C in the dark.

4. Pour solution B into the outflow chamber of the gradient former, and solution A into the other chamber.

5. Apply stirrer to the outflow chamber of the gradient former, and turn the magnetic stirrer on.
6. Open the inside valve of the gradient former, then the outside valve, and turn the peristaltic pump on.
7. Add the following regents to a 50 ml tube:

6% Acrylamide/0% Urea-Formamide	10 ml
TEMED	10 µl
APS	25 µl

8. Close the inside valve of the gradient former. Pour the solution into the outflow chamber of the gradient former, place the comb between the glass plates, and turn the peristaltic pump off.
9. Let the gel polymerize for at least 2 h. Remove the comb and rinse the slots with 0.5× TAE buffer to remove non-polymerized acrylamide. Attach the gel sandwich to the central core and immerse the core into the preheated aquarium. Attach the recirculating tube to the upper reservoir of the electrophoresis unit.
10. Mix the sample with a loading buffer (1:5) and load it onto the gel.
11. Electrophoresis is performed at constant voltage of 200 V for 6 h or 100 V during 12 h.
12. Incubate the gel after electrophoresis for 15 min in Milli-Q water containing ethidium bromide (0.5 µg ml^{-1}), rinse it for 10 min in Milli-Q water. Take a photograph of the gel using a UV transilluminator.

Following this procedure a fingerprint of the 16S rRNAs for different samples can be obtained. Figure 20.1 shows the spatial variation of bacterial 16S rRNA for samples obtained from different locations along the Tinto River.

DNA extraction from DGGE bands

The extraction of the DGGE bands allows more precise information about the type of microorganisms existing in the water column of the selected sampling sites to be obtained. Each band is excised with the help of a UV transilluminator and placed in an Eppendorf tube. Each band is incubated overnight with Milli-Q water at 37°C. Sometimes this elution is not enough for the biggest fragments, in which case other procedures can be applied, e.g. using 2-mercaptoethanol or dithiothreitol (Muyzer et al., 1996).

Sequencing of DGGE bands

The amount of DNA eluted from the DGGE band is not enough for sequencing. So a new PCR amplification for each band is necessary. The PCR conditions are the same as those described above. After cleaning the PCR products, sequencing can be performed using the same primers that were used for amplification.

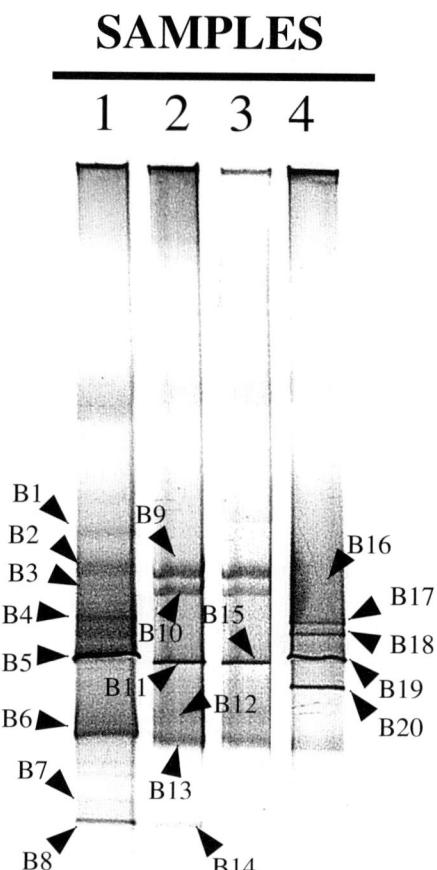

Figure 20.1. DGGE fingerprint of 16S rRNA genes present in different sampling sites of the Tinto River, using specific probes for Bacteria. Lane numbers correspond to sampling sites. Arrows indicate different bands.

Working with the sequences

As mentioned before, the sequences obtained with this methodology are short (up to 600 bp), so they are not appropriate for a real phylogenetic study (Stackebrandt and Rainey, 1998). On the other hand, they are very useful for comparing the retrieved sequences with a sequence database to identify the closest microorganisms or deposited orphan sequences. For example, partial 16S rRNA gene sequences from the bands resolved by DGGE can be compared with reference sequences contained in the EMBL Nucleotide Sequences Database using the BLAST program. In addition to obtaining information about 16S rRNA gene sequence similarities (identification), sequences could be subsequently aligned with 16S rRNA reference sequences in the ARB package (http://www.arb-home.de) (Ludwig et al., 2004) and inserted within a stable tree by using the ARB parsimony tool (Ludwig et al., 1998). Figure 20.2 shows the results obtained for retrieved samples

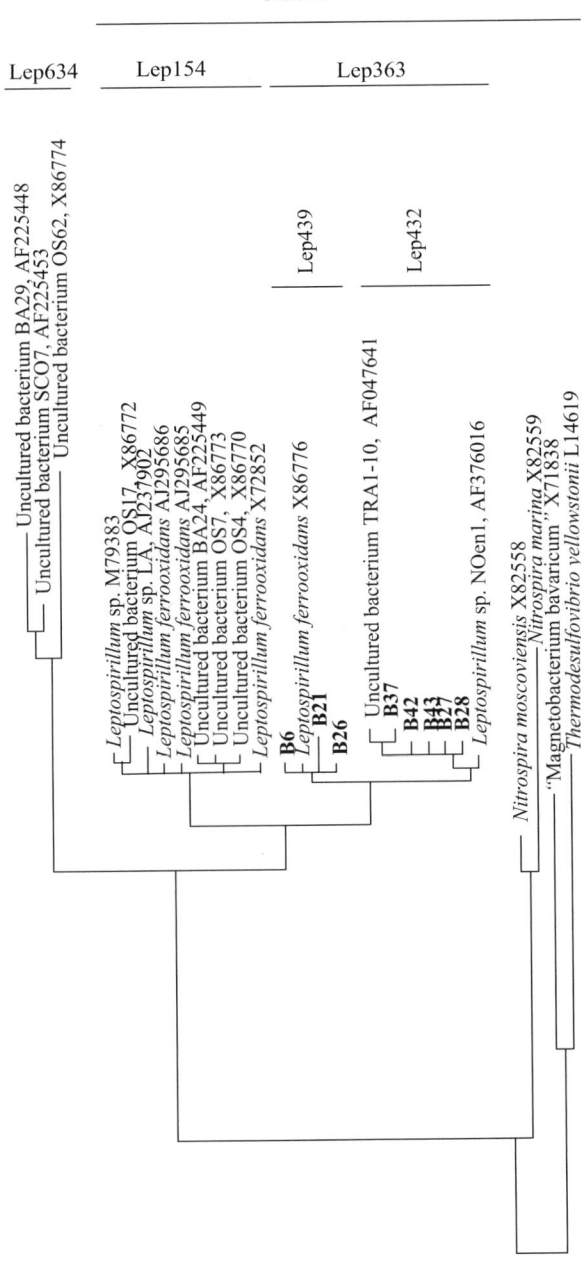

Figure 20.2. Phylogenetic affiliation of DGGE sequences with strong similarity to *Leptospirillum* spp. The phylogenetic tree was constructed as specified in the text. *Leptospirillum* spp. group in three clusters: *a*, *b* (*L. ferriphilum*) and *c* (*L. ferrooxidans*). Sequences from group *c* are distributed in two subgroups, named c_1 and c_2. All the DGGE sequences retrieved from the Tinto River correspond to *L. ferrooxidans* (group *c*). The specificity of the probes (LEP634, LEP154, LEP636, LEP439 and LEP432) designed to identify members of the different leptospirilli groups are shown. Note that probe NTR712 (described as a specific probe for members of the *Nitrospira* phylum) does not detect group *a* of leptospirilli, which have not been isolated in any acidic environment so far.

Acidophilic microorganisms

from the Tinto River corresponding to the genus *Leptospirillum* analyzed by DGGE.

Probe design

With DGGE we obtain information about the 16S rRNA gene sequences from the microorganisms present in the sample. Using these sequences specific hybridization probes for these microorganisms can be designed (Figure 20.2). An oligonucleotide probe can be designed using the PROBE_FUNCTION tool of the ARB package (Ludwig et al., 2004). The specificity of the newly designed probes should be evaluated with the PROBE_MATCH tool of the ARB package and by hybridization with references cultures (Amann, 1995).

Cloning of 16S rRNA Genes

For phylogenetic studies complete 16S rRNA gene sequences are needed. One way to generate these sequences is to clone the 16S rRNA genes present in the sample (Giovannoni et al., 1990; Ward et al., 1990). A complete 16S rRNA gene amplification can be obtained using universal primers for Bacteria or Archaea. After the amplification reaction we have a mixture of 16S rRNA genes from all Bacteria or Archaea present in the sample. To further analyze the sample a separation of the different genes is necessary. This can easily be achieved by cloning, inserting each amplified gene in a plasmid, and then into competent cells to facilitate their transformation. After plasmid extraction the inserted rRNA sequence can be sequenced (Giovannoni et al., 1990; Ward et al., 1990).

A standard protocol

The same considerations for samples obtained from an acidic ecosystem stated above, apply to this section. The main problem is to obtain intact DNA to facilitate the amplification reactions.

Sampling, DNA extraction and Polymerase Chain Reaction (PCR)

For sampling and DNA extraction, the same considerations given above for DGGE analysis apply. For PCR, primers to amplify the complete 16S rRNA genes (around 1492 bp) should be used. Depending on the origin of the sample it is sometimes difficult to obtain the complete 16S rRNA gene. Examples of universal primers for the archaeal and the bacterial domains are shown in Table 20.3.

1. Add to a PCR tube:
   a. 10×PCR buffer   5 µl
   b. dNTP mix (2.5 mM)   5 µl
   c. BSA (3 mg ml^{-1})   5 µl
   d. 50 pmol forward primer   0.5 µl

Table 20.3 Primers used for PCR amplification of 16S rRNA genes of the Bacteria and Archaea domains

Primer[a]	Target site	Sequence (5' to 3')	Specificity	Reference
8F	8–23 16S rRNA gene[b]	AGA GTT TGA TCM TGG C	Bacteria domain	Muyzer et al. (1995)
1507R	1492–1507 16S rRNA gene[b]	AAAC CTT GTT ACG ACT T	Bacteria domain	Muyzer et al. (1995)
25F	9–25 16S rRNA gene[b]	CYG GTT GAT CCT GCC RG	Archaea domain	Achenbach and Woese (1995)
1492R	1492–1513 16S rRNA gene[b]	TAC GGY TAC CTT GTT ACG ACT T	Archaea domain	Achenbach and Woese (1995)

[a]F (forward) and R (reverse) indicate the orientation of the primers in relation to the rRNA.
[b]*Escherichia coli* numbering of Brosius et al. (1981).

e. 50 pmol reverse primer   0.5 μl
   f. target DNA   1 μl (≈100 ng)
   g. Taq DNA polymerase (5 U μl^{-1})   0.5 μl
   h. Milli-Q water   to a final volume of 50 μl

2. PCR reaction:

   - I cycle

     95°C   5 min

   - 35 cycles

     95°C   1 min
     46°C (for Bacteria)
     52°C (for Archaea)   1 min
     72°C   3 min

   - I cycle

     95°C   1 min
     55°C   1 min
     72°C   10 min
     4°C   constant

3. Analyze 2 μl of the amplified products by conventional agarose gel electrophoresis to determine the size and concentration of the PCR product.

16S rRNA gene cloning

There are many different commercial kits for cloning genes of this size, such as pGEM-T and pGEM-T Easy Vector System (Promega) and TOPO-TA Cloning Kit (Invitrogen). The result is a clone library of the bacterial or archeal 16S rRNA genes from the extreme acidic sample.

Plasmid extraction (minipreps)

The extraction of the plasmid can be carried out using a traditional protocol (Sambrook and Russel, 2001) or a commercial kit, such as the Wizard Plus SV Minipreps DNA Purification System (Promega).

Screening of the clones

Before sequencing, a selection of the clones should be performed. In the case of acid mine drainage or the Tinto River the prokaryotic diversity is not very high, so the possibility of having more than one clone with the same 16S rRNA genes is high. One possibility is to screen the clone library by restriction fragment length polymorphism analysis (Nogales, 2001). The result of the restriction enzymatic digestion is tested with conventional electrophoresis. Plasmids with the same digestion patterns correspond to the same gene (Chandler et al., 1997). In this way the number of clones for sequencing is reduced. It is important to know the plasmid sequence in order to select which restriction enzymes to use.

Often the companies that sell the plasmids give indications on the most suitable restriction enzymes.

Sequencing

16S rRNA genes of selected clones can be sequenced using the recommended primers in the kit. The use of internal primers is convenient. To avoid ambiguities the sequence of both complementary strands is also advisable.

Working with the sequences

With longer sequences, a real phylogenetic study of the 16S rRNA genes is possible (Stackebrandt and Rainey, 1998). Sequences of 16S rRNA gene clones can be initially compared with reference sequences contained in the EMBL Nucleotide Sequences Database using the BLAST program to determine the affiliation of the most similar microorganism. In addition, a useful percentage of similarity will be obtained.

After the sequence generation and the identification of the different clones a more in-depth phylogenetic study can be undertaken. There are several programs especially designed for sequence analysis and the generation of phylogenetic trees. It is advisable to use all the available alternatives that are suited to the type of analysis that you might like to perform, in order to generate consensus trees. One possibility is the ARB package (Ludwig et al., 2004). Sequences can be aligned and then added to a general phylogenetic tree by parsimony (Ludwig et al., 1998). This tree is very useful because it contains all the sequences in the ARB database. Other trees can be generated using other methods and filters, and a final consensus phylogenetic tree can be designed. Some examples of the more useful algorithms are: parsimony, neighbor-joining and maximum likelihood. The position of a cloned *Acidiphilium* (ECT) from the Tinto River in a consensus tree obtained using the ARB program is shown in Figure 20.3.

Probe design

As with DGGE sequences, cloned 16S rRNA gene sequences can be used to design specific probes for the identification and quantification of the different microorganisms. Oligonucleotide probes can be designed using the ARB package, as described in the previous section (Ludwig et al., 2004).

Fluorescent *In situ* Hybridization (FISH)

Using DGGE and cloning we can obtain useful qualitative information about the diversity present in the samples, but these methodologies do not provide reliable quantitative data, key information for microbial ecology. To quantify the level of diversity detected by DGGE and cloning, fluorescent *in situ* hybridization (FISH) is a very powerful technique.

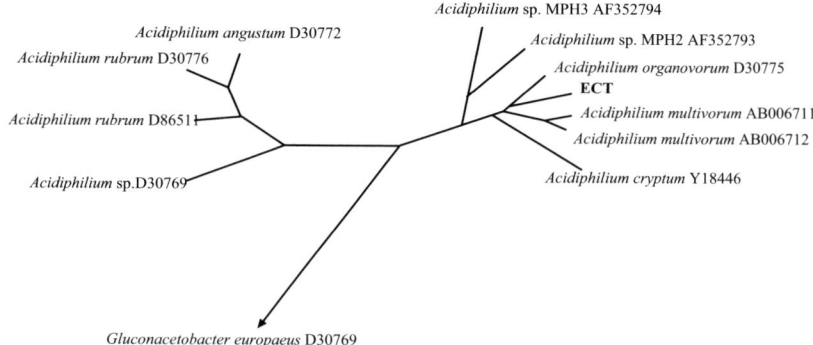

Figure 20.3. Phylogenetic tree showing the affiliation of 16S rRNA gene cloned sequence ECT retrieved from a Tinto River sample with sequences of members of the *Acidiphilium* genus. Phylogenetic trees were obtained using parsimony, neighbor-joining and maximum likelihood analysis with different sets of filters. A consensus tree was designed with all generated trees. *Gluconacetobacter europaeus* was used as outgroup.

FISH is a fast method for specific identification and quantification of intact cells in their natural environment. In this case, a fluorescent-labeled specific rRNA targeted oligonucleotide probe is used for the direct identification, independent of culture, and quantification of the microorganism (Amann et al., 2001). The use of *in situ* hybridization for counting and identifying organisms was proposed by Olsen et al. (1986), twenty years ago. The first assays were performed using radioactively labeled oligonucleotides (Giovannoni et al., 1988), and later by using fluorescent probes, which yield superb spatial resolution and can be detected very simply by using epifluorescent microscopy and fluorescent *in situ* hybridization (FISH) (Amann et al., 1990; DeLong et al., 1989).

First, samples must be fixed immediately after collection. Second, microorganisms are made permeable to oligonucleotide probes by fixation with aldehydes (formalin, paraformaldehyde, glutaraldehyde) or alcohols (methanol, ethanol) (Amann, 1995). When a fluorescent probe (labeled with a fluorescent dye, e.g. Cy3 or fluorescein) is placed in contact with permeable cells in fixed physico-chemical conditions, the fluorescent probe hybridizes with the specific intracellular target site in the ribosomes (Amann et al., 1995). Later a universal fluorochrome, like DAPI is used to evaluate the total number of viable cells in the sample. With epifluorescence microscopy the hybridized cells can be observed and counted, and after changing the filter, the total number of viable cells is counted.

A standard protocol

Specificity of probes

Initially, FISH is performed using available universal probes, even before starting DGGE and/or cloning. After the evaluation of the prokaryotic diversity using DGGE and/or cloned sequences, their sequences can be used to design new probes with different degrees of specificity.

Probes for FISH are 18 (±3)mer oligonucleotides complementary to specific zones of the 16S rRNA (Amann, 1995). The sequence of a probe can be obtained by a computer-assisted analysis using appropriate software, e.g. the PROBE_FUNCTION tool of the ARB package (Ludwig et al., 2004). The specificity of newly designed probes must be evaluated by comparison with sequences from a good database using the PROBE_MATCH tool of the ARB package. From time to time the comparison should be repeated after the incorporation of new sequences into the data bank. Furthermore, probe specificity should be checked by hybridization with reference systems with the target sequence (positive control) and with microorganisms from collections with sequences having some mismatch with the probe (negative control).

Once the probes have been designed they are ordered from an oligonucleotide synthesizing company such as Interactiva or Bonsai Technologies. The 18mer oligonucleotide is labeled with a fluorescent dye at the 5' end. The most commonly used fluorochromes for FISH are indocarbocyanine Cy3 (Cy3) and (6)-carboxyfluorescein-N-hydroxy-succinimide ester (fluorescein).

Some of the probes designed and used in the characterization of the microbial ecology of an extremely acid environment are shown in Table 20.4.

Fixation of cells

The protocols for hybridization in acidic conditions are the standard ones with few modifications. When working with acidic samples good fixation is extremely important. The protocols vary according to the type of sample:

- Water
    1. The sample must be fixed immediately. Paraformaldehyde dissolved in PBS (final concentration of fixed sample 4% v/v) is the compound more extensively used for fixation (Amann et al., 1995). However, in the case of water samples from acidic environments paraformaldehyde does not work properly. Heavy metals from the sample precipitate with the phosphate from the PBS buffer and interfere with the hybridization. The best results have been obtained using 37% formaldehyde. Formaldehyde is added to the water sample up to a final concentration of 4% (v/v). The amount of sample to be used will depend on the biomass. Normally with a cell density of 10^5 cells ml^{-1}, 1 ml of water sample is enough. Keep the reaction at 4°C for 2–4 h (never more than 14 h).
    2. Filter the fixed samples through a GTTP 025 Millipore filter (0.22 µm) using a filtration column.
    3. Without removing the filter from the filtration column, wash it with 10 ml of acidic water (pH 1.5) to remove excess formaldehyde and heavy metals. Wash again with 10 ml of PBS (pH 7.2) to neutralize the sample.
    4. Filters can be stored at −20°C prior to hybridization.

Table 20.4 Fluorescent-labeled oligonucleotide probes used for *in situ* hybridization experiments

Probe	Target	Sequence (5' to 3')	Specificity	Reference
EUB338	16S	GCT GCC TCC CGT AGG AGT	Bacteria domain	Amann et al. (1990)
EUB338-II	16S	GCA GCC ACC CGT AGG TGT	Planctomyces	Daims et al. (1999)
EUB338-III	16S	GCT GCC ACC CGT AGG TGT	Verrucomicrobiae (and others)	Daims et al. (1999)
ALF968	16S	GGT AAG GTT CTG CGC GTT	α-Proteobacteria	Neef, (1997)
ACD638	16S	CTC AAG ACA ACA CGT CTC	Acidiphilium spp.	González-Toril et al. (2003)
BET42a	23S	GCC TTC CCA CTT CGT TT	β-Proteobacteria	Manz et al. (1992)
GAM42a	23S	GCC TTC CCA CAT CGT TT	γ-Proteobacteria	Manz et al. (1992)
THIO1	16S	GCG CTT TCT GGG GTC TGC	Acidithiobacillus spp.	Stoffels (unpublished)
ACT465a	16S	GTC AAC AGC AGC TCG TAT	Group a At. ferrooxidans	González-Toril et al. (2003)
ACT465b	16S	GTC AAC AGC AGA TCG TAT	Group b At. ferrooxidans	González-Toril et al. (2003)
ACT465c	16S	GTC AAC AGC AGA TTG TAT	Group c At. ferrooxidans	González-Toril et al. (2003)
ACT465d	16S	GTC AAT AGC AGA TTG TAT	Group d At. ferrooxidans	González-Toril et al. (2003)
ATT985	16S	CCA GAC ATG TCA AGC CCA	At. thiooxidans, At. albertensis	González-Toril (unpublished)
THC642	16S	CAT ACT CCA GTC AGC CCG T	At. caldus	Edwards et al. (2000)
NTR712[a]	16S	CGC CTT CGC CAC CGG CCT TCC	Nitrospira group	Daims et al. (2001)
LEP154	16S	TTG CCC CCC CTT TCG GAG	Leptospirillum ferriphilum	González-Toril et al. (2003)
LEP634	16S	AGT CTC CCA GTC TCC TTG	Group a Leptospirillum spp.	González-Toril et al. (2003)
LEP636	16S	CCA GCC TGC CAG TCT CTT	Leptospirillum ferrooxidans	González-Toril et al. (2003)
LEP439	16S	CCT TTT TCG TCC CGT GCA	Group c1 Leptospirillum ferrooxidans	González-Toril et al. (2003)
LEP432	16S	CGT CCC GAG TAA AAG TGG	Group c2 Leptospirillum ferrooxidans	González-Toril et al. (2003)
SRB385	16S	CGG CGT CGC TGC GTC AGG	δ-Proteobacteria	Amann et al. (1990)
DSS658	16S	TCC ACT TCC CTC TCC CAT	Desulfosarcina, Desulfococcus, Desulfofrigus	Manz et al. (1998)
DSV698	16S	GTT CCT CCA GAT ATC TAC GG	Desulfovibrio spp.	Manz et al. (1998)
DSP648	16S	CTC TCC TGT CCT CAA GAT	Desulfosporosinus spp., Desulfitobacterium spp., Dehalobacter spp.	González-Toril (unpublished)
SUL228	16S	TAA TGG GCC GCG AGC TCC C	Sulfobacillus spp.	Hallberg and Johnson (2001)
ACM1160	16S	CCT CCG AAT TAA CTC CGG	Acidimicrobium spp.	González-Toril et al. (2003)
ARCH915	16S	GTG CTC CCC CGC CAA TTC CT	Archaea domain	Stahl and Amann (1991)
FER656	16S	CGT TTA ACC TCA CCC GAT C	Ferroplasma spp.	Edwards et al. (2000)
TMP654	16S	TTC AAC CTC ATT TGG TCC	Thermoplasma spp., Picrophilus spp.	González-Toril et al. (2003)
NON338	–	ACT CCT ACG GGA GGC AGC	Negative control	Amann et al. (1990)

[a]The name of this probe in the original publication is S-*-Ntspa-0712-a-A-21.

- Sediments and biofilms
    1. Sediments and biofilms can be fixed directly with formaldehyde diluted up to 4% in acid water (pH 1.5). The reaction is kept at 4°C during 4–6 h (never more than 14 h).
    2. Excess of formaldehyde and heavy metals are removed with an acid water wash, followed by PBS (pH 7.2). The elimination of the liquid phases from sediments can be done by centrifugation at low speed. In the case of biofilms it can be done by decantation. In the case of biofilms the wash should be repeated several times to make sure all the formaldehyde is removed.
    3. Samples can be stored in ethanol: PBS (50% v/v) at −20° C prior to hybridization.

Hybridization

Here we offer some protocols, with some differences depending on the type of sample, all of them based on Amann's original protocols (Amann, 1995), and adapted to the peculiarities of the environment from which the samples originated:

- Water
    1. Prepare 2 ml of hybridization buffer:

        0.9 M sodium chloride
        0.01% sodium dodecylsulfate
        20 mM Tris · HCl
        35% formamide
        Final pH 7.2

    The formamide percentage to be used may vary, depending on the probe (range between 0 and 60%).

    2. Cut a section of the GTTP 025 Millipore filter in which the sample has been fixed and place it on a clean microscope slide with the sample side up. The rest of the filter can be stored at −20°C for future hybridizations.
    3. Mix 8 μl of hybridization buffer with 1 μl of fluorescent probe (50 μM).
    4. Add the buffer/probe mix on the filter with the fixed sample.
    5. Place a slip of Whatman 3MM paper in a 50-ml tube and soak it with the rest of hybridization buffer (2 ml approximately).
    6. Place the slide with the filter and the buffer–probe mix into the tube and close the tube. Hybridize in horizontal position for 2 h at 46°C.
    7. In a 50-ml tube prepare 50 ml of washing buffer. Hybridization buffer without formamide can be used for this purpose. Preheat the buffer at 48°C.
    8. After hybridization (2 h) remove the slide under a hood and transfer the filter to the preheating washing buffer in the 50-ml tube.

9. Incubate for 20 min at 48°C.
10. Remove salts with a short wash of the filter with Milli-Q water and air dry.
11. Stain the cell with contrast fluorochrome: put a drop of 4′-6′diamidino-2 phenylindol (DAPI) (final concentration 1 µg ml^{-1}) over the filter and incubate for 30 s. Wash the filter briefly with Milli-Q water, followed by a wash in 80% ethanol. Air dry.
12. Place the filter with the fixed and hybridized cells on a new clean slide. Add a drop of a glycerol/PBS with a pH > 8.5 (e.g. Citifluor, Citifluor Ltd.). Cover with a cover slip.
13. View with an epifluorescence microscope equipped with suitable filters for probe fluorochrome and DAPI.

- Sediments

    1. Spread approximately 5 µl of the fixed sediments in a multiwell slide. Agarose can be used to inmobilize the sample (Llobet-Brossa *et al.*, 1998). Depending on the biomass, the sediment can be diluted. Sometimes sonication is convenient (Llobet-Brossa *et al.*, 1998).
    2. Air dry.
    3. Dehydrate the cells by successively washing with 50, 80 and 98% ethanol washes (3 min each). Slides can be stored dry at room temperature indefinitely.
    4. Hybridization is done the same way as described for a water sample (see above).

- Biofilm

    1. Depending on the characteristics of the biofilm, it could be treated in the same way as the sediments or it could be cut using a microtome.
    2. The immobilization could be done with agarose. From this point follow the sediment hybridization protocol.

An example of FISH from an acidic extreme sample is shown in Figure 20.4.

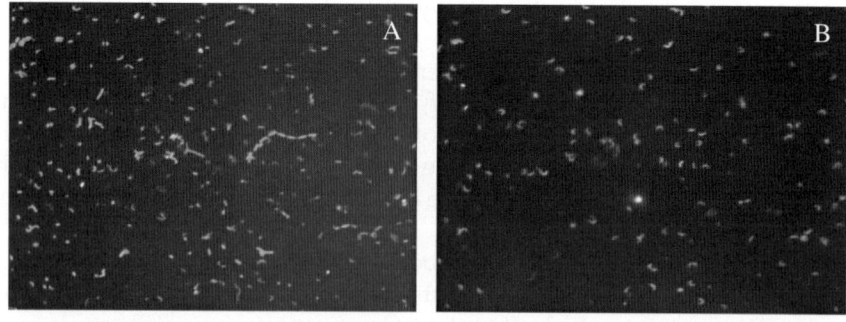

Figure 20.4. Epifluorescence micrographs of bacteria from the Tinto River. Sample obtained from the origin of the river (González-Toril *et al.*, 2003). (A) DAPI-stained cells; (B) same field showing hybridized cells with a specific probe for *L. ferrooxidans*. Probe: LEP636 (Cy3 labeled). See Plate 20.4 in Colour Plate Section.

Difficulties may be encountered when applying fluorescent *in situ* hybridization to environmental samples from highly eutrophic systems (Pernthaler *et al.*, 2002). Most bacteria in aquatic habitats are small, slow growing, or starving, and the intensity of the signal of hybridized cells is frequently below the detection limits or lost in high background fluorescence. This problem is common with sediments too. On the other hand, in acid mine drainage and in the Tinto River most of the microorganisms are strict chemolithotrophs, and they are usually associated with minerals and rocks. The study of the microorganism attached to solid substrates is very interesting, especially for biomining research, but in most of the cases high background fluorescence and low signal from the hybridized cells is obtained, lowering the efficiency of this powerful technique. To solve this problem an amplification of the signal can be introduced in the procedure. Pernthaler *et al.* (2002) applied the catalyzed reporter deposition (CARD) to FISH with excellent results, obtaining detection rates in marine sediments almost one order of magnitude higher than with conventional FISH.

CARD is based on the tyramide signal amplification. It was introduced more than a decade ago for immunoblotting and inmmunosorbent assays using horseradish peroxidase and haptenized tyramines. The basis and methodology for this technique is described in Pernthaler *et al.* (2002).

In our experience with acidophilic microorganisms we have applied this technique for the detection of microorganisms in direct contact with mineral substrates in bioleaching processes and in the characterization of subsurface microorganisms. Small pieces of pyrites were incubated with *At. ferrooxidans* and *L. ferrooxidans* strains. When active growth was observed, pyrites were removed from the culture and washed with acid water (pH 1.2). Later pyrites were incubated with 4% formaldehyde for 4 h to fix the cells. Excess formaldehyde was removed as described above and the piece of pyrite was immobilized with agarose. The CARD-FISH was performed on the pyrite following the protocol of Pernthaler *et al*. The distribution of hybridized fluorescent microbes can be seen in a confocal microscope. Figure 20.5 shows a view of iron-oxidizing bacteria attached on pyrite.

◆◆◆◆◆ ACIDOPHILIC EUKARYOTES

An important contribution of eukaryotic diversity has been reported to be associated with acidic environments, mainly Acid Mine Drainage (Amaral-Zettler *et al.*, 2003; Ehrlich, 2002; López-Archilla *et al.*, 2001). Unfortunately there is very little information about the nature of the eukaryotes that are able to deal with a proton gradient of five orders of magnitude across their membranes. Most of them lack any extracellular structure that could help to control the acidity of the habitat in which they develop. In addition, they are exposed to, and so must be resistant to, extremely high concentrations of toxic heavy metals normally present in

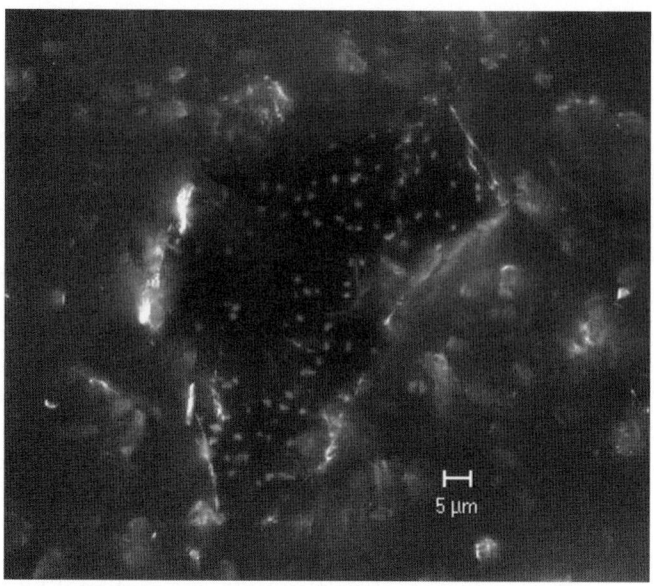

Figure 20.5. Confocal micrographs of bacteria adhered to pyrite. A massive sulfidic mineral sample from the Iberian Pyritic Belt (Tinto River) was exposed to iron-oxidizing bacterial (*L. ferrooxidans* and *At. ferrooxidans*). CARD-FISH using a universal probe for bacterial domain (EUB338 fluorescein labeled) was performed after removal of the solid sample for the incubation media. Attached bacteria, in green, can be observed on the surface of the mineral. See Plate 20.5 in Colour Plate Section.

acidic environments. In the Tinto River a high diversity of photosynthetic protists (algae) account for the greatest proportion of the biomass of the ecosystem (over 65%). An important number of filamentous fungi have been isolated from the acidic waters of the Tinto ecosystem, most of them able to grow in the extreme conditions of acidity and concentration of heavy metals of the habitat (López-Archilla et al., 2001). Also heterotrophic protists, important members of the consumers group of the food web, have been identified, mainly associated with biofilms. Most of the eukaryotic acidophilic microorganisms have been characterized by conventional phenotypic analysis (mainly morphology). It has to be pointed out, however, that except for filamentous fungi which are easily isolated using standard fungal media (López-Archilla et al., 2001), the remaining acidophilic eukaryotes thriving in the Tinto River with the exception of few algae (*Chlamydomonas* spp. and *Euglena* spp.) have not been isolated despite strenuous efforts. All the molecular techniques described in this chapter can in principle be applied to the characterization of the eukaryotic diversity of acidic environments, with the exception of FISH which is difficult to apply to the study of algae, due to the intrinsic strong fluorescence of their photosynthetic apparatus that interferes with the fluorescent signal of the probes. Obviously specific eukaryotic primers have to be used for amplification and cloning of small-subunit (18S) rRNA genes of eukaryotic microorganisms. The use of molecular ecology techniques, mainly cloning, has allowed a preliminary description of the eukaryotic diversity of the Tinto ecosystem, which basically agrees with the diversity described using conventional phenotypic techniques (Amaral-Zettler *et al.*, 2003; López-Archilla *et al.*, 2001).

List of suppliers

Boehringer Ingelheim GmbH
Corporate Headquarters, Binger Str. 173
55216 Ingelheim, Germany
Tel.: 49-6132-77-0
Fax: 49-6132-72-0
http://www.boehringer-ingelheim.com/

Pronase

Bio-Rad
1000 Alfred Nobel Drive
Hercules, CA 94547, USA
Tel.: 1-510-724-7000
Fax: 1-510-741-5817
http://www.bio-rad.com/

DGGE equipment, AG501-X8 mixed bed resin

Bonsai Technologies
Parc Technológic del Vallés
08290 Cerdanyola del Vallés
Barcelona, Spain
Tel.: +34-93-586-32-89
Fax: +34-93-580-43-96
http://www.bonsaitech.com/

Labeled oligonucleotides (probes) synthesis

CBS Scientific Co.
P.O. Box 856,
Del Mar, CA 92014, USA
Tel.: 1-858-755-4959, 1-800-243-4959
Fax: 1-858-755-0733
http://www.cbssci.com/

DGGE equipment

Citifluor Ltd.
18 Enfield Cloisters, Fanshaw Street,
London N1 6LD, UK
Tel.: +44-20-7739-6561
Fax: +44-20-7729-2936
http://www.citifluor.co.uk/

Citifluor

Ingeny International BV
Amundsenweg
174462 GP Goes, The Netherlands
Tel.: 31-113-222920
Fax: 31-113-222923
http://www.ingeny.com

DGGE equipment

Interactiva
Termo Hybaid GmbH, Sedanstraße 10,
D-89077 Ulm, Germany
Tel.: +49-731-935-79-290
Fax: +49-731-935-79-291
http://www.interactiva.de or http://www.thermohybaid.com

Labeled oligonucleotides (probes) synthesis

Invitrogen Ltd.
3 Fountain Drive
Inchinnan Business Park
Paisley PA4 9RF, UK
Tel.: 44-141-814 6100
Fax: 44-141- 814-6260
E-mail: euroinfo@invitrogen.com

TOPO-TA Cloning Kit

Millipore GmbH
Am Kronberger Hang 5
65824 Schwalbach, Germany
Tel.: +49-01805-045-645
Fax: +49-01805-045-644
http://www.millipore.com/

Gttp 025 Millipore filters, Millex-GS Millipore filters

Promega
High-Tech-Park, Schildkrötstraße 15
Mannheim D-68199, Germany
Tel.: 49-621-85010
Fax: 49-621-8501-222
http://www.promega.com

pGEM-T and pGEM-T Easy Vector System, Wizard Plus SV Minipreps DNA Purification System

Q-BIOgene
15 Morgan
Irvine, CA 92618-2005, USA
Tel.: 1-800-854-0530, 1-440-337-1200
Fax: 1-800-334-6999, 1-949-421-2675
http://www.bio101.com

Spin Kit for Soil, BIO 101

References

Achenbach, L. and Woese, C. (1995). 16S and 23S rRNA-like primers. In *Archaea. A Laboratory Manual* (K. R. Sowers and H. J. Schreier, eds), pp. 521–523. Cold Spring Harbor Laboratory Press, Cold Spring Harbor.

Akkermans, A. D. L., Mirza, M. S., Harmsen, H. J. M., Blok, H. J., Herron, P. R., Sessitsch, A. and Akkermans, W. M. (1994). Molecular ecology of microbes: A review of promises, pitfalls and true progress. *FEMS Microbiol. Rev.* **15**, 185–194.

Amann, R. I. (1995). In situ identification of microorganisms by whole cell hybridization with rRNA-targeted nucleic acid probes. In *Molecular Microbial Ecology Manual* (A. D. L. Akkermans, J. D. van Elsas and F. J. de Bruijn, eds), pp. 3.3.6.1–3.3.6.15. Kluwer Academic Publishers, Dordrecht.

Amann, R. I., Binder, B. J., Olson, R. J., Chisholm, S. W., Devereux, R. and Stahl, D. A. (1990). Combination of 16S rRNA-targeted oligonucleotide probes with flow cytometry for analyzing mixed microbial populations. *Appl. Environ. Microbiol.* **56**, 1919–1925.

Amann, R. I., Ludwig, W. and Schleifer, K. H. (1995). Phylogenetic identification and in situ detection of individual microbial cells without cultivation. *Microbiol. Rev.* **59**, 143–169.

Amann, R., Fuchs, B. M. and Behrens, S. (2001). The identification of microorganisms by fluorescence in situ hybridisation. *Curr. Opin. Biotechnol.* **12**, 231–236.

Amaral-Zettler, L. A., Gómez, F., Zettler, E., Keenan, B. G., Amils, R. and Sogin, M. L. (2003). Eukaryotic diversity in Spain's River of Fire. *Nature* **417**, 137.

Amils, R., González-Toril, E., Gómez, F., Fernández-Remolar, D., Rodríguez, N., Malkim, M., Aguilera, A. and Amaral-Zettler, L. A. (2004). Importance of chemolithotrophy for early life on earth: the Tinto River (Iberian Pyritic Belt) case. In *Origins* (J. Seckbach, ed.), pp. 463–480. Kluwer Academic Publishers, Dordrecht.

Benison, K. C. and Laclair, D. A. (2003). Modern and ancient extreme acid saline deposits: terrestrial analogs for Martian environments. *Astrobiology* **3**, 609–618.

Benz, M., Brune, A. and Schink, B. (1998). Anaerobic and aerobic oxidation of ferrous iron at neutral pH by chemoheterotrophic nitrate-reducing bacteria. *Arch. Microbiol.* **169**, 159–165.

Bond, P., Smriga, S. P. and Banfield, J. F. (2000). Phylogeny of microorganisms populating a thick, subaerial, predominantly lithotrophic biofilm at an extreme acid mine drainage site. *Appl. Environ. Microbiol.* **66**, 3842–3849.

Bridge, T. A. M. and Johnson, D. B. (1998). Reduction of soluble iron and reductive dissolution of ferric iron-containing minerals by moderately thermophilic iron-oxidizing bacteria. *Appl. Environ. Microbiol.* **64**, 2181–2186.

Brierley, C. L. and Brierley, J. A. (1973). A chemoautotrophic and thermophilic microorganism isolated from an acid hot spring. *Can. J. Microbiol.* **19**, 183–188.

Brierley, J. A. and Brierley, C. L. (1999). Present and future commercial applications of biohydrometallurgy. In *Biohydrometallurgy and the Environment toward the Mining of the 21st Century* (R. Amils and A. Ballester, eds), Part A, pp. 81–89. Elsevier, Amsterdam.

Brock, T. D. and Gustafson, J. (1976). Ferric iron reduction by sulfur- and iron-oxidizing bacteria. *Appl. Environ. Microbiol.* **32**, 567–571.

Brock, T. D., Brock, K. M., Belly, R. T. and Weiss, R. L. (1972). *Sulfolobus*: a new genus of sulfur-oxidizing bacteria living at low pH and high temperature. *Arch. Microbiol.* **84**, 54–68.

Brosius, J., Dull, T. J., Sleeter, D. D. and Noller, H. F. (1981). Gene organization and primary structure of a ribosomal RNA operon from *Escherichia coli*. *J. Mol. Biol.* **148**, 107–127.

Chandler, D. P., Fredrickson, J. K. and Brockman, F. J. (1997). Effect of the PCR template concentration on the composition and distribution of total community 16S rDNA clone libraries. *Mol. Ecol.* **6**, 475–482.

Colmer, A. R., Temple, K. L. and Hinkle, H. E. (1950). An iron-oxidizing bacterium from the acid drainage of some bituminous coal mines. *J. Bacteriol.* **59**, 317–328.

Corbett, C. M. and Ingledew, W. J. (1987). Is FeIII/FeII cycling an intermediate in sulphur oxidation by *Thiobacillus ferrooxidans*? *FEMS Microbiol Lett.* **41**, 1–6.

Daims, H., Bruhl, A., Amann, R., Schleifer, K. H. and Wagner, M. (1999). The domain-specific probe EUB338 is insufficient for the detection of all Bacteria: Development and evaluation of a more comprehensive probe set. *System. Appl. Microbiol.* **22**, 434–444.

Daims, H., Nielsen, P., Nielsen, J. L., Juretschko, S. and Wagner, M. (2001). Novel *Nitrospira*-like bacteria as dominant nitrite-oxidizers in biofilms from wastewater treatment plants: diversity and in situ physiology. *Water Sci. Technol.* **3**, 416–523.

DeLong, E. F., Wickham, G. S. and Pace, N. R. (1989). Phylogenetic strains: ribosomal RNA-based probes for the identification of single cells. *Science* **243**, 1360–1363.

Edwards, K. J., Bond, P. L., Gihring, T. M. and Banfield, J. F. (2000). An archaeal iron-oxidizing extreme acidophile important in acidic mine drainage. *Science* **287**, 1796–1798.

Ehrlich, H. L. (2002). *Geomicrobiology*, 4th edn., Marcel Dekker, New York.

Fernández-Remolar, D., Rodríguez, N., Gómez, F. and Amils, R. (2003). The geological record of an acidic environment driven by iron hydrochemistry: the Tinto River system. *J. Geophys. Res.* 108(2003)10.1029/2002JE001918.

Fernández-Remolar, D., Morris, R. V., Gruener, J. E., Amils, R. and Knoll, A. H. (2005). The Río Tinto basin, Spain: mineralogy, sedimentary geobiology and implications for interpretation of outcrop rocks at Meridiani Planum, Mars. *Earth Planet. Sci. Lett.* **240**, 179–189.

Gehrke, T., Hallmann, R. and Sand, W. (1995). Importance of exopolymers from *Thiobacillus ferrooxidans* and *Leptospirillum ferrooxidans* for bioleaching.

In *Biohydrometallurgical Processing* (T. Vargas, C. A. Jérez, K. W. Wiertz and H. Toledo, eds), vol. 1, pp. 1–11. Universidad de Chile, Santiago.

Giovannoni, J. G., DeLong, E. F., Olsen, G. J. and Pace, N. R. (1988). Phylogenetic group-specific oligodeoxynucleotide probes for identification of single microbial cells. *J. Bacteriol.* **170**, 720–726.

Giovannoni, S. J., Britschgi, T. B., Moyer, C. L. and Field, K. G. (1990). Genetic diversity in Sargasso Sea bacterioplankton. *Nature* **345**, 60–63.

Golyshina, O. V., Pivovarova, T. A., Karavaiko, G. I., Kondrateva, T. F., Moore, E. R., Abraham, W. R., Lunsdorf, H., Timmis, K. N., Yakimov, M. M. and Golyshina, P. N. (2000). *Ferroplasma acidiphilum* gen. nov., sp. nov., an acidophilic, autotrophic, ferrous-iron-oxidizing, cell-wall-lacking, mesophilic member of the *Ferroplasmaceae* fam. nov. comprising a distinct lineage of the Archaea. *Int. J. Syst. Evol. Microbiol.* **50**, 997–1006.

González-Toril, E., Gómez, F., Rodríguez, N., Fernández, D., Zuluaga, J., Marín, I. and Amils, R. (2001). Geomicrobiology of the Tinto river, a model of interest for biohydrometallurgy. In *Biohydrometallurgy: Fundamentals, Technology and Sustainable Development* (V. S. T. Cuminielly and O. García, eds), Part B, pp. 639–650. Elsevier, Amsterdam.

González-Toril, E., Llobet-Brossa, E., Casamayor, E. O., Amann, R. and Amils, R. (2003). Microbial ecology of an extreme acidic environment, the Tito River. *Appl. Environ. Microbiol.* **69**, 4853–4865.

González-Toril, E., García-Moyano, A. and Amils, R. (2006). Phylogeny of prokaryotic microorganisms from the Tinto River. In *Proceedings of the International Biohydrometallurgical Symposium – 2005* (D. Rawlings, ed.). Elsevier, Amsterdam, in press.

Hallberg, K. B. and Johnson, D. B. (2001). Biodiversity of acidophilic prokaryotes. *Adv. Appl. Microbiol.* **49**, 37–84.

Hallberg, K. B. and Johnson, D. B. (2003). Novel acidophiles isolated from moderately acidic mine drainage waters. *Hydrometallurgy* **71**, 139–148.

Hallberg, K. B. and Lindström, E. B. (1994). Characterization of *Thiobacillus caldus*, sp. nov., a moderately thermophilic acidophile. *Microbiology UK* **140**, 3451–3456.

Harrison, A. P. Jr. (1983). Genomic and physiological comparisons between heterotrophic thiobacilli and *Acidiphilium cryptum*, *Thiobacillus versutus* sp. nov., and *Thiobacillus acidophilus* nom. rev. *Int. J. Syst. Bacteriol.* **33**, 211–217.

Hiraishi, A. and Kitamura, H. (1984). Distribution of phototrophic purple nonsulfur bacteria in activated sludge systems and other aquatic environments. *Bull. Jpn. Soc. Sci. Fish.* **50**, 1929–1937.

Johnson, D. B. (1999). Importance of microbial ecology in the development of new mineral technologies. In *Biohydrometallurgy and the Environment Toward the Mining of the 21st Century* (R. Amils and A. Ballester, eds), Part A, pp. 645–656. Elsevier, Amsterdam.

Johnson, D. B. and Roberto, F. F. (1997). Heterotrophic acidophiles and their roles in the bioleaching of sulfide materials. In *Biomining: Theory, Microbes and Industrial Processes* (D. E. Rawlings, ed.), pp. 259–279. Springer, Berlin.

Kelly, D. P and Harrison, A. H. (1989). Genus *Thiobacillus*. In *Bergey's Manual of Systematic Bacteriology* (J. T. Staley, M. P. Bryant, N. Pfennig and J. G. Holt, eds), 1st edn., vol. 3, pp. 1842–1858. Williams & Wilkins, Baltimore.

Kelly, D. P. and Wood, A. P. (2000). Reclassification of some species of *Thiobacillus* to the newly designated genera *Acidithiobacillus* gen. nov.,

Halothiobacillus gen. nov. and *Thermithiobacillus* gen. nov. *Int. J. Syst. Evol. Microbiol.* **50**, 511–516.

Leathen, W. W., Kinsel, N. A. and Braley, S. A. Jr. (1956). *Ferrobacillus ferrooxidans*: a chemosynthetic autotrophic bacterium. *J. Bacteriol.* **72**, 700–704.

Llobet-Brossa, E., Rosselló-Mora, R. and Amann, R. (1998). Microbial community composition of Wadden Sea sediments as revealed by fluorescence in situ hybridization. *Appl. Environ. Microbiol.* **64**, 2691–2696.

López-Archilla, A. I., Marín, I. and Amils, R. (1993). Bioleaching and interrelated acidophilic microorganisms from Río Tinto, Spain. *Geomicrobiol. J.* **11**, 223–233.

López-Archilla, A. I., Marín, I. and Amils, R. (2001). Microbial community composition and ecology of an acidic aquatic environment: the Tinto River, Spain. *Microb. Ecol.* **41**, 20–35.

Ludwig, W., Strunk, O., Klugbauer, S., Klugbauer, N., Weizenegger, M., Neumaier, J., Bachleitner, M. and Schleifer, K. H. (1998). Bacterial phylogeny based on comparative sequence analysis. *Electrophoresis* **19**, 554–568.

Ludwig, W., Strunk, O., Westram, R., Richter, L., Meier, H., Yadhukumar, Buchner, A., Lai, T., Steppi, S., Jobb, G., Förster, W., Brettske, I., Gerber, S., Ginhart, A. W., Gross, O., Grumann, S., Hermann, S., Jost, R, König, A., Liss, T., Lüßmann, R., May, M., Nonhoff, B., Reichel, B., Strehlow, R., Stamatakis, A., Stuckmann, N., Vilbig, A., Lenke, M., Ludwig, T., Bode, A. and Schleifer, K. H. (2004). ARB: a software environment for sequence data. *Nucl. Acids Res.* **32**, 1363–1371.

Mackintosh, M. E. (1978). Nitrogen fixation by *Thiobacillus ferrooxidans*. *J. Gen. Microbiol.* **105**, 215–218.

Manz, W., Amann, R., Ludwig, W., Wagner, M. and Schleifer, K. H. (1992). Phylogenetic oligodeoxynucleotide probes for the major subclasses of *Proteobacteria*: problems and solutions. *System. Appl. Microbiol.* **15**, 593–600.

Manz, W., Eisenbrecher, M., Neu, T. R. and Szewzyk, U. (1998). Abundance and spatial organization of Gram-negative sulfate-reducing bacteria in activated sludge investigated by in situ probing with specific 16S rRNA targeted oligonucleotides. *FEMS Microbiol. Ecol.* **25**, 43–61.

Margesin, R. and Schinner, F. (2001). Potential of halotolerant and halophilic microorganisms for biotechnology. *Extremophiles* **5**, 73–83.

Marsh, R. M. and Norris, P. R. (1983). The isolation of some thermophilic, autotrophic, iron- and sulfur-oxidizing bacteria. *FEMS Microbiol. Lett.* **17**, 311–315.

Muyzer, G., de Bruyn, J. C., Schmedding, D. J. M., Bos, P., Westbroek, P. and Kuenen, J. G. (1987). A combined immunofluorescence-DNA-fluorescence staining technique for enumeration of *Thiobacillus ferrooxidans* in a population of acidophilic bacteria. *Appl. Environ. Microbiol.* **53**, 660–664.

Muyzer, G., de Waal, E. C. and Uitterlinden, A. G. (1993). Profiling of complex microbial populations by denaturing gradient gel electrophoresis analysis of polymerase chain reaction-amplified genes coding for 16S rRNA. *Appl. Environ. Microbiol.* **59**, 695–700.

Muyzer, G., Teske, A., Wirsen, C. O. and Jannasch, H. W. (1995). Phylogenetic relationships of *Thiomicrospira* species and their identification in deep-sea hydrothermal vent samples by denaturing gradient gel electrophoresis of 16S rDNA fragments. *Arch. Microbiol.* **164**, 165–172.

Muyzer, G., Hottenträger, S., Teske, A. and Wawer, C. (1996). Denaturing gradient gel electrophoresis of PCR-amplified 16S rDNA – A new molecular approach to analyse the genetic diversity of mixed microbial communities. In *Molecular Microbial Ecology Manual* (A. D. L. Akkermans, J. D. van Elsas and F. J. de Bruijn, eds), 2nd edn., pp. 3.4.4.1–3.4.4.22. Kluwer Academic Publishers, Dordrecht.

Myers, R. M., Maniatis, T. and Lerman, L. S. (1987). Detection and localization of single base changes by denaturing gradient gel electrophoresis. *Meth. Enzymol.* **155**, 501–527.

Neef, A. (1997). Anwendung der in situ-Einzelzell-Identifizierung von Bakterien zur Populationsanalyse in komplexen mikrobiellen Biozönosen. Ph.D. Thesis, Technical University of Munich.

Niehaus, F., Bertoldo, C., Kähler, M. and Antranikian, G. (1999). Extremophiles as a source of novel enzymes for industrial application. *Appl. Microbiol. Biotechnol.* **51**, 711–729.

Nogales, B., Moore, E. R., Llobet-Brossa, E., Rossello-Mora, R., Amann, R. and Timmis, K. N. (2001). Combined use of 16S ribosomal DNA and 16S rRNA to study the bacterial community of polychlorinated biphenyl-polluted soil. *Appl. Environ. Microbiol.* **67**, 1874–1884.

Norris, P. R., Clark, D. A., Owen, J. P. and Waterhouse, S. (1996). Characteristics of *Sulfobacillus acidophilus* sp. nov. and other moderately thermophilic mineral-sulphide-oxidizing bacteria. *Microbiology UK* **142**, 775–783.

Olsen, G. J., Lane, D. J., Giovannoni, S. J., Pace, N. R. and Stahl, D. A. (1986). Microbial ecology and evolution: a ribosomal RNA approach. *Annu. Rev. Microbiol.* **40**, 337–365.

Peng, J. B., Yan, W. M. and Boo, X. Z. (1994). Solid medium for the genetic manipulation of *Thiobacillus ferrooxidans*. *J. Gen. Appl. Microbiol.* **40**, 243–253.

Pernthaler, A., Pernthaler, J. and Amann, R. (2002). Fluorescence in situ hybridization and catalyzed reporter deposition for the identification of marine bacteria. *Appl. Environ. Microbiol.* **68**, 3094–3101.

Pivovarova, T. A., Markosyan, G. E. and Karavaiko, G. I. (1981). The auxotrophic growth of *Leptospirillum ferrooxidans*. *Microbiology* (English translation of *Mikrobiologiya*) **50**, 339–344.

Postgate, J. R. (1984). *The Sulphate-Reducing Bacteria*, 2nd edn., Cambridge University Press, Cambridge.

Pronk, J. T., Bruyn, J. C., Bos, P. and Kuenen, J. G. (1992). Anaerobic growth of *Thiobacillus ferrooxidans*. *Appl. Environ. Microbiol.* **58**, 2227–2230.

Raskin, L., Stromley, J. M., Rittmann, B. E. and Stahl, D. A. (1994). Group specific 16S rRNA hybridization probes to describe natural communities of methanogens. *Appl. Environ. Microbiol.* **60**, 1232–1240.

Rawlings, D. E. (1997). Restriction enzyme analysis of 16S rRNA genes for a rapid identification of *Thiobacillus ferrooxidans*, *Thiobacillus thiooxidans* and *Leptospirillum ferrooxidans* strains in leaching environments. In *Biohydrometallurgical Processing* (T. Vargas, C. A. Jérez, J. V. Wiertz and H. Toledo, eds), vol. II, pp. 9–18. University of Chile, Santiago.

Rawlings, D. E. (2002). Heavy metal mining using microbes. *Annu. Rev. Microbiol.* **56**, 65–91.

Russell, N. J. (2000). Toward a molecular understanding of cold activity of enzymes from psychrophiles. *Extremophiles* **4**, 83–90.

Sambrook, J. and Russel, D. W. (2001). Preparation of plasmid DNA by alkaline lysis with SDS: Minipreparation. In *Molecular Cloning. A Laboratory*

Manual (J. Sambrook and D. W. Russel, eds), 3rd edn., vol. 1, pp. 1.32–1.34. Cold Spring Harbor Laboratory Press, New York.

Sand, W., Rhode, K., Sobotke, B. and Zenneck, C. (1992). Evaluation of *Leptospirillum ferrooxidans* for leaching. *Appl. Environ. Microbiol.* **58**, 85–92.

Sand, W., Gehrke, T., Jozsa, P. G. and Schippers, A. (2001). (Bio)chemistry of bacterial leaching-direct vs. indirect bioleaching. *Hydrometallurgy* **59**, 159–175.

Schrenk, M. O., Edwards, K. J., Goodman, R. M., Hamers, R. J. and Bandfield, J. F. (1998). Distribution of *Thiobacillus ferrooxidans* and *Leptospirillum ferrooxidans*: Implications for generation of acidic mine drainage. *Science* **279**, 1519–1522.

Sheffield, V. D., Cox, D. R., Lerman, L. S. and Meyers, R. M. (1989). Attachment of a 40-base pair G+C-rich sequence (GC-clamp) to genomic DNA fragments by the polymerase chain reaction results in improved detection of single-base changes. *Proc. Natl. Acad. Sci. USA* **86**, 232–236.

Silverman, M. N. and Lundgren, D. G. (1959). Studies on chemoautotrophic iron bacterium *Ferrobacillus ferrooxidans*. *J. Bacteriol.* **77**, 642–647.

Stackebrandt, E. and Rainey, F. A. (1998). Partial and complete 16S rDNA sequences, their use in generation of 16S rDNA phylogenetic trees and their implications in molecular ecological studies. In *Molecular Microbial Ecology Manual* (A. D. L. Akkermans, J. D. van Elsas and F. J. de Bruijn, eds), pp. 3.1.1–3.1.17. Kluwer Academic Publishers, Dordrecht.

Stahl, D. A. and Amann, R. (1991). Development and application of nucleic acid probes. In *Nucleic Acid Techniques in Bacterial Systematics* (E. Stackebrandt and M. Goodfellow, eds), pp. 205–248. John Wiley & Sons, Chichester.

Suzina, N. E., Severina, L. O., Senyushkin, A. A., Karavaiko, G. I. and Duda, V. I. (1999). Ultrastructural organization of membrane system in *Sulfobacillus thermosulfidooxidans*. *Mikrobiologiya* **68**, 429–436.

Temple, K. L. and Colmer, A. R. (1951). The autotrophic oxidation of iron by a new bacterium *Thiobacillus ferrooxidans*. *J. Bacteriol.* **62**, 605–611.

Tuovinen, O. H. and Kelly, D. P. (1973). Studies on the growth of *Thiobacillus ferrooxidans*. 1. Use of membrane filters and ferrous iron agar to determine viable numbers and comparison with $^{14}CO_2$-fixation and iron oxidation as measures of growth. *Arch. Mikrobiol.* **88**, 285–298.

Ward, D. M., Weller, R. and Bateson, M. M. (1990). 16S rRNA sequences reveal numerous uncultured microorganisms in a natural community. *Nature* **345**, 63–65.

Widdel, F., Schnell, S., Heising, S., Ehrenreich, A., Assmus, B. and Schink, B. (1993). Ferrous iron oxidation by anoxygenic phototrophic bacteria. *Nature* **362**, 834–836.

Wintzingerode, F. V., Göbel, U. B. and Stackebrandt, E. (1997). Determination of microbial diversity in environmental samples: pitfalls of PCR-based rRNA analysis. *FEMS Microbiol. Rev.* **21**, 213–229.

Halophiles

◆◆

21 Characterization of Natural Communities of Halophilic Microorganisms
22 Cultivation of Haloarchaea
23 Extraction of Halophiles from Ancient Crystals
24 The Assessment of the Viability of Halophilic Microorganisms in Natural Communities
25 Characterization of Lipids of Halophilic Archaea
26 Characterization of Organic Compatible Solutes of Halotolerant and Halophilic Microorganisms
27 Genetic Systems for Halophilic Archaea
28 The Isolation and Study of Viruses of Halophilic Microorganisms
29 Detection, Quantification and Purification of Halocins: Peptide Antibiotics from Haloarchaeal Extremophiles
30 Storage of Halophilic Bacteria

21 Characterization of Natural Communities of Halophilic Microorganisms

Carol D Litchfield, Masoumeh Sikaroodi and Patrick M Gillevet
Department of Environmental Science and Policy, George Mason University, Manassas, VA 20110, USA

◆◆

CONTENTS

Introduction
Preparation for whole community analysis
Cloning using chemically competent cells
Sequencing
Analysis of clone data with custom PERL script
BIOLOG
Concluding remarks

◆◆◆◆◆ INTRODUCTION

Halophilic prokaryotes grow in elevated salt concentrations, from approximately 10% sodium chloride to saturation and some can even survive in salt crystals. Thus, their detection by cultivation or molecular markers requires special adjustments to the "standard" techniques. This applies to both the Bacteria and the Archaea. This chapter focuses on the application of methods for the detection and identification and potential metabolic activity of whole communities of halophiles from hypersaline environments, specifically the PCR procedure for amplification of 16S rRNA genes, the subsequent fingerprinting, cloning, sequencing, and analysis, and the metabolic technique of BIOLOG.

Molecular fingerprinting of microbial communities is a relatively easy and accurate technique to estimate the complexity and dynamics of microbes in any environment since many organisms resist cultivation (Ritchie *et al.*, 2000; Spring *et al.*, 2000). Recent advances in molecular techniques such as LH-PCR and RNA hybridization have enabled the identification of uncultivated taxa and provided insights into the expression of key functional genes in the environment and the microbial dynamics in those environments (Amann and Ludwig, 2000;

Dunbar *et al.*, 2001; Frischer *et al.*, 2000; Litchfield and Gillevet, 2002; Lydell *et al.*, 2004; Mills *et al.*, 2003; Pace, 1997; Rappé *et al.*, 1998; Seviour *et al.*, 2003; Suzuki *et al.*, 1998). The various methods are all based on sequence polymorphisms, or the slight differences that occur in the genomes of all organisms, and the amplification of the variable regions of specific genes. This process produces PCR amplicons of various lengths when examining a microbial community, and these amplicons can then be separated by several techniques. Usually, the PCR reaction procedure involves the amplification of a variable portion of the 16S ribosomal RNA gene with universal or specific primers. The size analysis of the PCR products can be determined by any of several procedures including Denaturing Gradient Gel Electrophoresis (DGGE), Terminal Restriction Fragment Length (T-RFLP), or Length Heterogeneity (LH).

The first molecular description of halophiles in salterns was reported by Benlloch *et al.* (1995). They performed Southern blots on selected clones after PCR and compared the clones with known cultures. This was a time consuming and not an especially sensitive way to analyze the community in a solar saltern. However, during sequencing of selected clones, they were able to detect two clones that did not match any known cultures. The next report on halophilic communities involved the use of restriction enzymes to cut the DNA, and the resulting products were separated by electrophoresis on denaturing polyacrylamide gels. The authors used *HinfI* alone or in combination with *MboI* or *AluI*. Based on the numbers of restriction fragments (RFLP), the authors concluded that the greatest diversity was detected when all three enzymes were used and that the diversity was greater among the Bacteria than the Archaea (Benlloch *et al.*, 1996).

This finding points out one of the inherent problems with restriction analyses that may be a special problem with respect to halophilic communities. With either RFLP or the more popular Terminal Restriction Fragment Length Polymorphisms (T-RFLP), it is essential that the 16S rRNA gene be susceptible to digestion by the selected enzymes. When examining a natural community, it is important to use multiple enzymes which can complicate the analysis of the resulting data by having different patterns and numbers of amplicons. With T-RFLP only the terminal amplicons are analyzed using an automatic sequencer, but even with this limited digestion not all of the 16S rRNA may have been digested (Mills *et al.*, 2003).

Further work on the saltern at Santa Pola was reported by the co-workers in Rodríguez-Valera's laboratory and this time the interspacer region of the 16–23S was analyzed (Benlloch *et al.*, 2001). The investigators examined both crystallizer water and cultures obtained from growth in broth by Ribosomal Internal Spacer Analysis (RISA) and then partially sequenced the clones. Again, the authors noted the presence of the SPhT phylotype frequently encountered in the Santa Pola ponds by the previous molecular analyses. The results (Figure 21.1) show the various molecular weights and the sources of the RISA bands. The authors concluded that the RISA bands from the crystallizer were different from those obtained from pure cultures.

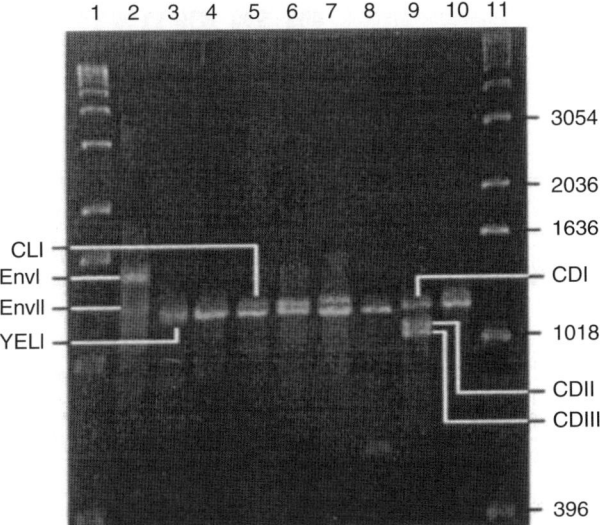

Figure 21.1. RISA analysis from environmental sample II total DNA and haloarchaeal cultures. Lanes 1 and 11: ladder 1 kb. Lane 2: environmental sample II. Lane 3: culture A-YE-L obtained in tubes with 5 ml of liquid media. Lane 4: culture A-YE-L obtained in 15 ml. Lane 5: culture A-C-L in 5 ml. Lane 6: culture A-C-L in 15 ml. Lane 7: culture A-YE-D in 5 ml. Lane 8: culture A-YE-D in 15 ml. Lane 9: culture A-C-D in 5 ml. Lane 10: culture A-C-D in 15 ml. Bands chosen for cloning and subsequent sequencing are labeled as EnvI, EnvII, YELI, CLI, CDI, CDII, and CDIII. Reproduced from figure 2, page 16 in Benlloch et al. (2001) with kind permission of Springer Science and Business Media.

In further studies that involved the same saltern, investigators from four laboratories evaluated the robustness of different molecular and cultivation techniques (Benlloch et al., 2002; Casamayor et al., 2002). Most of the emphasis in both of these articles centered on the molecular approach and compared the final results from the different laboratories. Regardless of which technique was used or which laboratory performed the analysis, the final results indicated similar major components in both the Bacteria and Archaea with one main prokaryotic group found in the 4–15% salt range and a second prokaryotic group in the 22–37% salinity range. Eukaryotes were also examined and found to be dominant in the lower salinity ponds. In a more detailed analysis of the genera, Benlloch et al. (2002) reported the presence of the *Cytophaga-Flavobacterium-Bacteroides* group throughout the ponds while for the Archaea, the marine Archaea group III was noted most frequently in the lower salt ponds. Clones in the higher salinity ponds were generally members of the *Halobacteriaceae* and again included the archaeal phylotype typical of this system, SPhT. Figure 21.2 shows the complexity of the Bacteria found throughout the saltern when the PCR products were analyzed by DGGE (Benlloch et al., 2002). Sixty-three bands were excised from the gel and sequenced with the majority being identified as members of the γ-Proteobacteria.

The complexity of the archaeal PCR products when the RISA technique was supplemented with RFLP is shown in Figure 21.3

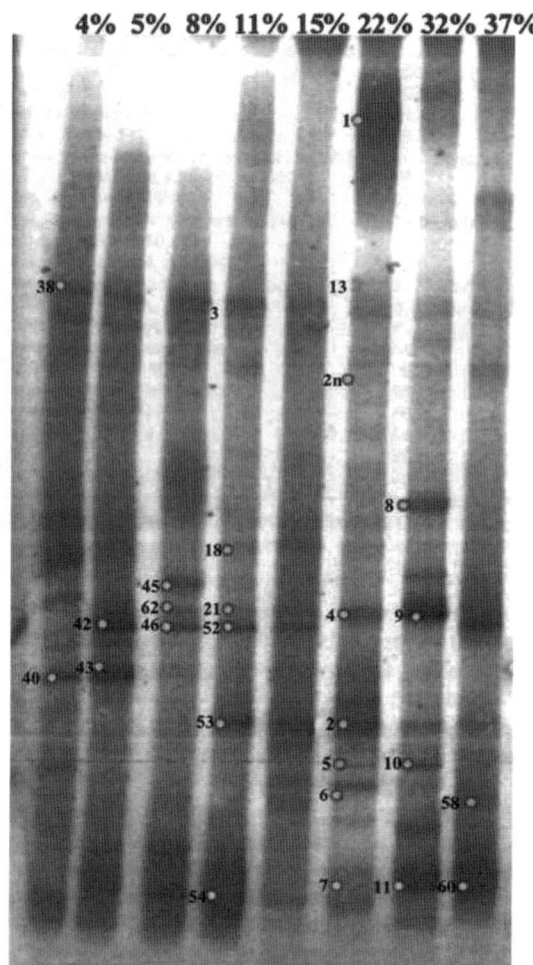

Figure 21.2. Bacterial 16S rRNA gene DGGE profile obtained by the UiB laboratory along the salinity gradient. Bands marked with a numbered dot were excised from the gel, reamplified and sequenced. The closest relatives for these sequences are reported in tables 1–3 in Benlloch et al. (2002) except for bands 18 and 21 (94.9% similarity to marine cyanobacterium Hstpl4), bands 58 and 60 (97.9% similarity to *Salinibacter ruber*) band 40 (74.8% similarity to *Psychroflexus torquis*), and band 43 (76.2% similarity to *Bacillus halophilus*). Reproduced from figure 1, page 350 in Benlloch et al. (2002) with permission of Blackwell Science Ltd. See Plate 21.2 in Colour Plate Section.

(Casamayor et al., 2002). The RFLP was added to the RISA because few amplicons were observed by RISA alone. This combined treatment gave band numbers similar to what had been observed in the DGGE for both the Bacteria and the Archaea. However, it prevented direct comparison with T-RFLP or the DGGE profiles. Analysis of all of the DGGE bands resulted in about 50% of the bands being heteroduplexes. This was not surprising considering each laboratory used different primers so that closely related sequences could easily form heteroduplexes. The common uncultivated square bacterium (SPhT) was again the predominant archaeon detected by these techniques.

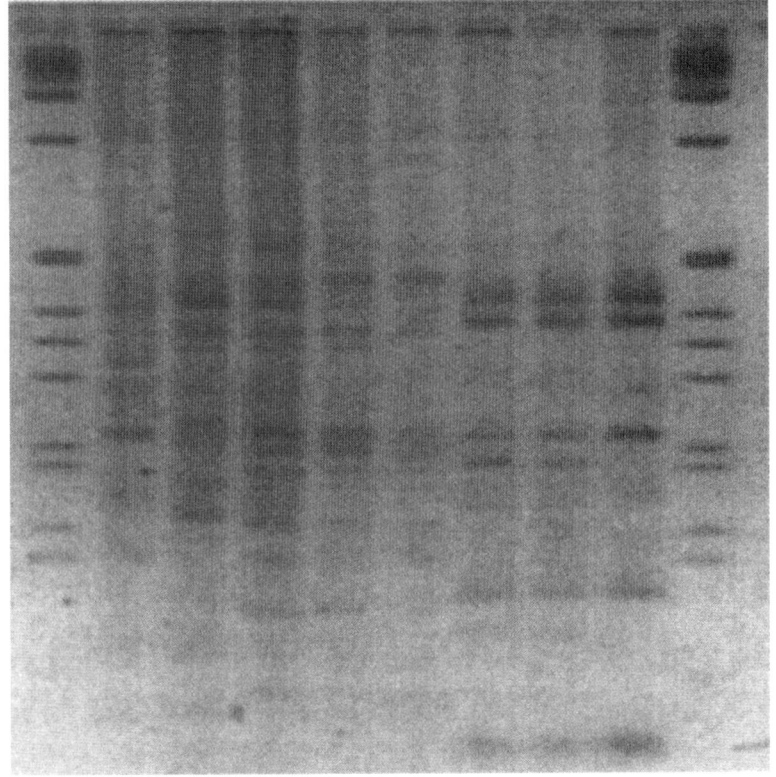

Figure 21.3. Example of a RISA(+RFLP (Restriction Fragment Length Polymorphism)) gel for Archaea. (A) PCR products from RISA were digested with the restriction enzyme *Hinf*I; M, 1 kb ladder. Reproduced from Figure 6A, page 344 in Casamayor *et al.* (2002) with permission of Blackwell Science Ltd. See Plate 21.3 in Colour Plate Section.

LH-PCR

None of the studies reported so far have used the Length Heterogeneity approach to fingerprinting the halophilic communities in salterns. This section describes how to fingerprint halobacterial communities using the LH-PCR procedure. In LH-PCR, the fingerprints of the variations in amplicon length are separated on a denaturing polyacrylamide gel or a capillary system. The peak area in the profile is proportional to the abundance of that amplicon in the community. The method was first used to estimate the diversity present in marine bacterioplankton (Suzuki *et al.*, 1998). Further studies have evaluated the robustness of this method and found it to be highly reproducible yielding good representations of communities that have been assessed by alternative methods such as DGGE and T-RFLP (Dunbar *et al.*, 2001; Mills *et al.*, 2003; Ritchie *et al.*, 2000). The reproducibility of the LH-PCR method is shown

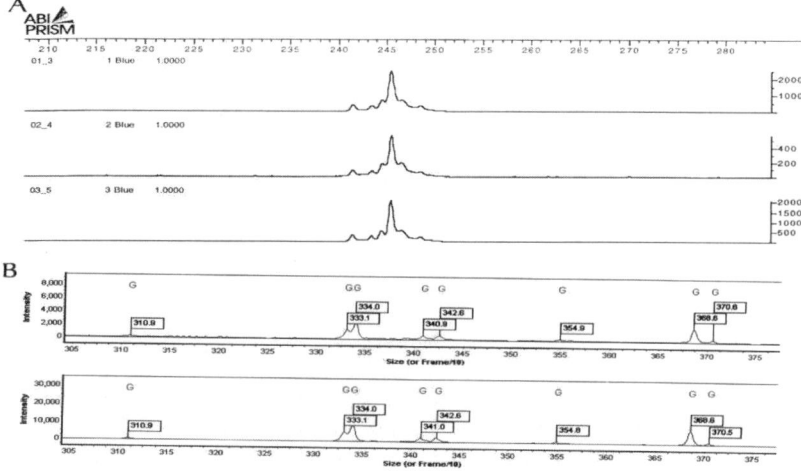

Figure 21.4. Reproducibility of the LH-PCR method using either the ABI 377 (A) or the SpectruMedix SCE9610 (B).

in Figure 21.4 where the top portion was performed with samples from the saltern in Eilat, Israel on an ABI 377 Genetic Analyzer while the bottom portion is a sample from Great Salt Lake, Utah, that was analyzed on SpectruMedix 9610 fluorescent capillary sequencer. In each case, the results are from different PCR amplifications (Litchfield et al., unpublished data).

As with all of the fingerprinting techniques, there may be more than one taxon represented within any one amplicon length or Operational Taxonomic Unit (OUT). Consequently, cloning and sequencing are essential to determining the specific taxonomic composition of the community. So, while LH-PCR is inexpensive and rapid, the subsequent cloning and sequencing are expensive. Additionally, attempts to relate experimental amplicon size with the size calculated from the clone sequence have shown that the clones need to be fingerprinted to correlate back to LH-PCR. Although there is the possibility of chimera formation during community fingerprinting because of the presence of multiple 16S rRNA genes, this has not been found to be a problem using our LH-PCR protocol. Specifically, we use short amplicon sizes and long extension times in the amplification cycles to minimize chimera formation (Grace and Wang, 1996, 1997).

◆◆◆◆◆ PREPARATION FOR WHOLE COMMUNITY ANALYSIS

Samples

Aqueous samples or sediments are collected aseptically from the environment of interest. The microorganisms in the aqueous samples must be concentrated prior to performing the molecular analyses. This concentration can be performed using tangential flow membrane filtration or by centrifuging at 12 000 × g for 15 to 20 min at 4°C. Depending on the source water, generally 3–4 l must be concentrated to obtain sufficient material for molecular analyses. However, for metabolic evaluation, aqueous samples may be used as received.

Genomic DNA Extraction

Normally, when analyzing pure cultures, they can be resuspended in Tris-EDTA (TE) buffer or diethylpyrocarbonate (DEPC)-treated water, and lysed at 96°C for 10 min in a thermal cycler. The lysed cultures can then be directly used in the PCR reactions. However, this is not a very successful technique for obtaining the 16S rRNA genes for whole community analyses. Consequently, the method most successful in our laboratories has used the BIO101 Fast-Spin® Kit for Soil (QBiogen). The following is a modification of the manufacturer's method. Note that all materials and reagents, with the exception of ethanol and microcentrifuge tubes, are provided in the kit.

Supplies needed: DNA Fast Spin Kit for Soil
RNA/DNA-free microcentrifuge tubes–2 ml
Micropipettors and appropriate tips
100% ethanol

- Add 978 µl sodium phosphate buffer and 122 µl MT buffer to a Multimix 2 tube.
- Add up to 500 mg or 350 µl of sample to the Multimix 2 Tissue Matrix tube. The volume to be added should depend on the estimated concentration of cells – more cells, less volume; but do not attempt to add more than 350 µl.
  NOTE: The tube should not be more than 7/8 full or there will be leakage during the bead-beating step.
- Secure the tubes in the FastPrep Instrument and process for 30 seconds at speed 5.5. Alternatively, if a bead beater system is not available, you may tape the tubes very securely onto a vortex mixer and shake the tubes at the highest setting for 10 min.
- Centrifuge the tubes at 14 000 × g for 10–12 min.
- Resuspend the samples and repeat Steps 3 and 4.
- Transfer the supernatants to clean 2.0 ml microfuge tube. Add 250 µl PPS reagent and mix by shaking by hand 10 times.
- Centrifuge at 14 000 × g for 5 min to pellet the debris.
- Transfer the supernatant to a clean 1.5 ml microfuge tube and add 750 µl of the resuspended Binding Matrix.
  NOTE: It is important to keep the Binding Matrix resuspended when you are doing several extractions.
- Place on a rotator or mix by hand for two min to allow binding of the DNA to the matrix.
- Place the tubes in a rack and allow to stand for at least 3 min.

- Discard the first 500 µl of the supernatant. Resuspend the DNA-Binding Matrix by gently mixing with a pipettor tip and transfer 600 µl to a clean Spin Filter.
- Centrifuge at 14 000 × g for 1 min; discard the filtrate; and repeat until all of the mixture has been centrifuged.
- Add 100 ml of 100% ethanol to the bottle containing SEWS; mix and label the bottle with the date the ethanol was added and the initials of the person who added the ethanol.
- Add 500 µl of the ethanol–SEWS mixture to the spin filter and centrifuge at 14 000 × g for 1 min and discard the filtrate.
- Repeat the previous step but centrifuge for 2 min.
- Place the Spin Filter in a fresh catch tube and air dry for 5 min with the lid open.
- Add 75–80 µl of 65°C DEPC water or DES water (supplied in the kit) and centrifuge for 1 min at 14 000 × g. This washes the DNA off of the filter and into the catch tube.
- Repeat the previous step. The DNA is now ready for testing and should be stored long term at −20°C or stored at 4°C for immediate use. Check presence of DNA and yield by 1% agarose gel electrophoresis.

Fingerprinting with Length Heterogeneity-PCR (LH-PCR)

The basic principle of LH-PCR fingerprinting is that the size of the amplification products over the first 2–3 variable regions in the 16S rRNA gene will be different for different groups of bacteria. Therefore, the different microbial members of the environmental sample will provide different molecular weight peaks when the fluorescent amplicons are separated on a sequencer. This technique was first developed by Suzuki et al. (1998) and Rappé et al. (1998).

We have found that the *Tfl* enzyme is more reliable at this stage of the study than other polymerases. The consensus primers we have developed for the halophilic Archaea are shown in Table 21.1. For the halophilic Bacteria, we have found that the universal primers 27F and 355R (Lane, 1991) can be used in this system. The fluorochromes necessary for detection of the individual amplicons are usually either 6-carboxyfluorescein (6-FAM) or NED^{TM} (Applied Biosystems Inc.). These are attached on the 5′ end of either the forward or reverse primer.

The additions for the *Tfl* polymerase reaction are shown in Table 21.2 while the thermal cycler conditions are listed in Table 21.3.

Preparing PCR products for fingerprinting

For LH-PCR fingerprinting, any sequencing instrument may be used. The principle is that the fluorescent amplicons will be detected by the instrument which results in a peak whose area under the peak is proportional to the amount of that amplicon. It should be noted that operation of a sequencer will be different for each instrument, and the conditions will vary slightly between laboratories using the same instrument. Data can be analyzed using the data analysis software specific to the sequencer/genetic analyzer used.

Table 21.1 Consensus forward and reverse primers for halophilic Archaea

Primer number	Sequence 5' to 3'
HK 1F	ATTCCGGTTGATCCTGCCGG
H 90F	TCCGATTTAGCTCCCCGCCA
H 287F	AGGTAGACGGTGGGGTAAC
H 571F	GCCTAAAGCGTCCGTAGCT
H 755F	CCACAGTGAGGGACGAAAG
H 926F	AAACTTAAAGGAATTGGCGG
H 1192F	CTACCGTTGCCCGTTCCTTC
H 1387F	GAATACGTCCCTGCTCCTTG
H 305R	GTTACCCCACCGTCTACCT
H 589R	AGCTACGGACGCTTTAGGC
H 774R	CTTTCGTCCCTCACTGTCGC
H 949R	GTTGTAGTGCTCCCCCGCA
H 1211R	GAAGGAACGGGCAACGGTAG
H 1392R	ACGGGCGGTGTGTRC
H 1406R	CAAGGAGCAGGGACGTATTC
H 1492R	GTTACCTTGCCCGTTCCTTC
H 1540R	AGGAGGTGATCCAGCCGCAG

Table 21.2 Protocol and reagents used in preparing samples for PCR

Reagent	Amount (µl)
DEPC water	11.9
10X Buffer*	2
Vortex	
MgCl$_2$* (25 mM)	2
Vortex	
dNTP (25 µM) (2.5 mM each)	2
Vortex	
Forward primer (10 µM)	1
Reverse primer (10 µM)	1
Vortex	
Tfl* (5 U µl^{-1})	0.1
DNA template	2 to 3

*Components of *Tfl* kit

Table 21.3 Thermal cycler operating conditions for halophilic Archaea

Step	Temperature (°C)	Time (min)
1	94	4
2	94	1
3	55	1
4	74	3
5	Back to step 2, 25 times	
6	74	10
7	4	4
8	END	

LH-PCR Fingerprint Data Analysis

Theory

In our laboratory, the LH-PCR products are separated using the SCE9610 capillary fluorescent sequencer (SpectruMedix LLC) and are analyzed with the GenoSpectrum™ software package (Version 2.01). The software package deconvolves the fluorescence data into electropherograms with the peaks of the electropherograms representing different populations of microflora with PCR amplicons of different length in base pairs (bp). All fingerprinting data are analyzed using custom software (Interleave 1.0) that combines data from several runs, interleaves the various profiles, normalizes the data, calculates the averages for each amplicon size, and determines diversity indices. The normalized peak areas are calculated by dividing an individual peak area by the total peak area in that profile. LH-PCR fingerprint patterns (i.e. presence or absence of certain amplicon peaks) associated with different experimental classes are identified by visual inspections of histograms of the LH-PCR amplicon normalized average abundances.

Analysis of LH-PCR Fingerprints

The LH-PCR fingerprints are quite complex and are difficult to compare visually. Multidimensional scaling techniques can be used to simplify the patterns in molecular fingerprints so that one can see global patterns (Johnson et al., 2003; Rees et al., 2004). Specifically, Eigen analysis of the LH-PCR fingerprints is done on individual normalized samples to determine global clustering of the LH-PCR fingerprints to see if there is a correlation between the fingerprints and the experimental classes. Principal Coordinate Analysis (PCO) is performed using the Multi Variate Statistical Package (MVSP) (Kovach, 1999). It should be noted that these techniques do not correlate experimental class with specific variables but instead correlate the classes with all the data.

We also compare the diversity of the samples using classic Shannon Diversity, Eveness, and Richness (Andreoni et al., 2004; De Lipthay et al., 2004; Gallagher et al., 2004; Haack et al., 2004; Ibekwe and Grieve, 2004). Our custom PERL script calculates the above indices and then performs a T-test to determine if they are significantly different from each other (Poole, 1974).

◆◆◆◆◆ CLONING USING CHEMICALLY COMPETENT CELLS

Although the fingerprints provide an indication of the diversity of the halophilic Archaea community, it is not possible to know from these fingerprints which taxa are represented by the different peaks or how many taxa might be forming the same amplicon length PCR product. Consequently, as in all of the fingerprinting efforts, cloning and sequencing of nonfluorescent PCR products provides a better picture of the community composition.

To do this, fresh PCR product without the fluorochrome is prepared as above except the final extension step is increased to 10–15 min. After verifying that PCR products are present on a 1% agarose gel, the ligation must be performed on the same day. Waiting overnight may result in the loss of any 3′A-overhangs which are required as the plasmid vector has a single 3′-thymidine overhang. Lost 3′A-overhangs can be replaced by following the TOPO TA^{TM} cloning manual instructions. There are three major steps in this cloning procedure. The first is the ligation stage, the second is the transformation step, and the third is the plating and selection of colonies for sequencing. The major modification that we perform is to add kanamycin (10 mg ml^{-1}) to the ImMedia plates (Invitrogen). This antibiotic reduces the number of satellite colonies.

◆◆◆◆◆ SEQUENCING

Again, either the SpectruMedix SCE9610 or an ABI 377 Genetic Analyzer may be used. However, unlike the fingerprinting where the LH-PCR products are essentially ready to be placed in the sequencer, the nonfluorescent PCR products must be specially prepared and then cleaned of excess primers and other reagents. Two methods are listed below for purification.

For sequencing, the PCR products must be free from any other chemicals than the product to be sequenced. We routinely use either the magnetic bead procedure or the Sephadex method for purification of the PCR products. In each case, we follow the manufacturer's protocol. Sequencing with the BigDye Terminator is then performed according to the manufacturer's instructions.

◆◆◆◆◆ ANALYSIS OF CLONE DATA WITH CUSTOM PERL SCRIPT

One of the main problems with the analysis of clones from environmental samples is the matching of each clone with its nearest named genus and species. Typically, GenBank is searched using BLAST (Altschul et al., 1997), but this usually identifies uncultured environmental samples with little or no phylogenetic information (Hagström et al., 2002). Although these GenBank identities can be used to correlate clones with other environmental communities, the construction of a phylogenetic tree consisting of only unnamed uncultivated microorganisms has little informational significance.

We have taken a novel approach to identify reference genera and species for the phylogenetic analysis and comparative ecological analysis of these clones. We perform BLAST analysis on the Ribosomal Database

(RDP) and associate the highest hit with a corresponding RDP number (Cole et al., 2003). These RDP numbers are derived from a hierarchical classification system based on the multiple alignment of all known Small Subunit ribosomal DNA data (SSU rDNA) in GenBank.

The clone sequence data are first trimmed off the cloning vector, and then the RDP database is searched using MEGABLAST. The hit table is parsed and the sequence with the highest bit score is chosen as the closest relative of the clone in question. The clone is then annotated with that RDP number. The clone data, the corresponding RDP hit, and the putative identification are stored in an associative array and written to a disk as a Genetic Data Environment (GDE) formatted database (Smith et al., 1994). The annotated clone data, as well as the GenBank identifications, can then be loaded back into GDE for informative phylogenetic and comparative analysis.

We currently use a version of GDE that has recently been ported to Mac OS X's UNIX environment. This algorithm uses MEGABLAST to compare the clone sequence data to the RDP database and compiles a table using the RDP numbers to correlate species identification within a hierarchical classification scheme. This custom PERL script annotates each clone with the corresponding RDP number obtained from the MEGABLAST search results and this number is used to classify the family/genus/species of each clone. To simplify the analysis, the data are generally filtered to report only those families/genera/species that constitute greater than 5% of the microbial community. Additionally, the PERL script calculates rarefaction curves on the clone data based on user selected family, genus or species criteria.

The clone libraries can then be analyzed using classic rarefaction analysis (Kemp and Aller, 2004) to determine the estimated diversity of the phylotypes in the community.

We next align the clone sequences from a library using Clustal X (Thompson et al., 1997) and construct phylogenetic trees using PAUP (Swofford, 2001). In addition, the libraries from different samples can be aligned and compared using tools such as Libshuff (Singleton et al., 2001) which statistically compares the phylotypic diversity in the samples.

Alternatively, the sequences of the clones may be compared directly with the sequences in GenBank. For this analysis, see Hall (2001), Altschul et al. (1990, 1997), Zhang and Madden (1997), and Zhang et al. (2000).

Figure 21.5 is an example of a cladogram comparing environmental isolates from Cargill Solar Salt Plant ponds with RDP references. The LH-PCR clone sequences were compared to sequences in the RDP database to assess patterns of the microflora using CloneID 1.0 (BioSpherex LLC). This algorithm uses MEGABLAST to compare the clone sequence data to the RDP database and compiles a table using the RDP numbers to correlate species identification within a hierarchical classification scheme. This custom PERL script annotates each clone with the corresponding RDP number obtained from the megablast search results and this number is used to classify the family/genus/species of each clone. We can see that

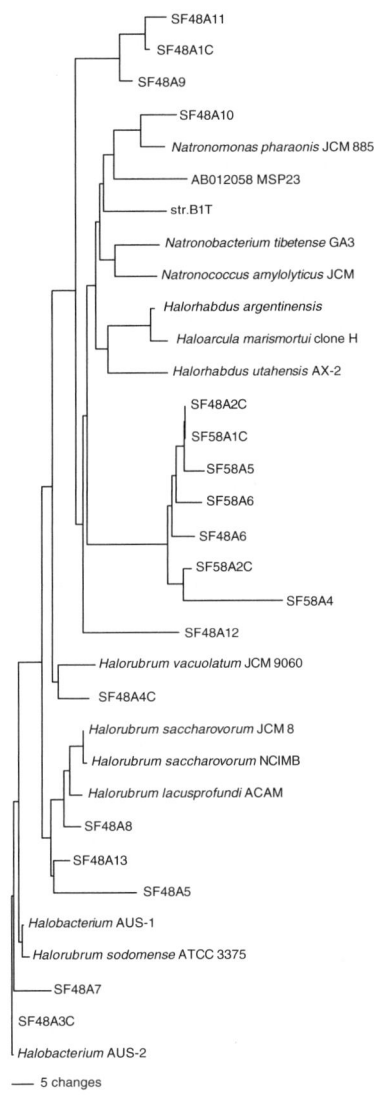

Figure 21.5. Example of a cladogram comparing environmental isolates from Cargill Solar Salt Plant ponds with RDP references.

some of the environmental isolates are similar to known characterized species (i.e. *Natronomonas* spp., *Haloarcula* spp. and *Halorubrum* spp.) but there is one novel clade that is not very similar to any named species in the current database (the clade defined by SF58A4).

◆◆◆◆◆ BIOLOG

The BIOLOG procedure was introduced in 1989 as a method for identifying bacteria (Bochner, 1989). This was accomplished by having a 96-well microtiter plate to which 95 different carbon sources were added along with proprietary nutrients. The A-1 well was the control and there were different plates for Gram-positive and Gram-negative cultures. If the wrong plate is used, the A-1 well will turn purple.

The principle of the BIOLOG system is to test for dehydrogenases using tetrazolium violet, a redox dye that competes for the hydrogens released by the dehydrogenases. Positive wells turn pink to dark purple within 18–24 h for pure cultures. There are a number of configurations that have been developed including plates specifically designed for Gram-positive bacteria, anaerobic microbes, and ECO plates. The plates used by most investigators have been the GN plates designed for Gram-negative bacteria.

It was quickly realized that this system can be applied to environmental samples thus allowing the scientist to estimate the metabolic potential of a community (Garland, 1996). Indeed, numerous studies showed the microbial metabolic diversity over time in various environments including grassland soils (Fleißbach and Mader, 1997), tropical soils (Sharma et al., 1997), the northern Chihuahuan Desert (Zak et al., 1994), and basaltic aquifers (Colwell and Lehman, 1997), to specifically name only a few of the varied environments in which this technique has been applied. Instead of a presence/absence analysis, the procedure was later modified to include the time to development of the color as an indication of the rate of metabolism of the substrate (Garland and Lehman, 1999).

The difficulty of interpretation of BIOLOG data with whole communities is that it is a cultivation procedure, albeit not a traditional one. Given the prolonged incubation times necessary because of low microbial biomass in environmental samples, the results reflect those organisms which can grow on a particular substrate. Hence it is essential to remember that the method is evaluating the potential for some members of a community to metabolize that substrate. As pointed out in 2002 (Litchfield and Gillevet, 2002), the results of respirometry on substrates that were positive on the BIOLOG plate are not always congruent.

Furthermore, there are limitations to the use of BIOLOG in hypersaline systems. It was noted that replicate GN plates inoculated with samples containing greater that 14–15% total salts were not reproducible (Litchfield et al., 2001). Therefore, it is advisable whenever using BIOLOG to inoculate replicate plates. Slight differences may be observed simply because of the heterogeneity of the microbial distribution in the sample, but anything greater than 5% (an arbitrary number used by us) in the three replicates could indicate salt inhibition.

Supplies needed: Sterile multichannel pipettor basins
Multi-channel pipettor and tips
GN 2™ (newer version of the plates)
Microtiter Plate reader or the BIOLOG System if interested in time to color development

Procedure

- Add approximately 25–30 ml of sample to the sterile sample basin.
- With an 8- or 12-channel pipettor, transfer 140 µl to each well on the plate.
  NOTE: The addition of only 140 µl to each well insures that there is no splashing from one well to another during the reading of the plates.
  NOTE: Do not touch the bottom of the wells when adding the samples. This will prevent carry over of the nutrients from one well to another which can lead to difficulty in interpreting the results.
- Cover the plates, insert into ziplock bags to reduce evaporation, and incubate at *in situ* temperatures. Plates may be incubated for up to 6 weeks or until color development has stabilized if the primary interest is in the overall metabolic potential. However, if the primary interest is in the rate of metabolic potential inherent in the community, then the plates should be read within 4–24 h.

Data Analysis

It is possible to simply record the data on a presence/absence basis and perform similarity analyses between the different samples. The MVSP program is used for these analyses (Kovach, 1999). An example of this is shown in Figure 21.6 where lower salinity samples from the saltern in Eilat, Israel, were tested using the GN plates (Irby, 2001). The analysis was done using nearest neighbor and the simple matching coefficient.

Alternatively, principal component analysis (also available on the MVSP program) or Niche Overlap Index (Wilson and Lindow, 1994) can provide more information about the effects of various nutrients on the microbial community structure over time (Irby, 2001). An example of such an analysis is shown in Figure 21.7 which compares the specific metabolic potential of the samples shown in Figure 21.6.

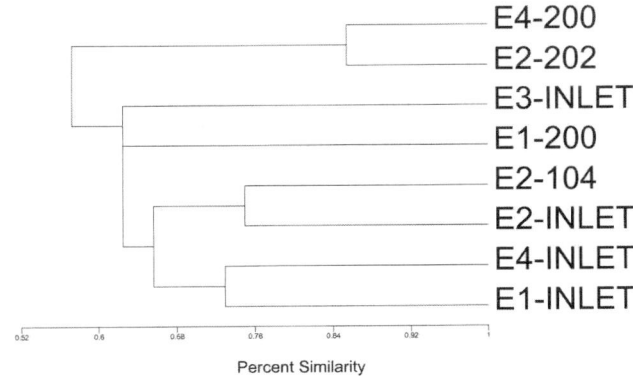

Figure 21.6. Similarity matrix for carbon utilization patterns by the whole community in samples from a solar saltern in Eilat, Israel.

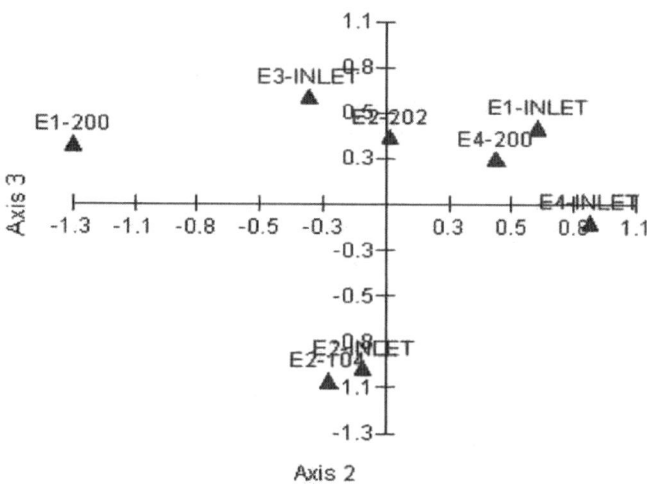

Figure 21.7. Principal component analysis of the BIOLOG data for the samples shown in Figure 15.6. See Plate 21.7 in Colour Plate Section.

Examination of the Eigen vectors of the axes allows one to determine the carbon sources most responsible for the placement of the samples in the respective quadrants, for example, the negative axis 2 factors were citric acid, *cis*-aconitic acid, L-threonine, D-glucuronic acid, L-asparagine and propionic acid, while the positive factors for axis 3 were gentiobiose, cellobiose, inosine, β-methyl-D-glucoside, N-acetyl-glucosamine, glucose-1-phosphate and glucose-6-phosphate. Again, it should be emphasized that these are only the metabolic potentials of the community over a prolonged incubation period, and they may not reflect short term utilization patterns as observed by Litchfield and Gillevet (2002) for Mono Lake.

◆◆◆◆◆ CONCLUDING REMARKS

Both molecular techniques and BIOLOG provide an indication of the biocomplexity in saline environments. In the one case, the fingerprints and subsequent cloning and sequencing provide insights into the numbers and types of taxa present, while BIOLOG provides insights into the potential metabolic capabilities of the community members. Each method is useful in increasing our understanding of these extreme environments, and each has its limitations which must be borne in mind when interpreting the data. During the last 10 years, great strides have been made in recognizing the microbial complexity of saline systems using both of these techniques. They have provided the stimulus to isolate new bacteria such as *Salinibacter* spp. and other organisms currently being studied in numerous laboratories.

List of suppliers

Applied Biosystems Inc.
850 Lincoln Centre Drive
Foster City, CA 94404, USA
Tel.: 1-800-327-3002
Fax: 1-650-638-5998
http://www.appliedbiosystems.com

Taq Big Dye Terminator chemistry; 6-carboxyfluorescein (6-FAM); HiDiFormamide

BIOLOG
21124 Cabot Blvd.
Hayward, CA 94545, USA
Tel.: 1-510-785-2564
Fax: 1-510-782-4639
http://www.biolog.com

Biolog Plates: Eco, GN and GP

BioSpherex LLC
15 Wiltshert Court
Sterling, VA, USA
Tel.: 1-703-430-8562
Fax: 1-801-881-7746

Interleave 1.0

Kovach Computing Services
85 Nant-y-Felin, Pentraeth, Isle of Angelesey
LL75 8UY, Wales, UK
Tel.: 44-1248-450414 (outside the UK)
Fax: 44-1248-450259 (outside the UK)
http://www.kovcomp.com

Multi Variate Statistical Package

MJ Reseach
590 Lincoln St.
Waltham, MA 02451, USA
Tel.: 1-888-735-8437
Fax: 1-888-923-8080
http://www.mjr.com

Thermal Cycler

Perkin Elmer
11642 Old Baltimore Pike
Beltsville, MD 20705, USA

Tel.: 1-301-937-8811
Fax: 1-301-937-6928
http://www.perkinelmer.com

Tfl DNA polymerase kit; SpectraCount Microtiter Plate reader

Promega Corp.
2800 Woods Hollow Road
Madison, WI 53711-5399, USA
Tel.: 1-800-356-9526
Fax: 1-800-356-1970
http://www.promega.com

pGEM-T Easy Vector System II

Q-Biogene, Inc.
2251 Rutherford Rd.
Carlsbad, CA 92008, USA
Tel.: 1-800-424-6101
Fax: 1-800-900-9224
http://www.qbiogene.com

BIO101 Fast-Spin® Kit for Soil; FastPrep Instrument

Spectrumedix LLC,
2124 Old Gatesburg Road
State College, PA 16803, USA
Tel.: 1-814-867-8600
Fax: 1-814-867-4513
http://www.spectrumedix.com

BaseSpectrum Version 2.0; Genospectrum Version 2.08

References

Altschul, S. F., Gish, W., Miller, E. W. and Lipman, D. J. (1990). Basic local alignment search tool. *J. Mol. Biol.* **215**, 403–410.
Altschul, S. F., Madden, T. L., Schäffer, A. A., Zhang, Z., Miller, W. and Lipman, D. J. (1997). Gapped BLAST and PSI-Blast: a new generation of protein database search programs. *Nucleic Acid Res.* **25**, 3389–3402.
Amann, R. and Ludwig, W. (2000). Ribosomal RNA-targeted nucleic acid probes for studies in microbial ecology. *FEMS Microbiol. Rev.* **24**, 555–565.
Andreoni, V., Cavalca, L., Rao, M. A., Nocerino, G., Bernasconi, S., Dell'Amico, E., Colombo, M. and Gianfreda, L. (2004). Bacterial communities and enzyme activities of PAHs polluted soils. *Chemosphere* **57**, 401–412.
Benlloch, S., Martínez-Murcia, A. and Rodríguez-Valera, F. (1995). Sequencing of bacterial and archaeal 16S rRNA genes directly amplified form a hypersaline environment. *Syst. Appl. Microbiol.* **18**, 574–581.

Benlloch, S., Acinas, S. G., Martínez-Murcia, A. and Rodríguez-Valera, F. (1996). Description of prokaryotic biodiversity along the salinity gradient of a multipond solar saltern by direct PCR amplification of 16S rDNA. *Hydrobiologia* **329**, 19–31.

Benlloch, S., Acinas, S. G., Antón, J., López-López, A., Luz, S. P. and Rodríguez-Valera, F. (2001). Archaeal biodiversity in crystallizer ponds from a solar saltern: culture versus PCR. *Microb. Ecol.* **41**, 12–19.

Benlloch, S., López-López, A., Casamayor, E. O., Øvreås, L., Goddard, V., Daae, F. L., Smerdon, G., Massana, R., Joint, I., Thingstad, F., Pedrós-Alió, C. and Rodríguez-Valera, F. (2002). Prokaryotic genetic diversity throughout the salinity gradient of a coastal solar saltern. *Environ. Microbiol.* **4**, 349–360.

Bochner, B. (1989). "Breathprints" at the microbial level. *ASM News* **55**, 536–539.

Casamayor, E. O., Massana, R., Benlloch, S., Øvreås, L., Díez, B., Goddard, V. J., Gasol, J. M., Joint, I., Rodríguez-Valera, F. and Pedrós-Alió, C. (2002). Changes in archaeal, bacterial and eukaryal assemblages along a salinity gradient by comparison of genetic fingerprinting methods in a multipond solar system. *Environ. Microbiol.* **4**, 338–348.

Cole, J. R., Chai, B., Marsh, T. L., Farris, R. J., Wang, Q., Kulam, S. A., Chandra, S., McGarrell, D. M., Schmidt, T. M., Garrity, G. M. and Tiedje, J. M. (2003). The Ribosomal Database Project (RDP-II): previewing a new autoaligner that allows regular updates and the new prokaryotic taxonomy. *Nucleic Acid Res.* **31**, 442–443.

Colwell, F. S. and Lehman, R. M. (1997). Carbon source utilization profiles for microbial communities from hydrologically distinct zones in a basalt aquifer. *Microb. Ecol.* **33**, 240–251.

De Lipthay, J. R., Johnsen, K., Albrechtsen, H. J., Rosenberg, P. and Aamand, J. (2004). Bacterial diversity and community structure of a subsurface aquifer exposed to realistic low herbicide concentrations. *FEMS Microbiol. Ecol.* **49**, 59–69.

Dunbar, J., Ticknor, L. O. and Kuske, C. R. (2001). Phylogenetic specificity and reproducibility and new method for analysis of terminal restriction fragment profiles of 16S rRNA genes from bacterial communities. *Appl. Environ. Microbiol.* **67**, 191–197.

Fleißbach, A. and Mader, P. (1997). Carbon source utilization by microbial communities in soils under organic and conventional farming practice. In *Microbial Communities: Functional versus Structural Approaches* (H. Insam and A. Rangger, eds), pp. 109–120. Springer-Verlag, Berlin.

Frischer, M. E., Danforth, J. M., Newton-Healy, M. A. and Saunders, F. M. (2000). Whole-cell versus total RNA extraction for analysis of microbial community structure with 16S rRNA-targeted oligonucleotide probes in salt marsh sediments. *Appl. Environ. Microbiol.* **66**, 3037–3043.

Gallagher, J. M., Carton, M. W., Eardly, D. F. and Patching, J. W. (2004). Spatio-temporal variability and diversity of water column prokaryotic communities in the eastern North Atlantic. *FEMS Microbiol. Ecol.* **47**, 249–262.

Garland, J. L. (1996). Analytical approaches to the characterization of samples of microbial communities using pattern of potential C source utilization. *Soil Biol. Biochem.* **28**, 213–221.

Garland, J. L. and Lehman, R. M. (1999). Dilution extinction of community phenotypic characters to estimate relative structural diversity in mixed communities. *FEMS Microbiol. Ecol.* **30**, 333–343.

Grace, C.-Y. and Wang, Y. (1996). The frequency of chimeric molecules as a consequence of PCR co-amplification of 16SrRNA genes from different bacterial species. *Microbiology* **142**, 1107–1117.

Grace, C.-Y. and Wang, Y. (1997). Frequency of formation of chimeric molecules as a consequence of PCR coamplification of 16S rRNA genes from mixed bacterial genomes. *Appl. Environ. Microbiol.* **63**, 4645–4650.

Haack, S. K., Fogarty, L. R., West, T. G., Alm, E. W., McGuire, J. T., Long, D. T., Hyndman, D. W. and Forney, L. J. (2004). Spatial and temporal changes in microbial community structure associated with recharge-influenced chemical gradients in a contaminated aquifer. *Environ. Microbiol* **6**, 438–448.

Hagström, A., Pommier, T., Rohwer, F., Simu, K., Stolte, W., Svensson, D. and Zweifel, U. L. (2002). Use of 16S ribosomal DNA for delineation of marine bacterioplankton species. *Appl. Environ. Microbiol.* **68**, 3628–3633.

Hall, B. G. (2001). *Phylogenetic Trees Made Easy.* Sinauer Associates, Inc., Sunderland.

Ibekwe, A. M. and Grieve, C. M. (2004). Changes in developing plant microbial community structure as affected by contaminated water. *FEMS Microbiol. Ecol.* **48**, 239–248.

Irby, A. (2001). *Microbial Diversity in Solar Salterns: Characterization of Whole Halophilic Communities by their Carbon Utilization Patterns and Examinations of Halophilic Bacteria that Grow at Elevated Magnesium Concentrations.* M.Sc. Thesis, George Mason University, Fairfax.

Johnson, M. J., Lee, K. Y. and Scow, K. M. (2003). DNA fingerprinting reveals links among agricultural crops, soil properties, and composition of soil microbial communities. *Geoderma* **114**, 279–303.

Kemp, P. F. and Aller, J. Y. (2004). Bacterial diversity in aquatic and other environments: what 16S rDNA libraries can tell us. *FEMS Microbiol. Ecol.* **47**, 161–177.

Kovach, W. L. (1999). *MVSP - A MultiVariate Statistical Package for Windows, ver. 3.1.* Kovach Computing Services, Pentraeth, Wales, U.K.

Lane, D. J. (1991). 16S/23S rRNA sequencing In *Nucleic Acid Techniques in Bacterial Systematics* (E. Stackebrandt and M. Goodfellow, eds), pp. 115–174. John Wiley & Sons, Chichester.

Litchfield, C. D. and Gillevet, P. M. (2002). Microbial diversity and complexity in hypersaline environments. *J. Indust. Microbiol. Biotechnol.* **28**, 48–56.

Litchfield, C. D., Irby, A., Kis-Papo, T. and Oren, A. (2001). Comparative metabolic diversity in two solar salterns. *Hydrobiologia* **466**, 73–80.

Lydell, C., Dowell, L., Sikaroodi, M., Gillevet, P. and Emerson, D. (2004). A population survey of members of the phylum *Bacteroidetes* isolated from salt marsh sediments along the east coast of the United States. *Microb. Ecol.* **48**, 263–274.

Mills, D. E., Fitzgerald, K., Litchfield, C. D. and Gillevet, P. M. (2003). A comparison of DNA profiling techniques for monitoring nutrient impact on microbial community composition during bioremediation of petroleum-contaminated soils. *J. Microbiol. Meth.* **54**, 54–74.

Pace, N. (1997). A molecular view of microbial diversity and the biosphere. *Science* **276**, 734–740.

Poole R. (1974). *An Introduction to Quantitative Ecology.* McGraw-Hill, New York.

Rappé, M. S. W., Suzuki, M. T., Vergin, K. L. and Giovannoni, S. (1998). Phylogenetic diversity of ultraplankton plastid small-subunit rRNA genes recovered in environmental nucleic acid samples for the Pacific and Atlantic coasts of the United States. *Appl. Environ. Microbiol.* **64**, 294–303.

Rees, G. N., Baldwin, D. S., Watson, G. O., Perryman, S. and Nielsen, D. L. (2004). Ordination and significance testing of microbial community composition derived from terminal restriction fragment length polymorphisms: application of multivariate statistics. *Antonie van Leeuwenhoek* **86**, 339–347.

Ritchie, N. J., Schutter, M. E., Dick, R. P. and Myrold, D. D. (2000). Use of length heterogeneity PCR and fatty acid methyl ester profiles to characterize microbial communities in soil. *Appl. Environ. Microbiol.* **66**, 1668–1675.

Seviour, R. J., Mino, T. and Onuki, M. (2003). The microbiology of biological phosphorus removal in activated sludge systems. *FEMS Microbiol. Rev.* **27**, 99–127.

Sharma, S., Piccolo, A. and Insam, H. (1997). Different carbon source utilization profiles of four tropical soils from Ethiopia. In *Microbial Communities: Functional versus Structural Approaches* (H. Insam and A. Rangger, eds), pp. 132–139. Springer-Verlag, Berlin.

Singleton, D. R., Furlong, M. A., Rathbun, S. L. and Whitman, W. B. (2001). Quantitative comparisons of 16S rRNA gene sequence libraries from environmental samples. *Appl. Environ. Microbiol.* **67**, 4374–4376.

Smith, S. W., Overbeek, R., Woese, C. R., Gilbert, W. and Gillevet, P. M. (1994). The genetic data environment (GDE): an expandable graphic interface for manipulating molecular information. *CABIOS* **10**, 671–675.

Spring, S., Schulze, R., Overmann, J. and Schleifer, K. (2000). Identification and characterization of ecologically significant prokaryotes in the sediment of freshwater lakes: molecular and cultivation studies. *FEMS Microbiol. Rev.* **24**, 573–590.

Suzuki, M., Rappé, M. S. and Giovannoni, S. J. (1998). Kinetic bias in estimates of coastal picoplankton community structure obtained by measurements of small-subunit rRNA gene PCR amplicon length heterogeneity. *Appl. Environ. Microbiol.* **64**, 4522–4529.

Swofford, D. L. (2001). PAUP*. *Phylogenetic Analysis Using Parsimony* (*and Other Methods). Sinauer Associates, Sunderland.

Thompson, J. D., Gibson, T. J., Plewniak, F., Jeanmougin, F. and Higgins, D. G. (1997). The Clustal X windows interface: flexible strategies for multiple sequence alignment aided by quality analysis tools. *Nucleic Acids Res.* **24**, 4876–4882.

Wilson, M. and Lindow, S. E. (1994). Ecological similarity and coexistence of epiphytic ice-nucleating (Ice$^+$) *Pseudomonas syringae* strains and a non-Ice-nucleating (Ice$^-$) biological control agent. *Appl. Environ. Microbiol.* **60**, 3128–3137.

Zak, J. C., Willig, M. R., Moorehead, D. L. and Wildman, H. G. (1994). Functional diversity of microbial communities: a quantitative approach. *Soil Biol. Biochem.* **26**, 1101–1108.

Zhang, Z. and Madden, T. L. (1997). PowerBLAST: A new network BLAST application for interactive or automated sequence analysis and annotation. *Genome Res.* **7**, 649–656.

Zhang, Z., Schwartz, S., Wagner, L. and Miller, W. (2000). A greedy algorithm for aligning DNA sequences. *J. Comput. Biol.* **7**, 203–214.

22 Cultivation of Haloarchaea

David Burns and Mike Dyall-Smith
Department of Microbiology and Immunology, University of Melbourne, Melbourne 3010, Australia

◆◆◆

CONTENTS

Introduction
Methods of isolation
Cultivation of salt lake microorganisms
Cultivation of water samples on solid media
Nutrients
Serial dilution/extinction culture
Summary

◆◆◆◆◆ INTRODUCTION

The isolation and laboratory cultivation of extremely halophilic Archaea (Order Halobacteriales) can present some significant technical challenges. The high salinities and incubation temperatures may cause salt precipitates to form, and limits the use of certain additives, such as gelling agents (e.g. gellan gum or Gelrite®) and indicators (e.g. skim milk powder for detecting protease activity, which coagulates rapidly in high salt media). Drying of media must be minimized or salinities will change significantly over the often lengthy incubation periods (up to 10 weeks), and this can lower the growth rate of organisms that are relatively slow-growing even at their optimum salt concentration. Nevertheless, a wide range of haloarchaea have been isolated from hypersaline waters, including members of almost all of the eighteen described genera within the family *Halobacteriaceae* (Grant *et al.*, 2001; Oren, 2001). Halophilic members of the domain Bacteria can comprise up to 25% of the prokaryotic community in these environments but they are usually at a much lower percentage (Antón *et al.*, 2000). At times, the flagellated green alga *Dunaliella* can also be present (Casamayor *et al.*, 2002).

Until recently, microbiological studies of hypersaline lakes have often reported viable counts far lower than the direct counts of the same waters, and it was not clear how accurately the laboratory isolates reflected the dominant organisms in the natural environment. We now know that in most cases, they did not. The identities and relative abundances of organisms in natural populations can now be determined with reasonable certainty using PCR-based strategies that target 16S small subunit ribosomal ribonucleic acid (16S rRNA) genes. While comparatively few studies of this type have been performed so far (Benlloch et al., 2001; Bowman et al., 2000; Burns et al., 2004a; Elshahed et al., 2004; Ochsenreiter et al., 2002), the results suggest that some of the most readily isolated and studied genera are only minor fractions of the *in situ* community. This is seen in cases such as the genera *Haloarcula* and *Haloferax*, which are estimated to make up only small fractions of the *in situ* community, but are over-represented in isolation studies (Antón et al., 1999; Burns et al., 2004a).

While the microbial diversity in hypersaline lakes is low, the productivity and cell densities are often high and, even in lake waters with saturated salt concentrations, microbial cell densities can reach up to 10^7–10^8 cells ml^{-1} (see review by Oren, 2001). These populations reflect a lack of predation and often quite high nutrient levels (Oren, 2002). The low total diversity makes hypersaline lakes an ideal candidate system for ecological studies. Understanding nutrient cycling and population dynamics are more manageable tasks with such relatively simple ecosystems compared to a high diversity environment like soil, which can contain as many as 12 000–18 000 species g^{-1} (Torsvik et al., 1996). In particular, multi-pond solar salterns represent ideal candidate model systems due to their managed nature, in which salt concentrations are kept relatively constant over time.

Isolation of the dominant organisms from hypersaline environments is a prerequisite for understanding their ecology, and for studying one of the most easily accessed groups of Archaea, the haloarchaea, but until recently, the dominant species, as identified by fluorescent *in situ* hybridization (Antón et al., 1999), clone library (Bowman et al., 2000) and microscopic analysis (Walsby, 1980), had not been grown (Burns et al., 2004a and literature cited within). Presumably responsible for a significant part of the nutrient cycling in this environment, the square, flat and gas-vacuolated, bacteriorhodopsin-containing archaeon first described by Walsby (1980) was cultivated in 2004 using relatively unconventional techniques (Bolhuis et al., 2004; Burns et al., 2004b). Another group, the Antarctic Deep Lake group, also represents a significant fraction of the microbial community, at least in Antarctic and Southern Hemisphere salt lakes, and was only cultivated last year (Burns et al., 2004a).

It has been shown in soil that reduced sample inoculum increases the cultivation success rate, presumably due to reduced competition (Davis et al., 2005; Olsen and Bakken, 1987). These studies also found that lower nutrient levels lead to greater cultivation success. Both of these approaches have been applied to various extents to saltern crystallizer ponds with varying but limited degrees of success (Benlloch et al., 2001).

Isolation of pure strains is an essential requirement to properly study organisms as it allows study of their metabolic capabilities and hence an insight into their role in the natural environment. In addition, pure cultures allow investigations into the interaction of specific strains and/or other organisms in their natural environment, including the isolation and study of their viruses. By combining molecular methods with newer cultivation techniques, it should be possible to gain considerable new insights into the community structure, dynamics and evolution of hypersaline environments.

♦♦♦♦♦ METHODS OF ISOLATION

Sampling of Salterns and Salt Lakes

There are a number of considerations that should be taken into account when sampling for cultivation studies:

1. The geographical co-ordinates of each site should be recorded, preferably using a GPS device with a low error.
2. Samples should be collected in sterile, water-tight containers with a capacity of at least a few hundred millilitre. While a few millilitre is usually enough for cultivation purposes, you may find if you want a chemical composition of the water, the analytical laboratory may ask for 100–500 millilitre.
3. Avoid contamination with flora from your body and clothing by holding the sample bottle away from you – either on a string or at the end of a stick.
4. Whether you take water or solid salt from the lake bed depends on the nature of the study, but the majority of haloarchaea are found suspended in the water column, and those with gas vesicles will be near the surface. Sometimes this is not an issue because the water depth is only a few centimetres, and the open end of the bottle must be skimmed rapidly in order to obtain sufficient liquid.
5. If you are sampling at some distance from the laboratory, storage of samples at 4°C is may be prudent, so an insulated container and ice should be taken. However, you should try to avoid this and simply use your sample as soon as possible.
6. Label the sample bottles clearly, with date, time, and location (GPS coordinates).
7. A digital camera to record the sample site is also useful documentation.

♦♦♦♦♦ CULTIVATION OF SALT LAKE MICROORGANISMS

Salt Water

For the cultivation of organisms from thalassohaline lakes, the base of all media is a salt solution that mimics concentrated seawater. Many haloarchaea isolated from such waters will tolerate a fairly generous salinity range and ionic composition (although growth rates will decrease away from their optima). Previously published salt-water compositions

(Oren et al., 2001) are likely to be acceptable. For athalassohaline waters, a chemical analysis may be useful to determine if there are significant deviations from the relative ionic ratios of seawater. For example, the Dead Sea has very high levels of Mg^{2+} and organisms isolated from these waters, such as *Halorubrum sodomense* (Oren, 1983) and *Halobaculum gomorrense* (Oren et al., 1995), require similarly high Mg^{2+} levels for growth. In general, you should try to match the salt composition to that of the environment being studied.

In specific instances, greater cultivation success may be achieved using sterilized water taken from the same environment as the isolation is occurring from. Centrifugation and/or filtration may be used to remove excess organic matter before sterilization (particularly important if a specific substrate is to be added). Hypersaline waters that are visibly turbid will block filters rapidly, so this is best performed after centrifugation (and perhaps also a pre-filtration step). Autoclaving should be considered essential as, even after 0.2-µm filtration, high concentrations of viruses are likely to be present.

Concentrated Seawater Solution – 30% Stock

First, it is important to note that haloarchaea are generally very sensitive to detergents (and will lyse!), so glassware should be rinsed well after cleaning so that no traces of soaps or detergents remain.

The composition of this stock solution is based upon that described by Rodriguez-Valera et al. (1980) and Torreblanca et al. (1986) and contains salts in approximately the same proportions as found in sea water but at a much higher concentration of total salt. The original formula of Rodriguez-Valera et al. contains low levels of $NaHCO_3$ and NaBr, but we have found them to be unnecessary. We have also slightly adjusted the masses to more closely give a total of 30% total salt, and to make weighing out simpler. Note that when using hydrated salts (e.g. $MgSO_4 \cdot 7H_2O$), the masses need to be adjusted to allow for the water content.

1. Add the following salts to a large beaker.

Salt	$g\,l^{-1}$
NaCl	240
*$MgCl_2 \cdot 6H_2O$	30
*$MgSO_2 \cdot 7H_2O$	35
KCl	7
**NaBr	0.8
**$NaHCO_3$	0.2

*Note the hydrated salts: adjust if using anhydrous salts
**We do not normally add these to our media.

2. Add double-distilled water to near the final required volume and dissolve the salts completely using a glass stirring rod or a magnetic stirrer (warming may help, but is not normally required).

3. Add 5 ml $CaCl_2 \cdot 2H_2O$ *slowly* from a 1 M sterile stock solution (final concentration = 0.5 g l^{-1}).
4. Adjust the pH up to 7.5 with a minimum volume of 1 M Tris·HCl buffer (pH 7.5). This is usually about 2 ml l^{-1}. Alternatively, a smaller volume of 1M Tris base can be used. You may have to lower the pH with a small amount of HCl in order to add buffer without overshooting the pH (it is often at or above 7.5 before buffer is added).
5. Transfer to a large, graduated cylinder, then top up with water to the exact final volume. Do not use graduated beakers as they can be very inaccurate.
6. Dispense into convenient volumes (e.g. 100–500 ml) in autoclavable glassware (e.g. Schott bottles).
7. Autoclave at 101 kPa (15 psi) for 15 min (100 ml volumes) or 30 min (500 ml volumes). Store at room temperature. There should be no precipitate after autoclaving. If there is, try adding the $CaCl_2$ solution after autoclaving and when the solution has cooled to room temperature. If the salt solution is to be used completely within 1–2 weeks, it is convenient to store it, unsterilized, in a large sealed flask at 4°C.
8. Dilute the SW stock solution using double-distilled water as required to make up your preferred concentration of salt water.

The salt-water stock can be used directly to prepare culture media for neutrophilic haloarchaea. However, for the isolation of haloarchaea from natural waters it may be prudent to include some additional trace elements, such as those in the SL10 solution, described below.

Trace Element Solution SL10

The components of the trace element solutions SL10 (Widdel *et al.*, 1983) are added and dissolved in the order listed.

SL10	
H_2O	1.0 l
25% HCl	10 ml
$FeCl_2 \cdot 4H_2O$	1.5 g
$CoCl_2 \cdot 6H_2O$	190 mg
$MnCl_2 \cdot 4H_2O$	100 mg
$ZnCl_2$	70 mg
H_3BO_3	6 mg
$Na_2MoO_4 \cdot 2H_2O$	36 mg
$NiCl_2 \cdot 6H_2O$	24 mg
$CuCl_2 \cdot 2H_2O$	2 mg

The trace element solution is autoclaved under air, in 25-ml aliquots in 50-ml screw-capped bottles. Large stocks do not need to be sterilized for storage. Use 1 ml l^{-1} of media.

♦♦♦♦♦ CULTIVATION OF WATER SAMPLES ON SOLID MEDIA

Agar is the most commonly used gelling agent for halophiles. However, for isolation, washed agar (or agarose) is recommended; it allows more specific formulation of the isolating media by removing impurities in the agar.

Washed Difco-Bacto Agar

In order to remove inhibitory components from this agar, it is washed with distilled water as follows:

1. Add 16.5 g agar (Difco-Bacto) (10% allowance for loss during washing) to distilled water.
2. Stir vigorously for 5 min and then allow the agar to settle for 30 min and decant the supernatant.
3. Repeat this process until the supernatant is clear (2–4 rinses.)
4. Add sufficient water to make the volume to 50 ml, stir the washed agar into a suspension, and mix with the desired medium (i.e. for 1.5% final concentration in 1 l of medium). Do not forget to take into account the dilution effect of the agar suspension on the final salt concentration.

As an alternative to washed agar, agarose or gellan (Gelrite®, Phytagel™) may be used. The expense of agarose makes this suitable only for small volumes. Gellan sets almost instantly when mixed with solutions of moderate to high salinity.

Once isolates have been cultivated for a few passages, it is often possible to use standard (unwashed) agar and media with higher nutrient levels than that used for isolation. These changes may give higher growth rates and cell densities. *This is a known phenomenon – but it is still important to be scrupulous in the isolation phase.*

Plating and Incubation

A range of dilutions should be made. If a statistically reliable viable count is required, you should aim to have colony numbers in the range 30–300. A higher viable count is usually obtained with minimal crowding on the plate – ten to fifty colonies.

Plates should be placed in sealed plastic containers to prevent moisture loss over the incubation period. This is especially important for longer incubations, which are more likely to yield novel or uncommon isolates (Burns et al., 2004a; Davis et al., 2005). To maintain humidity over extended incubations to reduce plate moisture loss, some exposed water agar plates or even an open vessel of water may be placed in the incubation container.

The container may be opened periodically to count colonies (e.g. at 2, 4 and 8 weeks) and check the progress of the experiment. This is also a good time to remove excess moisture from the lid (either replace the lid or simply flick out the water).

Growth of most isolates will be satisfactorily incubated at 37°C in the dark. However, for those seeking isolates that utilize bacteriorhodopsin, for instance, incubation under lights with reduced oxygen levels may be preferable. Remember that if algae such as *Dunaliella salina* are present in the water sample, these may also grow on the plates incubated under light, producing either green or red/orange colonies (depending upon the salt concentration).

Incubation at 37°C is suitable for the growth of most extremely halophilic Archaea (and *Salinibacter*), and is convenient because incubators at this temperature are often available. However, the optimum temperature for growth is usually about 10°C higher than this, and raising the temperature can significantly improve growth rates on solid and liquid media (Robinson et al., 2005). For example, the widely studied haloarchaeon *Haloferax volcanii* grows significantly faster at 45°C than at 37°C (Thorsten Allers, personal communication cited in the Halohandbook, 2004), as does *Hfx. mediterranei* (Lillo and Rodríguez-Valera, 1990). Growth near the optimum temperature may be useful in the isolation and maintenance of slow-growing haloarchaea.

It is imperative that strict aseptic technique be adhered to when isolating and passaging slow-growing organisms to avoid contaminants swamping the growth of the desired isolate. Increased diversity appears with time rather than dilution, and it is better to have crowded plates incubated longer than sparsely populated plates incubated only a few days. Plates should be incubated aerobically if attempting to isolate from the water column. For reduced oxygen conditions, commercially available sachets or containers may be used to lower the oxygen concentration to an appropriate level.

If cultures are passed into liquid media, some strains may exhibit reduced growth if shaken at rates that are commonly used for *E. coli* cultures (>150 rpm). In such cases, shaking at a reduced rate (50–80 rpm) or without shaking, may improve growth. We are not sure if this represents sensitivity to oxygen or physical damage due to turbulence.

◆◆◆◆◆ NUTRIENTS

Major Nutrients

Excluding haloalkaliphiles, most haloarchaea and halophilic bacteria (e.g. *Salinibacter*) will grow on Modified Growth Medium (MGM), a medium formulated around a salt water solution that mimics the composition of concentrated seawater, and a rich nutrient source (5 g l^{-1} peptone and 1 g l^{-1} yeast extract). However, some groups, while able to grow on MGM, will display optimum growth only when supplements are added, or different media used. The selection of nutrients used may vary considerably dependent on the target species. The majority of organisms may be isolated on rich media, such as MGM, and may actually require complex media for isolation (such as the Antarctic Deep Lake group (Burns et al., 2004a.) However, some specific organisms require a particular nutrient or substrate to grow, as demonstrated by the requirement for pyruvate in the isolation of the Square Haloarchaea of Walsby (Bolhuis et al., 2004; Burns et al., 2004b).

Minor Nutrients

Addition of vitamins after autoclaving media is a prudent step to ensure growth of fastidious organisms. This may be especially important if artificial media are used, rather than media prepared from natural water (i.e. water from the same lake as the sample).

Vitamin 10 stock

H_2O	1000 ml
4-aminobenzoate	13 mg
d-(+)-biotin	3 mg (store at 4°C)
nicotinic acid	33 mg
hemicalcium D-(+)-pantothenate	17 mg (store at 4°C)
pyridoxamine hydrochloride	50 mg (store at 4°C)
thiamine chloride hydrochloride	33 mg (store at 4°C)
cyanocobalamin	17 mg (store at 4°C)
D,L-6,8-thioctic acid	10 mg (store at 4°C)
riboflavin	10 mg (store at 4°C)
folic acid	4 mg (store at 4°C)

Vitamin 10 solution is sterilized by filtration into sterile bottles using 0.2-μm pore-size cellulose acetate filters. The bottles are wrapped in aluminium foil to protect against light, and stored at 4°C. Add aseptically to media at 3 ml l^{-1} (Janssen et al., 1997). Note that in the latter reference, these additives are denoted as Vitamin solution 1 and Vitamin solution 2. Vitamin 10 is simply a combination of these two with the same final concentrations of ingredients.

The media we have found to give good growth over the widest range of isolates is DBCM (below). It may be used without an additional carbon source, or, if a good carbon source for the isolate is known or suspected, with substrate included. Growth is definitely enhanced with additional substrate, but as listed here, growth is still reasonable. It is based on the Chemically Defined Medium (CDM) of Kauri et al. (1990) but with some differences. For easy comparison, the recipe for CDM is given in Section 2.5.1 of The Halohandbook (http://www.microbiol.unimelb.edu.au/staff/mds/HaloHandbook/).

DBCM (DB Characterization Media)

The basic salt-water stock is made to the desired concentration (here 25%), using the MDS salt water recipe (Section 2.2 in *The Halohandbook*) and dilute appropriately.

1. To autoclaved SW, add the following (filter-sterilized):
   Per litre

1 M NH$_4$Cl	5 ml
K$_2$HPO$_4$ buffer	2 ml
SL10 Trace Elements	1 ml
Vit10 vitamin solution	3 ml
Carbon source	See below

2. Carbon source: ~10 mM pyruvate, or 0.44 ml per 100 ml of a filter-sterilized, 25% w/v refrigerated stock solution. This works well for most strains.
3. It is not necessary to check the pH – it should still be around 7–7.5. We have found that pyruvate does not seem to substantially alter the pH (including end products, etc.). If using an acid-producing or acidic substrate, extra buffering may be required. NaHCO$_3$ (at 5 ml l^{-1}) is recommended for this purpose.
4. Add 50 ml of an appropriate MGM per litre of the medium. The basic formulation will give acceptable growth by itself, but most growth should come from an added substrate known to be utilized by the culture (e.g. glycerol or pyruate). If performing characterization or other tests, a lesser amount of MGM may be used to limit growth occurring from the MGM (i.e. lower background growth).

0.5 M Potassium Phosphate Buffer, pH 7.5

1. For 200 ml 0.5 M KPO$_4$ buffer (pH 7.5)
   1 M K$_2$HPO$_4$ 83.4 ml
   1 M KH$_2$PO$_4$ 16.6 ml
2. Check pH is close to 7.5
3. Add equal volume (100 ml) of pure H$_2$O
4. Autoclave

SERIAL DILUTION/EXTINCTION CULTURE

If the target organism is not being isolated on solid media, it is possible to isolate axenic cultures in liquid media. Microscopes fitted with laser-tweezers (Huber et al., 1995) or mechanical micromanipulators would probably work well for placing individual cells in culture tubes, but these are not widely available. A simpler and cheaper method is to use serial dilution (or extinction) cultures. The sample is serially diluted (in replicate tubes) so that, at a sufficiently high dilution, individual tubes are expected to contain only one cell. Cultures arising from such tubes should be pure. Dilutions can be performed in any suitable container, but for large experiments, plastic tissue culture trays (with many wells per tray) are useful. To maintain correct osmolarity over extended incubation periods, volumes >1 ml are preferable. This also allows sufficient material for screening (PCR, microscopy) and for passaging.

Cells should be diluted in the media being used for growth. The correct dilution factors will depend on the sample cell density. A direct count should be performed to estimate appropriate dilutions. However, as cultivation efficiency will vary, some small experiments should be performed to gain a more accurate picture of which dilutions should be targeted before large-scale isolation attempts are made.

Short Protocol for Extinction Cultures

1. Make a direct count of cell numbers. The acridine-orange stain is quick and convenient (see below).
2. Prepare 10-fold serial dilutions at least two orders of magnitude above and below the direct count. MacCartney tubes, tissue culture trays, or test tubes are among the most suitable containers to use for this step. Initial dilutions can be performed in single tubes, and expanded to multiple replicates once the desired dilution range is reached. It is important to serially dilute after this stage; do not just create each dilution from the one tube using appropriate volumes.
3. Incubate and screen cultures. The ideal dilution level is one that is positive for growth in about 30% of the samples. Lower positive percentages increase the chance of pure cultures, while much higher than 30% increases significantly the chance of obtaining a mixed culture. However, too low a growth percentage means you will have to screen an excessive number of cultures. It is better in this case to move up a dilution level, or move to step 4.
4. If necessary, prepare another dilution series using the results gained for cultivation efficiency to narrow down the dilution ranges (say 5-fold instead of 10-fold). At this stage, it may also be possible to vary incubation and/or media conditions. Do this by adding a series to the existing conditions, rather than simply replacing them.

If you wish to calculate a Most Probable Number, it is necessary to have dilutions that are entirely positive for growth, and dilutions that are entirely negative for growth (Cochran, 1950).

Enrichment Culture

As an alternative to serial dilutions, enrichment cultures can be performed. Enrichment may be preceded by serial dilutions, or other techniques for increasing the proportion of the target organism in the sample inoculum. A number of media have been developed for the selection or enrichment of specific microorganisms in hypersaline water samples. Advantage has been made of cell physiology, including anaerobic growth using nitrate as an electron acceptor (Hochstein and Tomlinson, 1985; Tomlinson et al., 1986); anaerobic growth on L-arginine (to enrich for *Hbt. salinarum*) (Oren and Litchfield, 1999), and phototrophy (Oesterhelt and Krippahl, 1983). *Halobacterium salinarum* has complex nutritional requirements whereas many other haloarchaea can grow in simple, defined media with single carbon/energy sources (Grant et al., 2001). While there is considerable variation in substrate utilization among the haloarchaeal taxa (Grant et al., 2001), there are relatively few examples in the literature where the nature of the carbon source has been used for the enrichment of specific groups. Haloarchaea were isolated by enrichment culture on the C_{20} hydrocarbon eicosane (Bertrand et al., 1990; Emerson et al., 1994), and the square haloarchaeon appears to require pyruvate for successful isolation from mixed cultures (Bolhuis et al., 2004). There is also a curious study by Allen Wais (Wais, 1988), which showed that the growth of haloarchaea from natural hypersaline waters could be significantly enhanced by the addition of an extract of *Halobacterium salinarum* cells. The components involved in this effect have yet to be identified.

Archaea and Bacteria show great differences in their sensitivity to antimicrobial drugs, and this information has also been used for formulating selective media that suppress either Bacteria or haloarchaea. For example, haloarchaea were recently isolated from a low salt environment (containing a high percentage of bacteria) using selective media containing ampicillin and kanamycin (Elshahed et al., 2004). On the other hand, the isolation of *Salinibacter ruber* (an extremely halophilic bacterium) utilized anisomycin (Hummel and Böck, 1987) and bacitracin (Mescher and Strominger, 1975) to suppress the growth haloarchaea (Oren et al., 2004). For a recent review of the antibiotic sensitivities of haloarchaea, see Oren (2001).

Monitoring Growth

If using sterilized water from the natural environment, or a medium with a very low substrate concentration, cell growth is unlikely to reach the point of becoming visibly turbid. In this case, more sensitive methods are required to monitor growth, and some alternatives are given below.

Direct Microscopy

If simply screening for growth, direct microscopy can be useful for densities above about 10^5 cells ml^{-1}. Phase contrast or fluorescent microscopy (e.g. Acridine Orange) of culture samples are very straightforward and generally give unambiguous results. If the target organism has a distinct morphology (such as the ADL or SHOW (square haloarchaea of Walsby) groups) or feature (such as gas vesicles), then microscopy will also provide evidence for identifying isolates, purity of cultures, or for the progress of an enrichment.

Acridine Orange Fluorescent Staining of Halobacteria

This is a rapid and easy method for visualizing unfixed haloarchaeal cells. Cell structure seems to be preserved well and the fluorescence is usually bright and covers the entire cell. The method is based on various sources. Antifade agents (e.g. DABCO, PPD) are necessary or the fluorescence fades dramatically within 30 s of UV illumination. DABCO precipitates in high salt, and cannot be used, but PPD (para-phenylenediamine) is satisfactory. We have not tried other agents.

Fluorescent Staining of Live Cells Using Acridine Orange

1. Depending upon the concentration of cells, the culture may need to be diluted 10 or more fold. This should be done using sterile media with the same salt concentration as the culture to be counted (otherwise cells will distort or lyse). For low cell densities, it may be necessary to concentrate the cells by centrifugation (e.g. 5 min, 6000 rpm, benchtop microfuge), then gently resuspend in a one-tenth volume (of the same medium). Note: Centrifugation may collapse gas vesicles or damage delicate cells (e.g. square haloarchaea).
2. To 100 µl of culture, add 5 µl of PPD stock solution (10% w/v and mix quickly by tapping the side of the tube with a finger.
3. Add Acridine Orange stain (1–5 µl of a 1 mg ml^{-1} AO stock, made up in water) and mix quickly (by tapping with a finger rather than vortexing). The sample should then show a faint orange colour. Leave for at least 5 min for the stain to penetrate cells. If detecting growth, go to step 4. If measuring growth, go to step 5.
4. To *detect growth* (but not measure it), put 5–10 µl of the stained sample on a clean glass slide, gently place a coverslip on top, and seal the edges with vaseline or a quick-drying slide mounting solution. (Go to step 6)

    a. We have found that agar-coated slides work poorly with a high salt culture sample. The dried agar (or agarose) does not absorb much of the water from the sample, and so does not immobilize cells well. There may also be a high background fluorescence with some agar-coated slides.

5. To *measure growth*, you need a counting chamber, preferably one that is designed to count bacteria (i.e. has a small depth). We use a Thoma counting chamber with a 0.1 µm distance between the coverslip and etched slide, and with metal clamps to hold the coverslip. The grid pattern has squares with sides of 50 µm. Add sufficient cell suspension to fill the gap between the coverslip and slide and to cover the grid area of the slide.

6. Examine microscopically by epifluorescence using the correct filter. Objectives at 40× and 100× will allow slides to be scanned rapidly at high power and for individual cells to be examined in detail with oil immersion. When using the counting chamber, you will need to move the focus up and down in order to see and count all the cells in the volume of each field of view. You will usually see the bright but blurred fluorescence of cells even when well out of focus. Count the cells in enough squares to attain between 50 and 200, and calculate the cells ml^{-1} in the original culture (taking the dilution or concentration steps into account).

Antifade Solution: Para-Phenylene Diamine (PPD) Stock Solution

1. 0.1 g PPD in 1 ml of pure water in a 1.5-ml plastic microfuge tube (Use a mask and gloves).
2. Mix and allow 5–10 min to dissolve (may need to warm slightly).
3. Cover the tube with metal foil and store at −20°C. PPD is light sensitive and if you leave it out on the bench for a day or so, the solution will go progressively black. If you store it at −20°C and keep it away from the light, it should last for weeks to months.

Acridine Orange (AO) Stain Solution

1. A stock solution of 1 mg ml^{-1} in pure water (Use a mask and gloves).

You need very little stain solution per sample (1–5 µl), so 1–10 ml should last quite a while. However, weighing out milligram quantities of powder can be difficult, so a two-stage approach is easier, i.e. make a 10 mg ml^{-1} solution (e.g. 0.1 g AO in 10 ml water) and dilute a small volume of this 10-fold to make 1 mg ml^{-1}.

Like PPD, Acridine Orange is light sensitive, so wrap the solutions in foil and do not expose to sunlight longer than is necessary. If a fine, orange precipitate forms, and will not redissolve (by warming), you can remove it by centrifugation (5 min, 12 000 rpm) and filtration (0.45-µm filter), although it may be best to simply make up another stock. Store both stock solutions at −20°C.

PCR Screening for Growth and Identification

PCR is an excellent method for screening small-volume cultures during isolations, as well as any culture that cannot reach visible turbidity. It provides a high level of sensitivity and has the added advantage of providing information that can identify specific organisms, including whether the culture is mixed or pure.

DNA should be extracted from the culture by any method that is preferred by the researcher, although particular care must be taken in the case of extreme halophiles to remove salts from the final preparation. The target would usually be the 16S rRNA gene (although other, highly conserved genes could be used if desired), and is amplified using consensus primers that are sufficiently broad to cover all possible organisms (Lane, 1991).

While the presence of microorganisms can be detected purely by the presence of an amplified target sequence, the PCR can include a third primer (multiplex PCR) chosen so that an additional product will be amplified if the culture contains a specific organism. Design of a suitable organism-specific primer may be performed using the ARB 16S sequence editing and database program (Ludwig et al., 2004), or gleaned from existing literature. The PCR product can also be sequenced or digested (using restriction enzymes) for confirmation of the culture identity and purity.

As an example, the protocol used to detect SHOW organisms in extinction culture tubes is given below.

Multiplex PCR Screening Protocol

1. DNA is extracted from cultures using any method that is preferred, and that will work for the organisms being studied. A relatively quick method for screening of cultures for SHOW organisms and most haloarchaea (those with weak cell walls consisting of only an S-layer), is given below. (For organisms with strong cell walls, more severe lysis conditions would be needed.)
   a. Remove a small volume from a culture tube (0.2–1 ml) to a clean plastic microfuge tube, centrifuge the cells to a pellet (2 min, 12 000 rpm), and remove all salt water and any pelleted crystals (with a fine-bore pipette tip). Removal of as much salt as possible is important in this method, as carry over of salt in the DNA extract may severely inhibit the polymerase used in PCR.
   b. Lyse the cells by quickly adding 200-µl distilled water to the cell pellet and mixing (by up-and-down pipetting). If no visible cell pellet is seen after step a, use 100-µl distilled water. Use 1 µl of lysate in PCR reactions.

2. 16S rRNA gene sequences of haloarchaea are amplified by PCR using consensus primers F1, ATTCCGGTTGATCCTGC (Ihara et al., 1997) and 1492Ra, ACGGHTACCTTGTTACGACTT (Grant et al., 1999). These primers will amplify most of the gene, giving an amplicon of around 1.5 kb.
   a. Addition of a third primer to the PCR, one that binds within the F1/1492Ra amplicon, can be used to detect the presence of a specific organism. In the example given here, we use the SHOW-specific (reverse-) primer SHOWprb, ACGGCACAACAG AGACGC (Burns et al., 2004b). In the presence of DNA from a SHOW organism, the SHOWprb primer, in combination with

the F1 primer, would give an amplicon of around 800 nt. Positive SHOW cultures would be expected to give two amplicons in this multiplex PCR, i.e. 1.5 and a 0.8 kb, while cultures of non-SHOW organisms would give just the 1.5-kb amplicon.

The multiplex PCR could include more primers, but then greater effort is required to optimize and troubleshoot the reaction. We would not recommend more than two or three specific primers.

3. PCR reactions contain final concentrations of 1.75 mM $MgCl_2$, 1x PCR buffer (Qiagen), 200 μM dNTPs, 37.5 pmol forward primer, 37.5 pmol organism-specific reverse primer, 60 pmol reverse primer and 2 units HotStarTaq DNA polymerase (Qiagen). The volume is made up to 49 μl with pure water.
4. 1 μl of cell lysate is added, and the solution mixed.
5. PCR cycling conditions are: 15 min/95°C; 30 × (1 min/95°C; 30 s/46°C; 90 s/72°C); 10 min/72°C.
6. The products of multiplex PCR are then examined by agarose gel electrophoresis (1.2% agarose gels). Negative control (e.g. no DNA added) and positive control reactions should be included in PCR assays (e.g. DNA extracted from a salt lake sample containing visible numbers of square haloarchaea, and that produces a SHOW-specific PCR amplicon with the F1 and SHOWprb primers).
7. Cultures that give SHOW-specific amplicons (0.8 kb) should be analysed further by sequencing the PCR products and comparing them with the sequence databases. A small amount of the culture can also be removed for microscopy (e.g. Acridine Orange stain, see above), to confirm that SHOW cells are present.

◆◆◆◆◆ SUMMARY

The authors hope that this chapter has clearly demonstrated that most extreme halophiles can be cultivated using inexpensive materials and relatively simple variations of traditional methods. A combination of factors, including advances in the understanding of the differences often required between isolation media and standard laboratory media, the need for greater patience in cultivation, and molecular screening techniques have radically improved the prospects for studying the dominant haloarchaea in natural hypersaline waters. This development, along with genomic and metagenomic studies (of both cells and haloviruses) that are underway, will no doubt lead to many new insights into life in these extreme environments.

List of suppliers

Greiner Bio-One GmbH
Maybachstrasse 2

D-72636 Frickenhausen, Germany
E-mail: info@de.gbo.com

24 well per plate tissue culture trays

Fluka Chemicals
c/- Sigma Aldrich
www.sigmaaldrich.com

Pyruvic acid, sodium salt

Oxoid Limited
Wade Road, Basingstoke
Hampshire, RG24 8PW
England
E-mail: oxoid@oxoid.com

Yeast extract (L21), Peptone (L37)

References

Antón, J., Llobet-Brossa, E., Rodríguez-Valera, F. and Amann, R. (1999). Fluorescence in situ hybridization analysis of the prokaryotic community inhabiting crystallizer ponds. *Environ. Microbiol.* **1**, 517–523.

Antón, J., Rosselló-Mora, R., Rodríguez-Valera, F. and Amann, R. (2000). Extremely halophilic bacteria in crystallizer ponds from solar salterns. *Appl. Environ. Microbiol.* **66**, 3052–3057.

Benlloch, S., Acinas, S. G., Antón, J., López-López, A., Luz, S. P. and Rodríguez-Valera, F. (2001). Archaeal biodiversity in crystallizer ponds from a solar saltern: culture versus PCR. *Microb. Ecol.* **41**, 12–19.

Bertrand, J. C., Almallah, M., Acquaviva, M. and Mille, G. (1990). Biodegradation of hydrocarbons by an extremely halophilic archaebacterium. *Lett. Appl. Microbiol.* **11**, 260–263.

Bolhuis, H., te Poele, E. M. and Rodríguez-Valera, F. (2004). Isolation and cultivation of Walsby's square archaea. *Environ. Microbiol.* **6**, 1287–1291.

Bowman, J. P., McCammon, S. A., Rea, S. M. and McMeekin, T. A. (2000). The microbial composition of three limnologically disparate hypersaline Antarctic lakes. *FEMS Microbiol. Lett.* **183**, 81–88.

Burns, D. G., Camakaris, H. M., Janssen, P. H. and Dyall-Smith, M. L. (2004a). Combined use of cultivation-dependent and cultivation-independent methods indicates that members of most haloarchaeal groups in an Australian crystallizer pond are cultivable. *Appl. Environ. Microbiol.* **70**, 5258–5265.

Burns, D. G., Camakaris, H. M., Janssen, P. H. and Dyall-Smith, M. L. (2004b). Cultivation of Walsby's square haloarchaeon. *FEMS Microbiol. Lett.* **238**, 469–473.

Casamayor, E. O., Massana, R., Benlloch, S., Øvreås, L., Díez, B., Goddard, V. J., Gasol, J. M., Joint, I., Rodríguez-Valera, F. and Pedrós-Alió, C. (2002). Changes in archaeal, bacterial and eukaryal assemblages along a salinity gradient by comparison of genetic fingerprinting methods in a multipond solar saltern. *Environ. Microbiol.* **4**, 338–348.

Cochran, W. G. (1950). Estimation of bacterial densities by means of the "most probable number". *Biometrics* **6**, 105–116.

Davis, K. E., Joseph, S. J. and Janssen, P. H. (2005). Effects of growth medium, inoculum size, and incubation time on culturability and isolation of soil bacteria. *Appl. Environ. Microbiol.* **71**, 826–834.

Elshahed, M. S., Najar, F. Z., Roe, B. A., Oren, A., Dewers, T. A. and Krumholz, L. R. (2004). Survey of archaeal diversity reveals an abundance of halophilic *Archaea* in a low-salt, sulfide- and sulfur-rich spring. *Appl. Environ. Microbiol.* **70**, 2230–2239.

Emerson, D., Chauhan, S., Oriel, P. and Breznak, J. A. (1994). *Haloferax* sp. D1227, a halophilic archaeon capable of growth on aromatic compounds. *Arch. Microbiol.* **161**, 445–452.

Grant, S., Grant, W. D., Jones, B. E., Kato, C. and Li, L. (1999). Novel archaeal phylotypes from an East African alkaline saltern. *Extremophiles* **3**, 139–145.

Grant, W. D., Kamekura, M., McGenity, T. J. and Ventosa, A. (2001). Class III. Halobacteria *class nov.* In *Bergey's Manual of Systematic Bacteriology* (D. R. Boone, R. W. Castenholz and G. M. Garrity, eds), pp. 294–334. Springer-Verlag, New York.

Hochstein, L. I. and Tomlinson, G. A. (1985). Denitrification by extremely halophilic bacteria. *FEMS Microbiol. Lett.* **27**, 329–331.

Huber, R., Burggraf, S., Mayer, T., Barns, S. M., Rossnagel, P. and Stetter, K. O. (1995). Isolation of a hyperthermophilic archaeum predicted by in situ RNA analysis. *Nature* **376**, 57–58.

Hummel, H. and Böck, A. (1987). 23S ribosomal RNA mutations in halobacteria conferring resistance to the anti-80S ribosome targeted antibiotic anisomycin. *Nucleic Acids Res.* **15**, 2431–2443.

Ihara, K., Watanabe, S. and Tamura, T. (1997). *Haloarcula argentinensis* sp. nov. and *Haloarcula mukohataei* sp. nov., two new extremely halophilic *Archaea* collected in Argentina. *Int. J. Syst. Bacteriol.* **47**, 73–77.

Janssen, P. H., Schuhmann, A., Morchel, E. and Rainey, F. A. (1997). Novel anaerobic ultramicrobacteria belonging to the *Verrucomicrobiales* lineage of bacterial descent isolated by dilution culture from anoxic rice paddy soil. *Appl. Environ. Microbiol.* **63**, 1382–1388.

Kauri, T., Wallace, R. and Kushner, D. J. (1990). Nutrition of the halophilic archaebacterium, *Haloferax volcanii*. *System. Appl. Microbiol.* **13**, 14–18.

Lane, D. J. (1991). 16/23s rRNA sequencing. In *Nucleic Acid Techniques in Bacterial Systematics* (E. Stackebrandt and M. Goodfellow, eds), pp. 115–175. John Wiley & Sons, New York.

Lillo, J. G. and Rodriguez-Valera, F. (1990). Effects of culture conditions on poly(β-hydroxybutyric acid) production by *Haloferax mediterranei*. *Appl. Environ. Microbiol.* **56**, 2517–2521.

Ludwig, W., Strunk, O., Westram, R., Richter, L., Meier, H., Yadhukumar, Buchner, A., Lai, T., Steppi, S., Jobb, G., Forster, W., Brettske, I., Gerber, S., Ginhart, A. W., Gross, O., Grumann, S., Hermann, S., Jost, R., Konig, A., Liss, T., Lussmann, R., May, M., Nonhoff, B., Reichel, B., Strehlow, R., Stamatakis, A., Stuckmann, N., Vilbig, A., Lenke, M., Ludwig, T., Bode, A. and Schleifer, K. H. (2004). ARB: A software environment for sequence data. *Nucleic Acids Res.* **32**, 1363–1371.

Mescher, M. F. and Strominger, J. L. (1975). Bacitracin induces sphere formation in *Halobacterium* species which lack a wall peptidoglycan. *J. Gen. Microbiol.* **89**, 375–378.

Ochsenreiter, T., Pfeifer, F. and Schleper, C. (2002). Diversity of Archaea in hypersaline environments characterized by molecular-phylogenetic and cultivation studies. *Extremophiles* **6**, 267–274.

Oesterhelt, D. and Krippahl, G. (1983). Phototrophic growth of halobacteria and its use for isolation of photosynthetically-deficient mutants. *Ann. Microbiol.* (Paris) **134B**, 137–150.

Olsen, R. A. and Bakken, L. R. (1987). Viability of soil bacteria: optimization of plate-counting technique and comparison between total counts and plate counts within different size groups. *Microb. Ecol.* **13**, 59–74.

Oren, A. (1983). *Halobacterium sodomense* sp. nov., a Dead Sea halobacterium with an extremely high magnesium requirement. *Int. J. Sys. Bacteriol.* **33**, 381–386.

Oren, A. (2001). The Order *Halobacteriales*. In *The Prokaryotes: An Evolving Electronic Resource for the Microbiological Community* (M. Dworkin, S. Falkow, E. Rosenberg, K.-H. Schleifer and E. Stackebrandt, eds), 3rd edn., release 3.2, 25 July 2001. Springer-Verlag, New York (http://link.springer-ny.com/link/service/books/10125).

Oren, A. (2002). Molecular ecology of extremely halophilic Archaea and Bacteria. *FEMS Microbiol. Ecol.* **39**, 1–7.

Oren, A. and Litchfield, C. D. (1999). A procedure for the enrichment and isolation of *Halobacterium*. *FEMS Microbiol. Lett.* **173**, 353–358.

Oren, A., Gurevich, P., Gemmell, R. T. and Teske, A. (1995). *Halobaculum gomorrense* gen. nov., sp. nov., a novel extremely halophilic archaeon from the Dead Sea. *Int. J. Syst. Bacteriol.* **45**, 747–754.

Oren, A., Elevi, R., Ionescu, D. and Lipski, A. (2004). Selective isolation of *Salinibacter*, an extremely halophilic member of the Bacteria, from saltern crystallizer ponds. *ASM 104th General Meeting.* New Orleans, LA.

Robinson, J. L., Pyzyna, B., Atrasz, R. G., Henderson, C. A., Morrill, K. L., Burd, A. M., Desoucy, E., Fogleman, R. E., 3rd, Naylor, J. B., Steele, S. M., Elliott, D. R., Leyva, K. J. and Shand, R. F. (2005). Growth kinetics of extremely halophilic Archaea (family *Halobacteriaceae*) as revealed by Arrhenius plots. *J. Bacteriol.* **187**, 923–929.

Rodriguez-Valera, F., Ruiz-Berraquero, F. and Ramos-Cormenzana, A. (1980). Isolation of extremely halophilic bacteria able to grow in defined inorganic media with single carbon sources. *J. Gen. Microbiol.* **119**, 535–538.

Tomlinson, G. A., Jahnke, L. L. and Hochstein, L. I. (1986). *Halobacterium denitrificans* sp. nov., an extremely halophilic denitrifying bacterium. *Int. J. Syst. Bacteriol.* **36**, 66–70.

Torreblanca, M., Rodríguez-Valera, F., Juez, G., Ventosa, A., Kamekura, M. and Kates, M. (1986). Classification of non-alkaliphilic halobacteria based on numerical taxonomy and polar lipid composition, and description of *Haloarcula* gen. nov. and *Haloferax* gen. nov. *Syst. Appl. Microbiol.* **8**, 89–99.

Torsvik, V., Sørheim, R. and Goksøyr, J. (1996). Total bacterial diversity in soil and sediment communities – a review. *J. Indust. Microbiol.* **17**, 170–178.

Wais, A. C. (1988). Recovery of halophilic archaebacteria from natural environments. *FEMS Microbiol. Ecol.* **53**, 211–216.

Walsby, A. E. (1980). A square bacterium. *Nature* **283**, 69–71.

Widdel, F., Kohring, G.-W. and Mayer, F. (1983). Studies on dissimilatory sulfate-reducing bacteria that decompose fatty acids. III. Characterization of the filamentous gliding *Desulfonema limicola* gen. nov. sp. nov., and *Desulfonema magnum* sp. nov. *Arch. Microbiol.* **134**, 286–294.

23 Extraction of Halophiles from Ancient Crystals

Russell H Vreeland
Department of Biology, West Chester University, West Chester, PA 19383, USA

◆◆◆

CONTENTS

Introduction
Steno's principles
Sample selection
Extraction methods
Final thoughts

◆◆◆◆◆ **INTRODUCTION**

Throughout much of the twentieth century, there were scattered reports describing the isolation of microorganisms from a variety of geologic materials (Cano and Borucki, 1995; Dombrowski, 1963; Mormile et al., 2003; Namyslowski, 1913; Reiser and Tasch, 1960; Tasch, 1963a,b; Vreeland et al., 2000). These reports were generally met either with skepticism or outright assault (negative references) (see e.g. Powers et al., 2001). Despite the negative attacks, these reports persisted and in the last decade of the century, they appeared with increasing frequency, and with greater acceptance by the scientific community.

The effort to find microorganisms in ancient salts and mines actually began early when Boleslaw Namyslowski conducted his survey of brine ponds within an Austrian salt mine (Namyslowski, 1913). While Namyslowski made no attempt to isolate or culture microbes from this source, his artistic renderings showed numerous microbes and even protozoa inhabiting the underground brine pools. These diagrams were of such high quality and detail that it is still possible to identify the genus and species of these organisms. It is also possible to find these same organisms inhabiting modern surface brine pools and salterns. In fact much of the current debate revolves around this latter point since many scientists feel this to be clear evidence of modern surface contamination and not long-term survival. However, in addition to providing a methodological description, this chapter will attempt to describe a series of steps and considerations to provide a reasonable level of evidence to support a conclusion that microbes do indeed survive within these marvelous salt rocks.

Aside from the Namyslowki article, the next most famous article describing long-term survival was that of Dombrowski (1963), a contemporary of two US researchers who also worked in the area (Reiser and Tasch, 1960; Tasch, 1963a). After these two articles appeared, very little development of the field occurred until the latter portion of the 1980s, all of the 1990s and into the current century. The research in this field began to increase following the recognition by Huval and Vreeland (1991) that most underground salt brines, when they reached earth's surface, actually contained viable halophilic Archaea that had not been present in the water before it became salt saturated. Norton and Grant (1988) subsequently used a series of wonderful micrographs to show that halophilic Archaea collected in the small fluid inclusions of laboratory grown crystals. Then Vreeland et al. (1998) conducted an extensive underground survey demonstrating that viable microbial populations were heterogeneously distributed within a Permian aged salt formation. This heterogeneity helps explain part of the difficulty faced when attempting to reproduce isolations. Since the presence of viable populations cannot presently be predicted by simple examination of the rock record, one individual could easily isolate a living microbe while another, using the same material from a different portion of the formation (or even the same core) could easily find nothing (Vreeland et al., 1998).

At the end of the twentieth century and continuing to the present, a flurry of reports have appeared describing positive isolations of living microbes and of intact DNA from ancient halite minerals. These reports began in 2000 with the isolation of four Permian aged microbes (Vreeland et al., 2000) from a surface sterilized salt crystal and have continued with isolations from crystals taken from Death Valley salts (Mormile et al., 2003), DNA from a variety of sites (Fish et al., 2002) and the description of at least two new archaeal species obtained from large salt rocks of the Zechstein basin (Denner et al., 1994; Gruber et al., 2004; Radax et al., 2001; Stan-Lotter et al., 1999, 2000, 2002). As such positive results continue to build, they are naturally being met with growing acceptance and at least some attention is turning toward understanding the mechanisms for this long-term survival. This will mean that many more laboratories will need to utilize the appropriate methods for selecting and sampling primary salt crystals in order to isolate enough microbes for detailed analyses and comparison.

In reality, the increasing acceptance appears to be due, at least in part, to continuing efforts to improve the isolation and sterilization techniques used. Credit for the acceptance is also attributable to an increased recognition, on the part of the biological community, of the need to carefully select samples that are demonstrably primary and datable, rather than simply retrieving a rock left on a mine floor. The purpose of this chapter is to provide a detailed review of the currently accepted methods for sampling ancient crystals in order to begin this very important investigation into long-term survival. Due to the absolute necessity for communicating with other disciplines in order to work in this field (primarily geologists and geochemists), the chapter begins with a brief review of basic geological principles. It then incorporates an additional brief description (on a microbiology level) of some geochemical analyses that might be applied and concludes with a description of each of the techniques that have been successfully used by various research groups.

◆◆◆◆◆ STENO'S PRINCIPLES

Few microbiologists have probably heard of the devout seventeenth century genius Nicolaus Steno (aka Niels Stensen; now the beatified Nicolai Stenonis), considered by many to be the father of modern geology. He was however more than a geologist, being one of the foremost anatomists of his day and later a catholic cardinal (Cutler, 2003). Prior to his geological discoveries and entry to priesthood, Nicolai Steno was perhaps the best known anatomist of his era discovering among other things, parotid gland ducts proving these to be the source of saliva, muscle geometry, brain structure, the pineal gland and tear ducts. Yet, it is his recognition of the basic principles of geology for which he is best known and which most impacts this particular chapter.

Steno's three stratigraphic principles are: first the *Principle of Superposition*; second the *Principle of Original Horizontality*; and third the *Principle of Original Lateral Continuity*. Each of these principles directly impact examination of ancient saline formations for trapped microbes and must be considered when selecting samples for use (Figure 23.1).

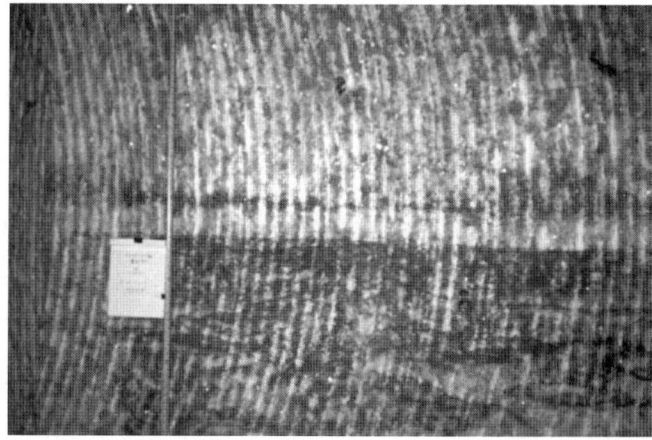

Figure 23.1. Halite layering within a well-preserved Permian aged salt formation. The layers provide a clear example of some of the Steno Principles. Each depositional cycle is clearly illustrated as a horizontal band of similar material. Older layers are toward the bottom of the slide. Several small dissolution pits are present in the middle layers. Note however, that these pits do not extend across multiple layers or out of the image. The ruler marks 2 meters. See Plate 23.1 in Colour Plate Section.

The principle of superposition states quite simply that within sedimentary rock layers, younger material will overlay older rocks. Consequently, in a well-preserved sedimentary (or evaporite) formation, layers on top were deposited after those below. This simple interpretation is complicated a bit by depositional cycles in which lower salinity water floods the pan dissolving some of the more accessible, previously deposited salts. This dissolution creates readily visible pits or dissolution pipes in the underlying layers. In this case, the material within the pipe is not the same age as the surrounding layer. Instead this material is actually equivalent, in age, to the layer in which the pipe or pit originates. Identifying salts from such features can only be accomplished while viewing the parent material and cannot be readily determined once a rock has been dislodged from the formation (Figure 23.1).

The horizontality principle is exactly what the name implies. All sedimentary or evaporate features are deposited under the force of gravity. Therefore they are horizontal (unless some underlying hard feature such as a boulder or rock gets in the way) when first deposited. In a very stable formation, this horizontal layering remains that way. However, if the formation is subjected to other geological forces the layers may buckle or deform. In the most severe instance (such as the salt domes of the southern United States) the layers become vertical. This type of deformation is a clear indicator that the rock has been subjected to rather severe forces that may easily have destroyed any preserved or living material. This may also result (as in salt domes) in the youngest salt layers being at the center of the formation while the oldest material is on the outside. In this case any living material collected can only be considered as dating from the time period when geologic forces stopped affecting the formation.

The principle of continuity is again relatively understandable. Basically, within the confines of a basin each layer is stacked in a continuous event and is therefore considered a single horizon. This continuity means that samples taken from anywhere within a continuous horizon are the same throughout the basin. Once again, the continuity can be interrupted by localized flooding or formation of pits and pipes and by intense geologic forces (Figure 23.1).

It should be clear that if the goal of a study is to find ancient biological materials or a live trapped microbe, samples must be carefully examined *in situ* before being removed from the surrounding rock. Alternatively, if the goal is to simply find an underground halophile or to collect a souvenir then any nice-looking crystal will suffice.

There is one final Steno discovery that is relevant to this discussion. During the development of his three principles, Steno also realized the significance of fossilized material contained within surrounding rocks. In the case of trapped shells and bones, Steno recognized that in order for the shell or bone to make an impression on the surrounding material, that shell or bone must have been present when the rock formed. A similar argument must be made for organisms trapped inside primary rocks. That is, if a formation meets all three of Steno's principles and if a crystal is sealed and primary, then any organisms (or biological materials) inside the crystal must have been present when the crystal formed. Ergo, that material is as old as the crystal and the formation layer from which it came. These are the essential data and arguments used by researchers such as Fish *et al.* (2002), Mormile *et al.* (2003) and Vreeland *et al.* (2000) in arguing that their isolates were hundreds of thousands to hundreds of millions of years in age.

◆◆◆◆◆ SAMPLE SELECTION

Ultimately, the quality of any analysis of ancient materials, whether for the purpose of isolating ancient materials, live microbes or even paleochemistry to learn about the environment that generated the formation, is totally dependent on the specific crystal selected. The reality is that after millions of years, even formations which eminently meet all of Steno's principles contain a significant number of non-primary or altered crystals. While these samples can reveal a great deal about the history of the formation, a complete description of all the cues and clues is beyond the scope of this chapter. It is important however, that researchers know the differences between primary and altered crystals in order to prevent embarrassing errors. Consequently, a brief description of some basic differences has been included in the next section. This short discussion should be considered only as a guide and not as a comprehensive description.

Primary vs. Non-primary Materials

When the goal of a study is to isolate either living microbes or remnant biological materials that are the remains of ancient life forms, it is critical that only the highest quality primary crystals be used. Distinguishing primary crystals from those that have been altered is not a trivial matter and is best done by scientists that have specialized in this type of research. Readers who wish to obtain more detailed information about these materials and how to examine them should refer to the classic research of Roedder (1984) and Lowenstein and Hardie (1985) before undertaking this effort. What should be said here is that most primary crystals (or primary areas within halite crystals) are very small, perhaps only millimeters in size and at most a few grams in mass. Very large rocks weighing one hundred grams or more cannot be considered primary. Therefore samples should be carefully cut to remove non-primary regions. Individual crystals should then be examined at least with a dissecting scope to look for evidence of cracks, misshapen inclusions (those that are rounded, elongated or which contain gas bubbles all being suspect). Extremely large individual crystals should also be discarded or used as gifts.

◆◆◆◆◆ EXTRACTION METHODS

Crystal Sterilization

Regardless of the overall goals of the project, some thought should be given to contamination control from the surface and from surrounding modern materials. Obviously, the more intense the goals imparted to a project (finding ancient life or examining a crystal from Mars for Martian life), the more stringent the levels of cleanliness. There are several procedures that can be used to sterilize salt crystals and more will undoubtedly be developed. All of the procedures that have been developed have some common properties. The sections below describe both the considerations and controls that have already been used as well as details of the methodology.

Considerations and controls

For obvious reasons, sterilization of the surface of any sample being used is extremely important. This aspect has actually been the downfall of several previous claims of the isolation of microorganisms due to the fact that those making the claims were unable to defend their sterilization procedures. The rationales for the considerations that should lead all sterilization protocol choices have been discussed in some detail in other publications (Vreeland and Powers, 1998; Vreeland and Rosenzweig, 1998, 2002). The basic considerations should, however, be re-stated here.

Probably the first and most important consideration that should be used is that the sterilization technique must be rigorous enough to be effective on the surfaces but gentle enough to not harm any material inside. This pretty effectively eliminates heating or flaming as a useful method. This is especially true since salt conducts heat rather well so even a brief exposure to sterilizing temperatures is quickly transmitted into the crystal.

Second, the sterilizing technique should not expose a new crystal surface when applied. This may not seem that important since it could be argued that one is exposing fresh surfaces, therefore anything now on the surface would be from the inside anyway. The problem with that argument is that it places one in an indefensible position responding to questions about the level of sterility assurance used in any experiment.

The third consideration is that the procedure should at least be quantifiable or controlled. Microbiologists readily recognize that stating that something is "sterilized" is, in reality, based upon probability. Basically something is considered sterile if there is only one chance in one million that the material or container contains a living microbe. Too many previous claims of isolation have been unable to provide such information. Even if that type of effectiveness of the technique has not been quantitated, any protocol should be defensible by some sort of control data showing (at the minimum) that the technique did destroy isolated organisms or biological material.

There are two protocols that meet the above criteria of gentleness, surface preservation and controlled and that have been successfully applied to salt crystals. These are described in some detail below. For completeness, a third technique successfully used with amber but not with salt has been included.

Ethanol

The use of ethanol as a sterilant is familiar to most microbiologists. The advantages of ethanol are that it is inexpensive and since it is non-corrosive and relatively easy to transport, it may even be used below ground in some areas. The disadvantage is that ethanol is not really a good sterilant, having a phenol coefficient of only 0.04. Still, ethanol is the sterilant used by many researchers. Denner et al. (1994), Dombrowski (1963), Gruber et al. (2004), Fish et al. (2002), Mormile et al. (2003), Radax et al. (2001) and Stan-Lotter et al. (1999) have all used undiluted ethanol to sterilize salt rock samples. When using ethanol, the salt should be completely immersed in the ethanol for at least 2 h (Denner et al., 1994; Radax et al., 2001; Stan-Lotter et al., 1999). Further, most authors use undiluted ethanol to prevent dissolution even though 70% (w/v) ethanol solutions make better disinfectants. Unfortunately, a 70% solution will dissolve some, or all of, the salt. Removal of the ethanol is accomplished either by simple evaporation or in many cases using a flame. While this may not seem to be the most efficient sterilization method, Norton et al. (1993) did utilize an orange halophilic coccus as an external control to show that the technique successfully eliminated the Archaea on the surface of the crystal without harming the red halophilic rods inside the crystal. This simple control had not been used by others and certainly provided a higher degree of acceptance for the resulting isolation.

Acids and bases

The most involved sterilization procedure that has been used within this field is that of Rosenzweig et al. (2000). It is also the only technique that has provided quantifiable sterility assurance. The technique takes advantage of the fact that NaCl is insoluble in concentrated solutions of HCl and NaOH. The overall procedure described by Rosenzweig et al. (2000) involves working within a Biosafety level 3 (US BSL-3) facility inside a laminar flow hood that has been treated for two hours with a UV germicidal lamp, followed by cleaning with any of several liquid sterilants.

Protocol

1. Select and clean crystals.
2. Examine the crystal for signs of damage. This includes cracks, gas bubbles, and misshapen inclusions. These should be excluded from sampling.
3. Sterilize each selected crystal in autoclaved 10 M NaOH for at least 5 min.

4. Wash the crystals in autoclaved saturated NaCl brine for 2 min.
5. Sterilize in filter sterilized 10 M HCl for at least 5 min.
6. Wash for 2 min in autoclaved Na_2CO_3 buffered NaCl saturated brine. The purpose for this final buffer is to neutralize residual acid, so it generally is not made to exacting standards. A solution of saturated NaCl is prepared, then excess sodium carbonate is added and the solution is sterilized. Crystals are placed into this neutralizing solution which naturally effervesces until the acid has been removed. This final wash should always be sampled for the presence of contaminating microbes.

For purposes of high level sterilization, these solutions are autoclaved for 45–60 min at 121°C. The HCl solution cannot be autoclaved and must be filter sterilized using 0.22-micron filters.

In this protocol, crystals being sampled must be fully immersed in the acid and base. If the material being sampled is very small it may float on the surfaces of these dense solutions, so researchers should make sure the material sinks into the sterilant. This technique lends itself well to large crystals that must be sterilized in individual containers, or to very small fragments that can be handled in groups of 6 using 24 well tissue culture plates. When using such small plates, the biggest difficulty is picking up the small crystals from the bottom of the wells. This is best accomplished using pointed plastic tweezers that can be kept sterile by immersion in disinfectant followed by HCl.

One liability inherent in the technique is the need to remove the base from the crystal to prevent neutralizing the acid, then removing the acid in order to maintain a proper medium pH. While neither of these washes are technically difficult, they do add steps and potential contamination (especially the final wash), so additional sterility controls are required.

Rosenzweig et al. (2000) originally tested the procedure against three different organisms, a halophilic archaeon and two different bacteria, one a Gram-negative halophile (*Halomonas elongata*), the other a Gram-positive spore former that was actually isolated from a salt crystal sterilized by the procedure. Kill curves showed that exposure to 10 M NaOH for only 1 min reduced the microbial population by a factor of 10^4, while exposure to 10 M HCl for 1 min resulted in a 10^8-fold reduction in the microbial population. Thus, successive runs through concentrated NaOH and HCl would yield a probability of contamination approaching 10^{-12}.

Extraction Techniques

Once the surface of a crystal has been adequately sterilized, any extraction technique can be used to extract the samples inside. The decision is made largely on the basis of the sizes of the inclusions and number of inclusions present. Regardless of the technique chosen, all sampling should be performed inside a treated class II or better laminar flow hood, inside a US BSL-3 facility.

Dissolution

The advantages of simply dissolving the crystal are that the entire crystal and all inclusion fluids contained are sampled. Second, there is less handling or manipulation of the material and therefore less chance for contamination. Finally, crystals being dissolved can be shaved to remove any questionable sample portions, leaving only the highest quality portions of the sample. Dissolution is really the only choice available when working with truly primary chevron crystals which contain large numbers of very small inclusions. The primary disadvantage is that the sample is destroyed, so each crystal must be photographed in order to provide documentation of sample quality. Methodologically there is little to describe for this extraction method. Depending on the size of the crystal being sampled, medium salt concentrations should be adjusted to compensate for the additional salt coming from the crystal.

Drilling

The technique for drilling a small hole in the crystal, then extracting the inclusion fluid is based on that used for chemical analyses by Das et al. (1990) and Horita et al. (1991, 1996). It is also the technique used with success by Vreeland et al. (2000). Drilling a crystal sample preserves the sample for future analyses or for making additional samples if the crystal contains more than one usable inclusion. The primary disadvantage is that only inclusions larger than the smallest usable drill bits can be sampled by this method. Unfortunately, experience has shown that the smallest carbon steel bits (0.25 mm) tend to bind and shear if the inclusion is deeper than half the length from the point to the shaft. This severely limits their use. Consequently, the smallest usable carbon steel bits are at least 0.5 mm in diameter. The gold-plated bits used by dentists are available in sizes as small as 0.1 mm but these have not yet been applied and they are expensive compared to the $9.00 (USD) carbon steel drill. One other problem with large inclusions is that they may be contained in non-primary materials meaning that additional proof of sample quality is required.

The overall drilling system requires use of a micromanipulator in order to provide slow accurate positioning and movement of the drill (Figure 23.2). Ideally, the manipulator should allow movement in three dimensions and should allow one to shift the drill only a few micrometers at a time. The actual drilling system can be a simple variable speed Dremel-type drill on a flexible shaft to accommodate the movement of the drill (Figure 23.2). The drill system can also be far more complex including special made wire drills encased in Lucite (Das et al., 1990; Horita et al., 1996; Lazar and Holland, 1988). This sampling can even reach

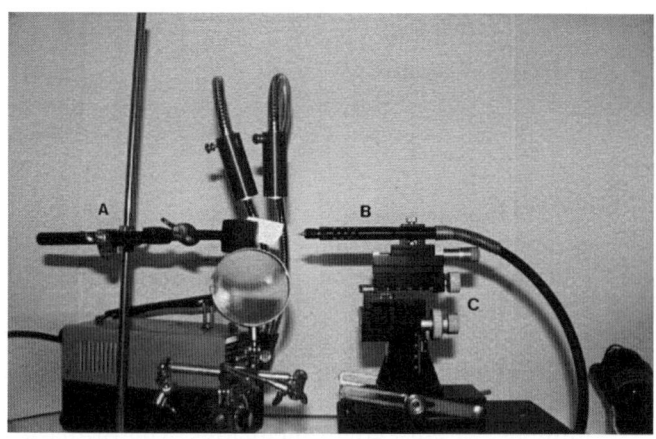

Figure 23.2. Typical homemade crystal drilling system. The drill is secured in the clamp of a micromanipulator which is then used to guide the spinning bit into the crystal. This particular micromanipulator allows alignment of the crystal in two dimensions before beginning the drilling process. Alignment of such a drill system requires being able to see the inclusion from two sides of the crystal. This can be difficult with some materials and with the crystal to be sampled placed behind the glass of a laminar flow hood. Gaining this vision requires smoothing and polishing two surfaces prior to sterilization. Smoothing is best done using a hand held drill (similar to that shown) fitted with a simple grinding bit. Salt crystals can be easily polished by rubbing the sides with a rag wetted with distilled water. See Plate 23.2 in Colour Plate Section.

a high level of sophistication using a computer controlled microscope with an attached drill such as that shown in Figure 23.3.

Regardless of the system used, all drill bits should be sterilized prior to use. The best way to sterilize and protect these small wire bits is to place them inside yellow micropipette tips prior to sterilization. The tapered tip protects the wire bit but the shaft of the bit is sufficiently exposed allowing attachment to the drills prior to removal of the plastic. When drilling with bits, this small fluid extraction must be done with a small bore needle (generally 20 gauge) which has been pre-sterilized. This gauge needle has proven to be most useful since smaller gauge wires quickly become clogged with crystals that may form by agitation as the brine is drawn into the tip.

◆◆◆◆◆ FINAL THOUGHTS

As is obvious from the foregoing discussion, many of the choices made in relation to the methods for extracting ancient materials from crystals are based upon the exact goals of the project. One aspect that must be mentioned and is true for any and all types of research with ancient materials is that no microbiologist should work alone. This type of research simply cannot be accomplished without close collaboration with professional geologists and or paleoclimatologists. Their opinions about samples, their knowledge of rock materials and how to "read" a formation cannot be duplicated or gleaned from a text, no matter how good it is. Finally, successfully extracting microbes from ancient materials requires time and patience. As might be expected from currently available information (Vreeland *et al.*, unpublished), the older the material being sampled, the more likely one is to find sterile samples. In addition, laboratory observations have shown that cells trapped in younger materials will recover much quicker. For instance, 100 000-year-old samples often require 2–3 weeks incubation while 250 000 000-year-old materials may have to be incubated for 3–5 months. The recommendation from this author is that if a reader samples a really old sample and a live

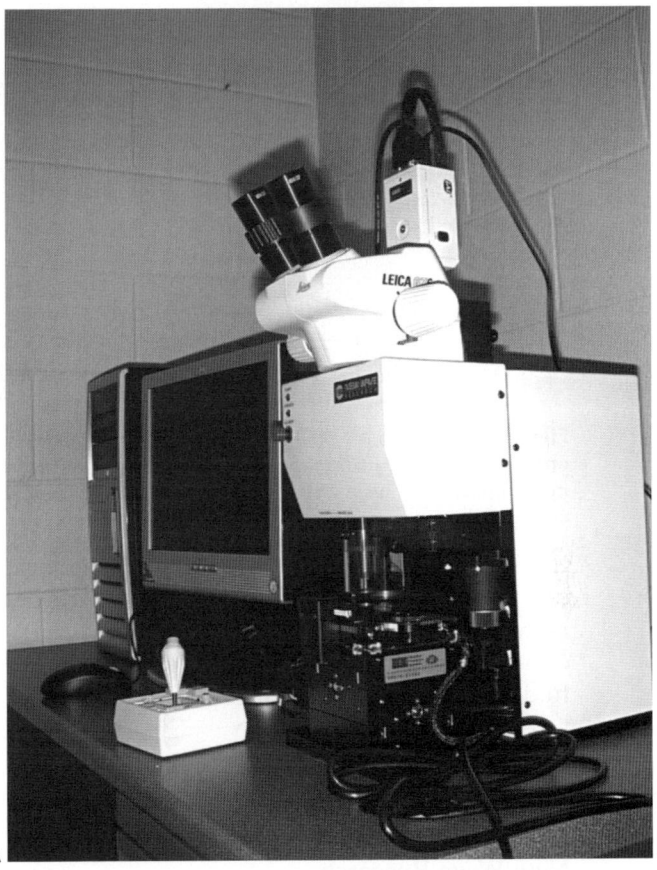

A

Figure 23.3. Computer controlled Micro-Mill system for sampling crystals and all ancient rock materials. (A) This unit is laser guided and can provide a close approximation of the depth of an inclusion. The unit provides a video image of the entire drilling process. A crystal to be sampled is placed beneath the objective lens where it is photographed and measured. Co-ordinates are then laid onto the sample, which are then used to guide the drilling process *via* the computer system seen behind the drill. Alternatively, the drill can be guided by the joystick shown in the foreground. The microscope can be separately placed into a sterile field and computer controlled from outside. See Plate 23.3 in Colour Plate Section.

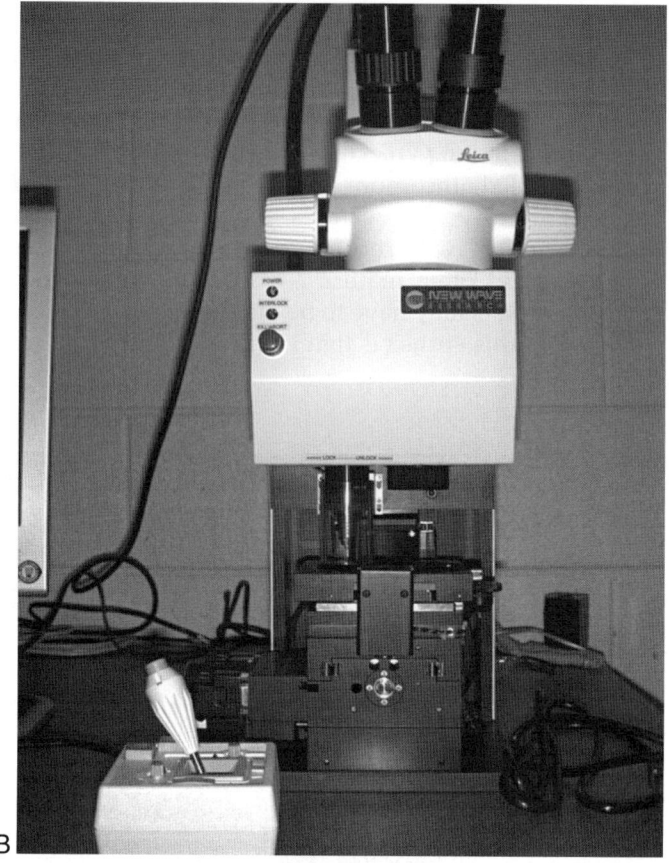

Figure 23.3. *Continued.* (B) A closer view of the microscope and controlling joystick. The objective lens is located on the right with the drill on the left. The precision controls on this instrument allow one to drill a 50-micron hole to a point near the inclusion, then, after a quick change of bits remove a 30-micron core sample from inside the crystal without touching the sides of the bore hole. See Plate 23.3 in Colour Plate Section.

organism appears in 24–48 h, it is probably a contaminant. So as a final bit of advice, sample a lot, be patient, be careful and you will undoubtedly find life in ancient materials. Good luck!

List of suppliers

New Wave Research
48660 Kato Road
Fremont, CA 94538, USA
Tel.: +1-510-249-1550
Fax: +1-510-249-1551
www.new-wave.com

Sampling MicroDrill

Drill bits the small drill bits described in this chapter may be purchased from a variety of tool makers. Two potential sources:

Braesselerusa
One Brassler Blvd.
Savannah, GA 31419, USA
Tel.: +1 800-841-4522
Fax: +1 888-610-1937
www.brasselerusa.com

Dental drills

McMaster – Carr Supply Company
473 Ridge Road
Dayton, NJ 08810-0317, USA
Tel.: +1 732-329-3200
Fax: +1 732-329-3772
www.mcmaster.com

Standard Carbide drills

References

Cano, R. and Borucki, M. K. (1995). Revival and identification of bacterial spores in 25 to 40 million year old Dominican amber. *Science* **268**, 1060–1064.

Cutler, A. (2003). *The Seashell on the Mountaintop – A Story of Science, Sainthood and the Humble Genius who Discovered a New History of the Earth.* Dutton, New York.

Das, N., Horita, J. and Holland, M. D. (1990). Chemistry of fluid inclusions in halite from the Salina group of the Michigan Basin: implications for late Silurian seawater and the origin of sedimentary groups. *Geochim. Cosmochim. Acta* **54**, 319–327.

Denner, E. B. M., McGenity, T. J., Busse, H.-J., Grant, W. D., Wanner, G. and Stan-Lotter, H. (1994). *Halococcus salifodinae* sp. nov., an archaeal isolate from an Austrian salt mine. *Int. J. Syst. Bacteriol.* **44**, 774–780.

Dombrowski, H. (1963). Bacteria from Paleozoic salt deposits. *Ann. New York Acad. Sci.* **108**, 453–460.

Fish, S., Shepherd, T. J., McGenity, T. J. and Grant, W. D. (2002). Recovery of 16S ribosomal RNA gene fragments from ancient halite. *Nature* **417**, 432–436.

Gruber, C., Legat, A., Pfaffenhuemer, M., Radax, C., Weidler, G., Busse, H.-J. and Stan-Lotter, H. (2004). *Halobacterium noricense* sp. nov., an archaeal isolate from a bore core on an alpine Permian salt deposit, classification of *Halobacterium* sp. NRC-1 as a strain of *H. salinarum* and emended description of *H. salinarum*. *Extremophiles* **8**, 431–441.

Horita, J. T., Friedman, J., Lazar, B. and Holland, D. H. (1991). The composition of Permian seawater. *Geochim. Cosmochim. Acta* **55**, 417–432.

Horita, J. T., Weinberg, A., Das, N. and Holland, D. H. (1996). Brine inclusions in halite and the origin of the middle Devonian prairie evaporates of western Canada. *J. Sedim. Res.* **66**, 956–964.

Huval, J. H. and Vreeland, R. H. (1991). Phenotypic characterization of halophilic bacteria isolated from groundwater sources in the United States. In *General and Applied Aspects of Halophilic Bacteria* (F. Rodriguez-Valera, ed.), pp. 53–62. Plenum Press, New York.

Lazar, B. and Holland, H. D. (1988). The analysis of fluid inclusions in halite. *Geochim. Cosmochim. Acta* **52**, 485–490.

Lowenstein, T. K. and Hardie, L. A. (1985). Criteria for the recognition of salt-pan evaporites. *Sedimentology* **32**, 627–644.

Mormile, M. R., Biesen, M. A., Gutierrez, M. C., Ventosa, A., Pavolvich, J. B., Onstott, T. C. and Frederickson, J. K. (2003). Isolation of *Halobacterium salinarum* retrieved directly from halite brine inclusions. *Environ. Microbiol.* **5**, 1094–1102.

Namyslowski, B. (1913). Über unbekannte halophile Mikroorganismen aus dem Innern des Salzbergwerkes Wieliczka. *Bull. Acad. Sci. Krakow B* **3/4**, 88–104.

Norton, C. F. and Grant, W. D. (1988). Survival of halobacteria within fluid inclusions in salt crystals. *J. Gen. Microbiol.* **134**, 1365–1373.

Norton, C. F., McGenity, T. J. and Grant, W. D. (1993). Archaeal halophiles (halobacteria) from two British Salt Mines. *J. Gen. Microbiol.* **139**, 1077–1081.

Powers, D. W., Vreeland, R. H. and Rosenzweig, W. D. (2001). Biogeology – how old are bacteria from the Permian age? – Reply. *Nature* **411**, 155–156.

Radax, C., Gruber, C. and Stan-Lotter, H. (2001). Novel haloarchaeal 16S rRNA gene sequences from Alpine Permo-Triassic rock salt. *Extremophiles* **5**, 221–228.

Reiser, R. and Tasch, P. (1960). Investigation of the vialbility of osmophile bacteria of great geological age. *Kans. Acad. Sci Trans.* **63**, 31–40.

Roedder, E. (1984). The fluids in salt. *Amer. Mineralogist* **69**, 413–439.

Rosenzweig, W. D., Woish, J., Petersen, J. and Vreeland, R. H. (2000). Development of a protocol to retrieve microorganisms from ancient salt crystals. *Geomicrobiology* **17**, 185–192.

Stan-Lotter, H., McGenity, T. J., Legat, A., Denner, E. B. M., Glaser, K., Stetter, K. O. and Wanner, G. (1999). Very similar strains of *Halococcus salifodinae* are found in geographically separated Permo-Triassic salt deposits. *Microbiology* **145**, 3565–3574.

Stan-Lotter, H., Radax, C., Gruber, C. McGenity, T. J., Legat, A., Wanner, G. and Denner, E. B. M. (2000). The distribution of viable microorganisms in Permo-Triassic rock salt. In *SALT 2000 – 8th World Salt Symposium* (R. M. Geertman, ed.), vol. 2, pp. 921–926. Elesvier Science, Amsterdam.

Stan-Lotter, H., Pfaffenhuemer, M., Legat, A., Busse, H.-J., Radax, C. and Gruber, C. (2002). *Halococcus dombrowskii* sp. nov., an archaeal isolate from a Permo-Triassic alpine salt deposit. *Int. J. Syst. Evol. Microbiol.* **52**, 1807–1814.

Tasch, P. (1963a). Dead and viable fossil salt bacteria. *Univ. of Wichita Bull.* **39**, 4–7.

Tasch, P. (1963b). Fossil content of salt and associated evaporites. In *Symposium on Salt* (A. C. Bersticker, K. E. Hoekstra and J. F. Hall, eds), pp. 96–101. Northern Ohio Geological Society, Cleveland.

Vreeland, R. H. and Powers, D. W. (1998). Microbiological considerations for sampling ancient salt formations. In *Biology and Geochemistry of Hypersaline Environments* (A. Oren, ed.), pp. 53–73. CRC Press, Boca Raton.

Vreeland, R. H. and Rosenzweig, W. D. (1998). Microorganisms in ancient salt formations: possibilities and potentials. In *Enigmatic Microorganisms and Life in Extreme Environments* (J. Seckbach, ed.), pp. 387–398. Kluwer Academic Publishers, Dordrecht.

Vreeland, R. H. and Rosenzweig, W. D. (2002). The question of uniqueness of ancient bacteria. *J. Indust. Microbiol. Biotechnol.* **28**, 32–41.

Vreeland, R. H., Piselli, A. F., McDonnough, S. and Meyers, S. (1998). Distribution and diversity of halophilic bacteria in a subsurface salt formation. *Extremophiles* **2**, 321–331.

Vreeland, R. H., Rosenzweig, W. D. and Powers, D. W. (2000). Isolation of a 250 million year old halotolerant bacterium from a primary salt crystal. *Nature* **407**, 897–900.

24 The Assessment of the Viability of Halophilic Microorganisms in Natural Communities

Helga Stan-Lotter[1], Stefan Leuko[2], Andrea Legat[1] and Sergiu Fendrihan[1]

[1] Division of Molecular Biology, Department of Microbiology, University of Salzburg, Hellbrunnerstr. 34, A-5020 Salzburg, Austria; [2] Australian Centre for Astrobiology, Department of Biological Science, Macquarie University, Sydney, NSW 2109, Australia

◆◆◆

CONTENTS

Introduction
Determination of colony forming units (CFUs) of extremely halophilic microorganisms
Staining haloarchaea with DAPI
LIVE/DEAD® BacLight™ bacterial viability kit
Conclusions

◆◆◆◆◆ **INTRODUCTION**

Understanding microbial processes in various environments depends on the knowledge about the presence and identity of those microorganisms. Improved methodology has greatly increased this knowledge; as pointed out by Paul (1993), the marine and freshwater environments have seen a particularly dramatic revision in the estimation of microbial abundances. While plate counting in the 1970s indicated numbers of 10^1–10^4 bacteria per ml and consequently, aquatic bacterial biomass was considered insignificant, the introduction of epifluorescence microscopy revealed bacterial abundances several orders of magnitude higher (for references, see Paul, 1993). The environments where halophilic microorganisms are found include aquatic habitats of varying salinity, salt marshes, surface salt lakes, subterranean salt lakes, the Dead Sea and some other places (Javor, 1989; Oren, 2002). Many of those sites can be considered as the remnants from ancient surface waters and oceans, which may also be true for their microbial populations. An increased awareness of the communities in hypersaline environments can be noted and recognized by an increase of publications and meetings. This chapter contains a contribution to the methods for identification of extremely halophilic microorganisms.

Hypersaline waters contain at least 10–12% salt, and concentrations of up to 35% salt, which is usually the point of onset of precipitation, are present in thalassohaline brines, which derive from sea water (see Rodríguez-Valera, 1993). Other hypersaline waters contain salts from the dissolution of continental minerals, and these are called athalassohaline (Rodríguez-Valera, 1993). The salts, which are deposited during evaporation of sea water or continental brines, are the origin of evaporitic rocks. Similarly as for other environments, progress in methodology has greatly increased our knowledge of microbial diversity in this type of niche. As recently as 1981, Larsen (1981) described mined rock salts as free from bacteria, although isolations of halophilic microorganisms from ancient salt sediments had occasionally been reported since the early decades of the twentieth century (for references, see Grant et al. (1998) and McGenity et al. (2000)). The first formally named strain from a Permian salt sediment was *Halococcus salifodinae* (Denner et al., 1994); several further haloarchaeal isolates from ancient rock salt were described (Gruber et al., 2004; Stan-Lotter et al., 1999, 2002; Vreeland et al., 2002). The application of molecular methods yielded amazingly large numbers of different 16S rRNA gene sequences or fragments (Fish et al., 2002; Radax et al., 2001), which suggested the presence of a sizable microbial community in rock salt. Are these microorganisms viable? Some of them certainly are, as the cultivation experiments proved; recently, a congruence between a 16S rRNA gene fragment and an isolated strain was found (Stan-Lotter et al., submitted). The classical determination of colony forming units (CFUs) is often very time consuming with extremely halophilic Archaea, since they can take months to produce visible colonies, whether isolated from ancient rock salt (Gruber et al., 2004) or crystallizer ponds (Burns et al., 2004). Additional techniques are therefore required, if an assessment of viability is wanted. Why may such knowledge be of importance? Several reasons are given in the following paragraphs:

The salt sediments can be viewed as remnants from ancient hypersaline oceans. The fluid inclusions of Permian rock salt were found to contain cations and anions in a similar composition as today's sea water (Horita et al., 1991). Dating of the salt deposits by sulfur-isotope analysis (ratios of $^{32}S/^{34}S$ as measured by mass spectrometry), in connection with information from stratigraphy, indicated a Permo-Triassic age for Alpine and Zechstein deposits (Holser and Kaplan, 1966). This estimate was independently confirmed by the identification of pollen grains, which possessed distinct morphological features and could be assigned to extinct plants, in the sediments (Klaus, 1974). We have stressed that the age of the microorganisms cannot simply be deduced from the age of the sediments from which they were isolated, since there are no methods yet to determine the age of a prokaryotic cell (Kunte et al., 2002; Stan-Lotter et al., 2004). However, the possibility of microbial survival for extreme periods of time cannot be ignored, since it would likewise be difficult to envisage a migration of large numbers of diverse halophiles into the rocks after deposition (McGenity *et al.*, 2000; Stan-Lotter *et al.*, 2004). Numerous subterranean environments have been found to contain microbial life, such as deep sub-seafloor sediments, crustal and sedimentary rocks, and ancient deposits (see Pedersen, 2000, for a review). The investigation of a halophilic community in an evaporitic salt sediment could give insights into relationships between halophiles and perhaps other ocean-dwelling microorganisms in past times, and may lead to an understanding of their usage of scarce energy sources and their mechanisms of long-term survival.

If halophilic prokaryotes on Earth can remain viable for very long periods of time, then it is reasonable to consider the possibility that viable microorganisms may exist – or may have existed in the past – in similar subsurface salt deposits on other planets or moons. This notion becomes all the more plausible in view of the detection of halite in extraterrestrial materials: the SCN meteorites (Gooding, 1992), which stem from Mars (Treiman *et al.*, 2000), were found to contain halite. Quite recently, the data from the US rovers Spirit (http://www.msnbc.msn.com/id/5166705/) and Opportunity (http://www.missionspace.info/news/merupdate/saltwater.html) suggested the formation of at least some Martian deposits from concentrated salt water. Traces of halite were detected in the Murchison and other carbonaceous meteorites (Barber, 1981); even macroscopic crystals of extraterrestrial halite, together with sylvite (KCl) and water inclusions, were found in the Monahans meteorite, which fell in Texas in 1998; the pieces were inspected days after being collected (Zolensky *et al.*, 1999). Another line of evidence for the existence of extraterrestrial salt was provided by the Galileo spacecraft; its onboard magnetometer detected fluctuations that are consistent with the magnetic effects of currents flowing in a salty ocean on the Jovian moon Europa (McCord *et al.*, 1998).

The search for life in the solar system and beyond is a goal of several space agencies in the twenty-first century (Foing, 2002). Extraterrestrial materials might be return samples, or well-characterized meteorites, which may contain halite, as described above. In any case, the development of methods for life detection in extraterrestrial samples will be crucial, and, due to the severe requirements of high sensitivity and reliability, will likely spawn applications to terrestrial samples as well.

The staining procedures using the fluorescent dyes DAPI (4′,6-diamidino-2-phenylindole) and the LIVE/DEAD® *Bac*Light™ Bacterial Viability kit (henceforth referred to as the LIVE/DEAD® kit), which we described recently (Leuko *et al.*, 2004), facilitate the enumeration of haloarchaeal cells and judgment about their viability. The usefulness of the LIVE/DEAD® kit can be extended by pre-staining cells before embedding in artificial halite, thus providing a model system for the exploration of microbial responses to extreme environmental conditions (Fendrihan and Stan-Lotter, 2004).

In this chapter, protocols for staining of haloarchaea with fluorescent dyes and correlation with CFUs, including improved media for growth of cells from environmental samples, are described.

◆◆◆◆◆ DETERMINATION OF COLONY FORMING UNITS (CFUs) OF EXTREMELY HALOPHILIC MICROORGANISMS

General Information

Extremely halophilic prokaryotes grow optimally in media containing 2.5–5.2 M NaCl (Kushner and Kamekura, 1988); they comprise both bacterial and archaeal genera (Kamekura, 1998); the latter are now often called haloarchaea. High concentrations of salt in the culture media reduce the solubility of other components; thus, published recipes should be followed carefully and recommended commercial sources, particularly those for less well-defined ingredients (e.g. casamino acid hydrolysates, yeast extracts), should be used. Generation times of haloarchaea range from several hours to several days for aerobes under optimum conditions (Burns et al., 2004; Gruber et al., 2004; Tindall, 1992). Agar plates are therefore poured at least 5–6 mm high, and during incubation, which is usually done between 37 and 42°C, they are wrapped in plastic bags to avoid desiccation. Supplementation of media with rock salt in addition to commercial NaCl (which was used here for some strains) increased growth rates and reduced generation times (Stan-Lotter et al., 2003).

Equipment and Reagents

Temperature-controlled shaker Innova 4080 (New Brunswick Scientific)
Glass sidearm flasks, 100 ml
Stereo zoom microscope Microzoom 1150 (Micros)
Sterile spatulas (Sigma)
Sterile dissolved rock salt
Sterile media for halophilic microorganisms:

1. Medium M2 was slightly modified from Hochstein (1987) and contains (in g l^{-1}): NaCl, 200; $MgCl_2 \cdot 6H_2O$, 20; KCl, 2; $CaCl_2 \cdot 2H_2O$, 0.2; yeast extract (Difco), 5; Hycase (casein hydrolysate; Sigma), 5; pH is adjusted to 7.4 with dilute HCl. Medium M2S (see below) was supplemented with dissolved and filtered rock salt.
2. ATCC medium no. 2185 contains (in g l^{-1}): NaCl, 250; $MgSO_4 \cdot 7H_2O$, 20; trisodium citrate $\cdot 2H_2O$, 3; KCl, 2; tryptone (Roth), 5; yeast extract (Difco), 3; 0.1 ml of a filter-sterilized trace metal solution (per 200 ml: $ZnSO_4 \cdot 7H_2O$, 1.32 g; $MnSO_4 \cdot H_2O$, 0.34 g; $Fe(NH_4)(SO_4)_2 \cdot 6H_2O$, 0.78 g; $CuSO_4 \cdot 5H_2O$, 0.14 g); pH is adjusted to 7.0.
3. Improved medium for recovery of isolates from rock salt and hypersaline waters (Stan-Lotter et al., 2003) contains (in g l^{-1}): NaCl, 200; KCl, 2; $CaCl_2 \cdot 2H_2O$, 0.1; $FeSO_4 \cdot 7H_2O$, 0.01; $MnCl_2 \cdot 4H_2O$, 0.00036; in addition, it contains 10 or 150 mM $MgCl_2$; 0, 1 or 10 g l^{-1}

glycerol; 0 or 0.5 g l^{-1} tryptone (Roth); 0 or 0.5 g l^{-1} Hycase (Sigma); 0, 0.1, 0.5 or 1 g l^{-1} yeast extract (Difco); 0 or 2 g l^{-1} soluble starch (Merck).

Agar (20 g l^{-1}) is added to media for solidification.

Haloarchaeal type strains

Halococcus dombrowskii H4 DSM 14522^T
Halobacterium salinarum DSM 3754^T
(both were obtained from the DSMZ)

Protocols

a. Preparation of growth media supplemented with rock salt.
   1. Dissolve 300 g portions of rock salt (bore cores from salt mines in Altaussee, Austria, which contained 92–95% halite; samples were a gift from Salinen Austria, Bad Ischl) in distilled water to saturation at ambient temperature (21–23°C) with slight shaking.
   2. Filter solution (approximately 32% dissolved solids) through Stericups (0.22 μm pores; Millipore).
   3. Mix M2 medium 1:1 (v/v) with saturated filtered rock salt solution and then sterilize for 15 min at 121°C. The rock-salt-supplemented medium is called M2S.
   4. Prepare liquid media and agar plates with M2S, ATCC medium 2185 (II), or improved recovery medium (III), respectively.

b. Dissolving rock salt for determination of microbial content.
   1. Surface-sterilize salt drill cores or rock salt lumps (weight about 300–400 g) by thorough flaming with a Bunsen burner flame (see Radax *et al.*, 2001).
   2. Add sterile distilled water, which had been adjusted to pH 7.0 with dilute NaOH, in increments of about 50 ml, for several days, to yield a solution saturated at ambient temperature.
   3. Analyze solution by making spread plates and/or stain with fluorescent dyes.

c. Growth of halophilic microorganisms; enumeration.
   1. Grow haloarchaeal type strains with shaking at 37°C in side arm flasks in 20-ml portions of medium M2S (*Hcc. dombrowskii* H4 DSM 14522^T) and M2 medium (*Hbt. salinarum* 3754^T), respectively.
   2. Spread portions of 100 μl of unknown samples (hypersaline brines, dissolved rock salt) on agar plates, made with the three media described above.
   3. Incubate at 37–42°C until colonies become visible (use stereo zoom microscope).
   4. Determine CFUs (colony forming units) by manual counting.

Comments

For use as a supplement, rock salt may be dissolved quickly at elevated temperatures. For use as source of unknown microorganisms, large amounts of rock salt are surface sterilized, such that the temperature of the interior will not be elevated and potential microorganisms will be preserved. Water for dissolution is added stepwise, in order to protect any haloarchaeal cells from low salt stress.

Although several automatic colony counters are on the market, the differences in pigmentation, size and shape of colonies apparently preclude the automated determination of CFUs of halophilic microorganisms from environmental samples; instead, manual counting, aided by 4–6-fold magnification of colonies, should be used.

◆◆◆◆◆ STAINING HALOARCHAEA WITH DAPI

General Information

DAPI (4′, 6-diamidino-2-phenylindole) is a fluorescent stain which binds preferentially to the AT-rich regions of dsDNA (Kubista et al., 1987). There have been suggestions that at higher salinity (more than a few %) DAPI reacts unspecifically with marine bacteria, or binds less effectively to DNA (Zweifel and Hagström, 1995). However, successful staining of extremely halophilic Archaea with DAPI has been reported (Antón et al., 1999, Leuko et al., 2004, 2005). Some special problems are encountered with haloarchaea: while halococci, which possess thick cell walls (Grant et al., 2001), can be stained with DAPI following fixation and permeabilization by ethanol, all other haloarchaea (e.g. members of the genera *Halobacterium, Haloarcula, Haloferax, Halorubrum*) will lyse during these procedures, due to the absence of a cell wall and the fragility of their outer surface layers (Grant et al., 2001). A detailed method, which involves embedding in agarose prior to fixing and staining with DAPI, has been developed by Antón et al. (1999) for these genera and was, slightly modified, used in our laboratory.

Equipment and Reagents

Epifluorescence microscope, filters and camera:

Zeiss Axioskop, equipped with a mercury lamp HBO 100; the filter set 01 (Zeiss) was used, which contains with BP 365, FT 395 and LP 397 (excitation at 365 nm, beam splitting mirror at 395 nm, excitation barrier filter at 397 nm). A charge-coupled device camera (DEI-750CE, Optronics) was used for recording of images, with the Zeiss KS-200 software package, version 3.0.

Humidified chamber (plastic dishes 14 × 14 cm) with paper towels lining the bottom, soaked with distilled water.
DAPI working solution: made from a frozen stock solution (1 mg DAPI, dissolved in 1 ml H_2O) by diluting 1:100 with H_2O.
Phosphate-buffered saline (150 mM NaCl, 1.5 mM KH_2PO_4, 9.5 mM Na_2HPO_4 buffer, pH 7.4).
TN buffer: 4 M NaCl, 50 mM Tris-HCl, pH 7.4.
Agarose solution: 0.5% (SeaKem LE, Biozym) in 22% NaCl.
Formaldehyde solution: 6% made up in 24% NaCl.
Glass microscope slides.
Dark glass slides with 6 wells (Marienfeld).
Glass cover slips (Roth; near UV transparent).

Protocols

a. DAPI staining of *Halococcus* strains:

1. Centrifuge 200 µl of a culture in late exponential phase at 12.000 × g for 3 min.
2. Add 30 µl of H_2O to the pellet, mix thoroughly.
3. Add 70 µl of ethanol (96%), incubate 1–2 min at room temperature.
4. Centrifuge cells again; take up pellet in 20 µl of sterile TN buffer.
5. Place 3 µl of the sample on a glass slide, mix with 3 µl DAPI working solution, incubate for 3–5 min in the dark.
6. View in the epifluorescence microscope, using filter set 01.
7. Record images with CCD camera.

b. DAPI staining of non-coccoid haloarchaea:

1. Warm 500 µl of agarose solution to 55°C; add 500 µl of sample.
2. Mix carefully; spot in portions of 15 µl on the wells of a six-well slide.
3. Dry samples at room temperature.
4. Overlay each sample with 100 µl of formaldehyde solution.
5. Incubate slides incubated for 16 h at 4°C in a humid chamber.
6. Wash slides extensively with phosphate-buffered saline; air dry.
7. Dehydrate in ethanol (Secco Solv, max 0.02% H_2O) by immersing for 3 min each in 50, 80 and 100% ethanol.
8. Stain each sample with 3 µl of DAPI working solution; incubate for 3–5 min in the dark.
9. View in epifluorescence microscope, using filter set 01.
10. Record images with CCD camera.

Examples of stained halococci (*Hcc. dombrowskii* H4 DSM 14522[T]) and non-coccoid haloarchaea (*Hbt. salinarum* DSM 3754[T]) are shown in Figure 24.1. Nucleoids can be discerned in both strains following staining with DAPI.

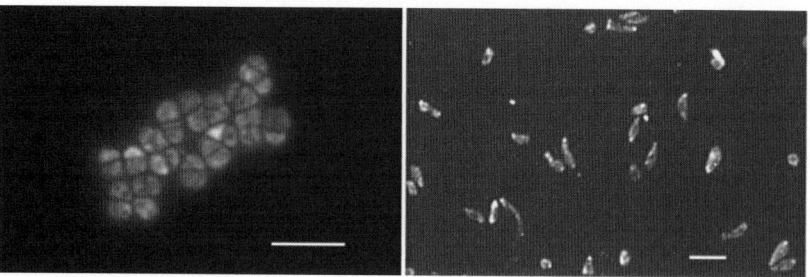

Figure 24.1. Staining of haloarchaea with DAPI. Left: *Halococcus dombrowskii* H4 DSM 14522^T; right: *Halobacterium salinarum* DSM 3754^T (Photograph by S. Rittmann). Bars, 2 μm. See Plate 24.1 in Colour Plate Section.

Advantages:

DAPI staining is a comparatively simple and rapid technique which can be used in the presence of high ionic strength (up to 4.2 M NaCl).
Nucleoids in cells are usually discernible.
Enumeration of total amounts of halococci is theoretically possible.

Disadvantages:

Non-coccoid haloarchaea require an elaborate protocol prior to staining.
Unable to discriminate viable from non-viable cells (see Leuko *et al.*, 2005).
DAPI in concentrations of 0.014 μM will not allow re-cultivation of cells (Leuko *et al.*, 2004, and references therein).

◆◆◆◆◆ LIVE/DEAD® *Bac*Light™ BACTERIAL VIABILITY KIT

Basic Procedures

General information

The LIVE/DEAD® kit consists of the nucleic acid stains SYTO 9 and propidium iodide (PI). Whereas SYTO 9 penetrates most membranes freely and has moderate affinity to nucleic acids, propidium iodide is highly charged and therefore normally cell-impermeant. It will, however, penetrate damaged membranes; since it possesses a high affinity for nucleic acids, it can displace weaker bound dyes, such as SYTO 9 (Haugland, 2002). This is the basis for the simultaneous application of both dyes for viability staining – viable cells with an intact membrane exhibit green fluorescence, whereas dead cells, due to a compromised membrane, show strong red fluorescence. Although the LIVE/DEAD® kit was developed for mesophilic, non-halophilic microorganisms, it was successfully applied to thermophilic and extremely halophilic Archaea (Leuko *et al.*, 2004). Final concentrations of 10 μM SYTO 9 and 30 μM PI are used for most staining applications (Bunthof *et al.*, 2001).

Equipment and reagents

Epifluorescence microscope, filters and camera:

Zeiss Axioskop, equipped with a mercury lamp HBO 100; the filter set 25 (Zeiss) was used, which contains with filters TBP 400/495/570, FT 410/505/585 and TBP 460/530/610. A charge-coupled device camera (DEI-750CE, Optronics) was used for recording of images, with the Zeiss KS-200 software package, version 3.0.

LIVE/DEAD® BacLight bacterial viability kit L-7012 (Molecular Probes, USA): stock solutions are 3.34 mM SYTO 9 and 20 mM propidium iodide (PI), both in dimethyl sulfoxide.
Working solutions: 0.5 mM SYTO 9, 1.5 mM propidium iodide, both prepared fresh daily with water.
Glass microscope slides.
Glass cover slips (Roth; near UV transparent).

Protocols

1. Make a 1:1 mixture of SYTO 9 and PI working solutions.
2. Mix 50 µl of salt-containing samples – taken directly from liquid cultures or cell suspensions – with 2 µl of a 1:1 mixture of SYTO 9 and PI.
3. Incubate for 15 min in darkness at room temperature.
4. View in epifluorescence microscope, using filter set 25.
5. Record images with CCD camera.

Detection of Halophilic Microorganisms in Environmental Samples

General information

A direct correlation between fluorescent particles obtained by staining with the LIVE/DEAD® kit and potential CFUs can be obtained using a confocal scanning laser microscope, since this instrument affords precise volumes of fields of view. Data obtained with haloarchaeal type strains showed a satifactory relationship between CFUs obtained by spread plates and numbers of viable cells following staining with the LIVE/DEAD® kit and viewing in a confocal scanning laser microscope (Leuko et al., 2004). In addition, the LIVE/DEAD® stains are, in contrast to DAPI, not growth-inhibitory in the concentrations used here (Leuko et al., 2004); thus, rare samples could be used for both enumeration and further analysis, following cultivation of resident microorganisms.

Equipment and samples

Epifluorescence microscope (as above) with filters for LIVE/DEAD® staining.
Control culture (Hcc. dombrowskii H4 DSM 14522^T).
Rock salt dissolved under sterile conditions.
Hypersaline waters (Dead Sea, subterranean salt lakes).

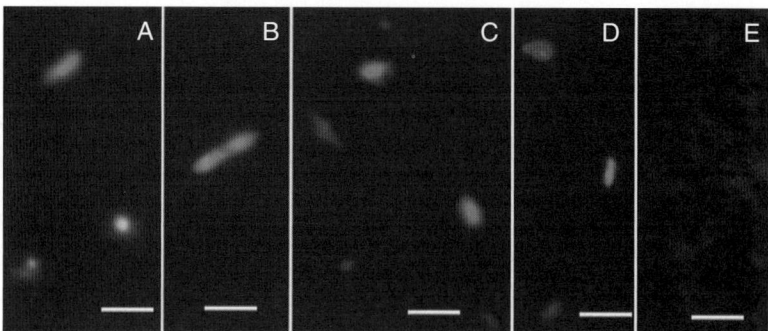

Figure 24.2. Fluorescent particles in environmental samples. Dead Sea water (A, B) and dissolved rock salt (C–E) from the Permian salt deposit at Altaussee, Austria, were treated with the LIVE/DEAD® kit without any prior preparation; (E) hazy staining of material in dissolved rock salt. Bars, 2 µm (reprinted with permission from the American Society for Microbiology, Washington D.C.). See Plate 24.2 in Colour Plate Section.

Protocol

1. Without further treatment, stain portions of 50 µl of the samples with the LIVE/DEAD® kit.
2. As a control, mix aliquots of dissolved rock salt with a small amount of a *Hcc. dombrowskii* H4 DSM 14522^T culture prior to staining.

Examples for visualizing particles in environmental hypersaline samples are shown in Figure 24.2. Several particles of about 1–2 µm length with green fluorescence in a sample of Dead Sea water (Figure 24.2A, B) and in dissolved rock salt (Figure 24.2C–E) from the salt mine in Altaussee, Austria, are visible. According to their sizes and shapes, they could represent viable halophilic prokaryotes. Some hazy appearing material was detected in dissolved rock salt, which produced weak green fluorescence (Figure 24.2E). In water from the Dead Sea, rod-shaped particles of green fluorescence, including an apparently dividing cell (Figure 24.2B) were present; some particles showed a yellowish fluorescence (Figure 24.2A). CFU from this sample were between 180 and 240 per ml on agar plates containing M2S medium, thus corroborating the presence of viable halophilic microorganisms. CFU from dissolved rock salt varied between 10 and 130 ± 20 per ml, similar as had been reported earlier (Stan-Lotter et al., 2000). No autofluorescence was detected in any of the environmental samples investigated so far (subterranean salt lakes; dissolved rock salt from Alpine Permian salt; Dead Sea water), but yellowish particles were occasionally observed (as in Figure 24.2A) also in rock salt samples (Fendrihan, unpublished).

Stability of Cell-bound LIVE/DEAD® Dyes

Fluorescent dyes are light sensitive to various extents and generally should be handled while avoiding exposure to light. For most

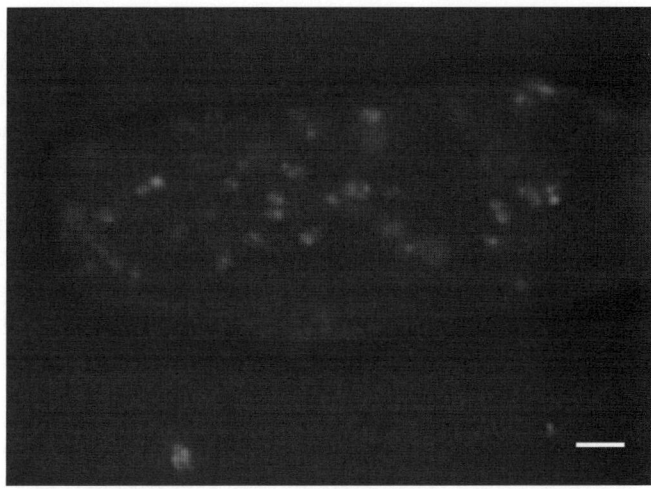

Figure 24.3. Cells of *Halococcus dombrowskii* H4 DSM 14522^T in fluid inclusions. Cells were stained with the LIVE/DEAD® kit prior to embedding in crystals of NaCl. Epifluorescence microscopy was performed following entrapment of cells for 3 days. Bar, 5 μm. See Plate 24.3 in Colour Plate Section.

applications, the stability of the stains, following reaction with cells, is of minor importance, since enumeration of microorganisms is the primary concern. We observed that haloarchaea stained with the LIVE/DEAD® kit kept their fluorescence for at least several days, when embedded in salt crystals on a microscopic slide and stored at room temperature in the dark. Figure 24.3 shows cells of *Hcc. dombrowskii* H4, which were stained with the LIVE/DEAD® kit as described above, and subsequently embedded in salt crystals by drying the sample on a glass slide for 3 days. The sample was inspected with the epifluorescence microscope and the cells were still fluorescent. They had accumulated preferentially in the fluid inclusions of the artificial halite.

Advantages

The procedure is technically simple.
No special equipment, except epifluorescence microscope, is necessary;
Staining at high ionic strength is possible.
Cell-bound dyes are stable, at least short-term.

Disadvantages

Measures only one physiological parameter.

◆◆◆◆◆ CONCLUSIONS

Staining of microorganisms with fluorescent dyes in the presence of high ionic strength (up to 4.2 M NaCl) is possible. Morphology, size and the presence of nucleoids (which is considered to be indicative of active cells; Howard-Jones *et al.*, 2001) can be detected in the epifluorescence microscope; an assessment of the intactness or damage of membranes, whether of bacterial or archaeal composition (Kates, 1993), can be made. Extremes of pH do not interfere with application of the LIVE/DEAD® kit (Leuko et al., 2004). Staining with DAPI does not distinguish between viable and dead bacterial cells (see Howard-Jones et al., 2001), which was also true for haloarchaea (Leuko et al., 2005). The LIVE/DEAD® kit is thought to permit a differentiation between active and dead cells. For information about the true status of microbial cells, determination of CFUs is still the most valuable approach, though not always feasible. A useful property of the dyes of the LIVE/DEAD® kit is their non-interference – when used at low concentrations, as for staining – with subsequent growth experiments.

The stability of cell-bound dyes from the LIVE/DEAD® kit was higher than expected (see Figure 24.3), since fluorescence was detectable for at least 3 days following staining. This property should allow further studies on the location of haloarchaea in salt crystals and might provide a model system for the study of microorganisms in future extraterrestrial samples.

Acknowledgments

This work was supported by the Austrian Science Foundation (FWF), projects P13995-MOB and P16260-B07. We thank Michael Mayr, Salinen Austria, and Christa Schaubmaier, State Hospital of Linz, for samples of rock salt and Dead Sea water, respectively.

List of suppliers

Biozym Biotech Trading GmbH
Wehlistraße 27 b
A - 1200 Wien, Austria
Tel: +43 1 334 0156 0

SeaKem LE agarose

DSMZ (Deutsche Sammlung von Mikroorganismen und Zellkulturen GmbH)
Mascheroder Weg 1b
D 38124 Braunschweig, Germany
Tel: +49 531 2616-0
http://www.dsmz.de

Cultures of haloarchaea, haloarchaeal type strains

ICN Biomedicals GmbH
Thüringerstr. 15
D-37269 Eschwege, Germany
Tel: +49 5651 921-0

Tris Ultra pure

Paul Marienfeld KG
c/o Klösch, Laborbedarf
Herbststr. 20
A-1160 Wien, Austria
Tel: +43 1 492 5185

Dark glass slides with wells

Millipore Ges.m.b.H.
Hietzinger Hauptstraße 145
A-1130 Wien, Austria
Tel: +43 1 877 89 26

Stericups

Molecular Probes Europe BV
PoortGebouw, Rijnsburgerweg 10,
2333 AA Leiden, The Netherlands
Tel: +31 71 5233378

LIVE/DEAD® BacLight™ kit

Roth c/o Lactan
Puchstr. 85
A-8020 Graz, Austria
Tel: +43 316 323 6920

Tryptone

Sigma Aldrich Chemical Company
Favoritner Gewerbering 10
A-1100 Wien, Austria
Tel: +43 1 605 81 20

DAPI

Carl Zeiss GmbH
Modecenterstraße 16
A-1034 Wien, Austria
Tel: +43 1 79518 0

Epifluorescence microscopes

Other supplies such as glassware, slides, cover slips, pipette tips, plastic tubes, plastic dishes can be obtained from any scientific supply company (e.g. VWR Scientific). Research grade chemicals (formaldehyde, Hycase) were obtained from Sigma Aldrich Chemical Company; agar, ethanol, microscope grade immersion oil, salts, trace elements were obtained from VWR Scientific.

References

Antón, J., Llobet-Brossa, E., Rodríguez-Valera, F. and Amann, R. (1999). Fluorescence *in situ* hybridization analysis of the prokaryotic community inhabiting crystallizer ponds. *Environ. Microbiol.* **1**, 517–523.

Barber, D. J. (1981). Matrix phyllosilicates and associated minerals in C2M carbonaceous chondrites. *Geochim. Cosmochim. Acta* **45**, 945–970.

Bunthof, C. J., van Schalkwijk, S., Meijer, W., Abee, T. and Hugenholtz, J. (2001). Fluorescent method for monitoring cheese starter permeabilization and lysis. *Appl. Environ. Microbiol.* **67**, 4264–4271.

Burns, D. G., Camakaris, H. M., Janssen, P. H. and Dyall-Smith, M. L. (2004). Combined use of cultivation-dependent and cultivation-independent methods indicates that members of most haloarchaeal groups in an Australian crystallizer pond are cultivable. *Appl. Environ. Microbiol.* **70**, 5258–5265.

Denner, E. B. M., McGenity, T. J., Busse, H.-J., Grant, W. D., Wanner, G. and Stan-Lotter, H. (1994). *Halococcus salifodinae* sp. nov., an archaeal isolate from an Austrian salt mine. *Int. J. Syst. Bacteriol.* **44**, 774–780.

Fendrihan, S. and Stan-Lotter, H. (2004). Survival of halobacteria in fluid inclusions as a model of possible biotic survival in Martian halite. In *Mars and Planetary Science and Technolgy. Selected Papers from EMC'04* (H. N. Teodorescu and H. S. Griebel, eds), pp. 9–18. Performantica Press, Iasi, Romania.

Fish, S. A., Shepherd, T. J., McGenity, T. J. and Grant, W. D. (2002). Recovery of 16S ribosomal RNA gene fragments from ancient halite. *Nature* **417**, 432–436.

Foing, B. (2002). Space activities in exo/astrobiology. In *Astrobiology. The Quest for the Conditions of Life* (G. Horneck and C. Baumstark-Khan, eds), pp. 389–398. Springer-Verlag, Berlin.

Gooding, J. L. (1992). Soil mineralogy and chemistry on Mars: possible clues from salts and clays in SNC meteorites. *Icarus* **99**, 28–41.

Grant, W. D., Gemmell, R. T. and McGenity, T. J. (1998). Halobacteria – the evidence for longevity. *Extremophiles* **2**, 279–288.

Grant, W. D., Kamekura, M., McGenity, T. J. and Ventosa, A. (2001). Class III. Halobacteria class. nov. In *Bergey's Manual of Systematic Bacteriology* (D. R. Boone, R. W. Castenholz and G. M. Garrity, eds), 2nd edn., vol. I, pp. 294–301. Springer-Verlag, New York.

Gruber, C., Legat, A., Pfaffenhuemer, M., Radax, C., Weidler, G., Busse, H. J. and Stan-Lotter, H. (2004). *Halobacterium noricense* sp. nov., an archaeal isolate from a bore core of an alpine Permo-Triassic salt deposit, classification of *Halobacterium* sp. NRC-1 as a strain of *Halobacterium salinarum* and emended description of *Halobacterium salinarum*. *Extremophiles* **8**, 431–439.

Haugland, R. P. (2002). LIVE/DEAD BacLight bacterial viability kits. In *Handbook of Fluorescent Probes and Research Products* (J. Gregory, ed.), 9th edn., pp. 626–628. Molecular Probes, Eugene.

Hochstein, L. I. (1987). The physiology and metabolism of the extremely halophilic bacteria. In *Halophilic Bacteria* (F. Rodriguez-Valera, ed.), vol. II, pp. 67–83. CRC Press Inc., Boca Raton.

Holser, W. T. and Kaplan, I. R. (1966). Isotope geochemistry of sedimentary sulfates. *Chem. Geol.* **1**, 93–135.

Horita, J., Friedman, T. J., Lazar, B. and Holland, H. D. (1991). The composition of Permian seawater. *Geochim. Cosmochim. Acta* **55**, 417–432.

Howard-Jones, M. H., Frischer, M. E. and Verity, P. G. (2001). Determining the physiological status of individual bacterial cells. In *Methods in Microbiology, Vol. 30, Marine Microbiology* (J. Paul, ed.), pp. 175–206. Academic Press, New York.

Kamekura, M. (1998). Diversity of extremely halophilic bacteria. *Extremophiles* **2**, 289–295.

Kates, M. (1993). Membrane lipids of Archaea. In *The Biochemistry of the Archaea* (M. Kates, D. J. Kushner and A. T. Matheson, eds), pp. 261–295. Elsevier Science Publishers, Amsterdam.

Klaus, W. (1974). Neue Beiträge zur Datierung von Evaporiten des Oberperm. *Carinthia II, 164, Jahrg* **84**, 79–85.

Kubista, M., Åkerman, B. and Nordén, B. (1987). Characterization of interaction between DNA and 4′, 6-diamidino-2-phenylindole by optical spectroscopy. *Biochemistry* **26**, 4545–4553.

Kunte, H. J., Trüper, H. and Stan-Lotter, H. (2002). Halophilic microorganisms. In *Astrobiology. The Quest for the Conditions of Life* (G. Horneck and C. Baumstark-Khan, eds), pp. 185–200. Springer-Verlag, Berlin.

Kushner, D. J. and Kamekura, M. (1988). Physiology of halophilic eubacteria. In *Halophilic Bacteria* (F. Rodriguez-Valera, ed.), vol. I, pp. 109–140. CRC Press, Boca Raton.

Javor, B. J. (1989). *Hypersaline Environments: Microbiology and Biogeochemistry.* Springer-Verlag, Berlin.

Larsen, H. (1981). The family *Halobacteriaceae*. In *The Prokaryotes. A Handbook on Habitat, Isolation and Identification of Bacteria* (M. P. Starr, H. Stolp, H. G. Trüper, A. Balows and H. G. Schlegel, eds), vol. I, pp. 985–994. Springer Verlag, Berlin.

Leuko, S., Legat, A., Fendrihan, S. and Stan-Lotter, H. (2004). Evaluation of the LIVE/DEAD BacLight kit for extremophilic Archaea and environmental hypersaline samples. *Appl. Environ. Microbiol.* **70**, 6884–6886.

Leuko, S., Legat, A., Fendrihan, S., Wieland, H., Radax, C., Gruber, C., Pfaffenhuemer, M., Weidler, G. and Stan-Lotter, H. (2005). Isolation of viable haloarchaea from ancient salt deposits and application of fluorescent stains for in situ detection of halophiles in hypersaline environmental samples and model fluid inclusions. In *Adaptations to Life at High Salt Concentrations in Archaea, Bacteria and Eukarya* (N. Gunde-Cimerman, A. Oren and A. Plemenitas, eds), Springer, Dordrecht, pp. 91–104.

McCord, T. B., Mansen, G. B., Fanale, F. P., Carlson, R. W., Matson, D. L., Johnson, T. V., Smythe, W. D., Crowley, J. K., Martin, P. D., Ocampo, A., Hibbitts, C. A. and Granahan, J. C. (1998). Salts on Europa's surface detected by Galileo's near Infrared Mapping Spectrometer. The NIMS team. *Science* **280**, 1242–1245.

McGenity, T. J., Gemmell, R. T., Grant, W. D. and Stan-Lotter, H. (2000). Origins of halophilic microorganisms in ancient salt deposits. *Environ. Microbiol.* **2**, 243–250.

Oren, A. (2002). *Halophilic Microorganisms and their Environments.* Kluwer Academic Publishers, Dordrecht.

Paul, J. H. (1993). The advances and limitations of methodology. In *Aquatic Microbiology, an Ecological Approach* (T. E. Ford, ed.), pp. 15–46. Blackwell Scientific, Boston.

Pedersen, K. (2000). Exploration of deep intraterrestrial microbial life: current perspectives. *FEMS Microbiol. Lett.* **185**, 9–16.

Radax, C., Gruber, C. and Stan-Lotter, H. (2001). Novel haloarchaeal 16S rRNA gene sequences from Alpine Permo-Triassic rock salt. *Extremophiles* **5**, 221–228.

Rodríguez-Valera, F. (1993). Introduction to saline environments. In *The Biology of Halophilic Bacteria* (R. H. Vreeland and L. I. Hochstein, eds), pp. 1–23. CRC Press, Boca Raton.

Stan-Lotter, H., McGenity, T. J., Legat, A., Denner, E. B. M., Glaser, K., Stetter, K. O. and Wanner, G. (1999). Closely related strains of *Halococcus salifodinae* are found in geographically separated Permo-Triassic salt deposits. *Microbiology* **145**, 3565–3574.

Stan-Lotter, H., Radax, C., Gruber, C., McGenity, T. J., Legat, A., Wanner, G. and Denner, E. B. M. (2000). The distribution of viable microorganisms in Permo-Triassic rock salt. In *SALT 2000. 8th World Salt Symposium* (R. M. Geertman, ed.), vol. 2, pp. 921–926. Elsevier Science, Amsterdam.

Stan-Lotter, H., Pfaffenhuemer, M., Legat, A., Busse, H. J., Radax, C. and Gruber, C. (2002). *Halococcus dombrowskii* sp. nov., an archaeal isolate from a Permo-Triassic alpine salt deposit. *Int. J. Syst. Evol. Microbiol.* **52**, 1807–1814.

Stan-Lotter, H., Radax, C., Gruber, C., Legat, A., Pfaffenhuemer, M., Wieland, H., Leuko, S., Weidler, G., Kömle, N. and Kargl, G. (2003). Astrobiology with haloarchaea from Permo-Triassic rock salt. *Int. J. Astrobiol.* **1**, 271–284.

Stan-Lotter, H., Radax, C., McGenity, T. J., Legat, A., Pfaffenhuemer, M., Wieland, H., Gruber, C. and Denner, E. B. M. (2004). From intraterrestrials to extraterrestrials – viable haloarchaea in ancient salt deposits. In *Halophilic Microorganisms* (A. Ventosa, ed.), pp. 89–102. Springer-Verlag, Berlin.

Tindall, B. J. (1992). The family Halobacteriaceae. In *The Prokaryotes* (A. Balows, H. G. Trüper, M. Dworkin, W. Harder and K. H. Schleifer, eds), pp. 768–808. Springer-Verlag, Berlin.

Treiman, A. H., Gleason, J. D. and Bogard, D. D. (2000). The SNC meteorites are from Mars. *Planet Space Science* **48**, 1213–1230.

Vreeland, R. H., Straight, S., Krammes, J., Dougherty, K., Rosenzweig, W. D. and Kamekura, M. (2002). *Halosimplex carlsbadense* gen. nov., sp. nov., a unique halophilic archaeon, with three 16S rRNA genes, that grows only in defined medium with glycerol and acetate or pyruvate. *Extremophiles* **6**, 445–452.

Zolensky, M. E., Bodnar, R. J., Gibson, E. K., Nyquist, L. E., Reese, Y., Shih, C. Y. and Wiesman, H. (1999). Asteroidal water within fluid inclusion-bearing halite in an H5 chondrite, Monahans (1998). *Science* **285**, 1377–1379.

Zweifel, U. L. and Hagström, Å. (1995). Total counts of marine bacteria include a large fraction of non-nucleoid-containing bacteria (ghosts). *Appl. Environ. Microbiol.* **61**, 2180–2185.

25 Characterization of Lipids of Halophilic Archaea

Angela Corcelli and Simona Lobasso
Dipartimento di Biochimica Medica, Biologia Medica e Fisica Medica, Università degli Studi di Bari, Piazza G. Cesare, 70124 Bari, Italy

◆◆◆

CONTENTS

Membrane lipids of halophilic Archaea
Techniques for extraction, separation and analysis
Functional role

◆◆◆◆◆ MEMBRANE LIPIDS OF HALOPHILIC ARCHAEA

The general architecture of the cell membrane is mainly determined by the interactions occurring between lipid molecules and proteins. It is now evident that lipids, besides being the main structural components of the cell membrane, play a critical role in important aspects of cellular functions. In addition, membrane lipid structures and patterns are particularly useful taxonomically to distinguish and recognize representatives of the different domains of the phylogenetic trees, as well as between different groups in the same domain.

In general, Archaea thrive in harsh or extreme environments: the halophiles in saturated salt, the methanogens in anoxic environments (and at elevated temperatures) and the extreme thermophiles at very high temperatures. The salt-loving archaeal microorganisms, which are typically found in the salt ponds along the coasts and hypersaline lakes, are grouped in the family *Halobacteriaceae*. Many members of the *Halobacteriaceae* are characterized by the presence of beautiful red-orange pigments, called bacterioruberins, in the cytoplasmic membrane, which help to screen out UV radiation and protect the cells from the harmful effects of sunlight.

In contrast with Eukarya and Bacteria containing largely diacylglycerol-derived membrane lipids and some monoacyl-monoalkyl-glycerol-derived lipids, archaeal membrane lipids are unique in containing diphytanylglycerol diether lipids. Interestingly, archaeal diether phospholipids are the mirror images of the analogous diester glycerophospholipid found in all other organisms (Figure 25.1); both glycerols in the

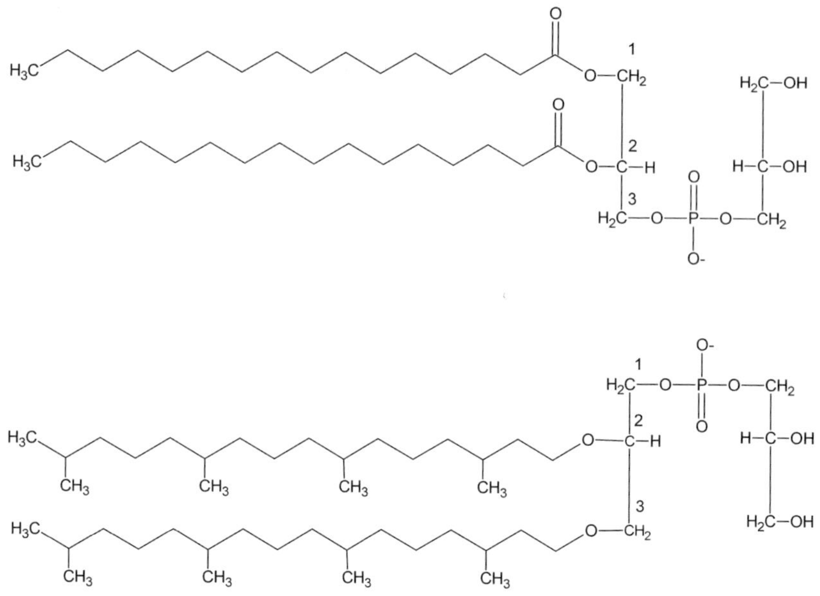

Figure 25.1. Comparison of the structures of diester glycerophospholipids (upper panel) and diphytanylglycerol diether analogs (lower panel). Archaeal lipids are glycerol diether isoprenoid derivatives, whereas "normal" lipids are glycerol diester fatty acid derivatives. Note that the glycerol ethers contain an *sn*-2,3 stereochemistry that is opposite to that of the naturally occurring *sn*-1,2 stereochemistry of glycerophospholipds of Bacteria and Eukarya.

archaeal glycerophospholipids have the opposite stereoconfiguration to those in other prokaryotes and in eukaryotes. This configuration of archaeal diether phospholipids would protect the extreme halophiles against hydrolysis of their membrane phospholipids by stereospecific phospholipases excreted by other organisms, thus helping these microorganisms to survive in competitive environments. Furthermore the branched isoprenoid chains would also ensure that the membrane lipids are in the liquid-crystalline phase at all ambient temperatures (Lindsey et al., 1979).

Most of the phospholipids and glycolipids of extreme halophiles are anionic, so that their negatively charged groups would impart a high negative charge density to the halophile membranes. No nitrogenous base-containing phospholipids, such as phosphatidylserine or phosphatidylethanolamine, are present in extreme halophiles, and this may be characteristic of the halophilic Archaea, in contrast to the methanogenic and thermophilic Archaea. Information on lipid biosynthesis in Archaea is available thanks to labelling studies with whole cells (Kates, 1993; Kamekura and Kates, 1988).

It has been an open question whether Archaea contain any fatty acids, along with the required fatty acid synthetase enzyme. Small amounts of straight-chain fatty acids have been found in the extreme halophiles (Corcelli et al., 1996; Pugh et al., 1971) and in methanogens (Pugh and Kates, 1994). The fatty acid synthetase (FAS) complex of *Halobacterium salinarum* was partially characterized more than 30 years ago;

FAS activities were low and the enzyme was inhibited by the high intracellular salt concentration (Pugh et al., 1971). The fatty acids were shown not to be incorporated into the membrane lipids, but to be used to acylate membrane proteins. Free palmitate has been found associated with halorhodopsin in *Halobacterium salinarum* at a ratio of 1–2 molecules per molecule of halorhodopsin (Colella et al., 1998; Corcelli et al., 1996). High-resolution crystallographic structures now available have confirmed the association of palmitate with halorhodopsin and described the specific mode of interactions between protein and fatty acid (Kolbe et al., 2000).

One- or two-dimensional thin-layer chromatography (TLC) is a useful technique for the rapid characterization of the lipids present in halophilic archaeal isolates. Furthermore mass spectrometry and NMR spectroscopic techniques are powerful analytical methods in studies of structural characterization of lipid mixtures, isolated lipid components or chemical degradation products of these lipids.

The study of lipid composition of archaeal extreme halophilic microorganisms is of particular interest not only for its relevance for taxonomy, but also to shed light on membrane assembly and function and for the presence of a large variety of unusual structures of potential biotechnology applications.

Neutral Lipids

Neutral lipids account for approximately 10% of the total lipid content of the halophilic Archaea of the family *Halobacteriaceae* (Kamekura and Kates, 1988); they consist almost exclusively of C_{20}–C_{50} isoprenoids and isoprenoid-derived compounds (Figure 25.2). The following types have

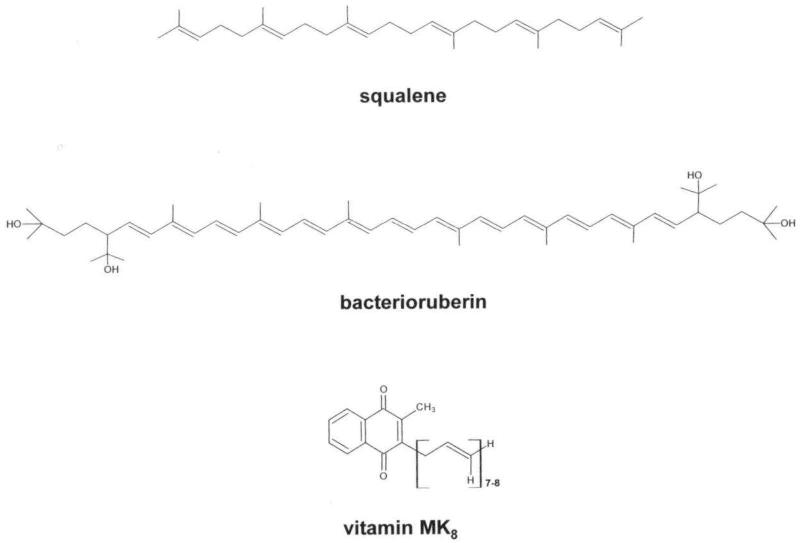

Figure 25.2. Structures of neutral lipids of extreme halophilic Archaea.

been reported: carotenoids (C_{40}-isoprenoid compounds), bacterioruberins (C_{50}-isoprenoid compounds), quinones, geranylgeraniol (C_{20}-isoprenoid lipid), neutral phytanyl ethers (Kushwaha and Kates, 1978), C_{30}-isoprenoid compounds (squalene, dihydrosqualene, tetrahydrosqualene, dehydrosqualene) (Kushwaha et al., 1972), and indole (Kushwaha et al., 1977). Quinones may account for about 9% of the total neutral lipid content of the cells (Kamekura and Kates, 1988). The major respiratory quinones in the *Halobacteriaceae* are MK-8 and MK-8(H_2), two menaquinones with eight isoprenoid units (Collins et al., 1981).

Polar Lipids

The main lipid core of archaeal extreme halophiles, which represents the hydrophobic portion of complex polar membrane lipids, is usually obtained from archaeal lipids by strong-acid methanolysis or acetolysis to remove polar groups such as phosphate esters or sugars (Kates, 1986).

The diether core lipid that forms the basis for most polar lipid structures present in the family *Halobacteriaceae* is 2,3-di-O-phytanyl-*sn*-glycerol (C_{20}, C_{20}), also called archaeol (Kates, 1978); in some haloalkaliphile and *Halococcus* species, C_{20},C_{25}- and C_{25},C_{25}-diether variants of the diphytanylglycerol diether lipid core were also identified (Kates, 1993). The diphytanylglycerol diether lipid core is considered as one of the most useful chemotaxonomic markers of the Archaea domain.

A great variety of polar lipids is encountered in the different representatives of the *Halobacteriaceae*, including phospholipids, sulfolipids and glycolipids.

Phospholipids

The phospholipid structures have been shown to be archaeol analogs of: phosphatidylglycerol (PG), methyl ester of phosphatidylglycerophosphate (PGP-Me), phosphatidylglycerosulfate (PGS) and phosphatidic acid (PA) (Figure 25.3). PGP-Me is the major phospholipid in all extreme halophiles and extreme haloalkaliphiles, having been identified by FAB-MS and TLC in several genera of extreme halophiles, including *Halobacterium*, *Haloarcula*, *Haloferax*, *Halococcus*, *Natronobacterium* and *Natronococcus* (Kates et al., 1993). It should be noted that PGP, as well as PA, may be detected only as minor components, probably present as biosynthetic intermediates. Another minor phospholipid, that has been identified in *Natronococcus occultus*, is the cyclic form of PGP (Lanzotti et al., 1989).

Glycolipids

Glycolipids have become useful taxonomic markers in the classification of different genera of halophilic Archaea. The structures of the glycolipids appear to be derived from a basic diglycosyl archaeol, mannosyl-glucosyl-diphytanylglycerol (DGD), by substitution of sugar or sulfate groups at the 3 or 6 position of the mannose residue. Detailed information

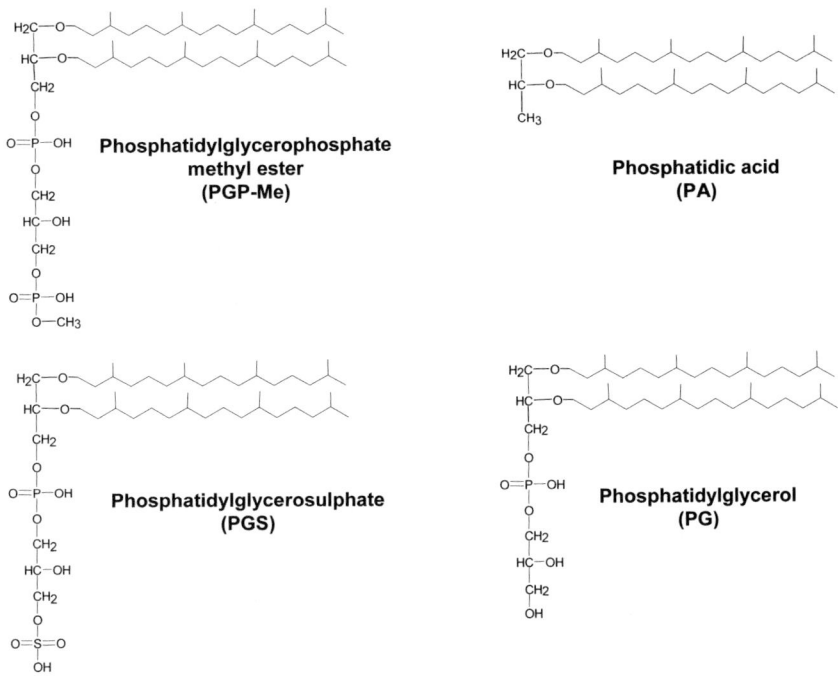

Figure 25.3. Structures of phospholipids of extreme halophilic Archaea.

on the occurrence of the specific archaeal glycolipids in the different genera of the *Halobacteriaceae* is available (Kates, 1986; Oren, 2002).

Here we report a few examples of the variants of glycolipid structures of *Halobacteriaceae*. The sulfated diglycosyl archaeol S-DGD-1 (1-O-[α-D-mannose- (6'-SO_3H)-(1'→2')-α-D-glucose]-2,3-di-O-phytanyl-*sn*-glycerol) is the major glycolipid in the genus *Haloferax* (Kushwaha et al., 1982); while S-DGD-3 (1-O-[α-D-mannose-(2'-SO_3H)-(1'→4')-α-D-glucose]-2,3-di-O-phytanyl-*sn*-glycerol) or S-DGD-5 (1-O-[α-D-mannose-(2'-SO_3H)-(1'→2')-α-D-glucose]-2,3-di-O-phytanyl-*sn*-glycerol) have been found in representatives of the genus *Halorubrum* (Tindall, 1990; Trincone et al., 1990, 1993). In the genus *Halobacterium*, the major glycolipid is the sulfated triglycosyl archaeol S-TGD-1(1-O-[β-D-galactose-(3'-SO_3H)-(1'→6')-α-D-mannose-(1'→2')-α-D-glucose]-2,3-di-O-phytanyl-*sn*-glycerol) (Kates, 1978); in addition, the sulfated tetraglycosyl archaeol S-TeGD (1-O-[β-D-galactose)- (3'-SO_3H) (1'→6')- α-D-mannose-(3←1')- galactofuranose-(1'→2')-α-D-glucose]-2,3-di-O-phytanyl-*sn*-glycerol) has been also found (Smallbone and Kates, 1981). The structures of S-TGD-1, S-DGD-1 and S-DGD-5 are reported in Figure 25.4.

Cardiolipins

Another interesting class of phospholipids consists of dimeric phospholipids having four chains in the hydrophobic tail. In Bacteria and Eukarya, the major representative of this class is diphosphatidylglycerol, also called cardiolipin.

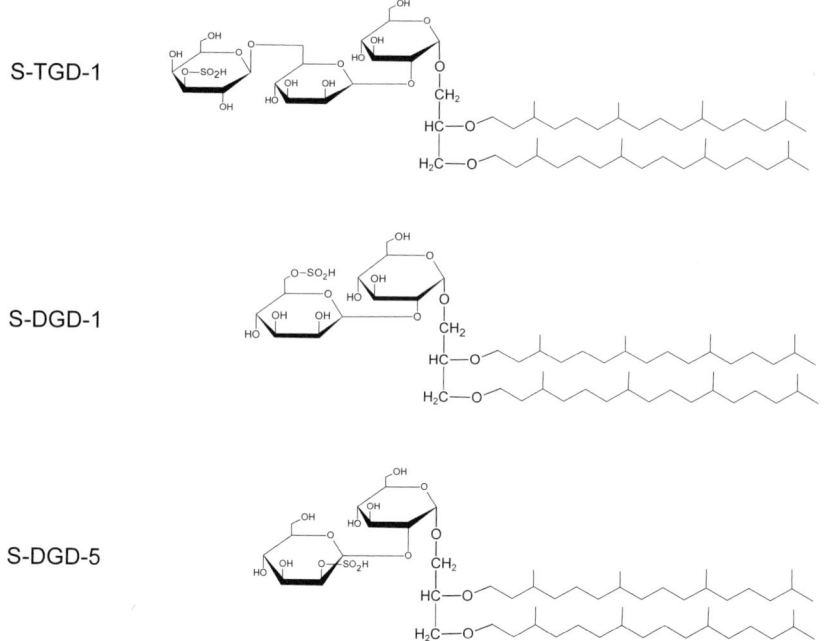

Figure 25.4. Structures of some glycolipids of extreme halophilic Archaea. S-TGD-1 is present in *Halobacterium salinarum*, S-DGD-1 in *Haloferax volcanii* and S-DGD-5 in *Halorubrum trapanicum*.

Reports of the occurrence of cardiolipin and derivatives within the archaeal domain are relatively scarce. Only recently two analogs of cardiolipins have been identified for the first time in the purple membrane of the archaeon, *Halobacterium salinarum* (Corcelli et al., 2000). They are the glycerol diether analog of bisphosphatidylglycerol (BPG or cardiolipin), sn-2,3-di-O-phytanyl-1-phosphoglycerol-3-phospho-sn-di-O-phytanylglycerol, and a novel complex phosphosulfoglycolipid, called glycocardiolipin or GlyC or S-TGD-1-PA (3'-SO$_3$H)-Galp-α1,6Manp-α1,2Glcpα-1-1-[sn-2,3-di-O-phytanylglycerol]-6-[phospho-sn-2,3-di-O-phytanylglycerol] (Figure 25.5). Both novel lipids have two diphytanylglycerol moieties in their molecule, making their structures analogous to that of eukaryal cardiolipin.

Figure 25.6 shows the lipid composition of the purple membrane of *Halobacterium salinarum* examined by TLC (c) and negative-ion ESI-MS mass spectra of purified GlyC (a) and BPG (b). Interestingly GlyC consists of a sulfotriglycosyl diphytanylglycerol esterified to the phosphate group of phosphatidic acid; in addition NMR analysis (Figure 25.7A) shows that its polar head group is composed of the same sugars in the same sequence and anomeric configuration as in S-TGD-1, the major membrane glycolipid of *Halobacterium*, namely: βGalp→αMan→αGlc. The sulfate group is also located on C-3 of the galactose residue, as in S-TGD-1 (Kates and Deroo, 1973).

Another glycocardiolipin was recently found in a membrane fraction isolated from an extreme halophilic archaeon, *Halorubrum* sp., isolated from the salt ponds in Margherita di Savoia, Italy (Lopalco et al., 2004).

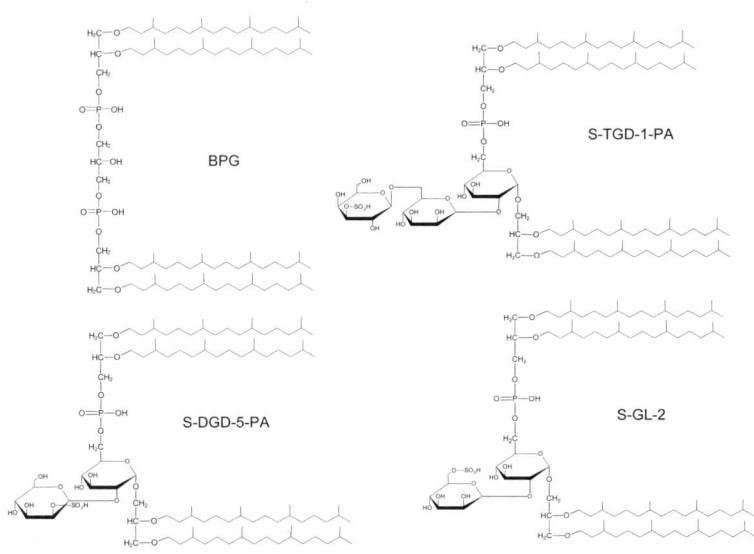

Figure 25.5. Structures of the archaeal cardiolipins. S-TGD-1-PA and BPG were discovered in the purple membrane of *Halobacterium salinarum* (from Lobasso et al., 2003); S-DGD-5-PA was discovered in the membranes of *Halorubrum trapanicum* (Lopalco et al., 2004) and S-GL-2 in *Haloferax volcanii* (Sprott et al., 2003).

By means of NMR (Figure 25.7B) and ESI-MS (Figure 25.8B) analyses it was shown that the glycocardiolipin of *Halorubrum* also has the structure of a sulfo-glyco-diether-phosphatidic acid, i.e. a phospholipid dimer consisting of the glycolipid S-DGD-5 esterified to phosphatidic acid, namely S-DGD-5-PA, (2'-SO$_3$H)-Man$p$-α1,2Glc$p$α-1-1-[sn-2,3-di-O-phytanylglycerol]-6-[phospho-sn-2,3-di-O-phytanylglycerol] (see structure in Figure 25.5). Again, NMR analysis shows that the sugars in the novel phospholipid are the same and in the same order as the glycolipid S-DGD-5 isolated from the *Halorubrum* membranes, in which the sulfated mannose is linked to glucose (Figure 25.7B). Clearly the glycolipids S-TGD-1 and S-DGD-5 are precursors of S-TGD-1-PA and S-DGD-5-PA, respectively (Corcelli *et al.*, 2000; Lopalco *et al.*, 2004).

Finally the glycosyl cardiolipin analog S-GL-2, (6'-SO$_3$H)-Man$p$-α1,2Glc$p$α-1-1-[sn-2,3-di-O-phytanylglycerol]-6-[phospho-sn-2,3-di-O-phytanylglycerol] was found in the membranes of *Haloferax volcanii* (Sprott *et al.*, 2003) (Figure 25.5).

◆◆◆◆◆ TECHNIQUES FOR EXTRACTION, SEPARATION AND ANALYSIS

Extraction of Total Lipids

Lipids that are largely in hydrophobically associated form may be extracted with relatively non-polar solvents such as ethyl ether,

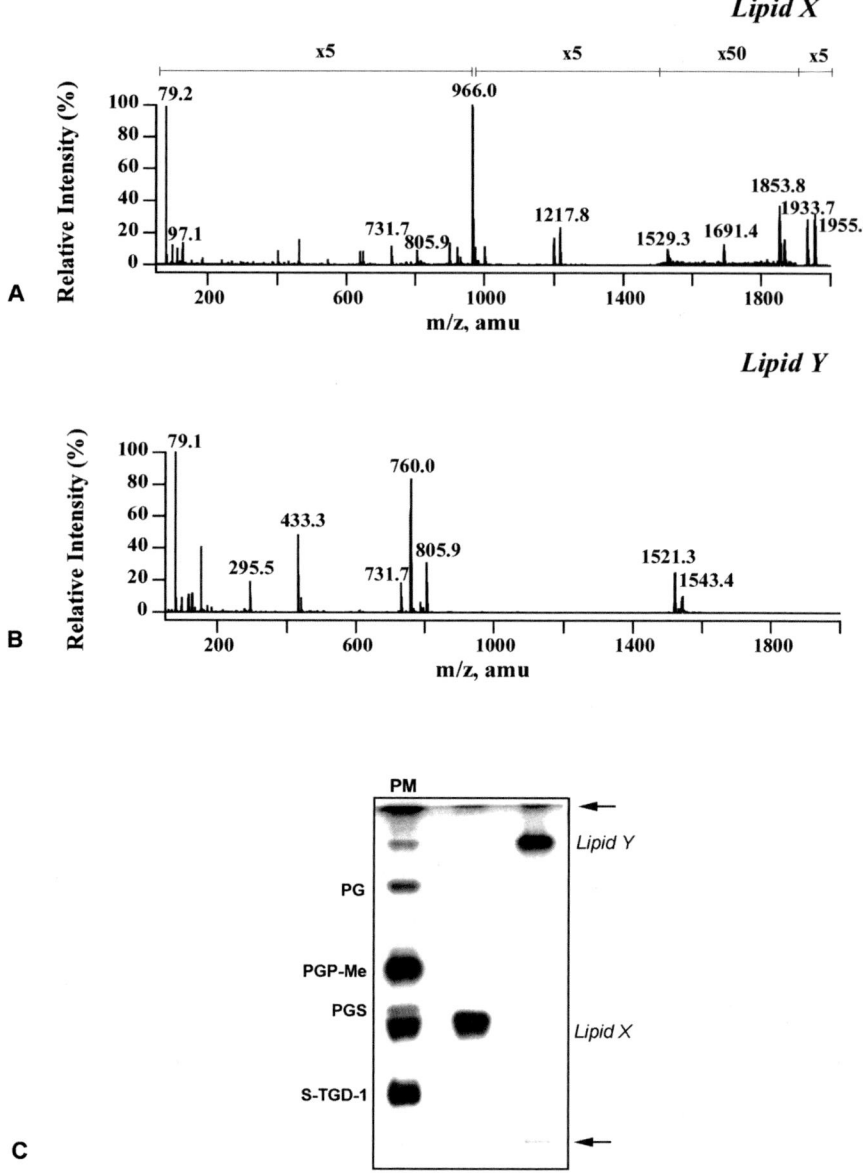

Figure 25.6. Negative-ion ESI-MS mass spectra (A,B) and TLC (C) of GlyC and BPG purified from the purple membrane of *Halobacterium salinarum* (from Corcelli et al., 2000). Lipids X and Y indicate GlyC and BPG, respectively. Note in the ESI mass spectrum of purified lipid X (A) a strong parent molecular ion peak at m/z 966.0 (doubly charged) plus a corresponding smaller one at m/z 1933.7 (singly charged), and in that of purified lipid Y (B) a parent ion at m/z 760.0 (doubly charged) and at m/z 1521.3 (singly charged). The plate (C) was developed with solvent chloroform–methanol–90% acetic acid (65:4:35, v/v) and charred with 0.5% sulfuric acid/ethanol. The individual lipid components of purple membrane can be identified by their R_f values relative to those of authentic standard markers, by their staining behaviour with specific reagents, by their P:lipid and sugar:lipid molar ratios and by their mass spectra. All of the components, except the two lipids named X and Y, had previously been reported for PM lipids of *Halobacterium cutirubrum* and *Halobacterium halobium* (Kushwaha et al., 1975, 1976; Kates et al., 1982).

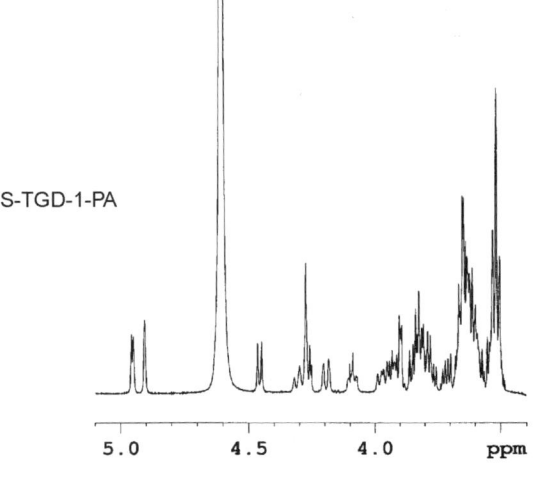
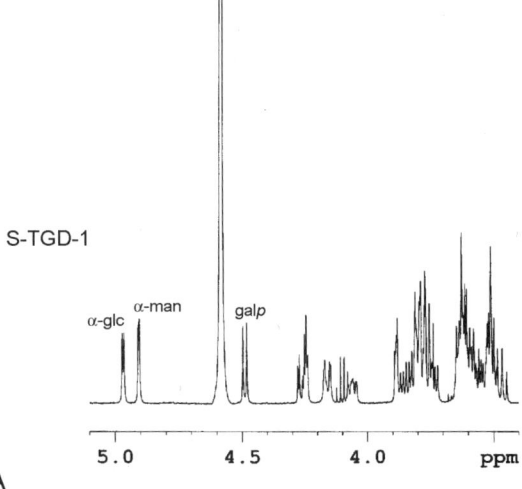

Figure 25.7. Identification of the sugars in the polar head of glycosyl cardiolipin analogs and of their precursor glycolipids by NMR. (A) ^{1}H-NMR spectra of the glycocardiolipin S-TGD-1-PA (upper panel) and the glycolipid S-TGD-1 (lower panel); (B) ^{1}H-NMR spectra of the other glycosyl-cardiolipin S-DGD-5-PA (upper panel) and the glycolipid S-DGD-5 (lower panel) (Corcelli et al., 2000; Lopalco et al., 2004).

chloroform or benzene. However, membrane-associated polar lipids require polar solvents such as ethanol or methanol to disrupt the hydrogen bonding or electrostatic forces between the lipids and the protein. Covalently bound lipids, by contrast, cannot be extracted directly by any solvents, but must first be cleaved from the complex by acid or alkaline hydrolysis.

One of the most versatile and effective lipid extraction procedures is that of Bligh and Dyer (1959), as modified for extreme halophiles (Kates, 1986; Kates and Kushwaha, 1995). Briefly, this procedure uses a one-phase alcoholic solvent system, namely methanol–chloroform–water

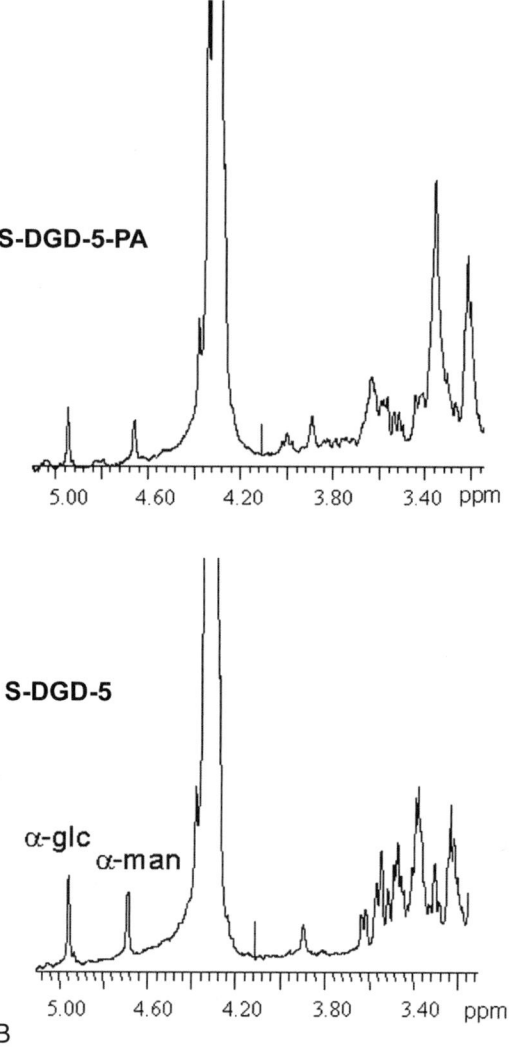

Figure 25.7. *Continued.*

(2:1:0.8, v/v) which rapidly and efficiently extracts the lipids; the extract is then diluted with one volume each of chloroform and methanol–water (1.0:0.9), any water–soluble contaminants are thus readily partitioned into the methanol–water phase, leaving the lipids relatively free of contaminants in the chloroform phase. Any emulsions formed can be broken by centrifugation and small-to-trace amounts of water can be removed from the chloroform phase by azeotropic distillation with benzene or toluene during concentration of the extract either in a vacuum or under a stream of nitrogen.

For example, lipids of the purple membrane (PM) of *Halobacterium salinarum* are extracted by adding 18.75 ml of methanol–chloroform (2:1, v/v) to a PM suspension containing about 8 mg of the protein bacteriorhodopsin in 5 ml of water. The mixture is gently shaken for

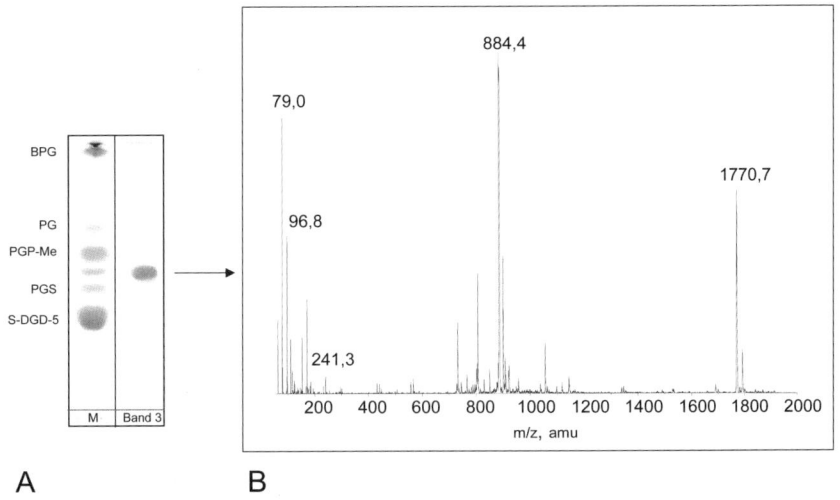

Figure 25.8. TLC (A) and negative-ion ESI-MS (B) spectrum of the glycosyl-cardiolipin analog S-DGD-5-PA isolated and purified from *Halorubrum* sp. membranes (modified from Lopalco et al., 2004).

several minutes until complete protein denaturation and bleaching is obtained. After centrifugation, the supernatant extract is decanted into a separatory funnel and the residue is resuspended in 23.75 ml of methanol–chloroform–water (2:1:0.8, v/v), the mixture is then shaken and centrifuged. Chloroform and water (6.25 ml each) are then added to the combined supernatant extracts to obtain a two-phase system. After complete separation of the two phases (requiring few hours at room temperature, in the dark), the chloroform phase, diluted with benzene, is brought to dryness under nitrogen; dried lipids are re-suspended in a small chloroform volume and stored at −20°C. In order to verify the completeness of lipid extraction, the whitish denatured material left after the lipid extraction is resuspended in methanol–chloroform–water (2:1:0.8) and residual lipids are re-extracted following the above procedure. Up to 10% of membrane lipids can be found in the re-extract. An enrichment of phospholipids and glycolipids tightly bound to membrane proteins can be found in the re-extract; in the case of bacteriorhodopsin, the major component of lipid re-extract is glyco-cardiolipin (Catucci et al., 2004).

The same procedure can be followed for lipid extraction from cells. However it is important to note that cells of extreme halophiles require high salt concentrations (3–4 M) to maintain their cell envelope structure intact; for this reason, these cells should always be kept in salt concentrations close to those of the growth medium (4 M). Washing or suspension of the cells in salt solution lower than 3 M or in water will result in cell lysis and the release of DNA and RNA, forming intractable gels, from which lipids cannot be readily extracted; furthermore it should be considered that, when halobacterial cells are under hypo-osmotic stress, changes in the lipid composition may occur (Lobasso et al., 2003; Lopalco et al., 2004).

Acetone Fractionation into Neutral and Polar Lipids

The simplest and often the most efficient procedure for separating polar lipids (phosphatides and glycolipids) from neutral lipids (including pigments) is the procedure of acetone precipitation (Kates and Kushwaha, 1995). The method depends on the general insolubility in cold acetone of most glycolipids and phosphatides and the fact that glycerides and other neutral lipids are soluble in cold acetone.

About 20 volumes of cold acetone are added to the total lipid extract dissolved in chloroform; then the tube is closed, well mixed on the vortexer and left overnight at $-20°C$. The polar lipids are collected by centrifuging at 2500 rpm for 5 min and the red-coloured acetone supernatant containing the neutral isoprenoid lipids and pigments is decanted into a round-bottom flask. The residual pellets are washed twice by suspending them in 10–15 portions of cold acetone and centrifuging until the washings are only faintly coloured. The precipitated phosphatides are freed of excess solvent in a stream of nitrogen and dried in a vacuum in a desiccator; the dried residue is weighed and dissolved to a known concentration in chloroform. The combined acetone supernatants are concentrated to dryness under nitrogen and the neutral lipids, weighed and dissolved in chloroform.

In general, the acetone-insoluble material will contain 95% or more of the lipid-P and only traces of neutral lipids material (e.g. pigments), whereas the acetone-soluble fraction will contain all of the neutral lipids and only traces of phosphatides.

Analyses of Lipid Extracts

Thin-layer chromatography (TLC)

TLC is one of the most effective and versatile techniques for separation of intact complex lipids (phospholipids, glycolipids, etc.) and their lipid moieties, as well as for neutral lipids; it offers many advantages that make it the method of choice in most instances: rapidity, detectability of all organic compounds by charring and preparative-scale operation. In order to increase the resolving power of chromatography, a two-dimensional development may be carried out in a relatively short time without involving any specialized equipment.

Preparative TLC of total lipid extracts of extreme halophilic Archaea can be carried out on silica gel plates (20 × 20 cm and a 0.5-mm thick layer, while for analytical purposes TLC plates 10 × 20 cm and a 0.25-mm thick layer can be used.

TLC plates should be run twice in the chromatographic chamber with chloroform–methanol (1:1) up to the top of the plate (to remove any traces of lipids), then dried in air and heated at 120°C before applying the sample.

A suitable aliquot of a chloroform solution of lipids is concentrated in a nitrogen stream to give a 2–5% solution. For standard analytical plates, 2–20 μl (containing 10–100 μg lipids) of the chloroform solution per spot (1.5 cm apart and 1.5 cm from the bottom of the plate) are applied using Hamilton microsyringes. The sample may also be applied as a row (0.5–1 cm long) of slightly overlapping very small spots using a fine capillary or a Hamilton syringe. For preparative plates, a 2–5% chloroform solution of lipids is used and an amount containing 10–15 mg of lipids per plate is applied to one 0.5-mm plate.

In all cases, the plates should be chromatographed immediately after the application of the sample. Usually 60–80 ml of solvent is added to glass tanks and allowed to reach equilibrium for 1–2 h; one or two plates are placed in the tank and the solvent is allowed to ascend to about 1–2 cm of the top of the plate (40–50 min); the plates are removed from the tank and the solvent is allowed to evaporate for a short period in the fume hood. Then the chromatogram should be stained immediately.

For two-dimensional chromatography, the spot is applied at one corner, the chromatogram developed in the first solvent, dried briefly, turned through 90° and developed in the second solvent; it is then dried and stained.

The following solvents are used for separation of the indicated lipid classes by TLC:

a. *Neutral lipids*: hexane or petroleum ether (b.p. 60–70°C)–ethyl ether–acetic acid (90:10:1 or 80:20:1 or 70:30:1, v/v) (Kates, 1986);
b. *Phospholipids and glycolipids*: chloroform–methanol–90% acetic acid (65:4:35, v/v) and chloroform–methanol–water (65:25:4, v/v) for *Halobacterium*, *Halorubrum*, *Haloferax* and *Haloarcula* lipids; chloroform–methanol–glacial acetic acid–water (85:22.5:10:4, v/v) for *Halococcus* and *Natronobacterium* (Kates and Kushawa, 1995).

The following stains have been found to be useful in detection of lipids on TLC plates:

a. 5% sulfuric acid in water or 0.5% in ethanol, followed by charring at 120°C for all lipids (Kates, 1986);
b. Iodine vapour for all lipids (Kates, 1986);
c. Molybdenum blue reagent specific for phospholipids (Corcelli *et al.*, 2000);
d. 0.5% α-naphtol in methanol/water (1:1, v/v) specific for glycolipids (Siakotos and Rouser, 1965);
e. 2% azure A in 1 mM sulfuric acid for sulfatides and sulfoglycolipids (Kean, 1968);
f. 0.25% ninhydrine in acetone/lutidine (9:1) for free amino groups (Marinetti, 1964).

Table 25.1 shows the TLC retention factor (R_f) in solvent chloroform–methanol–90% acetic acid (65:4:35, v/v) and staining behavior of various archaeal lipid components.

Table 25.1 TLC R_f and staining behaviour of various archaeal lipid components

Lipid	TLC R_f	Staining		
		Sugar	Phosphate	Sulfate
S-TGD-1	0.07	+	−	+
S-TGD-1-PA	0.21	+	+	+
S-DGD-5	0.31	+	−	+
PGS	0.40	−	+	+
S-DGD-5-PA	0.46	+	+	+
PGP-Me	0.54	−	+	−
PG	0.78	−	+	−
BPG	0.94	−	+	−
Neutral lipids + pigments	0.99–1			

Solvent: chloroform–methanol–90% acetic acid (65:4:35, v/v).

Isolation and purification of individual lipid components (Kates, 1986)

The components of the lipid extracts of extreme halophiles can be separated by preparative TLC in solvent chloroform–methanol–90% acetic acid (65:4:35, v/v). After scraping the silica band corresponding to each component from up to four large plates, about 30 ml of methanol–chloroform–water (2:1:0.8, v/v) can be added for lipid extraction from the silica (three times). After centrifugation, the supernatants are combined and by addition of the appropriate amount of chloroform–water (1:1, v/v) two phases are obtained. The chloroform layer is collected, diluted with benzene and brought to dryness under a stream of nitrogen. Each component can be further purified by re-chromatography in neutral solvent (chloroform–methanol–water, 65:25:4, v/v) and recovered from silica, as described above. The purity of the final material can be checked on silica gel HPTLC plates in the previously mentioned solvent chloroform–methanol–90% acetic acid (65:4:35, v/v).

Examples of isolated and purified lipids from *Halobacterium salinarum* and *Halorubrum* sp. are given in Figure 25.6C and Figure 25.8A.

Quantification of individual lipids of the total extracts

The quantitative analyses of lipid contents in a lipid extract can be performed by densitometric measurements. For quantitative determination, peak data of the unknowns are correlated with data from calibration standards chromatographed on the same TLC plate. The staining intensity can be evaluated by video densitometry using appropriate software. The standard curves of authentic standards are linear in the concentration range 1–10 µg.

Electrospray ionization mass spectrometry (ESI-MS)

Among the soft ionization methods developed for analysis of polar biomolecules, ESI-MS represents a major breakthrough for biological MS because the technique is directly applied to solutions, does not require derivatization reactions, is characterized by high sensitivity and moderate experimental complexity, and provides reproducible results. Using standard equipment, ESI-MS can be used for characterization and quantification of phospholipids and other polar lipids in an unprocessed total lipid extract from cells or sub-cellular structures. The use of a nano-ESI source allows the analysis of picomolar or even subpicomolar amounts of polar lipids. Furthermore the use of synthetic analogs as internal standards can also allow quantification of a particular class of polar lipid.

When operated in the single-stage MS mode, ESI mass spectra of cellular lipid extracts show almost exclusively molecular ion species of polar lipids. The choice of ion polarity sets the first level of specificity in phospholipid analysis by ESI-MS, since the possible charge states of a phospholipid class in solution determine the optimal ion polarity. A single molecular ion is present with a set of isotopic signals, and all m/z values generally reported refer to the monoisotopic molecular weight. On the basis of their monoisotopic m/z value, the major ion signals in the spectra can be correctly assigned to individual phospholipid molecular species. Due to different ionization efficiencies, the relative signal intensities of the ions of different phospholipid classes do not directly represent their molar abundances. The use of tandem mass spectrometric techniques, such as parent ion and neutral scanning (Brugger et al., 1997), together with the knowledge of phospholipid class-specific fragmentations in these tandem mass (MS/MS) analyses allows fine detailed structural studies of the phospholipids and glycolipids present in the lipid extract (Corcelli et al., 2004).

The samples for ESI-MS analyses can be prepared as follows: the lipid extracts of cells or membranes of extreme halophiles are taken to dryness under a gentle stream of nitrogen and re-dissolved in a small volume (20–200 µl) of chloroform/methanol (1:1, v/v) for negative ion measurements and of chloroform/methanol containing 1 mM ammonium acetate (1:3, v/v) for positive ion analyses.

Table 25.2 shows the ESI-MS diagnostic peaks (m/z) of various lipid components of archaeal halophiles and Table 25.3 lists the main peaks observed in the ESI-MS lipid profiles of various cultured halophilic microorganisms (Lattanzio et al., 2002). It should be noted that unsulfated diglycosyl, triglycosyl and tetraglycosyl lipids, when present, appear as minor peaks in the ESI-MS spectra only, as they are difficult to ionize.

ESI-MS analyses of unprocessed lipid extract of brine biomass have successfully been used to obtain information on the types of Archaea that dominate in the biomass collected from environments such as the Dead Sea or salt crystallizer ponds (Lattanzio et al., 2002; Oren, 1994; Oren and Gurevich, 1993). In particular, as glycolipids of extremely halophilic Archaea might serve as chemotaxonomic markers for classification of these organisms, the presence of certain glycolipid peaks in the total lipid extract of the biomass can be considered indicative of the presence of various genera of the *Halobacteriaceae* family (Oren, 2001; Oren and Gurevich, 1993). The major peak at m/z 1055 found in both

Table 25.2 Diagnostic peaks (m/z values) of the archaeal polar lipids present in halophilic microorganisms (modified from Lattanzio et al., 2002)

Lipid	m/z (amu)
Bisphosphatidylglycerol (BPG)	760*
Phosphatidylglycerol (PG)	806
Phosphatidylglycerosulfate (PGS)	886
Phosphatidylglycerosulfate methyl ester (PGP-Me)	900
Glycocardiolipin (GlyC or S-TGD-1-PA)	966*
S-DGD-5-PA	885*
Sulfated diglycosyl diphytanylglycerol (S-DGD-1 and S-DGD-5)	1056
Diglycosyl diphytanylglycerol (DGD-1)	976
Sulfated triglycosyl diphytanylglycerol (S-TGD-1)	1218
Triglycosyl diphytanylglycerol (TGD-1)	1138
Sulfated tetraglycosyl diphytanylglycerol (S-TeGD-1)	1381

*Double charged peaks.

Table 25.3 m/z values observed in the ESI-MS (negative ion) spectra of total lipid extracts from cultures of various halophilic archaeal microorganisms (Lattanzio et al., 2002)

Genus, Species	Phospholipids			Glycolipids					Cardiolipins	
	806	886	900	976	1056	1138	1218	1381	760	966
Hfx. gibbonsii ATCC 33959	x	x	x	x	x		x[b]		x	
Hfx. denitrificans ATCC 35960	x	x	x	x		x[a]	x[a]		x	
Hfx. volcanii ATCC 29605	x	x	x	x					x	
Hfx. mediterranei ATCC 33500	x	x	x	x			x[a]		x	
Hrr. saccharovorum JCM 8865	x	x	x	x			x[a]		x	
Har. vallismortis ATCC 29715	x	x	x		x		x		x	x[a]
Har. marismortui ATCC 43049	x	x	x		x		x		x	x
Halobacterium salinarum NRC 1	x	x	x				x		x	x
Halobacterium salinarum NRC 817	x	x	x				x		x	x

[a] present as a very small peak;
[b] minor peaks.

salts of Margherita di Savoia (Southern Italy) and Eilat (Israel) shows that monosulfated diglycosyl diether lipids occur in the genera such as *Haloferax*, *Halobaculum* and *Halorubrum* are far more abundant in the salt biomass than for example in *Halobacterium* (Lattanzio et al., 2002).

Nuclear magnetic resonance spectroscopy (NMR)

^{31}P-NMR

The common phospholipids from biological sources can be detected and quantified by phosphorus-31 nuclear magnetic resonance (^{31}P-NMR). However, because phospholipids tend to form aggregates in solution, fairly broad ^{31}P phospholipid resonance signals are generally observed that prevents the resolution and quantification of the constituent phospholipids. Detergents have been used in the presence of EDTA to dissolve phospholipids, resulting in a narrowing of ^{31}P-NMR resonances (London and Feigenson, 1979).

An increased resolution and narrowing of NMR resonances was obtained by the method of phospholipids analysis developed by Meneses and Glonek (1988); in particular this method is also based on the use of EDTA to bind the bivalent cations which induce phospholipid aggregations by charge–charge interactions with the phosphate groups. The sample for ^{31}P-NMR analysis can be prepared as follows: between 1 and 5 mg of individual purified phospholipids or total lipid extract is dissolved in 0.8 ml of deuterated chloroform. To this solution, 0.4 ml of methanol reagent containing Cs/EDTA prepared, as described in Meneses and Glonek (1988) is added and the mixture stirred gently. Two liquid phases are obtained, a major chloroform phase and a smaller water phase. By using a Pasteur pipette, the sample is placed in an NMR test tube where it separates within one minute. The sample tube turbine is adjusted so that only the chloroform phase is detected by the NMR spectrometer's receiver coil. Magnetic field stabilization is obtained through the deuterium resonance of deuterated chloroform. Usually the samples can be analyzed with proton broad–band decoupling to eliminate ^{1}H-^{31}P multiplets. Under these conditions each spectral resonance corresponds to a single phosphorus.

Table 25.4 presents the ^{31}P-NMR chemical shifts obtained from some isolated and purified archaeal phospholipids of *Halobacterium salinarum*, dissolved in the Cs/EDTA containing the analytical reagent, previously described. As an example, the ^{31}P-NMR spectrum of the total lipid extract of the purple membrane (PM) of *Halobacterium salinarum*, prepared as previously described, is shown in Figure 25.9. By comparing the chemical shifts in the spectrum in Figure 25.9 with those of the authentic standards, the major peaks were assigned to GlyC (1.96 ppm), PG (1.72 ppm) and PGP-Me (1.35 and 2.59 ppm). The assignment of the two different phosphorus signals of PGP-Me can be obtained by ^{31}P proton coupled NMR analysis. It appears quite difficult to distinguish PGS and cardiolipin in the spectrum, as their resonances almost overlap (at 1.37 and 1.4 ppm, respectively) and are very close to one of the two signals of PGP-Me (at 1.36 ppm). The proportions of the major lipids in the lipid

Table 25.4 ^{31}P chemical shifts of individual phospholipids of *H. salinarum*

Lipid	^{31}P δ (ppm)
GlyC	1.964
PGS	1.374
PGP-Me	1.357 and 2.594
PG	1.725
BPG	1.401

Each phospholipid is solubilized in the Cs/EDTA-containing reagent Meneses and Glonek, 1988.

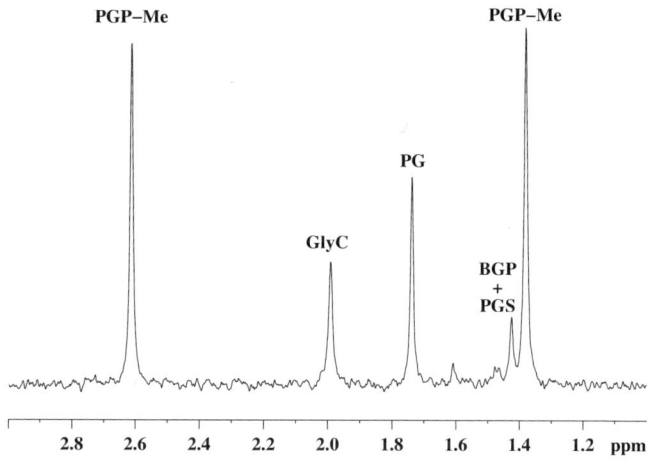

Figure 25.9. ^{31}P-NMR of the total lipid extract of the purple membrane of *Halobacterium salinarum* (from Corcelli et al., 2002a).

extract can also be obtained by comparing the areas of the peaks in the ^{31}P-NMR spectrum.

Despite its low sensitivity compared to ESI-MS analysis, ^{31}P-NMR analysis has the advantage of detecting all phospholipid species in a single experiment and with the same sensitivity. The above procedure is capable of resolving the ^{31}P resonance of various phosholipids sufficiently to permit accurate quantification; furthermore it is rapid, precise and involve a minimum of chemical manipulations of the extracts that may bias the results.

^{1}H-NMR

Archaeal lipids, having a diphytanylglycerol diether lipid core, exhibit a typical pattern of resonances in the ^{1}H-NMR spectrum. The presence of sharp methyl signals at approximately 0.8 ppm together with a broad methylene envelope at approximately 1.2 ppm is diagnostic for the presence of the archaeol core of the lipids of halophilic Archaea.

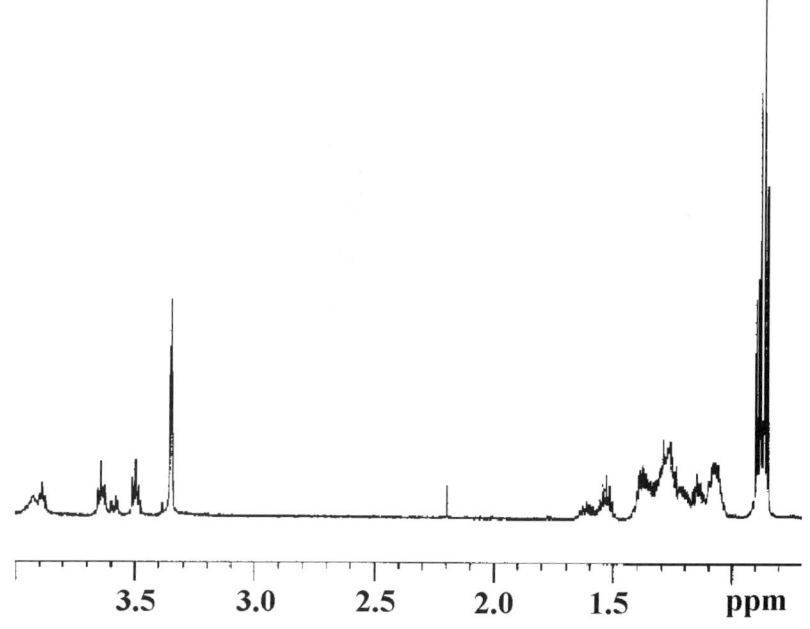

Figure 25.10. ^{1}H-NMR of the archaeal cardiolipin BPG, isolated and purified by TLC from *Halobacterium salinarum* (from Corcelli *et al.*, 2000).

As an example, the ^{1}H-NMR spectrum of bisphosphatidylglycerol (BPG), isolated from *Halobacterium salinarum*, is reported in Figure 25.10. In the case of glycolipids several signals are present in the region 3–5 ppm and among these the signals of anomeric protons of the sugar units are particularly useful (see below). ^{1}H-NMR spectroscopy can also be useful to analyze unprocessed total lipid extract.

The ^{1}H-NMR spectrum of the total lipid extract of the purple membrane (PM) of *Halobacterium salinarum* is shown in Figure 25.11. Although there is a considerable overlapping of resonances from the different lipid components, some of the lipids have a structure-specific resonance or set of resonances which allow their rapid identification and quantification. In particular in the 4.3–5.1 ppm region there are twin signals attributable to the anomeric protons of the two glycolipids S-TGD-1 and GlyC, which both have the same three sugars in the same sequence in the polar head (see molecular structures). In addition, the olefinic protons of squalene are also clearly distinct. The twin doublets around 4.4 ppm are characteristic of the anomeric proton of galactose, but are partially masked by the water peak, while the doublets of anomeric protons of glucose (around 4.85 ppm) and mannose (around 4.80 ppm) are isolated from other signals. ^{1}H-NMR spectra can be taken in CDCl$_3$/CD$_3$OD (4:3 v/v); approximately 2–5 mg of authentic standard lipid or total lipid extract can be dissolved in 700 μl forming one phase and TMS can be used as an internal standard to give relative ^{1}H chemical shifts.

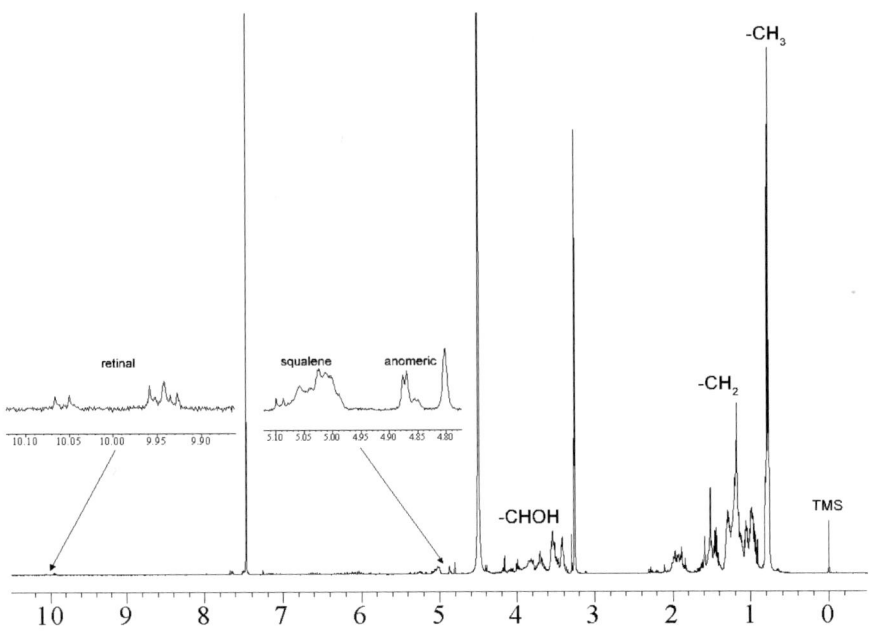

Figure 25.11. ^{1}H-NMR of the total lipid extract of the purple membrane of *Halobacterium salinarum* (from Corcelli *et al.*, 2002a). The two regions around 10 and 5 ppm, corresponding to aldehydic protons of retinal, and squalene and anomeric protons, respectively, are enlarged. Peaks at 7.26, 4.5, 3.3 ppm are due to residual protic solvents.

◆◆◆◆◆ FUNCTIONAL ROLE

While the chemical and physical properties and aggregation states of membrane lipids of Bacteria and Eukarya are well known, the knowledge of the physical properties of archaeal lipids is relatively limited. However, even from the limited knowledge of their properties and characteristics, it is clear that archaeal lipids are well adapted as membrane components in extreme environmental conditions. As ether lipids cannot be degraded easily, are temperature- and mechanically resistant, and highly salt tolerant, they are considered an ideal biomaterial to prepare lipid matrix for membrane protein reconstitution. In particular, membranes of halophiles are stable over a wide range of salt concentrations and at elevated pH values and are well adapted to the halophilic conditions.

It has been shown that, in contrast with liposomes formed from *Escherichia coli* lipids, liposomes formed from archaeal halophilic lipids remain stable in solutions containing up to 4 M of NaCl and KCl (van de Vossenberg *et al.*, 1999). The proton permeability of the liposomes from lipids of halophiles is independent of the salt concentration and was essentially constant between pH 7 and pH 9 (van de Vossenberg *et al.*, 1999). Recent comparative permeability studies of liposomes prepared from lipids of *Halobacterium salinarum* and *E. coli* show that the water permeability is not affected by the presence of an ether bond instead of ester bond or phytanyl chains instead of acyl chains (Mathai *et al.*, 2001).

The primary physical forces for organizing biological membranes are the hydrophobic interactions between the chains of lipid molecules. On the other hand, individual lipids might be selectively bound by hydrophobic interactions between their head groups and the membrane proteins. The lateral organization of phospholipids and their heads structures may play an important role in the so-called lateral proton conductance. The lateral proton conductance of phospholipids of extreme halophiles packed in monolayer has been measured (Teissie et al., 1990). It has been suggested that the polar heads of PGP-Me and PG may play an important role in coupling the light-activated proton pump of the purple membrane with the proton consuming ATP synthase of the red membrane of *Halobacterium salinarum* (Teissie et al., 1990). Therefore in studying the function of membrane lipids in cells, one major challenge is to elucidate the trans-membrane distribution of lipids and how it is regulated.

Most information on the cytoplasmic membrane structure of extreme halophiles, and in particular on the organization of archaeal lipids in the lipid bilayer, as well as on lipid–protein interactions, has been obtained from morphological, biophysical and biochemical studies on the specialized purple membrane patches of *Halobacterium salinarum* (Grigorieff et al., 1996; Krebs and Isenbarger, 2000; Oesterhelt and Stoeckenius, 1974; Weik et al., 1998). The purple membrane contains the only protein bacteriorhodopsin and a well-defined set of phospholipids and glycolipids. By electron microscopy and specific labelling, the glycolipids have been found exclusively in the outer leaflet of the membrane bilayer (Henderson et al., 1978). Localization of archaeal glycolipids in the purple membranes of *Halobacterium salinarum* has also been achieved by *in vivo* labelling and neutron diffraction (Weik et al., 1998). It has been suggested that the specific hydrophobic interactions between the tryptophan aromatic residues in the protein and carbohydrates in the lipids may play an important role in the stabilization and assembly of bacteriorhodopsin in the purple membrane (Weik et al., 1998). It is not clear whether both glycolipids S-TGD-1 and glycocardiolipin interact directly with bacteriorhodopsin. The available biochemical data indicate that the interactions of bacteriorhodopsin with glycocardiolipin are stronger than those with S-TGD-1. It is extremely difficult to remove glycocardiolipin from the annulus of bacteriorhodopsin in the course of solubilization with detergents and even after lipid extraction with denaturing organic solvents a residual amount of glycocardiolipin still remains associated with denatured bacteriorhodopsin (Catucci et al., 2004). Furthermore the glycocardiolipin seems to play an important role in stability and maintaining the trimeric structure of bacteriorhodopsin (Lopez et al., 1999).

The glycocardiolipin/bacteriorhodopsin stoichiometry in the purple membrane has been determined by analyzing an unprocessed lipid extract of the purple membrane with both of ^{31}P-NMR and ^{1}H-NMR spectroscopy (Corcelli et al., 2002a). The molar ratio of GlyC to BR is one and appears to be consistent with its possible location in the intra-trimer space.

Interestingly, until now glycocardiolipin has been found only in the extreme halophiles and in association with bacteriorhodopsin, while the archaeal bisphosphatidylglycerol is widespread in all genera of the family *Halobacteriaceae* (Lattanzio et al., 2002). The association of archaeal bisphosphatidylglycerol with an archaeal terminal oxidase has been reported (Corcelli et al., 2002b).

The steps of archaeal lipid biosynthesis were described in details by Morris Kates in his pioneering studies on lipids of extreme halophiles (Kates, 1993; Kates and Kushwaha, 1978; Kates et al., 1968; Kamekura and Kates, 1988). Recently, novel aspects regarding the regulation of the

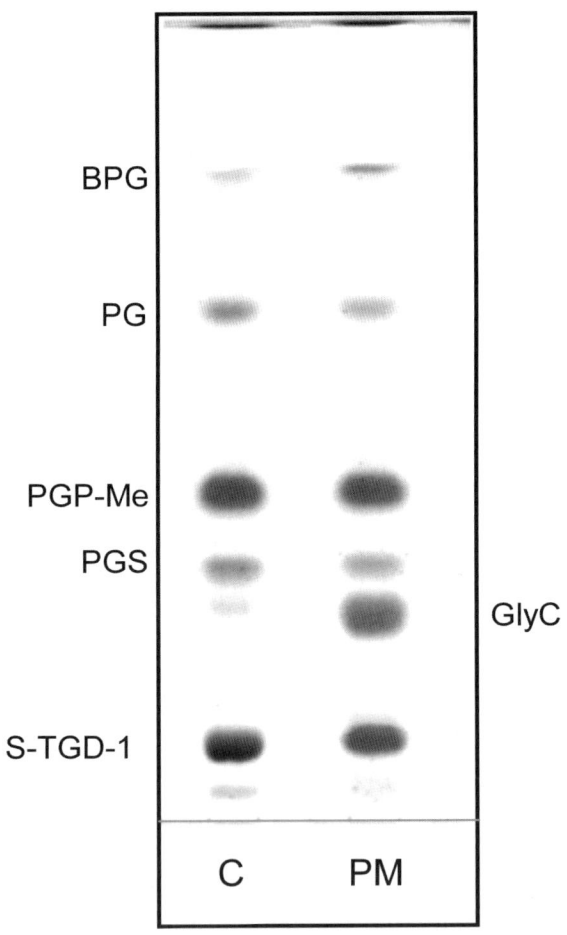

Figure 25.12. *De novo* synthesis of glycocardiolipin of *Halobacterium salinarum* under osmotic shock (from Lobasso et al., 2003). Lipids were extracted from whole cells (C) resuspended in 4 M NaCl and from purple membrane (PM) and analyzed by TLC. The plate was developed with solvent chloroform–methanol–90% acetic acid (65:4:35, v/v) and charred with 5% sulfuric acid. The abbreviated names of individual lipid components in the extract have been reported. Glycocardiolipin is only present in the lipid extract of membranes isolated after cell disruption by osmotic shock, following the standard procedure of Oesterhelt and Stoeckenius (1974).

lipid biosynthesis in the archaeal extreme halophiles have been described (Lobasso et al., 2003; Lopalco et al., 2004). Interestingly, when archaeal extreme halophiles are exposed to low salt conditions, relevant changes in the cell membrane lipid composition occur; during the adaptation of extreme halophilic microorganisms to hypo-osmolarity the membrane undergoes significant biochemical and structural changes, especially in terms of an increase in the content of cardiolipin analogs.

As stated above, *Halobacterium salinarum* cells possess a phospholipid dimer consisting of a glycolipid sulfo-triglycosyl-diether (S-TGD-1) esterified to the phosphatidic acid (PA), i.e. S-TGD-1-PA, called glyco-cardiolipin (GlyC). GlyC is almost absent in the lipid extract of cells of *Halobacterium salinarum*, while it is abundant in the purple membrane (PM), which is typically isolated after cell disruption by osmotic shock (Figure 25.12). It has been shown that osmotic stress specifically induces GlyC increase at the expense of S-TGD-1 in *Halobacterium salinarum* cells (Figure 25.13). Analogously, membranes isolated after cell disruption by osmotic shock from *Halorubrum* sp. are highly enriched in archaeal bisphosphatidylglycerol (BPG) and reveal the presence of the cardiolipin analog or phospholipid dimer (Figure 25.14), having the structure of a sulfo-diglycosyl-diether-phosphatidic acid (S-DGD-5-PA). By exposing cells to low salt medium, it has been clearly demonstrated that osmotic

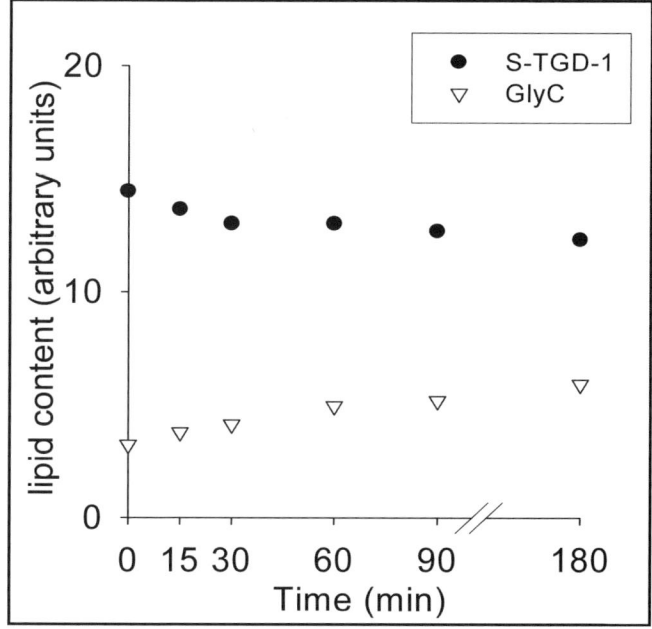

Figure 25.13. The time course of changes in GlyC and S-TGD-1 in *Halobacterium salinarum* NRC-1 cells after dilution in low salt medium (from Lobasso et al., 2003). Lipids were extracted from equivalent aliquots of cells undergoing hypotonic shock after different incubation times (0, 15, 30, 60, 90 and 120 min); 20 µg of different extracts have been loaded on a plate staining by spraying with 5% H_2SO_4 followed by incubation at 120°C for 13 min; then the lipid contents at the different incubation times were estimated by video densitometric analyses.

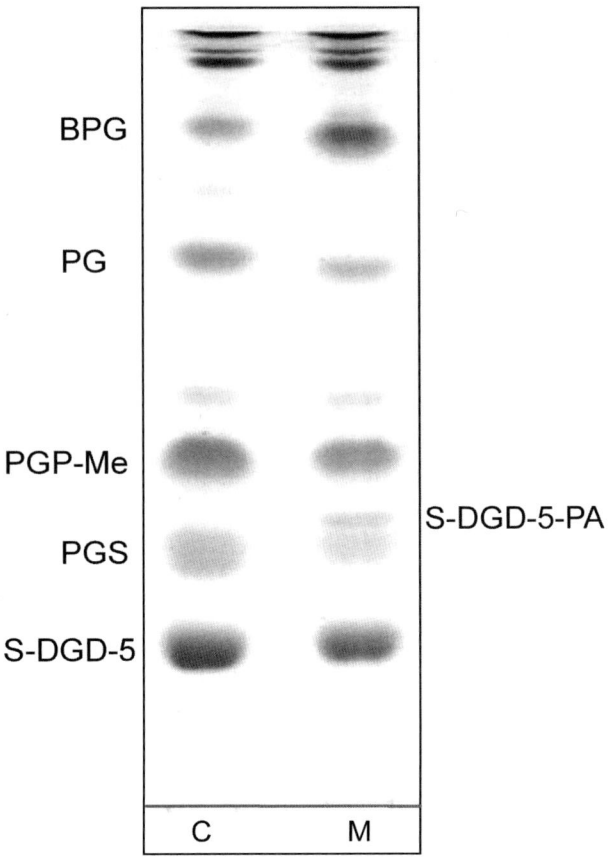

Figure 25.14. *De novo* synthesis of glycosyl-cardiolipin S-DGD-5-PA and archaeal BPG of *Halorubrum* sp. under osmotic shock (from Lopalco *et al.*, 2004). Lipids were extracted from whole cells resuspended in 4 M NaCl (C) and from membranes (M), isolated after cell disruption by osmotic shock, following the method described in Lopalco *et al.* (2004) and analyzed by TLC. The plate was developed with solvent chloroform–methanol–90% acetic acid (65:4:35, v/v) and charred with 5% sulfuric acid. The abbreviated names of individual lipid components in the extract have been reported. S-DGD-5-PA is only present in the lipid extract of membranes isolated after cell disruption by osmotic shock, while BPG content is higher than that of whole cells.

shock induces a specific increase in the content of archaeal BPG and S-DGD-5-PA in *Halorubrum* sp., at the expense of phosphatidylglycerol (Figure 25.15) and the glycolipid S-DGD-5, respectively (Lopalco *et al.*, 2004). In conclusion, osmotic shock stimulates the *de novo* synthesis of phospholipid dimers, i.e. cardiolipin analogs, in extreme halophilic Archaea.

It has been suggested that higher archaeal cardiolipin cell content could protect the cell from lysis. The increase in archaeal cardiolipin occurring in cells undergoing osmotic shock may represent the physiological response of the microorganisms to low external osmolarity. The modifications in lipid composition could affect the physical–chemical properties of the membrane, such as bilayer thickness, membrane fluidity and transport properties.

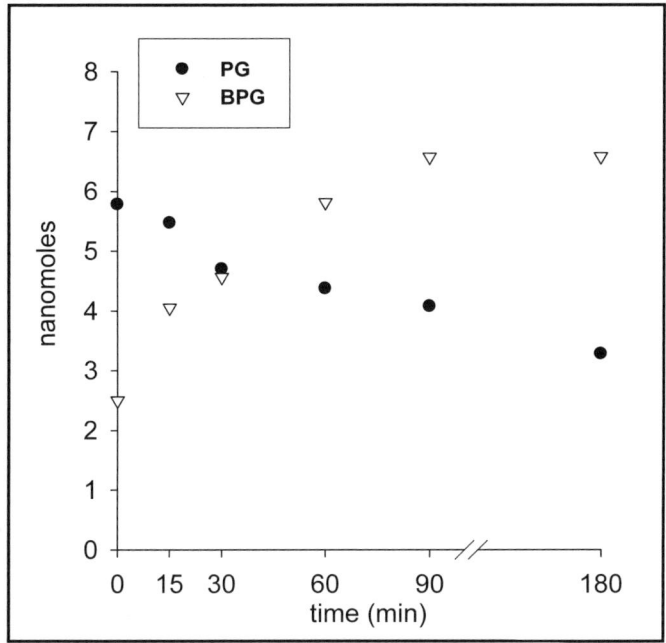

Figure 25.15. Time course of changes in BPG and PG contents occurring in swelling and disrupting *Halorubrum* sp. cells after dilution in low salt medium (from Lopalco et al., 2004). Lipids were extracted from equivalent cell aliquots taken at different incubation times (0, 15, 30, 60, 90 and 120 min) from the dilution; 40 µg of different extracts have been loaded on a plate staining by spraying with 5% H_2SO_4, followed by incubation at 120°C; values on the y axis (in nanomoles) were estimated by video densitometric analyses.

Much remains to be learned concerning the precise asymmetric arrangement of the lipids in the membrane bilayer, the interaction of the lipids with the membrane proteins, as well as the function of the membrane lipid asymmetry with respect to ion transport, permeability to nutrients, proton transport and conductance, and energy transduction.

References

Bligh, E. G. and Dyer, W. J. (1959). A rapid method of total lipid extration and purification. *Can. J. Biochem. Physiol.* **37**, 911–917.

Brugger, B., Erben, G., Sandhoff, R., Wieland, F. T. and Lehemann, W. D. (1997). Quantitative analysis of biological membrane lipids at the low picomole level by nano-electrospray ionization tandem mass spectrometry. *Proc. Natl. Acad. Sci. USA* **94**, 2339–2344.

Catucci, L., Lattanzio, V. M. T., Lobasso, S., Agostiano, A. and Corcelli, A. (2004). Role of endogenous lipids in the chromophore regeneration of bacteriorhodopsin. *Bioelectrochemistry* **63**, 111–115.

Colella, M., Lobasso, S., Babudri, F. and Corcelli, A. (1998). Palmitic acid is associated with halorhodopsin as a free fatty acid. Radiolabeling of halorhodopsin with ^{3}H-palmitic acid and chemical analysis of the reaction products of purified halorhodopsin with thiols and NaBH$_4$. *Biochim. Biophys. Acta* **1370**, 273–279.

Collins, M. D., Ross, H. N. M., Tindall, B. J. and Grant, W. D. (1981). Distribution of isoprenoid quinones in halophilic bacteria. *J. Appl. Bacteriol.* **50**, 559–565.

Corcelli, A., Lobasso, S., Coltella, M., Palmisano, F., Guerrieri, A. and Trotta, M. (1996). Role of palmitic acid on the isolation and properties of halorhodopsin. *Biochim. Biophys. Acta* **1281**, 173–181.

Corcelli, A., Colella, M., Mascolo, G., Fanizzi, F. P. and Kates, M. (2000). A novel glycolipid and phospholipid in the purple membrane. *Biochemistry* **39**, 3318–3326.

Corcelli, A., Lattanzio, V. M. T., Mascolo, G., Papadia, P. and Fanizzi, F. P. (2002a). Lipid-protein stoichiometries in a crystalline biological membrane: NMR quantitative analysis of the lipid extract of the purple membrane. *J. Lipid Res.* **43**, 132–140.

Corcelli, A., Palese, L. L., Lobasso, S., Lopalco, P., Orlando, A. and Papa, S. (2002b). Archaeal cytochrome c oxidase activity in cardiolipin-enriched membranes isolated from an extremely halophilic microorganism. Abstract, *4th International Congress "Extremophiles 2002"*. September, 22–26, Napoli, Italy.

Corcelli, A., Lattanzio, V. M. T., Mascolo, G., Babudri, F., Oren, A. and Kates, M. (2004). Novel sulfonolipid in the extremely halophilic bacterium *Salinibacter ruber*. *Appl. Environ. Microbiol.* **70**, 6678–6685.

Grigorieff, N., Ceska, T. A., Dowing, K. H., Baldwin, J. M. and Henderson, R. (1996). Electron-crystallographic refinement of the structure of bacteriorhodopsin. *J. Mol. Biol.* **259**, 393–421.

Henderson, R., Jubb, J. S. and Whytock, S. (1978). Specific labelling on the protein and lipid on the extracellular surface of purple membrane. *J. Mol. Biol.* **123**, 259–274.

Kamekura, M. and Kates, M. (1988). Lipids of halophilic archaebacteria. In *Halophilic Bacteria* (F. Rodriguez-Valera, ed.), vol. II, pp. 25–54. CRC Press, Boca Raton.

Kates, M. (1978). The phytanyl ether-linked polar lipids and isoprenoid neutral lipids of extremely halophilic bacteria. *Prog. Chem. Fats Lipids* **15**, 301–342.

Kates, M. (1986). *Techniques of Lipidology. Laboratory Techniques in Biochemistry and Molecular Biology*, vol. 3, Part 2, 2nd edn., Elsevier, Amsterdam.

Kates, M. (1993). Membrane lipids of Archaea. In *The Biochemistry of Archaea (Archaebacteria)* (M. Kates, D. J. Kushner and A. T. Matheson, eds), pp. 261–295. Elsevier, Amsterdam.

Kates, M. and Deroo, P. W. (1973). Structure determination of the glycolipid sulfate from the extreme halophile *Halobacterium cutirubrum*. *J. Lipid Res.* **14**, 438–445.

Kates, M. and Kushwaha, S. C. (1978). Biochemistry of lipids of extremely halophilic bacteria. In *Energetics and Structure of Halophilic Microrganisms* (S. R. Caplan and M. Ginzburg, eds), pp. 461–480. Elsiever, Amsterdam.

Kates, M. and Kushwaha, S. C. (1995). Isoprenoids and polar lipids of extreme halophiles. In *Archaea: A Laboratory Manual, Halophiles* (S. DasSarma and E. M. Fleischman, eds), pp. 35–54. Cold Spring Harbor Laboratory Press, New York, NY.

Kates, M., Wassef, M. K. and Kushner, D. J. (1968). Radioisotopic studies on the biosynthesis of the glyceryl diether lipids of *Halobacterium cutirubrum*. *Can J. Biochem.* **46**, 971–977.

Kates, M., Kushwaha, S. C. and Sprott, G. D. (1982). Lipids of purple membrane from extreme halophiles and of methanogenic bacteria. *Meth. Enzymol.* **88**, 98–111.

Kates, M., Moldoveanu, N. and Stewart, L. C. (1993). On the revised structure of the major phospholipid of *Halobacterium salinarium*. *Biochim. Biophys. Acta* **1169**, 46–53.

Kean, E. L. (1968). Rapid, sensitive spectrophotometric method for quantitative determination of sulphatides. *J. Lipid Res.* **9**, 319–327.

Kolbe, M., Besir, H., Essen, L. O. and Oesterhelt, D. (2000). Structure of the light-driven chloride pump halorhodopsin at 1.8 Å resolution. *Science* **288**, 1390–1396.

Krebs, M. P. and Isenbarger, T. A. (2000). Structural determinants of purple membrane assembly. *Biochim. Biophys. Acta* **1460**, 15–26.

Kushwaha, S. C. and Kates, M. (1978). 2,3-Di-*O*-phytanyl-*sn*-glycerol and prenols from extremely halophilic bacteria. *Phytochemistry* **17**, 2029–2030.

Kushwaha, S. C., Pugh, E. L., Kramer, J. K. G. and Kates, M. (1972). Isolation and identification of dehydrosqualene and C_{40} carotenoid pigments in *Halobacterium cutirubrum*. *Biochim. Biophys. Acta* **260**, 492–506.

Kushwaha, S. C., Kates, M. and Martin, W. G. (1975). Characterization and composition of the purple and red membrane from *Halobacterium cutirubrum*. *Can. J. Biochem.* **53**, 284–292.

Kushwaha, S. C., Kates, M. and Stoeckenius, W. (1976). Comparison of purple membrane from *Halobacterium cutirubrum* and *H. halobium*. *Biochim. Biophys. Acta* **426**, 703–710.

Kushwaha, S. C., Kates, M. and Kramer, J. K. G. (1977). Occurrence of indole in cells of extremely halophilic bacteria. *Can. J. Microbiol.* **23**, 826–828.

Kushwaha, S. C., Kates, M., Juez, G., Rodriguez-Valera, F. and Kushner, D. J. (1982). Polar lipids of an extremely halophilic bacterial strain (R-4) isolated from salt ponds in Spain. *Biochim. Biophys. Acta* **711**, 19–25.

Lanzotti, V., Nicolaus, B., Trincone, A., De Rosa, M., Grant, W. D. and Gambacorta, A. (1989). A complex lipid with a cyclic phosphate from the archaebacterium *Natronococcus occultus*. *Biochim. Biophys. Acta* **1001**, 31–34.

Lattanzio, V. M. T., Corcelli, A., Mascolo, G. and Oren, A. (2002). Presence of two novel cardiolipins in the halophilic archaeal community in the crystallizer brines from the salterns of Margherita di Savoia (Italy) and Eilat (Israel). *Extremophiles* **6**, 437–444.

Lindsey, H., Petersen, N. O. and Chan, S. I. (1979). Physicochemical characterization of 1,2-diphytanoyl-*sn*-glycero-3-phosphocholine in model membrane systems. *Biochim. Biophys. Acta* **555**, 147–167.

Lobasso, S., Lopalco, P. Lattanzio, V. M. T. and Corcelli, A. (2003). Osmotic shock induces the presence of glycocardiolipin in the purple membrane of *Halobacterium salinarum*. *J. Lipid Res.* **44**, 2120–2126.

London, E. and Feigenson, G. W. (1979). Phosphorus NMR analysis of phospholipids in detergents. *J. Lipid Res.* **20**, 408–412.

Lopalco, P., Lobasso, S., Babudri, F. and Corcelli, A. (2004). Osmotic shock stimulates de novo synthesis of two cardiolipins in an extreme halophilic archaeon. *J. Lipid Res.* **45**, 194–201.

Lopez, F., Lobasso, S., Colella, M., Agostiano, A. and Corcelli, A. (1999). Light-dependent and biochemical properties of two different bands of bacteriorhodopsin isolated on phenyl-sepharose CL-4B. *Photochem. Photobiol.* **69**, 599–604.

Marinetti, G. V. (1964). Chromatographic analysis of polar lipids on silicic acid impregnated paper. In *New Biochemical Separation* (A. T. James and L. J. Morris, eds), pp. 339–377. Van Nostrand, Princeton, NJ.

Mathai, J. C., Sprott, G. D. and Zeidel, M. L. (2001). Molecular mechanisms of water and solute transport across archaebacterial lipid membranes. *J. Biol. Chem.* **276**, 27266–27271.

Meneses, P. and Glonek, T. (1988). High resolution ^{31}P NMR of extracted phospholipids. *J. Lipid Res.* **29**, 679–689.

Oesterhelt, D. and Stoeckenius, W. (1974). Isolation of cell membrane of *Halobacterium halobium* and its fractionation into red and purple membrane. *Meth. Enzymol.* **31**, 667–678.

Oren, A. (1994). Characterization of the halophilic archaeal community in saltern crystallizer ponds by means of polar lipid analysis. *Int. J. Salt Lake Res.* **3**, 15–29.

Oren, A. (2001). The order Halobacteriales. In *The Prokaryotes. An Evolving Electronic Resource for the Microbiological Community* (M. Dworkin, S. Falkow, E. Rosenberg, K.-H. Schleifer and E. Stackebrandt, eds), 3rd edn., release 3.2, 25 July 2001. Springer-Verlag, New York, (http://link.springer-ny.com/link/service/books/10125/).

Oren, A. (2002). *Halophilic Microorganisms and their Environments*. Kluwer Academic Publishers, Dordrecht.

Oren, A. and Gurevich, P. (1993). Characterization of the dominant halophilic Archaea in a bacterial bloom in the Dead Sea. *FEMS Microbiol. Ecol.* **12**, 249–256.

Pugh, E. L. and Kates, M. (1994). Acylation of proteins of the archaebacteria *Halobacterium cutirubrum* and *Methanobacterium thermoautotrophicum*. *Biochim. Biophys. Acta* **1196**, 38–44.

Pugh, E. L., Wassef, M. K. and Kates, M. (1971). Inhibition of fatty acid synthetase in *Halobacterium cutirubrum* and *Escherichia coli* by high salt concentrations. *Can. J. Biochem.* **49**, 953–958.

Siakotos, A. N. and Rouser, G. (1965). Analytical separation of nonlipid water soluble substances and gangliosides from other lipids by dextran gel column chromatography. *J. Am. Oil Chemists' Soc.* **42**, 913–919.

Smallbone, B. W. and Kates, M. (1981). Structural identification of minor glycolipids in *Halobacterium cutirubrum*. *Biochim. Biophys. Acta* **655**, 551–558.

Sprott, G. D., Laroque, S., Cadotte, N., Dicaire, C. J., McGee, M. and Brisson, J. I. (2003). Novel polar lipids of halophilic eubacterium *Planococcus* H8 and archaeon *Haloferax volcanii*. *Biochim. Biophys. Acta* **1633**, 179–188.

Teissie, J., Prats, M., LeMassu, A., Stewart, L. C. and Kates, M. (1990). Lateral proton conduction in monolayers of phospholipids from extreme halophiles. *Biochemistry* **29**, 59–65.

Tindall, B. J. (1990). Lipid composition of *Halobacterium lacusprofundi*. *FEMS Microbiol. Lett.* **66**, 199–202.

Trincone, A., Nicolaus, B., Lama, L., De Rosa, M., Gambacorta, A. and Grant, W. D. (1990). The glycolipid of *Halobacterium sodomense*. *J. Gen. Microbiol.* **136**, 2327–2331.

Trincone, A., Trivellone, E., Nicolaus, B., Lama, L., Pagnotta, E., Grant, W. D. and Gambacorta, A. (1993). The glycolipid of *Halobacterium trapanicum*. *Biochim. Biophys. Acta* **1210**, 35–40.

van de Vossenberg, J. L., Driessen, A. J., Grant, W. D. and Konings, W. N. (1999). Lipid membranes from halophilic and alkali-halophilic Archaea have a low H^+ and Na^+ permeability at high salt concentration. *Extremophiles* **3**, 253–257.

Weik, M., Patzelt, H., Zaccai, G. and Oesterhelt, D. (1998). Localization of glycolipids in membrane by *in vivo* labeling and neutron diffraction. *Mol. Cell* **1**, 411–419.

26 Characterization of Organic Compatible Solutes of Halotolerant and Halophilic Microorganisms

Mary F Roberts
Merkert Chemistry Center, Boston College, Chestnut Hill, MA 02465, USA

◆◆◆

CONTENTS

Osmotic stress
Organic solutes in microorganisms
Identification of osmolytes by NMR spectroscopy
NMR methods to refine osmolyte biosynthetic pathways
NMR methods to monitor solute dynamics in cells
Other methods to monitor organic osmolytes

◆◆◆◆◆ OSMOTIC STRESS

One of the basic responses of cells is to accumulate intracellular solutes to balance external osmotic pressure produced by salts and other material that can not diffuse across the cell membrane. In the absence of suitable intracellular particles, water efflux to equilibrate water activity on both sides of the membrane will reduce the cell volume, which in turn can alter intracellular concentrations of metabolites and alter enzyme action as well as binding phenomena. Cells are often subjected to changes in the extracellular ionic strength and this alteration of external osmotic pressure must be counterbalanced in a timely fashion for cells to survive.

Halotolerant and halophilic microorganisms are particularly adept at countering high external osmotic pressure with appropriate concentrations of solutes. There are two broad classes of solutes accumulated in response to external osmotic pressure: (i) inorganic solutes, primarily cations such as K^+ and Na^+ (occasionally Mg^{2+}) and anions such as chloride, phosphate, carbonate and sulfate, and (ii) organic solutes, typically small, highly soluble molecules that may or may not be charged. The inorganic ions are specifically transported into cells – K^+ by active transport while Na^+ and chloride are internalized by facilitated diffusion since the extracellular concentrations of these two ions are usually quite high in media (Müller and Oren, 2003). Aerobic Archaea such as *Halobacterium* sp. (Oren, 2002), and *Natronococcus occultus* grown in rich media (Desmarais *et al.*, 1997), and the bacterium *Salinibacter ruber* (Oren *et al.*, 2002) use inorganic salts almost exclusively for osmotic balance. However, most microorganisms use organic solutes (for a review of solutes in Archaea, see Roberts, 2004). The organic solutes accumulated in the cell to balance external osmotic pressure, also termed osmolytes, may be transported from the media or synthesized from carbon sources in the cell. Of all the small organic solutes synthesized in cells, only a small number of solutes are actually used for this purpose. Part of the reason is that osmolyte concentrations may need to vary significantly during cell growth if the external medium exhibits large changes in salt/osmotic pressure. This suggests that a good osmolyte is one whose intracellular concentration can vary significantly with little effect on the overall metabolism and growth of the cell. Such molecules were termed "compatible solutes" by Brown (1976). The typical organic osmolyte is polar (it may be uncharged, zwitterionic or anionic) and, with only a few exceptions, not an intermediate in biochemical pathways. There must also be a mechanism for efficiently regulating the concentration of the osmolyte in response to external NaCl. Osmolytes also have other roles in cells – most notably they aid in stabilizing macromolecule structures (Bolen and Baskakov, 2001; Butler and Falke, 1996; Faria *et al.*, 2004; Foord and Leatherbarrow, 1998; Kim *et al.*, 2003; Liu and Bolen, 1995; Timasheff, 2002). Most of these interactions appear nonspecific, but more specific effects can occur. Some osmolytes may regulate molecular chaperones in cells (Diamant *et al.*, 2001). This further puts constraints on the types of molecules suitable for osmotic balance in cells.

◆◆◆◆◆ ORGANIC SOLUTES IN MICROORGANISMS

What are these unusual organic molecules accumulated by halotolerant and halophilic organisms? Until the advent of high-resolution NMR spectroscopy, little was known of the range of solutes used for osmotic balance in cells. However, in the last 25 years the library of organic osmolytes has increased significantly in large part due to diverse NMR experiments. The three types of organic osmolytes include uncharged solutes, zwitterions and organic anions. A brief overview of each type is presented along with spectroscopic parameters that identify the different solutes (Table 26.1).

Uncharged Solutes

Halophilic eukaryotes often accumulate glycerol or *myo*-inositol (Borowitzka and Brown, 1974). However, Bacteria and Archaea do not accumulate these solutes for osmotic balance. Instead, the uncharged

Table 26.1 Organic solutes synthesized and accumulated as osmolytes in halotolerant and halophilic microorganisms and their ^{13}C and ^{1}H chemical shifts

Solute	Chemical shifts, ppm ^{13}C	(^{1}H)
Uncharged solutes:		
α-glucosylglycerol	C1 98.9	
	C2 73.1	
	C3 74.2	
	C4 70.8	
	C5 72.7	
	C6 61.7	
	C1′ 62.6	
	C2′ 79.9	
	C3′ 61.6	
α-mannosylglyceramide	C1 100.1	(4.92)
	C2 70.4	(4.10)
	C3 70.7	(3.90)
	C4 67.1	(3.67)
	C5 73.9	(3.74)
	C6 61.4	(3.89/3.75)
	C1′ 175.5	
	C2′ 77.9	(4.32)
	C3′ 62.8	(3.87)
trehalose	C1 94.0	
	C2 73.0	
	C3 73.3	
	C4 70.5	
	C5 71.8	
	C6 61.4	
N-α-carbamoyl-L-glutamine 1-amide	C1 181.9	
	C2 57.9	(4.13)
	C3 31.9	(1.93/2.12)
	C4 35.8	(2.40)
	C5 182.5	
	C6 165.1	

(continued)

Table 26.1 Continued

Solute	Chemical shifts, ppm	
	^{13}C	(^{1}H)
N-acetylglutaminylglutamine amide	C1 179.1	
	179.1	
	177.0	
	175.6	
	174.9	
	C2 54.6	
	54.3	
	C3 28.0	
	C4 32.4	
	C5 23.0	
Zwitterions:		
betaine	C1 170.0	
	C2 67.2	(3.8)
	C3 54.4	(3.2)
ectoine	C1 177.5	
	C2 54.4	(4.0)
	C3 22.7	(2.0)
	C4 38.5	(3.2/3.4)
	C5 161.5	
	C6 19.5	(2.2)
hydroxyectoine	C1 176.0	
	C2 61.4	(4.0)
	C3 61.2	(4.5)
	C4 44.1	(3.2/3.4)
	C5 161.3	
	C6 19.5	(2.3)
Nγ-acetyldiaminobutyrate	C1	
	C2	(3.7)
	C3	(2.1)
	C4	(3.3)
	C5	
	C6	(2.0)

(continued)

Table 26.1 Continued

Solute	Chemical shifts, ppm ^{13}C	(^{1}H)
Nε-acetyl-β-lysine	C1 178.8	
	C2 39.2	(2.45)
	C3 50.0	(3.47)
	C4 30.3	(1.57)
	C5 25.1	(1.66)
	C6 39.5	(3.17)
	C7 174.8	
	C8 22.3	(1.90)
β-glutamine	C1 178.9	
	C2 39.0	(2.55)
	C3 47.3	(3.75)
	C4 37.2	(2.70)
	C5 175.3	
Anionic solutes:		
L-α-glutamate	C1 175.6	
	C2 55.4	(3.70)
	C3 27.7	(2.10)
	C4 34.1	(2.35)
	C5 182.3	
β-glutamate	C1 178.5	
	C2 39.2	(2.54)
	C3 47.7	(3.75)
	C4 39.2	(2.54)
	C5 178.5	
β-hydroxybutyrate	C2 46.9	
	C3 66.3	
	C4 22.7	
poly-β-hydroxybutyrate	C2 44.5	
	C3 71.4	
	C4 19.8	

(continued)

Table 26.1 Continued

Solute	Chemical shifts, ppm	
	^{13}C	(^{1}H)
α-glucosylglycerate	C1 99.0	(5.0)
	C2 73.2	(3.5)
	C3 74.2	(3.8)
	C4 70.5	(3.4)
	C5 72.5	(3.75)
	C6 62.0	(3.65)
	C1' 178.1	
	C2' 80.5	(4.15)
	C3' 64.6	(3.8)
α-mannosylglycerate	C1 99.1	(4.89)
	C2 70.5	(4.07)
	C3 70.8	(3.90)
	C4 67.2	(3.65)
	C5 73.5	(3.71)
	C6 61.4	(3.87/3.74)
	C1' 176.6	
	C2' 77.8	(4.27)
	C3' 63.3	(3.86/3.75)
α-diglycerophosphate	C1 66.8	
	C2 71.1	
	C3 62.5	
di-*myo*-inositol-1,1'-phosphate	C1 77.8	(3.98)
	C2 72.5	(4.22)
	C3 71.8	(3.51)
	C4 73.5	(3.59)
	C5 75.5	(3.27)
	C6 72.8	(3.71)

(*continued*)

Table 26.1 Continued

Solute	Chemical shifts, ppm ^{13}C (^{1}H)
di-2-O-β-mannosyl-DIP	Inositol: C9 62.8 (3.87) C1 76.4 (4.10) C2 80.0 (4.46) C3 70.5 (3.54) C4 73.0 (3.66) C5 74.3 (3.32) C6 72.0 (3.81) Mannose: C1 101.0 (4.93) C2 70.9 (4.15) C3 73.1 (3.69) C4 67.2 (3.56) C5 76.6 (3.42) C6 61.4 (3.92/3.73)
sulfotrehalose	C1 94.6 (5.28) C2 71.8 (3.75) C3 73.4 (3.92) C4 70.0 (3.62) C5 72.6 (4.00) C6 60.8 (3.90) C1' 92.3 (5.56) C2' 77.6 (4.35) C3' 71.3 (4.10) C4' 70.3 (3.68) C5' 72.9 (3.94) C6' 61.2 (3.88)

solutes they accumulate include carbohydrates modified so that their reducing end is no longer reactive. Trehalose, an α-1 to α-1 linked glucose dimer, is associated with many eukaryotic organisms subjected to water loss. It is also the major osmolyte in *Actinopolyspora halophila*, where it accounts for 15% w/v in these cells when they are grown in 24% NaCl (Nyyssölä and Leisola, 2001). More common are monosaccharides where a glycosidic bond is formed with either glycerol or glyceramide. For example, glucosylglycerol is an osmolyte in *Stenotrophomonas*, a member of the Proteobacteria (Roder et al., 2005), and α- and β-mannosylglyceramide are accumulated in *Rhodothermus marinus* (Silva et al., 1999). Only two other uncharged solutes have been detected as significant solutes in Bacteria and Archaea – an amino acid and a dipeptide, each modified to remove charged groups. N-α-carbamoyl-L-glutamine 1-amide has been detected in *Ectothiorhodospira marismortui* (Galinski and Oren, 1991). N-acetylglutaminylglutamine amide

(NAGGN) is accumulated by halophilic purple sulfur bacteria (Smith and Smith, 1989).

Zwitterionic Solutes

There are many more zwitterionic solutes than uncharged solutes found in halotolerant and halophilic Bacteria and Archaea (Martin et al., 1999). Most of these are derived from amino acids and modified so that they are relatively inert to metabolism. Glycine betaine (N,N,N-trimethylglycine) is found in all three domains and is found in many halophilic bacteria. In an early study, Imhoff and Rodriguez-Valera (1984) measured the average betaine concentration as a function of external NaCl in a variety of halophiles. Average betaine concentrations in 3, 10 and 20% NaCl were 0.21 ± 0.02, 0.65 ± 0.06 and 0.97 ± 0.09 M. In most cells where it is detected, the betaine is transported into the cells from the medium. Two basic types of betaine transporters have been detected in organisms including halophiles: (i) an ABC-type (Lai et al., 1999) (ii) members of the BCCT (betaine–choline–carnitine transporter) family (Vermeulen and Kunte, 2004). There are, however, a few organisms that can synthesize betaine either reductively (successive methylation of glycine using SAM as the methyl donor (Lai et al., 2005; Nyyssölä and Leisola, 2001; Nyyssölä et al., 2000, 2001)) or by oxidation of choline (conversion of the alcohol to aldehyde and then to carboxylate (Andresen et al., 1988; Canovas et al., 1998b; Nyyssölä and Leisola, 2001; Rozwadowski et al., 1991)).

Ectoine, 1,4,5,6-tetrahydro-2-methyl-4-pyrimidine carboxylic acid, is another very common zwitterionic osmolyte (Galinski et al., 1985). It is synthesized by a wide range of halotolerant and halophilic bacteria and is the major solute in aerobic chemoheterotrophic bacteria (Galinski, 1995). It has also been detected in methylotrophic bacteria (Doronina et al., 2000, 2003). A variant of ectoine, hydroxyectoine, has been detected in *Bacillus* spp. grown in high salt medium (Kuhlmann and Bremer, 2002). Ectoine levels respond to changes in extracellular NaCl, carbon source and degree of aeration (Onraedt et al., 2003). Under most conditions, ectoine only accumulates in exponentially growing cells (Regev et al., 1990). The biosynthesis of ectoine (Canovas et al., 1998a; Louis and Galinski, 1997; Peters et al., 1990) starts with an intermediate in amino acid metabolism, aspartate semialdehyde that is converted to L-2,4-diaminobutyric acid (DABA). DABA is a cationic molecule and it has no protective role in cells (Canovas et al., 1997). However, the next intermediate in the biosynthesis of ectoine, Nγ-acetyldiaminobutyric acid (NADA), is zwitterionic and it can act as a suitable osmolyte in *Halomonas elongata* strains selected for the loss of the ectoine synthase (Canovas et al., 1999).

Methanogens have not been documented to synthesize ectoine. Although many can use betaine for osmotic balance if it is supplied in the medium (Robertson et al., 1990a), they tend to synthesize two other unusual zwitterions as osmolytes: Nε-acetyl-β-lysine (Sowers et al., 1990;

Sowers and Gunsalus, 1995), and β-glutamine (Lai et al., 1991). Both of these solutes are preferentially accumulated at higher external NaCl concentrations. In several methanogens, Nε-acetyl-β-lysine is preferentially accumulated when the external NaCl is above 5%; β-glutamine accumulation requires higher NaCl. The use of β-amino acids would appear to be an excellent strategy of generating a true "compatible solute," since most enzymes that utilize the normal L-α-amino acid can not use the β-amino acid as a substrate. β-Glutamine has only been detected in *Methanohalophilus portucalensis* FDF1 grown with ≥ 2 M NaCl (Lai et al., 1991). It is synthesized from β-glutamate by an unusual glutamine synthetase (Roberts et al., 1992; Robinson et al., 2001). Biosynthesis of Nε-acetyl-β-lysine occurs in two steps: (i) isomerization of α-lysine to β-lysine, and (ii) acetylation of the ε-amino group (Roberts et al., 1992; Robertson et al., 1992b). Genes for the enzymes that catalyze these two transformations have been identified in *Methanosarcina mazei* Go 1 and other methanogens (Pflüger et al., 2003). If the genes are deleted in *Methanococcus maripaludis*, the cells can no longer grow in high external NaCl. This indicates that the synthesis of this zwitterion is absolutely critical to growth of these organisms at high NaCl.

Anionic Solutes

Anionic solutes in halotolerant and halophilic microorganisms are the most diverse group of osmolytes. Their prevalence in Archaea, in particular, correlates the high intracellular K^+ in these cells. In some cases, the organic anion is present in such quantities to act as a counterion for the K^+. The negative charge on the solute is provided by three functional groups: carboxylate, phosphate or sulfate.

L-α-glutamate is a ubiquitous solute and usually present at >20 mM in most cells. In medium containing moderate NaCl, this solute can be a contributor to the intracellular osmotic balance (meaning intracellular concentrations are >0.1 M). β-Glutamate is prevalent in many methanogens where it is often accumulated along with α-glutamate (Robertson et al., 1989, 1990b). The intracellular glutamates increase with increasing external NaCl, although some methanogens show an upper limit (~1 M NaCl) above which there is a switch to accumulation of the zwitterionic solute Nε-acetyl-β-lysine (Ciulla and Roberts, 1999; Robertson et al., 1992b). Interestingly, it is not directly derived from α-glutamate by a glutamate mutase. Rather it appears to be made from α-hydroxyglutarate (generated by reduction of α-ketoglutarate); the CoA derivative is dehydrated, then an amine is added to generate β-glutamate (Graupner et al., 2005). Although β-glutamate is mostly seen in Archaea, it has been detected in *Nocardiopsis halophila* (DasSarma and Arora, 2002).

Anionic versions of the uncharged carbohydrates are osmolytes in many different halotolerant and halophilic Bacteria and Archaea. At lower external NaCl, α-glucosylglycerate in *Methanohalophilus portucalensis* and α- or β-mannosylglycerate in *Rhodothermus* sp. act as osmolytes. In *R. marinus*, increasing the external NaCl causes a switch to accumulating the neutral mannosylglyceramide rather than mannosylglycerate (Silva et al., 1999).

Anionic solutes accumulated in response to external NaCl can also have other roles in Bacteria and Archaea. In Bacteria, poly-β-hydroxybutyrate is often used as a storage compound, but in the deep sea organism *Photobacterium profundum* SS9 β-hydroxybutyrate and a small polymer are preferentially synthesized in cells grown at high hydrostatic pressures (280 atm). They also increase with increasing external osmotic pressure, hence are termed piezolytes (Martin et al., 2002). Why these particular molecules are synthesized at high hydrostatic pressure is unclear. It is likely that the hydroxybutyrates aid in stabilizing macromolecules under these conditions.

Perhaps the most intriguing of the anionic solutes are di-*myo*-inositol-1,1'-phosphate, DIP (Ciulla et al., 1994; Scholz et al., 1992) and a variant with mannose attached to each inositol ring (Lamosa et al., 1998). These phosphodiesters are accumulated in hyperthermophilic Archaea at growth temperatures above 80°C (Ciulla et al., 1994; da Costa et al., 1998; Martins et al., 1996; Santos and da Costa, 2002). In *Archaeoglobus fulgidus*, a different phosphodiester, α-diglycerophosphate is the major osmolyte when then cells are grown below 80°C (Martins et al., 1997). However, above 80°C, there is a switch to DIP. α-diglycerophosphate looks like DIP with three carbons removed in each ring, so what is the advantage of the inositol ring at high temperatures? Perhaps the extensive hydroxyl network enhances water tension at high temperatures and this in turn stabilizes folded proteins.

The last of the unusual anionic solutes is sulfotrehalose – trehalose where a sulfate group is attached to one of the glucose rings on C(2). Sulfotrehalose is associated with the haloalkaliphilic Archaea *Natronococcus* and *Natronobacterium*, but only when the cells are grown in defined media (Desmarais et al., 1997). Under these conditions, the intracellular concentration of sulfotrehalose is comparable to the intracellular K^+.

◆◆◆◆◆ IDENTIFICATION OF OSMOLYTES BY NMR SPECTROSCOPY

Since osmolytes are present in cells at relatively high concentrations, they are well suited to be studied by NMR methods. Although they can be monitored inside the cells (Motta et al., 2004), it is often more useful to extract small molecules from the cell and then examine the extract by ^{13}C and/or ^1H methods. The extract usually has a mixture of solutes so that strategies for identification of compounds are needed. As will be illustrated with examples from several Archaea, there are a variety of two-dimensional NMR experiments that are suitable for this. Once the solutes are identified, they can also be quantified by comparing resonance intensities of the solutes to an added standard.

Cell Extraction Methods

There are two primary methods for lysing cells to prepare extracts for NMR spectroscopy that contain all the soluble small molecules but not the macromolecules or aggregates such as membranes: (i) ethanol extraction and (ii) perchloric acid extraction. When a cell suspension is mixed with 70% (v/v) ethanol, the ethanol partitions into the membrane causing it to swell and lyse. The amount of ethanol is critical for this procedure to work. Less than 50% ethanol will lyse most cells, but not precipitate all the proteins so that the spectrum could be complicated by resonances from small macromolecules. 80% or more ethanol can lower the solubility of some of these polar solutes so that the spectrum has a distribution of solutes different from what occurs *in vivo*. $HClO_4$ treatment alters the protonation state of proteins and lipids in cell wall or plasma membrane and this (along with the loss of pH gradients across membranes) causes cell lysis and precipitation of proteins, nucleic acids and cell wall components. While effective at cell lysis, $HClO_4$ treatment may not be appropriate if solutes are acid-labile.

Ethanol extracts

1. Cells are centrifuged and the pellet resuspended in 5–10 ml isotonic solution that does not contain any organic molecules (NaCl, KCl, etc., but no yeast extract, substrates, etc.). The washed resuspension is centrifuged again and the pellet used for extraction.
2. The cell pellet is mixed with 5–10 ml 70% (v/v) ethanol and centrifuged. After removal of the supernatant the pellet is treated with ethanol at least two more times.
3. The supernatants (containing soluble small molecules) are combined, frozen in liquid nitrogen and lyophilized.
4. The dry sample is solubilized with 0.5–1 ml D_2O (99.9%) if ^1H NMR spectra will be obtained, or with a H_2O/D_2O (4:1) mixture.
5. The residual pellet is resuspended in 1–5 ml H_2O for protein analysis (easily done using the Bradford protein assay with bovine serum albumin as the standard). This allows one to compare the total amount of different organic solutes with the total protein in the aliquot of cells that was extracted.

Perchloric acid extracts

1. Cells are harvested, washed and centrifuged as above.
2. Cell pellets are mixed with 1–2 ml cold 7% (v/v) $HClO_4$ and kept on ice for 20–30 min.
3. The solution is centrifuged to separate soluble small molecules from precipitated/denatured macromolecules.
4. The supernatant is neutralized with 1 M KOH; the $KClO_4$ precipitates reducing the ionic strength of the solution.
5. After lyophilization, the small molecule fraction is dissolved in D_2O as described above.

An additional step can be used in preparing the extracts. The combined supernatants can be passed over a Chelex-100 column (typically 0.5 ml) to remove any paramagnetic impurities. Alternatively, one can add 0.1–1 mM EDTA to the final H_2O/D_2O extracts.

Other extraction procedures

On occasion other extraction procedures can be used. For example, water-immiscible organic solvents (CH_2Cl_2, butanol) added to a cell pellet will solubilize membrane lipids and hydrophobic proteins in the organic phase (for example, see Galinksi and Herzog, 1990). The aqueous phases will be enriched in water-soluble small molecules and some proteins although many macromolecules are denatured by this treatment. One advantage for this method is that most degradative enzymes are inactivated by high concentrations of organic solvents. The disadvantage is that some small molecules may partition in both aqueous and organic phases making quantitation difficult.

One-dimensional NMR

Natural abundance ^{13}C NMR

If a sufficiently large sample of cells can be extracted, this is the easiest way of identifying and quantifying organic solutes. It was critical for detecting and identifying ectoine (Galinski et al., 1985) and several β-amino acids (Lai et al., 1991; Robertson et al., 1989; Sowers et al., 1990). The 1.1% natural abundance of ^{13}C means that $^{13}C-^{13}C$ coupling will not occur. The large chemical shift range for carbon is also advantageous. If protons are decoupled, then each carbon in a molecule will have a distinct resonance indicative of the type of carbon (e.g. $-CH_3$, $-CH_2-$, $-CH_2O-$, $=CH-$, etc.). ^{13}C chemical shifts (referenced to DSS (3-(trimethylsilyl)propanesulfonate) at 0.0 ppm) for known organic solutes in halotolerant and halophilic organisms are presented in Table 26.1. The absolute shifts may vary depending on extract ionic strength and pH, but the basic pattern of resonances and relative chemical shifts should be reasonably diagnostic for a given osmolyte. A reasonable sample would be an extract from 100 to 200 ml of a cell culture with an $OD_{660} \sim 1$. If one has no idea of the range of solutes in the extract, best conditions would be to use a 30 000 Hz sweep width centered around where CH_3OH resonates. The number of transients depends on the solute concentrations, but 400–2000 transients will usually detect resonances for the major solutes in the extract. To avoid saturation of protonated carbons, acquisition conditions should use ~70° pulse and a recycle time of 2–3 s. Carbonyl or quaternary carbons may not be relaxed by these conditions. The plus variable NOEs makes the integration of these resonances unreliable for quantifying solutes. However, the integrated intensity of CH_n groups is proportional to the amount of carbon in a given solute. One typically adds a known amount of a compound (formate (171.0 ppm) and dioxane (62.4 ppm) have been used for this purpose) that does not overlap with any other carbons in order to get an absolute amount of each solute. The amount of each solute in the extract can then be related to the total amount of protein that was in the cells that were extracted (step 5 of the ethanol extraction procedure).

^{1}H NMR

If fewer cells are available, this is the nucleus of choice for identifying solutes. Spectra are more complex than ^{13}C because of ^{1}H–^{1}H coupling, and in some cases the overlap of multiplets makes it difficult to integrate the resonances for individual solutes. In general, the higher the field, the more separated resonances will be and the easier to identify (Fan, 1997) and integrate the solutes. Table 26.1 lists ^{1}H chemical shifts for many of the solutes; again there is some variation depending on the extract conditions. Aside from increased sensitivity, another advantage of ^{1}H NMR is that connectivities of groups can be easily obtained with 2D experiments allowing one to trace out the backbone of the molecule. Useful acquisition parameters include a 6000 Hz sweep width centered on HDO (referenced as 4.75 ppm). Presaturation conditions are often useful to decrease the intensity of the residual water (HDO) resonance. For most samples, a saturation delay of either 2 or 3 s is adequate for suppressing the water resonance. Since ^{1}H is so sensitive, far fewer transients, typically, 16–128, are adequate for examining extracts with moderate concentrations of organic solutes. Free induction decays are processed with a 1 Hz line broadening. To quantify solutes, resonances whose identity has been established are integrated and compared with an added standard (formate at 8.5 ppm makes a particularly good choice since little is found in this region in extracts).

As an example, we have ethanol extracted 100 ml culture of *Methanocaldococcus jannaschii* grown in medium containing 5% NaCl to an OD$_{660}$ ~ 0.2 and obtained the ^{1}H spectrum at 500 MHz (Figure 26.1). Multiplets for α- and β-glutamate are the major resonances in the spectrum. However, there are other resonances as well that are not as easily identified in the 1D ^{1}H NMR spectrum. Other techniques are needed to identify these.

^{31}P NMR

Since some of the osmolytes detected in halotolerant organisms contain phosphorus (e.g. α-diglycerophosphate or DIP). ^{31}P is not as sensitive as ^{1}H but more sensitive than natural abundance ^{13}C. Obtaining spectra with a 8000-Hz sweep width centered around inorganic phosphate is adequate for detection of most biological phosphates. ^{31}P chemical shifts, referenced with respect to an external capillary of H$_3$PO$_4$ (at 0.00 ppm), are very sensitive to pH. This makes it easy to distinguish phosphodiesters from monophosphates since the latter will show changes in their chemical shifts over a pH range of 4–9, while the phosphodiesters will have invariant chemical shifts.

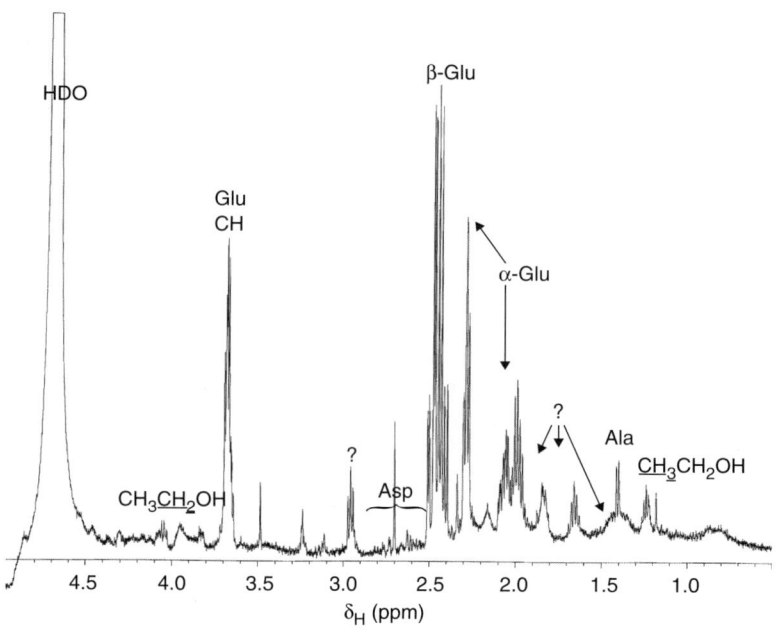

Figure 26.1. 500 MHz ^{1}H spectrum of an ethanol extract of *Methanocaldococcus jannaschii* cells grown in medium with 5% NaCl.

Two-dimensional NMR methods

Identification of mixtures of organic molecules in an aqueous solution can be aided by a suite of two-dimensional NMR methods. A combination of TOCSY, COSY and HMQC (or HSQC) should be adequate to sort out all the different solutes that contribute to osmotic balance in a particular microorganism. Several of these will be illustrated with extracts from *Methanocaldococcus jannaschii*.

^{1}H–^{1}H experiments

The homonuclear TOCSY (TOtal Correlated SpectroscopY) experiment is useful for determining which signals arise from protons within a spin system, yielding long range as well as short range correlations. As long as there are no heteroatoms separating the CH units, a stretch of CH$_n$ groups can be identified by this experiment. A TOCSY contour plot for the extract from *M. jannaschii* grown in 5% NaCl (Figure 26.2) shows the two spin system for β-glutamate and the four spin system for α-glutamate (the two protons on C3 are nonequivalent). There is also another prominent spin system consisting of five protons. This five peak pattern and specific ^{1}H chemical shifts strongly suggest that there is α-lysine in the extract of the *M. jannaschii* cells. However, the TOCSY reveals other interesting spin systems hidden under the glutamates. The ellipse highlights the α-CH correlations for a several α-glutamate-like spin systems whose shifts are suggestive of a small peptide that is detected in cells that are salt or nutrient stressed (D. Wagner and M. F. Roberts, unpublished results).

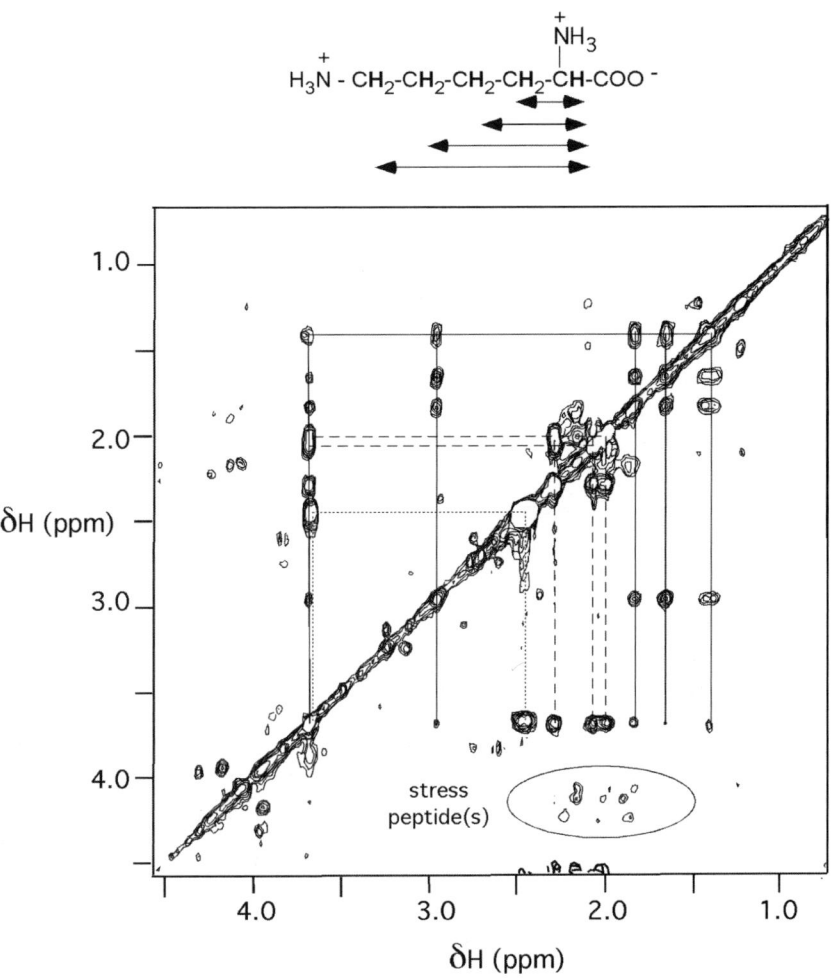

Figure 26.2. TOCSY contour plot showing the correlations of proton–proton interactions detected through the carbon backbone for an extract from *Methanocaldococcus jannaschii* grown in 5% NaCl. Spin systems for α-glutamate (- - -) and β-glutamate (·····) are shown as well as a third spin system with five spins that is consistent with α-lysine (——).

Although a TOCSY is a good screen for the number of different spin systems and for osmolytes this usually means the number of solutes, it does not link one CH unit to the next. A COrrelated SpectroscopY (COSY) experiment links protons that are directly coupled. Figure 26.3 shows the connectivities for the α-lysine in the *M. jannaschii* extract. By tracing out the connectivities in the spin system starting with the –CH_2NH_2 group (labeled ε) whch has a relatively unique chemical shift, one can identify the next CH_2 group in the sequence (ε) and the protons coupled to it until one reaches the α–CH group (which happens to overlap with that of the glutamates as well). Knowing the linkages and ^1H resonance assignment is also critical for interpreting heteronuclear experiments.

The last useful homonuclear experiment in a NOESY (Nuclear Overhauser Effect SpectroscopY). Rather than coupling interactions, this

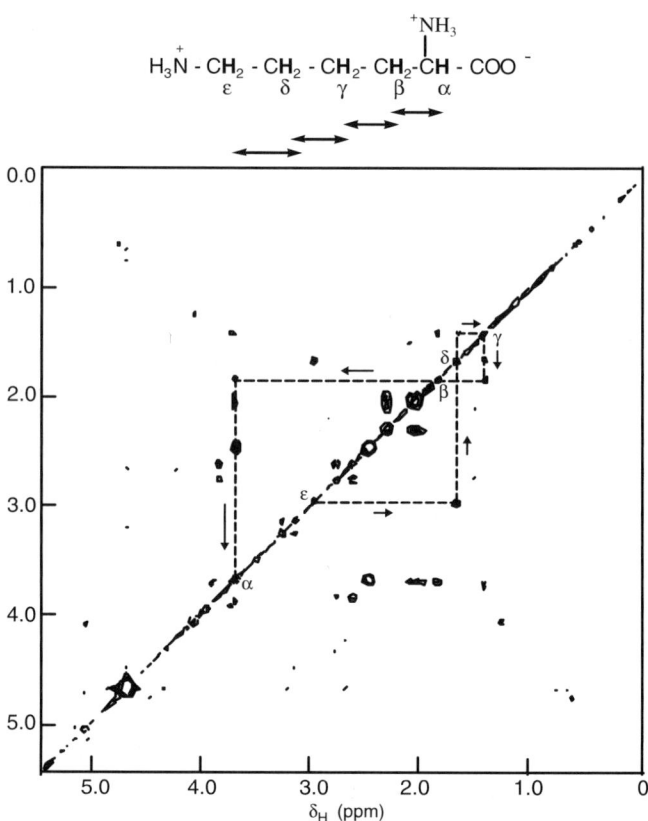

Figure 26.3. COSY contour plot illustrating the connectivities for α-lysine. The dashed line traces the connection starting with the ε–CH$_2$ through the five spins to the α–CH group.

identifies proton pairs that are spatially close. This dipolar interaction between spins is proportional to r^{-6} where r is the distance between the two protons. Key to the experiments is a "mixing time" that is usually 0.2–0.5 s for small molecules. This has been particularly useful in determining the linkages of carbohydrates. A good example is sulfotrehalose (Desmarais et al., 1997). The sulfate on the C(2) of one of the glucose rings alters its chemical shift in a way that could have suggested it was involved in a glycosidic bond. A NOESY experiment identified an off-diagonal element between the two protons connected to the anomeric carbons on each sugar ring (C(1), C(1′)), indicating they are spatially close (Figure 26.4B). This correlation was not detected in a TOCSY experiment (Figure 26.4A) so it can not represent through bond coupling but through a space interaction. The only way this can happen is, if the glycosidic bond between the two sugars is between the two C(1) carbons.

Heteronuclear NMR experiments

The ability to correlate ^{13}C and ^{1}H resonances is also an excellent diagnostic tool. Two types of these heteronuclear experiments are easily implemented and extremely useful as long as there is adequate sample

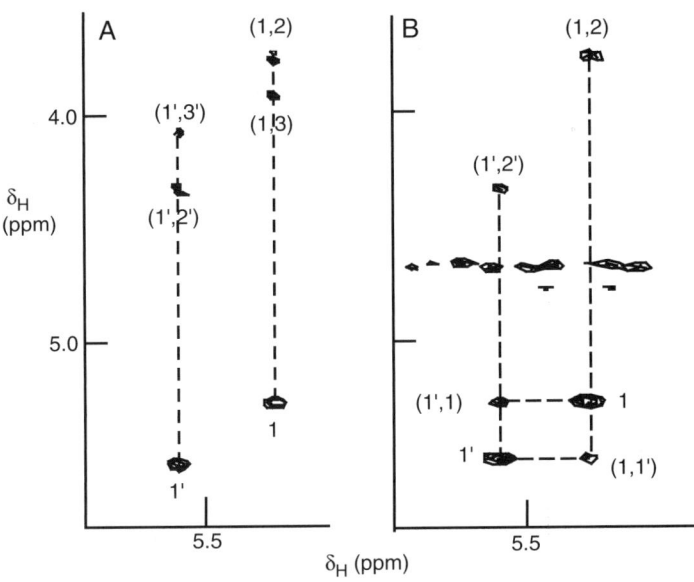

Figure 26.4. Comparison of (A) TOCSY and (B) NOESY slice for the C(1) protons on each glucose ring. The NOESY off-diagonal peaks between the protons on C(1) and C(1′) indicate that these carbons are linked *via* a glycosidic bond.

with natural abundance ^{13}C or the sample is enriched in ^{13}C. The HMQC (Heteronuclear Multiple Quantum Correlation) is a ^1H-detected experiment that correlates ^1H–^{13}C resonances using a spin echo difference experiment. The experiment exploits the coupling between carbons and their directly bonded protons ($^1J_{CH}$ is typically set to 140 Hz) to correlate each proton frequency with the frequency of the directly bonded carbon. An HMQC plot then has many peaks, each characterized by a ^1H and a ^{13}C chemical shift. Adequate sensitivity requires either moderately high extract concentrations or some degree of ^{13}C labeling. Figure 26.5 presents an HMQC plot for an *M. jannaschii* extract of cells grown in 3% NaCl with $^{13}CO_2$ present for 4 h. (^1H,^{13}C) pairs for the major solutes are easily detected. However, what is most striking with the HMQC are, many other peaks that are detected in this 2D experiment are not easily resolved in the homonuclear experiments. The HMQC is a particularly useful screen for changes in cell extracts either as a function of time or growth conditions. At a minimum it can be considered as a two-dimensional fingerprint of all the small molecules occurring at moderate concentrations in the cell. If one identifies all the components, then it can be used to quantify changes in the metabolome under different conditions.

An alternative to the HMQC is the HSQC (Heteronuclear Single Quantum Correlation) experiment. This experiment also correlates protons with directly bonded heteronuclei but uses a different pulse sequence (it is based on an INEPT (Insensitive Nuclei Enhanced by Polarization Transfer) experiment which monitors single quantum transitions). Either an HMQC or HSQC can be used, although there are small distinct advantages to each (Bax *et al.*, 1990; Norwood *et al.*, 1990).

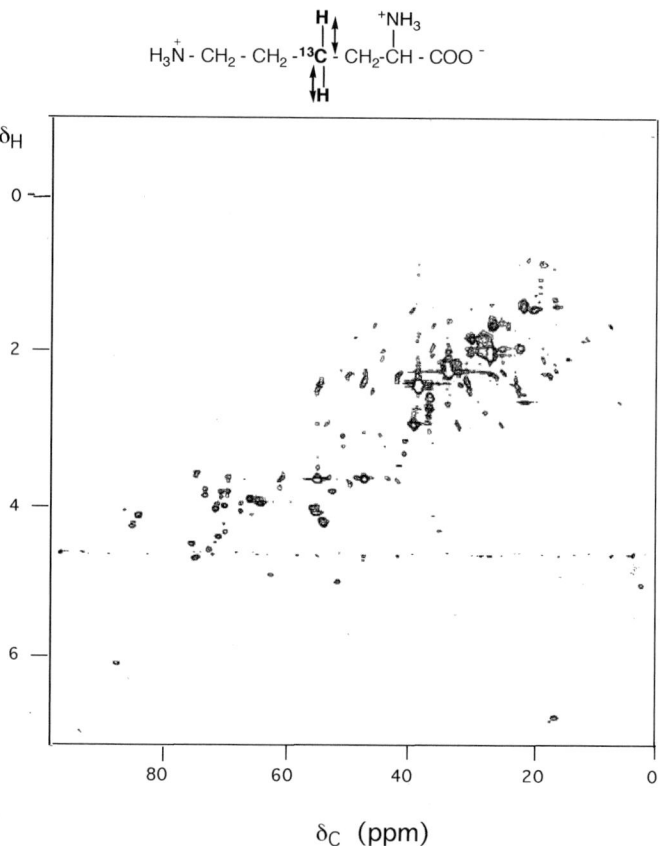

Figure 26.5. HMQC contour plot for an extract of *Methanocaldococcus jannaschii* grown in 3% NaCl with $^{13}CO_2$ present in the medium. The major solute ($^{13}C,^{1}H$) peaks are indicated as α (α-glutamate), β (β-glutamate) and L (L-α-lysine). Note all the other metabolites present in the extract and detected in this experiment.

A variant of the HMQC is the HMBC (Heteronuclear Multiple Bond Correlation) which is a ^{1}H-detected experiment that identifies proton nuclei coupled to ^{13}C that are separated by more than one bond, as shown in the scheme below. The key parameter in an HMBC experiment is a long range coupling constant for 2 or 3 bonds ($^{2}J_{CH}$ or $^{3}J_{CH}$ is usually set to 8 Hz). There is also a filter to suppress crosspeaks arising from $^{1}J_{CH}$ interactions, that has a delay that is matched to the inverse of $^{1}J_{CH}$. An HMBC experiment is particularly useful in connecting carbonyl/carboxyl groups with adjacent CH_n groups.

In halotolerant and halophilic organisms, solutes that contain phosphates include α-diglycerophosphate, DIP and mannosyl-DIP.

Connecting the ^{31}P in these solutes with nearby ^{1}H that are coupled to it (as indicated in the structure of α-glycerol phosphate below) can be done with a HETCOR (HETeronuclear CORrelation) experiment. This is a ^{31}P-detected experiment that identifies protons that are coupled to the phosphorus nuclei. It is not as sensitive as an HMQC or HSQC experiment. However, some spectrometer, indirect probes used for the heteronuclear ^{1}H-detected experiments can not be tuned to the higher phosphorus frequency. The ^{31}P-coupled protons can then be linked to other protons and carbons to determine/identify the structure of the solute. If the available spectrometer can carry out a ^{31}P–^{1}H HMQC or ^{31}P–^{1}HSQC, correlation of the phosphorus resonance with protons that are coupled can be achieved with much greater sensitivity than in the HETCOR experiment.

Good examples of ^{31}P–^{1}H correlation experiments are shown in the articles that identify α-diglycerol phosphate, DIP and mannosyl-DIP (Ciulla et al., 1994; Lamosa et al., 1998; Martins et al., 1996).

^{39}K NMR

K$^+$ is actively and specifically transported into cells (e.g. the Trk uptake system in *Halomonas elongata*; Kraegeloh et al., 2004). In halophiles, the concentration of this ion is often quite high. Intracellular K$^+$ is also often intertwined with the selection of intracellular organic solutes used for osmotic balance. While several analytical methods exist for quantifying K$^+$ (e.g. atomic absorption), ^{39}K NMR can also contribute to monitoring this ion in cells. This nucleus is quadrupolar (spin = 3/2, natural abundance 93.1%) and under normal conditions one can easily detect a resonance for only the aquo-K$^+$ complex (non-symmetric K$^+$ complexes have exceptionally broadened resonances and are not detected). Although ^{39}K NMR spectroscopy is not as sensitive as atomic absorption, it can be carried out on the same sample used for ^{1}H and other NMR analyses as long as the extraction procedure is slightly modified. The initial cell pellet should be washed with an isotonic solution that has no K$^+$ salts and replaces the total external K$^+$ concentration with Li$^+$ (usually LiCl). The washed cell pellet can either be examined directly for ^{39}K$^+$ content (usually at 4°C to stop any metabolism) or it can be osmotically lysed. For the latter, the cell pellet is mixed in 1–5 ml H$_2$O/D$_2$O (4:1). This allows the intracellular K$^+$ to equilibrate in the resuspension solution where it exists as the symmetric aquo-K$^+$ species. Since the relaxation time for ^{39}K is relatively short, the recycle delay can be short. However, if the relaxation is too short, much of the intensity will be lost after the pulse and before the acquisition of the free induction decay. In those cases, the ^{39}K resonance may be better monitored using a spin echo sequence (Robertson et al., 1992b). After obtaining an initial spectrum, a known amount of KCl is added to the sample and an additional spectrum is taken under identical conditions. Integration, using a fixed number of transients and integration scale before and after the exogenous K$^+$ was added, can be used to determine the concentration of potassium in the sample. Thus in turn can be normalized to protein in the extracted cells.

◆◆◆◆◆ NMR METHODS TO REFINE OSMOLYTE BIOSYNTHETIC PATHWAYS

In order to understand the regulation of these osmolyte concentrations in cells, one needs to first identify their biosynthetic pathways, then identify candidate enzymes. For many of the more unusual compatible solutes that do not occur in other cells and are not normal metabolites, can prove challenging. In the absence of homologous genes in other organisms, one usually proposes various pathways, then tries to assay for the indicated activities. Several generalizable NMR methods are useful because they do not select for a single reactant or product. Multiple reactions can be monitored at once and this information is used to unravel biosynthetic steps.

Isotopic Labeling

Once a series of reactions based on reasonable biochemistry has been proposed, it is often possible to design isotopically labeled precursors that can discriminate between pathways. The labeling can be done *in vivo* with intact cells or *in vitro* with enzyme extracts. Since many Archaea have finicky transport systems, the *in vivo* labeling can be difficult. A lack of label uptake into small molecules of the cell may reflect a lack of uptake of the precursor rather than a different biosynthetic pathway. Since the natural abundance of ^{13}C is 1.1%, this isotope is excellent as a biosynthetic tracer. Typically, one inoculates a culture containing the ^{13}C-labeled precursor in mM concentrations and allows the cells to grow for at least one or two doubling times. If aliquots are removed as a function of time, information of the time scale for solute synthesis can also be obtained.

A useful example is provided by the halophilic methanogen *Methanohalophilus portucalensis* FDF1 (Robinson and Roberts, 1997). When these methanogens are grown on a relatively nonspecific label source, such as ^{13}CH$_3$OH along with ^{12}CO$_2$ which will generate ^{13}CH$_3$^{12}CO-units, one obtains biosynthetic information on all the visible solutes extracted from the cells. In this case, there are many osmolytes

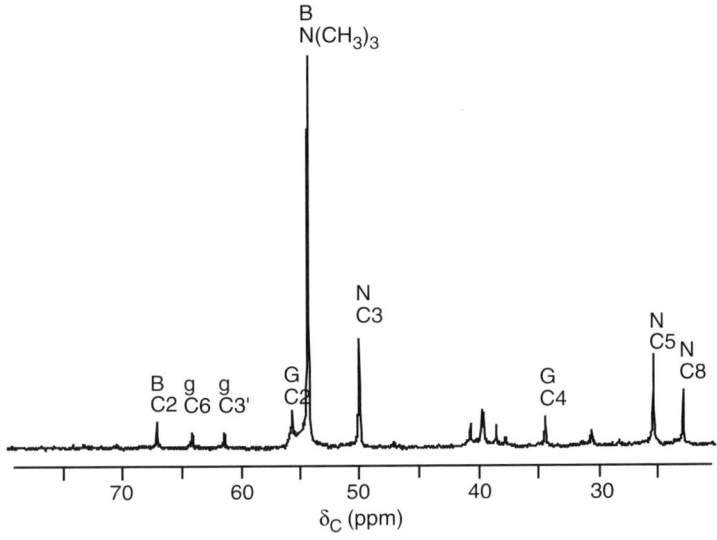

Figure 26.6. ^{1}H-decoupled ^{13}C spectrum (125.7 MHz) of an *Methanohalophilus portucalensis* extract grown with ^{13}CH$_3$OH and ^{12}CO$_2$: B, betaine; N, Nε-acetyl-β-lysine; G, α-glutamate; g, glucosylglycerate. Adapted from (Robinson and Roberts, 1997).

synthesized (betaine, Nε-acetyl-β-lysine, α-glutamate and glucosylglycerate (Robertson *et al.*, 1992a) and so the label uptake into specific carbons sheds light on many pathways in the organism (Figure 26.6). The observation that ^{13}C label is incorporated into the betaine N(CH$_3$)$_3$ peak clearly indicates these cells synthesize that molecule. Furthermore, labeling of only the methyl carbons is consistent with reductive methylation of glycine that is generated from serine (Roberts *et al.*, 1992). Under these growth conditions, α-glutamate but not β-glutamate was accumulated. Label uptake into glutamate C(2) and C(4) is consistent with a partial oxidative Krebs pathway to generate glutamate. ^{13}C incorporation into C(3) and C(5) of Nε-acetyl-β-lysine is consistent with the diaminopimelate pathway for lysine generation. As long as at least 2% ^{13}C is incorporated and an adequate volume of a cell suspension can be obtained, it is easy to directly detect the enrichment of specific carbons in the ^{13}C spectrum.

However, a more sensitive method to monitor ^{13}C uptake into small molecules is to use a 1D version of an HMQC experiment (Ciulla and Roberts, 1999). This generates a ^{1}H spectrum with only protons coupled to ^{13}C shown (Figure 26.7). Since the natural abundance of ^{13}C is 1.1%, small increases in ^{13}C labeling can be quantified. In this particular case, an aliquot (50 ml) of a low OD$_{660}$ culture of cells (~0.1) was grown with 10 mM H^{13}COO$^-$Na$^+$ added for 4 h (final OD$_{660}$ ~ 0.2). The resulting spectrum, which took about 1 h to acquire, has all the relevant resonances equivalently labeled since the formate is converted to CO$_2$ and H$_2$, and CO$_2$ is the precursor for all carbons. The resolution is lower, but the appropriate peaks are there and easily detected. This offers a sensitive way of monitoring ^{13}C content, either increases

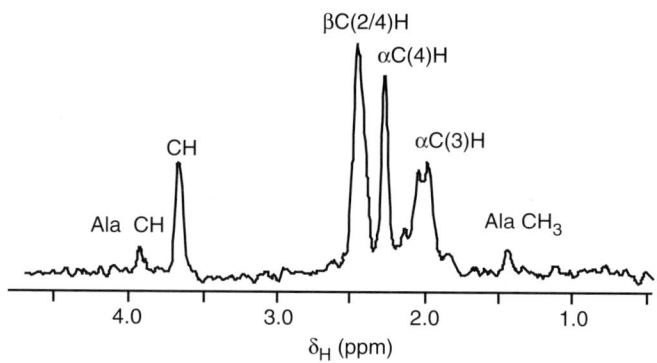

Figure 26.7. 1D HMQC spectrum of an extract of *Methanocaldococcus jannaschii* grown with $^{13}CO_2$ added to the medium. The resolution is not as good as in the regular 1D spectra, but all protons coupled to ^{13}C are visible.

or decreases in specific peaks, after a switch from the NMR-active to the NMR-inactive isotope.

Exploiting NMR to Monitor *in vitro* Enzyme Activities

A different approach to elucidating biosynthetic pathways for an osmolyte is to use NMR to monitor suspected reactions along the proposed biosynthetic pathway. This is a very useful procedure for osmolytes containing phosphate groups. One need not separate reactants and products, since the chemical shifts of different phosphorylated compounds are quite distinct and identifiable by 2D methods. Reaction mixtures can also have complex organic molecules, buffers, enzymes and cofactors present, and these do not contribute to the ^{31}P spectrum unless they contain phosphorus. An illustration of this methodology is the biosynthesis of DIP in different Archaea. A reasonable proposed pathway for generation of DIP (Chen et al., 1998) consists of four steps:

1. D-glucose-6-phosphate (G-6-P) is converted to L-inositol-1-phosphate (L-I-1-P) *via* inositol-1-phosphate synthase (IPS). This is the sole known pathway for synthesizing inositol in all other cells and likely to be functional here. To make a phosphodiester bond requires coupling of an activated phosphate ester with an alcohol. In the case of DIP, these are likely to be a nucleotide-diphosphate inositol ester as the activated precursor and *myo*-inositol as the free alcohol.
2. Coupling of a nucleotide triphosphate (NTP) to L-I-1-P would produce an NDP-inositol precursor.
3. Hydrolysis of the L-I-1-P by an inositol monophosphatase activity would generate *myo*-inositol, a molecule that is no longer chiral.
4. The final step in synthesizing DIP would be the generation of the phosphodiester linkage by condensing NDP-inositol with L-I-1-P (*via* a "DIP synthase" activity).

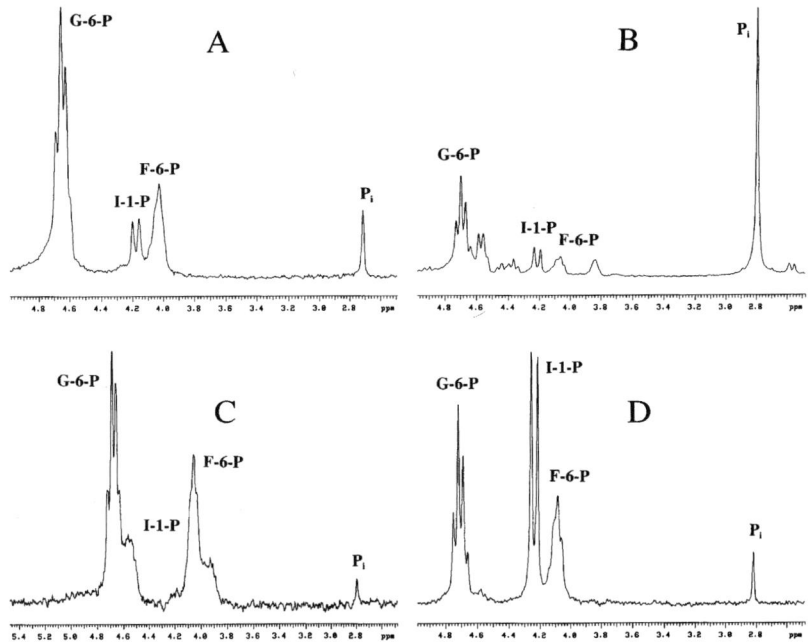

Figure 26.8. ^{1}H-coupled ^{31}P NMR spectra (202.3 MHz) illustrating I-1-P synthesis by crude enzyme extracts of (A) *Pyrococcus furiosus* with 1 mM EDTA (incubated at 85°C for 30 min), (B) *Thermotoga maritima* with 1 mM EDTA (incubated at 85°C for 10 min), and of *Archaeoglobus fulgidus* (C) with 1 mM EDTA or (D) 1 mM Mg^{2+} (incubated at 85°C for 30 min). The I-1-P doublets (4.2 ppm) with coupling constant of 8 Hz are shown as inserts in each spectrum. The assay conditions included 3-5 mg crude proteins, 5 mM G-6-P, 1 mM NAD$^+$ in 25 mM Tris-acetate, pH 8.0.

In vitro ^{31}P NMR spectra that monitor the IPS reaction in *Methanotorris igneus*, *Thermotoga maritima* and *Archaeoglobus fulgidus* are shown in Figure 26.8. The spectra are ^{1}H coupled because it is easy to identify I-1-P as a doublet amidst other phosphomonoesters such as G-6-P and fructose-6-phosphate (F-6-P), which appear as triplets or quartets. When 5-mM G-6-P is mixed with cell extract and NAD$^+$ (a cofactor of the enzyme), I-1-P is produced. In the case of the *A. fulgidus* extract, the synthesis of I-1-P requires divalent cation since in the presence of EDTA (Figure 26.8C) no I-1-P is visible while with 1 mM Mg^{2+}, the I-1-P is easily produced. However, the ^{31}P spectrum shows other reactions are occurring as well. The G-6-P is rapidly isomerized to F-6-P by glucose isomerase. There is also some limited phosphatase action since the resonance for inorganic phosphate is detected. Phosphatase activity is the highest in the *T. maritima* extract (other phosphorus resonances indicate other reactions are also occurring).

^{31}P spectra can also be used to monitor the final reaction in crude cell extracts (Figure 26.9). When 2 mM synthetic CDP-I (resonance ∼ −10 to −11 ppm) is mixed with protein extracts from *Methanotorris igneus*, *Pyrococcus furiosus* and 5 mM *myo*-inositol, DIP is produced (triplet ∼ −1 ppm). Again, additional resonances in the spectra attest to other

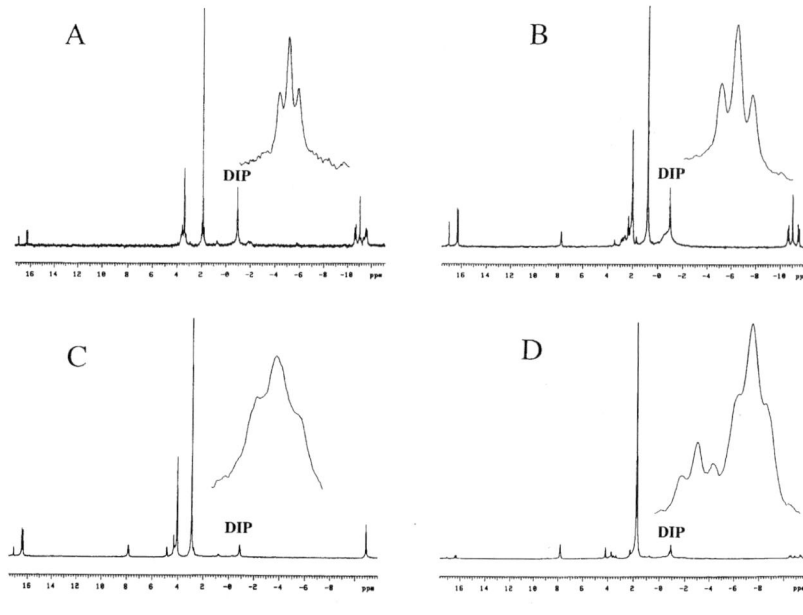

Figure 26.9. ^{1}H-coupled ^{31}P NMR spectra (202.3 MHz) illustrating DIP synthesis by crude enzyme extracts of (A) *Methanotorris igneus* (incubated at 85°C for 30 min), (B) *Pyrococcus furiosus* (incubated at 95°C for 30 min), (C) *Pyrococcus woesei* (incubated at 85°C for 4.5 h) and (D) *Thermotoga maritima* (incubated at 85°C for 10 min). The characteristic DIP triplets (−0.9 ppm) are shown as inserts in each spectrum. Assay conditions included 3–5 mg crude protein, 2–3 mM CDP-inositol, 5 mM *myo*-inositol, 6 mM MgCl$_2$ in 25 mM Tris acetate, pH 8.0. Note the two triplets for DIP isomers synthesized by *T. maritima* in the inset of (D).

reactions occurring with these molecules. The peaks around 16 ppm are likely to result from a cyclic form of I-1-P with a strained five-membered ring (cyclic-*myo*-inositol 1,2-phosphate (Volwerk et al., 1990)). However, the most interesting result from these *in vitro* studies is for *Thermotoga maritima*. Instead of a single DIP triplet, there are two: presumably one is for the DIP made with one L-I-1-P and the *myo*-inositol incorporated with the same stereochemistry, and the other for DIP with L-I-1-P and the *myo*-inositol incorporated as the other stereochemistry (a D-I-1-P unit, if one looks at the relation of this inositol ring compared to the phosphate in DIP).

◆◆◆◆◆ NMR METHODS TO MONITOR SOLUTE DYNAMICS IN CELLS

One would expect a true compatible solute to be relatively inert in the cell since it would not affect cell metabolism. How can this be measured? An easy way to monitor turnover of solute pools in cells is to grow cells with an NMR-active isotope (^{13}C or ^{15}N), then either remove or dilute the isotope with a non-NMR-active isotope (^{12}C, ^{14}N). These experiments are analogs to radioactive pulse/chase experiments. The latter have been used occasionally to measure the half-life of solutes in cells (e.g. fast turnover of glutamate and glutamine compared to α-glucosylglycerate in *Erwinia chrysanthemi* (Goude et al., 2004). Success in using the radioactivity approach requires good methods to separate and quantify solute, in the cited case high voltage electrophoresis coupled with paper chromatography. However, in the NMR version, one can examine many solutes at the same time for comparative isotope content (Roberts et al., 1990; Robertson et al., 1992b). The following example explores turnover of the amino group in osmolytes in *Methanothermococcus thermolithotrophicus* (Martin et al., 2001). Cells were grown to an $OD_{660} \sim 0.6$ in medium containing 3% NaCl and 4 mM ^{15}NH$_4$Cl (99% enriched), then harvested anaerobically and resuspended in medium containing sodium formate (a soluble substrate for methanogenesis) with 4 mM ^{14}NH$_4$Cl (and 15% D$_2$O). ^{1}H WALTZ decoupled ^{15}N NMR spectra (50.65 MHz on a 11.74 T magnet) were acquired at 50°C at various times after the start of the ^{14}N-chase. Acquisition conditions include a 7600 Hz sweep width, 90° pulse and 5 s recycle delay, and 32 transients. Offsets were chosen so that the ammonium resonance was in the center of the spectrum. As seen in Figure 26.10, the α-glutamate resonance

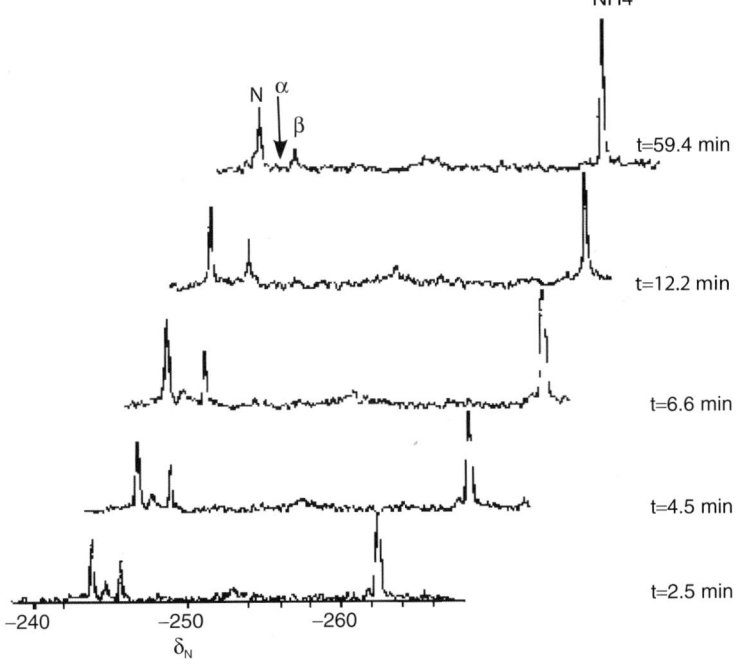

Figure 26.10. ^{1}H WALTZ decoupled ^{15}N NMR (50.65 MHz) spectra of intact *Methanothermococcus thermolithotrophicus* cell suspensions grown with 4 mM ^{15}NH$_4$Cl, anaeorbically harvested, and then resuspended in medium with formate and ^{14}NH$_4$Cl and incubated at 50°C. The time after the start of the chase is given to the right of each spectrum. The chemical shift region shown has resonances for the amino groups of Nε-acetyl-β-lysine (N), α-glutamate (α), β-glutamate (β), and ^{15}NH$_4^+$.

is rapidly lost with an estimated half-life of 5 min. In contrast, the half-lives for loss of ^{15}N from the β-glutamate and Nε-acetyl-β-lysine amino groups were much longer, 48 and >100 min, respectively. Transaminase activities could rapidly remove the ^{15}N while leaving the carbon backbone intact. The ^{15}N-pulse/^{14}N-chase experiments show that the amino groups of both β-amino acids were not transferred to other molecules and that the solute was not metabolized in the cell under these conditions, a characteristic of a compatible solute.

Similar experiments can be carried out with ^{13}C-labeled cells where the source of ^{13}C is either diluted 10-fold or removed (if a gas such as CO_2) and replaced with the ^{12}C-compound. For increased sensitivity, one can monitor ^{13}C content as a function of chase time using 1D HMQC acquisition conditions.

◆◆◆◆◆ OTHER METHODS TO MONITOR ORGANIC OSMOLYTES

While NMR is the most general method for identifying and quantifying osmolytes, other techniques can aid in analyzing the concentration of these solutes. Both mass spectrometry and HPLC (coupled with a sensitive method of solute detection) have been used. Solid-state NMR methods have not been specifically applied to monitoring osmolytes but could contribute to rapid analyses of labeled molecules in cells.

Mass Spectrometry

Mass spectrometry is the basis of metabolomics (where it is usually coupled with NMR, for example see Grivet *et al.* (2003), for a review of NMR and microorganism metabolomics) and proteomics (for review of use in proteomics see Joo and Kim, 2005). The technique offers high sensitivity although quantitation of small molecules is not easy (single amino acids can be difficult to detect with MALDI-TOF (Matrix Assisted Laser Desorption/Ionization-Time-Of-Flight) or ESI-MS (ElectroSpray Ionization Mass Spectrometry)). The major advantage of MS is to measure mass/charge for species; this should confirm the identities of solutes in an extract. Most analyses of novel solutes rely on MS to confirm the suspected identity of the solute.

Chromatography and Specific Detection

Depending on the types of organic osmolytes to be monitored, there are a variety of specific chromatographic separations and detection methods. Solutes with free amino groups, most notably free α- and β-amino acids, can be derivatized with a fluorophore either before or after chromatography, eluted on either hydrophobic or anion-exchange HPLC columns, and detected by fluorescence (Sowers et al., 1990; Lai et al., 1991). FMOC (9-fluorenylmethyl chloroformate) modification of amine-containing solutes has been shown to be highly sensitive and can be used to look for solute production from single bacterial colonies from agar plates (Kunte et al., 1993). This has also been adapted to monitoring ectoine after mild hydrolysis to release the free amino group from the ring. As long as one has standard solutes to determine elution times, HPLC with derivatization of amines is an excellent and rapid way to quantify this particular class of osmolytes in different extracts.

Quaternary amines cannot be detected with amine-reactive reagents and so they must be analyzed separately, if refractive index detection is not available. Solutes such as betaine can be precipitated as periodides (Grieve and Grattan, 1983) or reineckates (Greene, 1962), and then dissolved in organic solvent for measuring absorbance of the complex salt. These colorimetric assays are general assays for any quaternary amines and cannot discriminate between different water-soluble compounds (for example, betaine and choline would both react). However, if one is just monitoring betaine concentration in extracts, this is a useful protocol. The periodide salt method is easy and relatively rapid (Grieve and Grattan, 1983). Samples are acidified with 2 N H_2SO_4 (1:1 dilution) to which cold $KI-I_2$ is added (40 µl per µl acidified solution). After vigorous mixing, the sample is incubated at 4°C to precipitate the periodide crystals. After centrifugation and removal of the supernatant, the crystals are dissolved in dichloroethane and the absorbance measured at 365 nm. However, with both of these techniques, the detection method is very specific and will miss other classes of molecules (e.g. polyols, carbohydrates, solutes with modified amino groups).

Polyols as well as amino acids can be monitored using HPLC with refractive index detection (Kets et al., 1996). Again, the key is to have access to appropriate standards to calibrate the columns. For example, HPLC using a Lichrosper 100 NH_2 column with potassium phosphate (pH 7) and acetonitrile (20:80 v/v) was used to quantify mannitol and amino acids (glutamate and NAGGN) solutes in *Pseudomonas putida* (Kets et al., 1996). To identify individual polyols, Kets et al. (1994) ran a series of known polyols on an Animex HPX-87C column with distilled water as the eluent.

Recently, pulse amperometric detection has been coupled with anion exchange chromatography to yield a very sensitive method that can detect osmolytes such as ectoine (Riis et al., 2003). The sensitivity is high enough that the technique can be used to screen colonies on agar for solutes. The disadvantage is that the method has to be optimized for each particular molecule.

Solid-state NMR

Early investigations of N_2 assimilation in methanogens used solid-state ^{15}N NMR to monitor incorporation of this isotope in lyophilized cell pellets (Belay *et al.*, 1988). Taking a cue from current strategies in metabolomics (Griffin, 2003), one could use high-resolution magic angle sample spinning methods to observe ^{1}H spectra of intact cell pellets harvested under different conditions. Since osmolytes occur at such high intracellular concentrations, they should be visible along with all the macromolecules in the cells. Solid-state NMR analogs to COSY and NOESY exist, and could be used to confirm identities of discrete resonances in these extracts. Spectra could also be taken of just the pelleted debris from ethanol extraction (this might show interesting changes in other cell components). However, this is an area yet to be explored for microorganisms.

References

Andresen, P. A., Kaasen, I., Styrvold, O. B., Boulnois, G. and Strom, A. R. (1988). Molecular cloning, physical mapping and expression of the *bet* genes governing the osmoregulatory choline-glycine betaine pathway of *Escherichia coli*. *J. Gen. Microbiol.* **134**, 1737–1746.

Bax, A., Ikura, M., Kay, L. E., Torchia, D. A. and Tschudin, R. (1990). Comparison of different modes of two-dimensional reverse-correlation NMR for the study of proteins. *J. Magn. Reson.* **86**, 304–318.

Belay, N., Sparling, R., Choi, B. S., Roberts, M. F., Roberts, J. E. and Daniels, L. (1988). Physiological and ^{15}N NMR analysis of molecular nitrogen fixation by *Methanococcus thermolithotrophicus*, *Methanobacterium bryantii*, and *Methanospirillum hungatei*. *Biochim. Biophys. Acta* **971**, 233–245.

Bolen, D. W. and Baskakov, I. V. (2001). The osmophobic effect: natural selection of a thermodynamic force in protein folding. *J. Mol. Biol.* **310**, 955–963.

Borowitzka, L. J. and Brown, A. D. (1974). The salt relations of marine and halophilic species of the unicellular green alga, *Dunaliella*. The role of glycerol as a compatible solute. *Arch. Mikrobiol.* **96**, 37–52.

Brown, A. D. (1976). Microbial water stress. *Bacteriol. Rev.* **40**, 803–846.

Butler, S. L. and Falke, J. J. (1996). Effects of protein stabilizing agents on thermal backbone motions: a disulfide trapping study. *Biochemistry* **35**, 40595–40600.

Canovas, D., Vargas, C., Calderon, M. I., Ventosa, A. and Nieto, J. J. (1998a). Characterization of the genes for the biosynthesis of the compatible solute ectoine in the moderately halophilic bacterium *Halomonas elongata* DSM 3043. *System Appl. Microbiol.* **21**, 487–497.

Canovas, D., Vargas, C., Csonka, L. N., Ventosa, A. and Nieto, J. J. (1998b). Synthesis of glycine betaine from exogenous choline in the moderately halophilic bacterium *Halomonas elongata*. *Appl. Environ. Microbiol.* **64**, 4095–4097.

Canovas, D., Vargas, C., Iglesias-Guerra, F., Csonka, L. N., Rhodes, D., Ventosa, A. and Nieto, J. J. (1997). Isolation and characterization of salt-sensitive mutants of the moderate halophile *Halomonas elongata* and cloning of the ectoine synthesis genes. *J. Biol. Chem.* **272**, 25794–25801.

Canovas, D., Borges, N., Vargas, C., Ventosa, A., Nieto, J. J. and Santos, H. (1999). Role of Nγ-acetyldiaminobutyrate as an enzyme stabilizer and an intermediate in the biosynthesis of hydroxyectoine. *Appl. Environ. Microbiol.* **65**, 3774–3779.

Chen, L., Spiliotis, E. T. and Roberts, M. F. (1998). Biosynthesis of di-*myo*-inositol-1,1′-phosphate, a novel osmolyte in hyperthermophilic archaea. *J. Bacteriol.* **180**, 3785–3792.

Ciulla, R. A. and Roberts, M. F. (1999). Effects of osmotic stress on *Methanococcus thermolithotrophicus*: ^{13}C-edited ^{1}H-NMR studies of osmolyte turnover. *Biochim. Biophys. Acta* **1427**, 193–204.

Ciulla, R., Burggraf, S., Stetter, K. O. and Roberts, M. F. (1994). Occurrence and role of di-*myo*-inositol-1,1′-phosphate in *Methanococcus igneus*. *Appl. Environ. Microbiol.* **60**, 3660–3664.

da Costa, M. S., Santos, H. and Galinski, E. A. (1998). An overview of the role and diversity of compatible solutes in Bacteria and Archaea. *Adv. Biochem. Eng. Biotechnol.* **61**, 117–153.

DasSarma, S. and Arora, P. (2002). Halophiles. In *Encyclopedia of Life Sciences*. Vol. **8**, pp. 458–466. Nature Publishing Group, London.

Desmarais, D., Jablonski, P. E., Fedarko, N. S. and Roberts, M. F. (1997). 2-Sulfotrehalose, a novel osmolyte in haloalkaliphilic archaea. *J. Bacteriol.* **179**, 3146–3153.

Diamant, S., Eliahu, N., Rosenthal, D. and Goloubinoff, P. (2001). Chemical chaperones regulate molecular chaperones *in vitro* and in cells under combined salt and heat stresses. *J. Biol. Chem.* **276**, 39586–39591.

Doronina, N. V., Trotsenko, Y. A. and Tourova, T. P. (2000). *Methylarcula marina* gen. nov., sp. nov. and *Methylarcula terricola* sp. nov.: novel aerobic, moderately halophilic, facultatively methylotrophic bacteria from coastal saline environments. *Int. J. Syst. Evol. Microbiol.* **50**, 1849–1859.

Doronina, N. V., Darmaeva, T. D. and Trotsenko, Y. A. (2003). *Methylophaga alcalica* sp. nov., a novel alkaliphilic and moderately halophilic, obligately methylotrophic bacterium from an East Mongolian saline soda lake. *Int. J. Syst. Evol. Microbiol.* **53**, 223–229.

Fan, T. W.-M. (1997). Metabolite profiling by one and two-dimensional NMR analysis of complex mixtures. *Progr. Nucl. Magn. Res. Spec.* **28**, 161–219.

Faria, T. Q., Lima, J. C., Bastos, M., Macanita, A. L. and Santos, H. (2004). Protein stabilization by osmolytes from hyperthermophiles: effect of mannosylglycerate on the thermal unfolding of recombinant nuclease a from *Staphylococcus aureus* studied by picosecond time-resolved fluorescence and calorimetry. *J. Biol. Chem.* **279**, 48680–48691.

Foord, R. L. and Leatherbarrow, R. J. (1998). Effects of osmolytes on the exchange rates of backbone amide hydrogens in proteins. *Biochemistry* **37**, 2969–2978.

Galinski, E. A. (1995). Osmoadaptation in bacteria. *Adv. Microb. Physiol.* **37**, 272–328.

Galinski, E. A. and Herzog, R. M. (1990). The role of trehalose as a substitute for nitrogen-containing compatible solutes. *Arch. Microbiol.* **153**, 607–613.

Galinski, E. A. and Oren, A. (1991). Isolation and structure determination of a novel compatible solute from the moderately halophilic purple sulfur bacterium *Ectothiorhodospira marismortui*. *Eur. J. Biochem.* **198**, 593–598.

Galinski, E. A., Pfeiffer, H. P. and Trüper, H. G. (1985). 1,4,5,6-Tetrahydro-2-methyl-4-pyrimidinecarboxylic acid. A novel cyclic amino acid from halophilic phototrophic bacteria of the genus *Ectothiorhodospira*. *Eur. J. Biochem.* **149**, 135–139.

Goude, R., Renaud, S., Bonnassie, S., Bernard, T. and Blanco, C. (2004). Glutamine, glutamate, α-glucosylglycerate are the major osmotic solutes accumulated by *Erwinia chrysanthemi*. *Appl. Environ. Microbiol.* **70**, 6535–6541.

Graupner, M., Xu, H. and White, R. H. (2005). Biosynthesis of β-glutamate in *Methanocaldococcus jannaschii*. *J. Bacteriol.* Submitted for publication.

Greene, R. C. (1962). Biosynthesis of dimethyl-β-propiothetin. *J. Biol. Chem.* **237**, 2251–2254.

Grieve, C. M. and Grattan, S. R. (1983). Rapid assay for determination of water soluble quaternary ammonium compounds. *Plant and Soil* **70**, 303–307.

Griffin, J. L. (2003). Metabonomics: NMR spectroscopy and pattern recognitions analysis of body fluids and tissues for characterization of xenobiotic toxicity and disease diagnosis. *Curr. Opin. Chem. Biol.* **7**, 648–654.

Grivet, J.-P., Delort, A.-M. and Portais, J.-C. (2003). NMR and microbiology: from physiology to metabolomics. *Biochimie* **85**, 823–840.

Imhoff, J. F. and Rodriguez-Valera, F. (1984). Betaine is the main compatible solute of halophilic eubacteria. *J. Bacteriol.* **160**, 478–479.

Joo, W. A. and Kim, C. W. (2005). Proteomics of halophilic archaea. *J. Chromatogr. B* **815**, 237–250.

Kets, E. P. W., Galinski, E. A. and de Bont, J. A. M. (1994). Carnitine: a novel compatible solute in *Lactobacillus plantarum*. *Arch. Microbiol.* **162**, 243–248.

Kets, E. P. W., Galinski, E. A., de Wit, M., de Bont, J. A. M. and Heipieper, H. J. (1996). Mannitol, a novel bacterial compatible solute in *Pseudomonas putida* S12. *J. Bacteriol.* **178**, 6665–6670.

Kim, Y. S., Jones, L. S., Dong, A., Kendrick, B. S., Chang, B. S., Manning, M. C., Randolph, T. W. and Carpenter, J. F. (2003). Effects of sucrose on conformational equilibria and fluctuations within the native-site ensemble of proteins. *Prot. Sci.* **12**, 1252–1261.

Kraegeloh, A., Amendt, B. and Kunte, H. J. (2004). Potassium transport in a halophilic member of the *Bacteria* domain: identification and characterization of the K^+ uptake systems TrkH and TrkI from *Halomonas elongata* DSM 2581^T. *J. Bacteriol.* **187**, 1036–1043.

Kuhlmann, A. U. and Bremer, E. (2002). Osmotically regulated synthesis of the compatible solute ectoine in *Bacillus pasteurii* and related *Bacillus* spp. *Appl. Environ. Microbiol.* **68**, 772–783.

Kunte, H. J., Galinski, E. A. and Trüper, H. G. (1993). A modified FMOC-method for the detection of amino acid-type osmolytes and tetrahydropyrimidines (ectoines). *J. Microbiol. Meth.* **17**, 129–136.

Lai, M. C., Sowers, K. R., Robertson, D. E., Roberts, M. F. and Gunsalus, R. P. (1991). Distribution of compatible solutes in the halophilic methanogenic archaebacteria. *J. Bacteriol.* 5352–5358.

Lai, M.-C., Yang, D.-R. and Chuang, M.-J. (1999). Regulatory factors associated with synthesis of the osmolyte glycine betaine in the halophilic methanoarchaeon *Methanohalophilus portucalensis*. *Appl. Environ. Microbiol.* **65**, 828–833.

Lai, M.-C., Wang, C.-C., Chuang, M.-J., Wu, Y.-C. and Lee, Y.-C. (2005). Purification, partial characterization, substrate specificity, and regulation by potassium of a novel glycine sarcosine dimethylglycine N-methyltransferase from a halophilic methanoarchaeon. *Appl. Environ. Microbiol.*, submitted for publication.

Lamosa, P., Martins, L. O., da Costa, M. S. and Santos, H. (1998). Effects of temperature, salinity, and medium composition on compatible solute accumulation by *Thermococcus* spp. *Appl. Environ. Microbiol.* **64**, 3591–3598.

Liu, Y. and Bolen, D. W. (1995). The peptide backbone plays a dominant role in protein stabilization by naturally occurring osmolytes. *Biochemistry* **34**, 12884–12891.

Louis, P. and Galinski, E. A. (1997). Characterization of genes for the biosynthesis of the compatible solute ectoine from *Marinococcus halophilus* and osmoregulated expression in *Escherichia coli*. *Microbiology* **143**, 1141–1149.

Martin, D. D., Ciulla, R. A. and Roberts, M. F. (1999). Osmoadaptation in archaea. *Appl. Environ. Microbiol.* **65**, 1815–1825.

Martin, D. D., Ciulla, R. A., Robinson, P. M. and Roberts, M. F. (2001). Switching osmolyte strategies: response of *Methanococcus thermolithotrophicus* to changes in external NaCl. *Biochim. Biophys. Acta* **1524**, 1–10.

Martin, D. D., Bartlett, D. H. and Roberts, M. F. (2002). Solute accumulation in the deep-sea bacterium *Photobacterium profundum*. *Extremophiles* **6**, 507–514.

Martins, L. O., Carreto, L. S., da Costa, M. S. and Santos, H. (1996). New compatible solutes related to di-*myo*-inositol-phosphate in members of the order *Thermotogales*. *J. Bacteriol.* **178**, 5644–5651.

Martins, L. O., Huber, R., Huber, H., Stetter, K. O., da Costa, M. S. and Santos, H. (1997). Organic solutes in hyperthermophilic Archaea. *Appl. Environ. Microbiol.* **63**, 896–902.

Motta, A., Romano, I. and Gambacorta, A. (2004). Rapid and sensitive NMR method for osmolyte determination. *J. Microbiol. Meth.* **58**, 289–294.

Müller, V. and Oren, A. (2003). Metabolism of chloride in halophilic prokaryotes. *Extremophiles* **7**, 261–266.

Norwood, T. J., Boyd, J., Heritage, J. E., Soffe, N. and Campbell, I. D. (1990). Comparison of techniques for ^{1}H-detected heteronuclear ^{1}H-^{15}N spectroscopy. *J. Magn. Reson.* **87**, 488–501.

Nyyssölä, A., Kerovuo, J., Kaukinen, P., von Weymarn, N. and Reinikaiuem, T. (2000). Extreme halophiles synthesize betaine from glycine by methylation. *J. Biol. Chem.* **275**, 22196–22201.

Nyyssölä, A. and Leisola, M. (2001). *Actinopolyspora halophila* has two separate pathways for betaine synthesis. *Arch. Microbiol.* **176**, 294–300.

Nyyssölä, A., Reinikaiuem, T. and Leisola, M. (2001). Characterization of glycine sarcosine *N*-methyltransferase and sarcosine dimethylglycine *N*-methyltransferase. *Appl. Environ. Microbiol.* **67**, 2044–2050.

Onraedt, A., Walcarius, B., Soetaert, W. and Vandamme, E. J. (2003). Dynamics and optimal conditions of intracellular ectoine accumulation in *Brevibacterium* sp. *Commun. Agric. Appl. Biol. Sci.* **68**, 241–246.

Oren, A. (2002). Diversity of halophilic microorganisms: environments, phylogeny, physiology, and applications. *J. Ind. Microbiol. Biotechnol.* **28**, 56–63.

Oren, A., Heldal, M., Norland, S. and Galinski, E. A. (2002). Intracellular ion and organic solute concentrations of the extremely halophilic Bacterium *Salinibacter ruber*. *Extremophiles* **6**, 491–498.

Peters, P., Galinski, E. A. and Trüper, H. G. (1990). The biosynthesis of ectoine. *FEMS Microbiol. Lett.* **71**, 157–162.

Pflüger, K., Baumann, S., Gottschalk, G., Lin, W., Santos, H. and Müller, V. (2003). Lysine-2,3-aminomutase and β-lysine acetyltransferase genes of methanogenic archaea are salt induced and are essential for the biosynthesis Nε-acetyl-β-lysine and growth in high salinity. *Appl. Environ. Microbiol.* **69**, 6047–6055.

Regev, R., Peri, I., Gilboa, H. and Avi-Dor, Y. (1990). ^{13}C NMR study of the interrelation between synthesis and uptake of compatible solutes in two moderately halophilic eubacteria, Bacterium Ba1 and *Vibro costicola*. *Arch. Biochem. Biophys.* **278**, 106–112.

Riis, V., Maskow, T. and Babel, W. (2003). Highly sensitive determination of ectoine and other compatible solutes by anion-exchange chromatography and pulsed amperometric detection. *Anal. Bioanal. Chem.* **377**, 203–207.

Roberts, M. F. (2004). Osmoadaptation and osmoregulation in archaea: update 2004. *Front. Biosci.* **9**, 1999–2019.

Roberts, M. F., Choi, B. S., Robertson, D. E. and Lesage, S. (1990). Free amino acid turnover in methanogens measured by ^{15}N NMR spectroscopy. *J. Biol. Chem.* **265**, 18207–18212.

Roberts, M. F., Lai, M. C. and Gunsalus, R. P. (1992). Biosynthetic pathway of the osmolytes Nε-acetyl-β-lysine, β-glutamine, and betaine in *Methanohalophilus* strain FDF1 suggested by nuclear magnetic resonance analyses. *J. Bacteriol.* **174**, 6688–6693.

Robertson, D. E., Lesage, S. and Roberts, M. F. (1989). β-Aminoglutaric acid is a major soluble component of *Methanococcus thermolithotrophicus*. *Biochim. Biophys. Acta* **992**, 320–326.

Robertson, D. E., Noll, D., Roberts, M. F., Menaia, J. A. and Boone, D. R. (1990a). Detection of the osmoregulator betaine in methanogens. *Appl. Environ. Microbiol.* **56**, 563–565.

Robertson, D. E., Roberts, M. F., Belay, N., Stetter, K. O. and Boone, D. R. (1990b). Occurrence of β-glutamate, a novel osmolyte, in marine methanogenic bacteria. *Appl. Environ. Microbiol.* **56**, 1504–1508.

Robertson, D. E., Lai, M.-C., Gunsalus, R. P. and Roberts, M. F. (1992a). Composition, variation, and dynamics of major compatible solutes in *Methanohalophilus* strain FDF1. *Appl. Environ. Microbiol.* **58**, 2438–2443.

Robertson, D. E., Noll, D. and Roberts, M. F. (1992b). Free amino acid dynamics in marine methanogens. β-Amino acids as compatible solutes. *J. Biol. Chem.* **267**, 14893–14901.

Robinson, P. M. and Roberts, M. F. (1997). Effects of osmolyte precursors on the distribution of compatible solutes in *Methanohalophilus portucalensis*. *Appl. Environ. Microbiol.* **63**, 4032–4038.

Robinson, P., Neelon, K., Schreier, H. J. and Roberts, M. F. (2001). β-Glutamate as a substrate for glutamine synthetase. *Appl. Environ. Microbiol.* **67**, 4458–4463.

Roder, A., Hoffmann, E., Hagemann, M. and Berg, G. (2005). Synthesis of the compatible solutes glucosylglycerol and trehalose by salt-stressed cells of *Stenotrophomonas* strains. *FEMS Microbiol. Lett.* **243**, 219–226.

Rozwadowski, K. L., Khachatourians, G. G. and Selvaraj, G. (1991). Choline oxidase, a catabolic enzyme in *Arthrobacter pascens*, facilitates adaptation to osmotic stress in *Escherichia coli*. *J. Bacteriol.* **173**, 472–478.

Santos, H. and da Costa, M. S. (2002). Compatible solutes of organisms that live in hot saline environments. *Environ. Microbiol.* **4**, 501–509.

Scholz, S., Sonnenbichler, J., Schafer, W. and Hensel, R. (1992). Di-*myo*-inositol-1,1'-phosphate: a new inositol phosphate isolated from *Pyrococcus woesei*. *FEBS Lett.* **306**, 239–242.

Silva, Z., Borges, N., Martins, L. O., Wait, R., da Costa, M. S. and Santos, H. (1999). Combined effect of the growth temperature and salinity of the medium on the accumulation of compatible solutes by *Rhodothermus marinus* and *Rhodothermus obamensis*. *Extremophiles* **3**, 163–172.

Smith, L. T. and Smith, G. M. (1989). An osmoregulated dipeptide in stressed *Rhizobium meliloti*. *J. Bacteriol.* **171**, 4714–4717.

Sowers, K. R., Robertson, D. E., Noll, D., Gunsalus, R. P. and Roberts, M. F. (1990). Nε-acetyl-β-lysine: an osmolyte synthesized by methanogenic bacteria. *Proc. Natl. Acad. Sci. USA* **87**, 9083–9087.

Sowers, K. R. and Gunsalus, R. P. (1995). Halotolerance in *Methanosarcina* spp.: role of Nε-acetyl-β-lysine, β-glutamate, glycine betaine and K^+ as compatible solutes for osmotic adaptation. *Appl. Environ. Microbiol.* **61**, 4382–4388.

Timasheff, S. N. (2002). Protein-solvent preferential interactions, protein hydration, and the modulation of biochemical reactions by solvent components. *Proc. Natl. Acad. Sci. USA* **99**, 9721–9726.

Vermeulen, V. and Kunte, H. J. (2004). *Marinococcus halophilus* DSM 20408T encodes two transporters for compatible solutes belonging to the betaine-carnitine-choline transporter family: identification and characterization of ectoine transporter EctM and glycine betaine transporter BetM. *Extremophiles* **8**, 175–184.

Volwerk, J. J., Shashidhar, M. S., Kuppe, A. and Griffith, O. H. (1990). Phosphatidylinositol-specific phospholipase C from *Bacillus cereus* combines intrinsic phosphotransferase and cyclic phosphodiesterase activities: a ^{31}P NMR study. *Biochemistry* **29**, 8056–8062.

27 Genetic Systems for Halophilic Archaea

Brian R Berquist, Jochen A Müller and Shiladitya DasSarma
University of Maryland Biotechnology Institute, Center of Marine Biotechnology, Baltimore, MD 21202, USA

CONTENTS

Introduction
Haloarchaeal model systems
Genetic methods
Genomic approaches
Conclusions and future prospects

◆◆◆◆◆ INTRODUCTION

Halophilic archaea (haloarchaea) grow optimally in hypersaline environments containing 2–5 M NaCl, usually under aerobic conditions, with normal atmospheric temperature and pressure (DasSarma and Arora, 2002; Grant and Larsen, 1984). In contrast to other Archaea, which require specialized culturing equipment, haloarchaea can be grown in any microbiology laboratory using standard media supplemented with salt. As a result, many methods for genetic manipulation of haloarchaea have been developed and standardized over the last twenty years (see DasSarma and Fleischmann, 1995). With the recent availability of complete or nearly complete genome sequences for several haloarchaea (http://halo.umbi.umd.edu), these microorganisms have come to represent highly tractable model systems for genetic studies.

Haloarchaea are found in evaporatic brine pools in tropical, temperate and Antarctic regions of the world, usually at the surface, but also under the sea in submarine pools and underground in ancient salt deposits (DasSarma and Arora, 2002). Many hypersaline environments are dynamic with respect to temperature, ionic strength, pH, oxygen and sunlight. Not surprisingly, haloarchaea display a wide variety of biological responses (DasSarma, 2004; Kennedy *et al.*, 2001). To balance the high sodium chloride concentration in the medium, haloarchaea accumulate potassium chloride as a compatible solute, and all of their metabolic activities take place in a hypersaline cytoplasm. Haloarchaea also exhibit a high level of resistance to ultraviolet light as protection against the intense solar radiation usually found in their environment. Many species are highly motile, displaying phototaxis, chemotaxis and gas vesicle-mediated flotation. Some are capable of phototrophic growth, and most have facultative anaerobic capabilities.

All haloarchaea described thus far belong to the *Halobacteriaceae* family (Grant and Larsen, 1984). Of the 18 different genera of haloarchaea that have been described as of September 2004, genetic and genomic analysis has been concentrated on a few species and strains. The most intensively studied model systems, *Halobacterium* sp. NRC-1 and related strains, are well known for their production of purple membrane and gas vesicles, and their chemo- and phototactic capabilities. *Haloferax volcanii* is a system that has been of interest because it can grow on minimal medium. *Haloarcula marismortui* is also of growing interest as a result of the availability of the X-ray crystal structure of its large ribosomal subunit (Ban *et al.*, 2000). However, only *Halobacterium* spp. and *Hfx. volcanii* have well-developed genetic tools, including a facile transformation system, selectable markers, cloning and expression vectors, reporter genes, and gene replacement and knockout systems (DasSarma and Fleischmann, 1995).

Genome sequences are available for several common laboratory strains of haloarchaea. The first complete genome sequence was published in October 2000 for the model organism, *Halobacterium* sp. NRC-1 (Ng *et al.*, 1998, 2000). The genome sequences of *Har. marismortui* (Baliga *et al.*, 2004a) and *Hfx. volcanii* are also complete and nearly complete, respectively (Table 27.1). The genome of *Halobacterium* sp. NRC-1 is 2.57 Mbp in size, while the other two genomes are larger, over 4 Mbp, and harbor many paralogous genes. The genomes of these haloarchaea are relatively GC-rich (58–68%) and contain large and small extrachromosomal elements. The extrachromosomal replicons are rich in IS elements and rearrange at high frequency. This is in contrast to their chromosomes, which are quite stable, and in fact contain a smaller number of transposable IS elements than some other archaea, e.g. *Sulfolobus solfataricus* (DasSarma, 2004).

The combination of easy culturing, highly developed genetic and genomic tools, as well as interesting biology, makes haloarchaeal systems attractive for study. Past studies have been extremely fruitful, providing a deeper understanding of prokaryotic genome structure and dynamics,

Table 27.1 Haloarchaeal genomic sequences

Archaeal strain	Website	Reference
Halobacterium sp. NRC-1 (ATCC 700922)	http://halo.umbi.umd.edu/	Ng *et al.* (2000)
Haloferax volcanii DS2 (ATCC 29605)	http://www.tigr.org	Unpublished
Haloarcula marismortui (ATCC 43049)	http://halo.systemsbiology.net/	Baliga *et al.* (2004a)

lateral gene transfer, and DNA replication, transcription and translation. With the recent availability of complete genome sequences, systematic knockout strategies, DNA microarrays and proteomic approaches, haloarchaeal model systems now have the potential for greatly advancing microbiology, molecular biology and biotechnology.

◆◆◆◆◆ HALOARCHAEAL MODEL SYSTEMS

Halobacterium sp. NRC-1

Halobacterium species NRC-1 (ATCC 700922) is a rod-shaped strain, which grows well in complex media containing 4.3 M NaCl, with peptone, yeast extract, or casamino acids as carbon and energy sources (Figure 27.1). Optimal growth (~ 6 h generation time) is observed at 42°C. Growth on defined medium containing 15 amino acids and vitamins has also been reported (Grey and Fitt, 1976) and cell lysis is minimized in the presence of trace quantities of transition metal ions. Under low oxygen tension, *Halobacterium* sp. NRC-1 induces patches of purple membrane (containing bacteriorhodopsin) in the cell envelope and buoyant gas vesicles intracellularly. In combination, these features enable light

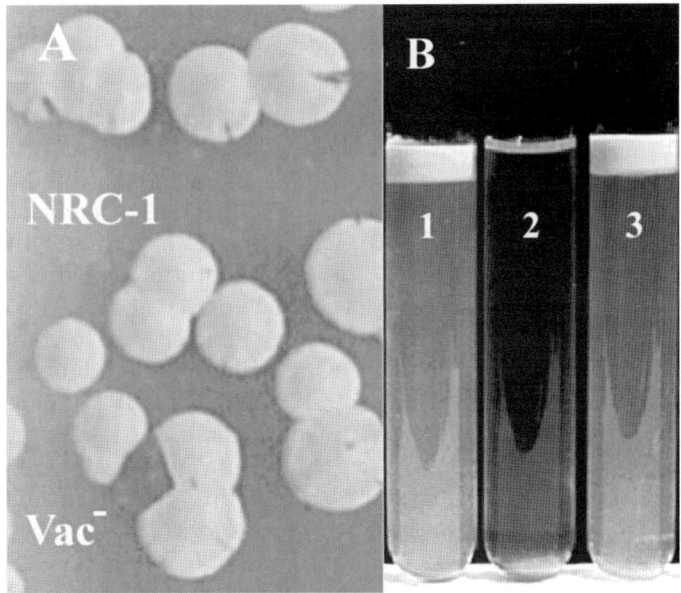

Figure 27.1. *Halobacterium* sp. NRC-1. (A) Colonies of the model haloarchaeal strain *Halobacterium* sp. NRC-1. Among pink wild-type (NRC-1) colonies, an orange gas vesicle-minus (Vac$^-$) mutant is visible. (B) Liquid cultures of NRC-1 (tube 1), Vac$^-$ mutant SD109 (tube 2) and a SD109 transformant containing the gas vesicle gene cluster on pFL2 (tube 3). Vac$^+$ floating cells at the top of tubes 1 and 3 and Vac$^-$ non-floating cells at the bottom of the tube 2 are visible. The meniscus is visible at the top of tube 2. See Plate 27.1 in Colour Plate Section.

harvesting for ATP synthesis in the photic zone of the water column (Baliga et al., 2001).

One of the most useful derivative strains of *Halobacterium* sp. NRC-1 is an isogenic in-frame deletant of the *ura3/pyrF* gene (NRC-1 (Δura3)), which is auxotrophic for uracil and 5-fluoroorotic acid-resistant (Ura$^-$, Foa^r) (Peck et al., 2000). The NRC-1 (Δura3) strain is used for generating gene knockouts using either selection for mevinolin-resistance (Mev^r) and counterselection against *ura3* (Foa^r) or both selection and counterselection for *ura3* (Lam and Doolittle, 1992; Peck et al., 2000). A number of single and multiple gene mutants have been successfully constructed using these procedures (e.g. Δbrp, Δblh, Δbrp/blh and ΔcrtY, retinal biosynthesis mutants (Peck et al., 2001, 2002); Δphr1, photolyase mutant (McCready and Marcello, 2003); ΔarsADRC, ΔarsB and ΔarsM, arsenic resistance mutants (Wang et al., 2004); and ΔdmsR, ΔdmsA and ΔdmsD, DMSO and TMAO utilization mutants (Müller and DasSarma, 2005), indicating that knockout of any non-essential gene is possible. A large number of studies have been conducted on the *bop* gene encoding the purple membrane protein, bacteriorhodopsin (BR). It should be noted that the commonly used BR overproducer (Pum^{+++}) strain S9 is not a derivative of or isogenic with NRC-1. It was generated from another natural *Halobacterium* sp., strain R1, by extensive chemical mutagenesis (Stoeckenius et al., 1979). Strain S9 likely contains dozens of mutations and allelic differences from NRC-1. The mutations responsible for the Pum^{+++} (deep purple) phenotype of S9 have been shown to be a double frame shift in the bacterio-opsin activator gene, *bat* (*bat** allele), which results in a 4 amino acid change in a putative oxygen/redox sensing domain in the encoded protein. Mutants of S9 lacking BR, e.g. SD20, which is purple membrane and bacterioruberin-deficient (Pum$^-$, Rub$^-$) as a result of an ISH1 insertion in *bat*, and SD23, which is Pum$^-$ as a result of an ISH1 insertion in *bop*, have been extensively characterized (Baliga et al., 2001; DasSarma, 1989). Other strains lacking BR have also been isolated and characterized (e.g. R1mR and L33) (DasSarma et al., 1983, 1984) (Table 27.2).

A large number of genetic studies have also been conducted on *gvp* genes using a gas vesicle-deficient (Vac$^-$) NRC-1 strain derivative deleted for the pNRC100 *gvp* gene cluster, in combination with a shuttle plasmid (pFL2) harboring either the wild type or mutated *gvp* gene cluster (DasSarma et al., 1994; Ng et al., 1994). Three classes of gas vesicle mutant strains were distinguishable: Class I mutants were extremely unstable, and gave rise to some progeny that were Vac$^+$ (1–5%) and others that were Vac$^-$. As result of this genetic instability, sectored colonies (Vac$^{\delta-}$) containing both Vac$^+$ and Vac$^-$ cells occurred. Class II mutants were either Vac$^{\delta-}$ or Vac$^-$ but stable with a reversion rate to Vac$^+$ at <1%. Class III mutants were stable Vac$^-$ derivatives (DasSarma et al., 1988). Class I strains characteristically contain heterogeneous populations of extrachromosomal replicons which include the wild-type pNRC100 containing all genes required for wild-type gas vesicle synthesis (*gvp* D, E, F, G, H, I, J, K, L, M, A, C, N, O), as well as at least one copy of a deleted variant of pNRC100 lacking the *gvp* gene cluster. Therefore Vac$^+$ and Vac$^-$ progeny strains are obtained through segregation of the heterogeneous

Table 27.2 Selected haloarchaeal strain characteristics and sources[a]

Haloarchaeal strains	Characteristics	Reference
Halobacterium sp. NRC-1	Wild-type ATCC 700922, JCM 11081	Ng et al. (2000)
Halobacterium sp. SK400/MPK414	NRC-1 $\Delta ura3$	Peck et al. (2000); Wang et al. (2004)
Halobacterium sp. S9	Purple membrane overproducer	Stoeckenius et al. (1979)
Halobacterium sp. R1	Gas vesicle-deficient mutant ATCC 19700	Stoeckenius et al. (1979)
Halobacterium sp. SD20	Purple membrane deficient mutant of S9 bat::ISH1	Baliga et al. (2001)
Halobacterium sp. SD23	Purple membrane deficient mutant of S9 bop::ISH1	Baliga et al. (2001)
Halobacterium sp. SD109	NRC-1 with 67 kb deletion of gvp gene cluster	Ng et al. (1994)
Halobacterium sp. SD112	NRC-1 with 59 kb deletion of gvp gene cluster	Ng et al. (1994)
Halobacterium sp. R1Mr	Purple membrane mutant of R1 bop::ISH2	DasSarma (1989)
Halobacterium sp. L33	Purple membrane mutant of S9 bop::ISH2	DasSarma (1989)
Haloferax volcanii DS2	Wild-type ATCC 29605, DSM 3757, JCM 8879	Charlebois et al. (1991)
Haloferax volcanii WFD11	DS2 cured of pHV2	Lam and Doolittle, (1989)
Haloferax volcanii WR479	WR341 $\Delta pyrE1 \Delta pyrE2$	Bitan-Banin et al. (2003)
Haloferax volcanii DS70	DS2 cured of pHV2	Wendoloski et al. (2001)
Haloferax volcanii H26	DS70 $\Delta pyrE2$	Allers and Ngo (2003)

[a]ATCC, American Type Culture Collection (http://www.atcc.org/); DSM = DSMZ, Deutsche Sammlung von Mikroorganismen und Zellkulturen GmbH (http://www.dsmz.de/); JCM, Japan Collection of Microorganisms (http://www.jcm.riken.go.jp/).

extrachromosomal replicon population. Class II mutant strains, including strain R1, appear to have lost gas vesicle production due to IS element insertion into regions of the *gvp* gene cluster. Class III mutant strains contain large deletions of pNRC100, including the *gvp* gene cluster, likely through intramolecular transposition of IS elements. Two class III mutants, strain SD109 and SD112, are well-characterized isogenic derivatives of strain NRC-1. They contain large (67 or 59 kbp, respectively) deletions of pNRC100, as a result of intramolecular transposition of ISH elements resulting in deletion of the entire *gvp* gene cluster (Ng et al., 1991, 1994).

Another series of genetic experiments have been carried out to identify the origins of replication in *Halobacterium* sp. NRC-1

(Berquist and DasSarma, 2003; Ng and DasSarma, 1993). Using a non-replicating plasmid containing the selectable marker *mev*, conferring resistance to mevinolin (Mev^r), two autonomous replication sequences were isolated, one from a common region of pNRC100 and 200, and another from the large chromosome. The pNRC replication region was analyzed using deletion and insertion mutagenesis, and showed the requirement of a gene, *rep*H, and its upstream region containing an AT-rich region with short repeats for autonomous replication. The chromosomal replication region required the *orc*7 gene, one of 10 *orc/cdc*6 eukaryotic-type DNA replication initiation genes in the genome, plus 750 bp upstream of the *orc*7 gene translational start. Within that 750 bp is a large inverted repeat which flanks an extremely AT-rich stretch of DNA. The large inverted repeat is highly conserved in nucleotide sequence upstream of *orc*7 orthologs in the genome sequences of *Hfx. volcanii* and *Har. marismortui*.

Haloferax volcanii

Haloferax volcanii DS2 (ATCC29605), isolated from shore mud of the Dead Sea, is disk-or cup-shaped, and sometimes pleiomorphic. It grows optimally, with a generation time of ~4 h in rich medium containing 1.5–2.5 M NaCl at 45°C. It requires at least 0.02 M Mg^{2+} and is tolerant of high Mg^{2+} (1.5 M) concentrations reflecting the Mg^{2+}-rich composition of its natural environment. Three media formulations have been created for *Hfx. volcanii*: a high salt-rich medium containing yeast extract and tryptone, a low salt-rich medium containing yeast extract and peptone, and a minimal medium containing glycerol and sodium succinate (DasSarma and Fleischmann, 1995). *Hfx. volcanii* can also grow in minimal medium, utilizing glucose as a sole carbon and energy source, although in practice the growth rate is extremely slow. The *Hfx. volcanii* physical map showed a 4.2 Mbp genome with a 3 Mbp chromosome and four extrachromosomal replicons pHV4 (690 kb), pHV3 (440 kb), pHV1 (86 kb) and pHV2 (6kb) (Charlebois *et al.*, 1989).

Useful derivatives of strain DS2 include strain *Hfx. volcanii* WFD11, which has been cured of the natural cryptic plasmid pHV2 by ethidium bromide treatment (Charlebois *et al.*, 1987). Strain DS2 has been used as a host and several pHV2 derivatives are popular cloning and shuttle vectors. However, this strain contains an unstable pHV3 replicon leading to slow growth, filamentation and cell lysis in about 10% of cells (Wendoloski *et al.*, 2001). Similarly, strain DS70 was cured of the small plasmid pHV2 without the use of potent mutagens (Wendoloski *et al.*, 2001). Strain WR479, another DS2 derivative with deletions of both the *pyr*E1 and *pyr*E2 genes and an Ura$^-$ Foa^r phenotype (Bitan-Banin *et al.*, 2003), is used as a host for gene knockouts employing a method similar to that developed for *Halobacterium* sp. strain NRC-1. Strain H26, a Δ*pyr*E2 derivative of strain DS70, was constructed to facilitate construction of gene knockouts by the same methodology (Allers and Ngo, 2003). Additional derivatives of strain H26 have recently been constructed, yielding $\Delta pyrE2\Delta leuB$ (Ura$^-$, Leu$^-$) leucine auxotrophs and $\Delta pyrE2$ $\Delta trpA$ (Ura$^-$, Trp$^-$) tryptophan auxotrophs (Allers et al., 2004).

Media and Growth

Generally, culturing of halophiles is relatively simple and conducted as for common aerobic bacteria. Liquid cultures are grown at 37–42°C in test tubes, Erlenmeyer flasks or Fernbach flasks with shaking at 200–300 rpm. For phototrophic strains fluorescent lighting may be used, but is not essential. Plates contain 2% agar, which is required for solidifying under hypersaline conditions, and are placed in airtight containers or a humidified environment to avoid crystallization of salt.

Media preparation

Descriptions are provided here for rich media for the two major model systems, Halobacterium sp. NRC-1 and Hfx. volcanii.

Halobacterium sp. NRC-1 CM$^+$ complex media containing, per liter:

NaCl	250 g
MgSO$_4$ · 7H$_2$O	20 g
KCl	2 g
Sodium citrate	3 g
Oxoid peptone	10 g
Trace metal stock (2000X)	0.5 ml

Adjust pH to 7.2 with NaOH

Trace metal stock (2000X):

FeSO$_4$ · 7H$_2$O	3.50 mg ml^{-1}
ZnSO$_4$ · 7H$_2$O	0.88 mg ml^{-1}
MnSO$_4$ · H$_2$O	0.66 mg ml^{-1}
CuSO$_4$ · 5H$_2$O	0.02 mg ml^{-1}

Dissolve in 0.1 N HCl

Oxoid peptone should be used for optimum purple membrane production. Peptone may be substituted with 7.5 g casamino acids and 10 g yeast extract. Trace metal stock is filter sterilized and added to cooled autoclaved medium.

Halobacterium sp. NRC-1 uracil-dropout medium containing, per liter:

NaCl	250 g
MgSO$_4$ · 7H$_2$O	20 g
Sodium citrate	3 g
KCl	2 g
Nitrogen base	10 g
Dropout formula	1.92 g

Adjust pH to 7.0 with NaOH

Note: A precipitate will form when the pH approaches neutrality. Trace metal stock is filter sterilized and added to media cooled to 65°C or lower.

Haloferax volcani complex medium containing, per liter:

NaCl	206 g
$MgSO_4 \cdot 7H_2O$	37 g
KCl	3.7 g
Yeast extract	3 g
Tryptone	5 g
$CaCl_2 \cdot 2H_2O$ (10 %)	5.0 ml
$MnCl_2$ (75 mg ml^{-1})	1.7 ml
1 M Tris-HCl (pH 7.2)	50 ml

Notes: For agar plates, 20 g of Bacto-Agar per liter is added to the liquid medium. Autoclaving is conducted at 15 lb in.$^{-2}$ for 20 min. Medium should be cooled to 65°C before addition of antibiotics and/or trace metals.

Salts inhibit gelling of agar, making a relatively high concentration essential to obtain a solid surface. Salts also reduce surface tension of the media, making removal of gas bubbles more difficult. Extra care must be taken to avoid formation of bubbles when pouring plates since their removal by flaming is less effective. Another problem in culturing various halophiles arises from the contamination of some batches of peptone with bile acids, leading to growth inhibition and cell lysis. Synthesis of purple membrane is especially sensitive to bile acids.

Cultures may be stored at 4°C for weeks to several months without extreme loss of viability; although extended storage may result in accumulation of mutants from IS element transpositions and other DNA rearrangements. For long-term storage, cultures may be frozen at –70°C after addition of glycerol to 15–25% (v/v) in liquid medium. Detailed culturing conditions and growth media for haloarchaea may be found in *Archaea, A Laboratory Manual: Halophiles* (DasSarma and Fleischmann, 1995).

◆◆◆◆◆ GENETIC METHODS

Transformation

Halobacterium and *Haloferax* spp., and several other haloarchaea can be transformed using a procedure involving spheroplast formation through divalent cation (Mg^{2+}) chelation with EDTA, generating competent cells with PEG treatment, cell regeneration after DNA uptake, and plating (Cline and Doolittle, 1992; Cline et al., 1989a,b; DasSarma and Fleischmann, 1995). The routine efficiency of the procedure is about 10^5–10^6 transformants per µg of common shuttle plasmid DNA. Main procedural variations for transformation of different halophiles reflect specific ionic concentrations of the media. An alternative procedure for *Halobacterium* sp. has been reported using freeze–thaw of cells, but the efficiency is relatively low in comparison to the EDTA–PEG procedure (Zibat, 2001).

Most of the haloarchaeal species tested for transformation have been found to encode restriction/modification systems, which reduce the efficiency of transformation with DNA from a heterologous host by a factor of 10^3 or greater (DasSarma and Fleischmann, 1995). DNA modification is evident by the inability of some restriction enzymes to digest chromosomal DNA from haloarchaea, and corresponding haloarchaeal restriction-modification activities have been reported (Lam and Doolittle, 1989; Ng et al., 1994; Patterson and Pauling, 1985). Purification of transforming DNA from an *Escherichia coli* Dam$^+$ strain can reduce or eliminate the restriction problem.

Transformation method

1. Grow a 50-ml culture of *Halobacterium* sp. NRC-1 at 42°C to an OD_{600} of 1.0 in a flask with good aeration.
2. Gently pellet cells by centrifugation at $6000 \times g$ for 10 min and decant supernatant.
3. Gently resuspend cells in 1/10 volume (4 ml) of spheroplasting solution (2 M NaCl, 27 mM KCl, 50 mM Tris-HCl (pH 8.75), 15% (w/v) sucrose).
4. Add 1/20 volume (200 µl) of 0.5 M EDTA (pH 8.0) and swirl gently.
5. After a 10 min incubation at room temperature, examine cells in phase-contrast microscope to verify spheroplast formation, which is characterized by perfectly spherical cell morphology.
6. Add 10 µl of DNA in spheroplasting solution or 10 mM Tris-HCl, 1 mM EDTA to 200 µl of spheroplast suspension. Examine cell morphology by phase-contrast microscopy to confirm absence of extensive cell lysis. Cells should appear spherical or pleomorphic and the number of visible cells should not change substantially after addition of DNA. Cell lysis results in increased viscosity and greatly reduces transformation efficiency.
7. Add 200 µl of PEG solution (6 ml of PEG600 in 4-ml spheroplasting solution) and mix gently by tapping.
8. After a 20 min incubation, add 1 ml of spheroplast dilution medium (4.3 M NaCl, 27 mM KCl, 80 mM $MgSO_4$, 10 mM sodium citrate, 1.4 mM $CaCl_2$, 50 mM Tris-HCl (pH 7.4), 15% (w/v) sucrose, 1% (w/v) Oxoid Peptone) to the mixture and gently invert tubes to mix.
9. Centrifuge at $6000 \times g$ for 4 min and remove supernatant. Gently resuspend cell pellets in 1 ml of spheroplast dilution media, transfer to sterile test tubes, and incubate the tubes with shaking at 42°C overnight.
10. After overnight incubation, add 1 ml of CM$^+$ medium and grow with shaking overnight at 42°C.
11. Check cultures by phase-contrast microscopy for regeneration. When >80% of cells have returned to rod shape (generally 1–2 days after addition of CM$^+$ medium), inoculate transformants on selective plates by spreading and incubate at 42°C. Visible colonies develop over 5–10 days.

For suspension of cells and transformation, the ionic strength of the medium is critical to maintain integrity of spheroplasts. Even a small amout of lysis will result in significant reduction in transformation efficiency due to the release and precipitation of DNA. The selections for transformation generally work well, although in rare cases cultures may contain large numbers of drug-resistant mutants due to Luria-Delbrück fluctuations. This is true for both mevinolin and 5-FOA resistance. The presence of homologous DNA will lead to the integration of transforming plasmids at high frequency, with a greater problem for larger regions of identity. Finally, the high mutation rate of some species and strains can lead to genetic variation among populations. Strict microbiological practice, including purification of strains and their long-term storage in frozen stocks, is essential for genetic studies.

Transfection and Natural Genetic Transfer

Halophilic archaea are susceptible to infection by a variety of halophages such as HF1, HF2, ΦCh1, His1 and ΦH (Dyall-Smith *et al.*, 2003). Early work done on ΦH indicated a linear 59 kb genome with ~2 kb of terminal repeats (Schnabel and Zillig, 1984; Schnabel *et al.*, 1982). ΦH has been used to transfect *Halobacterium* sp. using a spheroplast PEG protocol similar to the procedure used for transformations. Transfection efficiencies observed were between 10^6 and 10^7 per μg of DNA based upon plaque formation (Cline and Doolittle, 1987).

Natural genetic transfer has been observed in *Hfx. volcanii*. Initial observation of genetic transfer between *Hfx. volcanii* strains was based upon the observation that mixing of strains auxotrophic for an amino acid or nucleotide results in prototrophic strains (Mevarech and Werczberger, 1985). The generation of prototrophic recombinants required prolonged physical contact between cells and was not due to phage-mediated transduction. In addition, plasmids pHV2 and pHV11 (Rosenshine and Mevarech, 1989) could be transferred from a strain harboring those plasmids to a strain lacking them. These plasmids do not appear to be typical conjugative plasmids like those found in bacterial systems. Using electron microscopy, *Hfx. volcanii* cells were also found to be frequently interconnected by cytoplasmic bridges, providing a possible mechanism for genetic exchange (Rosenshine *et al.*, 1989).

Selectable Markers

One of the most commonly used selective markers for the haloarchaea confers resistance to mevinolin (Lam and Doolittle, 1992) (Table 27.3). Mevinolin is an inhibitor of 3-hydroxy-3-methylglutaryl coenzyme A (HMG-CoA) reductase, an enzyme involved in the production of isoprenoid side chains of lipids in archaea and sterols in eukaryotes. Selection of mevinolin-resistant (Mev^r) mutants of *Hfx. volcanii* resulted in the detection of an up-promoter mutation. The up-promoter allele was

Table 27.3 Commonly used selectable genetic markers for haloarchaea

Selection agent	Target species	Mode of action	Resistance gene	Gene product	Working concentration	Reference
Mevinolin	Halobacterium and Haloferax spp.	Inhibits lipid synthesis	hmg/mev	3-hydroxy-3-methylglutaryl-coenzyme A reductase	13–16 mg ml^{-1}	Lam and Doolittle (1992)
5-Fluoroorotic acid	Halobacterium and Haloferax spp.	Inhibits pyrimidine biosynthesis	ura3/pyrF and ura5/pyrE	orotidine-5′-monophosphate decarboxylase and orotate phosphoribosyl transferase	0.15–0.25 mg ml^{-1}	Peck et al. (2000); Bitan-Banin et al. (2003)
Novobiocin	Haloferax spp.	Inhibits DNA gyrase	gyrB	DNA gyrase subunit	0.1–0.2 mg ml^{-1}	Holmes and Dyall-Smith (1990)

cloned and has been used for selection in *Hfx. volcanii*, *Halobacterium* sp. and other haloarchaea. Additionally a second mevinolin resistance gene has been isolated for *Haloarcula hispanica*. This resistance marker also contains an up-promoter mutation for the HMG-CoA reductase gene, *hmg*, and this allele has been reported to recombine less frequently with the chromosomal copy of HMG-CoA reductase gene in *Hfx. volcanii* (Wendoloski *et al.*, 2001). Recombination between the *Hfx. volcanii* mevinolin resistance marker and the chromosomal *hmg* gene of *Halobacterium* sp. occurs at extremely low frequency and the background spontaneous mutation rate of wild-type *Halobacterium* sp. to Mev^r is $<10^{-7}$, making for an excellent selection system for haloarchaea.

Selectable markers for halophilic archaea allowing both selection and counterselection have been developed recently. These are based on the *Halobacterium* sp. NRC-1 *ura3* (or *pyrF*) gene, coding orotidine-5′-monophosphate decarboxylase, and *Hfx. volcanii pyrE2* genes, coding orotate phosphoribosyl transferase (Bitan-Banin *et al.*, 2003; Peck *et al.*, 2000). These markers are extremely useful for both gene replacements and gene knockouts. Forward selection can be carried out in a Δ*ura3* or Δ*pyrE2* background by selection for uracil prototrophy (Ura$^+$), using uracil-dropout medium for *Halobacterium* sp. and casamino acid medium for *Hfx. volcanii*. Counter selection can be carried out using resistance to 5-fluoroorotic acid (Foa^r), which is metabolized to a toxic intermediate (5-fluorouracil) in an Ura$^+$ background.

A variety of other selectable markers have also been described. The *gyrB* gene allele, containing mutations in the *N*-terminus of the coding sequence, has been used for selection of resistance to the type II topoisomerase inhibitor novobiocin (Holmes and Dyall-Smith, 1990, 1991; Holmes *et al.*, 1991). Several protein synthesis inhibitors, including anisomysin, thiostrepton and chloramphenicol, have also been used as selectable markers. These are particularly useful for replacements of rRNA genes, especially for *Halobacterium* spp. and other haloarchaea that have a single copy of the rRNA operon (Mankin *et al.*, 1992).

Shuttle and Cloning Vectors

Halophilic archaea are rich in plasmid diversity, and many recombinant vectors have been constructed (DasSarma and Fleischmann, 1995) (Table 27.4). Some of the most common vectors are derivatives of *Halobacterium* spp. extrachromosomal replicons, such as pNRC100 and pHH1, or miniplasmids pGRB, pHSB or pHGN, and *Hfx. volcanii* miniplasmid pHV2 or plasmid pHK2. These vectors generally replicate in all commonly used strains of halophilic Archaea tested.

A *Halobacterium* pNRC100 derived vector, pNG168 (8.9 kbp) (Figure 27.2), contains the pTZ18 vector, including the Amp^r determinant for selection in *E. coli*, multiple cloning region with unique *Apa*I, *Hind*III, *Eco*RI, *Pst*I, *Sma*I, *Bam*HI, *Spe*I and *Not*I restriction sites, the pNRC100 minimal replicon, pNG101, and the minimal *Hfx. volcanii* Mev^r marker for selection in halophiles. This plasmid is capable of replication in both

Table 27.4 Haloarchaeal genetic vectors

Plasmids	Features	Reference
pNG168	Haloarchaeal-*E. coli* shuttle plasmid for *Halobacterium* and *Haloferax* spp. Has multiple cloning sites and confers mevinolin resistance in haloarchaea and ampicillin resistance in *E. coli*. Based on pNRC100 replication origin.	DasSarma and Fleischmann (1995)
pUBP2	Haloarchaeal-*E. coli* shuttle plasmid for *Halobacterium* and *Haloferax* spp. Confers mevinolin resistance in haloarchaea and ampicillin resistance in *E. coli*. Based on pHH1 replication origin.	Blaseio and Pfeifer (1990)
pBB7G750	Low copy number haloarchaeal-*E. coli* shuttle plasmid for *Halobacterium* spp. Confers mevinolin resistance in haloarchaea and ampicillin resistance in *E. coli*. Based on *Halobacterium* sp. NRC-1 chromosomal replication origin.	Berquist and DasSarma (2003)
pWL104	Haloarchaeal-*E. coli* shuttle plasmid for *Halobacterium* and *Haloferax* spp. Has multiple cloning sites and confers mevinolin resistance in haloarchaea and ampicillin resistance in *E. coli*. Based on pHV2 replication origin.	Lam and Doolittle (1989)
pMDS99	Haloarchaeal-*E. coli* shuttle plasmid for *Halobacterium* and *Haloferax* spp. Has multiple cloning sites and confers mevinolin resistance in haloarchaea and kanamycin resistance in *E. coli*. Based on pHV2 replication origin.	Wendoloski et al. (2001)
pMDS20	Haloarchaeal-*E. coli* shuttle plasmid for *Haloferax* spp. Confers mevinolin resistance in haloarchaea and ampicillin resistance in *E. coli*. Based on pHK2 replication origin.	Holmes and Dyall-Smith (1990)
pHRZH	Haloarchaeal rRNA gene replacement vector for *Halobacterium* spp. Confers anisomycin and thiostrepton resistance.	Mankin et al. (1992)
pMPK408/ pSK400	Haloarchaeal *ura3* suicide vector for *Halobacterium* gene knockouts for *Halobacterium* spp. Integrates into genome and confers uracil-prototrophy and 5-FOA sensitivity in *Halobacterium* sp. and ampicillin resistance in *E. coli*.	Peck et al. (2000), Wang et al. (2004)

(*continued*)

Table 27.4 Continued

Plasmids	Features	Reference
pBB400	Haloarchaeal *ura3* suicide vector for *Halobacterium* gene knockouts for *Halobacterium* spp. Integrates into genome and confers uracil-prototrophy and 5-FOA sensitivity in *Halobacterium* sp. and ampicillin resistance in *E. coli*. Has a blue/white screen in *E. coli* and multiple cloning site.	Berquist and DasSarma (unpublished)
pGB70	Haloarchaeal *pyrE* suicide vector for for *Haloferax* spp. Integrates into genome and confers uracil-prototrophy and 5-FOA sensitivity.	Bitan-Banin et al. (2003)
pXLNov	Haloarchaeal *bop* gene expression vector for *Haloferax* spp. used for green fluorescent protein (GFP) fusion. Confers resistance to novobiocin.	Nomura and Harada (1998)
pSO7	Haloarchaeal sensory rhodopsin SRI (*sop1*) gene expression vector for *Halobacterium* spp. Confers resistance to mevinolin.	Krebs et al. (1993b)
pBBEV1	Haloarchaeal *bop* promoter expression vector for *Halobacterium* spp. Confers resistance to mevinolin. Includes *N*-terminal cleavable hexahistidine tag.	Berquist and DasSarma (unpublished)
pKJ408sfdx	Haloarchaeal HtrI expression vector using super *fdx* promoter for *Halobacterium* spp. Confers resistance to mevinolin.	Jung and Spudich (1998)
pSE1	Haloarchaeal *fdx*–DHFR fusion gene for promoter analysis in *Haloferax volcanii*. Confers resistance to trimethoprim	Danner and Soppa (1996)
pSD1	Haloarchaeal GPCR expression plasmid for *Haloferax* spp. Confers resistance to trimethoprim.	Patenge and Soppa (1999)
pENDS	Haloarchaeal *bop* promoter-based GPCR expression plasmid in *Halobacterium* spp. Confers resistance to mevinolin.	Bartus et al. (2003)
pFL2	Haloarchaeal gas vesicle expression plasmid for *Halobacterium* spp. Confers resistance to mevinolin	DasSarma et al. (1994)
pFM104D	Antigen display vector on gas vesicles for *Halobacterium* spp. Confers resistance to mevinolin	Stuart et al. (2001); (2004)
pWL204	Haloarchaeal tRNA gene expression vector for *Haloferax* spp. Confers resistance to mevinolin	Nieuwlandt and Daniels (1990)

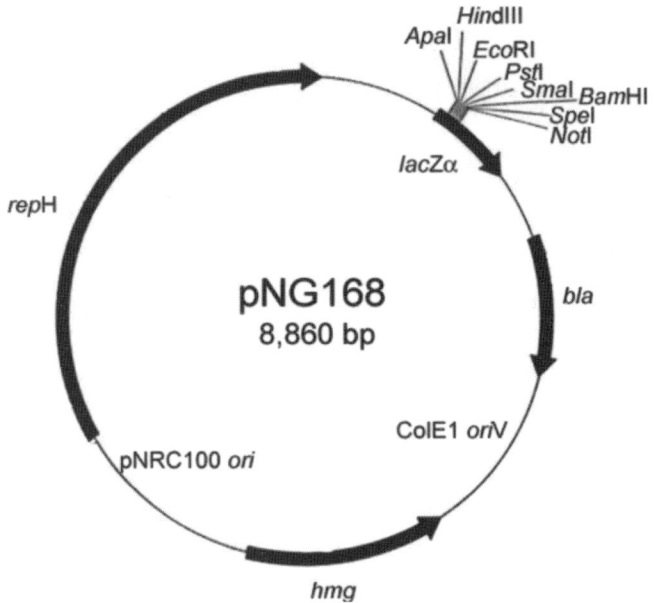

Figure 27.2. Haloarchaeal shuttle vector pNG168. This plasmid contains the *E. coli* pTZ19r replicon and the *Halobacterium* NRC-1 pNRC100 minimal replication region. The *bla* gene provides selection with ampicillin in *E. coli* and the *hmg* gene provides selection with mevinolin in haloarchaea. The multiple cloning site is located in the *lacZα* fragment gene and permits blue-white screening in *E. coli*. The plasmid is available from ATCC (Catalog no. MBA-77) and the sequence is available in GenBank (accession number AY291460).

Halobacterium sp. NRC-1 and *Hfx. volcanii* and has been used for a variety of genetic experiments (DasSarma *et al.*, 1994; Ng and DasSarma, 1993).

A similar vector to pNG168, pUBP2, is somewhat larger (12.3 kbp), and was constructed from the related *Halobacterium* sp. PHH1 plasmid pHH1 (Blaseio and Pfeifer, 1990; Pfeifer and Ghahraman, 1993). This vector contains the *E. coli* vector, pIBI, with the *Hfx. volcanii hmg* marker for Mev^r selection in halophiles, and pHH9, a natural deletion derivative of pHH1. Several unique restrictions sites, *Pst*I, *Bam*HI and *Sma*I, are available for cloning in this vector.

Three related *Halobacterium* spp. miniplasmids, pGRB1, pHSB1 and pHGN1 (~1700 bp), members of a large class of rolling circle type replication (RCR) replicons conserved in bacteria, archaea and eukaryotes (Ilyina and Koonin, 1992), have been used for vector construction by insertion of the Mev^r marker. Plasmid pHSB has been used to successfully transform a *Halobacterium* sp. not naturally harboring a RCR-type plasmid (Hackett and DasSarma, 1989). A recombinant shuttle vector, pHRZH, has been created utilizing pHSB, pBR322 *E. coli* plasmid vector, and a ribosomal RNA region containing a 16S gene, a 23S gene with mutations to provide resistance to thiostrepton and anisomycin, and 2 tRNA genes.

Shuttle vectors derived from the *Hfx. volcanii* natural miniplasmid pHV2 are in wide use and can be selected and maintained in *Haloferax* spp., *Halobacterium* spp. or *E. coli*. Initially, the Mev^r determinant was cloned in a pHV2 derivative to form pWL2. A variant was subsequently constructed (pWL104) containing the Mev^r determinant, portions of pHV2 and *E. coli* plasmid pAT153 conferring both tetracycline and ampicillin resistance, and a multiple cloning region with *Apa*I, *Eco*RI, *Pst*I, *Sma*I, *Bam*HI, *Spe*I and *Not*I restriction enzyme sites (Cline *et al.*, 1989a). Plasmid pHV2 was also used to create vector pMDS99 (7.3 kb), which uses the Mev^r determinant isolated from *Har. hispanica*, the pHV2 origin region, a pUC derivative containing *ori*P15A, and Kan^r.

Plasmid vectors for *Haloferax* sp. have also been based on the 10.5 kb RCR-type plasmid pHK2 (Holmes and Dyall-Smith, 1990; Holmes *et al.*, 1991, 1995). Initial recombinant plasmid construction contained the cloned novobiocin resistance element (the *gyr*AB operon containing a mutant *gyr*B allele). Shuttle vector pMDS1 was constructed by the addition of *E. coli* plasmid pBS$^+$ (Amp^r). Plasmids pMDS10 and pMDS11 were subsequently derived from pMDS1 by minimization of the pHK2 region, and pMDS20 (10.3 kbp) was derived from pMDS10 by deletion of the *gyr*A gene from the Novr cloned region (Holmes *et al.*, 1994).

Gene Replacements and Knockouts

A large number of gene replacements were constructed in the past involving manual screening. The largest number of mutants was characterized for the purple membrane *bop* gene, employing the mevinolin resistance marker (Krebs *et al.*, 1993a). In these cases, site-directed mutants of the *bop* gene were cloned into an *E. coli* plasmid vector containing the mevinolin resistance gene for selection in *Halobacterium*, but lacking an origin of replication of the haloarchaeon (a suicide plasmid vector). Integrants were selected after transformation of the host using Mev^r and excisants screened by loss of drug resistance. The resulting segregants were screened using allele-specific PCR primers. Alternatively, a mutant *bop* gene allele was introduced into a *bop* insertion mutant on a shuttle plasmid (Ni *et al.*, 1990). Another application of gene replacement was for the single rRNA operon of *Halobacterium* spp., where dominant mutations resulting in antibiotic resistance could be selected (Mankin *et al.*, 1992).

An exciting recent development is a new gene replacement and knockout method in the sequenced strain *Halobacterium* sp. NRC-1 using the *ura*3 gene, which is both selectable and counterselectable (Peck *et al.*, 2000). This approach is shown in Figure 27.3. A gene allele of interest (e.g. a deletion or point mutation) is first cloned into an *E. coli* plasmid containing the *ura*3 gene under the control of its own promoter, and the resulting suicide plasmid is introduced into the haloarchaeon host by transformation. Integrants are selected by uracil prototrophy (Ura$^+$, using commercially available uracil-dropout media components). Alternatively, mevinolin resistance (Mev^r) may also be used. Subsequently, excisants are selected by counterselecting for 5-fluoroorortic acid resistance (Foa^r), giving rise to both the original and replaced allele, which can be distinguished by PCR or phenotypic analysis.

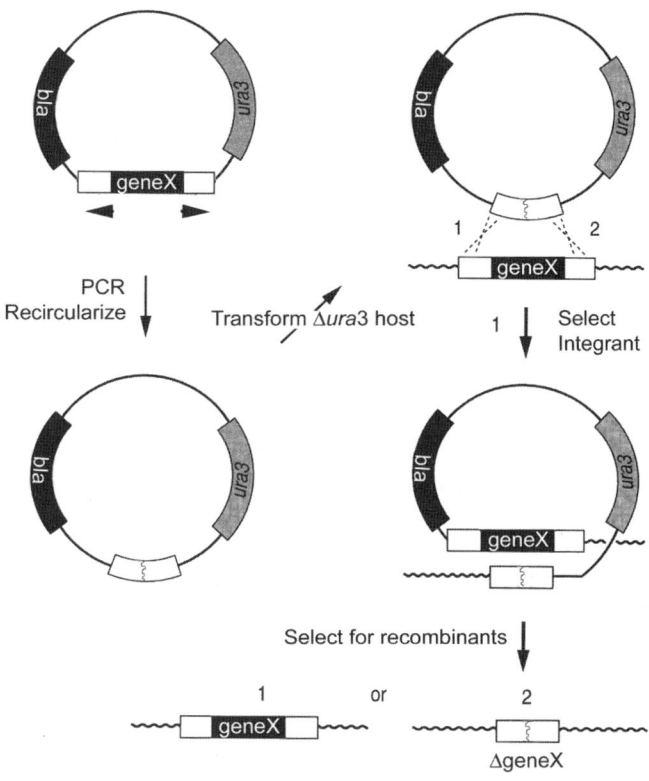

Figure 27.3. Gene knockout and replacement in the haloarchaea. The example shown is for selection and counterselection with *ura3*. A cloned haloarchaeal target gene (geneX) in a suicide plasmid vector (e.g. pMPK408 or pBB400, a second generation knockout vector which provides a blue-white screen in *E. coli* and a multiple cloning site), which do not replicate in haloarchaea, is used for PCR amplification (primers designated by arrowheads) and recircularization to provide for a precisely deleted gene. The plasmid is introduced into a Δ*ura3* haloarchaeon by transformation. Integrants are selected by uracil prototrophy using uracil dropout plates. Excisants are selected for by plating on plates containing 5-fluoroorotic acid. Depending on the site of the recombination (1 or 2), different outcomes are possible. Alternatively, mevinolin selection can also be used for integration and the *pyr*E2 gene can be used for selection and counterselection.

A similar approach has proved successful for *Hfx. volcanii*, but substituting the *pyr*E2 gene in place of *ura3*.

Insertion and Transposon Mutagenesis

Insertion mutagenesis is conducted by employing natural IS element insertions, gene cassettes, oligonucleotide linkers and transposons constructed from haloarchaeal IS elements. Natural phenotypic variants were used in early studies to identify *bop* and *gvp* regulatory genes and coordinately regulated genes. IS element insertions identified the *brp* and the *bat* genes necessary for purple membrane production and located upstream of *bop* (DasSarma, 1989). Similarly, IS element insertions identified *gvp*D, *gvp*E, and other genes necessary for wild-type gas vesicle synthesis (DasSarma et al., 1994). More recently, IS element insertions have been found in the *ura3* gene in Foa^r mutants (Peck et al., 2000). These studies represent a valuable natural transposon mutagenesis system in haloarchaea that has provided significant insights into gene function and regulation.

Gene disruptions made using gene cassettes and oligonucleotide linkers are also useful for genetic analysis. To study the functions of the *gvp* gene cluster on pNRC100 in gas vesicle formation, a kanamycin resistance (κ) gene cassette was used for scanning mutagenesis of the gene cluster cloned in a 24.5-kb *E. coli–Halobacterium* shuttle plasmid, pFL2 (DasSarma et al., 1994). Transformation of *Halobacterium* sp. SD109, an NRC-1 derivative deleted for the entire *gvp* gene cluster, with pFL2 and mutated pFL2 derivatives showed that while the unmutated gene cluster successfully programmed gas vesicle formation (Figure 27.1), derivatives with insertion of the κ cassette in nearly all of the *gvp* genes lacked normal gas vesicles. In most cases, the block in gas vesicle synthesis did not result from polar effects, since similar results were obtained for derivatives of the insertion mutants in which most of the internal portion of the κ cassette was deleted and only small (15–54 bp) insertions remained.

Similarly, one pNRC100 minireplicon, pNG11Δ12, was analyzed by linker scanning mutagenesis using a short oligonucleotide (Ng and DasSarma, 1993). Insertions in the *repH* gene knocked out the capability of the minireplicon for autonomous replication, while a second insertion at the same site restoring the *repH* reading frame, resulted in reversion to replication proficiency.

Some effort has been directed toward construction of recombinant transposons with selectable markers from natural IS elements. Synthetic transposons were constructed consisting of haloarchaeal ISH elements (ISH2, ISH26 or ISH28) flanking a mevinolin resistance determinant (Dyall-Smith and Doolittle, 1994). A subsequent construct ThD73 was created and contains the terminal inverted repeats (TIRs) of ISH28, the Mev^r determinant, pUC19 *E. coli* plasmid (allowing for recovery of transposons and sequencing of insertion sites), and the ISH28 transposase gene placed outside of the TIRs. Transformation of *Har. hispanica* with this transposon suicide plasmid construct and subsequent recovery of insertion site plasmids in *E. coli* revealed little target sequence specificity but with insertion site preference for lower GC% DNA (Woods et al., 1999). However, when transposons were introduced into haloarchaeal strains with perfectly homologous DNA (e.g. *Hfx. volcanii*), plasmids integrated at homologous sites through high-frequency homologous recombination.

Gene Reporters

Several reporter genes are available for haloarchaea. Commonly used gene reporters include the *bop* gene, a halophilic β-galactosidase and green fluorescent protein, plus its engineered derivatives, from the jellyfish *Aequorea victoria*. The bacterio-opsin gene, *bop*, has been used extensively for analyses of its own promoter elements. In *Halobacterium* sp., use of the *bop* gene as a reporter allows for a simple assay for promoter activity. Using a recipient *bop*-deficient *Halobacterium* strain (e.g. strain SD23), it is possible to clone mutated promoters upstream of the *bop* gene carried on a plasmid pNG168 derivative, transform the *bop*-deficient strain, and assay for production of purple membrane. Transformant colonies can be assayed qualitatively based upon phenotypic colony coloring. Pum$^-$ strains have an orange appearance, while Pum$^+$ to Pum^{+++} colonies appear purple on plates. This reporter gene system has been used to elucidate and characterize elements of the *bop* promoter, including the TATA box, the RY box and UAS, upstream activator sequence (Baliga and DasSarma, 1999, 2000).

A halophilic β-galactosidase (*bga*H) gene has been isolated and used as a reporter gene for promoter analyses in *Halobacterium* sp. and *Hfx. volcanii* (Holmes and Dyall-Smith, 2000; Holmes et al., 1997). When introduced into these organisms, which do not possess a β-galactosidase gene, β-galactosidase activity can be detected in cell lysates by colorimetric assays and colonies can be screened for activity by plating on solid media containing X-gal as a substrate. Studies using this reporter gene have examined production utilizing the *bop* gene promoter, mutagenized derivatives of the *fdx* (ferredoxin) gene promoter, and the *arc*B (ornithine carbamoyltransferase) gene promoter (Patenge et al., 2000). Additional studies have utilized the *bga*H reporter gene to study activity of promoter for genes involved in gas vesicle synthesis (Gregor and Pfeifer, 2001).

The commonly used reporter gene *gfp* from *A. victoria* can also be used as a reporter in the hypersaline intracellular milieu of haloarchaea. Initial functional analysis examined *bop* promoter-driven BR–GFP fusion protein production in *Halobacterium* sp. The BR–GFP fusions could be localized to the plasma membrane and fluorescence of the BR–GFP fusions indicated that they were bifunctional proteins that retained the two intrinsic functions of their components (Nomura and Harada, 1998). For *Hfx. volcanii*, GFP reporter gene expression has been investigated for detecting new genes which may be involved in the proteasome pathway of archaea. It has been observed that a soluble modified derivative of GFP with a red shifted mutation (Davis and Vierstra, 1998) can be expressed to high levels (~1% of total cell protein) and visualized in *Hfx. volcanii* cells (Reuter and Maupin-Furlow, 2004). These results indicate that GFP can be used as a versatile reporter in both model haloarchaea.

Promoter Analysis and Expression Vectors

The two main promoters commonly utilized for heterologous or homologous gene expression in haloarchaea are the ferredoxin (*fdx*) promoter, mainly for soluble proteins, and the *bop* promoter, for membrane proteins. Both promoters have been characterized by saturation mutagenesis and up-promoter mutations reported (Baliga and DasSarma, 1999, 2000; Danner and Soppa, 1996). Vectors containing 200 bp upstream of the *fdx* coding region are sufficient for constitutive expression. For *bop*, only 53 bp upstream of the coding region are required, but regulation is complex, and expression is highly dependent on the specific gene and construct (Baliga and DasSarma, 1999, 2000; Bartus et al., 2003).

Fusions of the *bop* promoter and N-terminal coding region have been used successfully for overproduction of BR mutants, halorhodopsin (the light-driven chloride pump), several sensory rhodopsins (the phototactic receptors), and two halotransducers (e.g. Heymann et al., 1993; Jung and Spudich, 1998; Krebs et al., 1993b; Ni et al., 1990). Additionally, foreign membrane proteins, e.g. mammalian G-protein coupled receptors (GPCRs), such as the human muscarinic acetylcholine (M_1) and adrenergic (A2b, β_2) receptors, have been expressed in haloarchaea (Bartus et al., 2003; Patenge and Soppa, 1999). However, both proteolysis and uncharacterized determinants at the 3′ end of the *bop* gene were found to limit the usefulness of this heterologous gene expression system.

The *fdx* promoter has been used to express a variety of soluble proteins (Jung and Spudich, 1998; Piatribratov et al., 2000). These expression systems have sometimes been coupled to His-tagged proteins, which have been successfully purified using metal affinity chromatography in a variety of systems. However, difficulties have also been encountered in haloarchaea due to binding of endogenous proteins with metal affinity (e.g. *Halobacterium* sp. NRC-1 protein VNG2021 is a frequent contaminant; Berquist and DasSarma, unpublished).

In addition to the *fdx* and *bop* promoter vectors, a few other specialized expression systems have also been used successfully. The gas vesicle gene cluster has been modified to incorporate a cloning site in the *gvp*C gene in the pFM104D shuttle vector and used to produce fusion proteins attached to the surface of floating vesicles (Stuart et al., 2001, 2004). This system has been used for antigen expression and delivery, and development and testing of vaccine candidates. Also, pWL204, a derivative of pWL102 with a *Hfx. volcanii* tRNA^{Lys} gene promoter cloned into it, has been used for expression of tRNA genes to study RNA processing (Nieuwlandt and Daniels, 1990).

◆◆◆◆◆ GENOMIC APPROACHES

Systematic Gene Knockouts

Recently a facile gene knockout methodology was developed for *Halobacterium* sp. strain NRC-1, and subsequently mimicked for *Hfx. volcanii*, using the selectable and counterselectable *ura3* gene (Figure 27.3). This development is a step toward systematic knockout of genes in these model strains. This approach allows, in principle, analysis of any gene allele of interest, with either a precise deletion or site-directed mutation. For a gene deletion, the gene of interest plus ~500 bp of 5′ and 3′ flanking DNA is cloned into an *E. coli* plasmid which also harbors the *ura3* gene plus its own promoter. Inverse PCR is used to generate an in-frame gene deletion. Subsequent ligation of the inverse PCR product then yields a final knockout vector for the gene of interest. Alternatively, the 5′ and

3′ flanking regions can be amplified separately, and a triple ligation into the *E. coli* vector containing *ura3* can be performed to yield a final vector containing only DNA flanking the gene of interest. This vector is introduced into the host *via* transformation. The suicide vector integrates through homologous recombination into either the 5′ or 3′ flanking DNA region of the gene of interest, forming a merodiploid strain that is selected for by uracil prototrophy (Ura$^+$) using uracil dropout media (Wang *et al.*, 2004).

In an alternate approach, an *E. coli* plasmid with both the *ura3* gene and the Mev^r determinant was used, and Mev^r selected in *Halobacterium* sp. Excisants are then counterselected by using 5-fluororotic acid resistance (Ura$^-$, Foa^r). The resulting colonies have either a wild-type copy of the gene of interest or the wild-type gene copy has been replaced with the deletion allele, yielding a gene knockout (Peck *et al.*, 2000). The colonies are then screened by PCR to determine which allele they encode.

An additional advantage to these strategies is that the resulting knockout strain can be used in further transformations to knockout other genes of interest, resulting in double or triple knockout strains. The approach for directed gene knockouts or replacements in *Halobacterium* sp. strain NRC-1 Δ*ura3* is shown in Figure 27.3. A similar approach is possible in *Hfx. volcanii* utilizing the *pyrE2* gene in place of *ura3* (Bitan-Banin *et al.*, 2003). This methodology is a vast improvement over previous gene deletions and replacements, which required extensive manual screening, and it sets the stage for systematic knockout of genes in haloarchaea.

DNA Microarrays

In recent years, a variety of microarray platforms with differences in probe type and probe application to the microarray slide have been employed for transcriptome analysis of *Halobacterium* sp. NRC-1 as well as *Hfx. volcanii*. The first microarrays used for *Halobacterium* sp. harbored PCR-product-based probes. Purified PCR products for 2413 open reading frames (ORFs) from the *Halobacterium* sp. NRC-1 genome were spotted in quadruplicate onto polyamine glass slides. This type of microarray was used for a transcriptome comparison of the *Halobacterium* sp. strains NRC-1, S9, SD20 and SD23 (Baliga *et al.*, 2002). Also reported was a 70-mer oligonucleotide microarray in a study testing the effect of high UV irradiation on *Halobacterium* sp. NRC-1 (Baliga *et al.*, 2004b). Pre-synthesized oligonucleotides representing 2400 ORFs were spotted onto glass slides, similar to the production of PCR-product-based microarrays.

Most recently, an *in situ* synthesized oligonucleotide array, using ink-jet technology (Hughes *et al.*, 2001) have been used successfully for transcriptome analysis of DMSO/TMAO respiration and response to UV irradiation in *Halobacterium* sp. NRC-1 (Müller and DasSarma, 2005; McCready *et al.*, 2005; Figure 27.4). Oligomer (60-mer) probes were designed for 2474 ORFs utilizing the program OligoPicker (Wang and

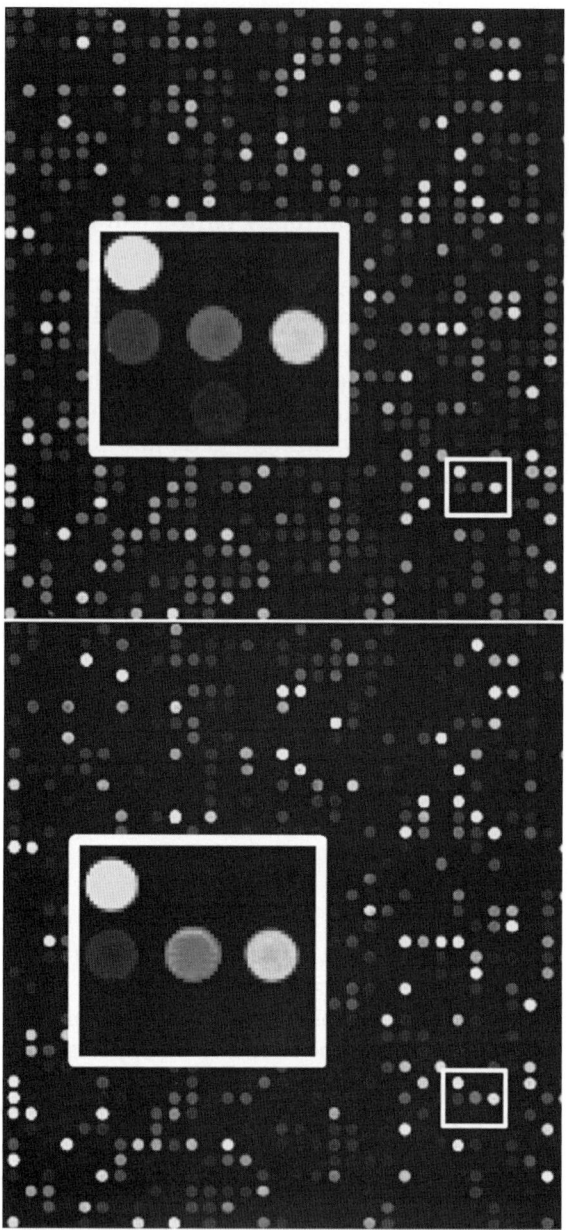

Figure 27.4. Oligonucleotide microarray analysis of the genetically tractable halophilic archaeon *Halobacterium sp.* NRC-1. The upper panel represents an environmental perturbation experiment comparing anaerobic growth *via* TMAO respiration *versus* aerobic growth. The lower panel denotes a genetic perturbation experiment in which a $\Delta dmsR$ strain is compared with a $dmsR^+$ strain. The inserts highlight the expression changes of *dms*A, coding for the DMSO/TMAO reductase in this extremophile. See Plate 27.4 in Colour Plate Section.

Seed, 2003). Neither the relatively high GC content of the chromosome, nor the difference in proportion of GC content between the three replicons of *Halobacterium* sp. NRC-1 were problematic in the probe design. Up to three probes were designed per ORF with a mean T_m of 81°C and a T_m range of 3°C. The microarray slides harbor both gene-probes (~8000 features per array) as well as ~400 negative and positive control spots to test hybridization conditions. Unused microarray slides can be stored for several months under a dry atmosphere at ambient temperatures and in the dark. A hallmark of *in situ* synthesized oligonucleotide arrays is the low occurrence of data outliers due to non-uniform spot morphology or background noise. Features replicated within a single array show low differences in absolute processed signal intensities and spot-to-spot variation for replicate experiments. Green and red fluorescent signal intensities have a dynamic range in excess of three orders of magnitude allowing simultaneous analysis of low- and high-intensity features.

A different microarray approach was taken to study metabolism in *Hfx. volcanii*. Since the unfinished status of the genome sequence of *Hfx. volcanii* did not allow for the construction of an ORF-specific microarray, a shotgun microarray was constructed (Zaigler et al., 2003). A genomic DNA clone library of *Hfx. volcanii* was generated and converted into a PCR product library. A fragment size of 1.4–1.6 kbp was chosen to limit representation of more than one gene per clone. Aliquots of the PCR products were then spotted onto glass slides to obtain a onefold coverage microarray. After hybridization, the identity of the regulated genes was determined by sequencing the respective clones. This microarray type performed well with very low variation coefficients for most spots in replicate experiments.

Protocols for RNA isolation, cDNA preparation and labeling, hybridization, and signal detection used for haloarchaeal microarrays are similar to the protocols published by Eisen and Brown (1999). Slide-image processing may be carried out using one of the freely or commercially available software. Typically, spot signal intensity extraction, subtraction of background intensities and assessment of spot signal significance, normalization of fluorescence intensities, and calculation of signal ratios are carried out for each feature. In order to identify differentially expressed genes, further statistical analysis of the transcript data may be carried out by employing programs performing Analysis Of Variance (ANOVA, Sharov et al., 2005), or programs that use Bayesian statistics (Baldi and Long, 2001; Hatfield et al., 2003). A detailed description on microarray database implementation and management may be found in Bowtell and Sambrook, 2003. The same reference also contains valuable information on further analyses of gene expression patterns like Cluster analysis.

Proteomics

Another approach that has been applied to haloarchaeal model systems is proteomics. In this approach, proteins from whole cells or subcellular components are analyzed by mass spectrometry (MS), after fractionation using either two-dimensional polyacrylamide gels (2D-GE) or liquid chromatography. Extensive efforts have been made on analysis of membrane proteins of *Halobacterium* species and the development of 2D-GE system for both *Halobacterium* spp. and *Hfx. volcanii* (Evans *et al.*, 2003; Goo *et al.*, 2003; Joo and Kim, 2005; Klein *et al.*, 2005; Karadzic and Maupin-Furlow, 2005; Tebbe *et al.*, 2005). One study has also examined changes in proteomic profiles of cells grown under differential conditions (Choi *et al.*, 2005).

Whole cell and soluble proteins have been analyzed using both 2D-GE matric-assisted laser desorption/ionization time of flight (MALDI-TOF) and liquid chromatography/tandem mass spectrometry (LC/MS/MS) (Baliga *et al.*, 2002; Choi *et al.*, 2005; Goo *et al.*, 2003; Tebbe *et al.*, 2005). In an initial study of the model strain, *Halobacterium* sp. NRC-1, proteins were isolated by ultracentrifugation of whole cell lysates and analyzed by LC/MS/MS. A total of 426 proteins were identified, about one-fifth of the proteome, including 232 soluble proteins, 165 insoluble proteins and 29 that were found in both fractions. More recently, the soluble protein fraction from a similar strain, R1, was analyzed by 2D-GE and MALDI-TOF MS analysis of tryptic fragments. A reference map was established for most proteins which distribute in the narrow pI range between 3.5 and 5.5, and 661 proteins were identified. Evidence was obtained for posttranslational modification of a significant percentage of proteins (94 proteins). Similar approaches have been used to generate 2D GE maps for *Hfx. volcanii* (Evans *et al.*, 2003; Karadzic and Maupin-Furlow, 2005).

Two reported studies are directed at membrane components of *Halobacterium* sp. In one study (Blonder *et al.*, 2004), the purple membrane of *Halobacterium halobium*, a closely related strain to NRC-1, was used as a model for integral membrane proteins. A method using buffered methanol was used to solubilize and extract memrane proteins which were analyzed by microcapillary liquid chromatography and tandem mass spectrometry after tryptic digestion. This method avoided the use of detergents, strong acids or cyanogen bromide, which can interfere with mass spectrometry. A second study of 2D GE demonstrated several problems associated with integral membrane proteins of another closely related *Halobacterium* strain (Klein *et al.*, 2005). Not surprisingly, irreversible denaturation of integral membrane proteins was observed in the isoelectric focusing gel, making them unsuitable for analysis by this method. In addition, mass fingerprinting of tryptic peptides by MALDI-TOF is not favorable. Nevertheless, it was possible to identify 114 integral membrane proteins using LC/MS/MS.

◆◆◆◆◆ CONCLUSIONS AND FUTURE PROSPECTS

Haloarchaea represent excellent experimental models for studies of archaeal, extremophilic, and aspects of eukaryotic biology, such as DNA replication and transcription. *Halobacterium* sp. NRC-1 and *Hfx. volcanii* represent excellent systems for traditional genetic as well as genomic studies. The availability of the complete genome sequence for *Halobacterium* sp. NRC-1, since 2000, has allowed significant advances in the development of postgenomic methodology of this organism.

This includes gene knockout capability, DNA microarrays and proteomic approaches. As a result, many interesting biological studies are possible in the future, and several have already been reported.

In the near future, some other haloarchaeal systems will also become attractive for detailed analysis and biotechnological applications. The genome of *Har. marismortui* coupled with the high-resolution structure of the ribosomal large subunit will provide excellent opportunities for development of new antibiotics. Full benefits will be realized after further development of genetic methods for this organism. The complete genome sequence of *Hfx. volcanii* will permit a more detailed view of the metabolic capabilities of this prototrophic archaeon. Other haloarchaea, which expand our understanding of the environmental niches and physiological capabilities, are sure to be found and will stimulate more research on this interesting, useful, and tractable class of microrganisms.

Acknowledgments

We are grateful to the National Science Foundation for support of research on halophilic archaea in the authors' laboratory (MCB-0450695 MCB-0296017, MCB-0296016 and MCB-0135595). We also thank Priya DasSarma for careful reading and editing of the manuscript.

List of suppliers

Sigma-Aldrich
P.O. Box 14508
St. Louis, MO 63178, USA
Tel: 1-800-325-3010
http:/www.sigmaaldrich.com/

(Sigma Y0626) Nitrogen base
(Sigma Y1501) Dropout formula

Oxoid Inc.
800 Proctor Avenue
Ogdensburg, NY 13669, USA
Tel: 1-613-226-1318
http:/www.oxoid.com

(Oxoid L34) Neutralized Peptone

Merck & Co., Inc.
P.O. Box 100
One Merck Drive
Whitehouse Station, NJ 08889-0100, USA
Tel: 1-908-423-1000
http:/www.merck.com

Mevinolin/Lovastatin

References

Allers, T. and Ngo, H. P. (2003). Genetic analysis of homologous recombination in Archaea: *Haloferax volcanii* as a model organism. *Biochem. Soc. Trans.* **31**, 706–710.

Allers, T., Ngo, H. P., Mevarech, M. and Lloyd, R. G. (2004). Development of additional selectable markers for the halophilic archaeon *Haloferax volcanii* based on the *leuB* and *trpA* genes. *Appl. Environ. Microbiol.* **70**, 943–953.

Baldi, P. and Long, A. D. (2001). A Bayesian framework for the analysis of microarray expression data: regularized t-test and statistical inferences of gene changes. *Bioinformatics* **17**, 509–519.

Baliga, N. S. and DasSarma, S. (1999). Saturation mutagenesis of the TATA box and upstream activator sequence in the haloarchaeal *bop* gene promoter. *J. Bacteriol.* **181**, 2513–2518.

Baliga, N. S. and DasSarma, S. (2000). Saturation mutagenesis of the haloarchaeal *bop* gene promoter: identification of DNA supercoiling sensitivity sites and absence of TFB recognition element and UAS enhancer activity. *Mol. Microbiol.* **36**, 1175–1183.

Baliga, N. S., Kennedy, S. P., Ng, W. V., Hood, L. and DasSarma, S. (2001). Genomic and genetic dissection of an archaeal regulon. *Proc. Natl. Acad. Sci. USA* **98**, 2521–2525.

Baliga, N. S., Pan, M., Goo, Y. A., Yi, E. C., Goodlett, D. R., Dimitrov, K., Shannon, P., Aebersold, R., Ng, W. V. and Hood, L. (2002). Coordinate regulation of energy transduction modules in *Halobacterium* sp. analyzed by a global systems approach. *Proc. Natl. Acad. Sci. USA* **99**, 14913–14918.

Baliga, N. S., Bonneau, R., Facciotti, M. T., Pan, M., Glusman, G., Deutsch, E. W., Shannon, P., Chiu, Y., Weng, R. S., Gan, R. R., Hung, P., Date, S. V., Marcotte, E., Hood, L. and Ng, W. V. (2004a). Genome sequence of *Haloarcula marismortui*: a halophilic archaeon from the Dead Sea. *Genome Res.* **14**, 2221–2234.

Baliga, N. S., Bjork, S. J., Bonneau, R., Pan, M., Iloanusi, C., Kottemann, M. C., Hood, L. and DiRuggiero, J. (2004b). Systems level insights into the stress response to UV radiation in the halophilic archaeon *Halobacterium* NRC-1. *Genome Res.* **14**, 1025–1035.

Ban, N., Nissen, P., Hansen, J., Moore, P. B. and Steitz, T. A. (2000). The complete atomic structure of the large ribosomal subunit at 2.4 Å resolution. *Science* **289**, 905–920.

Bartus, C. L., Jaakola, V. P., Reusch, R., Valentine, H. H., Heikinheimo, P., Levay, A., Potter, L. T., Heimo, H., Goldman, A. and Turner, G. J. (2003). Downstream coding region determinants of bacterio-opsin, muscarinic acetylcholine receptor and adrenergic receptor expression in *Halobacterium salinarum*. *Biochim. Biophys. Acta* **1610**, 109–123.

Berquist, B. R. and DasSarma, S. (2003). An archaeal chromosomal autonomously replicating sequence element from an extreme halophile, *Halobacterium* sp. strain NRC-1. *J. Bacteriol.* **185**, 5959–5966.

Bitan-Banin, G., Ortenberg, R. and Mevarech, M. (2003). Development of a gene knockout system for the halophilic archaeon *Haloferax volcanii* by use of the *pyrE* gene. *J. Bacteriol.* **185**, 772–778.

Blaseio, U. and Pfeifer, F. (1990). Transformation of *Halobacterium halobium*: development of vectors and investigation of gas vesicle synthesis. *Proc. Natl. Acad. Sci. USA* **87**, 6772–6776.

Blonder, J., Conrads, T. P., Yu, L. R., Terunuma, A., Janini, G. M., Issaq, H. J., Vogel, J. C. and Veenstra, T. D. (2004). A detergent- and cyanogen bromide-free method for integral membrane proteomics: application to *Halobacterium* purple membranes and the human epidermal membrane proteome. *Proteomics* **4**, 31–45.

Bowtell, D. and Sambrook, J. (eds) (2003). *DNA Microarrays: A Molecular Cloning Manual*, Cold Spring Harbor Laboratory Press, Plainview, N.Y.

Charlebois, R. L., Lam, W. L., Cline, S. W. and Doolittle, W. F. (1987). Characterization of pHV2 from *Halobacterium volcanii* and its use in demonstrating transformation of an archaebacterium. *Proc. Natl. Acad. Sci. USA* **84**, 8530–8534.

Charlebois, R. L., Hofman, J. D., Schalkwyk, L. C., Lam, W. L. and Doolittle, W. F. (1989). Genome mapping in halobacteria. *Can. J. Microbiol.* **35**, 21–29.

Charlebois, R. L., Schalkwyk, L. C., Hofman, J. D. and Doolittle, W. F. (1991). Detailed physical map and set of overlapping clones covering the genome of the archaebacterium *Haloferax volcanii* DS2. *J. Mol. Biol.* **222**, 509–524.

Choi, J., Joo, W. A., Park, S. J., Lee, S. H. and Kim, C. W. (2005). An efficient proteomics based strategy for the functional characterization of a novel halophilic enzyme from *Halobacterium salinarum*. *Proteomics* **5**, 907–917.

Cline, S. W. and Doolittle, W. F. (1987). Efficient transfection of the archaebacterium *Halobacterium halobium*. *J. Bacteriol.* **169**, 1341–1344.

Cline, S. W. and Doolittle, W. F. (1992). Transformation of members of the genus *Haloarcula* with shuttle vectors based on *Halobacterium halobium* and *Haloferax volcanii* plasmid replicons. *J. Bacteriol.* **174**, 1076–1080.

Cline, S. W., Schalkwyk, L. C. and Doolittle, W. F. (1989a). Transformation of the archaebacterium *Halobacterium volcanii* with genomic DNA. *J. Bacteriol.* **171**, 4987–4991.

Cline, S. W., Lam, W. L., Charlebois, R. L., Schalkwyk, L. C. and Doolittle, W. F. (1989b). Transformation methods for halophilic archaebacteria. *Can. J. Microbiol.* **35**, 148–152.

Danner, S. and Soppa, J. (1996). Characterization of the distal promoter element of halobacteria *in vivo* using saturation mutagenesis and selection. *Mol. Microbiol.* **19**, 1265–1276.

DasSarma, S. (1989). Mechanisms of genetic variability in *Halobacterium halobium*: the purple membrane and gas vesicle mutations. *Can. J. Microbiol.* **35**, 65–72.

DasSarma, S. (1993). Identification and analysis of the gas vesicle gene cluster on an unstable plasmid of *Halobacterium halobium*. *Experientia* **49**, 482–486.

DasSarma, S. (2004). Genome sequence of an extremely halophilic archaeon. In *Microbial Genomes* (C. M. Fraser and K. E. Nelson, eds), Humana Press, Inc. Totowa, N.J.

DasSarma, S. and Arora, P. (2002). Halophiles. In *Encyclopedia of Life Sciences*, pp. 458–466. Macmillan Press, London.

DasSarma, S. and Fleischmann, E. M. (eds) (1995). *Archaea: Laboratory Manual, Halophiles*, Cold Spring Harbor Laboratory Press, Plainview, N.Y.

DasSarma, S., RajBhandary, U. L. and Khorana, H. G. (1983). High-frequency spontaneous mutation in the bacterio-opsin gene in *Halobacterium halobium* is mediated by transposable elements. *Proc. Natl. Acad. Sci. USA* **80**, 2201–2205.

DasSarma, S., RajBhandary, U. L. and Khorana, H. G. (1984). Bacterio-opsin mRNA in wild-type and bacterio-opsin deficient *Halobacterium halobium* strains. *Proc. Natl. Acad. Sci. USA* **81**, 125–129.

DasSarma, S., Halladay, J. T., Jones, J. Donovan, J. W., Giannasca, P. J. and Tandeau de Marsac, N. (1988). High-frequency mutations in a plasmid-encoded gas vesicle gene in *Halobacterium halobium*. *Proc. Natl. Acad. Sci. USA* **85**, 6861–6865.

DasSarma, S., Arora, P., Lin, F., Molinari, E. and Yin, L. R. (1994). Wild-type gas vesicle formation requires at least ten genes in the *gvp* gene cluster of *Halobacterium halobium* plasmid pNRC100. *J. Bacteriol.* **176**, 7646–7652.

Davis, S. J. and Vierstra, R. D. (1998). Soluble, highly fluorescent variants of green fluorescent protein (GFP) for use in higher plants. *Plant Mol. Biol.* **36**, 521–528.

Dyall-Smith, M. L. and Doolittle, W. F. (1994). Construction of composite transposons for halophilic Archaea. *Can. J. Microbiol.* **40**, 922–929.

Dyall-Smith, M., Tang, S. L. and Bath, C. (2003). Haloarchaeal viruses: how diverse are they? *Res. Microbiol.* **154**, 309–313.

Eisen, M. B. and Brown, P. O. (1999). DNA arrays for analysis of gene expression. *Meth. Enzymol.* **303**, 179–205.

Evans, E. C., Horn, T., Wagner, M. A., Eschenbrenner, M., Mujer, C. V. and DelVecchio, V. G. (2003). Isolation protocol for two-dimensional-polyacrylamide gel electrophoresis analysis of *Haloferax volcanii* proteome. *Biotechniques* **35**, 478–480.

Goo, Y. A., Yi, E. C., Baliga, N. S., Tao, W. A., Pan, M., Aebersold, R., Goodlett, D. R., Hood, L. and Ng, W. V. (2003). Proteomic analysis of an extreme halophilic archaeon, *Halobacterium* sp. NRC-1. *Mol. Cell. Proteomics* **2**, 506–524.

Grant, W. D. and Larsen, H. (1984). Extremely halophilic archaeobacteria. In *Bergey's Manual of Systematic Bacteriology* (J. Holt, ed.), pp. 2216–2233. Williams & Wilkins, Baltimore.

Gregor, D. and Pfeifer, F. (2001). Use of a halobacterial *bga*H reporter gene to analyse the regulation of gene expression in halophilic archaea. *Microbiology* **147**, 1745–1754.

Grey, V. L. and Fitt, P. S. (1976). An improved synthetic growth medium for *Halobacterium cutirubrum*. *Can. J. Microbiol.* **22**, 440–442.

Hackett, N. R. and DasSarma, S. (1989). Characterization of the small endogenous plasmid of *Halobacterium* strain SB3 and its use in transformation of *H. halobium*. *Can. J. Microbiol.* **35**, 86–91.

Hatfield, G. W., Hung, S. P. and Baldi, P. (2003). Differential analysis of DNA microarray gene expression data. *Mol. Microbiol.* **47**, 871–877.

Heymann, J. A., Havelka, W. A. and Oesterhelt, D. (1993). Homologous overexpression of a light-driven anion pump in an archaebacterium. *Mol. Microbiol.* **7**, 623–630.

Holmes, M. L. and Dyall-Smith, M. L. (1990). A plasmid vector with a selectable marker for halophilic archaebacteria. *J. Bacteriol.* **172**, 756–761.

Holmes, M. L. and Dyall-Smith, M. L. (1991). Mutations in DNA gyrase result in novobiocin resistance in halophilic archaebacteria. *J. Bacteriol.* **173**, 642–648.

Holmes, M. L. and Dyall-Smith, M. L. (2000). Sequence and expression of a halobacterial β-galactosidase gene. *Mol. Microbiol.* **36**, 114–122.

Holmes, M. L., Nuttall, S. D. and Dyall-Smith, M. L. (1991). Construction and use of halobacterial shuttle vectors and further studies on *Haloferax* DNA gyrase. *J. Bacteriol.* **173**, 3807–3813.

Holmes, M., Pfeifer, F. and Dyall-Smith, M. (1994). Improved shuttle vectors for *Haloferax volcanii* including a dual-resistance plasmid. *Gene* **146**, 117–121.

Holmes, M. L., Pfeifer, F. and Dyall-Smith, M. L. (1995). Analysis of the halobacterial plasmid pHK2 minimal replicon. *Gene* **153**, 117–121.

Holmes, M. L., Scopes, R. K., Moritz, R. L., Simpson, R. J., Englert, C., Pfeifer, F. and Dyall-Smith, M. L. (1997). Purification and analysis of an extremely halophilic β-galactosidase from *Haloferax alicantei*. *Biochim. Biophys. Acta* **1337**, 276–286.

Hughes, T. R., Mao, M., Jones, A. R., Burchard, J., Marton, M. J., Shannon, K. W., Lefkowitz, S. M., Ziman, M., Schelter, J. M., Meyer, M. R., Kobayashi, S., Davis, C., Dai, H., He, Y. D., Stephaniants, S. B., Cavet, G., Walker, W. L., West, A., Coffey, E., Shoemaker, D. D., Stoughton, R., Blanchard, A. P., Friend, S. H. and Linsley, P. S. (2001). Expression profiling using microarrays fabricated by an ink-jet oligonucleotide synthesizer. *Nature Biotechnol.* **19**, 342–347.

Ilyina, T. V. and Koonin, E. V. (1992). Conserved sequence motifs in the initiator proteins for rolling circle DNA replication encoded by diverse replicons from eubacteria, eucaryotes and archaebacteria. *Nucleic Acids Res.* **20**, 3279–3285.

Joo, W. A. and Kim, C. W. (2005). Proteomics of halophilic archaea. *J. Chromatogr. B Analyt. Technol. Biomed. Life Sci.* **815**, 237–250.

Jung, K. H. and Spudich, J. L. (1998). Suppressor mutation analysis of the sensory rhodopsin I-transducer complex: insights into the color-sensing mechanism. *J. Bacteriol.* **180**, 2033–2042.

Karadzic, I. M. and Maupin-Furlow, J. A. (2005). Improvement of two-dimensional gel electrophoresis proteome maps of the haloarchaeon *Haloferax volcanii*. *Proteomics* **5**, 354–359.

Kennedy, S. P., Ng, W. V., Salzberg, S. L., Hood, L. and DasSarma, S. (2001). Understanding the adaptation of *Halobacterium* species NRC-1 to its extreme environment through computational analysis of its genome sequence. *Genome Res.* **11**, 1641–1650.

Klein, C., Garcia-Rizo, C., Bisle, B., Scheffer, B., Zischka, H., Pfeiffer, F., Siedler, F. and Oesterhelt, D. (2005). The membrane proteome of *Halobacterium salinarum*. *Proteomics* **5**, 180–197.

Krebs, M. P., Mollaaghababa, R. and Khorana, H. G. (1993a). Gene replacement in *Halobacterium halobium* and expression of bacteriorhodopsin mutants. *Proc. Natl. Acad. Sci. USA* **90**, 1987–1991.

Krebs, M. P., Spudich, E. N., Khorana, H. G. and Spudich, J. L. (1993b). Synthesis of a gene for sensory rhodopsin I and its functional expression in *Halobacterium halobium*. *Proc. Natl. Acad. Sci. USA* **90**, 3486–3490.

Lam, W. L. and Doolittle, W. F. (1989). Shuttle vectors for the archaebacterium *Halobacterium volcanii*. *Proc. Natl. Acad. Sci. USA* **86**, 5478–5482.

Lam, W. L. and Doolittle, W. F. (1992). Mevinolin-resistant mutations identify a promoter and the gene for a eukaryote-like 3-hydroxy-3-methylglutaryl-coenzyme A reductase in the archaebacterium *Haloferax volcanii*. *J. Biol. Chem.* **267**, 5829–5834.

Mankin, A. S., Zyrianova, I. M., Kagramanova, V. K. and Garrett, R. A. (1992). Introducing mutations into the single-copy chromosomal 23S rRNA gene of the archaeon *Halobacterium halobium* by using an rRNA operon-based transformation system. *Proc. Natl. Acad. Sci. USA* **89**, 6535–6539.

McCready, S. and Marcello, L. (2003). Repair of UV damage in *Halobacterium salinarum*. *Biochem. Soc. Trans.* **31**, 694–698.

McCready, S., Müller, J. A., Boubriak, I., Berquist, B. R., Ng, W. L., and DasSarma, S. (2005). UV irridation induces homologous recombination genes in a model archaeon. *Saline Systems* **1**, 3.

Mevarech, M. and Werczberger, R. (1985). Genetic transfer in *Halobacterium volcanii*. *J. Bacteriol.* **162**, 461–462.

Müller, J. A. and DasSarma, S. (2005). Genomic analysis of anaerobic respiration in the archaeon *Halobacterium* sp. strain NRC-1: dimethyl sulfoxide and trimethylamine N-oxide as terminal electron acceptors. *J. Bacteriol.* **187**, 1659–1667.

Ng, W. L. and DasSarma, S. (1993). Minimal replication origin of the 200-kilobase *Halobacterium* plasmid pNRC100. *J. Bacteriol.* **175**, 4584–4596.

Ng, W. L., Kothakota, S. and DasSarma, S. (1991). Structure of the gas vesicle plasmid in *Halobacterium halobium*: inversion isomers, inverted repeats, and insertion sequences. *J. Bacteriol.* **173**, 1958–1964.

Ng, W. L., Arora, P. and DasSarma, S. (1994). Large deletions in class III gas vesicle deficient mutants of *Halobacterium halobium*. *System. Appl. Microbiol.* **16**, 560–568.

Ng, W. V., Ciufo, S. A., Smith, T. M., Bumgarner, R. E., Baskin, D., Faust, J., Hall, B., Loretz, C., Seto, J., Slagel, J., Hood, L. and DasSarma, S. (1998). Snapshot of a large dynamic replicon in a halophilic archaeon: megaplasmid or minichromosome? *Genome Res.* **8**, 1131–1141.

Ng, W. V., Kennedy, S. P., Mahairas, G. G., Berquist, B., Pan, M., Shukla, H. D., Lasky, S. R., Baliga, N. S., Thorsson, V., Sbrogna, J., Swartzell, S., Weir, D., Hall, J., Dahl, T. A., Welti, R., Goo, Y. A., Leithauser, B., Keller, K., Cruz, R., Danson, M. J., Hough, D. W., Maddocks, D. G., Jablonski, P. E., Krebs, M. P., Angevine, C. M., Dale, H., Isenbarger, T. A., Peck, R. F., Pohlschröder, M., Spudich, J. L., Jung, K. W., Alam, M., Freitas, T., Hou, S., Daniels, C. J., Dennis, P. P., Omer, A. D., Ebhardt, H., Lowe, T. M., Liang, P., Riley, M., Hood, L. and DasSarma, S. (2000). Genome sequence of *Halobacterium* species NRC-1. *Proc. Natl. Acad. Sci. USA* **97**, 12176–12181.

Ni, B. F., Chang, M., Duschl, A., Lanyi, J. and Needleman, R. (1990). An efficient system for the synthesis of bacteriorhodopsin in *Halobacterium halobium*. *Gene* **90**, 169–172.

Nieuwlandt, D. T. and Daniels, C. J. (1990). An expression vector for the archaebacterium *Haloferax volcanii*. *J. Bacteriol.* **172**, 7104–7110.

Nomura, S. and Harada, Y. (1998). Functional expression of green fluorescent protein derivatives in *Halobacterium salinarum*. *FEMS Microbiol. Lett.* **167**, 287–293.

Patenge, N. and Soppa, J. (1999). Extensive proteolysis inhibits high-level production of eukaryal G protein-coupled receptors in the archaeon *Haloferax volcanii*. *FEMS Microbiol. Lett.* **171**, 27–35.

Patenge, N., Haase, A., Bolhuis, H. and Oesterhelt, D. (2000). The gene for a halophilic β-galactosidase (bgaH) of *Haloferax alicantei* as a reporter gene for promoter analyses in *Halobacterium salinarum*. *Mol. Microbiol.* **36**, 105–113.

Patterson, N. H. and Pauling, C. (1985). Evidence for two restriction-modification systems in *Halobacterium cutirubrum*. *J. Bacteriol.* **163**, 783–784.

Peck, R. F., DasSarma, S. and Krebs, M. P. (2000). Homologous gene knockout in the archaeon *Halobacterium salinarum* with ura3 as a counterselectable marker. *Mol. Microbiol.* **35**, 667–676.

Peck, R. F., Echavarri-Erasun, C., Johnson, E. A., Ng, W. V., Kennedy, S. P., Hood, L., DasSarma, S. and Krebs, M. P. (2001). *brp* and *blh* are required for synthesis of the retinal cofactor of bacteriorhodopsin in *Halobacterium salinarum*. *J. Biol. Chem.* **276**, 5739–5744.

Peck, R. F., Johnson, E. A. and Krebs, M. P. (2002). Identification of a lycopene β-cyclase required for bacteriorhodopsin biogenesis in the archaeon *Halobacterium salinarum*. *J. Bacteriol.* **184**, 2889–2897.

Pfeifer, F. and Ghahraman, P. (1993). Plasmid pHH1 of *Halobacterium salinarium*: characterization of the replicon region, the gas vesicle gene cluster and insertion elements. *Mol. Gen. Genet.* **238**, 193–200.

Piatribratov, M., Hou, S., Brooun, A., Yang, J., Chen, H. and Alam, M. (2000). Expression and fast-flow purification of a polyhistidine-tagged myoglobin-like aerotaxis transducer. *Biochim. Biophys. Acta* **1524**, 149–154.

Reuter, C. J. and Maupin-Furlow, J. A. (2004). Analysis of proteasome-dependent proteolysis in *Haloferax volcanii* cells, using short-lived green fluorescent proteins. *Appl. Environ. Microbiol.* **70**, 7530–7538.

Rosenshine, I. and Mevarech, M. (1989). Isolation and partial characterization of plasmids found in three *Halobacterium volcanii* isolates. *Can. J. Microbiol.* **35**, 92–95.

Rosenshine, I., Tchelet, R. and Mevarech, M. (1989). The mechanism of DNA transfer in the mating system of an archaebacterium. *Science* **245**, 1387–1389.

Schnabel, H. and Zillig, W. (1984). Circular structure of the genome of phage ΦH in a lysogenic *Halobacterium halobium*. *Mol. Gen. Genet.* **193**, 422–426.

Schnabel, H., Zillig, W., Pfaffle, M., Schnabel, R., Michel, H. and Delius, H. (1982). *Halobacterium halobium* phage ΦH. *EMBO J.* **1**, 87–92.

Sharov, A. A., Dudekula, D. B. and Ko, M. S. (2005). A web-based tool for principal component and significance analysis of microarray data. *Bioinformatics*. Bioinformatics Advance Access published on February 25, 2005, DOI 10.1093/bioinformatics/bti343.

Stoeckenius, W., Lozier, R. H. and Bogomolni, R. A. (1979). Bacteriorhodopsin and the purple membrane of halobacteria. *Biochim. Biophys. Acta* **505**, 215–278.

Stuart, E. S., Morshed, F., Sremac, M. and DasSarma, S. (2001). Antigen presentation using novel particulate organelles from halophilic archaea. *J. Biotechnol.* **88**, 119–128.

Stuart, E. S., Morshed, F., Sremac, M. and Dassarma, S. (2004). Cassette-based presentation of SIV epitopes with recombinant gas vesicles from halophilic archaea. *J. Biotechnol.* **114**, 225–237.

Tebbe, A., Klein, C., Bisle, B., Siedler, F., Scheffer, B., Garcia-Rizo, C., Wolfertz, J., Hickmann, V., Pfeiffer, F. and Oesterhelt, D. (2005). Analysis of the cytosolic proteome of *Halobacterium salinarum* and its implication for genome annotation. *Proteomics* **5**, 168–179.

Wang, X. and Seed, B. (2003). Selection of oligonucleotide probes for protein coding sequences. *Bioinformatics* **19**, 796–802.

Wang, G., Kennedy, S. P., Fasiludeen, S., Rensing, C. and DasSarma, S. (2004). Arsenic resistance in *Halobacterium* sp. strain NRC-1 examined by using an improved gene knockout system. *J. Bacteriol.* **186**, 3187–3194.

Wendoloski, D., Ferrer, C. and Dyall-Smith, M. L. (2001). A new simvastatin (mevinolin)-resistance marker from *Haloarcula hispanica* and a new *Haloferax volcanii* strain cured of plasmid pHV2. *Microbiology* **147**, 959–964.

Woods, W. G., Ngui, K. and Dyall-Smith, M. L. (1999). An improved transposon for the halophilic archaeon *Haloarcula hispanica*. *J. Bacteriol.* **181**, 7140–7142.

Zaigler, A., Schuster, S. C. and Soppa, J. (2003). Construction and usage of a onefold-coverage shotgun DNA microarray to characterize the metabolism of the archaeon *Haloferax volcanii*. *Mol. Microbiol.* **48**, 1089–1105.

Zibat, A. (2001). Efficient transformation of *Halobacterium salinarum* by a "freeze and thaw" technique. *Biotechniques* **31**, 1010–1012.

28 The Isolation and Study of Viruses of Halophilic Microorganisms

Kate Porter and Mike Dyall-Smith
Department of Microbiology and Immunology, University of Melbourne, Melbourne 3010, Australia

◆◆

CONTENTS

Introduction
Cultivation methods for haloarchaea
Isolation and cultivation of haloviruses
Characterization methods for haloviruses
Transfection of haloarchaea by halovirus DNA

◆◆◆◆◆ INTRODUCTION

This chapter will focus exclusively on viruses of the extremely halophilic Archaea (Family *Halobacteriaceae*), since members of the latter group make up the great majority of microbes living in the salt-saturated waters of salt lakes and saltern crystallizer ponds (Benlloch *et al.*, 2001; Burns *et al.*, 2004a). For recent, comprehensive reviews on halobacteria, see Oren (2001) and Grant *et al.* (2001). Haloarchaeal viruses (haloviruses) have been studied for some decades, but much of the early work was severely hampered, as in other areas of environmental microbiology, by the inability to culture the dominant haloarchaea present in salt lakes. Care must be taken in interpreting the earlier literature, which deals exclusively with haloviruses of the head–tail morphotype that infect strains of a single genus, *Halobacterium*. In addition, the stocks of most of these viruses are now probably lost. However, we are now at an exciting stage in haloviral study, as the cultivation barrier has recently been overcome, and a much wider diversity of viruses are being discovered and studied (Dyall-Smith *et al.*, 2003). Table 28.1 lists the haloviruses published in the readily accessible literature, and those currently under study are indicated. The references can be used to track the publication history relating to each virus.

The ecological importance of haloviruses has been demonstrated by direct electron-microscopy of salt lake waters (Guixa-Boixareu *et al.*, 1996;

Table 28.1 Summary of haloviruses

Name	Host[a]	Source	Temperate/Lytic	Genome size (kb)[b]	% G+C (mol%)	Morphology	Size (head; tail) (nm)	Comments	References
φCh1[f]	*Nab. magadii*	*Nab. magadii*	Temperate, integrates into host genome	58.5	62	Isometric head and contractile tail. Density 1.38 g cm^{-3}	70; 130	RNA in mature virus particles. Partial adenine methylation. Similar to ΦH. Fully sequenced	Baranyi et al. (2000), Klein et al. (2000, 2002), Witte et al. (1997)
ΦH	*Hbt. salinarum* ATCC 29341 (+ 4 other strains)	*Hbt. salinarum*	Temperate	59	65	Isometric head and contractile tail	64; 170	Well studied and partly sequenced. Forms plasmid in lysogens. Similar to Hs1 and φCh1	Blaseio and Pfeifer (1990), Gropp et al. (1992), Ken and Hackett (1991), Schnabel and Zillig (1984), Schnabel et al. (1982), Stolt and Zillig (1992, 1994), Torsvik (1982)
ΦN	*Hbt. salinarum* NRL/JW (+ 1 other strain)	*Hbt. salinarum*	ND[c]	56	70	Isometric head and non-contractile tail	55; 80	5-methyl-cytosine replaces cytosine	Vogelsang-Wenke and Oesterhelt (1988)
B10	*Halobacterium* sp. B10	ND[c]	ND[c]	ND[c]	ND[c]	Isometric head and non-contractile tail	ND[c]		Torsvik (1982)
HF1[f]	*Hfx. lucentense* NCIMB 13854 (+ 4 other genera)	Saltern, Australia	Lytic	77.7	55.8	Isometric head and contractile tail	58; 94	Broad host range, including *Hfx. volcanii*. Fully sequenced. Similar to HF2. Few restriction sites	Nuttall and Dyall-Smith (1993), Dyall-Smith et al. (2003)

Name	Host	Source	Lifestyle	Genome (kb)	G+C%	Morphology	Dimensions	Notes	References
HF2	*Hrr. coriense* ACAM 3911 (+ 1 other species of same genus)	Saltern, Australia	Lytic	75.9	55.8	Isometric head and contractile tail	58; 94	Host range restricted to members of *Halorubrum*. Fully sequenced. Similar to HF1. Few restriction sites	Dyall-Smith et al. (2003), Nuttall and Dyall-Smith (1993, 1995), Tang et al. (2002, 2004)
Hh-1	*Hbt. salinarum* ATCC 29341 (+ 1 other strain)	Fish sauce	Temperate (?) (immune to super-infection)	37.6	67	Bradley Group B	60; 100	Virus released without cell lysis	Pauling (1982), Rohrmann et al. (1983)
Hh-3	*Hbt. salinarum* ATCC 29341 (+ 1 other strain)	Fish sauce	Temperate (?) (immune to super-infection)	29.6	62	Bradley Group B	75; 50	Differs from Hh-1 in restriction pattern of its genome, and in the sizes of structural proteins	Pauling (1982), Patterson and Pauling (1985), Rohrmann et al. (1983)
Hs1	*Hbt. salinarum* stn. 1	*Hbt. salinarum*	Lytic	ND[c]	ND[c]	Bradley Group A	50; 120	Restriction map similar to ΦH	Torsvik (1982), Torsvik and Dundas (1974, 1980)
His1	*Har. hispanica* ATCC 33960	Saltern, Australia	Lytic/carrier	14.9	40	Spindle-shaped and small tail	74 × 44; 7	Distantly related to His2. Fully sequenced	Bath and Dyall-Smith (1998), Dyall-Smith et al. (2003), Unpublished[a]
His2	*Har. hispanica* ATCC 33960	Saltern, Australia	Lytic/carrier	16.2	ND[c]	Spindle-shaped and pleomorphic particles.	62 × 62; 0	Distantly related to His1. Fully sequenced	Dyall-Smith et al. (2003), Unpublished[a]
Ja.1	*Hbt. salinarum* NRC 34001 (+ 2 other strains)	Salt ponds, Jamaica	ND[c]	230[e]	ND[c]	Bradley Group A	90; 150	Broad range of *Hbt. salinarum* strains. Particle density 1.55 g ml^{-1}	Torsvik (1982), Wais et al. (1975)

(*continued*)

Table 28.1 Continued

Name	Host[a]	Source	Temperate/Lytic	Genome size (kb)[b]	% G+C (mol%)	Morphology	Size (head; tail) (nm)	Comments	References
PH1	Har. hispanica ATCC 33960 (+ 1 other strain)	Salt lake, Australia	Lytic	ND[c]	ND[c]	Spherical, lipid layer	51	Related to SH1	Unpublished
S45	Hbt. salinarum NRC 34001 (+ 3 other strains)	Salt ponds, Jamaica	Lytic	ND	ND	Bradley Group B1	40; 70	Persistent infection. Cells survive virus release	Daniels and Wais (1984)
S5100	Hbt. salinarum NRC 34001	Salt ponds, Jamaica	Lytic	65	76	Isometric head and contractile tail	65; 76		Daniels and Wais (1990, 1998)
SH1	Har. hispanica ATCC 33960 (+ 1 other strain)	Salt lake, Australia	Lytic	30.9	68.4	Spherical, lipid layer	70	Related to PH1. Few restriction enzymes will cut. Fully sequenced	Bamford et al. (2005), Porter et al. (2005)

[a] Hbt. cutirubrum and Hbt. halobium strains originally described as host cells are now designated Hbt. salinarum. Only culture collection numbers or strains are given for these strains. Hosts from other genera are named. Har = Haloarcula; Hbt = Halobacterium; Hfx = Haloferax; Hrr = Halorubrum; Nab = Natrialba.
[b] All haloviruses (to date) have linear double-stranded DNA genomes.
[c] ND, not determined.
[d] C. Bath, B.Sc. (honours) thesis, Department of Microbiology and Immunology, The University of Melbourne, 1995.
[e] Approximate value only.
[f] Haloviruses being studied currently are in bold letters.

Oren et al., 1997), where high levels of virus-like particles were observed, many with unusual morphologies (lemon-shaped or round particles). The spectrum of genome sizes associated with populations of virus-like particles has also been examined (Diez et al., 2000). Ecological studies of haloviruses have been reported by Wais and Daniels (Daniels and Wais, 1990; Wais and Daniels, 1985; Wais et al., 1975). The latter authors isolated *Halobacterium* spp. and their viruses directly from Jamaican salt lakes, and noted the significant effect of salt concentration on virulence. More recent molecular analyses of several haloviruses, including full genome sequences of some, have been presented in separate publications, e.g. HF1/HF2 (Tang et al., 2002, 2004), and φCh1 (Klein et al., 2002). The ΦH1 genome remains incomplete but is remarkably similar to that of φCh1 (Zillig et al., 1988).

◆◆◆◆◆ CULTIVATION METHODS FOR HALOARCHAEA

The cultivation of host cells is critical for the isolation and study of haloviruses, and the reader is directed to the relevant chapters on this elsewhere in this book and recent publications on the subject (e.g. Burns et al., 2004a), as well as the on-line (and free) handbook on haloarchaeal methods ("The HaloHandbook"), which may be found at http://www.microbiol.unimelb.edu.au/staff/mds/resources.html. We will assume this background information in describing the halovirus methods below.

◆◆◆◆◆ ISOLATION AND CULTIVATION OF HALOVIRUSES

Many of the haloviruses that have been studied intensively at the molecular level are temperate and were isolated (often by accident) from lysogenic laboratory strains of haloarchaea (see Table 28.1). These include φCh1, ΦH and ΦN (Schnabel et al., 1982; Vogelsang-Wenke and Oesterhelt, 1988; Witte et al., 1997). Others are natural isolates from hypersaline waters, and these viruses are often lytic, for example S45, HF2 and His1 (Bath and Dyall-Smith, 1998; Daniels and Wais, 1984; Nuttall and Dyall-Smith, 1993). In both cases, simple overlay plaquing methods adapted from conventional bacteriophage studies were used for isolation and determining virus concentrations. For example, Allen C. Wais and colleagues isolated viruses from the salt ponds of Yallahs in Jamaica, using enrichment cultures of *Halobacterium salinarum* strains inoculated with water samples (Daniels and Wais, 1984, 1990; Wais et al., 1975). An important discovery by the Wais group was the observation that salt concentration affects halovirus virulence, being increased at lower salt levels (Daniels and Wais, 1990, 1998). This has been put to great use in our laboratory by adjusting the salt concentration of plaque assays and growth culture media to maximize virus growth. We have isolated viruses on hosts such as *Haloferax*, *Haloarcula* and *Halorubrum*, and while enrichment cultures did not appear to be useful in increasing the levels of viruses before isolation, the direct plaque assay of water samples has provided adequate numbers of plaques (Bath and Dyall-Smith, 1998; Nuttall and Dyall-Smith, 1993; Porter et al., 2005). The difference in success of enrichment cultures may reflect the dominant hosts in the salt lakes studied in Jamaica and Australia. Numbers of plaques vary widely between water samples and the isolating hosts used in the plaque assay. One memorable Australian sample had a direct titre of 10^5 PFU ml^{-1} on *Haloarcula hispanica*, but generally our plaque numbers are around 1–2 per plate on hosts such as *Har. hispanica*, *Haloferax lucentense* and *Halorubrum coriense*. We have rarely recovered haloviruses from Australian salt lake samples on *Hfx. volcanii* DS2, perhaps reflecting the nature of this host. *Hfx. volcanii* DS2 was isolated from the Dead Sea, the water of which has levels of magnesium much higher than thalassohaline lakes.

Sometimes cleared zones that mimic plaques result from particulates in the water sample. Such particles should be largely removed by a low-speed centrifugation prior to plaque assay. The removal of all cells from the sample is also important because many haloarchaea produce halocins (Haseltine et al., 2001; Torreblanca et al., 1994), which could produce zones of inhibition in the overlay strain. In both cases, careful inspection of the centre of the plaque using a hand lens (which will show up particulates or microcolonies of cells clearly in the exact centre) and a further round of plaque titration should identify true virus plaques from false ones. While most haloarchaea appear to be able to form overlay layers suitable for plaque formation, some strains are more problematic, such as the recently isolated square haloarchaea (Burns et al., 2004b) and the members of the Antarctic Deep Lake group (Burns et al., 2004a). We are currently developing conditions for virus isolation on such strains.

Isolation of Haloviruses by Direct Plating

In recent years, halovirus isolation by direct plating of saltern water has yielded several viruses of diverse morphotypes (Bath and Dyall-Smith, 1998; Nuttall and Dyall-Smith, 1993; Porter et al., 2005). The technique is achieved through the use of simple overlay plating, adapted from the method of Adams (1959) used from plaque assay of bacteriophages. The isolation of haloviruses may be improved by screening on a range of potential hosts, as well as using a range of salt concentrations and incubation temperatures to stimulate plaquing. As identified by the Wais group, lower salt concentrations often stimulate better plaquing (Daniels and Wais, 1990, 1998). The general method is as follows:

1. Collect a water sample aseptically from a hypersaline pond.
2. Remove cells and cellular debris from 5 to 10 ml of water by centrifugation (5000 *g*, 10 min, room temperature). Retain the supernatant.

3. Mix the cleared water sample, host cells and molten agar medium to prepare overlay plates (described below). Incubate aerobically for visible plaques.
4. Pick an agar plug from an arising plaque with a glass Pasteur pipette or a sterile plastic micropipette tip.
5. Resuspend the agar plug in 500 µl of halovirus diluent (HVD; 2.47 M NaCl, 90 mM $MgCl_2$, 90 mM $MgSO_4$, 60 mM KCl, 3 mM $CaCl_2$, 10 mM Tris-HCl, pH 7.2).

Haloviral Plaque Assay

Thus far, standard plaque assays have proven to be the easiest method for isolating and estimating viral titers and we use overlay plaquing methods adapted from Adams' methods for plaque assay of bacteriophages (Adams, 1959). Typically, each halovirus has a defined host range, generally with one host producing the best growth under a set of specific plaque assay conditions (see Table 28.2). In general, the most efficient plaquing can be obtained through the use of salt concentrations and temperatures below the optimum for host growth. For example, *Har. hispanica* grows optimally at 37°C on modified growth medium (MGM) (Nuttall and Dyall-Smith, 1993) containing 23% (w/v) salt water, but plaques (for a number of different viruses) are generally best at 30°C

Table 28.2 Plaque assay conditions for haloviruses

Virus	Haloarchaeal host	Optimal plaque assay conditions	References
φCh1	*Nab. magadii*	37°C, 1–2 weeks, 24% salt	Witte *et al.* (1997)
ΦH	*Hbt. salinarum*	42°C, 3 days, 27% salt	Schnabel *et al.* (1982)
HF1	*Har. hispanica, Har. marismortui, Hfx. volcanii, Hfx. lucentense, Hbt. salinarum, Hrr. trapanicum*	*Haloterrigena* stain DS11, 37°C, 2 days, 12% salt	Nuttall and Dyall-Smith (1993)
HF2	*Hrr. coriense, Hrr. saccharovorum*	*Hrr. coriense*, 37°C. 2 days, 18% salt	Nuttall and Dyall-Smith (1993)
His1	*Har. hispanica*	30°C, 3 days, 18% salt	Bath and Dyall-Smith (1998)
SH1	*Har. hispanica, Halorubrum.* sp. (natural isolate; see Porter *et al.*, 2005)	*Har. hispanica*, 30°C, 3 days, 18% salt	Porter *et al.* (2005)

and 18% (w/v) salt water for this host. The procedure, in detail is as follows:

1. Prepare serial dilutions of the virus sample in MGM or HVD.
2. To duplicate 10 ml centrifuge tubes, add 100 µl of each dilution to 150 µl of host culture (in early exponential phase) and 3–4 ml of top-layer MGM (0.7 % (w/v) agar; maintained at 50°C). Mix gently for 2 s.
3. Pour the mixture over a solid MGM 90 mm plate (1.5% (w/v) agar).
   - The plate should be at room temperature or slightly warmer. It is best to hold the plate at a slight angle when first pouring the overlay, so the molten agar runs quickly across the plate. Then, rapidly move the overlay around to cover the plate and place on a perfectly horizontal surface to set (5–10 min).
4. After plates have set, invert and incubate aerobically for visible plaques.
   - Due to the high salt content, agar plates will dry very quickly, with the formation of salt crystals over the surface. To prevent this, plates should be incubated in a sealed bag or container. In some cases, the inclusion of a small dish of water within the container during incubation, or storage of the plates overnight at 4°C after incubation, may improve the clarity of plaques.

Preparation of Halovirus Lysates

Cultures of halovirus may be obtained by several methods; however, the liquid cultivation of haloviruses appears to be the most efficient method of achieving large volumes of virus at high titers. As in plaque assays, each halovirus has an optimum set of growth conditions and, again, more efficient virus production can often be achieved using salt concentrations and temperatures below the optimum for host growth.

1. Inoculate MGM with a 10% volume of host culture in early exponential phase.
   - The MGM salt concentration should be optimal for viral growth, rather than that of the host. Most viruses will grow well in a volume ratio of 1:4 (liquid to flask).
2. Inoculate the culture with virus.
   - The optimal multiplicity of infection (MOI) varies amongst the haloviruses, but tends to be relatively low. For example, good growth of HisI may be achieved using a MOI of ~0.1 (Bath and Dyall-Smith, 1998), whilst SH1 grows well with a MOI of ~0.05 (Porter et al., 2005), ΦH and HF1 at a MOI of ~0.01 (Nuttall and Dyall-Smith, 1993; Schnabel et al., 1982), and HF2 with a MOI of ~0.001 (Nuttall and Dyall-Smith, 1993).

3. Incubate aerobically with agitation for cleared cultures, indicating cell lysis (usually 2–4 days, 30 or 37°C)

   - Gentle agitation appears to be important for obtaining lysates of high titer. We use 100 rpm here.

4. For new viruses, sample at daily intervals and follow the turbidity and virus titer.

   - Some viruses or culture conditions do not produce obvious culture lysis, particularly for large culture volumes. However, the titer may still reach high levels. In some cases, the titer may also drop after the peak, so knowing the optimum incubation period is important.

Purification of Head–Tail Haloviruses

For sturdy haloviruses like the head–tail viruses (for example, ΦH and HF1), conventional bacteriophage purification strategies, with some modifications, have proven satisfactory. A separate method is also described for more delicate viruses. Most haloviruses require high salt levels for stability, so this must be maintained throughout the purification process. The use of chloroform to ensure the release of virus particles from cells and lyse any remaining cells should only be incorporated if it does not reduce virus titer (see below). Some isolates, such as HF2, are sensitive to chloroform (Nuttall and Dyall-Smith, 1993). Many haloviruses are stable at temperatures of 40°C or above, but it is safer to maintain temperatures at around 10°C during the purification. For example, SH1 is stable in HVD at 50°C, but loses infectivity at 25°C in the high concentrations of caesium chloride used for density gradient centrifugation. This deterioration is prevented at 10°C.

1. Harvest virus-infected cultures when clearing occurs, indicating cell lysis.

   - For some haloviruses, complete cell lysis never occurs. In these cases, it is best to harvest when OD_{550} is at a minimum.

2. Add DNase I (1 µg ml^{-1}) and RNase A (1 µg ml^{-1}). Incubate at 37°C for 1 h.
3. Pellet unlysed cells and cell debris by low-speed centrifugation (Sorvall GSA; 9000 rpm, 30 min, 10°C).
4. If the virus is known to be insensitive to chloroform, add 100 µl chloroform. Mix.

   - Omit this step if the organic solvent sensitivity of the virus is unknown. Alternatively, filtration through a 0.22 µm filter will remove any unlysed cells.

5. Precipitate virus from the cleared lysate by the addition of 10% (w/v) polyethylene glycol 6000 MW (PEG$_{6000}$). Incubate at 4°C for 24 h.

6. Concentrate the virus by centrifugation (Sorvall GSA; 9000 rpm, 30 min, 4°C).
7. Carefully decant the supernatant and very carefully wash the surface of the visible pellet with a slow stream of 1–2 ml of cold HVD (using a Pasteur pipette), to remove residual PEG_{6000}. Discard this wash, sucking out all the liquid from the tube. Removing the PEG_{6000} will make it much easier to resuspend the precipitated virus particles into a homogeneous suspension.
8. Resuspend the pellet in 3–4 ml HVD.
9. Layer the virus on caesium chloride (0.6 g ml^{-1} in HVD).
10. Centrifuge to equilibrium (Beckman 50Ti; 30 000 rpm, 20 h, 10°C)
11. Collect the light-scattering virus band.

- The virus band is most clearly seen if viewed in a darkened room with a small lamp shining directly above the centrifuge tube. A syringe with a long needle (preferably bent at 90° a few mm from the tip) is useful to harvest specific regions of the gradient.

12. Remove the CsCl from the harvested material by dialysis against HVD or by centrifugation and resuspension in HVD. For dialysis, the harvested material is dialysed against HVD (2×3 h/($500 \times$ volume)) and stored at 4°C.

- The titre and A_{260} (or the protein concentration) should be measured for each purification. In this way, virus concentrations (PFU ml^{-1}), and batch-to-batch quality (PFU A_{260}^{-1} or PFU μg^{-1}) can be assessed. The presence of host cell contaminants, such as flagella or membrane fragments, can be ascertained by negative-stain EM.

Purification of Spindle and Round Haloviruses

Some haloviruses, such as the lemon-shaped His1 and the round SH1, have particles that are structurally weaker than those with head–tail morphology, and require alternate purification procedures to minimize damage (Bath and Dyall-Smith, 1998; Porter et al., 2005). These viruses are also less dense, making them more difficult to free from cell debris, and they contain lipids, making them sensitive to organic solvents. In addition, the use of PEG precipitation to concentrate virus often causes aggregates of virus particles that are difficult to disperse. Described below are our purification methods for our most sensitive viruses (His1 and SH1).

1. Harvest virus-infected cultures when clearing occurs, indicating cell lysis.

- For some haloviruses, complete cell lysis never occurs. In these cases, it is best to harvest when OD_{550} is at a minimum. For new viruses, this will require some preliminary experiments to determine when the minimum occurs.

2. Pellet unlysed cells and cell debris by low-speed centrifugation (Sorvall GSA; 6000 rpm, 30 min, 10°C).

   - We used to use an Arklone (freon 113) extraction step here to minimize virus aggregates caused by host membrane lipids, but this solvent is no longer commercially available. Alternative solvents include Vertrel XF® (decafluoropentane) and HFE-7100® (methoxy-nonafluorobutane).

3. Pellet virus from the supernatant (Beckman SW28; 23 000 rpm, 13 h, 10°C) onto a 4 ml cushion of 30% (w/v) sucrose (made in HVD).

   - Each bucket holds 36 ml and there are 6 per rotor, for a total volume of 216 ml per spin. For larger volumes, the PEG precipitation method can be used (described in the previous section), but unless pellets are carefully washed to remove traces of PEG, they are often difficult to resuspend to fully dispersed virus particles (rather than clumps).
   - The use of a sucrose cushion was found to greatly alleviate the difficulty in resuspending virus pellets to homogeneous suspensions.

4. Resuspend the pellet in 3–4 ml HVD using a micropipette with a 1 ml tip to disperse the pellet by rapidly sucking and squirting. Remove the mixture to a plastic tube and vortex mix at room temperature to fully resuspend the virus. Any residual aggregates can be removed by a low-speed centrifugation. Store at 4°C until the next step.

5. Purify the virus preparation in a linear 10–70% (w/v) sucrose (in HVD) gradient (Beckman SW28; 23 000 rpm, 2 h, 10°C).

   - If a gradient maker is not available, a 10–70% sucrose gradient can be made by gently layering 4–5 ml each of the following sucrose solutions, starting with the most dense: 70, 60, 50, 40, 30, 20 and 10%. Store overnight at 4°C, to allow a continuous gradient to form, then load sufficient virus sample on top to fill the tube, and centrifuge.

6. Collect the light-scattering virus band.

   - The virus band is most clearly seen if viewed in a darkened room with a small lamp shining directly above the centrifuge tube.

7. Dialyse the virus against HVD (as above) and store at 4–8°C.

   - The titre and A_{260} (or the protein concentration) should be measured for each purification. In this way, virus concentrations (PFU ml^{-1}), and batch-to-batch quality (PFU A_{260}^{-1} or PFU μg^{-1}) can be assessed. The presence of host cell contaminants, such as flagella or membrane fragments, can be ascertained by negative-stain EM.

◆◆◆◆◆ CHARACTERIZATION METHODS FOR HALOVIRUSES

For accurate taxonomic assignments, the characteristics of newly isolated viruses need to be determined. These are also useful to know when formulating or refining purification regimes (e.g. sensitivity to chloroform and CsCl). Here we describe methods for some of the more basic, virological parameters, particularly those that are more specific for haloviruses. As a guide, the recent description of halovirus SH1 (Porter *et al.*, 2005) contained a number of typical characterization tests (growth curves, stabilities, genome size, structural proteins).

Host-Range Determinations

One of the most important tests is to determine whether isolates can infect one (monovalent) or more than a single species of host (divalent, polyvalent). When reading the halovirus literature, it should be remembered that the number of haloarchaeal species available for laboratory study has increased dramatically over the last 20 years (Grant *et al.*, 2001; Oren, 2001), and we now have a much greater understanding of their numerical significance in the natural environment. In addition, the isolating host of a virus may not produce the best plaques or highest titres. For example, HF1 was isolated on *Hfx. lucentense* but plaques much better on *Halorubrum trapanicum* DS66 (a derivative clone of NCIMB 784). Consequently, screening a wide range of haloarchaea for their ability to host viruses can improve the ease of characterization. Of the recently isolated haloviruses, there are examples that are monovalent (e.g. His1), divalent (HF2, SH1) and polyvalent (HF1) (Bath and Dyall-Smith, 1998; Nuttall and Dyall-Smith, 1993; Porter *et al.*, 2005). The valencies of many others are uncertain because most of the currently described haloarchaeal genera were not known at the time the early head–tail haloviruses were isolated and studied.

It is important for host-range studies to have high titred virus stocks to overcome restriction barriers (Daniels and Wais, 1984) and, as mentioned earlier, virus plaquing may be stimulated through the use of salt concentrations and temperatures lower than the host optima. The method below is convenient when testing the host ranges of a number of halovirus isolates at the same time.

1. In a 10 ml centrifuge tube, add 150 µl of host culture in mid-exponential phase ($OD_{550} = 0.8$–1.0) and 3–4 ml of top-layer MGM (0.7% (w/v) agar; maintained at 50°C). Mix gently for 2 s.
2. Pour the mixture over a solid MGM plate (1.5% (w/v) agar).
3. After plates have set, spot 5 µl of a virus sample onto the top of the agar.
    - In order to overcome any potential restriction barriers, virus stocks of high titer ($\geq 10^{11}$ PFU ml^{-1}) should be used for spot tests. This may require concentration of the cell lysate to achieve a sufficiently high titred stock.

- If you have a number of viruses to test, the underside of each layer agar plate can be marked with a grid pattern, and each square labelled.

4. After virus samples are dry, invert the plate and incubate aerobically for visible plaques (1–3 days).

- It is sometimes observed that there is inhibition of growth of the lawn culture, rather than true plaque formation. This is likely to be due to halocin production by the host used to grow the virus. The halocin can be removed by dialysing the lysate, or by using semi-purified or purified virus preparations (which should also provide higher titres).
- Susceptible hosts should then be tested quantitatively using the standard plaque assay to determine the efficiency of plaquing relative to the isolating host. Restriction barriers can then be assessed.

Virus Particle Stability Determinations

These are no longer critical for formal taxonomic proposals of new virus groups but provide useful information for formulating purification regimes, storage conditions and for studying the structural characteristics of haloviruses. For example, treatments that break virus particles into distinct components, such as removing an outer layer, can aid in determining the locations and interactions between subunits. After the treatments described below, virus titres are determined by plaque assay (as described above).

Caesium chloride stability

Equilibrium density centrifugation in CsCl gradients is commonly used to purify viruses, but this method exposes the virus to high concentrations of this salt (≥ 3.5 M), and it is important to know the stability of a new virus in CsCl before using this method. All the haloviruses we have worked with are stable in such gradients at 4–10°C, but SH1 is unstable in CsCl at 25°C, so care must be taken to maintain cold conditions during exposure to CsCl.

1. Dilute a virus sample in caesium chloride in a volume ratio of 1:100 (virus to caesium chloride)

- Use a CsCl stock solution (in HVD) at a concentration similar to the expected density of the virus, or the method you wish to use. For example, 0.6 g ml^{-1} (3.57 M) CsCl is used in the virus purification protocol given above, but other methods can use concentrations of 1–1.7 g ml^{-1}.

2. Incubate at appropriate temperatures (e.g. 10 and 25°C). After 24 h, remove a sample and dilute in HVD.

Chloroform sensitivity

Chloroform is commonly used in purification regimes to free virus particles from host cell membranes and to lyse host cells, examples include ΦH (Vogelsang-Wenke and Oesterhelt, 1988) and φCh1 (Witte *et al.*, 1997). Sensitivity to chloroform was also used widely in classical virology to discriminate between viruses with or without lipid envelopes. As might be expected, all the lipid-containing viruses (SH1, His1) are chloroform sensitive. Head–tail haloviruses differ in sensitivity, even between related viruses; for example HF1 is resistant whereas HF2 titres drop to negligible values after 5 min of exposure (Nuttall and Dyall-Smith, 1993).

1. In a 1.5 ml plastic microfuge tube, mix a virus sample in chloroform in a volume ratio of 1:4 (chloroform to virus).
2. Incubate at 25°C, with constant agitation in a shaker. At appropriate time points, allow the mix to settle, remove a sample from the upper layer (without any residual chloroform) and dilute in HVD.

Salt concentration

Like their hosts, haloviruses are often very sensitive to a low ionic environment. This has many implications for purification, storage and downstream manipulation of halovirus particles. Some haloviruses are resistant to low salt, such as His1 and Hh-1 (Bath and Dyall-Smith, 1998; Pauling, 1982) and ΦN (Vogelsang-Wenke and Oesterhelt, 1988) retains infectivity even in distilled water.

1. Dilute a virus sample in pure water in a volume ratio of 1:1000 (virus to pure water) and mix quickly.
2. Incubate at room temperature, with constant, gentle agitation. At appropriate time points, remove a sample and dilute in HVD.

Salt stabilization

Viruses that are sensitive to depleted ionic environments may be stabilized by the presence of certain ions, such as Mg^{2+} (e.g. HF1, HF2 and SH1) (Nuttall and Dyall-Smith, 1993; Porter *et al.*, 2005).

1. Pellet virus from the supernatant (Beckman SW28; 23 000 rpm, 13 h, 10°C).
2. Rinse the pellet with the appropriate solution.
   - We make up mixtures based on the recipe of HVD (see above), but with various concentrations of NaCl, KCl, $MgCl_2$ and $MgSO_4$.
3. Resuspend the pellet in the appropriate solution.
4. Incubate at room temperature, with constant, gentle agitation. At appropriate time points, remove a sample and dilute in HVD.

pH stability

Stability to changes in pH is an important characteristic to know for the storage and handling of haloviruses, particularly for haloalkiliphilic viruses such as ɸCh1. The pH of thalassohaline salt lakes, at around saturation, is about 8, and haloviruses isolated from these environments are stable at pH values ranging from 7.5 to 8. pH stability curves for haloviruses have rarely been reported.

1. Dilute a virus sample in the appropriate pH buffer in a volume ratio of 1:100 (virus to buffer).

    - The buffer is HVD, buffered with appropriate Tris·HCl (pH 6–10) or potassium phosphate (pH 5–8).

2. Incubate at room temperature for 30 min. Remove a sample and dilute in HVD.

Temperature stability

Storage and daily manipulation of haloviruses are affected by the temperature stability of haloviruses. Most haloviruses are relatively stable at room temperature, at least in the short term, and can be kept for several years at 4°C.

1. Dispense equal volumes (e.g. 50–100 µl) of virus preparation to a number of plastic tubes (to be incubated at different temperatures).

    - Microcentrifuge tubes (1.5 ml) or PCR tubes (0.2 ml) are convenient for this purpose as their caps provide liquid-tight seals.

2. Incubate each tube at the appropriate temperature for 30 min. Remove a sample and dilute in HVD.

    - A temperature range of 4–80°C provides a good coverage, as many haloviruses are stable over a wide range of temperatures. A convenient way of incubating at a range of temperatures is to use a PCR machine that allows a temperature gradient across the sample block. This can be done at a wider range at first, and then narrower ranges can be used.

Negative-Stain Transmission Electron Microscopy

Examination of halovirus preparations by negative-stain transmission electron microscopy (TEM) is widely used to determine particle morphology, purity and quality. However, for viruses that require high salt for stability, the drying of samples onto grids produces salt crystals that may occlude large regions of the grid. Alternate methods, such as cryo-electron microscopy, can circumvent this but are often not readily available and TEM may be the only practical option. For some viruses, such as His1, virion particles may be relatively insensitive to ionic environment and dilution of virus preparations with pure water will improve TEM visualization. In other cases, it is necessary to rely on dilution of the salt in the stain. Our method for negative-stain TEM is adapted from that of Tarasov *et al.* (2000) and is relatively simple.

If particle disruption is a problem, fixation with gluteraldehyde (Nuttall and Dyall-Smith, 1993) can be tried (e.g. 6%, 1 h on ice) but solutions should be buffered with 4-morpholinoethane sulfonic acid (MES) instead of Tris-HCl, as the latter compound will react with the fixative.

1. Place a 20 µl drop of the virus sample on a clean surface (such as a square of Parafilm® plastic wrap, fixed to the bench top by tape or by pressing down on the corners with a fingernail).
2. Using tweezers, carefully place the formvar-coated grid on top of the virus sample (coated side down) and allow virus particles to adsorb to the grid for 1.5–2 min.
    - The best results are obtained with new grids. We generally purchase formvar-coated or carbon-coated copper grids (ProSciTech).
3. Place a 20 µl drop of 2% uranyl acetate near the virus sample.
    - We have found that the best results are obtained with uranyl acetate, whilst poorer results are obtained with ammonium molybdate. The uranyl acetate solution should be filtered through a 0.22 µm filter just prior to use.
4. Lift the grid off from the virus sample and remove most of the attached liquid by a light touch of the edge to a piece of filter paper (Whatman No. 1 or equivalent).
5. Place the grid, formvar side down, onto the uranyl acetate solution and leave for 1–2 min. While staining, this step also dilutes the salts in the sample.
6. Remove the grid from the stain and absorb as much excess liquid as possible by touching the edges with filter paper.
7. Allow the grid to air dry (5–10 min).

Extraction of DNA from Halovirus Preparations

Extraction of DNA from haloviruses is an integral part of virus characterization. Our methods for haloviral DNA extraction have been modified from those of Ausubel et al. (1994) for bacteriophages. The removal of salts is an important factor in good quality DNA preparations.

1. In a 1.5 ml microcentrifuge tube, mix 200 µl of purified virus with 300 µl of pure water.
    - For reference, for haloviruses such as SH1, $\sim 10^{11}$ PFU yields ~ 0.5–1 µg DNA.
2. Add proteinase K (to 50 µl ml^{-1}) and SDS (to 0.1%). Vortex for 2 s and incubate at 37°C for 1 h. (We use stock solutions of 20% SDS and 17 mg ml^{-1} of proteinase K.)

- Non-protease treated DNA may be obtained by omitting the proteinase K. However, in the absence of proteinase K treatment, halovirus SHI must be incubated at 56°C or above (in step I), in order to prevent DNA degradation by a co-purifying nuclease.

3. Extract the DNA using phenol–chloroform–*iso*-amyl-alcohol.
   - Although one phenol–chloroform extraction is adequate for viruses such as HFI, some viruses, including SHI, require at least three phenol–chloroform extractions in order to eliminate activity by a co-purifying nuclease.
4. Precipitate the DNA by adding a 1/10-volume of 3 M sodium acetate (50 µl, final concentration of 0.3 M) and two volumes (1 ml) of 100% ethanol. Mix well and incubate on ice for 15 min.
5. Pellet the DNA by centrifugation (16 000 g, 15 min, room temperature).
6. Wash the DNA pellet twice with 1 ml of 70% (w/v) ethanol.
7. Dry the DNA pellet. Resuspend in 20–50 µl of pure water.

Extraction of Protein from Halovirus Preparations

The extraction and purification of viral proteins is a necessary step before protein identification methods can be applied. Often the genome sequence has been completed before the virus structural proteins are analysed. The methods below have been modified from those described by Ausubel *et al.* (1994), Schägger and von Jagow (1987) and Laemmli (1970).

1. Disrupt the particles and precipitate proteins from purified virus by the addition of cold 10% trichloroacetic acid (TCA) and incubate on ice for 15 min.
   - As a guide, for SHI virus, 2×10^{12} PFU represents about I mg of protein.
   - The precipitation step is necessary to eliminate high levels of salt that will interfere with electrophoresis.
2. Centrifuge to pellet the proteins (12 000 g, 15 min, 4°C)
3. Wash protein pellet 3 × in cold acetone (12 000 g, 10 min, 4°C). Drain thoroughly after the last wash to remove the acetone. Note that acetone is flammable.
4. Dry the pellet in a vacuum desiccator (15 min, room temperature).
5. Resuspend in a sample buffer that is appropriate for the gel system to be used for protein separation and heat treat, immediately to destroy any proteases. If not to be used immediately, store the protein preparation at −20°C.

◆◆◆◆◆ TRANSFECTION OF HALOARCHAEA BY HALOVIRUS DNA

The PEG-mediated introduction of viral DNA into spheroplasts of susceptible haloarchaeal cells will initiate halovirus infection and produce viable virus particles. This allows the genetic study of halovirus genomes by direct manipulation of their DNA and also overcomes the cell-receptor barrier that restricts haloviral host range. Our methods for haloarchaeal transfection have been adapted from those of Cline and colleagues who transfected *Hfx. volcanii* with DNA from the *Hbt. salinarum*-specific halovirus ΦH (Cline and Doolittle, 1987).

1. To a 1.5 ml microcentrifuge tube, add 1 ml of a haloarchaeal culture (OD_{550} of 0.8–1.0).
2. Pellet the cells by centrifugation (3400 g, 15 min, room temperature).
3. Gently resuspend the cells in 500 μl of buffered spheroplasting solution (1 M NaCl, 27 mM KCl, 15% sucrose, 50 mM Tris-HCl, pH 8.75).
4. Pellet the cells by centrifugation (2300 g, 10 min, room temperature).
5. Gently resuspend the cells in 100 μl of buffered spheroplasting solution, with glycerol (1 M NaCl, 27 mM KCl, 15% sucrose, 15% glycerol, 50 mM Tris-HCl, pH 8.75).
6. To form spheroplasts, add EDTA (pH 8.0; 50 mM) to the cells and mix by tapping the tube. Incubate at room temperature for 10 min.
7. To 1.5 ml microcentrifuge tubes containing 1–10 μl viral DNA, add 100 μl of spheroplasts. Mix gently (by tapping) and incubate at room temperature for 2–5 min.

   - Add no more than 1 μg of DNA, as higher levels of DNA will precipitate with the addition of PEG.

8. Add 100 μl of 60% PEG MW 600 (PEG_{600}), made up in 40% unbuffered spheroplasting solution (1 M NaCl, 27 mM KCl, 15% sucrose). Mix gently and incubate at room temperature for 20 min.
9. Add 1 ml of MGM.

   - Use MGM with the optimum salt concentration for growth of the haloarchaeal strain. For example, for transfection of *Har. hispanica* cells, use 23% MGM, whilst for *Hfx. lucentense*, use 18% MGM.

10. Pellet the cells by centrifugation (4000 g, 5 min, room temperature). Gently resuspend the cells in 1 ml of MGM and allow recovery by incubating at 37°C for 2 h (shaking, 180 rpm).
11. Make serial 10-fold dilutions of the cell suspension in MGM and assay by plaque titration using a lawn culture of a susceptible host, as described earlier.

    - If plating directly after the 2 h recovery period, then you will be counting infectious foci. The eclipse period for most haloviruses is at least 4 h.
    - Transfection efficiencies vary markedly between different viruses and cells so a number of serial dilutions may need to be plated. Some viruses have yet to be shown to transfect at all (e.g. HFl, HF2) while

others show high activity e.g. ΦH DNA into *Hfx. volcanii* gave $\sim 10^6$–10^7 transfectants μg^{-1} (Cline and Doolittle, 1987).
- If very small numbers of transfectants are expected, and the lawn culture is different to the transfectant cells, then the zero dilution plates of the plaque assay may show turbid plaques because of mixed cell population.

List of suppliers

3M Corporation
Corporate Headquarters
3M Center
St. Paul, MN 55144-1000, USA
Tel: +1-888-364-3577
http://www.3m.com/

HFE-7100 3M™ Novec™ Engineered Fluid

ProSciTech
P.O. Box 111
Thuringowa, Queensland, Australia
Tel: +61 7 4773 2244
http://www.proscitech.com.au

Formvar film/400 mesh copper grids (#GSCu400F-50), Strong Carbon Film on 400 Mesh Cu (#GSCu400C-50), Uranyl acetate (#CO79)

Sigma-Aldrich
P.O. Box 14508
St. Louis, MO 63178, USA
Tel: +1-800-325-3010
http://www.sigmaaldrich.com/

Vertrel XF® (cat. #94884)

References

Adams, M. H. (1959). *Bacteriophages*. Interscience, New York.
Ausubel, F. M., Brent, R., Kingston, R. E., Moore, D. D., Seidman, J. G., Smith, J. A. and Struhl, K. (1994). *Current Protocols in Molecular Biology*. John Wiley and Sons, New York.
Bamford, D. H., Ravantii, J. J., Rönnholm, G., Laurinavičius, S., Kukkaro, P., Dyall-Smith, M., Somerharju, P., Kalkkinen, N. and Bamford, J. K. (2005). Constituents of SH1, a novel membrane-containing virus infecting the halophilic euryarchaeon *Haloarcula hispanica*. *J. Virol.* **79**, 9097–9107.
Baranyi, U., Klein, R., Lubitz, W., Kruger, D. H. and Witte, A. (2000). The archaeal halophilic virus-encoded Dam-like methyltransferase M.

φCh1-I methylates adenine residues and complements dam mutants in the low salt environment of *Escherichia coli*. *Mol. Microbiol.* **35**, 1168–1179.

Bath, C. and Dyall-Smith, M. L. (1998). His1, an archaeal virus of the Fuselloviridae family that infects *Haloarcula hispanica*. *J. Virol.* **72**, 9392–9395.

Benlloch, S., Acinas, S. G., Antón, J., López-López, A., Luz, S. P. and Rodríguez-Valera, F. (2001). Archaeal biodiversity in crystallizer ponds from a solar saltern: culture versus PCR. *Microb. Ecol.* **41**, 12–19.

Blaseio, U. and Pfeifer, F. (1990). Transformation of *Halobacterium halobium*: development of vectors and investigation of gas vesicle synthesis. *Proc. Natl. Acad. Sci. USA* **87**, 6772–6776.

Burns, D. G., Camakaris, H. M., Janssen, P. H. and Dyall-Smith, M. L. (2004a). Combined use of cultivation-dependent and cultivation-independent methods indicates that members of most haloarchaeal groups in an Australian crystallizer pond are cultivable. *Appl. Environ. Microbiol.* **70**, 5258–5265.

Burns, D. G., Camakaris, H. M., Janssen, P. H. and Dyall-Smith, M. L. (2004b). Cultivation of Walsby's square haloarchaeon. *FEMS Microbiol. Lett.* **238**, 469–473.

Cline, S. W. and Doolittle, W. F. (1987). Efficient transfection of the archaebacterium *Halobacterium halobium*. *J. Bacteriol.* **169**, 1341–1344.

Daniels, L. L. and Wais, A. C. (1984). Restriction and modification of halophage S45 in *Halobacterium*. *Curr. Microbiol.* **10**, 133–136.

Daniels, L. L. and Wais, A. C. (1990). Ecophysiology of bacteriophage S5100 infecting *Halobacterium cutirubrum*. *Appl. Environ. Microbiol.* **56**, 3605–3608.

Daniels, L. L. and Wais, A. C. (1998). Virulence in phage populations infecting *Halobacterium cutirubrum*. *FEMS Microbiol. Ecol.* **25**, 129–134.

Diez, B., Antón, J., Guixa-Boixereu, N., Pedrós-Alió, C. and Rodríguez-Valera, F. (2000). Pulsed-field gel electrophoresis analysis of virus assemblages present in a hypersaline environment. *Int. Microbiol.* **3**, 159–164.

Dyall-Smith, M., Tang, S. L. and Bath, C. (2003). Haloarchaeal viruses: how diverse are they? *Res. Microbiol.* **154**, 309–313.

Grant, W. D., Kamekura, M., McGenity, T. J. and Ventosa, A. (2001). Class III. Halobacteria class nov. In *Bergey's Manual of Systematic Bacteriology* (D. R. Boone, R. W. Castenholz and G. M. Garrity, eds), pp. 294–334. Springer-Verlag, New York.

Gropp, F., Grampp, B., Stolt, P., Palm, P. and Zillig, W. (1992). The immunity-conferring plasmid pFHL from the *Halobacterium salinarium* phage FH: nucleotide sequence and transcription. *Virology* **190**, 45–54.

Guixa-Boixareu, N., Calderón-Paz, J. I., Heldal, M., Bratbak, G. and Pedrós-Alió, C. (1996). Viral lysis and bacterivory as prokaryotic loss factors along a salinity gradient. *Aquat. Microb. Ecol.* **11**, 215–227.

Haseltine, C., Hill, T., Montalvo-Rodriguez, R., Kemper, S. K., Shand, R. F. and Blum, P. (2001). Secreted euryarchaeal microhalocins kill hyperthermophilic Crenarchaea. *J. Bacteriol.* **183**, 287–291.

Ken, R. and Hackett, N. R. (1991). *Halobacterium halobium* strains lysogenic for phage ΦH contain a protein resembling coliphage repressors. *J. Bacteriol.* **173**, 955–960.

Klein, R., Greineder, B., Baranyi, U. and Witte, A. (2000). The structural protein E of the archaeal virus φCh1: evidence for processing in *Natrialba magadii* during virus maturation. *Virology* **276**, 376–387.

Klein, R., Baranyi, U., Rössler, N., Greineder, B., Scholz, H. and Witte, A. (2002). *Natrialba magadii* virus φCh1: first complete nucleotide sequence and functional organization of a virus infecting a haloalkaliphilic archaeon. *Mol. Microbiol.* **45**, 851–863.

Laemmli, U. K. (1970). Cleavage of structural proteins during the assembly of the head of the bacteriophage T4. *Nature* **227**, 680–685.

Nuttall, S. D. and Dyall-Smith, M. L. (1993). HF1 and HF2: Novel bacteriophages of halophilic archaea. *Virology* **197**, 678–684.

Nuttall, S. D. and Dyall-Smith, M. L. (1995). Halophage HF2: genome organization and replication strategy. *J. Virol.* **69**, 2322–2327.

Oren, A. (2001). The Order *Halobacteriales*. In *The Prokaryotes: An Evolving Electronic Resource for the Microbiological Community* (M. Dworkin, S. Falkow, E. Rosenberg, K.-H. Schleifer and E. Stackebrandt, eds), 3rd edn., Springer-Verlag, New York (http://link.springer-ny.com/link/service/books/10125). release 3.2, 25 July 2001.

Oren, A., Bratbak, G. and Heldal, M. (1997). Occurrence of virus-like particles in the Dead Sea. *Extremophiles* **1**, 143–149.

Patterson, N. H. and Pauling, C. (1985). Evidence for two restriction-modification systems in *Halobacterium cutirubrum*. *J. Bacteriol.* **163**, 783–784.

Pauling, C. (1982). Bacteriophages of *Halobacterium halobium*: isolated from fermented fish sauce and primary characterization. *Can. J. Microbiol.* **28**, 916–921.

Porter, K., Kukkaro, P., Bamford, J. K., Bath, C., Kivelä, H. M., Dyall-Smith, M. L. and Bamford, D. H. (2005). SH1: A novel, spherical halovirus isolated from an Australian hypersaline lake. *Virology* **335**, 22–33.

Rohrmann, G. F., Cheney, R. and Pauling, C. (1983). Bacteriophages of *Halobacterium halobium*: Virion DNA's and proteins. *Can. J. Microbiol.* **29**, 627–629.

Schnabel, H. and Zillig, W. (1984). Circular structure of the genome of phage ΦH in a lysogenic *Halobacterium halobium*. *Mol. Gen. Genet.* **193**, 422–426.

Schnabel, H., Zillig, W., Pfaffle, M., Schnabel, R., Michel, H. and Delius, H. (1982). *Halobacterium halobium* phage ΦH. *EMBO J.* **1**, 87–92.

Schägger, H. and von Jagow, G. (1987). Tricine-sodium dodecyl sulfate-polyacrylamide gel electrophoresis for the separation of proteins in the range from 1 to 100 kDa. *Anal. Biochem.* **166**, 368–379.

Stolt, P. and Zillig, W. (1992). *In vivo* studies on the effects of immunity genes on early lytic transcription in the *Halobacterium salinarium* phage ΦH. *Mol. Gen. Genet.* **235**, 197–204.

Stolt, P. and Zillig, W. (1994). Gene regulation in halophage ΦH – more than promoters. *Syst. Appl. Microbiol.* **16**, 591–596.

Tang, S. L., Nuttall, S., Ngui, K., Fisher, C., Lopez, P. and Dyall-Smith, M. (2002). HF2: a double-stranded DNA tailed haloarchaeal virus with a mosaic genome. *Mol. Microbiol.* **44**, 283–296.

Tang, S. L., Nuttall, S. and Dyall-Smith, M. (2004). Haloviruses HF1 and HF2: evidence for a recent and large recombination event. *J. Bacteriol.* **186**, 2810–2817.

Tarasov, V. Y., Pyatibratov, N. G., Tang, S. L., Dyall, S. M. and Fedorov, O. V. (2000). Role of flagellins from A and B loci in flagella formation of *Halobacterium salinarum*. *Mol. Microbiol.* **35**, 69–78.

Torreblanca, M., Meseguer, I. and Ventosa, A. (1994). Production of halocin is a practically universal feature of archaeal halophilic rods. *Lett. Appl. Microbiol.* **19**, 201–205.

Torsvik, T. (1982). Characterization of four bacteriophages for *Halobacterium*, with special emphasis on phage Hs1. In *Archaebacteria* (O. Kandler, ed.), p. 351, Gustav Fischer, Stuttgart.

Torsvik, T. and Dundas, I. D. (1974). Bacteriophage of *Halobacterium salinarium*. *Nature* **248**, 680–681.

Torsvik, T. and Dundas, I. D. (1980). Persisting phage infection in *Halobacterium salinarium* str.1. *J. Gen. Virol.* **47**, 29–36.

Vogelsang-Wenke, H. and Oesterhelt, D. (1988). Isolation of a halobacterial phage with a fully cytosine-methylated genome. *Mol. Gen. Genet.* **211**, 407–414.

Wais, A. C. and Daniels, L. L. (1985). Populations of bacteriophage infecting *Halobacterium* in a transient brine pool. *FEMS Microbiol. Ecol.* **31**, 323–326.

Wais, A. C., Kon, M., MacDonald, R. E. and Stollar, B. D. (1975). Salt-dependent bacteriophage infecting *Halobacterium cutirubrum* and *H. halobium*. *Nature* **256**, 314–315.

Witte, A., Baranyi, U., Klein, R., Sulzner, M., Luo, C., Wanner, G., Krüger, D. H. and Lubitz, W. (1997). Characterization of *Natronobacterium magadii* phage ϕCh1, a unique archaeal phage containing DNA and RNA. *Mol. Microbiol.* **23**, 603–616.

Zillig, W., Reiter, W.-D., Palm, P., Gropp, F., Neumann, H. and Rettenberger, M. (1988). Viruses of Archaebacteria. In *The Bacteriophages* (R. Calendar, ed.), pp. 517–555. Plenum Publishing Corporation, New York.

29 Detection, Quantification and Purification of Halocins: Peptide Antibiotics from Haloarchaeal Extremophiles

Richard F Shand
Department of Biological Sciences, Northern Arizona University, Flagstaff, AZ 86011, USA

◆◆

CONTENTS

Introduction
Detecting and quantifying halocin activity
Halocin purification

◆◆◆◆◆ **INTRODUCTION**

Production of peptide (<5–10 kDa) or protein (>10 kDa) antibiotics is a near universal feature of life. There are hundreds of examples from Bacteria (bacteriocins) as well as from a wide variety of Eucarya (eucaryocins) including mammals, frogs, insects, protozoans and plants (see references in O'Connor and Shand, 2002; a constantly growing database currently containing some 880 eucaryocin sequences is at http://www.bbcm.univ.trieste.it/~tossi/pag1.htm). In animals, peptide antibiotic production is considered part of the animal's innate immune system. The scope of protein antibiotic production in the Archaea (archaeocins) is just emerging (O'Connor and Shand, 2002). In extremely halophilic members of the domain Archaea (haloarchaea), protein antibiotic production (halocins) is a near universal feature of haloarchaeal rods (Torreblanca *et al.*, 1994). However, the only other genus of the domain Archaea to show archaeocin production is *Sulfolobus* (sulfolobicins produced by the provisionally named "Sulfolobus islandicus"; Prangishvili *et al.*, 2000). The production of archaeocins in other members of the domain Archaea has yet to be reported, most likely because investigators have yet to look for them.

Halocin Characteristics

Table 29.1 summarizes the characteristics of halocins described so far (a complete bibliography of the halocin literature can be found at http://jan.ucc.nau.edu/~shand/bibliography.html). The microhalocins tend to be thermostable, with some withstanding boiling (e.g. halocins A4, C8 and S8), while the larger halocins are thermolabile (e.g. halocins H1 and H4). Similarly, the microhalocins can be desalted and will retain activity while the larger halocins are salt-dependent.

Activity spectra

Halocin activity is usually confined to other haloarchaea, with some inhibiting many species (a broad spectrum) and others just a few (a narrow spectrum; see Table 29.1). However, halocins A4, R1 and S8, all inhibit three species of *Sulfolobus* (a hyperthermophilic crenarchaeote), with halocin R1 also inhibiting *Methanosarcina thermophila* (Hazeltine et al., 2001). Halocins are not active against Bacteria or Eucarya for at least two reasons. First, the larger halocins lose activity when desalted – a prerequisite for testing against non-extreme halophiles. Although microhalocins can withstand desalting, the way this experiment is performed is to desalt the microhalocin and then place it onto a lawn of sensitive haloarchaeal cells to determine activity. Consequently, the desalted microhalocin is placed back into a hypersaline environment, so it is not clear if the microhalocin is still active in a desalted state, or if it is inactive but simply refolds and reacquires activity when it is placed back into a hypersaline environment. Second, most eucaryocins and many bacteriocins (but by no means all) carry a positive net charge and interact with the negatively charged cell membrane lipids of the target cell, either to gain access to the interior or to disrupt the cell membrane. In contrast, microhalocins are neutral, so there is no way to interact with the membrane lipids of the target cell.

Secretion, processing, mechanism of action and halocin immunity

Halocin preproteins (e.g. halocin H4) or preproproteins (e.g. halocin S8) are secreted using the twin-arginine translocation (Tat) pathway. This is the preferred pathway for translocating proteins in the haloarchaea (Rose et al., 2002). The Tat signal sequence is removed in halocin H4 (Cheung et al., 1997), but it is unknown if it is removed after the halocin S8 preproprotein is externalized. Halocin S8 (36 amino acids) is processed from the interior of the HalS8 preproprotein, leaving a 230 amino acid amino terminal protein and a 45 amino acid carboxy terminal peptide (Price and Shand, 2000). The mechanism of action is only known for halocin H6/H7 (see Table 29.1), where it inhibits the Na^+/H^+ antiporter (Meseguer et al., 1995). Interestingly, this halocin also inhibits mammalian Na^+/H^+ antiporters (Alberola et al., 1998) indicating that this halocin can be desalted and can retain function. Immunity genes and factors have not been characterized for any halocin, but we speculate that one possible role for the 230 amino acid amino terminal protein released

Table 29.1 Halocin characteristics. From O'Connor and Shand (2002); reproduced with permission

Halocin	Producer (source)	Size	GenBank Accession #	Thermal Stability	Salt Dependent	Activity Spectrum[a]	Mechanism	References
A4	Strain TuA4 (solar saltern, Tunisia)	7435 Da	—	≥1 week at boiling[b]	No	Broad *Sulfolobus* spp.	ND	Hazeltine et al. (2001), Kemper, S. and Shand, R. (unpublished), Robinson (2004)
C8	*Halobacterium* strain AS7092 (Great Chaidan Salt Lake, China)	6.3 kDa	AY310321	>60 min at 100°C	No	Broad	ND	Li et al. (2003)
G1	*Halobacterium* strain GRB (solar saltern, France)	ND[c]	—	ND	ND	Broad	ND	Soppa and Oesterhelt (1989)
H1	*Haloferax mediterranei* Xai3 (solar saltern, Spain)	31 kDa	—	<50°C	Yes	Broad	Membrane permeability	Platas (1995), Platas et al. (1996), Rodriguez-Valera et al. (1982)
H2	Strain GLA22 (solar saltern, Spain)	ND	—	ND	ND	Broad	ND	Rodriguez-Valera et al. (1982)
H3	Strain GAA12 (solar saltern, Spain)	ND	—	ND	ND	Broad	ND	Rodriguez-Valera et al. (1982)

(*continued*)

Table 29.1 Halocin characteristics. From O'Connor and Shand (2002); reproduced with permission

Halocin	Producer (source)	Size	GenBank Accession #	Thermal Stability	Salt Dependent	Activity Spectrum[a]	Mechanism	References
H4	*Haloferax mediterranei* R4 (solar saltern, Spain)	39.6 kDa (preprotein) 34.9 (mature)	U16389	<60°C	Partially[e]	Narrow	Proton flux?	Cheung et al. (1997), Meseguer and Rodriguez-Valera (1985, 1986), Meseguer et al. (1995), Perez (2000), Rodriguez-Valera et al. (1982), Shand et al. (1999)
H5	Strain MA220 (solar saltern, Spain)	ND	—	ND	ND	Narrow	ND	Rodriguez-Valera et al. (1982)
H6/H7	*Haloferax gibbonsii* Ma2.39[d] (solar saltern, Spain)	32 kDa	—	≤90°C	No	Narrow	Na^+/H^+ antiporter inhibitor	Alberola et al. (1998), Meseguer et al. (1995), Rodriguez-Valera et al. (1982), Torreblanca et al. (1989)
R1	*Halobacterium* strain GN101 (solar saltern, Mexico)	3.8 kDa	—	60°C	No	Broad *Sulfolobus* spp., *Methanosarcina thermophila*	ND	Hazeltine et al. (2001), O'Connor (2002), Rdest and Sturm (1987), Shand et al. (1999)
S8	Strain S8a (Great Salt Lake, UT)	33.9 kDa (preprotein) 3.6 kDa (mature)	AF276080	≥24 h at boiling[b]	No	Broad *Sulfolobus* spp.	ND	Hazeltine et al. (2001), Price and Shand (2000), Shand et al. (1999)

[a] Activity spectrum refers to inhibition of haloarchaea unless otherwise indicated.
[b] This study was done at 2113 m (7000 ft) water boils at 93°C at this elevation.
[c] ND: not determined.
[d] Halocin H6 is produced by *Haloferax gibbonsii* Ma2.39. This strain is proprietary and should not be confused with a different halocin-producing strain, *Hfx. gibbonsii* Ma2.38 (ATCC 33595). Halocin H7 is halocin H6, but is produced by a halocin overproducing mutant of *Hfx. gibbonsii* Ma2.39 called *Hfx. gibbonsii* Alicante SPH7.
[e] See text.

after processing of the halocin S8 preproprotein is to provide immunity to the producer (O'Connor and Shand, 2002).

◆◆◆◆◆ DETECTING AND QUANTIFYING HALOCIN ACTIVITY

Detecting Halocin Activity

Sampling cultures

For most halocins, and for most antibiotics in general, activity first appears in the culture supernatant either during the transition between exponential growth and stationary phase, or in early- to mid-stationary phase. However, there are variations:

(i) halocin H1 activity first appears during mid-exponential phase (Platas et al., 1996);
(ii) halocin H4 activity first appears and spikes during the transition into stationary phase, but then declines rapidly from peak levels (Cheung et al., 1997); and
(iii) halocin R1 appears relatively late in stationary phase (O'Connor, 2002). Consequently, cultures should be sampled during each of the following stages of growth: exponential phase, the transition into stationary phase, early- to mid-stationary phase and late stationary phase. Culture samples need not be large (250–500 µl). The cells should be removed by centrifugation ($10\,000 \times g$ for 5 min) and the culture supernatant transferred to a new microcentrifuge tube, being careful not to aspirate any cells from the cell pellet. Culture supernatants are stored at 4°C. Note that the microhalocins (e.g. C8, R1 and S8) can be stored for years at 4°C without loss of activity. However, halocin H4 (34.9 kDa) has a shelf life at 4°C of about four months (Perez, 2000).

Preparing top agar overlays (lawns) of halocin-sensitive cells

Since no one haloarchaeon is sensitive to all halocins, a battery of haloarchaea should be used when surveying for halocin activity as exemplified in Kis-Papo and Oren (2000) (in this study, 11 strains were used). However, *Halobacterium salinarum* NRC817 (nearly genetically identical to *Halobacterium* spp. NRC-1) is sensitive to most halocins (in the Kis-Papo and Oren study, NRC-1 was sensitive to 26 of the 29 halocin producing isolates) and is a good place to begin the survey.

Strain NRC817 will be used as a model for this methodology. NRC817 medium contains per liter: 245 g NaCl, 20 g $MgSO_4 \cdot 7H_2O$, 3 g tri-Na-citrate $\cdot 2H_2O$, 2 g KCl, 0.5 g tryptone, 0.3 g yeast extract and 43.75 mM Tris-HCl (pH 7.2). The medium is titrated to pH 7.2, filter sterilized and stored at room temperature in the dark. Trace elements are added to the medium to a final concentration of 1× just prior to inoculation; a 1000X stock solution contains, per 100 ml: 5 mg $CuSO_4 \cdot 5H_2O$, 455 mg $Fe(NH_4)_2(SO_4)_2 \cdot 6H_2O$, 30 mg $MnSO_4 \cdot H_2O$ and 44 mg $ZnSO_4 \cdot 7H_2O$. The stock solution is filter sterilized and stored in small aliquots at $-20°C$ for short-term storage, or $-80°C$ for long-term storage. Aliquots are thawed once and then discarded. Typically, an NRC817 lawn cell culture is prepared in 25 ml of medium and grown in a 125 ml baffled flask at $41°C$ in a shaking water bath.

To ensure reproducibility, lawn cells are prepared by placing the culture into balanced growth (see Robinson *et al.*, 2005 for details):

(1) From an exponentially growing culture, cells are diluted into warmed, aerated media to an OD_{600} of 0.01.
(2) The culture is grown to an OD_{600} of ≤ 0.2 and then diluted again into warmed, aerated media to an OD_{600} of 0.01.
(3) Cells in the culture in step 2 are grown to an OD_{600} of 0.6–0.8 for use in lawns.

Bottom agar plates contain 1.5% (w/v) Difco agar in the NRC817 medium described above. The flask or bottle containing the agar medium is weighed, then brought to a boil before autoclaving and autoclaved for 20 min. The agar is cooled to $60°C$ in a water bath, reweighed and sterile water added to replace any water lost during autoclaving. Trace elements are added and bottom agar plates are poured into 15×100 mm Petri dishes. Top agar contains 0.75% (w/v) Difco agar and the same autoclaving procedure is followed as described for bottom agar plates. When the top agar is at $60°C$, trace elements are added and the medium (4 ml) is dispensed into 13×100 mm culture tubes held at $60°C$ (ensure that the level of the water in the water bath will be above the level of the medium and the empty tubes are at $60°C$ before adding the top agar). Bottom agar plates are warmed to $41°C$ before adding the top agar overlays. Lawn cell culture (100 µl; OD_{600} between 0.6 and 0.8) is added to the $60°C$ top agar, one at a time (to minimize exposure to a non-permissive growth temperature), mixed thoroughly by vortexing, and then poured onto the surface of the warmed bottom agar. Since 4 ml is just sufficient to cover the surface of the bottom agar, use of cool bottom agar plates will result in premature solidification of the top agar and incomplete coverage of the plate surface. Lawns are allowed to cool and then stored at $4°C$ in a sealed plastic bag if not used at that time.

Spotting supernatants onto lawns

A common misconception is that the lawn cells should be allowed to grow and develop prior to spotting of supernatants. This approach fails to produce zones of inhibition. The supernatant is spotted onto the newly prepared, undeveloped lawn and the lawn is incubated at $41°C$. In this way, zones of inhibition are easily visualized. Ten microliters of supernatant is sufficient to detect halocin activity. A positive control of a supernatant known to contain halocin activity and to produce a zone of inhibition on the lawn cells, and a negative control of uninoculated culture medium should also be spotted. A 15×100 mm Petri dish will accommodate 12, 10 μl spots, plus the two controls. Once the supernatant has been spotted, the lawns can be moved carefully to the incubator and placed right side up to allow the spots to dry. Once dry, the plates can be turned upside down and placed in a plastic bag to prevent desiccation. On fresh lawns, zones become visible after 24 h, but usually 36–48 h of development are needed to visualize fainter (more turbid) zones.

General considerations for preparing and using lawns

There are several factors that affect the sensitivity of the lawn cells to a halocin:

(1) *Density of the lawn cells.* Increasing the number of cells in the lawn (e.g. from 100 to 200 μl) will result in the lawns developing faster; but for some halocins it decreases the sensitivity of the assay.
(2) *Growth state of the cells when the lawn is prepared.* Using cells from early exponential phase (e.g. an OD_{600} of 0.2), or cells from stationary phase results in delayed appearance of the lawn. In addition, stationary phase cells have a very different physiology than exponentially growing cells, and respond differently when exposed to a halocin.
(3) *Length of storage of lawns.* Typically, lawns are stored for no more than two weeks at 4°C which ensures a fairly uniform sensitivity from any particular batch of lawns. The longer the lawns are stored, the longer they will take to develop. Interestingly, strain NRC817 shows increased sensitivity to some halocins upon prolonged storage at 4°C.

An alternative that is more rapid but less sensitive than spotting culture supernatants to screen for halocin production is to transfer cells from freshly grown colonies onto the lawns as long as the cells will grow on the lawn cell medium and grow at about the same rate as the lawn cells. If the colonies being screened grow much slower, the lawn cells will out grow the colonies and zones may not appear. This is an efficient first step in screening many colonies recovered from the environment on a single lawn. However, colonies that arise slowly relative to the lawn cells should be screened by spotting culture supernatants.

Quantifying Halocin Activity

Figure 29.1A shows quantification of halocin activity by two-fold serial dilutions to extinction. The first dilution in which no zone appears is the extinction dilution, and the activity is the reciprocal of the extinction dilution, reported in arbitrary units (AU). Dilution series need not involve large volumes of either diluent (typically sterile culture medium) or sample. A typical dilution series uses 60 μl of undiluted material followed by 30 μl of diluent in a series of tubes or wells (96-well microtiter plates are excellent for this). Thirty microliters of the undiluted sample are transferred to and mixed with the 30 μl of diluent in the first tube. The pipette tip is discarded and using a new tip, the process is repeated for as many dilutions as needed. Since a 15 × 100 mm Petri dish will hold

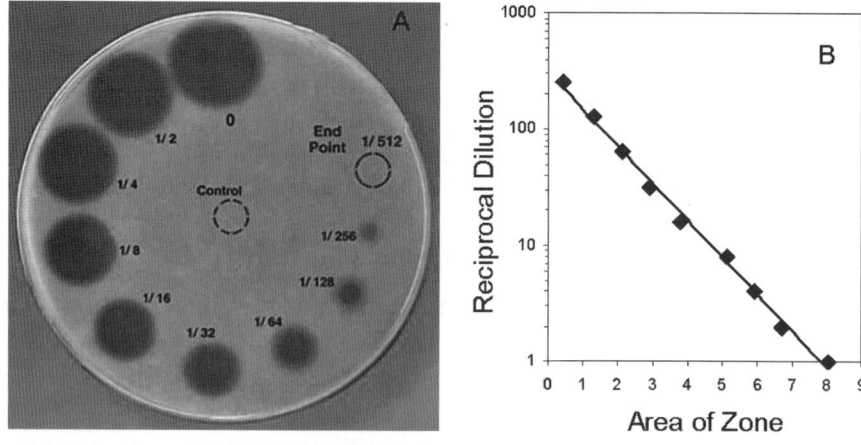

Figure 29.1. (A) Serial two-fold dilutions to extinction of a concentrated solution of halocin R1. The lawn cells are *Halobacterium salinarum* strain NRC817. The negative control is uninoculated medium. The halocin activity of this sample is 512 AU. (B) Correlation between the area of a zone of inhibition (from Figure 1A) and the logarithm of the activity ($R^2 = 0.995$). Note that the "0" dilution was assigned a value of 1 in order to be plotted logarithmically. See Plate 29.1 in Colour Plate Section.

12, 10 µl samples [plus two controls], we typically spot the undiluted sample and 11 dilutions for samples in which the activity is unknown. Note that while the pipette tip must be discarded between each dilution, a single pipette tip can be used to spot all of the dilutions onto the lawn if one begins the spotting with the most dilute sample and works their way backwards to the more concentrated samples.

This assay is independent of the volume spotted because it is dependent on the *concentration* of halocin, not the volume of the sample. Whether one spots 10 µl or 50 µl, the extinction dilution will be the same (hence the use of arbitrary units). Note that the diameter of the zone is not quantitative; for example, in Figure 29.1A the diameter of the ½ dilution is not one-half of the diameter of the zone of the undiluted sample. However, the area of the zone is proportional to the logarithm of the dilution as shown in Figure 29.1B. This graph could be used to estimate halocin activity based upon a single zone. However, this correlation is only relevant for a specific halocin; new graphs would have to be made for each new halocin, and all subsequent lawns would have to be identical (OD_{600}, length of storage, etc.) to the lawn from which the plot was created.

◆◆◆◆◆ HALOCIN PURIFICATION

Determining Physiological and Physicochemical Characteristics of Halocin-laden Supernatants

Since halocins vary widely in size and physicochemical properties, there is no one purification scheme for all halocins. However, the first place to start is to determine various physiological and physicochemical characteristics of the halocin of interest.

Halocin activity profile

Since not all halocins remain at maximal levels in stationary phase (e.g. halocin H4; Cheung *et al.*, 1997), a halocin activity profile *versus* growth phase experiment is carried out to determine where in the growth phase halocin activity first appears, where halocin activity peaks, and if halocin activity remains at peak levels in stationary phase. It is important to collect culture supernatants for subsequent purification steps when halocin activity is highest. In addition, it may be beneficial to optimize growth conditions with respect to temperature, salt concentration, and especially nutrient source to determine which condition produces the highest halocin activity levels (for example, see Platas *et al.*, 1996).

Thermal stability

If the halocin can withstand heat without losing activity, a simple heat precipitation step of the culture supernatant removes a significant amount of contaminating proteins. Since optimal growth temperatures for many haloarchaea are $\geq 50°C$ (Robinson *et al.*, 2005), the initial thermal stability experiment exposes culture supernatants to temperatures between 60°C and boiling in 10°C increments. The temperature at which the halocin supernatant was collected is used as a control. Subsequent experiments using smaller temperature increments can be conducted to refine the thermal stability profile. Although sampling times need to be determined empirically, a good starting point is to collect samples every 15 min over the first hour, and then at 30-min intervals over the next 3 h. The samples are then quantified for activity levels and compared to the control. In addition, the supernatants are examined for the presence of precipitates. Withstanding high temperatures for several hours without losing activity is typical for many microhalocins, as larger halocins heat-denature under these conditions. Note also that stability at modest temperatures (e.g. 60°C) for a few hours can be beneficial as heating culture supernatants at this temperature results in significant precipitation of contaminating proteins that are easily removed by filtration.

Resistance to desalting

This is a particularly important characteristic for halocin purification as halocins that denature in low salt concentrations can not be subjected to ion exchange chromatography. However, Perez (2000) found that halocin H4 (34.9 kDa) could be desalted to 150 mM NaCl and still retained detectable activity for a sufficient period of time to allow separation on an ion exchange column by HPLC. A desalting/resalting experiment is carried out using recursive dilution and reconcentration in a spin filtration device. The Millipore Ultrafree® Biomax spin filter units are excellent for this. The filter membranes are made of polysulfone – a low protein binding material to which halocins do not stick; if non-polysulfone membrane devices are used, the membranes should be checked to make sure that halocin activity is not lost due to the binding of the halocin to the membrane. They also have a tangential flow configuration (as opposed to a radial configuration) which minimizes clogging of the membrane, can be spun in either a fixed-angle or swinging bucket rotor, and come in a variety of volumes and nominal molecular weight cutoffs (NMWCO). For this experiment, a 5 kDa cutoff will retain nearly any halocin. One milliliter of halocin-laden supernatant is concentrated 10-fold to 100 µl in the spin filter. The filtrate is removed from the collection tube and saved. The retentate (in the same filtration device) is diluted with 900 µl of 10 mM Tris-HCl, pH 7.5. This recursive concentration/dilution procedure is performed three times, and each time, the filtrate is removed and saved. Ten microliters of the desalted supernatant is removed (the remaining desalted material [~90 µl] is left in the spin filter) and assayed along with the filtrates (the filtrates should show no halocin activity). A negative control using 10 mM Tris-HCl, pH 7.5 should be spotted on the assay plate. Resalting is accomplished in the same fashion using basal salts from the medium (basal salts are just the salt components from the medium – no carbon sources or other nutrients are included). Should the sample not withstand desalting with 10 mM Tris-HCl, modest amounts of NaCl (up to ~200 mM, which still permits ion exchange chromatography) can be used as the desalting medium. Should the halocin require low levels of NaCl to retain activity, the shelf life of the halocin in the lower salt concentration should be compared to that in the optimal salt concentration.

Resistance to acetone

Just as thermostability can be used to precipitate contaminating proteins using heat, resistance to acetone allows contaminating proteins to be removed by acetone precipitation. In a chemical fume hood, an equal volume of acetone is mixed thoroughly with the halocin-laden aqueous supernatant. The emulsion is then transferred to a separatory funnel and the phases allowed to separate. The high salt aqueous phase (containing the halocin) is at the bottom overlaid by acetone (with precipitated proteins). The aqueous phase is drained off, but this still contains some residual acetone. The acetone is driven from the aqueous solution by stirring and heating the solution at 60°C while blowing a stream of nitrogen gas over the surface until no acetone odor is detected.

Preliminary size determination

Frequently, NMWCO filters are used to estimate the size of a protein to be purified. However, determining the approximate size of a halocin using NMWCO filters is unreliable. For example, halocin R1 appears to be about 29 kDa on a gel filtration column. However, after heating the supernatant at 60°C and reapplication to the gel filtration column, the activity elutes at 3 kDa. This difference is due to the HalR1 protein binding to a "carrier protein"; the two dissociate upon heating (O'Connor, 2002). A second example is halocin S8. This halocin is 3.6 kDa, but partitions equally across a filter with a NMWCO of 30 kDa (Price and Shand, 2000).

A General Procedure for Purifying Thermostable, Salt-independent, Acetone-resistant Microhalocins

If the physicochemical parameters described above show that the halocin is resistant to heat, acetone and desalting, then the following procedure can be used to purify the microhalocin. This procedure involves generating a large volume of culture supernatant, concentrating the supernatant, heat and acetone precipitation, application of the supernatant to a gel filtration column, and then to an HPLC reversed-phase column under two different conditions.

Generation of cell-free supernatants

(1) We begin with 16 l of culture (grown in 8, 4-l baffled flasks) to the appropriate stage of the growth curve (as determined from previous physiology experiments; see above). A 1 ml sample is removed from each flask and pooled to determine initial halocin activity levels.
(2) Cells can be removed from the culture either by centrifugation or by tangential flow filtration using a large surface area 0.45 μm tangential flow filter (e.g. in a Millipore Pellicon™ tangential flow filtration device).

Concentration of the cell-free supernatant

The cell-free supernatant (or filtrate) is processed through a series of filters with progressively smaller NMWCO: 100 kDa, 30 kDa and 10 kDa. Each filtrate and each retentate is assayed to determine the amount of halocin activity. In this system, 16 l of supernatant can be reduced to 200–250 ml of retentate. At each step, a small amount of retentate and filtrate are saved for halocin activity assays. The retentate with the highest activity is used for subsequent steps.

Heat and acetone precipitation

(1) The 200–250 ml of retentate is heated (temperature and duration are determined previously; see thermal stability experiment above). Precipitated proteins are removed by filtration through a 0.22 μm filter equipped with a prefilter. The heat-treated material must be allowed to return to room temperature before filtering; vacuum filtration of hot material will result in the material boiling.
(2) The filtered, heat-treated material is then subjected to acetone precipitation (see acetone precipitation experiment above).

Concentration of the heat-treated, acetone-precipitated supernatant

(1) The heat-treated, acetone-precipitated supernatant is concentrated using Millipore Ultrafree® Biomax spin filters with a 5 kDa cutoff that hold 15 ml of sample (see desalting experiment above). Several filter units are employed and the retentates pooled (there should be no halocin activity in the filtrates). The 200–250 ml of retentates are reduced to about 2–4 ml.

(2) This material is then subjected to a Bradford assay (or other protein assay), to determine the amount of protein in the sample. This is an important parameter so as not to overload the gel filtration column.

Gel filtration column chromatography in high salt concentration

(1) Bio-Gel®P (polyacrylamide) gel filtration matrices from Bio-Rad come in variety of fractionation ranges and are compatible with high salt buffers. BioGel P-10M or P-30M (medium particle size) are a good place to start. Columns should be long (e.g. Bio-Rad Econo-columns, 2.5 cm × 120 cm) and packed to 112–114 cm. The hydration and elution buffers are basal salts from the growth medium (sodium azide is not needed in the column due to the high salt concentration).

(2) Due to the high viscosity of the running buffer, columns should not be run at greater than 0.04 ml min^{-1}. This is very slow, but resolution is lost above 0.04 ml min^{-1}. Suitable proteins for characterizing the column are ferritin (440 kDa), ovalbumin (40 kDa), carbonic anhydrase (29 kDa) and cytochrome C (12.5 kDa). Protein standards with molecular masses below 12.5 kDa are not soluble in high salt running buffers.

(3) The concentrated halocin sample is loaded onto the column in a volume of no more than 5 ml. Fractions are collected in 0.5 ml aliquots and every fifth fraction assayed for activity. The activity is then quantified in those fractions and a plot of elution volume *versus* halocin activity constructed.

(4) Fractions with "high" levels of halocin activity are pooled and concentrated in Millipore Ultrafree® Biomax spin filters with 5 kDa cutoffs. The term "high" is relative and which fractions are pooled is based on the graph of halocin activity *versus* elution volume.

Reversed-phase high performance liquid chromatography

The microhalocins that have been purified to date (see Table 29.1) are uniformly hydrophobic and can be purified using reversed-phase HPLC without losing activity.

(1) We have found POROS® 10 R2/H matrices (Applied Biosystems) to be superior to C8/C18 matrices for purifying microhalocins. Although C8/C18 columns traditionally are used to purify hydrophobic peptides, they are silane-linked and are labile in strong

base; POROS® 10 R2/H matrices are made of styrene and are stable in base.

(2) A typical reversed-phase HPLC run using a POROS 10 R2/H column begins with 2.2 mM trisodium phosphate (TSP; highly basic) as the aqueous phase. The base run is done first as TSP is not volatile and interferes with Edman degradation amino acid sequencing reactions. The reversed-phase column is washed with 10 column volumes (CV) of TSP. The sample is loaded onto the column and washed again with 5 CV of TSP. Protein is eluted with an isocratic wash of 10 CV of 20% (v/v) acetonitrile (ACN) in 2.2 mM TSP, followed by an ACN gradient from 20% (v/v) ACN to 100% ACN in 2.2 mM TSP in 15 CV. Fractions (1 ml) are collected and assayed for halocin activity. Note that, in fractions with higher ACN concentrations, small, distinct zones of inhibition appear on the assay plate due to the ACN, but these are easily distinguished from zones produced by halocin activity. Negative controls containing several concentrations of ACN in 2.2 mM TSP should be spotted on the assay plates. Fractions containing halocin activity are pooled and lyophilized in a SpeedVac® (Thermo Electron Corporation). The dried protein is resuspended in 18 MΩ water and neutralized with a small amount of HCl if needed until all of the protein is in solution.

(3) The protein from the base run is then subjected to an acid run using 0.1% (v/v) trifluoroacetic acid (TFA) and ACN instead of TSP and ACN. The procedure is the same as described for the base run except that TFA is used in place of TSP. Fractions are assayed for halocin activity and pooled. Since both TFA and ACN are volatile, the pooled material is concentrated easily in a SpeedVac® in preparation for amino acid sequencing (note that the material is not completely dried, but concentrated to a volume of 5-10 µl).

(4) Amino acid sequencing of the purified halocin can now be done, and the mass of the protein is determined by mass spectrometry. Once part (or all) of the amino acid sequence of the halocin is obtained, degenerate deoxyoligonucleotide probes can be designed to clone the gene.

General considerations while purifying halocins

(1) Halocin activity should be quantified at each step of the purification in order to follow the progress of the purification and to determine the loss of halocin activity along the way.

(2) Samples should be set aside at each step of the purification and assayed by SDS-PAGE in order to visualize the purity of the halocin at the final step. It is critical to know the purity of the halocin prior to sequencing by Edman degradation. Since microhalocins are small, tricine SDS-PAGE gels should be used as described in Schägger and von Jagow (1987). In addition, unlike glycine gels, tricine gels can handle high amounts of salt and so high salt fractions from gel filtration columns can also be run.

List of Suppliers

Applied Biosystems
850 Lincoln Centre Drive
Foster City, CA 94404,USA
Tel: 1+800-327-3002
Fax: 1+650-638-5998
www.appliedbiosystems.com

POROS® Perfusion Chromatography™ media and columns

Bio-Rad Laboratories (Life Science Research Group)
2000 Alfred Nobel Drive
Hercules, CA 94547, USA
Tel: +1-800-424-6723
Fax: +1-800-879-2289
www.bio-rad.com

Gel filtration matrices, Gel filtration Econo-columns

Millipore
290 Concord Road
Billerica, MA 01821,USA
Tel: +1-800-645-5476
Fax: +1-800-645-5439
www.millipore.com

Ultrafree® Biomax spin filters, Pellicon™ tangential flow filters

Thermo Electron Corporation
450 Fortune Boulevard
Milford, MA 01757, USA
Tel: 1+800-522-7763
Fax: 1+508-634-2127
www.thermo.com

SpeedVac®

References

Alberola, A., Meseguer, I., Torreblanca, M., Moya, A., Sancho, S., Polo, B., Soria, B. and Such, L. (1998). Halocin H7 decreases infarct size and ectopic beats after mycardial reperfusion in dogs. *J. Physiol.* **509.P**, 148P.

Cheung, J., Danna, K. J., O'Connor, E. M., Price, L. B. and Shand, R. F. (1997). Isolation, sequence, and expression of the gene encoding halocin H4, a bacteriocin from the halophilic archaeon *Haloferax mediterranei* R4. *J. Bacteriol.* **179**, 548–551.

Hazeltine, C., Hill, T., Montalvo-Rodriguez, R., Kemper, S. K., Shand, R. F. and Blum, P. (2001). Secreted euryarchaeal microhalocins kill hyperthermophilic crenarchaea. *J. Bacteriol.* **183**, 287–291.

Kis-Papo, T. and Oren, A. (2000). Halocins: are they involved in the competition between halobacteria in saltern ponds? *Extremophiles* **4**, 35–41

Li, Y., Xiang, H., Liu, J., Zhou, M. and Tan, H. (2003). Purification and biological characterization of halocin C8, a novel peptide antibiotic from *Halobacterium* strain AS7092. *Extremophiles* **7**, 401–407.

Meseguer, I. and Rodriguez-Valera, F. (1985). Production and purification of halocin H4. *FEMS Microbiol. Lett.* **28**, 177–182.

Meseguer, I. and Rodriguez-Valera, F. (1986). Effect of halocin H4 on cells of *Halobacterium halobium*. *J. Gen. Microbiol.* **132**, 3061–3068.

Meseguer, I., Torreblanca, M. and Konishi, T. (1995). Specific inhibition of the halobacterial Na^+/H^+ antiporter by halocin H6. *J. Biol. Chem.* **270**, 6450–6455.

O'Connor, E. M. (2002). *Purification and Characterization of Microhalocin R1 from Halobacterium salinarum GN101*. Doctoral dissertation, Northern Arizona University, Flagstaff.

O'Connor, E. M. and Shand, R. F. (2002). Halocins and sulfolobicins: The emerging story of archaeal protein and peptide antibiotics. *J. Indust. Microbiol. Biotechnol.* **28**, 23–31.

Perez, A. M. (2000). *Growth Physiology of Haloferax mediterranei R4 and Purification of halocin H4*. Masters thesis, Northern Arizona University, Flagstaff.

Platas, G. (1995). Characterization de la actividad antimicrobiana de la haloarquea *Haloferax mediterranei* Xia3. Tesis doctoral, Universidad Autónoma de Madrid, Madrid.

Platas, G., Meseguer, I. and Amils, R. (1996). Optimization of the production of a bacteriocin from *Haloferax mediterranei* Xia3. *Microbiología* **12**, 75–84.

Prangishvili, D., Holz, I., Stieger, E., Nickell, S., Kristjansson, J. K. and Zillig, W. (2000). Sulfolobicins, specific proteinaceous toxins produced by strains of the extremely thermophilic archaeal genus *Sulfolobus*. *J. Bacteriol.* **182**, 2985–2988.

Price, L. B. and Shand, R. F. (2000). Halocin S8: a 36-amino-acid microhalocin from the haloarchaeal strain S8a. *J. Bacteriol.* **182**, 4951–4958.

Rdest, U. and Sturm, M. (1987). Bacteriocins from Halobacteria. In *Protein Purification: Micro to Macro* (R. Burgess, ed.), pp. 271–278. Alan R. Liss, New York.

Robinson, J. L. (2004). *Haloarchaeal Growth Physiology, Characterization of Halocin A4, and Cloning tbp and tfb Genes*. Masters thesis, Northern Arizona University, Flagstaff.

Robinson, J. L., Pyzyna, B., Atrasz, R. G., Henderson, C. A., Morrill, K. L., Burd, A. M., DeSoucy, E., Fogelman, R. E., Naylor, J. B., Steele, S. M., Elliott, D. R., Leyva, K. J. and Shand, R. F. (2005). Growth kinetics of extremely halophilic *Archaea* (family *Halobacteriaceae*) as revealed by Arrhenius plots. *J. Bacteriol.* **187**, 923–929.

Rodriguez-Valera, F., Juez, G. and Kushner, D. J. (1982). Halocins: salt-dependent bacteriocins produced by extremely halophilic rods. *Can. J. Microbiol.* **28**, 151–154.

Rose, R. W., Bruser, T., Kissinger, J. C. and Pohlschröder, M. (2002). Adaptation of protein secretion to extremely high-salt conditions by extensive use of the twin-arginine translocation pathway. *Mol. Microbiol.* **45**, 943–950.

Schägger, H. and von Jagow, G. (1987). Tricine-sodium dodecyl sulfate-polyacrylamide gel electrophoresis for the separation of proteins in the range from 1 to 100 kDa. *Anal. Biochem.* **66**, 368–379.

Shand, R. F., Price, L. B. and O'Connor, E. M. (1999). Halocins: protein antibiotics from hypersaline environments. In *Microbiology and Biogeochemistry of Hypersaline Environments* (A. Oren, ed.), pp. 295–306. CRC Press, Boca Raton.

Soppa, J. and Oesterhelt, D. (1989). *Halobacterium* sp. GRB: a species to work with? *Can J. Microbiol.* **35**, 205–209.

Torreblanca, M., Meseguer, I. and Rodríguez-Valera, F. (1989). Halocin H6, a bacteriocin from *Haloferax gibbonsii. J. Gen. Microbiol.* **135**, 2655–2661.

Torreblanca, M., Meseguer, I. and Ventosa, A. (1994). Production of halocin is a practically universal feature of archaeal halophilic rods. *Letts. Appl. Microbiol.* **19**, 201–205.

30 Storage of Halophilic Bacteria

Jessica DiFerdinando[1] and Russell H Vreeland[1,2]
[1] Southeast Applied Research Inc., 1423 Candlewood Dr., Fredericksburg, VA 22407, USA;
[2] Department of Biology, West Chester University, West Chester, PA 19383, USA

CONTENTS

Introduction
Culture tracking
Storage systems
Reviving cultures

◆◆◆◆◆ INTRODUCTION

One problem faced by all dedicated microbiologists is that it is easier to collect microbes than it is to carry out extensive analyses on each one. Consequently, safe dependable storage methods must be instituted in every laboratory. Bacterial cultures in test tubes, plates or even microcentrifuge tubes are, of course easier to store than, say elephants or hippos, largely because hundreds to thousands will easily fit within a single refrigerator or freezer. This naturally leads to the proliferation of numerous duplicates and even instances of poorly marked tubes containing useless strains. Those whose careers are dedicated to halophilic microbes often face the problem of over-collection more so than others.

Let us face facts here, cultures of *Escherichia coli*, *Salmonella* or *Bacillus*, for all of their great microbiological value are really rather mundane when present on slants or plates. Not so with halophilic and halotolerant microbes, which present themselves in a wonderful, sometimes dizzying array of pigments, smells, shapes and properties. Also microbiologists studying halophiles face the fact that the diversity of the hypersaline world is horribly underappreciated and little studied. Couple that with the irritating propensity of hypersaline habitats to either appear without warning along a roadside or in some out of the way, horribly forsaken place to which one may never return and you have all the makings of complete storage chaos. Despite this, most halophilic researchers have tried to solve their overall storage problems using the rather standard techniques arising from the care and cultivation of non-halophilic microbes. In our experience, these techniques simply do not work that well, yet there have been no discussions or manuals that deal with this important aspect of hypersaline microbiology. Therefore, this chapter will attempt to redress this problem based upon experience within our particular laboratory. Some of the techniques have arisen from experimentation, while others have resulted from improvements following unexpected loss of interesting cultures that have been improperly stored. Some techniques have even arisen following conscious decisions to approach the problem in a more systematic manner. This text attempts to deal with both halophilic Archaea and halophilic Bacteria. We will use these terms throughout the text while reserving the term "halophiles" as a general descriptor for both groups. In addition, this chapter deals exclusively with those microbes that either require, or have optimal, sodium chloride concentrations at or above 6% (w/v). Obviously, some aspects of the storage of halophiles are similar to the storage of non-halophilic Bacteria. Some of the major differences are the result of the salinity of the medium in which the halophiles grow and the metabolic byproducts they produce.

◆◆◆◆◆ CULTURE TRACKING

Whenever a collection of organisms reaches a number greater than can be easily committed to memory a culture tracking system should be implemented. The best systems assign unique identifiers to each organism, are easily communicated to others and ultimately provide a jargon within the laboratory. There are many useful examples of systems found within the large permanent culture collections. These include the sequential numbering of acquisitions used initially at the American Type Culture Collection (ATCC) and still used by institutions such as the Deutsche Sammlung von Mikroorganismen und Zellkulturen (DSMZ) and the Japanese National Collection. Alternatively a system can be the somewhat more complex, but useful, letter–number system currently being used by the ATCC. This latter system is actually worth a brief description since it illustrates the potential need to reconstruct tracking systems over time. The initial tracking system used by ATCC was simply a sequential numbering system with each culture being assigned its number as it was acquired. This system worked quite well for most of the twentieth century. However, with the explosion of patented organisms and genomic depositions that occurred during the last two decades before the end of the millennium, the sequential numbering required nearly eight digits and became rather cumbersome. In response ATCC devised its current letter–number sequence. This system will almost certainly carry the collection for many more years. It works as follows. All prokaryotic cultures arriving at ATCC are assigned a designator beginning with the letter "B" followed by two more letters (currently AA) followed by a sequential number from 1 to 99 999, the next submission will be assigned number BAB – 1. This system means there are 99 999 "BAA" cultures and provides a total of over 67 million acquisitions before a new first letter is needed! Clearly very few private collections will need this level of tracking, however many small laboratories would benefit greatly by moving away from some of the inscrutable systems visible in the literature.

Our own laboratory is a case in point. Initially the collection began with a number–letter–number designator designed largely for internal use. In this case the first number indicated a particular field trip, the letter was supposed to indicate something about the culture (surprisingly H = halophile) with the last number indicating the isolate (e.g. 1H9 = *Halomonas elongata* (Vreeland *et al.*, 1980)). That quickly changed to a system where the letters indicated the sample sites (Ed = Enid, Oklahoma USA) which also did not work once others began depositing cultures with scant isolation information. Ultimately, we were forced not only to change the designators for new cultures, but it became necessary to convert the entire collection to an entirely new system. This was accomplished in a manner that lends itself to a simple explanation and rapid understanding by the masters and undergraduate level students populating the laboratory. Essentially, the system uses common laboratory notebooks and Excel-type spread sheets. Each book is numbered 1, 2 etc. (we are currently finishing book 2). The first number in a culture designator is simply the book number followed by a hyphen, the following number defines the page within the book followed by another hyphen, the third number designates the sequential entry on the page. Experience has also shown that such entries should only occur on odd pages reserving the even numbered pages for information, pictures, descriptions or other data on each culture. Consequently, all information on culture 2-9-3 is found in book 2, page 9, entry 3. The system works well, everyone becomes easily conversant and there are rarely foul-ups (we do have a set of 2–10s for instance) but overall this makes our collection easy to maintain as long as others accessing the collection remember keep track of our numbers when we provide a culture.

◆◆◆◆◆ STORAGE SYSTEMS

Within any working laboratory there are always different storage needs ranging from short-term to long-term systems. The sections below describe methods that can be applied in each situation.

Short-term Storage

Short-term storage includes everything from refrigerating broth and slants to simply tightening a cap and letting a tube sit on the bench top. It is, by a wide margin, the most frequently performed storage and also most often taken for granted. This is unfortunate since even the best long-term storage system begins with cultures that have been grown and stored for short periods of time. With halophilic microorganisms short-term storage periods can reasonably last between 6 and 9 months or as long as the medium does not begin precipitating salt crystals. This type of storage may be easily accomplished using the cells' normal growth medium prepared as either agar slope or even plates. Agar slope cultures are relatively easy to store; however, screw caps should be loosened by one quarter turn to allow the cultures to breathe. While the edges of the slopes will rapidly form crystals the cultures will generally remain viable at the bottom of the slope. Due to the hygroscopic nature of salt media these tubes will often collect a small puddle of liquid at the bottom following several weeks at 4°C. Since this water has a salt content nearly equivalent to that of the medium cultures usually survive. In addition, we have found that while all growth media can be useful for short-term storage, some growth media preserve cultures better than others. Halophiles that utilize carbohydrates and produce acid *via* a fermentation or incomplete oxidation reaction should, if possible, be stored on media lacking carbohydrates. If this is not possible the storage medium should be strongly buffered using something like 3-(*N*-Morpholino)propanesulfonic acid (MOPS) buffer. Without these precautions halophilic organisms will often lower the medium pH to lethal levels. This same precaution holds when reviving carbohydrate utilizers. For some reason that has not yet been defined, these halophiles seem to release even more acid upon re-growth than they do prior to storage.

Halophilic microbes can also be stored successfully on agar in Petri plates for up to two years provided the plates are inside plastic storage bags containing a wet paper towel placed beneath the plates. To ensure adequate survival, the bags are best sealed using zip-lock-type bags rather than with ties or heat since the former allows sufficient airflow for the cells. Inevitably, stored plates, broth and tubes will begin to dry out and crystals will form across the agar surface or at the bottom of the flask. While it is best to try to avoid this occurrence, it does not necessarily mean that the culture has been lost since most halophiles will survive crystallization and are easily recovered using the techniques described in the section on recovery and revival at the end of this chapter.

One final word on short-term storage of halophilic isolates (especially fresh archaeal isolates). While the methods described above are useful, they are really not recommended. In our experience, fresh isolates are especially problematic and a laboratory can lose as much as 50% of the fresh isolates during the first year in culture. The exact reason for such losses is not known; however, at least some losses might be caused by lysogenic viruses in the cells. For this reason fresh halophilic isolates should be placed into medium or long-term storage as soon as practical.

Medium Term Storage

We consider any culture storage up to about 5 years as medium term storage. Medium term storage of halophilic cultures is simply accomplished by freezing the cells in a deep freeze or even at −80°C. In reality, any of the long-term storage methods described below will protect the cultures for medium lengths of time without problems so there is no real reason to dedicate technical time to establishing such storage. In our experience, the biggest problem with cultures stored at −20 or −80°C appears to be a slow deterioration in viability, probably caused by oxidation. This has proven to be especially true of cultures placed onto ceramic beads or even frozen in ice brine. These losses are ultimately complete and affect every halophilic culture we attempted to store under these conditions. For this reason we believe that storage of halophilic microbes in freezers should not be considered for use with these organisms unless nothing else is available. If this method must be used each culture should be revived and re-stored at 2–3 year intervals in order to maintain viability.

Long-term Storage

As might be expected setting up a long-term storage system in any laboratory is basically worth the time, cost and effort. Long-term storage of halophilic microbes maintains viability, genetic purity and eases the overall workload for culture maintenance. The organisms are readily stored and in our experience recover very rapidly from most common storage methods.

Storage under liquid nitrogen

Long-term storage mainly consists of placing cultures under liquid nitrogen. They are easily revived when needed for study. A specific laboratory spreadsheet must be used to track all of the cultures by number, nitrogen storage location, date stored, medium preference, colony morphology and history. Culture storage locations should also be recorded in handwritten lab notebooks. These books also serve as a backup reference source if the electronic version fails or becomes corrupted. One final general comment, when using liquid nitrogen freezers we recommend obtaining a pair of very long forceps (the type used by ground keepers to lift trash without bending over work well) in order to retrieve the inevitable dislodged vial or cryostraw that sinks to the bottom of the tank.

Lyophilization

The techniques for freeze drying bacterial cells are well known and readily available in most biology departments. They are generally effective for long-term storage provided the vials can be adequately sealed to prevent intrusion of oxygen. That being said the technique is not widely used (and is often problematic) with halophilic microbes due to the need for several salts in order to maintain cellular integrity. This creates a problem since the high salt causes a lower melting temperature of the medium in which the cells are suspended. This sometimes results in the cell suspension melting, especially near the end of the process, rather than sublimation as is needed for lyophilization. This is not to indicate that halophilic Bacteria and Archaea cannot be lyophilized: the procedure is commonly used at the major culture collections.

At the ATCC, halophilic microorganisms are lyophilized according to the following procedure:

- Prepare 2.0 ml of freshly grown cultures at a density of approximately 10^9 viable cells per vial.
- Concentrate the cells further via centrifugation (generally 6000–9000 × g for 10 min being adequate) and suspend the cells in 0.2 ml of a solution containing ½ Reagent #20 plus 20% w/v NaCl for archaeal halophiles plus ½ volume of the growth medium. For non-acid producing halophiles a solution of glycine betaine (10%) + 20% w/v NaCl can also be used.
- Following lyophilization the vials are flame sealed to protect the cells from oxygen.
- Store the cultures under liquid nitrogen, or frozen at −80°C.

Reagent #20 contains 10% bovine serum albumin and 20% sucrose in distilled water; the solution is filter sterilized (Simione and Brown, 1991; J. Tang, ATCC, personal communication).

Studies on this type of storage of halophiles (using a simulated time of 25 years) have indicated that halophilic cultures freeze dried as described here will have a loss of viability of about two orders of magnitude (P. Krader, personal communication).

Storage on beads

One method for storing halophiles under liquid nitrogen simply uses screw-capped cryovials containing a concentrated solution of cells in media. This is then frozen under liquid nitrogen. As one needs additional culture small amounts of the ice are scrapped off with a sterile loop and used as inoculum (L. I. Hochstein, personal communication).

An improvement on this method utilized the same cryovials (1.5–2.0 ml), only these were now filled half way with glass or ceramic beads. Prior to use, these donut-shaped beads should have a dull finish or be gently shaken with fine grit abrasives to provide attachment sites for the cells. The vials and beads can then be sterilized at 121°C for 25 min. Pure cultures grown in liquid media are aseptically transferred to fill the cryovial and the cap is replaced. Glycerol is not generally added to the cryovial since we have found that many halophiles will produce an overabundance of acidic end products as a result of its fermentation when they are revived. The culture should be allowed to sit for one to two hours in order to allow cells to settle and attach to the beads. The cryovial may also be centrifuged gently to settle the cells, not to form a pellet. Any liquid settling above the level of the glass beads should be aseptically removed prior to storage. Cultures that settle rapidly should be shaken to redistribute the cells throughout the beads.

Storing the cryovials in liquid nitrogen requires some organization. Cryovials must be secured in aluminum canes that hold five vials in a vertical fashion (Figure 30.1). We normally designate the top vial in the cane as vial 1, with the last vial at the bottom of the cane designated

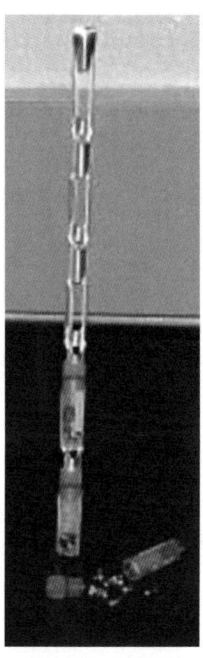

Figure 30.1. An aluminum cane holds five individual cryovials. Each sterilized cryovial contains up to 25 glass beads. Individual beads may be retrieved when it is necessary to revive the culture.

as vial 5. Each aluminum cane itself requires a designation by a letter of the alphabet. The first aluminum cane is designated "A" and the last is designated "Z". Each cane should also be secured in a plastic sheath and stored vertically inside a cylindrical receptacle. The cylinder has a number designation that is determined by the position around the liquid nitrogen tank. For example, the very first culture placed in a cryovial and in the tank would have a location designated as 1A1, or cylinder "1", aluminum cane "A" and position "1" on the cane.

The use of beads in cryovials has its advantages. They are an inexpensive means to store culture and the culture may be revived time after time as long as there are glass beads inside the vial. On average, we generally place 20–25 beads inside the vial to fill it half way so the culture can be accessed up to 24 times without needing to be replenished.

However, we have recently realized that the disadvantages of the beads in cryovials systems may outweigh the advantages. First, the vials are no longer sold pre-sterilized in dry form so all vials will contain Tryptic Soy Broth without salt and with 20% glycerol covering the glass beads. Consequently, the glass beads and vials must be purchased separately and then assembled and sterilized. This is merely just a limitation of time. A more serious limitation has proven to be a significant risk of contamination. When the cryovials and beads are purchased separately, the screw caps allow liquid nitrogen to fill the vial. While this does prevent harmful gases, especially oxygen, contacting the cells there is nothing preventing cells from escaping their vial and contaminating the liquid nitrogen in the tank. We should note that our experience has shown that there is little agitation in the tank; however, the possibility exists and if it occurred the result would be devastating. Consequently, a prudent practice would be to seal the tubes but this has proven to be technically difficult. After revival, a streak plate should be made to ensure purity of the culture. Most of the contamination we have experienced has arisen from an organism forming bright white colonies which were easily distinguished from the red halophile studied. The original organism is then re-isolated and the cryovial must be replaced. Another significant disadvantage has recently appeared in that some cryovials will split nearly in half after repeated freeze thaw cycles. This is not only dangerous to workers as it seriously increases contamination or even causes loss of all beads into the tank.

Storage using cryostraws

One of the newest developments is to store the halophilic organisms in cryostraws (Figure 30.2), such as those used to store animal sperm and embryos. The main advantage to cryostraws is that they are heat sealed so no liquid nitrogen comes in contact with the culture. This virtually eliminates the risk of cross-contamination from microbes floating in the nitrogen. One disadvantage is that the straws are a one-time use only method and are therefore more expensive. The straws are purchased pre-sterilized in packs of 20. Equipment required to heat seal and label the straws must be purchased from individual suppliers. Naturally, each supplier has its own specific straw, label maker, liquid nitrogen organizer

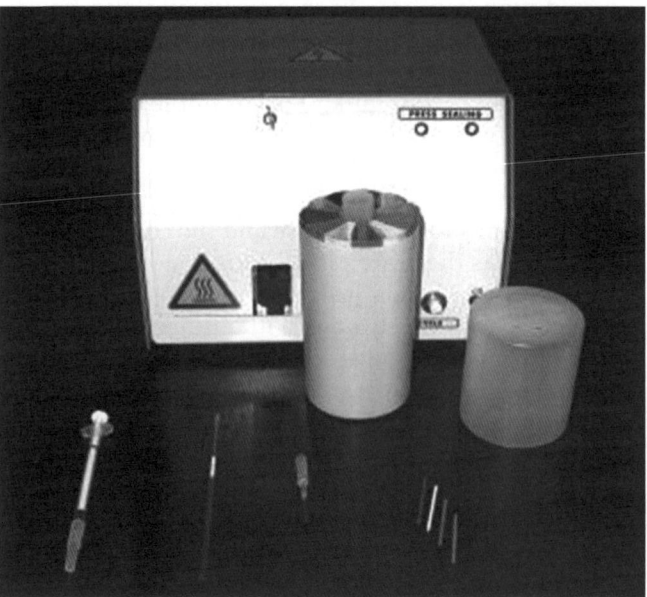

Figure 30.2. Supplies needed to fill cryostraws. From left to right: syringe fitted with nozzle adjustment, cryostraw, sterile filling nozzle and colored labeling rods. The welding machine is seen in the background along with the colored cylinder that holds the straws upright in the liquid nitrogen tank. See Plate 30.2 in Colour Plate Section.

and sealing unit. Since the straws can only be used once, multiple straws should be prepared for each culture.

Procedure

- Grow cultures to be stored in broth medium.
- Place a single sterile straw on the end of a syringe. Each straw has a cotton plug placed approximately two-thirds up one end. The syringe must be placed on the end nearest this plug.
- Insert the bottom of the straw into a sterile filling nozzle. This sterile, disposable nozzle separates the culture from the end of the straw avoiding contamination.
- Aspirate the culture directly into the straw leaving about two millimeters space between the culture and the cotton plug. Another 2 mm should be left between the culture and the end of the straw.
- Remove the straw from the broth and detach the filling nozzle. The syringe should remain attached to the opposite end of the straw, to prevent the liquid from flowing out of the straw.
- Insert the open end of the straw into the welding machine to seal the straw end as described by the manufacturer.
- Once sealed, the syringe can be removed, a labeling rod inserted and the open end of the straw welded shut. Even when sealed the straw should be handled gently so as not to disturb the air bubble on either end of the liquid culture. Bumping the straw may cause the culture to shift into the cotton plug or to move to the other end. Since these bubbles allow the culture to expand or contract upon freezing, shifting to one end or the other may damage the straw.
- Seal the label and culture inside the straw and immerse in liquid nitrogen. The labeling rods may be purchased in different colors and so each lab may devise a separate labeling system by color if they wish.

Since individual straws are only 20-mm long by about 5-mm wide, a single liquid nitrogen container will hold more straws than cryovials. Cryostraws are stored in special plastic cylinders sold by the manufacturers. One large cylinder of a liquid nitrogen tank will hold two plastic cylinders each containing over 120 straws. Once again, our system designates the top plastic cylinder "1" and the bottom cylinder "2". Each plastic cylinder is divided into eleven uniquely colored sections, like pieces of a pie. These sections retain the liquid nitrogen and may be removed from their cylinder and taken to a work station for use. This keeps most cultures frozen while the one that is needed may be removed. Naturally, we recommend that all straws for each culture be placed in the same colored section. This color location should be recorded.

Storage of cultures in crystals

Halophilic microorganisms, especially the halophilic Archaea and the spore-forming halotolerant Bacteria, offer one very special storage mechanism not available with other microbes. That is storage inside a salt crystal. This particular mechanism is relatively easy to set up. Using crystals is advantageous since they are sealed, very inexpensive, easy to produce in quantity and can maintain culture purity. The disadvantages associated with their use are that they must be handled carefully, they do absorb water and will therefore dissolve, they can be fragile and finally identification of which crystal holds which microbe must be done indirectly by labeling the vial holding the crystal. Nevertheless we have successfully used this method for several microbes. The brines used to produce the crystals may be simply saturated NaCl or may be a solution containing a small amount of potassium phosphate and magnesium chloride. Encasing cells in crystals produced from the growth media or from brines containing organic additives is not recommended since these materials cause defects in the crystal lattice, making the crystal more permeable. In addition, our experience has been that cells encased in crystals produced from high nutrient media actually have a lower recovery rate than those encased in pure halite. This result is presumably caused by the cells attempting to metabolize the nutrients producing toxic wastes within the confines of the small brine inclusions in which they become trapped (Norton and Grant, 1988). Producing storage crystals is extremely easy and is done using sterile microtiter or tissue culture plates of any size (24, 48 or 96 well plates have all been used) depending upon the number and size of crystals desired.

Procedure

- Grow cells to be stored in crystals in broth media, harvest by centrifugation and wash with sterile brine.
- Prepare a dense cell suspension in saturated salt brine, and distribute the suspension into wells of 24, 48 or 96 well flat bottom tissue culture plates provided with a lid.
- In order to trap the most cells per crystal, evaporate the brine over silica gel or calcium chloride for about two weeks. The lid of the plate should remain on to prevent contamination.
- Once crystals are visible remove the desiccant to allow the crystals to seal by growing slowly. This technique generally yields single crystals at the bottom of each well. The cells should be optically dense enough to be visible in the center of the crystal.

One advantage of this technique is that cells trapped in salt crystals are not considered dangerous for shipment and can be readily dispensed to other laboratories without problems. In addition, the evaporation within the wells yields very uniform crystals that are easily handled. Prior to reviving the cells, the crystals should be surface sterilized using either the NaOH/HCl system of Rosenzweig *et al.* (2000) or the ethanol technique of Norton *et al.* (1993) if the crystal has not contacted (or does not contain) spore-forming organisms.

◆◆◆◆◆ REVIVING CULTURES

In a situation where a slant or Petri plate has dried or been stored for a long period we find that the culture can often be revived by adding a small amount of sterile medium to the surface of the agar and then transferring the liquid into a flask. However, we find that adding the medium and then allowing the culture to rest undisturbed at room temperature for approximately 2 h before incubating and or shaking results in somewhat better recovery.

Reviving a culture from any of the above long-term storage methods is best accomplished using the medium in which the stored culture was originally grown. Broth suspensions generally are best, but solid medium in a Petri plate may also be used. When dealing with cultures stored on beads, the vial should be placed on ice and maintained as cold as possible. Individual beads are best retrieved using an inoculating needle with the tip bent to form a 90 degree angle. The sterile inoculating needle should be cooled well since sudden heat will rapidly kill the stored cultures. A single bead is retrieved by inserting the tip of the needle through the bead's center and carefully pulling the bead out of the vial. The bead is then dropped into the liquid medium. If using a Petri plate, the bead should be dropped onto the agar, the lid replaced, and the plate shaken vigorously to cause the bead to roll around on the surface. Newly revived halophiles recover best when allowed to rest in or on the medium, at room temperature, for at least 2 h before being placed into the incubation temperature.

To revive cultures in straws, the section where the straw is located should be removed from the goblet, placed on ice and the desired straw located. The straw should be defrosted on ice to avoid sudden temperature shock. The outside of the straw and scissors should be sterilized using a laboratory disinfectant according to manufacturer's instructions. When both straw and scissors are sterile, hold the straw at a 45 degree angle over a flask or Petri plate of the same medium in which the culture was originally grown. First cut the culture end of the straw, then the opposite end and allow the liquid culture to flow out of the straw and into the new media. Allow the culture to sit at room temperature for at least two hours before incubating.

When reviving cultures from a crystal the crystal should be surface sterilized, then simply placed into media (20% w/v NaCl for Archaea, 8% w/v NaCl for halotolerant Bacteria) and allowed to dissolve slowly. Once again allowing the culture to rest a bit before intense shaking and incubation seems to improve recovery.

List of suppliers

The suppliers listed below are those that have provided materials used with our culture collection. Readers should be aware that this is far from an exhaustive list and is not provided as an endorsement of either the company or the products. There are undoubtedly specific suppliers for different countries and/or even regions for some countries. This list is therefore being provided as a guide for those wishing to search for some of these specialized supplies.

Cryovials
Cryovials may be obtained from most common suppliers, including:

VWR Scientific

Fisher Scientific

Vials should be autoclavable, with screw cap closures, 2.0 ml capacity.

Pro-Lab Diagnostics
20-Mural Street, Unit 4
Richmond Hill, Ontario
Canada, L4B 1K3

The complete Microbank System (includes beads but with non-salt supplemented media)

Cryo-Bio System (I.M.V Division)
10 Rue Clemenceau
B.P. 81 – 61302
L'Aigle CEDEX
France

Cryostraws and all supplies

References

Norton, C. F. and Grant, W. D. (1988). Survival of halobacteria within fluid inclusions in salt crystals. *J. Gen. Microbiol.* **134**, 1365–1373.

Norton, C. F., McGenity, T. J. and Grant, W. D. (1993). Archaeal halophiles (halobacteria) from two British salt mines. *J. Gen. Microbiol.* **139**, 1077–1081.

Rosenzweig, W. D., Woish, J., Petersen, J., and Vreeland, R. H. (2000). Development of a protocol to retrieve microorganisms from ancient salt crystals. *Geomicrobiology* **17**, 185–192.

Simione, F. P. and Brown, E.M. (1991). *ATCC Preservation Methods: Freezing and Freeze Drying.* American Type Culture Collection, Rockville, MD.

Vreeland, R. H., Litchfield, C. D., Martin, E. L. and Elliot, E. (1980). *Halomonas elongata*, a new genus as species of extremely salt tolerant bacteria. *Int. J. Syst. Bacteriol.* **30**, 485–495.

Barophiles

◆◆

31 Handling of Piezophilic Microorganisms

31 Handling of Piezophilic Microorganisms

Chiaki Kato
Research Program for Marine Biology and Ecology, Extremobiosphere Research Center, Japan Agency for Marine–Earth Science and Technology, 2–15 Natsushima-cho, Yokosuka 237-0061, Japan

CONTENTS

Introduction
Traditional handling of piezophiles
The DEEPBATH system

◆◆◆◆◆ INTRODUCTION

Some extremophiles, living under the deep-ocean floor, are microorganisms that have adapted to a high-pressure environment and can grow more easily under high hydrostatic pressure conditions than at atmospheric pressure. We call such deep-sea extremophiles "piezophiles," meaning pressure- (piezo- in Greek) loving (-phile) organisms (Yayanos, 1995). In the laboratory, piezophiles must be isolated and cultivated under high-pressure conditions, and therefore special high-pressure equipment, such as pressure vessels, hydrostatic pumps, etc., is necessary. When sampling deep-sea sediments or seawater to obtain living piezophiles, we also need a pressure-retaining sampler to maintain the environmental pressure and temperature, because some piezophiles are sensitive to drastic pressure and temperature changes. The Japan Agency for Marine-Earth Science and Technology (JAMSTEC) has been developing a "*deep*-sea *ba*ro-piezophile and *th*ermophile isolation and cultivation system," referred to as the "DEEPBATH" system, for handling piezophiles under study (Kyo *et al.*, 1991). In this chapter, the traditional methods for the handling of piezophilic microorganisms are described along with novel systems.

♦♦♦♦♦ TRADITIONAL HANDLING OF PIEZOPHILES

In 1957, Zobell and Morita were first considering how to handle piezophilic microorganisms living under the deep-sea floor. They developed a titanium pressure vessels that could produce a pressure of up to 100 MPa for the handling and cultivation of such microbes (Zobell and Morita, 1957; see Figure 31.1A). They attempted several times to isolate piezophiles from deep-sea samples, but they were able to isolate only piezotolerant deep-sea microbes, which showed better growth under atmospheric pressure conditions but also grew at high pressure. One reason for their inability to isolate piezophiles was that such extremophiles can be sensitive to drastic changes in pressure and temperature, and it may be very difficult to maintain the microbes at atmospheric pressure. However, Zobell and Morita made a great contribution to the handling of microbes in high-pressure microbiology, and many researchers began to study biological physiology under pressure conditions (for a review, see Marquis, 1976).

In 1979, Yayanos and co-workers succeeded in isolating piezophilic microorganisms from amphipods recovered from a depth of 5782 m in the Philippine Trench using pressure-retaining traps. This was the first report on the isolation of a piezophilic microbe (Yayanos et al., 1979). They modified their pressure vessels for easier handling, using a pin system to retain the pressure instead of the original screw system (see Figure 31.1B). This modification solved the problem of pressure leakage due to wearing and metal fatigue of the screw, and the new system did not require much power for operation. Using those pressure vessels, the author's JAMSTEC group isolated numerous piezophiles during the past 10 years (Kato et al., 2000, 2004). The procedures for the sampling of deep-sea sediment to the isolation of piezophiles are described below.

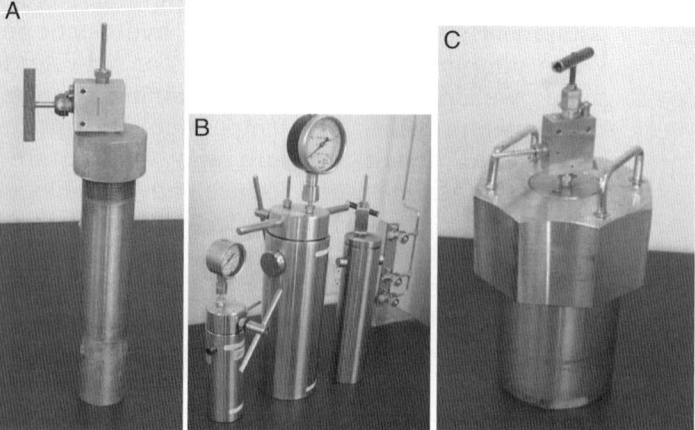

Figure 31.1. Pressure vessels in JAMSTEC projects. (A) Zobell-Morita-type pressure vessel, kindly provided by Prof. Richard Y. Morita. (B) Yayanos-type pressure vessels. Yayanos modified the pressure-retaining system using a pin for easier handling. (C) Large-type (1000 ml) pressure vessel. See Plate 31.1 in Colour Plate Section.

Sampling of Deep-sea Sediments

Deep-sea submersible systems are available in JAMSTEC. The manned submersible *Shinkai 6500* (maximum depth 6500 m, accommodating one scientist and two pilots during operation), and the unmanned submersible *Kaiko* (maximum depth 11 000 m, containing launcher and vehicle systems) are used for sampling from the deepest ocean (see the JAMSTEC home page, http://www.jamstec.go.jp). As shown in Figure 31.2, the sterilized sediment sampler and/or core sampler are controlled by the manipulators of the submersible, and the desired samples can be obtained. Then the samples are carried to the mother ship laboratories, where they are placed under environmental pressure and temperature conditions as soon as possible.

Mixed Cultivation of Microorganisms Under Environmental Pressure and Temperature Conditions

Deep-sea sediment samples obtained using the submersible are mixed with culture medium (e.g. Marine broth medium 2216, Difco Co.), and high-pressure cultivation initiated using pressure vessels under environmental pressure and temperature conditions. After several days of cultivation, the mixed cultivations are transferred to fresh media to continue culture under different conditions, if necessary.

Isolation of Piezophilic Microorganisms Using the Pressure Bag Method

The mixed cultivation is diluted and mixed with low melting point agar (0.7% agar) medium in sterilized plastic bags (10 cm in diameter, approximately 20 ml of agar medium; see Figure 31.3A) at 20°C. Then the bags are sealed and transferred to ice-cold water for solidification of the medium. After solidification, the bags are placed into large pressure vessels (see Figure 31.1C), pressurized to atmospheric pressure, and kept at room temperature for several days. After pressure cultivation, several microbial colonies can be identified in the bag, as shown in Figure 31.3B. Each colony is placed into a separate disposable syringe under a stereoscopic microscope, and cultivation under both high-pressure and atmospheric pressure conditions is initiated. If the isolates grow better under elevated pressure conditions than at atmospheric pressure, they are termed piezophiles.

Characterization of Isolates

To determine the optimal growth conditions of the isolates, multiple (20–30) pressure vessels are used to avoid the effects of repeated pressurization and decompression on the cells. If the pressure of

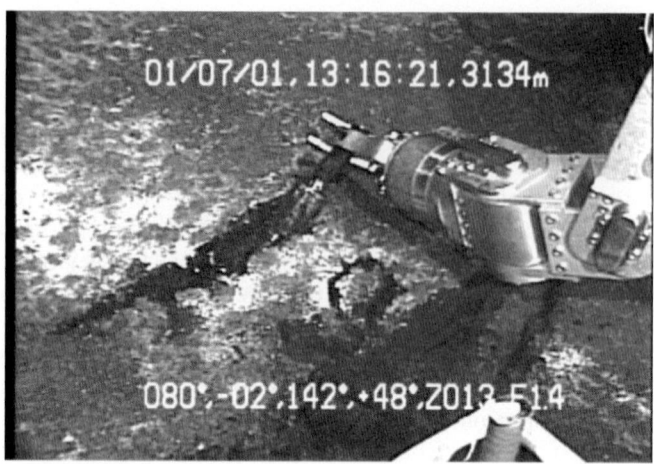

Figure 31.2. Sampling sediment from the deep-sea floor using a sterilized sediment sampler with the Shinkai 6500 manipulator. This sampling was performed at a bacterial mat site in the northeast Sea of Japan at a depth of 3134 m on July 1, 2001, during dive 6K#624 (observer, C. Kato). See Plate 31.2 in Colour Plate Section.

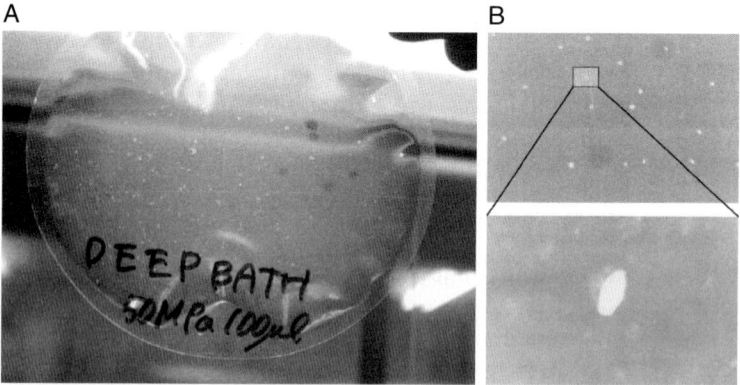

Figure 31.3. Sterilized plastic bag. (A) Plastic bag after high-pressure cultivation. Several microbial colonies formed at a pressure of 50 MPa. (B) Piezophile colonies in the bag. A single colony can be suctioned into a sterilized disposable syringe. See Plate 31.3 in Colour Plate Section.

a vessel is released, the same vessel is not repressurized to ensure that the results are reproducible. In the case of cultivation under aerobic conditions, oxygenated fluorinert (FC-72, Sumitomo-3M Co., Japan) is added to the cultures (25% of total volume; Kato et al., 1995). To study the physiological and molecular properties of the isolates, 200-ml scale cultivation is carried out using the large type of pressure vessel (Figure 31.1C).

Using the above methods, we have performed several studies on the taxonomy of piezophiles, gene expression controlled by pressure conditions and genomic analysis of piezophilic microorganisms (Kato et al., 2000, 2004; Nakasone et al., 1998, 2002). However, the 200-ml culture scale is sometimes insufficient to yield the desired amounts of protein, DNA, lipids, etc. because the maximum cell density of piezophile cultures is not very high (ca. 10^8–10^9 cells ml^{-1}). Therefore we developed a large high-pressure cultivation system, the DEEPBATH system, as described the next section.

◆◆◆◆◆ THE DEEPBATH SYSTEM

The DEEPBATH system consists of four separate devices: (1) a pressure-retaining sampling device, (2) dilution device under pressure conditions, (3) isolation device and (4) cultivation device (Kyo *et al.*, 1991). The system is controlled by central regulation systems, and the pressure and temperature ranges of the devices are from 0.1 to 65 MPa and from 0 to 150°C, respectively. The scale of the cultivation devices (2 sets) is 1.5 l each, and therefore cultures of up to 3 l can be obtained. The construction of the system and the sample stream are shown in Figure 31.4. The principle of the isolation of piezophiles is the dilution-to-extinction procedure, although in some cases microbes may be attached so firmly to the sediment that it is difficult to obtain a pure culture using the dilution-to-extinction procedure. Thus we combined the traditional procedure explained above and the DEEPBATH system protocol for isolation studies. Two experimental examples using the DEEPBATH system are described below.

Changes in the microbial community in Japan Trench sediment from a depth of 6292 m during cultivation without decompression (Yanagibayashi *et al.*, 1999)

A deep-sea sediment sample (about 5 cm^3) was obtained from the Mannequin Valley in the Japan Trench from a depth of 6292 m using the pressure-retaining sampling device of the *Shinkai 6500* submersible (dive #373, June 1997, observer C. Kato) under conditions of 65 MPa and 2°C. The obtained sample was diluted almost 100-fold with sterilized seawater with the dilution device at a pressure of 65 MPa and temperature of 2°C. Then 10 ml of the dilution solution was inoculated into the cultivation device (containing 1.5 l of Marine broth 2216 medium) with no change in pressure and temperature. Cultivation was repeated five times at 65 MPa and 10°C without decompression, and each time cultivation continued for approximately 4 days until the early stationary phase had been reached, as confirmed with a laser beam microbe density measurement system (Kyo *et al.*, 1991). Each cultivated bacterial mixture was centrifuged, and DNA was purified from the mixture to study microbial diversity. The changes in the microbial community are shown in Table 31.1. At the beginning stage, 15 bacterial clones with different 16S rRNA gene sequences were identified in the samples. Only two groups of the bacterial genera *Shewanella* and *Moritella* were identified after cultivation at 65 MPa. Interestingly, the genus *Moritella* became abundant during consecutive cultivation without decompression, and thus *Moritella* may be better adapted to growth under limiting oxygen conditions under high pressure than the piezophilic *Shewanella*.

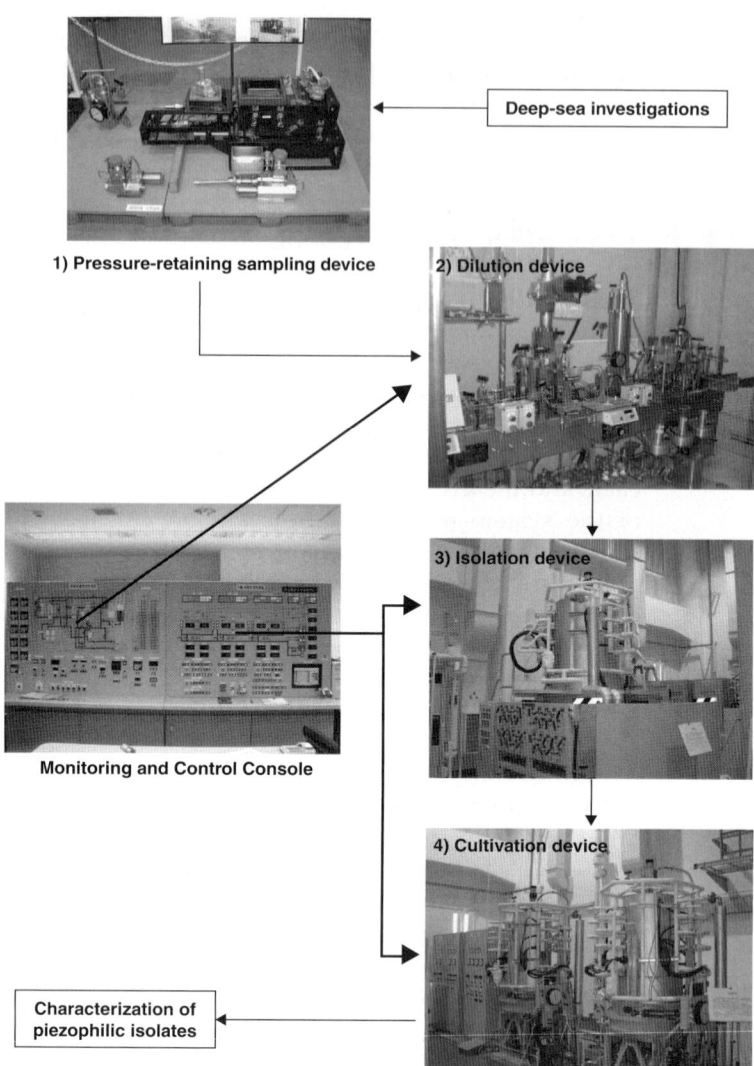

Figure 31.4. The DEEPBATH system. The system is composed of four devices: (1) pressure-retaining sampling device, (2) dilution device under pressure conditions, (3) isolation device and (4) cultivation device. The system is controlled by Monitoring and Control Console. See Plate 31.4 in Colour Plate Section.

Table 31.1 Number of clones in different groups from the sequence-amplified 16S rRNA genes from the DNA of cells (lst–5th cultures) consecutively cultivated at 65 MPa. The genera *Shewanella* and *Moritella* are typical piezophilic groups in the deep-sea environment (Kato et al., 2000, 2004)

Bacterial group	lst	2nd	3rd	4th	5th	Total clones
Genus *Shewanella*	63%	33%	10%	0%	0%	9
Genus *Moritella*	37%	67%	90%	100%	100%	37
Total clones	8	9	10	10	9	46

High-pressure Cultivation of Hyperthermophilic Microorganisms
(Canganella et al., 1997; Kato, 1997)

The "DEEPBATH" system can be used at temperatures up to 150°C, allowing high-pressure cultivation studies using hyperthermophilic micro-organisms, which can grow at around 100°C. Erauso et al. (1993) reported the cultivation of the hyperthermophilic archaeon "Pyrococcus abysii", isolated from a deep-sea hydrothermal vent, under elevated pressure conditions. They indicated that there were problems associated with the effects of repeated decompression and repressurization processes, which affected the quality of the growth results. DEEPBATH is currently the only system available which permits sampling without any change in pressure and temperature. Using this system, we have performed high-pressure and high-temperature (HP–HT) cultivation studies of two novel hyperthermophilic Archaea isolated from submarine hydrothermal vents, *Thermococcus peptonophilus* (isolated from the Izu-Bonin Trough at a depth of 1400 m; Gonzalez et al., 1995) and *Pyrococcus horikoshii* (isolated from the Okinawa Trough at a depth of 1300 m; Gonzalez et al., 1998). The optimal growth temperatures of *T. peptonophilus* and *P. horikoshii* are 85°C and 95°C, respectively. The results of HP–HT growth studies are shown in Figure 31.5. It was interesting that the temperature for growth of both hyperthermophilic Archaea could be increased under higher pressure conditions, and their growth profiles appeared piezophilic. In particular, the growth profile of *P. horikoshii* at 103°C was the same as that of an obligatory piezophilic microorganism (Figure 31.5B). Such piezophilic growth profiles in deep-sea hyperthermophiles are common, as indicated by Deming and Baross (1993), who suggested that these microbes might originate from deep subsurface environments. It is interesting to consider that the origin of life might have come from the subsurface. The DEEPBATH system should be useful in the field of subsurface microbiology in future investigations.

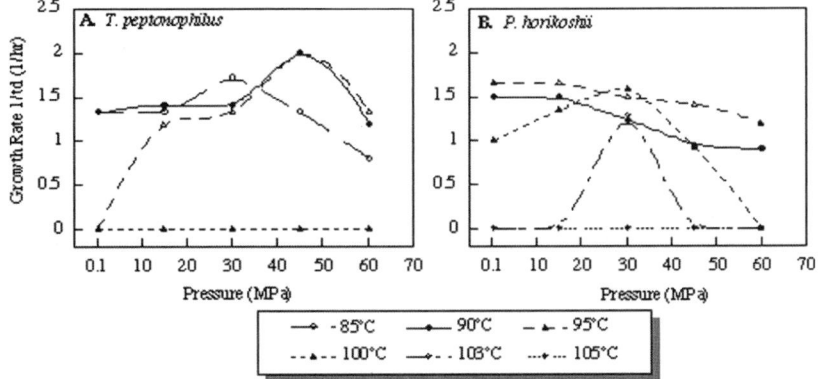

Figure 31.5. Effects of pressure on the growth rate of *Thermococcus peptonophilus* (A) and *Pyrococcus horikoshii* (B) at various temperatures. The growth rate is shown as $1/t_d$, where t_d represents the doubling time in hours.

Acknowledgements

I especially thank Prof. Richard Y. Morita for providing the valuable traditional pressure vessels to study piezophiles, and for his encouragement during the course of high-pressure research. I am very grateful to the submersible operation teams and the DEEPBATH operation staff for making it possible to undertake piezophilic studies. I also thank my colleagues who collaborated in JAMSTEC projects, and Ms. Cynthia Yenches for assistance in editing the manuscript.

List of Suppliers

Sumitomo-3M Co.
2-33-1 Tamagawadai, Setagaya-Ku
Tokyo 158-8583, Japan.
Tel: +81-3-3709-8111
http://www.mmm.co.jp/index.html

Fluorinert

References

Canganella, F., Gonzalez, J. M., Yanagibayashi, M., Kato, C. and Horikoshi, K. (1997). Pressure and temperature effects on growth and viability of the hyperthermophilic archaeon *Thermococcus peptonophilus*. *Arch. Microbiol.* **168**, 1–7.

Deming, J. W. and Baross, J. A. (1993). Deep-sea smokers: Windows to a subsurface biosphere? *Geochim. Cosmochim. Acta* **57**, 3219–3230.

Erauso, G., Reysenbach, A.-L., Godfroy, A., Meunier, J. R., Crump, B., Partensky, F., Baross, J. A., Marteinsson, V., Barbier, G., Pace, N. R. and Prieur, D. (1993). *Pyrococcus abyssi* sp. nov., a new hyperthermophilic archaeon isolated from a deep-sea hydrothermal vent. *Arch. Microbiol.* **160**, 338–349.

Gonzalez, J. M., Kato, C. and Horikoshi, K. (1995). *Thermococcus peptonophilus* sp. nov., a fast-growing, extremely thermophilic archaebacterium isolated from deep-sea hydrothermal vents. *Arch. Microbiol.* **164**, 159–164.

Gonzalez, J. M., Masuchi, Y., Robb, F. T., Ammerman, J. W., Maeder, D. L., Yanagibayashi, M., Tamaoka, J. and Kato, C. (1998). *Pyrococcus horikoshii* sp. nov., a hyperthermophilic archaeon isolated from a hydrothermal vent at the Okinawa Trough. *Extremophiles* **2**, 123–130.

Kato, C., Sato, T. and Horikoshi, K. (1995). Isolation and properties of barophilic and barotolerant bacteria from deep-sea mud samples. *Biodivers. Conserv.* **4**, 1–9.

Kato, C. (1997). Genomic analyses of hyperthermophilic archaea. *J. Japan. Oil Chem. Soc.* **46**, 517–523 (in Japanese).

Kato, C., Nakasone, K., Qureshi, M. H. and Horikoshi, K. (2000). How do deep-sea microorganisms respond to the environmental pressure? In *Cell and Molecular Response to Stress. Environmental Stressors and Gene Responses*

(K. B. Storey and J. M. Storey, eds), vol. 1, pp. 277–291. Elsevier Science, Amsterdam.

Kato, C., Sato, T., Nogi, Y. and Nakasone, K. (2004). Piezophiles: High pressure-adapted marine bacteria. *Mar. Biotechnol.* **6**, S195–S201.

Kyo, M., Tuji, T., Usui, H. and Itoh, T. (1991). Collection, isolation and cultivation system for deep-sea microbes study: concept and design. *Oceans* **1**, 419–423.

Marquis, R. E. (1976). High pressure microbial physiology. *Adv. Microb. Physiol.* **14**, 159–241.

Nakasone, K., Ikegami, A., Kato, C., Usami, R. and Horikoshi, K. (1998). Mechanisms of gene expression controlled by pressure in deep-sea microorganisms. *Extremophiles* **2**, 149–154.

Nakasone, K., Ikegami, A., Kawano, H., Usami, R., Kato, C. and Horikoshi, K. (2002). Transcriptional regulation under pressure conditions by the RNA polymerase σ^{54} factor with a two component regulatory system in *Shewanella violacea*. *Extremophiles* **6**, 89–95.

Yanagibayashi, M., Nogi, Y., Li, L. and Kato, C. (1999). Changes in the microbial community in Japan Trench sediment from a depth of 6292 m during cultivation without decompression. *FEMS Microbiol. Lett.* **170**, 271–279.

Yayanos, A. A. (1995). Microbiology to 10 500 meters in the deep sea. *Ann. Rev. Microbiol.* **49**, 777–805.

Yayanos, A. A., Dietz, A. S. and Boxtel, R. V. (1979). Isolation of a deep-sea barophilic bacterium and some of its growth characteristics. *Science* **205**, 808–810.

Zobell, C. E. and Morita, R. Y. (1957). Barophilic bacteria in some deep-sea sediments. *J. Bacteriol.* **73**, 563–568.

Radiation-Resistant Microorganisms

◆◆◆

32 Measuring Survival in Microbial Populations Following Exposure to Ionizing Radiation

32 Measuring Survival in Microbial Populations Following Exposure to Ionizing Radiation

Julie M Zimmerman and John R Battista
Biological Sciences, Louisiana State University and A & M College, Baton Rouge, LA 70803, USA

◆◆

CONTENTS

Preface
Background
Factors that influence the effect of ionizing radiation
Generating a survival curve

◆◆◆◆◆ PREFACE

Characterizing a cell's ability to tolerate ionizing radiation experimentally is straightforward. An investigator must generate a dose–response curve and use that curve to compare irradiated populations. In this chapter, when we refer to a dose–response (or survival) curve, we are describing a graph that portrays the relationship between exposure to a toxic agent and the viability of a microbial population. We focus on a method for assessing viability following exposure to ionizing radiation, but with slight modification these protocols may be used to define the lethal effects of any chemical or physical agent on most cultured microbes. Since the dose–response curve delineates a mathematical relationship between the extent of exposure (the dose applied) and viability, it provides a means of directly comparing the resistance of populations and estimating the statistical relevance of that comparison.

◆◆◆◆◆ BACKGROUND

Within the domains Archaea and Bacteria there are species capable of inhabiting environments inhospitable to all other forms of life (Horikoshi and Grant, 1998). These species have been isolated from hot springs and abyssal hydrothermal vents (Brock and Freeze, 1969; Burggraf et al., 1990), surviving at temperatures as high as 130°C (Kashefi and Lovley, 2003), at extremely acidic and alkaline pH (Norton et al., 1992; Sorokin et al., 2002), and in exceedingly dry and/or cold environments (Billi and Potts, 2002; Billi et al., 2000). Since the deleterious physical effects of these

Table 32.1 Species of prokaryotes described as ionizing radiation resistant

Species	Phylum	Reference
Bacteria		
Methylobacterium radiotolerans	α-Proteobacteria	Green and Bousfield (1983), Ito and Iizuka (1971)
Kocuria rosea	Actinobacteria	Brooks and Murray (1981), Rainey et al. (1997)
Acinetobacter radioresistens	γ-Proteobacteria	Nishimura et al. (1994)
Kineococcus radiotolerans	Actinobacteria	Phillips et al. (2002)
Hymenobacter actinosclerus	Flexibacter-Cytophaga-Bacteroides	Collins et al. (2000)
Chroococcidiopsis spp.	Cyanobacteria	Billi et al. (2000)
Rubrobacter xylanophilus	Actinobacteria	Ferreira et al. (1999)
Rubrobacter radiotolerans	Actinobacteria	Ferreira et al. (1999)
Deinococcus radiodurans	Deinococcus-Thermus	Battista and Rainey (2001)
Deinococcus proteolyticus	Deinococcus-Thermus	Battista and Rainey (2001)
Deinococcus radiophilus	Deinococcus-Thermus	Battista and Rainey (2001)
Deinococcus radiopugnans	Deinococcus-Thermus	Battista and Rainey (2001)
Deinococcus grandis	Deinococcus-Thermus	Battista and Rainey (2001)
Deinococcus geothemalis	Deinococcus-Thermus	Battista and Rainey (2001)
Deinococcus murrayi	Deinococcus-Thermus	Battista and Rainey (2001)
Deinococcus indicus	Deinococcus-Thermus	Suresh et al. (2004)
"*Deinococcus frigens*"	Deinococcus-Thermus	Hirsch et al. (2004)
"*Deinococcus marmoris*"	Deinococcus-Thermus	Hirsch et al. (2004)
"*Deinococcus saxicola*"	Deinococcus-Thermus	Hirsch et al. (2004)
Deinococcus hohokamensis	Deinococcus-Thermus	Rainey et al. (2005)
Deinococcus navajonensis	Deinococcus-Thermus	Rainey et al. (2005)
Deinococcus hopiensis	Deinococcus-Thermus	Rainey et al. (2005)
Deinococcus apachensis	Deinococcus-Thermus	Rainey et al. (2005)
Deinococcus maricopensis	Deinococcus-Thermus	Rainey et al. (2005)
Deinococcus pimensis	Deinococcus-Thermus	Rainey et al. (2005)
Deinococcus yavapaiensis	Deinococcus-Thermus	Rainey et al. (2005)
Deinococcus papagonensis	Deinococcus-Thermus	Rainey et al. (2005)
Deinococcus sonorensis	Deinococcus-Thermus	Rainey et al. (2005)
Truepera radiovictrix	Deinococcus-Thermus	Albuquerque et al. (2005)
Archaea		
"*Pyrococcus abyssi*"	Euryarchaeota	Jolivet et al. (2003b)
Pyrococcus furiosus	Euryarchaeota	DiRuggiero et al. (1997)
Thermococcus gammtolerans	Euryarchaeota	Jolivet et al. (2003a)
Thermococcus radiotolerans	Euryarchaeota	Jolivet et al. (2004)
Thermococcus marinus	Euryarchaeota	Jolivet et al. (2004)

conditions on isolated DNA are well documented, it is not unreasonable to assume that the genome of species in these environments is under constant assault, and that their capacity to endure DNA damage exceeds that of mesophilic neutrophiles such as *Escherichia coli*.

Ionizing radiation resistance, the ability to survive exposure to high doses of X- and gamma rays, is a characteristic shared by a relatively small number of prokaryotes that presumably have heightened ability to withstand the lethal effects of that exposure. Table 32.1 lists bacterial and archaeal species reported to exhibit ionizing radiation resistance. All of the species listed exhibit D_{10} doses – the dose needed to inactivate 90% of a population – in excess of 1000 Gy. The *Deinococcaceae* are the best-known family on this list (Battista and Rainey, 2001), and their resistance appears to arise from efficient repair of ionizing radiation-induced DNA damage (Battista, 1997; Battista *et al.*, 1999; Minton, 1996). The reasons for enhanced radioresistance among the other species listed are unknown, but it is assumed that these mechanisms involve enhanced DNA repair capacity and/or the ability to passively prevent radiation-induced damage from occurring *in vivo*.

Since there are no natural environments capable of generating greater than 400 mGy of gamma radiation annually, it is suspected that ionizing radiation resistance is an incidental character, the result of an evolutionary process that built an enhanced facility to repair DNA damage caused by another stress. For example, it has been reported that desiccation introduces extensive DNA damage, lesions similar to those sustained subsequent to exposure to ionizing radiation, into the genomes of *Deinococcus radiodurans* (Mattimore and Battista, 1996) and cyanobacteria of the genus *Chroococcidiopsis* (Billi *et al.*, 2000) as they dry. These organisms are highly tolerant of desiccation and at the same time extremely resistant to the lethal effects of ionizing radiation, suggesting that the ability to tolerate desiccation-induced DNA damage provides protection against ionizing radiation induced damage. Similarly, many of the species listed in Table 32.1 are thermophilic; elevated temperatures promote depurination (Lindahl, 1993), a condition that can lead to increases in DNA damage (especially DNA double strand breaks) that are associated with ionizing radiation-induced DNA damage. Thus, ionizing radiation resistance appears to be indicative of a species capacity to survive desiccation or extreme temperatures.

◆◆◆◆◆ FACTORS THAT INFLUENCE THE EFFECT OF IONIZING RADIATION ON MICROBIAL POPULATIONS

X- and gamma rays are forms of electromagnetic radiation characterized by very short wavelengths and sufficient kinetic energy to disrupt the atomic structure of molecules. When these high-energy photons are absorbed by living cells, they initiate a cascade of biological damage that may ultimately kill the cell (Ward, 1975). The effectiveness of ionizing radiation in bringing about cell death in a microbial population is influenced by a number of factors.

1. The dose administered: While it is intuitive that the numbers of organisms killed in an irradiated population increases with increasing dose, for many it is more difficult to appreciate that successive increments of the same dose do not kill equivalent numbers of organisms. To a first approximation the generation of damage is random within an irradiated system, and for a microbial culture this means that not every member of the population receives an equivalent amount of damage (Iwanami and Oda, 1985). If a 100 Gy dose of gamma radiation kills 90% of a population comprised of 10^7 cells, 9×10^6 cells die and 1×10^6 remain viable. If this surviving population is subjected to a second 100 Gy dose, 9×10^5 cells or 90% of the residual population die. In other words, a given dose will kill the same proportion of a specific organism in culture regardless of population size.

2. Target size: In general, the smaller the species being irradiated (the target), the more difficult it is to inactivate (Osborne et al., 2000). Prokaryotes are more radioresistant than eukaryotes, and viruses are more radioresistant than bacteria. Assuming all microbial targets are spherical, there is an inverse relationship between the D_{37} dose – the dose that is necessary to inactivate a single cell – and the target volume.

3. The presence of water: The effects of energy deposition in cells exposed to ionizing radiation are classified as direct – when energy transfer results in the ionization of a biologically important molecule – and indirect – when water is ionized and the products of that event, which include hydroxyl radical, modify biologically important molecules (Ward, 1988). Water is the most abundant molecule in living systems and indirect effects reportedly account for as much as 70% of the biological damage generated post-irradiation. Thus, ionizing radiation is much less effective when desiccated populations are irradiated. This is especially important when considering the irradiation of spore-forming species. Most of the water present in the vegetative cell is extruded during sporulation (Setlow, 1992), increasing the radioresistance of the spore and the overall radioresistance of a culture comprised of mixed vegetative and sporulating population.

4. The presence of oxygen: Anoxic cultures are more resistant to ionizing radiation than are cultures replete with oxygen. Oxygen readily reacts with carbon-centered radicals generated by the interaction of hydroxyl radicals with macromolecules, forming peroxyl radicals and propagating the damage caused by ionizing radiation (von Sonntag, 1987).

5. The stage of growth: In general, microbial cultures are most resistant to ionizing radiation in stationary phase. Two features of life in stationary phase are believed to contribute to this phenomenon (Kolter, 1992). First, stationary phase cultures are anoxic and as indicated above lower oxygen levels will limit the extent of damage caused by ionizing radiation. Second, the genomes of cells in stationary phase are not actively replicating. In E. coli, extensive DNA damage will block the

movement of the replication fork and this event may have catastrophic consequences. When the genome is not being replicated, this problem is avoided.

6. Species specific damage tolerance: Among prokaryotes, there are species that are very sensitive to ionizing radiation and species that are quite radioresistant. Most vegetative cells succumb to doses in the 30–80 Gy range, but species like *Deinococcus radiodurans* (Battista, 1997) and *Rubrobacter radiotolerans* (Ferreira *et al.*, 1999) survive exposure to doses 100-fold higher (Figure 32.1). In many instances the reasons for increased radiotolerance are not known, but for the best understood radioresistant organism, *D. radiodurans*, a substantial body of evidence indicates that this species has an enhanced DNA repair capacity. Loss of that capacity sensitizes this species to ionizing radiation (Battista, 1997; Battista *et al.*, 1999; Minton, 1996). Thus, an organism's radioresistance may increase for reasons that cannot be predicted based on the factors listed above.

◆◆◆◆◆ GENERATING A SURVIVAL CURVE

The method described below may be applied to a culture of any microorganism at any growth phase. In our laboratory, generation of a survival curve begins by conducting a preliminary study in which we determine the dose range in which some portion of the population remains viable. This analysis is followed by a detailed study in which the surviving fraction of the population is calculated after exposure of a series of

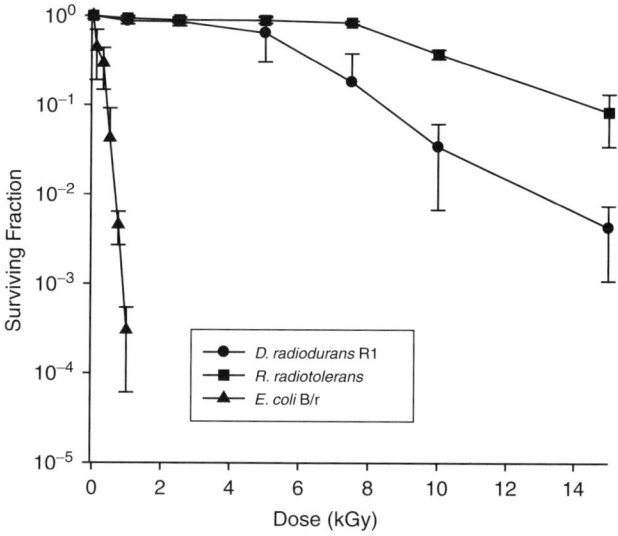

Figure 32.1. A typical survival curve comparing the ionizing radiation resistance of *Deinococcus radiodurans* R1, *Rubrobacter radiotolerans* and *Escherichia coli* B/r. The surviving fraction is plotted as a common log *versus* the dose of gamma radiation in kiloGray administered.

doses within this range. As indicated above, a cell's susceptibility to a given radiation dose is influenced by a number of parameters. Obtaining an accurate survival curve requires that each replicate study be carefully reproduced.

In the methods described, the investigator ultimately "counts colonies" and it is assumed that there is a linear relationship between the dose administered, the inactivation of an individual cell, and the inactivation of a colony forming unit. In the simplest situation, where a single cell gives rise to a single colony, this relationship is easily seen. The survival curve represents the survival of the individual cell. For species whose colony forming unit is composed of multiple cells (e.g. diplococci), the survival curve represents the survival of the colony forming unit unless the data are adjusted to reflect the number of individual cells involved. (See Daly *et al.* (2004) for an example of how to apply a correction.) If the species being investigated exhibits substantial variation in the numbers of cells that can make up a colony forming unit, these assumptions do not hold true. For example, if the cells of a species clump during growth in liquid, an individual colony could be formed by a single cell or hundreds of cells. Obtaining a reproducible titer before or after irradiation under such circumstances is only achieved if a method for evenly dispersing the cells can be identified.

Equipment and Reagents

- culture medium, both broth and solid media
- spectrophotometer
- gamma radiation source

Preliminary Studies

We recommend conducting a preliminary study under the growth conditions chosen for analysis. This is done to determine an appropriate dose range and to establish how susceptible the culture is to a given exposure before investing in the time and materials necessary to conduct a more thorough study.

1. Inoculate appropriate broth medium with the organism of interest, grow under well-defined conditions until the population reaches the desired growth phase.
2. Record the optical density (OD_{600}) of the culture.
3. Aliquot 1 ml of the culture into each of six containers. (The type of container used is not critical provided it is sterile and can be sealed to prevent contamination. We use sterile microcentrifuge tubes.)
4. Apply the desired radiation dose to each sample. Typically, we choose a series of five doses; each sample is subjected to a different dose. The sixth sample serves as a control and is not irradiated. The doses chosen vary with the organism and its expected radioresistance. When unsure of the species radioresistance, dose at 100, 250, 500,

1000, 2000 and 3000 Gy. If resistance is high at 3000 Gy, initiate a second preliminary study in which higher doses are applied. It is unusual to find a species that exhibits high levels of resistance following exposure to doses in excess of 3000 Gy, particularly if they are in exponential phase growth.

5. Following irradiation, serially dilute each sample using six 10-fold dilutions. The medium for dilution should be iso-osmotic relative to the growth medium. (Growth medium can be used, but typically we dilute using a solution that does not permit the cells to grow e.g. physiological saline.)
6. A 10–50 µl aliquot of each dilution in each series is placed in a defined pattern on a single agar plate, allowed to dry, and then grown under appropriate conditions. This pattern is repeated on at least three plates for each dose.
7. Observe the plates for growth. (Growth on the plates estimates the titer of the irradiated culture, and comparison of pattern of growth on plates corresponding to the un-irradiated control with the irradiated samples provides an indication of radioresistance at a given dose.)
8. If there is growth after the maximum dose, repeat the preliminary study using higher doses. Otherwise, determine the dilution for each dose (and the control) that will most likely result in a countable spread plate (30–300 colonies per plate) when the study is repeated in a manner appropriate for statistical analysis (see below).

The Survival Curve

The goal of this protocol is to generate a statistically valid survival curve suitable for comparing the radioresistance of different bacterial strains.

1. Using the information provided in the preliminary study, estimate the species' survival at each dose to be used in generating the survival curve. For example, if you estimate that a 1000-fold dilution of a 1000 Gy dose will result in a countable plate, prepare to generate an identical dilution for that dose when repeating the study. Since the preliminary results produce a rough prediction of survival, plan to also plate dilutions that are 10-fold higher and 10-fold lower, bracketing your estimate (For this example, plate 100-fold and 10 000-fold dilutions).
2. Repeat steps 1–5 of the preliminary study. It is critical that the culture be grown to approximately the same optical density to ensure reproducibility.
3. Spread 100 µl of the chosen dilutions on solid media for each dose. (Minimally, each dilution of each sample should be plated in triplicate.)
4. Incubate the plates under optimal growth conditions and allow colonies to form. Determine the number of surviving cells for each dose. The length of time necessary for growth is determined by the

organism examined but be aware that individuals within a population may recover at different rates and that the incubation time post-irradiation can influence the estimate of survival. Determining the number of survivors too quickly may underestimate the population's resistance to ionizing radiation. We recommend counting the plates more than once, separating the first and second counts by several days to avoid this problem.

5. Once counts are complete, calculate the titer (number of survivors ml^{-1}) at each dose. Use these numbers to determine the surviving fraction at each dose by dividing the titer post-irradiation with the titer of the un-irradiated population.
6. Repeat the study at least three times. Combine the data and plot surviving fraction *versus* dose on a semi-log plot (Figure 32.1). Include a graphic representation of standard deviation of the mean for each point on each graph plotted. We recommend including the standard deviation because it quantifies variability in the population being irradiated.

References

Albuquerque, L., Simões, C., Nobre, M. F., Pino, N. M., Battista, J. R., Silva, M. T., Rainey, F. A. and da Costa, M. S. (2005). *Truepera radiovictrix*, a new radiation resistant member of the Order *Deinococcales*. FEMS Microbiol. Lett. **247**, 161–169.

Battista, J. R. (1997). Against all odds: the survival strategies of *Deinococcus radiodurans*. Annu. Rev. Microbiol. **51**, 203–224.

Battista, J. R. and Rainey, F. A. (2001). Phylum BIV. "Deinococcus-Thermus" Family 1. *Deinococcaceae* Brooks and Murray 1981, 356^{VP}, emend. Rainey, Nobre, Schumann, Stackebrandt and da Costa 1997, 513. In *Bergey's Manual of Systematic Bacteriology* (D. R. Boone and R. W. Castenholz, eds), pp. 395–414. Springer, New York.

Battista, J. R., Earl, A. M. and Park, M. J. (1999). Why is *Deinococcus radiodurans* so resistant to ionizing radiation? Trends Microbiol. **7**, 362–365.

Billi, D. and Potts, M. (2002). Life and death of dried prokaryotes. Res. Microbiol. **153**, 7–12.

Billi, D., Friedmann, E. I., Hofer, K. G., Caiola, M. G. and Ocampo-Friedmann, R. (2000). Ionizing-radiation resistance in the desiccation-tolerant cyanobacterium *Chroococcidiopsis*. Appl. Environ. Microbiol. **66**, 1489–1492.

Brock, T. D. and Freeze, H. (1969). *Thermus aquaticus* gen. n. and sp. n., a nonsporulating extreme thermophile. J. Bacteriol. **98**, 289–297.

Brooks, B. W. and Murray, R. G. E. (1981). Nomenclature for "*Micrococcus radiodurans*" and other radiation-resistant cocci: *Deinococcaceae* fam. nov. and *Deinococcus* gen. nov., including five species. Int. J. Syst. Bacteriol. **31**, 353–360.

Burggraf, S., Fricke, H., Neuner, A., Kristjansson, J., Rouvière, P., Mandelco, L., Woese, C. R. and Stetter, K. O. (1990). *Methanococcus igneus* sp. nov., a novel hyperthermophilic methanogen from a shallow submarine hydrothermal system. Syst. Appl. Microbiol. **13**, 263–269.

Collins, M. D., Hutson, R. A., Grant, I. R. and Patterson, M. F. (2000). Phylogenetic characterization of a novel radiation-resistant bacterium from irradiated pork: description of *Hymenobacter actinosclerus* sp. nov. *Int. J. Syst. Evol. Microbiol.* **50**, 731–734.

Daly, M. J., Gaidamakova, E. K., Matrosova, V. Y., Vasilenko, A., Zhai, M., Venkateswaran, A., Hess, M., Omelchenko, M. V., Kostandarithes, H. M., Makarova, K. S., Wackett, L. P., Fredrickson, J. K. and Ghosal, D. (2004). Accumulation of Mn(II) in *Deinococcus radiodurans* facilitates gamma-radiation resistance. *Science* **306**, 1025–1028.

DiRuggiero, J., Santangelo, N., Nackerdien, Z., Ravel, J. and Robb, F. T. (1997). Repair of extensive ionizing-radiation DNA damage at 95°C in the hyperthermophilic archaeon *Pyrococcus furiosus*. *J. Bacteriol.* **179**, 4643–4645.

Ferreira, A. C., Nobre, M. F., Moore, E., Rainey, F. A., Battista, J. R. and da Costa, M. S. (1999). Characterization and radiation resistance of new isolates of *Rubrobacter radiotolerans* and *Rubrobacter xylanophilus*. *Extremophiles* **3**, 235–238.

Green, P. N. and Bousfield, I. J. (1983). Emendation of *Methylobacterium* Patt, Cole, and Hanson, 1976; *Methylobacterium rhodinum* (Heumann, 1962) comb. nov. corrig.; *Methylobacterium radiotolerans* (Ito and Iizuka, 1971) comb. nov. corrig.; and *Methylobacterium mesophilicum* (Austin and Goodfellow, 1979) comb. nov. *Int. J. Syst. Bacteriol.* **33**, 875–877.

Hirsch, P., Gallikowski, C. A., Siebert, J., Peissl, K., Kroppenstedt, R., Schumann, P., Stackebrandt, E. and Anderson, R. (2004). *Deinococcus frigens* sp. nov., *Deinococcus saxicola* sp. nov., and *Deinococcus marmoris* sp. nov., low temperature and drought-tolerating, UV-resistant bacteria from continental Antarctica. *Syst. Appl. Microbiol.* **27**, 636–645.

Horikoshi, K. and Grant, W. D. (eds) (1998). *Extremophiles: Microbial Life in Extreme Environments*. Wiley-Liss, New York.

Ito, H. and Iizuka, H. (1971). Taxonomic studies on a radio-resistant *Pseudomonas*. XII. Studies on the microorganisms of cereal grain. *Agric. Biol. Chem.* **35**, 1566–1571.

Iwanami, S. and Oda, N. (1985). Theory of survival of bacteria exposed to ionizing radiation. I. X and gamma rays. *Radiat. Res.* **102**, 46–58.

Jolivet, E., L' Haridon, S., Corre, E., Forterre, P. and Prieur, D. (2003a). *Thermococcus gammatolerans* sp. nov., a hyperthermophilic archaeon from a deep-sea hydrothermal vent that resists ionizing radiation. *Int. J. Syst. Evol. Microbiol.* **53**, 847–851.

Jolivet, E., Matsunaga, F., Ishino, Y., Forterre, P., Prieur, D. and Myllykallio, H. (2003b). Physiological responses of the hyperthermophilic archaeon "*Pyrococcus abyssi*" to DNA damage caused by ionizing radiation. *J. Bacteriol.* **185**, 3958–3961.

Jolivet, E., Corre, E., L' Haridon, S., Forterre, P. and Prieur, D. (2004). *Thermococcus marinus* sp. nov. and *Thermococcus radiotolerans* sp. nov., two hyperthermophilic archaea from deep-sea hydrothermal vents that resist ionizing radiation. *Extremophiles* **8**, 219–227.

Kashefi, K. and Lovley, D. R. (2003). Extending the upper temperature limit for life. *Science* **301**, 934.

Kolter, R. (1992). Life and death in stationary phase. *ASM News* **58**, 75–79.

Lindahl, T. (1993). Instability and decay of the primary structure of DNA. *Nature* **362**, 709–715.

Mattimore, V. and Battista, J. R. (1996). Radioresistance of *Deinococcus radiodurans*: functions necessary to survive ionizing radiation are also necessary to survive prolonged desiccation. *J. Bacteriol.* **178**, 633–637.

Minton, K. W. (1996). Repair of ionizing-radiation damage in the radiation resistant bacterium *Deinococcus radiodurans*. *Mutat. Res.* **363**, 1–7.

Nishimura, Y., Uchida, K., Tanaka, K., Ino, T. and Ito, H. (1994). Radiation sensitivities of *Acinetobacter* strains isolated from clinical sources. *J. Basic Microbiol.* **34**, 357–360.

Norton, S. A., Brownlee, J. C. and Kahl, J. S. (1992). Artificial acidification of a non-acidic and an acidic headwater stream in Maine, USA. *Environ. Pollut.* **77**, 123–128.

Osborne, J. C. Jr., Miller, J. H. and Kempner, E. S. (2000). Molecular mass and volume in radiation target theory. *Biophys. J.* **78**, 1698–1702.

Phillips, R. W., Wiegel, J., Berry, C. J., Fliermans, C., Peacock, A. D., White, D. C. and Shimkets, L. J. (2002). *Kineococcus radiotolerans* sp. nov., a radiation-resistant, gram-positive bacterium. *Int. J. Syst. Evol. Microbiol.* **52**, 933–938.

Rainey, F. A., Nobre, M. F., Schumann, P., Stackebrandt, E. and da Costa, M. S. (1997). Phylogenetic diversity of the deinococci as determined by 16S ribosomal DNA sequence comparison. *Int. J. Syst. Bacteriol.* **47**, 510–514.

Rainey, F. A., Ray, K., Ferreira, M., Gatz, B. Z., Nobre, M. F., Bagaley, D., Rash, B. A., Park, M.-J., Earl, A. M., Shank, N. C., Small, A. M., Henk, M. C., Battista, J. R., Kämpfer, P. and da Costa M. S. (2005). Extensive diversity of ionizing radiation-resistant bacteria recovered from a Sonoran Desert soil and the description of 9 new species of the genus *Deinococcus* from a single soil sample. *Appl. Environ. Microbiol.* **71**, 5225–5235.

Setlow, P. (1992). I will survive: protecting and repairing spore DNA. *J. Bacteriol.* **174**, 2737–2741.

Sorokin, D. Y., Gorlenko, V. M., Tourova, T. P., Tsapin, A. I., Nealson, K. H. and Kuenen, G. J. (2002). *Thioalkalimicrobium cyclicum* sp. nov. and *Thioalkalivibrio jannaschii* sp. nov., novel species of haloalkaliphilic, obligately chemolithoautotrophic sulfur-oxidizing bacteria from hypersaline alkaline Mono Lake (California). *Int. J. Syst. Evol. Microbiol.* **52**, 913–920.

Suresh, K., Reddy, G. S., Sengupta, S. and Shivaji, S. (2004). *Deinococcus indicus* sp. nov., an arsenic-resistant bacterium from an aquifer in West Bengal, India. *Int. J. Syst. Evol. Microbiol.* **54**, 457–461.

von Sonntag, C. (1987). *The Chemical Basis of Radiation Biology*. Taylor and Francis, Philadelphia.

Ward, J. F. (1975). Molecular mechanisms of radiation-induced damage to nucleic acids. *Adv. Rad. Biol.* **5**, 181–239.

Ward, J. F. (1988). DNA damage produced by ionizing radiation in mammalian cells: identities, mechanisms of formation, and reparability. *Prog. Nucleic Acid Res. Mol. Biol.* **35**, 95–125.

Strict Anaerobes

33 The Study of Strictly Anaerobic Microorganisms

33 The Study of Strictly Anaerobic Microorganisms

Kevin R Sowers[1] and Joy EM Watts[2]

[1] Center of Marine Biotechnology, University of Maryland Biotechnology Institute, Baltimore, MD 21202, USA; [2] Department of Biological Sciences, Towson University, Towson, MD 21252, USA

◆◆

CONTENTS

Introduction
Apparatus
Media preparation
Enrichment, isolation and maintenance of anaerobes
Anaerobic biochemistry

◆◆◆◆◆ **INTRODUCTION**

Most anaerobic extremophiles are obligate anaerobes that require strictly anoxic environment, and an extracellular milieu that is poised at a low redox potential. Despite their requirement for fastidiously anoxic conditions, these anaerobic microorganisms are ubiquitous in the environment and occur in a large range of "extreme" habitats that include anoxic sewage digestors, mammalian, ruminant and termite digestive tracts, polar lakes and tundra, geothermal submarine vents, calderas and hot springs, deep sea sediments, and deep subsurface rock. The combined processes of aerobic and anaerobic degradation ensures that CO_2 "fixed" as cell carbon by photosynthetic organisms and consumed by heterotrophs, is eventually restored to the atmosphere as CO_2, thus completing the global carbon cycle in even the most antianthropomorphic environments. Geological hydrogen-utilizing, CO_2-reducing extremophiles such as the methanogenic Archaea can also serve as primary producers of reduced carbon compounds in the absence of photosynthesis, which has significant implications for research in astrobiology and the search for extraterrestrial life.

Unlike facultative anaerobes and oxygen-tolerant anaerobes, many obligate anaerobes require stringent anaerobic conditions to maintain viability. This is achieved by using oxygen-free gases to prepare anoxic medium with the addition of chemical reducing agents to achieve a low redox potential (Hungate, 1950). Specialized glassware and vessels are utilized to maintain reduced anoxic conditions during growth. This chapter describes the apparatus and methodology for growth, isolation, scale-up of the most stringent anaerobes and the isolation of oxygen-labile biomolecules. Methods described herein are applicable to methanogenic Euryarchaeota, non-methanogenic hyperthermophiles within both the Euryarchaeota and Crenarchaeota, and hyperthemophilic bacteria within the *Thermotogales*. These methods are also applicable to "non-extremophiles" such as the iron-reducing and sulfate-reducing bacteria as well as other obligate anaerobes within the Bacteria.

◆◆◆◆◆ APPARATUS

Gassing Manifold for Anaerobic Media Preparation

To prepare anaerobic medium at the very minimum, a source of oxygen-free anaerobic gas, gas flow regulator, barbed tubing connector, rubber tubing and gassing cannula configured from syringes, are required. However, depending on circumstances described below a gassing manifold for medium preparation may require up to four primary components: gas supply, gas proportioner, oxygen trap and cannula.

Gas supply

Gases commonly used for growth of anaerobes include nitrogen, argon, hydrogen and carbon dioxide or mixtures of these gases, which are readily available from local gas distributors. A standard two-stage regulator connected to rubber tubing with a barbed tubing fitting or to copper refrigerator tubing (0.635-cm outer diameter) with tubing compression fittings (e.g. Swagelok™) is used to distribute the gas to a medium work station (Figure 33.1). When using copper tubing, it is advisable to use flexible reinforced tubing between the regulator and

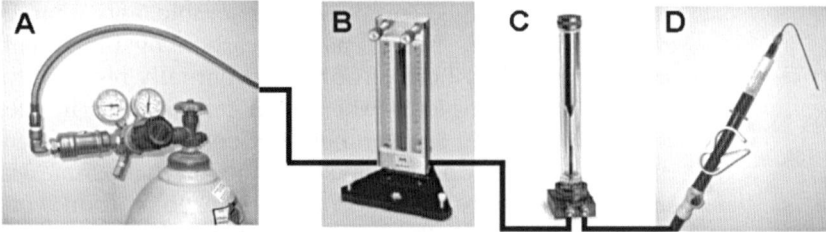

Figure 33.1. Gassing manifold for preparing media. An anaerobic gas tank with regulator is connected to system with flexible tubing (A) or copper tubing. When using multiple gas tanks instead of a single tank of pre-mixed gases, the tank lines are connected to a gas proportioner (B) to control the ratio of gases. An in-line oxygen scrubber (C) reduces traces of oxygen, then a rubber gas line is connected from the scrubber to a cannula (D) or to multiple cannulas with T-connectors. Construction of the apparatus is described in the text.

a bulkhead. This configuration makes it more convenient to remove the regulator when changing tanks while minimizing the risk of developing leaks at the junction of the tubing and regulator. Many media recipes use a carbonate buffer system and require a mixture of gases to be present in the headspace. The most common mixtures are nitrogen and carbon dioxide in 70/30 or 80/20% ratios and hydrogen and carbon dioxide in an 80/20% ratio in the case of hydrogen-oxidizing bacteria.

Gas proportioner

If gas mixtures are required for a specific medium, gases can be ordered pre-mixed in the required ratios from most gas distributors. However, if gas mixtures will be used at high rates, a more economical alternative is to buy individual gases and mix them to the appropriate ratio with a flow meter (Figure 33.1). The requisite multiple gas regulators and flow meter create a higher initial investment, but the reduced costs from gas charges will compensate for this expense over time.

Oxygen trap

Gas with an oxygen concentration of <5 ppm ("zero" grade or purer) is suitable for growth of even the most stringent anaerobes such as the methanogenic Archaea. In cases where oxygen levels are excessive, the gas purity is unreliable or just as an added precaution, a gas purifier should be included in the line. High-capacity oxygen traps that fit directly in-line with compression fittings (e.g. Swagelok^{TM}) are available from most gas chromatography companies. Alternative oxygen traps include the use of heated copper turning or catalytic pellets in glass columns, which are described in detail elsewhere (Sowers and Noll, 1995). Unlike the commercial gas purifiers, the latter systems can be regenerated indefinitely by purging with hydrogen, but the user must construct them and the initial cost is the same as the commercial systems.

Gassing cannula

The final components of the anaerobic gassing system are the gassing cannulas for distributing the gas. Cannulas consist of syringe barrels, either 1-ml glass tuberculin syringes with the flange cut off or 5-ml glass syringes with tubing inserted into the barrel, as shown in Figure 33.1. The barrels are filled with non-absorbent cotton and attached to the gas outlet with tubing that has low gas permeability (e.g. butyl, Viton^{TM}). Alternatively, a gassing manifold can be constructed from Teflon^{TM} tubing and compression fittings by inserting the tubing through a rubber stopper in the end of the syringe barrel. The cannula needles are prepared by heating a 19 G blunt-end needle approximately 1 cm from the hub to red hot with a Bunsen burner and bending the cannula needle to 45°. A quick-disconnect fitting attached to the other end of the tubing is paired with its mate, which is attached to the gas supply. The quick-disconnect enables the user to detach the cannula units from the gas lines for periodic autoclaving. Thereafter, the metal cannula is heated with a Bunsen burner before placing in a tube of sterile medium and the absorbent cotton prevents contaminants in the non-sterile gas line from entering the medium. A tubing clamp is placed in the gas line below the cannula so that each line can be closed individually when not in use. The cannula lines are all connected to a single gas feed line by glass T-shaped connectors. The feed line is in turn connected to a gas purifier, gas proportioner (optional) and gas source as shown in Figure 33.1. All tubing is secured in place with tubing clamps. A minimum of four cannulas should be prepared; two lines for gassing the receiving vessels, one line for gassing the prepared medium, and one line attached to the air inlet of a 3-valve pipet bulb for dispensing the medium (Sowers and Noll, 1995). An additional cannula fitted with a 13 G × 4" blunt end needle is convenient for degassing syringes prior to culture transfers.

All tubing connections and gas regulators should be periodically checked for leaks with a liquid leak detector or with an electronic gas leak detector. As a note of caution, hydrogen is an odorless, highly flammable gas that is lighter than air and forms explosive mixtures with air at concentrations above 4%. Installation of a flash arrestor and a normally-off electric solenoid valve in the hydrogen line immediately after the regulator is recommended in the event of accidental ignition or power failure, respectively.

Anaerobic Glove Box

The anaerobic glove box offers several advantages over gassing cannula for conducting research with anaerobes. Liquid medium can be dispensed without the need to manipulate gassing cannula and colony isolation and mutant screening can be conducted using standard Petri plates rather than more laborious roll tubes or bottle plates. In general, any conventional technique normally performed at the bench-top can be performed with the anaerobic extremophiles in the oxygen-free environment of the anaerobic glove box.

Glove box configurations

Both flexible and rigid anaerobic glove boxes have been described for experiments with even the most fastidious anaerobes. The Freter-type flexible anaerobic glove box first described by Aranki et al. (1969) is widely used but both types of glove box operate by similar principles and both are suitable for experiments with anaerobic extremophiles. However, gloveless systems that employ baffled sleeves or arm ports should be avoided because they tend to introduce levels of oxygen that are inhibitory to more fastidious anaerobes such as the methanogenic Archaea. The typical anaerobic glove box (Figure 33.2) consists of a sealed chamber with one or more sets of flexible gloves for access into the chamber. An airlock mounted on one end is used to transfer items into the glove box. The airlock is purged of air by three vacuum cycles

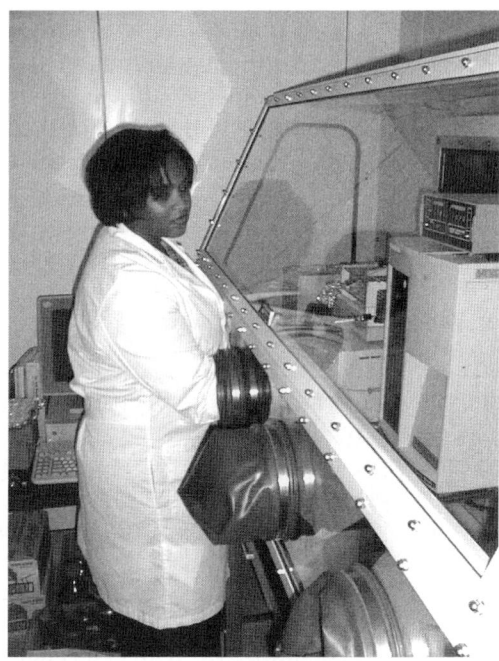

Figure 33.2. Anaerobic glove box capable of maintaining oxygen concentrations below 1 ppm. Materials are introduced into the glove box through the airlock located at the side.

and three backfilling cycles; the first two backfilling cycles contain only nitrogen while the third cycle contains the same gas mixture used in the glove box. Both manual and automated versions of the airlock are available. The internal glove box contains the desired anaerobic gases supplemented with 5% v/v hydrogen which functions in the oxygen removal system described below. The system requires a tank of nitrogen and a tank of premixed nitrogen and hydrogen in a 95/5% ratio.

Oxygen reduction

Since polymers used to construct flexible glove boxes (PVC) or the windows of rigid glove boxes (polycarbonate) and gloves (Neoprene) are slightly permeable to oxygen, there is a gradual buildup of oxygen that can exceed 1–2 ppm min^{-1} in the absence of an active oxygen removal system. In addition, residual oxygen is introduced into the glove box each time the airlock is opened after cycling. The oxygen removal system developed for the anaerobic glove box consists of reusable palladium-coated alumina pellets enclosed in a stainless steel mesh frame and mounted in front of a recirculating fan. The palladium catalyzes the reduction of any oxygen that enters the glove box in the presence of hydrogen contained in the glove box while the alumina support absorbs water formed by the reaction. The catalyst is regenerated by holding it at 115°C dry heat for 2–4 h, then cooling to room temperature before returning it to the glove box. Oxygen absorbed by the pellets during regeneration will cause the catalyst to heat up rapidly as soon as they are reintroduced to hydrogen in the glove box. Therefore, freshly regenerated catalyst packets should never be placed against the vinyl walls or base of the glove box since the heat that they generate will melt the PVC and create leaks. The catalyst packs should be immediately transferred to the catalyst holder as soon as they are removed from the airlock. Care must be taken when handling the catalysts to not vigorously shake the palladium-coated pellets as this reduces the life of the catalysts.

Controlling water vapor and hydrogen sulfide

Because the glove box is a closed, water-impermeable environment, water vapor generated by the catalytic reduction of oxygen and by evaporation of liquids will accumulate and eventually condense. This accumulation can be minimized by maintaining media in sealed vessels whenever possible, although introduction of some water vapor is inevitable whenever media containers are opened in the glove box. Excess water vapor can be removed by placing trays of silica gel, alumina, anhydrous calcium sulfate (Drierite^{TM}) or calcium chloride into the glove box. Alternatively the absorbents, with the exception of calcium chloride, which liquefies, can be enclosed in stainless steel mesh frames such as those used for the catalyst and placed in front of the circulating fan(s). Another contaminant of concern in anaerobic glove boxes is hydrogen sulfide, which is generated from sodium sulfide reductant contained in pre-reduced media and as an end product of sulfur-reducing microorganisms. The level of hydrogen sulfide, which irreversibly binds and inactivates the catalyst, can be minimized by placing activated carbon in the glove box as described above for water absorbents. Hydrogen sulfide levels can also be controlled with a solution of 0.25% (w/v) Ag_2SO_4 and 0.005 N H_2SO_4 in 50% (v/v) glycerol. Gas is sparged through the solution in a gas-washing bottle with a common aquarium pump. The water and hydrogen sulfide absorbents should be replaced whenever the catalyst packs are regenerated.

Maintaining gas ratios

The glove box, airlock and catalyst along with gases, vacuum pump and gas regulators are the basic components of a functional anaerobic glove box. The glove box atmosphere should be monitored for hydrogen in order to maintain an optimal level for oxygen reduction and oxygen should be monitored in order to determine the effectiveness of the catalyst. Electronic gas monitors offer convenient real-time assessment of gas concentrations. Gases can also be monitored using a gas chromatograph equipped with a TCD detector. Examples of suitable columns include Carboxen 1000 (60/80 mesh) packed in a 460 × 0.32 cm stainless steel column (Supelco, Inc) or a 30 m × 0.53 mm internal diameter capillary column containing GS-Molesieve (Agilent J&W Scientific). Carbon dioxide concentration can be tested with a Fyrite^{TM} combustion gas analyzer. Glove box accessories that may be useful include shelving to provide more usable space and an incandescent flaming device for sterilizing inoculation loops. It is advisable to have available spare sets of gloves and a mending kit to repair small punctures if a flexible glove box is used.

Specialized Culture Vessels for Growing Anaerobes

Growth in liquid medium

Hungate (1950) introduced the first effective technique for growing the most fastidious anaerobes which employed stoppered culture tubes as individual anaerobic culture chambers. Pre-reduced media prepared in culture tubes are sealed with anaerobic gas in the headspace providing a low oxidation–reduction potential in an oxygen-free atmosphere. Moore (1966) later introduced specialized anaerobe tubes with reinforced tapered necks that minimized the risk of breakage when sealing with a stopper (Figure 33.3). In the basic procedure, the tube is filled with oxygen-free gas with a gassing cannula. This cannula is then kept in the tube for continuous gassing as the medium is anaerobically transferred to the tube. The stopper is then placed in the tube opening with the cannula still in place; the tube is sealed by pulling the cannula up as pressure is applied to the top of the stopper. Once the cannula is removed, the stopper is completely seated with a slight twisting motion. The tubes containing medium are then autoclaved by clamping them in a tube press (Bellco Glass), which prevents extrusion of the stoppers during autoclaving. The tubes can be repeatedly opened and resealed for inoculation, assays, and transfers by the same procedure.

A more convenient alternative to the beveled top tubes and stoppers is the crimp-style anaerobe tube introduced by Macy *et al.* (1972) and serum vials (Figure 33.3). The tube consists of an 18×150 mm culture tube with serum vial-style neck. Both tubes and vials are sealed with thick butyl rubber septum stoppers designed to minimize diffusion of

Figure 33.3. Culture vessels used for anaerobic growth medium. The anaerobe tube (A) with the reinforced tapered neck is sealed with a conical butyl rubber stopper and the crimp-style anaerobe tube (B) and serum vial (C) are sealed with a butyl rubber septum secured in place with an aluminum crimp seal.

oxygen into the medium that is secured in place with an aluminum crimp seal. The latter feature eliminates the need for a tube press during autoclaving and minimizes the risk of accidental stopper extrusion as a result of an increase in gas pressure. After preparation, medium is inoculated with a syringe eliminating the need to open the vessel.

Colony growth on solidified medium

Most anaerobic extremophiles can be grown on solid media in Petri dishes. However, since even brief exposure to oxygen is lethal for many of these microorganisms, an anaerobic glove box is required for medium preparation and inoculation. Anaerobic medium containing agar is melted in sealed serum vials outside the glove box, then transferred into the glove box and poured into plates. Conventional polystyrene disposable Petri plates are suitable for mesophiles, but glass- or high-temperature-resistant plastic plates should be used for thermophiles. Although anaerobic extremophiles will tolerate the low partial pressure of oxygen associated with anaerobic glove boxes, they will only grow in an atmosphere with a low redox potential. Therefore, in order to grow most of these microorganisms after inoculation, an anaerobe jar must be employed that is capable of maintaining a reduced atmosphere and in the case of thermophiles, it must tolerate incubation at high temperatures. The medium is inoculated by spreading the culture on the surface in molten agar or directly with a sterile rod, then incubated in an anaerobic jar (Figure 33.4) under an atmosphere of nitrogen, carbon dioxide and hydrogen sulfide. The long incubation

Figure 33.4. Stainless steel anaerobe jar for incubating cultures grown in Petri plates on solidified medium.

periods of 2–14 days requires that the jars selected are not wholly composed of plastics or other polymers, which will permit the gradual permeation of oxygen. Metal culture jars are available from commercial sources and they can be constructed from modified pressure cookers, paint pressure canisters, glove box airlocks (Balch and Wolfe, 1976; Metcalf et al., 1998; Tumbula et al., 1995a), or custom-made by a machine shop (Balch et al., 1979). Inexpensive glass containers can be constructed from a modified canning jar (Apolinario and Sowers, 1996). Several commercially manufactured glass and/or metal anaerobe jars suitable for growing anaerobes from TORBAL, Oxoid and BBL are no longer available, but often can be found in storage or thorough used-equipment distributors. Since the plates are incubated in a sealed environment, the plates must be pre-dried to a critical water content; too much water causes confluence of colonies and excessive drying inhibits growth.

If an anaerobic glove box is not available, an alternative vessel for colony isolation is the agar roll tube used in conjunction with a roll tube spinner (Hungate, 1969). Molten agar is poured into an anaerobe tube, sealed under anaerobic gas, and placed on the spinner to form a monolayer of solidified medium on the inner surface of the tube. The large diameter 25×150 mm stoppered anaerobe tube (Figure 33.3) is most convenient for isolating colonies and provides more surface area than the standard 18×150 mm anaerobe tubes and crimp-style tubes. The inoculated roll tubes are then incubated upright in a conventional incubator. Colonies are transferred by opening the stopper, placing a sterile gassing cannula inside to maintain an anaerobic atmosphere and transferring the colony with an inoculating needle.

◆◆◆◆◆ MEDIA PREPARATION

Anaerobic pre-reduced medium introduced by Hungate (1950) consists of standard media components with four basic modifications: 1) oxygen-free water is prepared by boiling and or by sparging with anaerobic gas to displace dissolved oxygen; 2) medium is prepared under an anaerobic gas phase to exclude further introduction of oxygen; and 3) a reducing agent is added to lower the redox potential to below -300 mV; 4) media are dispensed into oxygen impermeable vessels and sealed under anaerobic gas to maintain these conditions during growth and storage.

Preparing Liquid Medium

Media recipes for extreme anaerobes are described in original publications, culture collection catalogs and in the Archaea: A Laboratory Manual, Methanogens (Sowers and Schreier, 1995) and Thermophiles (Robb and Place, 1995). Generally media for extremophiles consists of different concentrations of Na^+, Mg^{2+}, K^+ and Ca^{2+} salts depending on the optimal osmolarity of the strain, NH_4^+ salt as a nitrogen source, PO_4^{3-} salt as a phosphorus source, cysteine and S^{2-} as sulfur sources and reducing agents, $NaHCO_3$ or Na_2CO_3 as a carbon source and pH buffer, resazurin as a redox indicator, trace minerals and additional growth factor requirements such as vitamins, yeast extract, etc. Some of the most common energy sources include sugars or cell hydrolysates for heterotrophs, H_2 and CO_2 or S^0 for autotrophs, methylated amines and methylated sulfides for methylotrophs, acetic acid for aceticlastic species.

Medium is prepared in the following sequence to minimize precipitation:

1. Add approximately 90% of the total volume of deoxygenated H_2O to a flask containing a stirring bar. Water can be boiled and then cooled under a stream of anaerobic gas or sparged with a fritted glass sparging stone for 30–60 min while stirring to remove dissolved oxygen.
2. While sparging the vessel with a gassing cannula, add and completely dissolve all ingredients except phosphate, carbonate, cysteine and sulfide.
3. Add phosphate and allow to dissolve completely.
4. Add carbonate and allow to dissolve completely.
5. Add cysteine and allow to dissolve completely.
6. Adjust pH as required.
7. Use the remaining H_2O to wash salts off the walls of the flask.
8. Continue to sparge with anaerobic gas for 15 min while stirring to displace remaining dissolved oxygen.
9. Dispense medium with a pipette or syringe into tubes or bottles under N_2–CO_2.

After dispensing medium into tubes or bottles, secure septa with aluminum crimp seals. Autoclave media for 20 min under fast exhaust (gravity). Bottles should be autoclaved in wire baskets to protect from risk of explosion upon removal from the autoclave and safety glasses should be worn.

Note: Precipitates will form during autoclaving as dissolution of CO_2 increases the pH but may be redissolved by briefly shaking the medium after it has cooled and allowing it to stand for a period of time. Media can be stored without the addition of Na_2S for up to 6 months. For hydrogen-utilizing autotrophs, purge and pressurize the medium gas phase to 125 kPa with H_2–CO_2 (4:1 v/v) prior to inoculation. The medium will commonly be pink before autoclaving but will turn colorless after autoclaving. If the medium remains pink after autoclaving or becomes pink with shaking after it is cooled, this indicates that the medium contains traces of oxygen and should be discarded.

Preparing solidified medium

For growth of mesophilic species on solidified medium, prepare media as described above and dispense up to 100 ml into 160 ml serum vials containing premeasured quantities of agar (1–1.5% w/v) (Apolinario and

Sowers, 1996; Jones *et al.*, 1983; Tumbula *et al.*, 1995a). After sterilization by autoclaving, transfer media pre-cooled to 60°C into an anaerobic glove box and pour into Petri plates that have been equilibrated in the glove box for 16 h to remove adsorbed oxygen. For preparing large quantities of media, solidification can be prevented by holding medium in the glove box at 55°C in an incubator or water bath filled with glass beads until ready for use. For growth of thermophiles and hypertherophiles, thermostable gellum gum (Gelrite^{TM}) is substituted for agar as the solidifying agent (Fiala and Stetter, 1986; Erauso *et al.*, 1993, 1995). Since Gelrite^{TM} polymerizes with the divalent cations Mg^{2+} and Ca^{2+}, the polymer is autoclaved separately from the medium and combined immediately before dispensing. After solidification of the medium, dry plates by storing in an anaerobic glove box at ca. 40% relative humidity at room temperature for 48 h (Apolinario and Sowers, 1996). Thereafter, plates can be stored inverted in a sealed container such as an anaerobe jar or canning jar until ready for use. Inoculate media by spreading inocula onto the surface with a sterile inoculation loop or spreading rod. The rod should be stored in the anaerobic glove box overnight before use, to ensure that all O_2 is desorbed from it. The inoculation loop can be repeatedly sterilized with an electric heating coil in the glove box, but spreading rods must be sterilized prior to transfer into the glove box. Transfer inoculated inverted plates into an anaerobe jar. H_2S is introduced into the gas phase by including a small vial or empty Petri plate containing ca. 0.5 ml Na_2S solution. As the Na_2S solution is acidified by CO_2 in the gas phase, H_2S is released into the vessel. If CO_2 is not included in the gas phase, then acid must be added to the Na_2S solution vial immediately before sealing the anaerobe jar. Remove sealed anaerobe jars from the glove box and incubate at the recommended temperature. For hydrogen utilizing autotrophs, purge and pressurize the gas phase of the jar to 200 kPa with H_2–CO_2 (4:1) prior to incubation. Alternatively, H_2S can be added to the jars directly as a gas phase by purging and pressurizing the vessels with pre-mixed 79.9% N_2, 20% CO_2, 0.1% H_2S or 79.9% H_2, 20% CO_2, 0.1% H_2S. Pressurization should not exceed value recommended for canister.

Note: Take precautions necessary for protection from accidental explosion when pressurizing vessels.

◆◆◆◆◆ ENRICHMENT, ISOLATION AND MAINTENANCE OF ANAEROBES

Sample Collection

Although anaerobic extremophiles require highly reduced, anaerobic conditions for growth, they are ubiquitous in the environment and occur in most habitats where oxygen has been depleted. Examples of these habitats include sewage digestors, waste landfills, marine and freshwater sediments, Antarctic lakes, tundra, algal mats, hot springs, calderas and deep-sea geothermal vents. Most anaerobes reside in lower sediment layer samples and can be collected with coring devices or grab samplers. Since anaerobes are usually embedded in highly reduced sediments, they can be transferred from sampler to container without additional precautions to keep them anaerobic. The sediment is transferred to a glass container such as a canning jar, filled to top to exclude air and sealed with a metal screw top. The rim should be free of sediment to ensure an airtight seal around the rubber gasket. If samples are too small to fill the container, a portable lecture bottle equipped with a regulator, short gas line and cannula can be used to displace air in the sample container prior to sealing. The sample can then be transported to the lab or stored.

Enrichment and Isolation

Open an anaerobe tube or serum vial of sterile anaerobic medium inside an anaerobic glove box. If working on a bench, immediately insert a gassing cannula after opening the container. Inoculate medium with approximately 10–20% v/v of sediment. A convenient means of inoculating isolation or enrichment media is to use a syringe small enough in width to fit into the top of the medium container. The needle end is cut off and the syringe barrel is used to core the sample, then it is extruded into the medium with the plunger. Remove the cannula and seal the culture vessel. Incubate the culture at the ambient temperature of the sample source. If it is a hydrogen utilizing autotroph displace the gas phase with H_2 (or H_2–CO_2 mixture if required) prior to sealing the culture vessel. Enrichments containing hydrogen should be shaken during incubation to promote dissolution of the hydrogen. Hyperthermophiles are incubated in shaking water baths that contain polyethylene glycol or mineral oil to minimize evaporation since most dry shaking incubators do not heat above 60°C. Growth can be monitored by standard techniques such as direct microscopic counting of DAPI stained cells. However, the physiological activities of some anaerobic extremophiles can also be used for monitoring. For example, growth of methanogens can be monitored by gas chromatographic analysis of methane production and S^0 reducing species can be colorimetrically monitored by sulfide analysis. Once there is evidence of growth the 10–20% v/v of the culture is transferred to a fresh medium two or three times to ensure enrichment of the target microorganisms. Anaerobes are isolated by serially diluting the enrichment culture in liquid medium and inoculating onto medium solidified with agar for mesophiles or Gelrite for thermophiles requiring growth temperatures greater than 50°C. Since some anaerobes do not form colonies on solidified media an alternative approach is to incubate the serial dilutions directly and continue to sequentially transfer the highest growing dilution until the culture is axenic. The disadvantage of the latter approach is that it will only isolate the most abundant strain in the enrichment culture.

Large-Scale Growth of Anaerobes

Volumes up to 250 ml

Anaerobic cultures can be scaled up to virtually any volume for the production of cell material. Generally cultures grown to late exponential growth are sequentially transferred (5–20% v/v) to increasing larger volumes until the maximum culture volume is obtained. For example, a 10 ml culture started in an anaerobe tube is transferred to 100 ml of medium in a serum vial, 1000 ml in a septum-capped bottle, then 10 l in a fermentor or glass carboy. For cultures up to 250 ml, the medium can be prepared directly in media bottles (Figure 33.5A) and sealed under anaerobic gas with septum screw caps (Wheaton). Media should not exceed 50% of the total volume of the container. These bottles are composed of thick borosilicate glass, which is resistant to shattering caused by the pressure created by autoclaving, and coated with autoclavable plastic to contain the glass and the medium if the integrity of the vessel should fail. Bottles should always be autoclaved in an enclosed mesh basket and allowed to cool to near room temperature in the autoclave before handling to minimize the risk of injury from explosion of superheated liquids. The bottles can be inoculated with a syringe and placed in an incubator at the desired temperature. When growing hydrogen-utilizing microorganisms, the bottle is placed on a shaker as described above and the hydrogen periodically replaced in the gas phase of the culture.

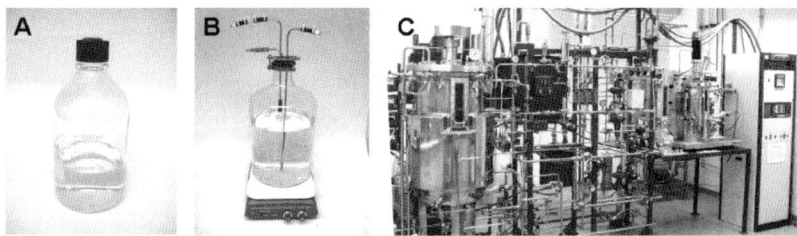

Figure 33.5. Scale up of anaerobic cultures. The 500 ml plastic-coated borosilicate medium bottle (A), which is inoculated with a syringe through a septum in the cap, can be used for culture volumes up to 250 ml. A 20 liter carboy for culture volumes up to 10 liters (B) is configured with a gas inlet with fritted glass sparger, a gas outlet with stainless steel check valve to prevent backflow of air into the medium and a liquid inlet line for inoculation and sampling of the culture during growth. The culture is agitated with a stir bar. A modified pilot scale fermentation unit (C) with 20 liter and 250 liter vessels is equipped with gas proportioner for controlling inlet anaerobic gas ratios and modified temperature controller and software for maintaining growth temperatures of up to 105°C.

Volumes greater than 250 ml

To prepare larger volumes, media can be prepared in glass carboys or fermentors (Figure 33.5B–C) (Adams, 1995; Boone et al., 1987; Daniels, 1995; Daniels and Belay, 1984; Mukhopadhyay et al., 1999; Sowers et al., 1984; Worthington et al., 2003). Media in carboys should not exceed 50% of the total volume and the carboy should always be vented when autoclaved to prevent potential explosion from pressure created by superheated liquid during cooling. As an additional precaution, the carboy should be autoclaved inside of a polypropylene tank (Nalgene^{TM}) without a top to allow penetration of the steam around the container. The carboy is cooled to room temperature in the autoclave, the air is then displaced with anaerobic gas by inserting a sterile cannula or connecting a gas line with sterile in-line filter to a sparging stone inserted through the stopper. After displacing the air the stopper is secured in place with a stopper clamp. Because of the volume the carboy is inoculated by inserting a sterile length of tubing equipped with Luer Lok fittings and 18 G hypodermic needles on each end. The needles are inserted into the septum of the inoculum bottle and the carboy stopper. The inoculum is transferred by activating the peristaltic pump. The carboy is placed in an incubator or climate-controlled room at the desired temperature. When growing hydrogen-utilizing microorganisms, a large stirring bar should be placed in the carboy and the culture agitated on a stir plate to promote the dissolution of the gas. The hydrogen is either periodically replaced in the gas phase of the culture or a gas mixture containing 5% hydrogen (v/v) can be continually passed through the medium with a sparging tube. When sparging the carboy should be vented to a fume hood to prevent a build up of potentially explosive hydrogen and the vent should be unobstructed at all time to prevent a build up pressure. The stopper should include a stainless steel check valve rated at 2–5 kPa to prevent accidental buildup of pressure.

Most standard fermentors of any volume can be used for growth of anaerobes with only slight modifications (Figure 33.5C). Firstly, all silicone o-rings and tubing should be replaced with less oxygen permeable polymers such as EPDM, Neoprene^{TM}, Viton^{TM} or butyl rubber (Sowers and Noll, 1995). Secondly, the air line connected to the instrument inlet is replaced with a gas line to tank(s) of anaerobic gas. If the fermentor is sterilized in an autoclave it should be vented, then sparged with an anaerobic gas once it is cooled and connected to the fermentation unit. If the fermentor is sterilized in-place, it is sparged with anaerobic gas prior to sterilization and set to continue sparging after the sterilization process. Fermentors are inoculated as described for carboys. When inoculating between fermentors the vessel containing the inoculum source is pressurized below the pressure rating of the vessel and inoculum transferred via a standard transfer line. Because of the large volumes of pre-reduced medium it is generally not necessary to purge the transfer line with anaerobic gas prior to transfer. Stainless steel fermentor vessels cannot be used to grow extreme acidophiles, which require custom glass-lined or titanium vessels to prevent metal corrosion. A cost-effective alternative for the scale-up of acidophiles is the use of a modified glass-lined water heater described by Worthington et al. (2003).

Anaerobic harvesting

Small volume cultures in anaerobe tubes can be directly harvested in many clinical centrifuges at low speeds of up to $4000 \times g$. The supernatant is then decanted with a syringe or unstoppered and decanted in an anaerobic glove box. Alternatively, cultures can be transferred to sterile deoxygenated Oak Ridge centrifuge tubes with o-ring seals inside an anaerobic glove box. The tubes are sealed, centrifuged in a standard high-speed centrifuge outside of the glove box, then returned to the glove box and decanted. For larger volumes, Oak Ridge bottles with o-ring seals and centrifuge rotors are available in volumes of up to 500 ml (3 liter total rotor volume). Continuous flow centrifugation is necessary for harvesting large volumes from carboys and fermentors. Either a standard high-speed centrifuge equipped with a continuous flow rotor (e.g. Beckman, Sorvall) or a dedicated vertical bowl continuous flow centrifuge (e.g. CEPA, Sharples) is suitable. Harvesting lines are connected between the carboy or fermentor vessels and the culture is transferred with a pump or by pressurizing the vessel. The flow of anaerobic medium through the rotor will keep the cell pellet anaerobic during the harvesting process. Once harvested, the cell pellet is rapidly transferred to a serum vial and sealed under anaerobic gas or transferred directly to liquid nitrogen for long-term storage. Generally the reducing potential of the microorganisms and the reduced medium in the pellet will protect the cells from oxygen for a brief period during the transfer.

Storage

Most anaerobic mesophiles and thermophiles can be stored for 2 to 3 months in anaerobic medium at room temperature, except for psychrophiles, which must be stored below their T_{max}. For long-term storage as frozen stocks prepare medium with the addition of 50% glycerol and dispense 2.5 ml into small (approx. 5–10 ml) serum vials. Seal under anaerobic gas and sterilize by autoclaving. To prepare cell material for storage, grow approximately to mid or late exponential growth approximately 0.5–1.0 g wet weight of centrifuged cell pellet. Cells can be harvest anaerobically by two means. If cultures are grown in 20 ml septum top anaerobe tube, these tubes can be centrifuged in a clinical centrifuge at $5000 \times g$, anaerobically decanted in an anaerobic glove box or with a syringe, and the pellets combined in 2.5 ml of supernatant. Alternatively cultures are opened in a glove box, decanted into Oak Ridge centrifuge tubes or bottles and removed from the glove box for centrifugation. The cultures are then decanted in the glove box and the pellets combined in 2.5 ml of supernatant with a pipette.

The cell suspension is then added to the glycerol medium and the vial is stored at –80°C. To revive a culture the vial is removed from the freezer and the cell suspension is then slowly warmed by exposure to room temperature. The septum is heated with a flame to both sterilize and soften the rubber. The contents of the vial are then transferred to fresh medium with a degassed syringe. Since glycerol can inhibit some microorganisms, the transfer should not exceed 10% v/v. Additional information on the long-term storage is discussed by Hippe (1984), Boone (1995) and Tumbula *et al*. (1995b).

♦♦♦♦♦ ANAEROBIC BIOCHEMISTRY

Many enzymes and cofactors in anaerobic microorganisms are poised at negative redox potentials. As a result, the exposure of many enzymes and cofactors to air will result in inactivation either by denaturation of the protein, or oxidation of an active site or requisite cofactor. Not all proteins in anaerobic microorganisms are oxygen labile, but those that are must be maintained under highly reduced conditions throughout the purification process in order to maintain activity. General guidelines for performing anaerobic biochemistry described below are based on principles described by Burgess *et al*. (1980), Cooper *et al*. (1979), Shriver (1969), and Sowers (1995).

Preparation of Anaerobic Buffers

The most convenient means of deoxygenating buffers is to prepare the buffer and store it to an anaerobic glove box long enough for the oxygen to diffuse from the liquid. Small volumes of up to 100 ml will become deoxygenated overnight with static incubation; larger volumes will require agitation overnight on a stir plate. If an anaerobic glove box is unavailable buffers and solutions can be deoxygenated with a gassing manifold used for preparing medium. The prepared solution is agitated overnight on a stir plate with under a flow of nitrogen or argon provided by a gassing cannula. A more rapid means of deoxygenating solutions is to use a pressure/vacuum-gassing manifold (Figure 33.6). The apparatus can consist of either a glass or metal manifold connected to several high vacuum valves. The manifold inlet is connected to a vacuum pump and an anaerobic gas tank *via* a three-way valve. The manifold valves are equipped with a length of tubing and 1 ml tuberculin syringe barrel with the top flange removed. Hypodermic needles (22–23 G) attached to the syringe barrels are used to perforate small serum vials or cuvettes through septum tops. For volumes greater than 50 ml, a length of tubing is used to connect a side arm flask containing the buffer solution directly to the manifold valve. The stopper on the flask should be secured with a stopper clamp to prevent it from blowing off under pressure. The flask is stirred on a stir plate during the deoxygenating process. The three-way valve is positioned in

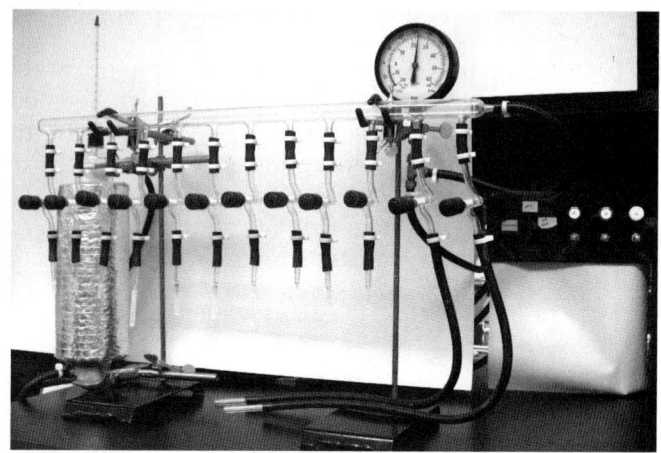

Figure 33.6. Manifold for the preparation of degassed solutions and containers.

the vacuum position until no more bubbles are seen rising from the solution. The three-way valve is then positioned in the pressure position and the vessel or cuvette is pressurized to no more than 5 kPa for 5 min. This procedure is repeated 5 more times. Once in the final pressure cycle, the manifold valve is closed and the septated vials and cuvettes are removed from the manifold. Flasks are removed after sealing the tubing with a tubing clamp or hemostat. Reducing agents such as dithiothreitol and β-mercaptoethanol are added after degassing to prevent their oxidation or volatilization during the vacuum cycles. A detailed description on the construction of a gassing manifold can be found earlier in this chapter.

Preparation of Anaerobic Cell-free Extracts

Cell-free extracts can be prepared with either a sonicator or French pressure cell. Cells are suspended in a recommended buffer (1:2 v/v) and cell lysates can be prepared by standard procedures with the entire sonication unit inside an anaerobic glove box. Alternatively, if the controller is too large to transfer through the airlock, the unit can be operated with only the probe within the glove box and the leads connected to the controller through an accessory port. In the absence of an anaerobic glove box, the probe can be inserted into a tube while purging with a gassing cannula as described above in *Anaerobic Medium Preparation*. The tube and cannula should be secured with clamps to prevent them from contacting the probe. The extract is then transferred to a deoxygenated Oak Ridge centrifuge tube, sealed under an N_2 atmosphere and centrifuged at $45\,000 \times g$ for 30 min.

For the preparation of cell lysates with a French pressure cell, the cell is modified by the addition of a length of 0.8 mm internal diameter flexible tubing to the exit port (Figure 33.7). Insert the un-beveled end of an 18 G bleeding needle into the other end of the tubing.

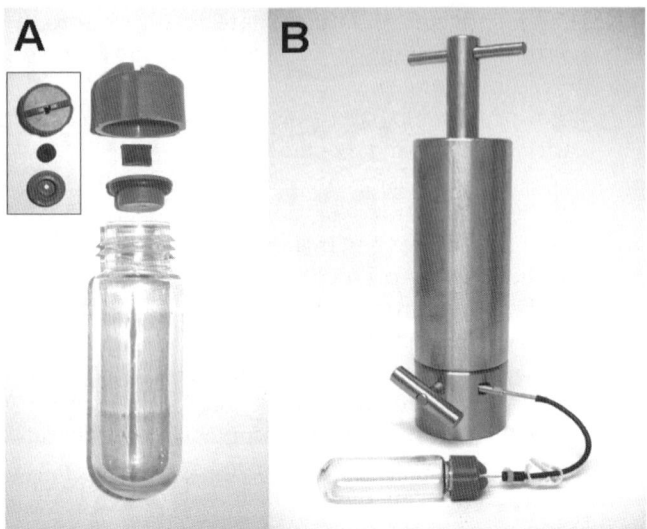

Figure 33.7. French pressure cell configured for preparing anaerobic cell-free extract. An Oak Ridge centrifuge tube (50 ml) with modified top assembly (A) is used for anaerobic centrifugation outside of an anaerobic glove box. Top assembly (Inset) is modified by drilling a 2.5 mm hole though the outer and inner tops and inserting a cored rubber septum between the tops to seal the hole. A needle attached to Viton^{TM} tubing is pushed through the septum and attached to the outlet of the French pressure cell. The cell is loaded in an anaerobic glove box and the lysate is collected into the modified Oak Ridge centrifuge tube (B).

Pre-equilibrate the disassembled cell in an anaerobic glove box for at least 1 hour to allow oxygen to desorb from surfaces. The cell is inverted on a stand and loaded by conventional techniques. The cell is then sealed with the closure plug assembly. While inverted, open the valve and slowly push the piston assembly until a drop of extract appears at the outlet needle. Close the valve and insert the outlet needle into an Oak Ridge centrifuge tube modified with a septum insert (Figure 33.7). Remove the pressure cell from the glove box and insert into a press. Lyse the cells at the recommended pressure and collect the lysate into the centrifuge tube. Remove the exit port needle from the centrifuge tube and centrifuge the lysate as described above. Return the centrifuge tube to the glove box and decant the supernatant into the desired container. Cell-free lysates can be used immediately or frozen. A convenient mean of storing cell-free extract anaerobically is to pellet the extract in liquid nitrogen in the anaerobic glove box, then transfer the pellets to a Dewar flask.

Anaerobic Precipitation of Proteins

Proteins may be precipitated by conventional means within an anaerobic glove box. Since volatile precipitating agents such as acetone will damage materials in the anaerobic glove box, the use of a non-volatile precipitating agent such as ammonium sulfate is recommended.

Pre-weigh ammonium sulfate in a covered, vented container and equilibrate overnight in an anaerobic glove box. The moisture content of the glove box should be minimized with desiccant to prevent absorption of water by the precipitating agent. To precipitate proteins, slowly add ammonium sulfate to the extract with constant stirring in an Erlenmeyer flask. Once the desired concentration of precipitating agent is dissolved, transfer the extract to an Oak Ridge centrifuge tube. Seal the tube, remove it from the glove box and centrifuge as recommended in the specific protocol.

If an anaerobic glove box is unavailable, anaerobic precipitations can be conducted with a gassing manifold and side-arm flask, although these conditions may not be as stringently anaerobic. A side-arm flask with a volume four times that of the protein solution is fitted with a stopper that has a butyl rubber septum inserted into a 12 mm bore hole in the center. A pre-measured quantity of ammonium sulfate is added to the side-arm of the flask, a stir bar is placed inside and a stopper is secured with steel wire or strapping tape. The flask is deoxygenated with a gassing manifold for six cycles. The flow of gas into the vessel should be adjusted to a rate that does not disturb the ammonium sulfate in the side arm of the flask. Cell-free extract is added to the flask with a degassed syringe and a 23 G needle inserted through the septum in the stopper. Aliquots of ammonium sulfate are added to the extract by tipping the flask at an angle to allow a portion of the reagent to fall from the side arm to the liquid. The reagent is dissolved completely on a stir plate before the next aliquot is added. The process is repeated until the entire reagent is dissolved in the extract. The precipitated extract is transferred by syringe to a degassed centrifuge tube modified with a septum insert (Figure 33.7). Centrifuge the extract as recommended in the specific protocol. After centrifugation, remove the supernatant fraction with a syringe and transfer to another degassed container. The precipitate can be redissolved in buffer and also transferred to another degassed container.

Protein Separations by Column Chromatography

Anaerobic liquid column chromatography, FPLC or HPLC can be performed with or without an anaerobic glove box. In all cases, the system is pre-reduced with anaerobic buffer prior to loading the sample. The advantage of the anaerobic glove box is that there is virtually no chance of oxygen diffusion through the system components and so fractions can be collected using a conventional fraction collector. The procedures for chromatography with and without an anaerobic glove box are also described.

Chromatography in an anaerobic glove box

Buffer is filtered through a 0.2 μm membrane filter into a vacuum flask and deoxygenated with a pressure-vacuum manifold or by equilibrating an open flask with stirring overnight in the glove box. The buffer volume should be no more than 1/4 that of the flask. If the buffer is deoxygenated with a gassing manifold, secure the stopper with a stopper clamp or steel wire and transfer the sealed flask into the glove box. After deoxygenating, add the appropriate reducing agent solution with a syringe pre-rinsed with deoxygenated water. Commonly used reducing agents include dithiothreitol, β-mercaptoethanol, and thioglycolate, usually in concentrations of 1 to 2 mM. Dithionite and sodium sulfide are sometimes used as reducing agents but can often inhibit enzyme activity. Refer to the individual protocol for recommended reducing agents. The buffer is transferred to a buffer bottle or a chromatography line is inserted directly into the flask. Accumulation of dissolved gases can be minimized by sealing the buffer container when it is not in use or by substituting helium for nitrogen in the glove box atmosphere. Transfer the cell-free extract into the glove box. If the sample is stored in liquid nitrogen, it can be transferred into the glove box in a Dewar flask. Although some liquid nitrogen will boil off during evacuation of the airlock, most of the nitrogen will remain. Equilibrate the column with at least 6 void volumes of anaerobic buffer before loading the sample. Elute the sample from the column using a recommended protocol and collect the fractions with a fraction collector. Fractions are sealed in glass serum vials and removed from the glove box for assaying or freezing. Alternatively, fractions can be transferred into vented plastic cryovials and stored under liquid nitrogen.

Chromatography without an anaerobic glove box

Buffer bottle tops are prepared with gas inlet and liquid outlet tubes submerged in the buffer and a gas outlet above the liquid level. Prepare buffers with a pressure-vacuum manifold as described above or by purging with helium through the gas inlet with stirring. After deoxygenating, add the appropriate reducing agent solution as described above. A slow flow of helium should be maintained when operating the system to prevent a vacuum from developing in the buffer bottle and the bottle should be sealed when not in use. Equilibrate the system and column with at least 6 void volumes of anaerobic buffer. If the sample is frozen, transfer it to a serum vial and degas using a gassing manifold. The vial containing the sample may be kept cold with ice water slurry. Degas the sample for 6 cycles. If the sample begins to foam, immediately release the vacuum to prevent denaturing of the protein. The samples may be diluted by adding anaerobic buffer with a syringe. Load the sample onto the column using a gas-tight syringe that has been purged 6 times with degassed water or buffer. Elute the sample from the column by the recommended protocol. Eluted fractions may be collected into degassed serum vials by attaching an 18 G needle to the column or UV monitor outlets and manually inserting the needle into the vials. Alternatively, if the eluate fractions must be frozen immediately, they may be transferred directly into cryovials submerged in a Dewar flask containing liquid nitrogen.

Activity Assays

Spectrophotometric assays can be performed inside or outside an anaerobic glove box. If the spectrophotometer is located within the glove box, the assays are conducted according to standard procedures. The following protocol describes techniques for anaerobic activity assays with a spectrophotometer located outside of an anaerobic glove box. The procedure employs "semimicro" (1 ml) quartz or glass cuvettes that accept round-stoppered tops (Figure 33.8). The PTFE polymer tops are replaced with a rubber septum. Transfer components to the cuvette with a gas-tight syringe. Before each use, rinse the syringe 6 times with degassed buffer or water-containing reducing agent. Equilibrate cuvettes and stoppers overnight in an anaerobic glove box. Add all ingredients to the cuvette except for the protein and seal with the stopper. Remove the cuvette containing the reaction mixture from the glove box and insert it into the spectrophotometer. Begin the reaction by adding the protein with a degassed syringe. If an anaerobic glove box is not available, the cuvettes are sealed with stoppers and deoxygenated with a pressure-vacuum manifold for 6 cycles as described above. Inject the deoxygenated reaction components into the cuvettes from serum vials with a gas-tight syringe. One serum vial should contain anaerobic buffer or water for purging the syringe.

Figure 33.8. Cuvette with round top fitted sealed with a rubber septum for degassing and additions by syringe.

Acknowledgments

This work was supported in part by grants to KS from the Department of Energy, Energy Biosciences Program (DE-FG02-93ER20106) and the National Science Foundation, Division of Molecular & Cellular Biosciences (MCB0110762).

List of suppliers

Agilent J&W Scientific
395 Page Mill Rd.
Palo Alto, CA 94306, USA
Tel: +1-800-227-9770
Fax: +1-302-633-8901
http://www.chem.agilent.com

Gas chromatography columns and packing

Alltech Associates, Inc.
2051 Waukegan Road
Deerfield, IL 60015-1899, USA
Tel: +1-708-948-8600
Fax: +1-708-948-1078
http://www.alltechweb.com

Oxygen scrubbers

Bel-Art Products, Inc.
6 Industrial Road
Pequannock, NJ 07440, USA
Tel: +1-973-694-0500
Fax: +1-973-694-7199
http://service.belart.com

Tubing clamps

Bellco Glass, Inc.
340 Edrudo Road
P.O. Box B
Vineland, NJ 08360-0017, USA
Tel: +1-609-691-1075
Fax: +1-609-691-3247
http://www.bellcoglass.com

Anaerobic culture tubes, stoppers, crimp seals, crimper, roll tube and roll tube spinner

Cole-Parmer Instrument Company
625 East Bunker Court

Vernon Hills, IL 60061, USA
Tel: +1-800-323-4340
Fax: +1-847-247-2929
http://www.coleparmer.com/

Gas regulators, electronic gas monitors, solenoid valves, flash arresters, metal and polymer tubing, quick disconnect tubing fittings

Coy Laboratory Products, Inc.
14500 Coy Drive
Grass Lake, MI 49240, USA
Tel: 313-475-2200
Fax: 313-475-1846
http://www.coylab.com

Anaerobic glove boxes, electronic gas monitor, Fyrite CO_2 detector, palladium catalyst, incandescent loop sterilizer

Don Whitley Scientific Limited
14 Otley Road
Shipley, West Yorkshire, BD17 7SE, UK
Tel: + 44-1274 595728
email: info@dwscientific.co.uk

Anaerobic glove boxes, stainless steel anaerobe jars

Kimble/Kontes Glass
Spruce Street
P.O. Box 729
Vineland, NJ 08360, USA
Tel: +1-856-692-3600
Fax: +1-856-794-9762
http://www.kimble-kontes.com

Glass carboys, vacuum valves, manifolds, oxygen scrubbing towers

Nalge Nunc International
75 Panorama Creek Drive
P.O. Box 20365
Rochester, NY 14602-0365, USA
Tel: +1-716-264-3898
Fax: +1-716-586-3294
http://www.nalgenelabware.com/

Tanks for autoclaving carboys

N. J. International Corporation
29/30, Ashok Chambers, 2nd Floor,
Devji Ratanshi Marg,
Mumbai 400 009, India

Tel: +91 22 5631 4651/52/53
Fax: +91 22 2370 4892
http://www.njintlindia.com

Stainless steel anaerobe jars

Popper & Sons, Inc.
300 Denton Avenue
New Hyde Park, NY 11040, USA
Tel: +1-516-248-0300
Fax: +1-516-747-1188
http://www.popperandsons.com

Syringe barrels and blunt end needles for cannula

Supelco, Inc.
Supelco Park
Bellefonte, PA 16823, USA
Tel: +1-814-359-3441
Fax: +1-814-359-5459
http://www.signa-aldrich.com/supelco

Gas chromatography columns and packing

Swagelok Co.
31400 Aurora Road
Solon, OH 44139, USA
Tel: +1-216-349-5934
Fax: +1-216-349-5843
http://www.swagelok.com

Compression fittings, metal tubing, flexible gas tubing, flow valves

Wheaton Science Products
1501 N. 10th Street
Millville, NJ 08332-2093, USA
Tel: +1-856-825-1100
Fax: +1-856-825-1368
http://www.wheatonsci.com

Glass serum vials, crimp seals, crimper, septum media bottles

References

Adams, M. W. W. (1995). Large-scale growth of hyperthermophiles. In *Archaea: A Laboratory Manual – Thermophiles* (F. T. Robb, K. R. Sowers, S. DasSarma, A. R. Place, H. J. Schreier and E. M. Fleischmann, eds), pp. 47–49. Cold Spring Harbor Laboratory Press, Plainview.

Apolinario, E. A. and Sowers, K. R. (1996). Plate colonization of *Methanococcus maripaludis* and *Methanosarcina thermophila* in a modified canning jar. *FEMS Microbiol. Lett.* **145**, 131–137.

Aranki, A. S., Salam, S. A., Kenny, E. B. and Freter, R. (1969). Isolation of anaerobic bacteria from human gingiva and mouse cecum by means of a simplified glove box procedure. *Appl. Microbiol.* **17**, 568–576.

Balch, W. E. and Wolfe, R. S. (1976). New approach to the cultivation of methanogenic bacteria: 2-mercaptoethanesulfonic acid (HS-CoM)-dependent growth of *Methanobacterium ruminantium* in a pressurized atmosphere. *Appl. Environ. Microbiol.* **32**, 781–791.

Balch, W. E., Fox, G. E., Magrum, L. J., Woese, C. R. and Wolfe, R. S. (1979). Methanogens: reevaluation of a unique biological group. *Microbiol. Rev.* **43**, 260–296.

Boone, D. R. (1995). Short- and long-term manitenance of methanogen stock cultures. In *Archaea: A Laboratory Manual – Thermophiles* (F. T. Robb, K. R. Sowers, S. DasSarma, A. R. Place, H. J. Schreier and E. M. Fleischmann, eds), pp. 79–83. Cold Spring Harbor Laboratory Press, Plainview.

Boone, D. R., Mathrani, I. M. and Mah, R. A. (1987). H_2–CO_2 recirculation and pH control for growth of methanogens in mass culture. *Appl. Environ. Microbiol.* **53**, 946–948.

Burgess, B. K., Jacobs, D. B. and Stiefel, E. I. (1980). Large-scale purification of high activity *Azotobacter vinelandii* nitrogenase. *Biochim. Biophys. Acta* **614**, 196–209.

Cooper, D. Y., Schleyer, H., Novack, B. G. and Rosenthal, O. (1979). Anaerobic techniques for spectrophotometric studies of oxidative enzymes with low oxidation-reduction potentials. *Pharmacol. Therap.* **8**, 339–358.

Daniels, L. (1995). Large-scale culture techniques for methanogenic Archaea. In *Archaea: A Laboratory Manual – Thermophiles* (F. T. Robb, K. R. Sowers, S. DasSarma, A. R. Place, H. J. Schreier and E. M. Fleischmann, eds), pp. 63–74. Cold Spring Harbor Laboratory Press, Plainview.

Daniels, L. and Belay, N. M. B. (1984). Considerations for the use and large-scale growth of methanogenic bacteria. *Biotechnol. Bioeng. Symp.* **14**, 199–213.

Erauso, G., Reysenbach, A. L., Godfroy, A., Meunier, J.-R., Crump, B. C., Partensky, F., Baross, J. A., Marteinsson, V. T., Barbier, G., Pace, N. R. and Prieur, D. (1993). *Pyrococcus abyssi* sp. nov., a new hyperthermophilic archaeon isolated from a deep-sea hydrothermal vent. *Arch. Microbiol.* **160**, 338–349.

Erauso, G., Prieur, D., Godfroy, A. and Raguenes, G. (1995). Plate cultivation techniques for strictly anaerobic, thermophilic, sulfur-metabolizing Archaea. In *Archaea: A Laboratory Manual – Thermophiles* (F. T. Robb, K. R. Sowers, S. DasSarma, A. R. Place, H. J. Schreier and E. M. Fleischmann, eds), pp. 25–29. Cold Spring Harbor Laboratory Press, Plainview.

Fiala, G. and Stetter, K. O. (1986). *Pyrococcus furiosus* sp. nov. represents a novel genus of marine heterotrophic archaebacteria growing optimally at 100°C. *Arch. Microbiol.* **145**, 56–61.

Hippe, H. (1984). Maintenance of methanogenic bacteria. In *Maintenance of Microorganisms: A Manual of Laboratory Methods* (B. E. Kirsop and J. J. S. Snell, eds), pp. 69–81. Academic Press, London.

Hungate, R. E. (1950). The anaerobic mesophilic celluolytic bacteria. *Bacteriol. Rev.* **14**, 1–49.

Hungate, R. E. (1969). A roll tube method for cultivation of strict anaerobes. In *Methods in Microbiology* (J. R. Norris and D. W. Ribbons, eds), pp. 117–132. Academic Press, New York.

Jones, W. J., Whitman, W. B., Fields, F. D. and Wolfe, R. S. (1983). Growth and plating efficiency of methanococci on agar media. *Appl. Environ. Microbiol.* **46**, 220–226.

Macy, J. M., Snellen, J. E. and Hungate, R. E. (1972). Use of syringe methods for anaerobiosis. *Am. J. Clin. Nutr.* **25**, 1318–1323.

Metcalf, W. W., Zhang, J. K. and Wolfe, R. S. (1998). An anaerobic, intrachamber incubator for growth of *Methanosarcina* spp. on methanol-containing solid media. *Appl. Environ. Microbiol.* **64**, 768–770.

Moore, W. E. C. (1966). Techniques for routine culture of fastidious anaerobes. *Int. J. Syst. Bacteriol.* **16**, 173–190.

Mukhopadhyay, B., Johnson, E. F. and Wolfe, R. S. (1999). Reactor-scale cultivation of the hyperthermophilic methanarchaeon *Methanococcus jannaschii* to high cell densities. *Appl. Environ. Microbiol.* **65**, 5059–5065.

Robb, F. T. and Place, A. R. (eds) (1995). *Thermophiles*, Cold Spring Harbor Laboratory Press, Plainview.

Shriver, D. F. (1969). *The Manipulation of Air-Sensitive Compounds*. McGraw-Hill, New York.

Sowers, K. R. (1995). Techniques for anaerobic biochemistry. In *Archaea: A Laboratory Manual – Thermophiles* (F. T. Robb, K. R. Sowers, S. DasSarma, A. R. Place, H. J. Schreier and E. M. Fleischmann, eds), pp. 15–47. Cold Spring Harbor Laboratory Press, Plainview.

Sowers, K. R. and Noll, K. M. (1995). Techniques for anaerobic growth. In *Archaea: A Laboratory Manual – Methanogens* (F. T. Robb, K. R. Sowers, S. DasSarma, A. R. Place, H. J. Schreier and E. M. Fleischmann, eds), pp. 15–47. Cold Spring Harbor Laboratory Press, Plainview.

Sowers, K. R. and Schreier, H. J. (eds) (1995). *Methanogens*, Cold Spring Harbor Laboratory Press, Cold Spring Harbor.

Sowers, K. R., Nelson, M. J. and Ferry, J. G. (1984). Growth of acetotrophic, methane-producing bacteria in a pH auxostat. *Curr. Microbiol.* **11**, 227–230.

Tumbula, D. L., Bowen, T. L. and Whitman, W. B. (1995a). Growth of methanogens on solidified medium. In *Archaea: A Laboratory Manual – Thermophiles* (F. T. Robb, K. R. Sowers, S. DasSarma, A. R. Place, H. J. Schreier and E. M. Fleischmann, eds), pp. 49–55. Cold Spring Harbor Laboratory Press, Plainview.

Tumbula, D. L., Keswani, J., Shieh, J. and Whitman, W. B. (1995b). Long-term maintenance of methanogen stock cultures in glycerol. In *Archaea: A Laboratory Manual – Thermophiles* (F. T. Robb, K. R. Sowers, S. DasSarma, A. R. Place, H. J. Schreier and E. M. Fleischmann, eds), pp. 85–87. Cold Spring Harbor Laboratory Press, Plainview.

Worthington, P., Blum, P., Perez-Pomares, F. and Elthon, T. (2003). Large scale cultivation of acidophilic hyperthermophiles for recovery of secreted proteins. *Appl. Environ. Microbiol.* **69**, 252–257.

Applications of Extremophiles

◆◆◆

34 Applications of Extremophiles: The Industrial Screening of Extremophiles for Valuable Biomolecules

34 Applications of Extremophiles: The Industrial Screening of Extremophiles for Valuable Biomolecules

Gilles Ravot, Jean-Michel Masson and Fabrice Lefèvre

Protéus SA, 70 allée Graham Bell, Parc Georges Besse, 30000 Nîmes, France

◆◆

CONTENTS

Introduction
Enzyme screening
Production of metabolites
Conclusions

◆◆◆◆◆ **INTRODUCTION**

Although the percentage of microorganisms isolated and cultured today represents only a small proportion of the existing microbes in the different biotopes on Earth – estimates vary between 0.001 and 15% depending on the considered ecosystem (Amann *et al.*, 1995) – many industrial applications based on microbial material have been developed. Microorganisms or biomolecules from microbial sources such as enzymes, metabolites or other cell components have now been used for years in industrial processes such as the enzymatic synthesis of fine chemicals, the production of antibiotics by means of fermentation, or the production of ingredients for cosmetic creams by simple cell extraction.

With the discovery of extremophilic microorganisms which are able to grow in a broad variety of conditions, and more particularly since the isolation during the 1980s of many novel hyperthermophilic species (Blöchl *et al.*, 1995), microbiologists quickly understood the strong potential of these microorganisms for industrial applications especially in the field of biocatalysis. Since biomolecules coming from extremophiles are in many cases more robust than the ones coming from their non-extremophilic counterparts, they can successfully replace existing biomolecules in several processes. In addition, extremophilic microorganisms and their biomolecules may present novel industrial opportunities for the production of novel and useful compounds such as compatible organic solutes or for the synthesis of products under novel physico-chemical conditions. Nevertheless, even with the characterization of a variety of interesting enzymes (called extremozymes) or metabolites, few processes using molecules isolated from extremophiles have been industrialized to date.

The cultivation of extremophilic microorganism at industrial scale, in many cases results in low biomass yield (except in the case of halophilic microorganisms). Several innovative fermentation processes have been developed to face this problem (Schiraldi and De Rosa, 2002), but these have a limited industrial efficiency. Therefore, the use of these organisms as efficient enzyme or metabolite producers is generally not economically viable, which explains the difficulties to bring extremophilic biomolecules on the market in the past. Nowadays, several mature and efficient technologies in microbiology, chemistry and molecular biology which speed up the discovery and development of valuable biomolecules are available. In addition, these technologies assist the production of such biomolecules under conditions compatible with economical industrial constraints. By filling the gap between academic discovery and industry, these technologies offer fascinating opportunities for the application of products originating from extremophiles.

In this chapter, we explore, in a non-exhaustive manner, the various screening approaches for the discovery of molecules with industrial interest, with a special focus on the screening of extremozymes.

◆◆◆◆◆ ENZYME SCREENING

So far, many different markets have been successfully supplied with enzymes generated by biotechnology. According to a report on enzymes for industrial applications (Business Communications Company, Inc., 2004), the global market for industrial enzymes, estimated at $1.9 billion in 2003, is expected to reach nearly $2.4 billion in 2009.

This market is divided into three application segments: food enzymes, animal feed enzymes and technical enzymes, which includes uses in cosmetics and personal care, agrochemistry, and the manufacturing of pharmaceuticals and fine chemicals. These last two segments show an increasing demand for new enzymes to address unmet needs in the field of chiral synthesis of increasingly complex molecules.

The use of enzymes provides benefits to the industry, making it possible to:

- develop new applications or new processes,
- improve the competitiveness of existing manufacturing processes by reducing the production costs,
- improve quality and reduce the amount of side-products,
- reduce the cost related to waste disposal,
- deliver substances that are difficult or even impossible to produce under commercially viable conditions using only chemical processes,
- develop competitive "green" chemistry.

The discovery of extremophiles opened a new area for industrial applications, by providing a novel source of more robust enzymes that can better fulfill the harsh specifications of industrial processes. Resulting from academic and industrial research programs, several enzymes originating from extremophiles are already on the market, such as pullulanases, lipases, DNA polymerases and ligases, xylanases, amylases, protease and cellulases (van den Burg, 2003; Vieille and Zeikus, 2001; Vieille et al., 1996).

Each enzyme market faces different challenges. However, within all segments timeline for enzyme discovery is one of the main critical considerations.

With the recent development of microbiology, chemistry and molecular biology technologies, the field of enzyme screening and optimization is now experiencing major R&D initiatives and concrete results. Provided that the industrial constraints have been properly defined, the technologies available and the know-how generated by academic research on life in extreme conditions enable to discover and design tailor-made enzymes from extremophiles within an appropriate timeline.

How to Detect Enzyme Activities in Extremophiles?

Tips and hints on screening assays

The search for new enzymes requires the development of suitable methods and assays for screening the catalytic activities in environmental samples, in pure cultures, or for identifying an enzyme during a molecular cloning step (Wahler and Reymond, 2001). High-throughput is often required to enable the testing of the highest number of microorganisms or clones, while meeting the industrial R&D deadlines. The screening process should also closely mimic the targeted conditions to rapidly result in the selection of an enzyme that fulfils the specifications defined by the industry. The screening assays must also feature a suitable dynamic range to be used for the selection of enzymes generated by directed evolution.

Within the context of chiral chemistry, high-throughput screening assays are often not only aimed at activity, but also at stereoselectivity and enantioselectivity. Although traditional analytical techniques (GC, HPLC or TLC) are the reference methods for the final assessment of enzymes characteristics in terms of enantioselectivity or conversion rate, these proven technologies are generally far too expensive and time-consuming for high-throughput screening purposes.

When it comes to screening for enzymes active under extreme conditions, the challenge is to have suitable screening assays to detect activities under these conditions with a satisfactory signal-to-noise ratio.

The best approach, when possible, consists in using the natural substrate to detect the product(s) generated by the enzyme activity or the substrate consumption. Examples are the detection of phosphate release for phytase screening, or detection of ammonium for nitrilase. The monitoring of the cofactor for cofactor-dependent enzymes such as alcohol dehydrogenases is also possible (direct measurement at UV wavelengths or indirect chemical or enzymatic measurements).

The following protocol has been successfully used to detect cofactor consumption during the screening of thermoactive alcohol dehydrogenases from thermophiles. This assay is based on a formazan precipitation reaction as described by Mayer and Arnold (2002), with a mixture of NAD and NADP as cofactors, and a pool of seven different alcohols (primary and secondary alcohols with short or long chains) as substrates:

1. 5 µl of the extract to be tested are incubated for 5–20 min at 80°C with 30 µl of buffer A (50 mM Tris-HCl, pH 8.0, 0.13% gelatin), 20 µl of a mixture of equal parts of methanol, ethanol, isopropanol, butanol, decanol, benzyl alcohol and octanol, 5 µl of 20 mM NAD, 5 µl of 20 mM NADP and 140 µl of buffer B (30 µM of phenazine methosulfate and 300 µM of nitroblue tetrazolium).
2. Reaction mixtures with an active alcohol dehydrogenase become dark blue within a few minutes at 80°C, while the negative control (without enzyme) stays clear.

An alternative is the direct monitoring of the consumption of either NAD or NADP with a pool of alcohol substrates. It can be achieved by following UV absorbance at 340 nm in UV compatible microtiter plates (Corning UV Plates #3635). For thermophilic enzymes, evaporation must be controlled by sealing plates with adhesive films or overlaying liquid in wells with mineral oil (SIGMA #M3516).

Many detection tests based on natural substrates have been developed to characterize pure enzymes from non-extremophilic microorganisms. Such tests may not be compatible with high- or even low-screening throughput because of lack of specificity or sensitivity, interference with compounds used in the culture media, etc.

When detection using natural substrates can not be considered, convenient protocols to test enzymatic activities at high-throughput are those involving fluorogenic or chromogenic substrates. For this purpose, reporter substrates are commercially available, such as umbelliferyl or nitrophenyl derivatives. Chromogenic substrates are generally more convenient than fluorogenic substrates as their reactions can be detected visually, implying that they are suitable for routine activity monitoring without instruments. Nitrophenyl derivatives require control of the final pH above pH 8.0 to enable the detection of nitrophenol at 414 nm. However, due to their chemical activation, all these reporter substrates generate a high level of background in extreme conditions, thus hindering the differentiation of positive strains from false positives.

Recent efforts in chemistry have been undertaken to make more chemical reactions compatible with high-throughput screening in extreme conditions. We have developed one such approach, called CLIPS-O^{TM} (for **CataLsyts Identification Per Substrate Oxidation**) jointly with the team led by Prof. Jean-Louis Reymond at the University of Bern, Switzerland (Reymond et al., 2001). The principle behind this technology is the use of a situation-appropriate "spacer" molecule between the substrate and

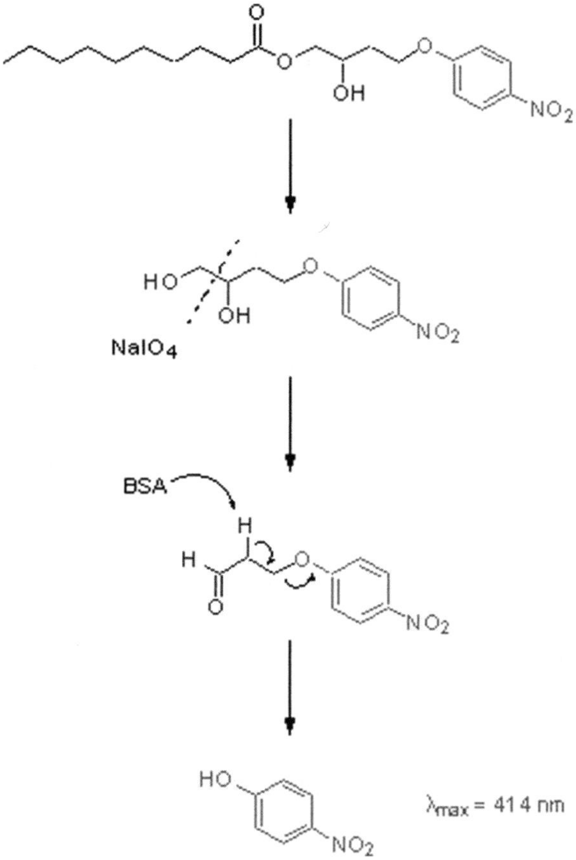

Figure 34.1. Detection of thermostable esterases with a CLIPS-O^{TM} substrate. After incubation with the enzyme, the ester CLIPS-O^{TM} substrate is cleaved into a hydrolyzed intermediate. Both the ester CLIPS-O^{TM} substrate and the hydrolyzed intermediates are extremely stable under a wide range of physico-chemical conditions, including high temperatures and extreme pH. The signal is released solely after a secondary chemical oxidization with sodium periodate, a reaction that only takes place if the intermediate diol has been produced. See Plate 34.1 in Colour Plate Section.

reporter moieties, followed by an enzymatic reaction to create a stable intermediate. A secondary reaction liberates the signal for measurement. The CLIPS-O^{TM} chemistry enables the synthesis of tailor-made substrates that mimic both the structure and the energetic state of many different industrial substrates. With such fluorogenic and chromogenic substrates, enzyme activity profiles can be rapidly characterized. An example of this chemistry technology for the screening of esterases or lipases is given in Figure 34.1.

These novel substrates combine high sensitivity, selectivity and the possibility to test enantioselectivity within a simple format suitable for high-throughput screening. The stability of these substrates enables to achieve excellent signal-to-noise ratio when screening under the specific conditions of the process targeted (e.g. under high temperature or under acidic or highly alkaline conditions). As an example, the screening for thermostable esterases and lipases from thermophilic microorganisms has been successfully implemented using such CLIPS-O^{TM} assays (Lagarde et al., 2002).

Given below is one procedure to detect thermophilic strains expressing lipases or esterases:

1. Mix 10 µl of sample with 8 µl of a CLIPS-O^{TM} substrate solution (20 mM in 50% acetonitrile, 50% pure water) and 74 µl of 0.2 M PIPES buffer, pH 7.0;
2. Incubate at the appropriate temperature during 40 min;
3. Add 40 µl of freshly prepared 100 mM NaIO$_4$ solution;
4. Add 4 µl of 50 mg ml^{-1} bovine serum albumin (BSA);
5. Add 40 µl of 0.2 M Na$_2$CO$_3$;
6. Leave at room temperature during 5–10 min;
7. Spin the sample 5 min;
8. Read in microtiter plate at 414 nm.

Fingerprinting enzymatic activities

To rapidly characterize enzymatic activities of an extremophile microorganism, wells of a microtiter plate can be filled with different substrates. Activity level is then monitored in each well after addition of cells. Activity values provide a fingerprint of the enzymatic diversity of the strain. These values can be converted into an array of colors depending on activity level to enable a rapid visual comparison of the fingerprints.

This type of measurement gives a much more complete and rapid assessment of enzyme activities than a single determination. It can be used to directly assess several enzymatic activities of a strain, to characterize the activity profiles of one particular enzyme, or to compare activity patterns of enzymes or phylogenetically related strains. Such a fingerprinting protocol is obviously made easier when all the activities are revealed under the same conditions (Wahler et al., 2001).

Examples of enzymes fingerprinting have been achieved with CLIPS-O^{TM} substrates. For this purpose, a library of stereochemically and structurally diverse fluorogenic and chromogenic CLIPS-O^{TM} substrates has been synthesized, including substrates for esterases, lipases, proteases, peptidases and epoxide hydrolases, and has been employed to characterize enzyme activity profiles of biocatalysts to be used in chiral chemistry (Wahler et al., 2002; see Figure 34.2).

Plate screening

As for other microorganisms, traditional plate screening (either with natural or artificial substrates) can be used for the screening of extremophilic microorganisms with slight modifications, especially regarding the solidifying agent. For pH values above 3, Gelrite® Gellan Gum (SIGMA G1910) or Phytagel^{TM} (SIGMA P8169) are used in order to prepare plates (or roll tubes) instead of agar which is unsuitable for growth of hyperthermophiles due to its low melting point. The concentration of solidifying agent has to be adjusted depending on the desired pH as mentioned by Reysenbach and Götz (2001). In the case of highly

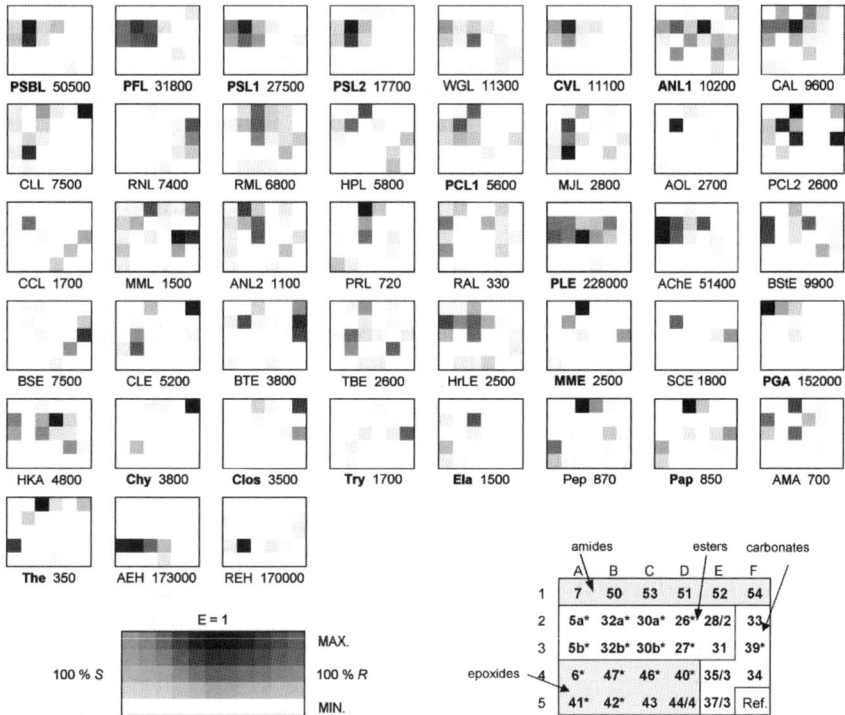

Figure 34.2. Enzymes fingerprinting with a set of CLIPS-O™ substrates (modified from Wahler *et al.*, 2002). Substrates marked * in the layout (bottom right) are enantiomeric pairs. Below each array: Enzyme code, and apparent maximum rate in pM s^{-1}, which appears as the darkest (gray scale or color) square in each array. Conditions: 0.1 mg ml^{-1} enzyme, 100 μM substrate, 20 mM borate pH 8.8 with 2.5% v/v DMF, 26°C, 2 mg ml^{-1} BSA, 1 mM NaIO$_4$. Enzymes marked in bold appear homogeneous (>90%) by SDS-PAGE analysis (PSBL – *Pseudomonas* sp. B lipoprotein lipase; PFL – *Pseudomonas fluorescens* lipase; PSL1 – *Pseudomonas* sp. lipoprotein lipase; PSL2 – *Pseudomonas* sp. lipoprotein lipase; WGL – wheat germ lipase; CVL – *Chromobacterium* visc. lipoprotein lipase; ANL1 – *Aspergillus niger* lipase; CAL – *Candida antarctica* lipase; CLL – *Candida lipolytica* lipase; RNL – *Rhizopus niveus* lipase; RML – *Rhizomucor miehei* lipase; HPL – hog pancreatic lipase; PCL1 – *Burkholderia cepacia* (basonym: *Pseudomonas cepacia*) lipase; MJL – *Mucor javanicus* lipase; AOL – *Aspergillus oryzae* lipase; PCL2 – *Burkholderia cepacia* (basonym: *Pseudomonas cepacia*) lipase; CCL – *Candida cylindracea* lipase; MML – *Mucor miehei* lipase; ANL2 – *Aspergillus niger* lipase; PRL – *Penicillium roqueforti* lipase; RAL – *Rhizopus arrhizus* lipase; PLE – pig liver esterase; AChE – *Electrophorus electricus* acetylcholine esterase; BStE – *Geobacillus stearothermophilus* (basonym: *Bacillus stearothermophilus*) esterase; BSE – *Bacillus* sp. esterase; CLE – *Candida lipolytica* esterase; BTE – *Geobacillus thermoglucosidasius* (basonym: *Bacillus thermoglucosidasius*) esterase; TBE – *Thermoanaerobacter brockii* esterase; HrLE – horse liver esterase; MME – *Mucor miehei* esterase; SCE – *Saccharomyces cerevisiae* esterase; PGA – *Escherichia coli* penicillin G acylase; HKA – hog kidney acylase I; Chy – bovine pancreatic chymotrypsin; Clos – *Clostridium histolyticum* clostripain; Try – hog pancreatic trypsin; Ela – porcine pancreatic elastase; Pep – porcine stomach mucosa pepsin; Pap – *Papaya latex* papain; AMA – *Aspergillus melleus* acylase I; The – "Bacillus thermoproteolyticus" thermolysin; AEH – *Aspergillus niger* epoxide hydrolase; REH – *Rhodotorula glutinis* epoxide hydrolase). Fingerprints are presented as arrays of squares with a two-color scale combining stereoselectivity and activity. The patterns observed consistently appear indistinguishable from one measurement to the next. See Plate 34.2 in Colour Plate Section.

acidophilic microorganisms the starch method described by Schleper et al. (1995) can be used. We currently use the following protocol as an almost universal media preparation method, with Phytagel™ as a solidifying agent for plate screening (as well as for isolation of extremophiles):

1. Prepare 2× concentrated medium without the solidifying agent and complement, if necessary, the medium in order to obtain a concentration in $MgCl_2 \cdot 6H_2O$ at least equal to 2 g l^{-1};
2. Prepare in a separated bottle a solution of 16 g l^{-1} Phytagel™;
3. Sterilize both solutions in an autoclave;
4. Before use, melt the Phytagel™ (using a microwave oven) and mix (vol/vol) with the media previously heated at approximately 50°C.

Additional equipment for plate screening

- Bellco tubes or glass Petri dishes for anaerobes or aerobes, respectively
- Gas impermeable butyl rubber stoppers
- Chemicals for media preparation
- Gases and gas mixtures for the culture of anaerobes
- Incubators
- Needle syringes
- Anaerobic chamber and anaerobic jars

Automation

Liquid handling automates enable the automation of a broad diversity of screening tests on cell extracts or on genomic clones. Solvent compatibility should be checked before implementing a screening in microtiter plates (see as an example http://www.corning.com/Lifesciences/technical_information/techDocs/chemcompplast.asp?region=ge%32 language=en). Pipeting organic solvents with liquid handlers requires use of several "up and down" pipeting steps to ensure solvent saturation of the tips before the final pipeting, otherwise wrong volumes may be distributed due to the surface tension coefficient of solvents.

Particular attention should be paid to the precise determination of intra-plate-test deviation: the mean as well as the standard deviation must be calculated on the results of the assay achieved in all the wells of one single plate. Assays with a standard deviation higher than 10–15% should not be validated. On the other hand, inter-plate deviation must be investigated by calculating mean and standard deviation on the results of the assay obtained in all the wells of several plates. Finally, positive and negative controls should be introduced in each plate (preferably not near the edge of the plate), as internal quality standards for the screening procedure.

When screening for thermophilic enzymes, evaporation in microtiter plates concentrates the signal response in the wells on the edge of the plates, thus potentially leading to false positives. To lower this effect, 96-well plates should be preferred and volumes above 150 µl should be used. One of the best ways to avoid evaporation is to use polystyrene plates (COSTAR #3599) sealed with either aluminum adhesive films

(COSTAR #6570), or by a thermal plate sealer (such as ALPS300™ from Abgene, or PlateLoc® from Velocity 11).

Microbial Strategies for Enzyme Screening of Extremophiles

Culture collection

Historically, because cell culture was the only way to obtain enzymes, the screening of microorganisms from culture collections was the method of choice to discover biocatalysts. Today, even if most microorganisms cannot be cultured using current culture techniques, this remains a strategy of choice in the industry in order to obtain novel enzymes.

In addition to the low diversity available (as compared to the existing one) even when dealing with large collections, the maintenance of the strains and the time required for the culture of the strains represent the major drawbacks of this method. Therefore, we prepare microtiter plates of extracts of the strains, compatible with high-throughput automation in order to speed up the discovery process. For that purpose, we have developed a simple and efficient procedure summarized below:

1. Grow the strains in appropriate medium to the end of the exponential growth phase (before entering the stationary phase);
2. Centrifuge the culture for 15 min at $8000 \times g$ (at $4°C$);
3. Resuspend the pellets in fresh culture medium (whether or not supplemented with 10% glycerol) in order to concentrate the cells 50 times (or 100 times if the growth of the strain is weak);
4. Transfer 100 µl portions to microtiter plates sorted according to the growth pH and temperature of the microorganisms so that same screening test can be run for all the strains on the plate. A portion of 100 µl of fresh culture medium is transferred to the microtiter plate and used as control;
5. Store at $-80°C$ (the microtiter plates can be stored for at least 2 years before thawing);
6. Thaw the microtiter plates gently on ice before use.

Unfortunately, when induction is required for the expression of the targeted enzyme, these microtiter plates cannot be used and a specific culture program must be set up prior to the screening of the strains. In such a case, the use of a computerized database for collection management is a key tool, especially when dealing with thousands of microorganisms growing in many different conditions. There is no commercial tool available fully devoted to the management of strain collections and recommended screening procedures. Therefore, the development of a dedicated database is necessary in order to provide the information required for microorganism screening and culture collection management. To be really efficient, this database must contain all available information: (i) on the strain – from the sampling conditions to the phenotypic and phylogenetic characterization including the isolation procedure, (ii) on the maintenance, as well as (iii) on previous screening results.

Environmental samples

In some cases, the expression of an enzyme may give a selective advantage under particular growth conditions that can be used for the selection of microbes expressing such enzymes. One of the currently used approaches in industry is based on the assimilation of the carbon or the nitrogen source in minimal medium, following expression of the targeted enzyme. The detoxification of a growth inhibitor was also successfully used in a strategy based on a selection approach. In both cases, an environmental sample – chosen according to the targeted biocatalyst and its application – is generally used to inoculate the culture broth.

After growth of the enrichment culture and measurement of the targeted enzyme activity when possible, the broth can be used for isolation as well as for cloning purposes. For the isolation of strains, the use of a plate screening assay, if available, is the fastest method to recover strains of interest. Since the growth and induction protocols must obviously be defined according to the targeted enzyme, its properties and industrial application, no universal method can be described. However, a non-exhaustive list of enzymes of interest and strategies applied for the selection of microorganisms expressing such activities is given in Table 34.1.

Molecular Strategies for Gene Cloning

Various approaches of molecular biology can be used to clone genes coding for enzymes from cultivable or non-cultivable extremophilic microorganisms:

– construction of a genomic library for molecular screening,
– creation of a genomic library for recombinant expression,
– bioinformatic analysis and PCR cloning.

Whatever the method used, the starting point is the genomic material extracted from the samples.

Table 34.1 Examples of enzymes of industrial interest for which useful enrichment strategies have been developed

Enzymes	Enrichment strategy	Substrates
Amylase	Carbon and energy source	Starch
Cellulase	Carbon and energy source	Cellulose
Xylanase	Carbon and energy source	Xylan
Nitrilase/Nitrile hydratase	Carbon and/or nitrogen source	Nitrile
Amidase	Carbon and/or nitrogen source	Amide
Protease	Carbon and energy source	Proteins
Chitinase	Carbon and energy source	Chitin
Lipase	Carbon and energy source	Triolein

DNA extraction and construction of genomic libraries

Obtaining quality genomic DNA is a critical step as many molecular biology procedures can be inhibited by residual salts, metals, ions or any other organic compounds that are present in the strain, culture medium or environmental sample.

When possible, one should try to centrifuge cells and wash them three times in an appropriate buffer to eliminate potential interfering compounds present in the samples.

Genomic DNA extraction from a pure culture

The QIAamp^{TM} DNA mini kit protocol (QIAGEN) is a chemical and enzymatic extraction-based protocol that is currently used for DNA extraction from pure cultures of extremophiles, without any need to modify the instructions of the manufacturer.

Due to the special structure of the outer envelopes of many extremophiles, such as specific pseudopeptidoglycan of many Archaea, the method described above may not always be the most appropriate. In that case a mechanical protocol should be preferred (see below).

Extraction of genomic DNA from enrichment/environmental samples

Extraction of genomic material from enrichment cultures or environmental samples can be carried out by an initial mechanical disruption of cells. Depending on the composition and viscosity of the sample, direct disruption at high pressure using a French press or dedicated cell disrupters (Cell D from Constant Cell Disruption Systems), grinding with mortar and pestle (automated grinder from Restch) or breakage by shaking with ceramic beads are the most efficient procedures. As far as possible we avoid using French press or Cell D to achieve DNA extraction, as these steps liberate nucleases and may shear genomic DNA. We usually apply a protocol from BIO101 Systems (FastDNA® SPIN kit for soil, BIO101 Systems, QBIOGENE) based on disruption with beads in buffers.

To recover genomic DNA from samples containing sulfur, a specific protocol has been described by Reysenbach and Götz (2001):

1. Suspend the sample in 500 µl of a pre-heated extraction buffer (100 mM Tris-HCl, pH 8.0, 1.4 M NaCl, 20 mM EDTA, 0.4% (vol/vol) 2-mercaptoethanol, 2% (wt/vol) cetyltrimethylammonium (CTAB), 1% polyvinyl-pyrrolidone (PVP 360));
2. Incubate 15 min at 65°C;
3. Add 500 µl of chloroform:isoamyl alcohol (24:1) and vortex for 2 min;
4. Spin for 15 min and transfer the aqueous layer to a fresh tube;
5. Precipitate DNA with an equal volume of isopropanol and 0.5 volume of 5 M NaCl at room temperature for 15 min;
6. Spin for 30 min and wash with 70% ethanol;
7. Resuspend in 100 µl of TE (10 mM Tris, pH 8.0 + 1 mM EDTA) and add 1 µl of RNAse (1 mg ml^{-1});

8. Incubate at 37°C for 30 min;
9. Precipitate DNA with 2 volumes of freezer-cold ethanol and 0.1 volume of 3 M sodium acetate at 4°C for 1 h,
10. Spin for 30 min at 4°C, wash with freezer-cold ethanol and dry the pellet.

Given below is an additional method to extract genomic DNA from soil samples. This method enables removal of humic acids from DNA with thiocyanate guanidine and polyvinyl-pyrrolidone (PVP). Volumes indicated are given for a starting sample of 2 g of solid sample (for example compost).

1. Resuspend the sample in 20 ml of a freshly made extraction solution prepared by mixing 17 ml of 1 mM sodium phosphate buffer pH 8.0 with 3 ml of a 6 M solution of guanidine thiocyanate in 0.1 M Tris-HCl, pH 8.0. Add a pinch of dithiothreitol and 0.4 g of PVP (weigh in a hood, and wear a mask!). The sodium phosphate buffer and guanidine thiocyanate solutions can be prepared in advance and stored at room temperature;
2. Vortex during 1 min and leave the tube on ice during 1 min. Repeat this step twice;
3. Centrifuge at $1000 \times g$ for 15 min at 4°C;
4. Transfer the supernatant in a new tube and add 1/10th of volume of 10% sodium dodecylsulfate (SDS);
5. Vortex, then incubate for 30 min at 60°C;
6. Leave on ice for 15 min;
7. Centrifuge at $4000 \times g$ for 15 min at 4°C;
8. Transfer the supernatant in a new tube and add 1/10th of volume of 3 M sodium acetate, pH 4.8 and 1 volume of isopropanol;
9. Invert the tube once or twice, then centrifuge at $10\,000 \times g$ for 30 min at 4°C;
10. Resuspend the pellet (not always visible) in 300 µl of TE,
11. Extract with 300 µl of an phenol–chloroform–isoamyl alcohol mixture (25/24/1, vol/vol/vol). Vortex, then spin for 2 min;
12. Transfer the (top) aqueous phase to a new tube. Add 1/10th of volume of 3 M sodium acetate, pH 4.8 and 2 volumes of cold ethanol;
13. Spin for 15 min. Discard the supernatant;
14. Add 500 µl of 70% ethanol to wash the pellet, and spin for 5 min;
15. Resuspend the dried pellet in 100 µl of TE.

RNA can be removed by a short treatment with 1/100th volume of RNAse at 1 mg ml^{-1} before phenol–chlorofom extraction.

Quality control of genomic DNA

Quality and concentration of all DNA samples are determined (after RNAse treatment) by recording a UV absorption spectrum between 220 and 320 nm in a UV-VIS spectrophotometer using a quartz glass microcuvette. The purity is deemed acceptable when the ratio A260:A280 is between 1.8 and 1.9. One A260 unit is assumed to correspond to a DNA concentration of 50 µg ml^{-1}.

Genomic library construction

Cloning genomic DNA from extremophiles does not require any specific protocol. Several commercially available kits can be used, depending on the final screening that will be achieved.

For molecular screening based on hybridization, partial DNA fragmentation with the restriction enzyme *Sau*3A I (or total DNA digestion with an appropriate enzyme) and further cloning in usual vectors such as pBR322 or pUC18 is the most obvious approach (Moore, 1995). The only key parameter is to make sure that the whole genomic DNA is statistically represented in the final library. As genomes of a number of extremophiles have already been sequenced and characterized (http://www.tigr.org/tigr-scripts/CMR2/CMRGenomes.spl), the genome size of a particular strain can be assessed by (i) phylogenetic positioning of this microorganism (by sequencing its 16S rRNA gene or using the ARDRA method (Vaneechoutte *et al.*, 1992), and (ii) comparison with the phylogenetically closest related known genome.

When such an assumption cannot be made, genome size can be determined by other methods that may be more time consuming, such as pulsed field gel electrophoresis, real time PCR (Wilhelm *et al.*, 2003) or flow cytometry (Vinogradov, 1994).

Using a partial digestion strategy, the following formula is used to determine the size of the library for a given probability that at least one copy of each gene has been cloned:

$$\text{Number of clones in the final library} = \ln(1 - P)/\ln(1 - I/G)$$

where P is the probability (usually set at 99%), I is the average insert size of the cloned genomic fragments and G is the whole genome size in base pairs.

Creation of genomic libraries for functional screening requires cloning of genomic fragments in vectors bearing a transcription promoter to drive the *in vivo* expression of the heterologous genes. Well-known vectors are the pET series (Novagen), pBSKS (Stratagene), pBAD (Invitrogen) for *E. coli*, pPIC for *Pichia pastoris* (Invitrogen) or pCOM for *Pseudomonas* sp. Leaky promoters should be preferred when the enzymatic activity is not supposed to be toxic, as this avoids the induction step that can be time consuming, difficult to automate, and may lead to overexpression of misfolded proteins (inclusion bodies). Since screening assays will have been calibrated to be as sensitive as possible, even a slight recombinant expression of the enzyme should be detectable.

Construction of genomic libraries with bacteriophages is a good alternative to plasmid vectors for both molecular and functional approaches, as it enables the cloning of larger genomic inserts. The capability of phages to package size-standardized DNA fragments and their efficiency of *E. coli* infection leads to high-quality and high-redundancy genomic libraries of extremophile genomes. These bacteriophage protocols are available as commercial kits such as Lambda ZAP® from Stratagene.

Functional screening *versus* molecular screening of libraries

Heterologous expression for functional screening

Expression cloning is a technology of choice for enzyme screening: genomic fragments are expressed in a recombinant host or *in vitro* under cell-free conditions, and enzyme activities are assayed at high-throughput with enzymatic tests. The choice of the recombinant host depends on the enzymatic activity which is looked for (to avoid endogenous activity of the host) and on genetic compatibility between this host and the genomic DNA from the extremophile. *E. coli* is the most preferred host, whereas *Pseudomonas* should be preferred for the expression of GC-rich DNA, and yeasts for proteins that could require glycosylation. A complete set of *E. coli* strains and vectors is now available to deal with a vast number of situations: strains expressing rare codons, strains that favor disulfide bridge formation, and strains that co-express molecular chaperones to improve protein folding (see Novagen website: http://www.emdbiosciences.com/html/NVG/compcellsfamily2.html).

From time to time, genetic deletion of an endogenous gene may be required to avoid background during screening. For example the deletion of the *E.coli appA* gene is required for the screening of new phytases from extremophiles. This can be achieved either by direct transformation of a DNA construct containing the targeted gene, the open reading frame of which has been interrupted by a selection marker (Miller, 1992), or by phage P1 transduction of *E. coli*, starting from a strain that has already lost this gene (Cherepanov and Wackernagel, 1995).

The highest throughput for screening recombinant expression libraries is obtained by using selection or plate assays. For selection assays, recombinant hosts are spread on plates containing either a substrate that is toxic for colonies which do not express the desired enzyme (for example, plating on arsenic for arsenate detoxification enzymes (Crameri *et al.*, 1997), or that only enables growth of hosts that had acquired the desired gene (for example, using auxotrophy complementation). Strains with specific deletions are useful for such a selection approach of recombinant clones. Available *E. coli* strains, with various genetic backgrounds, can be found at the *E. coli* Stock Center (http://cgsc.biology.yale.edu/cgsc.html).

For plate assays, a reporter molecule is included in the plate medium to enable the visual localization of hosts that had gained the enzyme of interest. Such reporters can be natural substrates themselves (for example, casein or tributyrin generates clear halos around colonies expressing proteases and lipases, respectively), or synthetic reporters (Remazol Brilliant Blue R-D-Xylan SIGMA #M5019 for the screening of xylanases, or Red Gluc #1365C from Research Organics for β-glucuronidases), or pH indicators.

As mentioned above, a key difficulty for screening of extremophile genes is the proper calibration of such plate or selection assays. High temperatures, extreme pH or high salt concentrations that can be required in some cases are not always compatible with the growth of the recombinant host or the plate screening itself. In such cases, colonies can be grown under appropriate conditions and then overlaid by a gel containing the substrate. A similar strategy can be used when substrates cannot directly be poured in the plate medium.

We have developed an assay for the plate screening of acylases based on the hydrolysis of lauryl-glutamate. A product (lauric acid) of the reaction creates a white "snow" precipitate around positive colonies.

1. Prepare 50 ml of 5× M9 medium without NH_4Cl (30 g l^{-1} Na_2HPO_4, 15 g l^{-1} KH_2PO_4, 2.5 g l^{-1} NaCl and 15 g l^{-1} agar; autoclave for 20 min at 121°C);
2. Add 200 ml of sterile water, 62.4 µl of 50 g l^{-1} $MgSO_4 \cdot 7H_2O$, 22.5 µl of 100 g l^{-1} $CaCl_2 \cdot 2H_2O$, 2 ml of 50% (wt/vol) glucose, 25 µl of 10 mg ml^{-1} thiamine and 6 ml of 250 g l^{-1} lauryl glutamate;
3. Add the appropriate selection antibiotic(s) for the strain and plasmid(s);
4. Plate the strain (we use E. coli BL21 DE3) on these plates;
5. Clones expressing an acylase hydrolyzing the lauryl-glutamate appear with a white "snow" precipitate around the colonies.

Such a strategy based on precipitation of long-chain organic acids can be applied to the detection of various activities such as for instance lipases/esterases, protease, amidase, nitrilase and others.

Microtiter plate assays (MTPA) are used when plate assays cannot be implemented due to the previously described drawbacks. Each colony is picked and cultivated in a well of a microtiter plate (optionally with an induction step for recombinant expression). The main advantages of these MTPA are:

– the easy level of automation with standard liquid handlers (provided that all necessary quality controls of aseptic technique during picking and liquid handling have been achieved),
– the possibility to add any kind of reagent and/or buffer, and the option to incubate, shake, seal plates and run multi-step assays,
– the possibility to duplicate a microtiter plate to keep a "mother" copy of each plate, while "daughter" copies undergo reaction assays,
– the possibility to store the genomic library sorted in MTP by adding 10% glycerol, sealing plates and freezing them at −80°C for future screenings.

When *in vivo* expression is not successful due to cytotoxicity, incorrect protein folding or more simply genetic incompatibility between the expression host and the genomic fragment to be expressed, cell-free *in vitro* expression approaches can be considered.

Cell-free expression is a technology that allows, solely on the basis of the genomic information, to produce all proteins encoded by a specific genome in microtiter plates and then to detect the targeted enzymes using a functional assay. We have developed a specific patented embodiment of this technology brand-named Phenomics^{TM}. Genomic DNA fragments are cloned in a vector between two transcription controlling sequences (T7 and T3 promoters). The technology therefore circumvents all technical pitfalls linked to regulation of gene transcription. After a colony picking step, each genomic fragment is transcribed in both directions in two separate wells. *In vitro* transcribed mRNAs are then translated into proteins using a specific *in vitro* cell-free translation reagent extract brand-named PheMix^{TM}. Several PheMix^{TM} extracts containing ribosomes from a phylogenetically related strain have been prepared to optimize gene expression according to the origin of the genomic DNA. This cell-free approach is tailored to eliminate all known obstacles that interfere with the compatibility between expression vectors and DNA fragments on the one hand and cellular host selected for expression on the other hand.

Given below is a protocol to achieve cell-free expression of extremophile genomic DNA:

1. Mix in a tube:

    Between 100 and 200 ng of mRNA (prepared by *in vitro* transcription and desalted);
    7.5 µl MM-T buffer (270 mM Tris-acetate pH 8.2, 12 mM DTT, 6.5 mM ATP, 5.4 mM GTP, 0.07 mg ml^{-1} folic acid and 6.5 mM EDTA);
    2 µl of 4 mg ml^{-1} *E. coli* tRNA;
    5 µl of amino acids mixture (20 amino acids) at 2.5 mM each;
    40–80 mM of acetyl phosphate (*);
    100–180 mM of potassium acetate (*);
    8–12 mM of magnesium acetate (*);
    15–20 µl of *in vitro* translation extract prepared as described by Zubay (1973) and thawed on ice;
    Add RNAse free sterile H$_2$O up to 50 µl.

    Be careful to use RNAse-free buffers, tubes and tips.

2. Incubate 1–3 h at the appropriate temperature (usually the growth temperature of the strain used for the preparation of the *in vitro* translation extract).

(*) salt concentrations must be optimized for each batch of *in vitro* translation extract and for each batch of mRNAs.

Molecular screening

Molecular screening of genomic libraries is achieved using well-known hybridization approaches (Brown, 1995). The protocol requires that DNA probes should have a minimum level of homology with the searched sequences, and the method is thus limited to the detection of genes homologous to already described ones. The probes can also be designed from the results of protein micro-sequencing.

Once a genomic fragment has been detected with a probe, it is sequenced. Another round of hybridization screening or a further step of reverse PCR amplification (Silver, 1991) are generally required to unravel the full length sequence of the gene. Molecular screening is more time-consuming but less expensive than functional screening. It is usually an alternative to expression cloning when no functional heterologous expression could be obtained.

Table 34.2 Available complete genomes (November 2004) sorted by their extremophilic origin, available from The Institute of Genomic Research (TIGR, Rockville, MD) and the National Center for Biotechnology Information (NCBI)

	Available sequences	Extremophiles				Non-Extremophiles
		(Hyper) Thermophiles	Thermo-acidophiles	Halophiles	Alkaliphiles	
Archaea	19	10	5	1	0	3
Bacteria	153	4			1	148
Total	172	14	5	1	1	151

Bioinformatics and genomes from extremophiles

As shown in Table 34.2, a significant number of genomes from extremophiles have been sequenced. These sequences, available on the web, can be "screened" using bioinformatics tools to identify and discover novel useful enzymes.

By expressing open reading frames (ORFs) of potential interest, either *in vivo* or *in vitro*, in order to validate their functions or to characterize their activities, the main difficulties of screening can be overcome. Such an approach was recently published for the screening for thermophilic alcohol dehydrogenases (Ravot *et al.*, 2003).

Bioinformatic alignments of extremophile sequences enable the design of degenerated probes to fish out new genes in genomes from pure culture collections, or directly in enrichment cultures or environmental samples.

We currently use a protocol based on specific degenerated primers called CODEHOP (**CO**nsensus-**DE**generate **H**ybrid **O**ligonucleotide **P**rimer) as described by Rose *et al.* (2003), to fish out new genes from the genomes of our extremophiles.

Enzymes from Extremophiles and Directed Evolution

Natural biodiversity provides backbones of new enzymes, and the discovery of extremophile proteins offers molecular structures that are very different from those known to date and may be associated with extreme performances. However, to cope with the industrial constraints, there is a need to rapidly engineer the natural enzymes to fine-tune their characteristics so that they fit exactly with these constraints. The term "directed evolution" covers a group of molecular biology techniques that allow the creation and selection of genes coding for improved proteins from an initial pool of parental genes, by mimicking *in vitro* the natural evolution processes. Using directed evolution, it is possible, for example, to create or increase resistance to solvents or to other toxic compounds, or to increase the enantioselectivity of an enzyme.

The first directed evolution protocols (DNA-Shuffling – Stemmer (1994) and Step – Zhao et al. (1998)) have relied on DNA-polymerase-based recombination. However, during cycles of recombination, additional mutations are introduced in genes – even when using high-fidelity polymerases – thus creating deleterious point mutations in the final recombinant genes. To fix that point we have developed and patented a ligation based protocol brand-named L-Shuffling™ (for Ligation Shuffling, Dupret et al., 2002). L-Shuffling™ enables the conservative recombination of parental genes to generate new improved proteins by avoiding the creation of point mutations during recombination. This generates a higher percentage of active clones in the final library, so that the desired recombinants can be found quicker during screening. The technical principle of L-Shuffling™ is as follows: genes are cleaved by restriction enzymes or by a DNAse I. These fragments are denatured and annealed on a DNA template (called the assembly matrix). This matrix enables the adjacent hybridization of the fragments ends. Parental genes, oligonucleotides, synthetic sequences or even fragments of the parental genes can be used as assembly matrix. After hybridization, a thermostable ligase catalyzes the ligation of the adjacent ends. Then cycles of denaturation–hybridization–ligation are repeated, before the purification of the recombined fragments on an agarose gel.

Here is one embodiment of the patented L-Shuffling™ protocol to recombine a family of homologous genes:

1. Align the parental genes to select a set of restriction enzymes that have restriction sites in the same places in all (or in at least two) genes;
2. Select an assembly matrix that will enable the adjacent hybridization of the fragments generated by this set of restriction enzymes (one particular embodiment consists in selecting two pools of restriction enzymes; each pool of corresponding fragments is used as assembly matrices for the other one);
3. Prepare the different genes to be shuffled either by a PCR amplification of the full length ORFs, or by a digestion with restriction enzymes from a plasmid;
4. Purify these genes on an agarose gel and recover the DNA with the QiaKit PCR products purification kit (QIAGEN);
5. Digest the genes with the set of selected restriction enzymes in order to obtain "recombination compatible" fragments between 100 and 500 base pairs long,
6. Purify these fragments with the QiaKit PCR products purification kit (QIAGEN) and recover them in pure sterile water;
7. Mix 500 ng of the prepared fragments with 30 ng of the assembly matrix (when using two pools of restriction sized fragments, mix 500 ng of fragments from pool #1 with 500 ng of fragments from pool #2) in a PCR tube;
8. Add 15 U of Ampligase (Epicentre #A3202K), 3 µl of enzyme buffer, and add water to a final volume of 30 µl;

9. Include a negative control reaction that includes everything but the ligase;
10. Overlay with mineral oil, and run 40 cycles: (94°C 5 min – 91°C 1 min – 65°C 4 min), then store at 4°C;
11. Check the quality of the recombination by analyzing 2 µl of the positive and negative reactions on an agarose gel (a clear band of recombined products should appear in the positive reaction, whereas a low molecular weight smear appears in the negative reaction);
12. Purify the recombined products on an agarose gel and clone them for a further screening step.

The whole process can be repeated several times with increasing screening stringency until the protein meets the goals that have been set. Once again, a good calibration of the screening assays is crucial when performing screening of shuffled libraries, as one should be able to discriminate slightly improved clones from wild type ones.

Expression, Purification and Immobilization of Extremozymes

Expression

Eventually, the key point for the industrialization of a biocatalyst remains the possibility to produce this protein at the industrial scale at reasonable cost. To overcome the difficulties to grow extremophilic microorganisms such as the hyperthermophilic ones at very high cell densities, overexpression of their proteins in mesophilic hosts such as *Escherichia coli*, *Bacillus subtilis* or yeasts is the preferred alternative. This strategy allows the use of conventional fermentors, which is not always possible with extremophiles due to their growth requirements (in particular temperature of growth and the corrosive effects of the media used).

Fortunately, in many cases such simple cloning in classical hosts leads to the functional expression of proteins. Moreover, these proteins, when expressed in mesophilic hosts, retain all the properties of the native enzymes, such as for instance, thermostablity or optimal activity at high temperature. This observation suggests that only few enzymes need extrinsic factors (salts or polyamines), post-transcriptional modifications (glycosylation) or specific chaperones (such as heat shock proteins) to be properly folded (Vieille and Zeikus, 2001). "Heterologous expression of Thermophiles Proteins" and "Heat Shock Proteins in Hyperthermophiles" are described in more detail in Chapter 11 of this book.

However, even when the expression of extremozymes is often possible, the level of expression can be very low. This difficulty, due mainly to a significantly different codon usage between the expressed gene and the optimal codon usage of the mesophilic host, can be overcome using *E. coli* strains containing tRNA genes for the complementation of rare codons or by the design of a synthetic gene optimized according to the codon usage of the host.

Purification

Depending on the final application, the level of expression and the production mode (intracellular or extracellular), purification of the enzyme may be required. The desired cost of the enzyme solution may also influence the purification procedure. When possible and when the extremozyme is produced in a traditional host such as *E. coli*, the difference in stability under extreme conditions between the protein's host and the recombinant protein can simplify the purification process. In the case of enzymes coming from hyperthermophilic microorganisms produced in *E. coli*, the following protocol (that can be scaled up at the industrial level) can be tested to achieve a thermoprecipitation for the recovery and the purification of the protein:

- after fermentation, the cells are concentrated (by centrifugation or diafiltration) and washed twice in 0.1 M phosphate buffer, pH 7.5,
- the cell are then suspended in the same phosphate buffer to obtain an optical density of 150–200 at 600 nm,
- TritonX 100 is added to the washed cell suspension to a final concentration of 0.5% (v/v) to improve recovery of the protein (this step can be omitted in some cases),
- The temperature of the suspension is then raised to 85°C under stirring conditions (300 rpm) and maintained at this value for 30 min,
- after cooling at 25°C, polyethylenimine (SIGMA P3143) is added to a final concentration of 0.3% (v/v) for flocculation and the cell suspension is then centrifuged at $4500 \times g$,
- finally, the supernatant is filtered using dead-end filtration (Millipore^{TM}, Polygard^{TM} CR Opticap^{TM} XL 5) according to the supplier's protocol,
- when required, the protein suspension may be concentrated by cross-flow ultrafiltration.

Immobilization

There are numerous techniques known for immobilization. Usually they involve the attachment of the enzyme onto a solid support by adsorptive means or by covalent binding. There are other techniques which use cross-linking of the enzyme in free or in crystalline form. Confining the enzyme into a restricted area by entrapment into a solid matrix or a membrane-restricted compartment is also possible, and this method is frequently used. Depending on the immobilization technique, the properties of the biocatalyst such as stability, mechanical resistance, selectivity, binding properties for substrates, pH and temperature features can be changed.

Cross-linking of enzymes involves the attachment of enzyme molecules with other enzyme molecules by covalent bonds, which results in insoluble high-molecular aggregates. The free enzyme molecules can also be cross-linked with other inactive "filler" proteins such as albumins. The most widely used reagent for immobilization by cross-linking is α,ω-glutaraldehyde, sometimes in combination with other cross-linkers like polyazetidine. The advantage of this method is its simplicity. Depending on the final industrial process constraints, the appropriate enzyme immobilization procedure has to be selected and optimized to produce the biocatalyst.

Here is a simple protocol to achieve cross-linking of proteins of extremophiles to formulate them as a powder amenable for fine chemistry reactions in organic solvents:

– Dilute the protein of interest in phosphate buffer with a concentration between 0.2 and 2 M at a pH that is compatible with the protein stability. Determine the protein concentration using, for example, the Bradford assay;
– Add 20% (w/w of protein) of maltitol and 10% (w/w of protein) of gluteraldehyde under vigorous agitation at room temperature;
– Shake the suspension for 30 min;
– Add 10% (w/w of protein) of Polymer KYMENE™ 617 (Hercules) and shake for another 15 min;
– The total suspension is then laid out on metal trays and dried until constant weight (depending on the protein, the drying temperature can be chosen between 37°C and 60°C);
– The reticulate mass is recovered and roughly ground to obtain a biocatalytic powder.

◆◆◆◆◆ PRODUCTION OF METABOLITES

In addition to enzymes, extremophiles are a source for a large variety of metabolites that are of interest for various biotechnological applications. Some of these compounds are already known, but can be advantageously produced in extremophiles, while others are new products. For example, the compatible solutes, ectoine and hydroxyectoine are efficiently produced in halophiles, where they can be harvested by "bacterial milking", e.g. hyposmotic shock with cell recycling (Sauer and Galinski, 1997). They are used for enzyme stabilization and as skin moisturizers (Motitschke *et al.*, 2001). Slightly halophilic hyperthermophiles like *Archaeoglobus fulgidus* produce new compatible solutes such as diglycerol phosphate, a compound that shows a real potential as enzyme stabilizer (Lamosa *et al.*, 2000). The characterization of these compatible solutes is discussed elsewhere (H. Santos; M. Roberts; Chapters 8 and 26, respectively in this book). While there is no specific screening process for these molecules, their production is generally highly dependent on the culture conditions, usually salt concentration and temperature. This implies that various culture conditions have to be tested (Martins *et al.*, 1997).

Archaeal extremophiles also produce unique ether-linked lipids that could be used in pharmaceutical applications (Sprott *et al.*, 1999), as well as extracellular or intracellular biopolymers such as poly(γ-D-glutamic acid) and polyhydroxyalkanoates such as poly-β-hydroxybutyrate that can be used as thickeners in the food industry or for production of biodegradable plastics (Heyazen *et al.*, 2000).

Anaerobic fermentation of cellulose or starch to produce ethanol can be performed under thermophilic conditions using a *Thermoanaerobacter ethanolicus* strain in pure or mixed culture (Ljungdahl and Wiegel, 1981). Organic acids, especially L(+) lactic acid, can also be produced under thermophilic conditions with thermotolerant bacteria (Combet-Blanc *et al.*, 2000; Fardeau *et al.*, 2004).

It has been claimed that extremophiles should represent a new and promising source for novel antibiotics. However, only a small fraction of extremophile biodiversity has been explored to date for antibiotics production. Screening for such compounds involves the preparation of extracts from both culture broth and cells in various solvents, followed by the assaying of these extracts for antimicrobial activity. Again, the major difficulty arises from the culture yields that are generally low, and the harshness of culture conditions that can be unfavorable for antibiotic stability (Horikoshi, 1999). Apart from the discovery of halocins in halophilic bacteria (see RF Shand, Chapter 29 in this book), the isolation of an antifungal agent from a moderately thermophilic *Pseudomonas* sp. has been described (Phoebe *et al.*, 2001).

The production of biosurfactants from psychrophilic marine strains grown on hydrocarbons (Yakimov *et al.*, 1999) could help accelerate oil spill remediation. A number of halophilic or halotolerant microorganisms, especially *Haloferax* and *Halomonas* species, produce highly sulfated exopolysaccharides that have the required properties for assisted oil recovery from oil wells. This is an interesting prospect since these microorganisms are relatively easier to cultivate on a large scale than other extremophiles. The use of (hyper)thermophilic microorganisms for *in situ* Microbial Enhanced Oil Recovery has also been evaluated in hot oil fields.

The tremendous diversity of extremophiles has just begun to be tapped for biotechnological applications, and there can be no doubt that numerous exotic metabolites await discovery.

◆◆◆◆◆ CONCLUSIONS

Biomolecules from extremophiles are still a largely unexplored source of new products such as enzymes for industrial or R&D applications, materials to create new reagents providing new functionality and new kits for research (new promoters, tags, fusion partners, chaperones), or new bioactive molecules. Study of life in extreme conditions provides fundamental information on these microorganisms, on means for isolating new strains, on their metabolism and their genetics. Based on genome sequences, one can estimate that between one to five thousand enzymes are produced by any single microorganism. A worldwide collection of extremophiles thus represents a tremendous potential for the discovery of new enzymes or metabolites for industrial applications.

From our past experience, we have observed that the most rewarding approach to capture all the benefits of extremophile biotechnology is to fully integrate this technology within the industrial technologies. This integration enables:

- speeding up the design and the development of new enzymes to address unmet needs,
- extending the existing product lines with proprietary products having improved performance,
- designing new procedures for refining raw material (such as renewable feed stock for agriculture),
- driving the discovery of new biomolecules of industrial interest (such as, for instance, developing polysaccharides from extremophiles as anti-fouling agents).

Limitations due to stringent industrial R&D deadlines are overcome by the emergence of new proven screening technologies that speed up the discovery of valuable molecules from these organisms, and enable to create tailor-made molecules that perfectly fit with harsh industrial, economic and environmental conditions.

Acknowledgements

We thank Dr. Bernard Ollivier, Dr. Lyuba Ryabova and Prof. Jean-Louis Reymond for their careful proofreading of this chapter. We thank Prof. Jean-Louis Reymond for providing figures on enzymes fingerprinting. And we thank all Protéus' staff for their useful contribution to the development of some of the mentioned protocols.

Useful Webpages

The Institute for Genomic Research (TIGR, Rockville, MD):
http://www.tigr.org/
National Center for Biotechnology Information (NCBI):
http://www.ncbi.nlm.nih.gov/
E. coli Stock Center: http://cgsc.biology.yale.edu/cgsc.html/

List of suppliers

Abgene
565 Blossom Road, Rochester
New York 14610, USA
Tel: +1-585-654-4800; +1-800-445-2812
Fax: +1-585-654-4810
http://www.abgene.com/

ALPS300™ Thermal plate sealers

Constant Systems Ltd
Cell-D
20 Bis, Rue du Chapitre
30150 Roquemaure, France
Tel: +33 (0) 4 66 82 82 60
Fax: +33 (0) 4 66 90 21 10
http://www.constantsystems.com

Cell D

Corning Inc. – Life Sciences
45 Nagog Park
Acton, MA, USA
Tel: +1-978-635-2200
Fax: +1-978-635-2476
http://www.corning.com/

UV Microtiter plates #3635

Epicentre
726 Post Road
Madison, WI 53713, USA
Tel: 1-608-258-3080
Fax: 1-608-258-3088
http://www.epibio.com/main.asp

Ampligase (#A3202K)

Hercules Inc.
1313 North Market Street
Wilmington, DE 19894-0001, USA
Tel: 1-302-594-5000
Fax: 1- 302-594-5400
http://www.herc.com/

Kymene^{TM} 617

Invitrogen Corporation
1600 Faraday Avenue
PO Box 6482
Carlsbad, California 92008, USA
Tel: 1-760-603-7200
Fax: 1-760 602-6500

U.S. Academic and Industrial Orders
Tel: 800-955-6288, Option 1
Fax: 800-331-2286
http://www.invitrogen.com/content.cfm?pageid=13

pBAD and pPIC vectors

Millipore
*290 Concord Rd.
Billerica, MA 01821, USA
Tel: 978-715-4321
http://www.millipore.com/msds.nsf/docs/69zptg*

Polygard™ CR Opticap™ XL5

Novagen
*EMD Biosciences, Inc.
10394 Pacific Center Court
San Diego, CA 92121, USA
Tel: 1-800-854-3417
Fax: 1-800-776-0999
http://www.emdbiosciences.com/html/NVG/home.html/*

pET series vectors

Qbiogene
*Parc d'Innovation
BP 50067
67402 Illkirch Cedex
Tel: +33 (0)3 88 67 54 25
Fax: +33 (0)3 88 67 19 45
http://www.qbiogene.com/*

BIO101 Systems (Fast DNA SPIN kit for soil)

QIAGEN Inc.
*27220 Turnberry Lane
Valencia, CA 91355, USA
Tel: 1-800-426-8157
Fax: 1-800-718-2056
http://www1.qiagen.com/*

QIAamp™ DNA mini kit and QiaKit PCR products purification kit

Research Organics
*4353 East 49th Street
Cleveland, OH. 44125, USA
Customer Service
Tel: 800-321-0570
International Tel: 216-883-8025
Fax: 216-883-1576
Technical Service Tel: 800-334-0144
http://resorg.com/*

Red Gluc #1365C

Retsch
Distributor: VERDER S.A.R.L.
Parc des Bellevues
Rue du gros chêne
95610 Eragny sur Oise, France
Tel: +33 (0)1 34643111
Fax: +33 (0)1 34644450
http://www.retsch-us.com/en/db/overview.php?AKZ=C11

Mortar Grinder RM 100

Sigma-Aldrich
P.O. Box 14508
St. Louis, MO 63178, USA
Tel: +1-800-325-3010
http://www.sigmaaldrich.com/

Mineral oil (SIGMA #M3516), Gelrite® Gellan Gum (SIGMA #G1910), Phytagel™ (SIGMA #P8169), Remazol Brilliant Blue R-D-Xylan (SIGMA #M5019), Polyethylenimine (SIGMA#P3143), Polystyrene plates (COSTAR #3599), Aluminum adhesive films (COSTAR #6570)

Stratagene
11011 N. Torrey Pines Road
La Jolla, CA 92037, USA
Tel: +1-512-321-3321
Fax: +1-512-321-3128
http://www.stratagene.com/homepage/

pBSKS vectors; Lambda ZAP® bacteriophage cloning kits

Velocity11
3565 Haven Avenue
Menlo Park, CA 94025, USA
Tel: 1-650-846-6600
Fax: 1-650-846-6620
http://www.velocity11.com/

PlateLoc® thermal plate sealers

References

Amann, R. I., Ludwig, W. and Schleifer, K. H. (1995). Phylogenetic identification and in situ detection of individual microbial cells without cultivation. *Microbiol. Rev.* **59**, 143–169.

Blöchl, E., Burggraf, S., Fiala, G., Lauerer, G., Huber, G., Huber, R., Rachel, R., Segerer, A., Stetter, K. O. and Völkl, P. (1995). Isolation, taxonomy and phylogeny of hyperthermophilic microorganisms. *World J. Microbiol. Biotechnol.* **11**, 9–16.

Brown, T. (1995). Southern blotting and hybridization. In *Current Protocols in Molecular Biology* (F. M. Ausubel, R. Brent, R. E. Kingston, D. D. Moore, J. G. Seidman, J. A. Smith, K. Struhl and V. B. Chanda, eds), pp. 2.9.1–2.9.15. John Wiley & Sons, Inc, New York.

Business Communications Company, Inc. (2004). *Enzymes for Industrial Applications, RC-147U*, (www.bccresearch.com).

Cherepanov, P. P. and Wackernagel, W. (1995). Gene disruption in *Escherichia coli*: Tc^R and Km^R cassettes with the option of Flp-catalyzed excision of the antibiotic-resistance determinant. *Gene* **158**, 9–14.

Combet-Blanc, Y., Ollivier, B. and Garcia J.-L. (2000). Bacterial strains of the genus *Bacillus*, closely phenotypically related to the genus *Lactobacillus*, culture method and use. US Patent #6,022,537.

Crameri, A., Dawes, G., Rodriguez, E., Silver, S. and Stemmer, W. P. (1997). Molecular evolution of an arsenate detoxification pathway by DNA shuffling. *Nature Biotechnol.* **15**, 436–438.

Dupret, D., Masson, J.-M. and Lefèvre, F. (2002). Method for obtaining in vitro recombined polynucleotides sequences, sequence banks and resulting sequences. Patent WO0009679.

Fardeau, M.-L., Combet-Blanc, Y. and Ollivier, B. (2004). Bacterial strains of genus *Exiguobacterium*, culture method and uses. Patent WO2004055173.

Heyazen, F. F., Rehm, B. H. A., Eberhardt, R. and Steinbüchel, A. (2000). Polymer production by two newly isolated extremely halophilic archaea: application of a novel corrosion-resistant bioreactor. *Appl. Microbiol. Biotechnol.* **54**, 319–325.

Horikoshi, K. (1999). Alkaliphiles: some applications of their products for biotechnology. *Microbiol. Mol. Biol. Rev.* **63**, 735–750.

Lagarde, D., Nguyen, H.-K., Ravot, G., Wahler, D., Reymond, J.-L., Hills, G., Veit, T. and Lefèvre, F. (2002). High-throughput screening of thermostable esterases for industrial bioconversions. *Org. Process Res. Dev.* **6**, 441–445.

Lamosa, P., Burke, A., Peist, R., Huber, R., Liu, M. Y., Silva, G., Rodrigues-Pousada, C., LeGall, J., Maycock, C. and Santos, H. (2000). Thermostabilization of proteins by diglycerol phosphate, a new compatible solute from the hyperthermophile *Archaeoglobus fulgidus*. *Appl. Environ. Microbiol.* **66**, 1974–1979.

Ljungdahl, L. G. and Wiegel, J. K. W. (1981). Anaerobic thermophilic culture. US Patent #4,292,407.

Martins, L. O., Huber, R., Huber, H., Stetter, K. O., da Costa, M. S. and Santos, H. (1997). Organic solutes in hyperthermophilic archaea. *Appl. Environ. Microbiol.* **63**, 896–902.

Mayer, K. M. and Arnold, F. H. (2002). A colorimetric assay to quantify dehydrogenase activity in crude cell lysates. *J. Biomol. Screen.* **7**, 135–140.

Miller, J. H. (1992). Experiments with transposable elements. In *A Short Course in Bacterial Genetics* (J. H. Miller, ed.), pp. 309–321. Cold Spring Harbor Laboratory Press, Cold Spring Harbor, New York.

Moore, D. D. (1995). Construction of recombinant DNA libraries. In *Current Protocols in Molecular Biology* (F. M. Ausubel, R. Brent, R. E. Kingston, D. D. Moore, J. G. Seidman, J. A. Smith, K. Struhl and V. B. Chanda, eds), pp. 5.0.3–5.4.4. John Wiley & Sons, Inc, New York.

Motitschke, L., Driller, H. and Galinski, E. (2001). Ectoin and ectoin derivatives as moisturizers in cosmetics. US Patent #6,403,112.

Phoebe, C. H., Jr, Combie, J., Albert, F. G., Van Tran, K., Cabrera, J., Correira, H. J., Guo, Y., Lindermuth, J., Rauert, N., Galbraith, W.,

Galbraith, V. and Selitrennikoff, C. P. (2001). Extremophilic organisms as an unexplored source for antifungal compounds. *J. Antibiotics* (Tokyo) **54**, 56–65.

Ravot, G., Wahler, D., Favre-Bulle, O., Cilia, V. and Lefèvre, F. (2003). High throughput discovery of alcohol dehydrogenases for industrial biocatalysis. *Adv. Synth. Catal.* **345**, 691–694.

Reymond, J.-L., Wahler, D., Badalassi, F., and Nguyen, H.-K. (2001). Method for releasing a product comprising chemical oxidation, method for detecting said product and uses thereof. Patent WO0192563.

Reysenbach, A.-L. and Götz, D. (2001). Methods for the study of hydrothermal vent microbes. In *Marine Microbiology – Methods in Microbiology*, (J. Paul, ed.), vol. 30, pp. 639–656. Academic Press, London.

Rose, T. M., Henikoff, J. G. and Henikoff, S. (2003). CODEHOP (COnsensus-DEgenerate Hybrid Oligonucleotide Primer) PCR primer design. *Nucleic Acids Res.* **31**, 3763–3766.

Sauer, T. and Galinski, E. A. (1997). Bacterial milking: a novel bioprocess for production of compatible solutes. *Biotechnol. Bioeng.* **57**, 306–313.

Schiraldi, C. and De Rosa, M. (2002). The production of biocatalysts and biomolecules from extremophiles. *Trends Biotechnol.* **20**, 515–521.

Schleper, C., Puehler, G., Holz, I., Gambacorta, A., Janekovic, D. Santarius, U., Klenk, H.-P. and Zillig, W. (1995). *Picrophilus* gen. nov., fam. nov.: a novel aerobic, heterotrophic, thermoacidophilic genus and family comprising *Archaea* capable of growing around pH 0. *J. Bacteriol.* **177**, 7050–7059.

Silver, J. (1991). Inverse polymerase chain reaction. In *PCR1, A Practical Approach* (M. J. McPherson, P. Quirke and G. R. Taylor, eds), pp. 137–146. Oxford University Press, Oxford.

Sprott, G.D., Patel G.B., Choquet, C.G. and Ekiel, I. (1999). Formation of stable liposomes from lipid extracts of *archaeobacteria* (archaea). US Patent #5,989,587.

Stemmer, W. P. (1994). Rapid evolution of a protein in vitro by DNA shuffling. *Nature* **370**, 389–391.

van den Burg, B. (2003). Extremophiles as a source for novel enzymes. *Curr. Opin. Microbiol.* **6**, 213–218.

Vaneechoutte, M., Rossau, R., De Vos, P., Gillis, M., Janssens, D., Paepe, N., De Rouck, A., Fiers, T., Claeys, G. and Kersters, K. (1992). Rapid identification of bacteria of the *Comamonadaceae* with amplified ribosomal DNA-restriction analysis (ARDRA). *FEMS Microbiol. Lett.* **72**, 227–233.

Vieille, C. and Zeikus, J. G. (2001). Hyperthermophilic enzymes: sources, uses, and molecular mechanisms for thermostability. *Microbiol. Mol. Biol. Rev.* **65**, 1–43.

Vieille, C., Burdette, D. S. and Zeikus, J. G. (1996). Thermozymes. *Biotechnol. Annu. Rev.* **2**, 1–83.

Vinogradov, A. E. (1994). Measurement by flow cytometry of genomic AT/GC ratio and genome size. *Cytometry* **16**, 34–40.

Wahler, D. and Reymond, J.-L. (2001). High-throughput screening for biocatalysts. *Curr. Opin. Biotechnol.* **12**, 535–544.

Wahler, D., Badalassi, F., Crotti, P. and Reymond, J. L. (2001). Enzyme fingerprints by fluorogenic and chromogenic substrate arrays. *Angew. Chem.* **113**, 4589–4592.

Wahler, D., Badalassi, F., Crotti, P. and Reymond, J. L. (2002). Enzyme fingerprints of activity, and stereo- and enantioselectivity from fluorogenic and chromogenic substrate arrays. *Chem. Eur. J.* **8**, 3211–3228.

Wilhelm, J., Pingoud, A. and Hahn, M. (2003). Real-time PCR-based method for the estimation of genome sizes. *Nucleic Acids Res.* **31**, e56.

Yakimov, M. M., Guiliano, L., Bruni, V., Scarfi, S. and Golyshin, P. N. (1999). Characterization of Antarctic hydrocarbon-degrading bacteria capable of producing bioemulsifiers. *Microbiologica* (Pavia) **22**, 249–256.

Zhao, H., Giver, L., Shao, Z., Affholter, J. A. and Arnold, F. H. (1998). Molecular evolution by staggered extension process (StEP) in vitro recombination. *Nature Biotechnol.* **16**, 258–261.

Zubay, G. (1973). In vitro synthesis of protein in microbial systems. *Annu. Rev. Genet.* **7**, 267–287.

Index

acidophilic microorganisms
  domain Archaea 480–1
    Acidianus brierleyi 481
    Ferroplasma sp. 481
    Sulfolobus acidocaldarius 481
  eukaryotes 501–3
  fluorescent in situ hybridization (FISH) 495–6
    catalyzed reporter deposition (CARD) 501
    fixation 497–9
    hybridization 499–500
    probe specificity 496–7
    standard protocol 496
  isolation 473–4
  microbial ecology analysis 481
    cloning of 16S rRNA genes 492
    denaturing gradient gel electrophoresis methodology (DGGE) 483–92
    gene cloning 494
    ribosomal RNAs 482
    study flow chart 482–3
  prokaryotes 474–81
    Acidimicrobium ferrooxidans 479
    Acidiphilium spp 479–80
    Acidithiobacillus caldus 477
    Acidithiobacillus thiooxidans 477
    actinobacteria 479
    Desulfosporosinus sp. 480
    domain bacteria 474–7
    Ferrimicrobium acidiphilum 479
    Leptospirillum sp. 478
    Sulfobacillus sp. 478
aerobic alkaliphiles 439
  media and culture conditions 443–5
  source environments 439
    high Ca^{2+} environments 440
    Na_2CO_3 environments 440
  types 441–3
    alkaliphiles 441
    Cellulomonas bogoriensis 441
    haloalkaliphiles 441
    Halomonas magadiensis 441
alkalithermophiles
  alkalithermophilic cultures 453
  antibiotic susceptibility 464
  characterization 462
  considerations 452
  cultivation methods 454
    media selection 454
    organic buffer solutions 455
  DNA extraction 464–5
    Gelrite 458
    Pseudomonas elodea 458
  incubation and culture equipment 452–3
  intracellular pH range determination 463–4
  isolation methods 460
    soft agar approach 460–2
  media, sterilization of 457
  medium pH, adjustment 455
  solid media preparation 458
    agar 458
    Gelrite, advantages 458–9
    silica gel 459
  temperature range determination 462–3
anaerobic microorganisms
  anaerobe maintenance 767–72
    anaerobes, growth of 769–71
    isolation and enrichment 768–9
    sample collection 767–8
    storage 771
  anaerobic biochemistry 772–7
    activity assays 777
    anaerobic buffers 772–3
    anaerobic cell-free extracts 773–4
    column chromatography 775–6
    protein anaerobic precipitation 774–5
  anaerobic glove box 760
    glove box configurations 760
    maintainance gas ratio 762–3
    oxygen reduction 761–2
    water vapor and hydrogen sulfide control 762
  culture vessels 763–5
    liquid medium 763–4
    solidified medium 764–5
  media preparation apparatus 758–63
    gas proportioner 759
    gas supply 758–9
    gassing cannula 759–60
    oxygen trap 759
  media preparation 765–7
anaerobic thermophiles 309–25
  electrotransformation (ET)
    efficiencies 315

anaerobic thermophiles (*continued*)
  high-efficiency protocol 316–17
    Thermoanaerobacterium saccharolyticum
      JW/SL-YS 485 314
    Thermoanaerobacterium thermosaccharolyticum 314
  exo- and endonuclease activities 311
  genome sequence information 310–11
  recombinant thermophile selection 318–20
  replicative shuttle vectors 311–12
  restriction endonuclease activity
    detection 322
    examination 322–3
  saccarolytic thermophiles, plating 317–18
  total DNA isolation 320–1
Archaea/Bacteria, thermophilic 204
Archaea, membrane lipids
  characterization techniques 587
  diphytanylglycerol diether lipids, role 585–6
  extraction technique 591, 593–5
  fatty acid synthetase (FAS) complex 586–7
  lipid extract analysis
    electrospray ionization mass spectrometry (ESI-MS) 598–601
    nuclear magnetic resonance spectroscopy (NMR) 601–3
    thin-layer chromatography (TLC) 596–7
  neutral lipid types 587–8
  polar lipids
    cardiolipins 589–91
    glycolipids 588–90
    phospholipids 588
  purple membrane (PM) lipid composition 590–5
  role, functional 604–9
  separation technique 596
athalassohaline 538, 570

BLAST analysis 523

cell growth 175–6
cell harvesting 177
  biomass quantification, methods 182–5
    cell counting 184
    cell lysis 183
    DNA quantification 185
    dry weight determination 183–4
    protein quantification 184–5
  extraction 177–9
    ethanol–chloroform 176–7
    perchloric/trichloroacetic acid 178–9

intracellular volume, determination 186
  NMR detection, identification, quantification 179–82
  purification methods
    adsorption chromatography 191–2
    gel filtration chromatography 190–1
    ion-exchange chromatography 189–90
    thin layer chromatography 187–9
  unknown solute, identification 192–5
    $^{1}H/^{13}C$ correlation spectra 193–4
    ethanol, ^{13}C NMR spectrum 193–4
Clostridium thermocellum 309
compatible solutes 174–95
  molecular representation 175
  roles and types 174
Cytophaga/Flavobacterium/Bacteriodes 12, 15, 151, 515

extremophile, handling methods 1–19
  Cyanidium caldarium 4
  Dunaliella acidophila 4
  mechanisms and behavior 14–17
    Deinococcus radiodurans 17
    Dunaliella 16
    Halanaerobiales 16
    Halobacteriaceae 15
    Polaromonas vacuolata 17
    Psychroflexus 17
    Salinibacter 15
  overview 5–14
    acidophiles 9–10
    alkaliphiles 10–11
    halophiles 11–12
    piezophiles 12–13
    psychrophiles 9
    radiation resistant 13–14
    thermophiles 7–8
  study 17–19
  Thermus aquaticus 4
extremophilic bacteria, lipid analysis 127–55
  Deinococcus/Thermus 128
  fatty acids 141–4
    cell cultivation/harvesting 144–5
    mass spectrometry analysis 146–51
    preparation 145
  isoprenoid quinones 151–4
    extraction 152
    growth of organisms 152
    HPLC 153
    identification 153
    purification 152
    separation 152

polar lipids 129–41
  characterization 137–40
  detection 135–6
  microbial growth and extraction 130
  TLC 131–5
Salinibacter ruber 128
Thermomicrobium roseum 128
Thermotogales 127

genomic DNA extraction 519–20
  BIO101 Fast-Spin® Kit (QBiogen) 519

haloarchaea cultivation 685–91
  challenges 535–6
  haloviruses, cultivation 692
    DNA extraction 696–7
    host-range determination 692–3
    negative-stain TEM 695–7
    protein extraction 697
    virus particle stability 693
  head–tail, purification 689–90
  isolation, direct plating 686–7
  isolation methods, saltern and sampling considerations 537
  lysate preparation 688–9
  plaque assay 687–8
  spindle/round halovirus, purification 690–1
haloarchaea cultivation, growth monitoring
  direct microscopy
    acridine orange (AO) stain solution 547
    acridine orange fluorescent staining 546–7
    antifade solution/para-phenylene diamine (PPD) stock solution 547
  PCR screening
    multiplex PCR screening protocol 548–9
haloarchaea, genetic systems 649–73
  genetic methods 656–8
    cloning vectors 660–4
    expression vectors, 667–8
    gene replacement/knockouts 664–5
    gene reporters 666–7
    genetic transfer 658
    insertion mutagenesis 665
    promoter analysis 667–8
    selectable markers 658–60
    transfection 658
    transformation 656–8
    transposon mutagenesis 666
  genomic approaches 668–72
    DNA microarrays 669–71

gene knockouts, systematic 668
    proteomics 671–72
  model systems 651–6
    Halobacterium sp. 651–3
    Haloferax volcanii 654–6
  haloarchaea transfection 698–9
Halobacteriaceae 2, 4, 11, 15, 442, 515, 535, 585, 587–9, 599, 606, 650, 681
  bacterioruberins, role 585
halocin study 703–15
  activity, detecting and quantifying 707–10
  characteristics 704
    activity spectra 704
    immunity 704–7
    mechanism of action 704–7
    processing 704–7
    secretion 704–7
  purification 710–15
    microhalocin purification 713–15
halophilic bacterial storage 719–29
  culture revival 728–9
  culture tracking 720–1
  long-term culture storage 723–8
    beads 724–6
    cryostraws 726–7
    crystals 727–8
    liquid nitrogen atmosphere 723
    lyophilization 723–4
  storage systems 721–8
    long-term storage 723
    medium term storage 722–3
    short-term storage 721–2
halophile extraction, ancient salt crystals
  sample selection
    primary vs. non-primary materials 557
  Steno's stratigraphic principles
  studies, early 553–4
halophile extraction methods
  crystal sterilization methodology 558–60
  extraction techniques
    dissolution 561
    drilling 561–2
halophile, hypersaline environments/salterns
  BIOLOG 525–8
  characterization techniques, basis 514
  clone data analysis, custom PERL script 523–5
    ribosomal database (RDP) 523–4
    genetic data environment (GDE) 524
    cladogram 524–5
  16S rRNA gene DGGE profile, bacteria 515–6
  restriction analysis, problems 514

halophile, hypersaline environments/
    salterns (*continued*)
  ribosomal internal spacer analysis (RISA)
    514–5
  RISA + RFLP (restriction fragment length
    polymorphism) gel, Archaea 515–17
  sequencing 523
  whole community analysis, preparation
    samples 518–9
    genomic DNA extraction 519–20
halophilic microorganisms
  colony forming units (CFUs), determination
    572–3
  DAPI (4′, 6-diamidino-2-phenylindole)
    staining 574–6
  LIVE/DEAD® BacLight™ bacterial
    viability kit 576
    detection, environmental samples 577–8
    stability, cell-bound LIVE/DEAD dyes
      578–80
  habitats 569–70
  viability assessment techniques 570–1
halotolerant/halophilic microorganisms
    characterization
  biosynthetic pathways 634–8
    isotopic labeling 634
    NMR monitoring 636–7
  cell solute dynamics 638–40
  organic osmolyte monitoring 640–2
    chromatography 640
    detection methods 640
    mass spectrometry 640
    solid-state NMR 641
  organic solutes 616–24
    anionic solutes 623–4
    uncharged solutes 616–22
    zwitterionic solutes 622–3
  osmolyte identification 624–34
    cell extraction methods 625
    NMR spectroscopical analysis 626–34
  osmotic stress 615–6
halovirus study 681–99
heat shock proteins (HSP) 233–48
  gene regulation 234–7
    array construction 235–6
    heat shock response, transcriptional
      analysis 234–5
    labeling and hybridization 236–7
    real time RT-PCR 237
HSP characterization 237–40
  biophysical and biochemical methods
    circular dichroism (CD) 243
    classical/static light scattering 242

differential scanning calorimetry
  (DSC) 243–4
dynamic light scattering (DLS) 241–2
fluorescence 242–3
Fourier transform infrared (FT-IR)
  244–5
peptide mapping, LC-MS 246–7
post-translational modifications site,
  identification 247
protein analysis, RP-HPLC 246
protein/peptide, N- and C-terminal
  sequencing 245
SDS-PAGE laser densitometry 245–6
recombinant expression 237–8
  expression vectors, types 238–40
HSP, metabolic engineering in 247–8
hyperthermophile cultivation 109–22
  bioreactor 112–16
    culture sampling 116
    high temperature–pressure (high T–P)
      bioreactor 112–16
    inoculum supply 115
    maintenance 116
    temperature/pressure control 115
    tubing and connectors 114
  Methanocaldococcus jannaschii cultivation
    116–20
    cell lysis 120–1
    gas–liquid mass transfer
      limitations 121
hyperthermophile studies 93–103
  glass gas-lift bioreactor 96–103
    description 96–8
    fermentation protocol 98–102
    growth parameters 102
    metabolic product analysis 102
  sulphur-reducing microorganisms 94–6
    approaches 95–6
    cultivation 94
    Thermococcales 94
hyperthermophiles 233, 323–4
  liposome-mediated DNA uptake 324
  plasmid preparation 323
  spheroplast formation 323–4
  transformed spheroplast plating 324
hyperthermophilic organisms 173
  carbon chemical shift values 181
hyperthermophilic virus
  Archaea, viruses of 340–1
  detection 337–9, 342–3
    host genome copies 343–4
    infectivity 339, 342
    plaque assays 339

transmission electron microscopy
  (TEM) 337–8
enumeration 332
hot habitat virus-like particles (VLPs) 332–6
  sampling and enrichment culture 332–3
morphotypes
  transmission electron micrographs 334–5
purification
  host strains 338–9
virus–host isolation 336–8
  host strain 338–9

ionizing radiation-resistant microorganisms
  745–51
  background 745–7
  effects of ionizing radiation 747–9
  prokaryotic species 746
  survival curve generation 749–52
    dose–response survival curve 749, 751–2
    equipment and reagents 750
    preliminary studies 750–1

length heterogeneity (LH)-PCR, halophiles
  517–8, 520–1
  data analysis 522
    GenoSpectrum™ software 522
    Interleave 1.0 software 522
  molecular fingerprinting 517–8
LIVE/DEAD stains
  propidium iodide (PI) 576
  SYTO 9, 576

MEGABLAST 524

PCR product analysis
  denaturing gradient gel electrophoresis
    (DGGE) 514
  length heterogeneity (LH) 514
  terminal restricted fragment length
    polymorphisms (T-RFLP) 514
piezophiles
  DEEPBATH system 737–9
    clone numbers 738
    devices 737–8
    effects of pressure 739
    hyperthermophile cultivation 739
    Shewanella and *Moritella* 737–8
  traditional handling 734–7
    deep-sea sediment sampling 735–6
    isolate characterization 735–7
    isolation 735–6
    JAMSTEC project pressure vessels 734
    mixed cultivation 735

proteomics 218–25
pschyrophile, biotechnological applications
  428
psychrophile culture media 381–2
  cyanobacteria, media for 383–4
  enrichment strategies 385–6
  low temperature culturing 386
  microalgae, media for 384–5
  saline habitats, media for 382–3
  storage of cultures 386–7
  yeasts and fungi, media for 384
psychrophile, field laboratory culturing
  community DNA, isolation of 387–8
  psychrotrophic communities 388–9
psychrophile genomics and proteomics 426–7
psychrophile habitat sampling 374
  aquatic sources 378–9
  ice 379–81
    ice cores 380
  soil sediments 376–7
    phylogenetic analyses 377–8
  snow 381
  thermal considerations 374–5
psychrophile protein generation 398
  analysing DSC thermograms 419–21
    Arrhenius plot 421
  chemical modification 401–4
    protein modification methodologies
      402–4
  differential scanning calorimetry 417–19
    DSC thermograms 418–21
  directed evolution 400–1
  dynamic fluorescence quenching 415–17
  protein stability 408
  protein structure from psychrophiles 404–7
    conformational stability 408
    crystallographic data 406–7
    kinetic stability 408
  site-directed mutagenesis 398–9
  spectrophotometric methods 408–10
    circular dichroism (CD) spectroscopy 410
    fluorescence and CD spectra 409
    fluorescence spectroscopy 410
    unfolding curve generation 410–12
  transverse urea gradient gel (TUG)–gel
    electrophoresis 421
    analysis 423
    principle 421–3
    procedure 424–6
  unfolding curves, conformational stability
    and 412–15
    chemical denaturation 412
    thermal denaturation 412–15

psychrophile protein generation (*continued*)
psychrophile proteins 395
   heterologous expression 397–8
      cold-adapted hosts 398
      protein production 396
psychrophilic microorganisms, handling of 371
   cold planet 371–3
      habitats 372
   psychrophiles and psychrotolerants 373–4

salt lake haloarchaea, cultivation
   base media
      salt water 537–9
      trace element solution SL10 539
   nutrients
      DB characterization media (DBCM) 543
      modified growth medium (MGM) 541–2
      serial dilution/extinction culture methodology 544–9
   solid media, agar 540
Sulfolobus solfataricus, functional genomics of thermophiles 201–26
   DNA microarray analysis 206
      cDNA synthesis and labelling 215–16
      cell growth and harvest 212–13
      cell lysis and RNA isolation 212, 214–15
      gene amplification and printing 212
      microarray hybridization 216–17
      scanning, data extraction and normalization 217–18
   medium 213
      defined 212–13
      rich 213
   proteomics 221
      2D electrophoresis, protein visualization and image analysis 222–3
      cell lysis 222
      cultivation and metabolic labelling 221–2
      peptide quantification 224–5
      protein isolation and identification 224–5

thalassohaline 537, 570, 686, 695
Thermoanaerobacterium saccharolyticum 309, 320
Thermoanaerobacterium thermolyticum 309
Thermococcales 253–5
   genetic tool, design 255–8
      pYS2 map 256
      transformation procedures 256–8
   virus-like particles 265–70
      DNA manipulation 269–70
      enrichment cultures 265–8
      host, relationship with 269
      novel VLP 268
      nucleotide sequence determination 270
      PAV1, genome of 269
      sequence analysis and annotation 270
      structure 269
      transmission electron micrographs 267
Thermococcales, plasmids of 258–71
   cloning and sequencing 264–5
   diversity 263
   isolation and purification 258–9, 264
   pGT5, DNA topological analysis of 259–60
      electrophoresis, types 260
   pGT5, sequence of 260–3
      isolation and sequencing 261
      ssDNA, identification 261–3
thermophile membranes 161–70
   lipid isolation 164–7
      assay, Soxhlet extraction 166
      equipment and reagents 165–6
      principle and applications 164–5
   liposome preparation, proton permeability measurements 167–9
      assay 169
      equipment and reagents 169
      principle and applications 167–9
thermophiles 202
   DNA microarrays 206–11
   genomics 203
   proteomics 220–1
thermophiles, hydrothermal environments 55–79
   classification, 60
      Aquificales 60
      Archaeoglobales 60
      Methanococcales 60
      Thermococcales 60
   cultivation conditions 77–8
   isolation, 78
      FISH technique 78
      optical tweezers 78
   media
      anaerobic techniques 63
      chemolithoautotrophic thermophiles 72–5
      cosmopolitan thermophiles 60
      heterotrophic, S^0-utilizing 70–2
      hydrogenotrophic methanogens 75–7
      modified anaerobic techniques 63
   sample 56–60
      collection 56–9
      preservation 60
      processing 60

thermophilic microorganisms, preservation 349–66
  deep freezing 360–6
    aliquots and freezing distribution 361–4
    cell suspension preparation of 361
    equipment 361
    glass capillary tube method 362–4
    storage and recovery 365–6
  double-vial preparation 353
  long-term methods
    deep freezing 351–2
    freeze-drying/lyophilization 352
    liquid-drying (L-drying) 352
  subculturing 350
  vacuum-drying 353–60
    centrifugal freeze-drying 353–60
    equipment 353
    liquid-drying 356–9
    secondary vacuum-drying 356–7
    storage and recovery 359–60
thermozymes 279
Thermus, gene transfer 280–3
  conjugation 282–3
  electroporation 283
  natural transformation 280–2
Thermus, genetic tools 283–304
  antisense expression vector 293–4
  E. coli/*Thermus* shuttle vectors 289–93
    pLUI and pMY series 291–2
    pMKI8 292–3
  expression vectors 298–303
  mutants, isolation 283–8
    bleomycin resistance marker 286–7
    insertional mutagenesis 285
    kanamycin resistance marker 284–6
    media 284
    orotate phosphoribosyltransferase marker 287–8
  promoter probe vectors 294–8
    pKANPROII, pTGTII 295–6
    pPP11 294–5
    pSIJ7mdhA 297–8
    pTEXImdh 296–7
    pTPROI 298–9
  protein thermostabilization, from mesophiles 303–4
  transposon mutagenesis 288–9
Thermus, genetic systems 279–304
transcriptomics 205